CISM COURSES AND LECTURES

The series presents lecture notes, monographs, edited works and proceedings in the field of Mechanics, Engineering, Computer Science and Applied Mathematics.
Purpose of the series is to make known in the international scientific and technical community results obtained in some of the activities organized by CISM, the International Centre for Mechanical Sciences.

INTERNATIONAL CENTRE FOR MECHANICAL SCIENCES

COURSES AND LECTURES - No. 402

BASICS OF ROBOTICS

THEORY AND COMPONENTS OF MANIPULATORS AND ROBOTS

EDITED BY

ADAM MORECKI
WARSAW UNIVERSITY OF TECHNOLOGY

AND

JÓZEF KNAPCZYK
CRACOW UNIVERSITY OF TECHNOLOGY

Springer-Verlag Wien GmbH

This volume contains 413 illustrations

Originally published by Springer-Verlag Wien New York in 1999

SPIN 10730568

In order to make this volume available as economically and as rapidly as possible the authors' typescripts have been reproduced in their original forms. This method unfortunately has its typographical limitations but it is hoped that they in no way distract the reader.

ISBN 978-3-211-83150-2 ISBN 978-3-7091-2532-8 (eBook)
DOI 10.1007/978-3-7091-2532-8

CONTENTS

PREFACE

Over thirty years of research and teaching experience have gone into preparation of this book. Various authors are responsible for the text:
Adam Morecki, for chapters 1, 2, 8, 9, 14, 15, 16 and 18;
Józef Knapczyk, for chapters 3, 4, 5, 6 and 7;
Mirosław Galicki, for chapters 11 and 12,
Antoni Wiliński, for chapter 13,
Teresa Zielińska, for chapter 17.

The authors express their thanks to the other persons who collaborated in these efforts: Prof. B. Fiałkowski, who revieved chapter 9, and Dr M. Olszewski and Dr J. Barczyk who assisted in the preparation chapters 8 and 9.

This text has been translated from revised 1994 Polish edition by: Sławomir Dzierżek, Jacek Kapanowski, Józef Knapczyk and Adam Morecki. The figures were made by Marcin Knapczyk.

Warsaw - Cracow, April 1999

Foreword

Robotics is a field of science and technology dealing with the issues of mechanical analysis, design, control, measurements, applications and maintenance of robots. Robots are now used widely in scientific research, manufacturing technology, civil engineering, transportation and agriculture, as well as in medicine, underwater investigations and space research. The theory and technology of robots and manipulators is an interdisciplinary field of study requiring the cooperation of specialists in various different scientific disciplines.

Recently we have seen the rapid development of industrial robots used in machine-building and electromechanical industries, in operations such as welding, painting and assembly, as well as with presses and finishing machine tools, such as grinding and polishing machines. The main purpose of using robots is to increase work quality and to release the operator from hard, or tedious labour, and especially from hazardous tasks. The largest purchaser of industrial robots is still the automotive industry, but electrical and electronic industries are also a sizeable clients, and more. recently, there has been a growing demand for robots from the service sector.

The number of produced and installed robots continues to grow. The level of complexity of these machines is increasing as well, especially in terms of artificial intelligence, i.e. sensing and vision (image recognition).

Robotics can be divided into the following branches:

- theoretical robotics (the theory of robots and manipulators),
- general robotics (the economic and social aspects, terminology, classification, education, development, and future trends),
- metrological robotics (robots for measurement, calibration, inspection, diagnostics, and testing),
- robotics of locomotion (one-, two-, four-, six-, eight- and multi-legged machines, as well as combined wheeled-and-legged machines capable of walking, running, jumping and crawling),
- robotics for medical applications and rehabilitation (manipulators and robots for surgical, therapeutical, orthotical and prosthetical applications, rehabilitation robots and man-machine systems),
- industrial robotics (robots and manipulators for applications in the electrical and mechanical industries, in food-processing, paper-making, glass-making, in the chemical, power engineering, and mining sectors),
- service robotics (robots used in hospital transport, for window cleaning, etc.)

- microrobotics (medical, inspection, and mobile microrobots),
- robots for other non-industrial applications (underwater investigation, space research, scientific study, military uses, fire-fighting, disaster-relief, civil engineering, agriculture and transportation).

Industrial robotics is closely connected with the fields of Flexible Manufacturing Systems (FMS), and Computer Integrated Manufacturing (CIM), while Annmaned Factories (AF) is a currently development.

Scenarios for the near future show rapid development of new generations of "intelligent" robots generated by advances in electronic microprocessor technology (the new scale of microchip integration, with over one billion components in one chip, the anticipated development of bioprocessors) and in mechanical processors (several thousand mechanical systems housed in one cubic inch). These robots will be better equipped to operate in changing environmental conditions than the previous generations.

We decided to write this book several years ago, to fill a gap in Polish scientific literature presenting the bases and scope of modern robotics. The contents of the book are divided into four parts. The first deals with the issues of topology, kinematic analysis, and static and dynamic force analysis, as well as geometrical and functional characteristics and grippers of manipulators and robots. The second part presents robot drives, motion transmission mechanisms, and the sensors and transducers used. The third part of the book is devoted to the issues of robot and manipulator control, and task planning. The last part discussess selected issues, such as robots and manipulators for underwater investigation, industrial robots, robots for rehabilitation, and flexible robots. The concluding sections deal with mobile robots and with walking machines, and microrobotics in particular.

Part One

THE MECHANICS OF MANIPULATORS AND ROBOTS

1. INTRODUCTION

1.1. Subjects of research and investigation

Theoretical robotics deals with the mechanical analysis and synthesis of open kinematic chains of manipulators and robots, taking into account friction, play in joints, compliances, vibration, positioning accuracy and orientation accuracy. Research the field has focused on the design or selection of drives and control systems, sensors, transducers, and components of artificial and machine intelligence.

Recent studies have addressed in particular: positioning and orientation accuracy, collision avoidance (following a spatial trajectory in workspace with obstacles), active force control, computer aided design of robots, voice communication and flexible manipulators. Solving the problems raised by these issues requires the application of a whole series of methods drawing on mathematics, mechanics, theory of machines and mechanisms, control engineering, computer science, cybernetics, theory of systems, metrology, diagnostics and theory of maintenance.

The modeling of flexible manipulators and investigation of their dynamic properties by means of a class of finite element methods is now the subject of extensive research. Another interesting field of study is that of human-spine type and elephant-trunk type manipulators. Parallel manipulator structures permitting higher velocities are a recent development.

Advances in material engineering have made new, light, composite materials available for robot construction. Computer aided-design is now an important tool in the design of robots. There are computer programs that can support the process of designing the robot structure as well as the selection of joint drives. The development of computer graphics has brought advances in computer simulation and animation applied to the off-line design of robots and robotized stations. New materials featuring shape memory have influenced the design of robot actuators and grippers. Other studies regard sensing and machine intelligence. New types of tactile, vision and force sensors have been developed, including fiber optic sensors. Many studies address the issue of force control in closed-loop feedback systems [1.9].

General robotics deals with the economic, social and educational aspects of robotics, including problems of safety. Studies on the standardization of terminology and symbols have an important part here. The discipline still does not have specific and reliable methods for estimating the economic and social consequences of robotization. Studies conducted by The International Standardization Organization (ISO) TC 184/SC2 Group (Robots for the Manufacturing Environment) and The International Federation of Robotics have focused on these issues.

Measurement and diagnostic methods play an important role in robotics. Measurements of kinematic and dynamic quantities, parameters describing the robot's accuracy and degree of wear require the development of new methods and special transducers. The measurement of the trajectory accuracy of the robot end-effector at high velocities is another example of the difficulties encountered. Calibration is also an important issue.

The use of mobile robots for inspection and testing (for example inspection of welded joints of large pressure vessels) is a separate field of study related to both mobile machines and metrology [1.4].

The robotics of walking machines involves the mechanical analysis, design and control of one-, two-, four-, six- and multi-legged walking machines, together with combined wheeled-and-legged machines. Various features of the body structure and motion of insects and mammals are utilized in the construction of these machines. Limb control mechanisms, as well as the manner in which these creatures walk, run, or jump over various terrains, are applied in machine design.

Considering the many types of legged locomotion found in the nature, selection of the optimum machine gait is one of the main research tasks, others are the choice of machine model, leg structure, number of legs, their configuration, the sensor and control systems for the legs and for the machine as a whole. The issues of terrain profile, situation recognition, machine stability, etc. are additional problems rendering the analysis and design process even more complex. The difficulty of designing a walking, running, or jumping machine is greater when the number of legs is lower: one- and two-legged machines are a suitable example. Two-legged machines featuring a high degree of anthropomorphism (anthropoid machines) constitute a special class, where the most advanced design is a robot-musician [1.1]. Suitably designed multi-stage control systems with integrated machine intelligence systems seem to be a requisite for the further development of these machines, allowing them to switch from slow walking (also called crawling) to fast motion and running. An interesting trend in the development of mobile machines is a snake-like machine, where the technical solution is a combined wheeled-and-legged design. The most advanced constructions of mobile machines are the six-legged walking machine built at The Ohio State University, and the running and jumping machines of the Carnegie-Mellon University [1.6].

Research and applications in the field of medical and rehabilitation robotics have concentrated on the application of manipulators in surgery, therapy, replacement and assistance in performing functions of the upper and lower limbs, and patient care. An interesting surgical application is the use of the PUMA robot to remove tissues from inside of the patient's scull. Another class of these machines is the rehabilitation manipulator, usually located on a wheelchair and controlled by voice or movements of the head, chin, tongue, or eyeball. There are also manipulator-equipped workstands for tetraplegics, to help them perform many functions (eating, reading, etc.).

Industrial robotics deals with the issues of industrial robots and the application of manipulators for the robotization of basic manufacturing processes, such as casting, welding, painting, coating, assembly, disassembly, press operation, and many other tasks requiring significant physical effort, or hazardous for man.

Recent years have seen the development of non-industrial robotics: manipulators installed on submarines for the investigation and future utilization of underwater resources, and on space shuttles for the maintenance of telecommunication satellites, or for taking ground samples on other planets; and a class of robots adapted for fire-fighting, or for removal of the effects of environmental disasters where the presence of men is undesirable. Various services, such as window cleaning, serving customers, hospital transport are seen as interesting applications for new generations of robots with specialized manipulation, or locomotion functions, and increased machine intelligence.

The degree of robotization of other non-industrial applications in civil engineering, agriculture, forestry, transportation and administration is also growing.

Agricultural applications include grain drying, fertilization, irrigation, weed and pest control, cultivation, tobacco, fruit and vegetable harvesting, branch cutting, fruit sorting and packing, and in supervisory tasks in sheep and poultry farming and during harvesting. In civil engineering robots will be employed increasingly in earth works, and construction works. Other fields of expanding robotized and computer aided systems are office work and housekeeping [1.9]. Microrobotics is also a promising field [1.11].

1.2. Industrial robotics

Studies in the field of industrial robotics have concentrated on mechanical structures, drive systems, locomotion, end effectors, sensor systems, controllers, programming systems, and machine intelligence.

The investigation of mechanical structures involves new structures and the issues of increasing positioning and orientation accuracy. The problem of reaching high accuracy in comparison with other heavy machines arises from the static and dynamic compliance of robot and manipulator components and systems. The most accurate constructions operate in the Cartesian coordinate system, but this does not ensure sufficient velocity, calling for study on new configurations and drive system arrangements. An interesting solution recently presented is the Delta robot with in-parallel geometry. Direct-drive solutions (SCARA-type robots), which are expected to enhance the positioning accuracy of industrial robots, are also finding wider application.

The use of composite materials in the construction of robots permits higher velocities, but causes dynamic compliance, which must be taken into account in the control process.

New solutions of pneumatic and hydraulic drives have cut construction expenses. Robots traveling on rails are used on various welding and assembly stands. Robots of the Automatic Guided Vehicle type (AGV) are used to transport materials.

Locomotion on legs is still in the experimental stage in many countries. State-of-the-art designs of grippers are still rudimentary compared with the abilities of the human hand. The application of electrostatic phenomena in gripper construction is a recent innovation, together with the adaptation of the gripper to object shape. It appears that grippers will continue to be driven by pneumatic, hydraulic and electric actuators.

Research and development in multi-task and multi-fingered grippers is carried out in many university laboratories. An important issue is the miniaturization of drive systems.

To date only three of the five human senses (vision, touch, taste, hearing and smell), namely touch, hearing and vision, have found applications in robotics. Voice control is a subject of intensive study all over the world, as are vision and distance sensors.

Studies in the field of vision have focused on object identification and position detection, inspection, quality control, navigation, and image analysis. Three-dimensional systems are now available on the market. Distance measurement by means of ultrasonic, infrared and laser sensors is another subject of research. Obstacle avoidance may be an interesting application for these techniques.

The main issue concerning tactile sensors is the evaluation of the forces between the gripper and its surrounding. A typical application is in assembly operations. Various types of tactile sensors are used to measure the forces:

- in the joints, or actuators (mainly in master-slave manipulators),
- between the robot arm and the gripper (measurement of force and torque components in the radiocarpal joint),
- in the contact area between the fingers and the grasped object (sensors of the artificial skin type), and
- in the robot workspace.

At the current level of development control systems are usually position controllers with limited capabilities of cooperation with external sensors. Efforts are being made to increase the capabilities of robot controllers by using multiprocessor systems that can provide smooth cooperation with the environment. Generally accepted interface standards are still lacking in this field.

One must also note that the current level of development in programming systems is not satisfactory. Robot manufacturers are offering off-line programming languages, but these are not generally accepted by users. Graphical programming interfaces and simulation are being extensively developed. Off-line programming languages with CAD-type data base access are especially important.

A significant area of study in machine intelligence (also called Artificial Intelligence - AI) is devoted to the support provided to the user in the processing of information for the task to be executed. This support, known as Knowledge Engineering, has much to offer to robotics. An expert system would be very useful when designing flexible manufacturing systems or flexible assembly systems. The expert system that could, for example, help select the optimum sequence of operations in the assembly of a given product. Knowledge presentation is also important in the development of programming systems and in the interpretation of information received from the sensors (image understanding).

In the field of integrated systems, the development of high-level languages supported by real-time multi-tasking systems seems to be an important trend. Mechatronic systems developed during the last decade and virtual reality methods are playing an increasingly important role in various fields of robotics.

Much attention is devoted to the issues of safety, economics, social aspects, robot applications, and the training of specialists [1.9, 1.10, 1.12].

As robotics is developing quickly in all these directions, recent symposia have focused on the separate issues of machine vision, robotized assembly, manipulator kinematics, flexible manufacturing systems, computer integrated manufacturing, and microrobotics. [1.11]

1.3. References

[1.1] Basis of Robotics: Theory and elements of manipulators and robots (in Polish). Ed. by A. Morecki and J. Knapczyk. Second edition, WNT Warszawa 1994.

[1.2] *Morecki A., Buć J.*: Practical experience with robot applications in Poland in 1985-1987; 18-th Int. Symp. on Industrial Robots. Proceedings. Ed. by C.W. Burckhardt, Lausanne 1988.

[1.3] Catalogue of robotized stands in domestic industry (in Polish). WPM WEMA, Warszawa 1986 and 1987 (in Polish).

[1.4] *Morecki A.*: Research problems in up-to-date robotics (in Polish). Proc. of 3-rd Domestic Conf. on Textile Machines and Cranes, Bielsko-Biała 1988.

[1.5] *Morecki A., Kaczanowski S.*: Central Programme of Applied Investigations - Industrial Robots. Proc. of 22-nd Int. Symp. on Industrial Robots, Detroit, Michigan, USA, 1991.

[1.6] *Morecki A.*: Basis of Robotics: Theory and elements of manipulators and robots (in Polish). 4-th Domestic Conference on Robotics, Wrocław 1993.

[1.7] Scientific projects and research grants in robotics. PIAP, Warszawa 1994.

[1.8] RO.MAN.SY. 6, Proc. of 11-th CISM-IFToMM Symp. on Theory and Practice of Robots and Manipulators. Ed. by A. Morecki, G. Bianchi and K. Jaworek. Springer-Verlag, Wien-New York 1995.

[1.9] Proc. of 5-th Domestic Conference on Robotics (in Polish), vol. 1 and 2. Technical Univ. of Wrocław, Wrocław 1996.

[1.10] *Morecki A.*: Modern Robotics: Investigations, applications, education (in Polish). Proc. of 15-th Domestic Conference on TMM, Białystok 1996.

[1.11] *Morecki A.*: Microrobotics and its applications (in Polish). Proc. of 5-th Domestic Conference on Robotics, Wrocław 1996.

2. BASIC CONCEPTS, DEFINITIONS AND SYSTEMATIZATION OF MANIPULATORS AND ROBOTS

2.1. Definitions

The majority of machines manufactured since the 1950 can be classified as working machines: engines, processors and transporters. But the classical definition of the machine, as proposed in 1875 by F. Reuleaux, has had to be changed as a new group of machines called cybernetic machines appeared on the scene. These machines include systems modeling biological and physiological processes operating in living nature, including men and animals.

In 1963 I. Artobolevski* proposed a new definition: an artificial device designed to replace energetic, physiological and intellectual functions of man. By energetic functions we mean the replacement of physical work; by physiological functions, the replacement of organs, such as the upper or lower limb; and by intellectual functions, the adaptation to and cooperation with the environment.

A cybernetic mechanism can be defined as a part of the cybernetic machinery replacing human movements, such as manipulation or locomotion [2.1].

A manipulator (*manus* - hand) can be defined as a cybernetic mechanism used to perform some of the human upper limb functions: manipulation, performed by a manipulator gripper; and reaching movements performed by an arm.

Today manipulators are usually composed of a single or double open kinematic chain having 5 to 9 degrees of freedom, a drive system (arrangement of motors and actuators), a control system, sensors, and a power supply system. Figure 2.1 shows the flowchart of the manipulator.

The anthropomorphic manipulator (*anthropos* - human, *morphe* - shape) resembles the human limb in shape (anatomy) and operation (physiology).

A robot is a technical device designed to perform some of the manipulation and locomotion functions of man, and featuring a certain level of power management, information processing, and machine intelligence (some autonomy of operation in a specific environment). Figure 2.2 shows a flowchart of the robot.

A special case of the robot is the **industrial robot**.

The industrial robot is a multi-function manipulator designed to manage materials, parts, or specialized tools in a programmable manner in order to perform various tasks.

* of the Russian Academy of Sciences

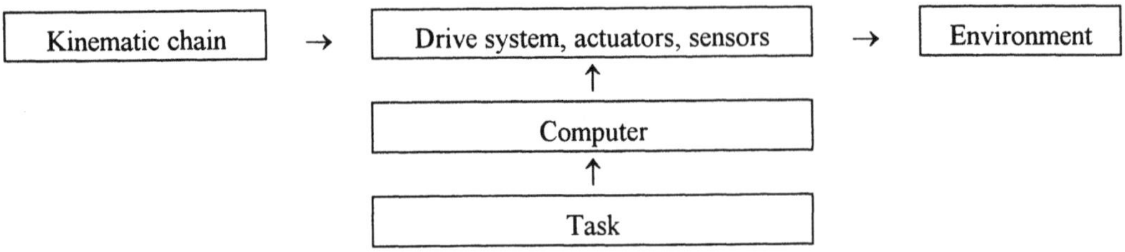

Fig. 2.1. Flowchart of the manipulator

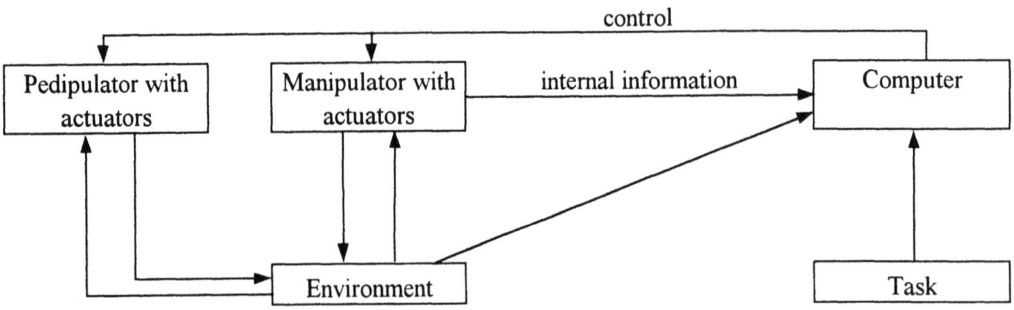

Fig. 2.2. Flowchart of the robot

Walking machines constitute a special class of robots.

A walking machine can be defined as a technical device designed to perform selected functions resembling locomotion on limbs in vertebrates and legs in insects. The locomotion of the walking machine is discrete and can be performed by using one, two, three, four, six, eight, or more "legs" to walk, run, or jump over a hard surface.

The pedipulator, or the "leg" of the walking machine may be a one-, two-, or three-link system. Fig. 2.3 shows a six-legged walking machine.

A robot, as can be seen in Fig. 2.2, consists of four main subassemblies:

1. **The mechanical system** containing suitable actuators and designed to perform the various functions of manipulation and locomotion. As has been said, the manipulator itself is usually a single, or double open kinematic chain having several degrees of freedom. The links of this chain are usually connected by rotational, or prismatic joints. If all the joints are rotational, the system is of the **anthropomorphic type**.

A 6-degree-of-freedom kinematic chain is needed to move objects in three-dimensional space. The first three degrees of freedom direct the manipulator gripper toward the desired position, while the remaining three are used to provide suitable orientation in space (Fig 2.4). The movement of the links are usually controlled by electric, or pneumatic actuators (when the actuating forces are small, not exceeding a few daN), or by hydraulic actuators (when the forces reach the level of several hundred daN).

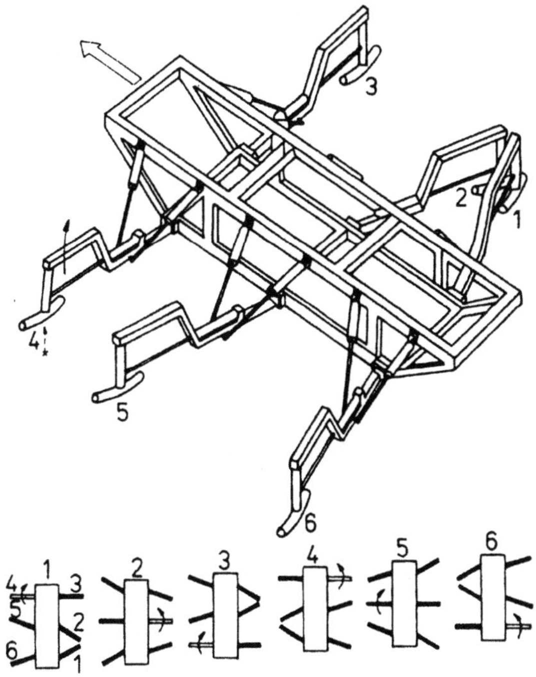

Fig. 2.3. Six-legged walking machine made at The Carnegie-Melon University (USA).

The locomotive part of the robot may present wheels, caterpillars, or legs, or a combination of wheels and legs. In the case of leg locomotion, the system normally consists of four, or six legs, each usually with 3 degrees of freedom.

2. **The working environment** surrounding the robot. If the robot is on a fixed base, the environment is limited to the workspace in which the gripper moves. But the environment is not defined purely by the robot's geometry; it also includes the physical properties of the environment and its contents, e.g. obstacles. Consequently, the robot must be able to take these elements into account.

3. **The task,** defined as the difference between the initial state of the environment and the final state, after completion of the programmed task. The task is usually described in a language appropriate for computer-control.

4. **The control system,** a computer responsible for generation of the control signals sent to the actuators moving the manipulator, according to *a priori* information (knowledge of the task to be executed) and *a posteriori* knowledge about current and former states of the robot and environment. The system also contains two significant closed feedback loops, providing internal sensory information and external sensory information. The use of this information in the control algorithm allows the robot to perform the programmed tasks.

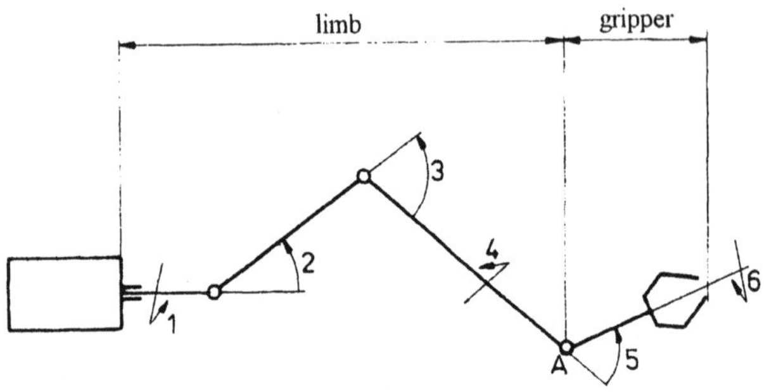

Fig. 2.4. 6-DOF kinematic chain of a robot

2.2 Kinematic chains of manipulators and robots

2.2.1. Degrees of freedom (DOF) and types of joints

As we have seen, the manipulator is usually an open kinematic chain containing links[1] connected by kinematic pairs of specific classes. In three-dimensional space, a rigid link has six degrees of freedom (Figure 2.5). Three are usually used to position the manipulator's end effector in space, while the other three provide the desired orientation of the end-effector with respect to the object of interest. An example of a 12-DOF manipulator installed on a spaceship is shown in Figure 2.6. In this arrangement the spaceship has six DOF which are the possible translatory motions P_1, P_2 and P_3, and the possible angular motions θ_1, θ_2 and θ_3. The manipulator arm has three DOF - θ'_1, θ'_2 and θ'_3, as does the gripper - θ'_4, θ'_5 and θ'_6.

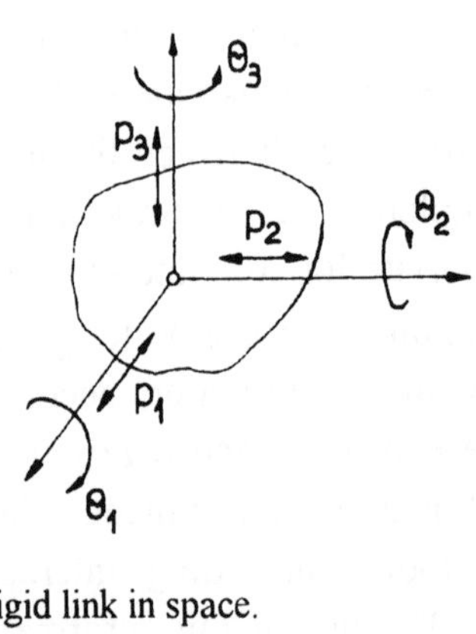

Fig. 2.5. Degrees of freedom of a rigid link in space.

[1] These links, which may be rigid or flexible, shall be considered rigid unless otherwise stated.

Table 2.1 shows possible types of joints, the relative motion provided by these joints, the DOF and a symbolic representation of joints. As can be seen, there are nine possible joints (nos. *2* to *10*) with 1 to 5 DOF for rotation, translation, and combinations of these two. The most common joints are types *2* and *3*.

Table 2.1. Joints used in kinematic chains of manipulators

No.	Type of joint	Relative motion	Number of DOF	Symbolic representation
1	Fixed connection	0 rotations 0 translations	0	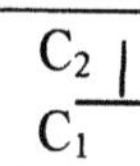 C_1 = link 1 C_2 = link 2
2	Revolute joint	1 rotation 0 translations	1	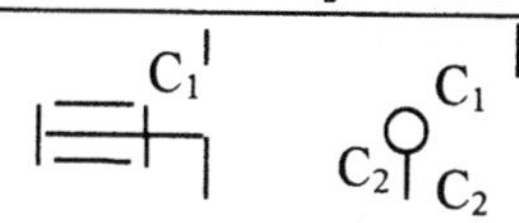
3	Prismatic joint	0 rotations 1 translation	1	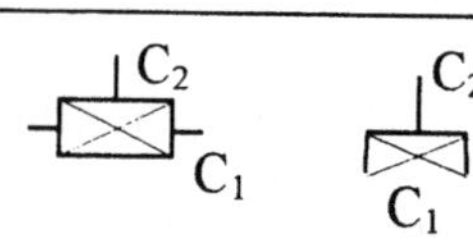
4	Screw joint	1 rotations 1 translation linked together	1	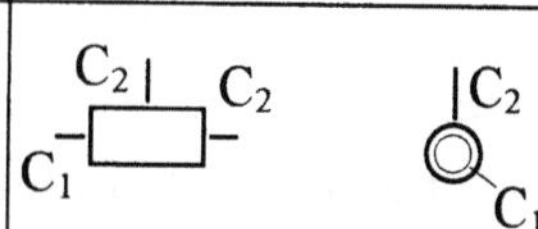
5	Cylindrical joint	1 rotation 2 translations	2	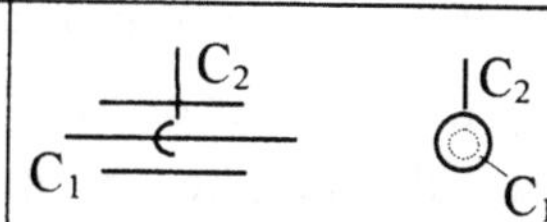
6	Joint providing rotation and two translations	3 rotations 0 translations	3	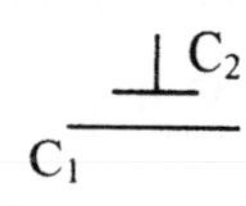
7	Spherical joint	2 rotations 2 translations	3	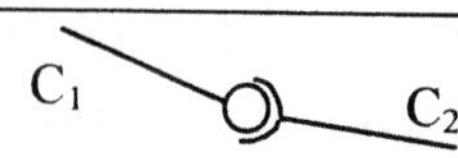
8	Linear contact joint	3 rotations 1 translation	4	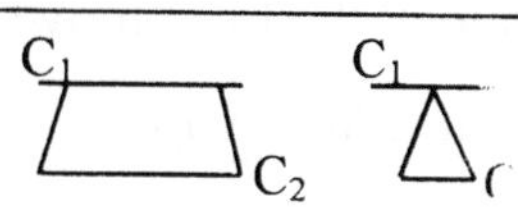
9		3 rotations 2 translations	4	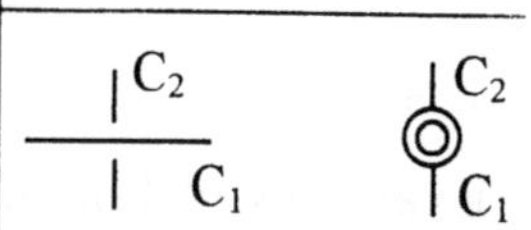
10	Point contact joint	3 rotations 3 translations	5	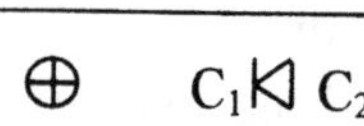
11	Free joint	3 rotations 3 translations	6	Unconstrained bodies

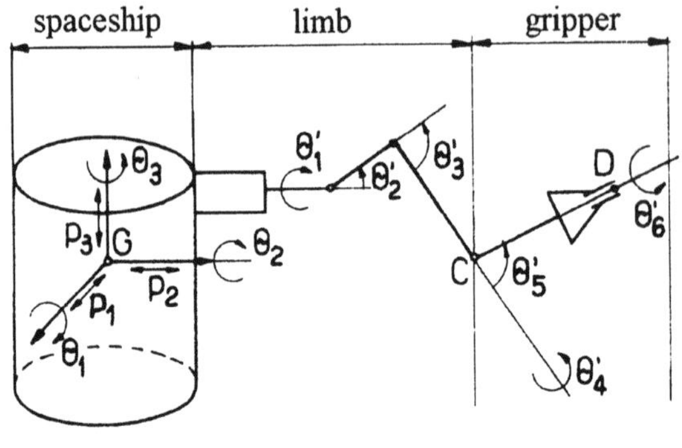

Fig. 2.6. Example of a 12-DOF manipulator

2.2.2. Mobility of open kinematic chains of manipulators

In the case of open kinematic chains the general mobility formula is used:

$$w = 6n - \sum_{i=1}^{n} ip_i \tag{2.1}$$

where:

w is the number of independent motions of the movable links relative to the base,

n, the number of movable links, and

p_i, the number of *i-th* class joints (see Table 2.1, items 2 to 10).

Since the number of movable links here equals the number of kinematic pairs, relationship (2.1) can also be written in the form (2.2)

$$w = 6\left(\sum_{i=1}^{n} p_i\right) - \sum_{i=1}^{n} ip_i = p_5 + 2p_4 + 3p_3 + 4p_2 + 5p_1 \tag{2.2}$$

meaning that the mobility of an open kinematic chain equals the sum of the number of degrees of freedom provided by its joints (kinematic pairs). In the special case when all the links move in one plane, relationships (2.1) and (2.2) can be written as

$$w = 3n - \sum_{i=1}^{n} ip_i \tag{2.3}$$

and

$$w = p_5 + 2p_4 \tag{2.4}$$

Examples of kinematic chains of planar and spatial manipulators are shown in Figure 2.7: a) is a wrist with intersecting joint axes (at C), while b) is a wrist with three axes of rotation coincident with a straight line through the gripper's center point; c) illustrates planar chains of the *PP* type (c_1), *PR* type (c_2), *RP* type (c_3), and *RR* type (c_4); d) shows a mechanism with two closed chains (loops); e), a paralellogram linkage; and f), a pantograph linkage. In g) and h) we have a mechanism that provides the displacement of the end link, while maintaining its orientation constant.

Figure 2.8 gives some examples of the calculation of manipulator mobility: a) shows a manipulator operating in a Cartesian coordinate system. The positioning chain can perform three motions (*1, 2* and *3*) and the gripper, two (*4* and *5*). In this case joints *1* and *2* of the chain are prismatic, *3* and *4* are cylindrical, and *5* is a revolute joint. Applying formula (2.2) we obtain $w = 3 + 2 = 5$

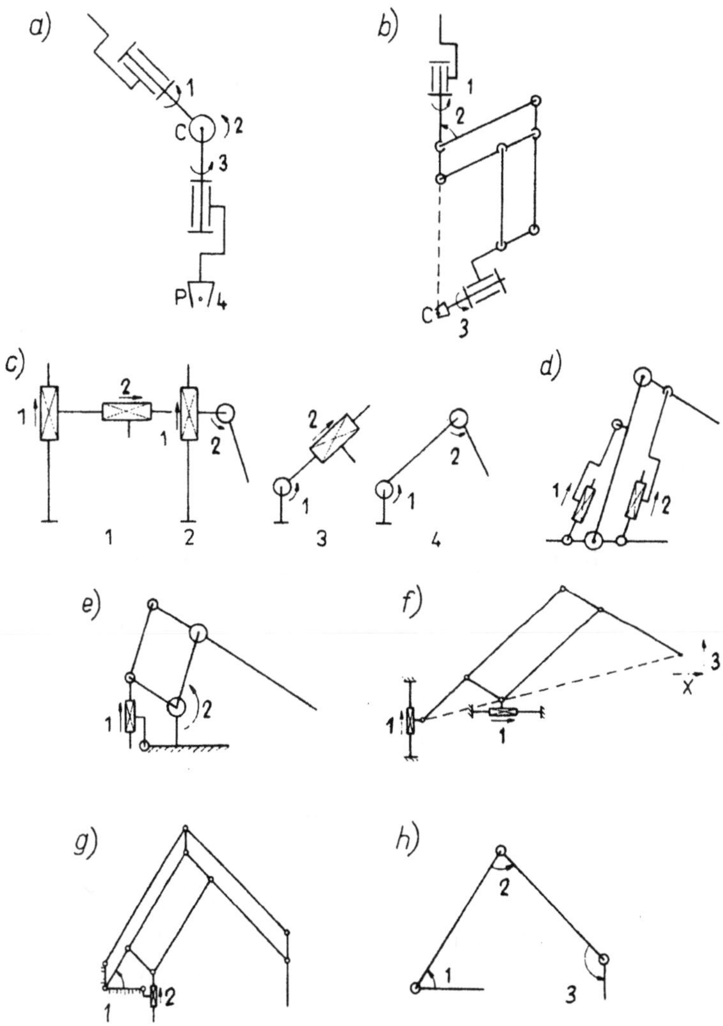

Fig. 2.7. Various mechanisms used in manipulators: a) wrist with intersecting axes, b) three axes of rotation coincident with a straight line, c) planar chains, d) mechanism with two closed kinematic chains (loops), e) paralellogram linkage, f) pantograph linkage, g), h) constant orientation mechanisms

The drive system of this type of manipulator requires five actuators.

Figure 2.8b shows a kinematic chain used in Versatran manipulators. The arrows indicate the possibilities of relative motion. This kinematic chain features four revolute joints (*1, 4, 5* and *6*) and two prismatic joints (*2* and *3*). Applying relationship (2.2) we obtain $w = 6$. Figure 2.8c shows a chain with 8 DOF , and Figure 2.8d, a chain with seven revolute joints ($w = 7$) [2.2].

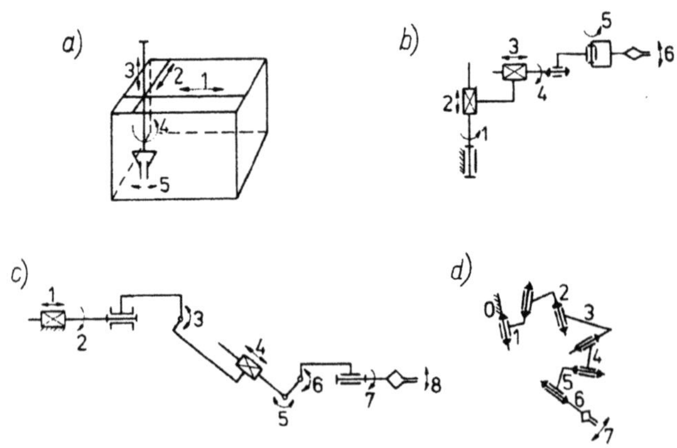

Fig. 2.8. Determining the mobility of kinematic chains used in manipulators: a) manipulator operating in the Cartesian coordinate system, b) Versatran manipulator, c) 8-DOF manipulator, d) manipulator with revolute joints only.

2.2.3. Variations of kinematic chains used in manipulators

Let us consider modifications of kinematic chains containing only revolute (R) and prismatic (*P*) joints, permitting only rotational or translatory motion. The theory of combinations tells us that the number of possible variations of kinematic chains consisting of two links can be found using the formula

$$L_k = 2^k \tag{2.5}$$

where k is the number of joints, or the mobility of the kinematic chain.

When $k = 1$, there are two possible variations, one with a revolute and the other with a prismatic joint. When $k = 2$, four variations can be obtained involving revolute (R) and prismatic (*P*) joints (Fig. 2.9a): *PR, RR, RP* and *PP*.

By changing the orientation of the axes of kinematic pairs (the orientation can be changed by $\pi/2$), additional variations can be obtained (Fig. 2.9b). When k =3, there are eight possible variations (Fig. 2.10a), from *PPP* to *OOO*. Figure 2.10b gives examples of 18 further variations of the basic *OOO* kinematic chain, obtained by changing the location of the axes.

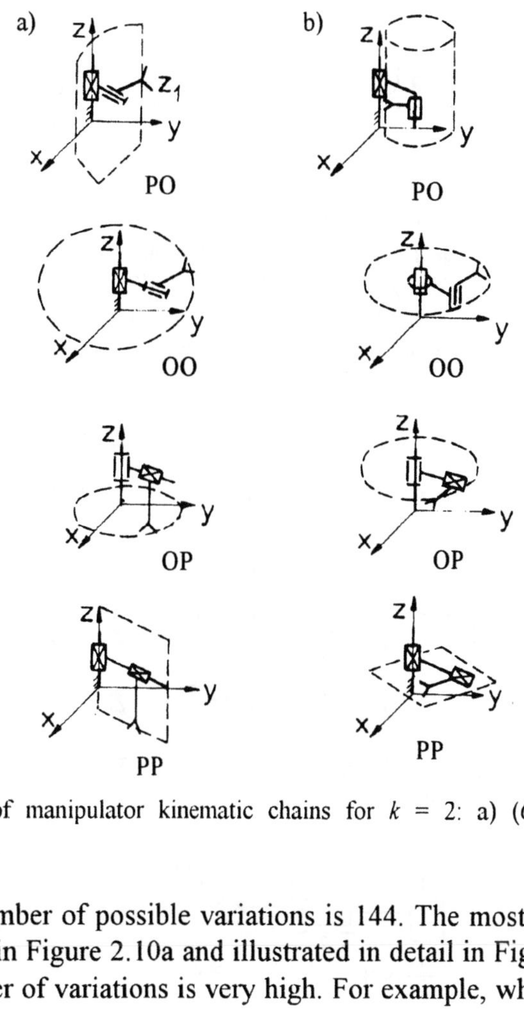

Fig. 2.9. Variations of manipulator kinematic chains for k = 2: a) (O) and (P) variations, b) additional variations.

The total number of possible variations is 144. The most common variations are indicated by *) in Figure 2.10a and illustrated in detail in Figure 2.11. When k = 4, 5, 6..., the number of variations is very high. For example, when k = 4, thirty-two variations can be found, and this number can be expanded to more than one thousand by changing the location of the axes. Linking the variations shown in Figure 2.11 with the number of their applications, these variations may be put in the order shown in Table 2.2.

Table 2.2. Application of different variations (in percents)

Variation	*POP*	*OOO*	*PPP*	*OOP*	*POO*
%	47	25	14	13	1

The variations shown in Table 2.2 have been analyzed for 195 different industrial robots. The *POO* variation is undergoing rapid development (the SCARA robot is based on this variation).

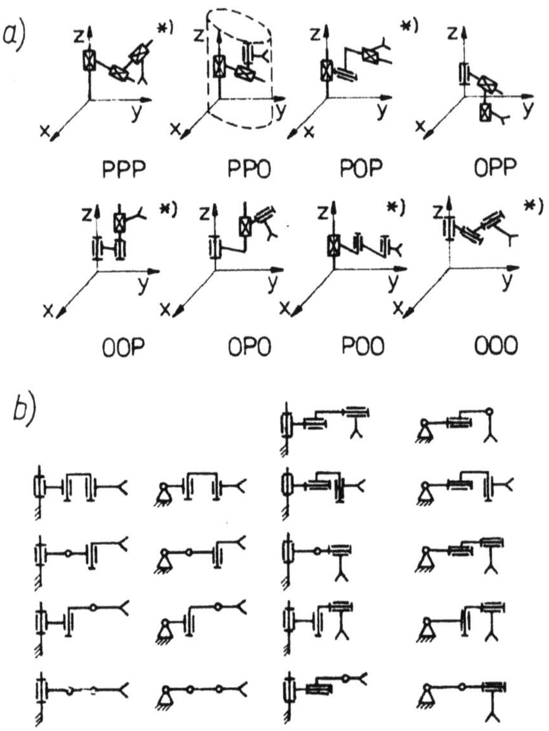

Fig. 2.10. Variations of manipulator kinematic chains for k = 3: a) variations from *PPP* to *OOO*, b) 18 further variations of the *OOO* kinematic chain.

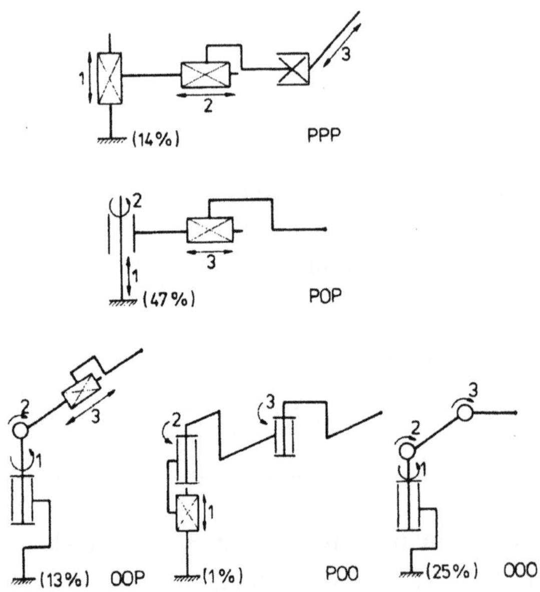

Fig. 2.11. The most common variations of kinematic chains used in manipulators.

2.2.4. The matrix approach to various kinematic chains of manipulators

2.2.4.1. Gripper degrees of freedom, configurations and mobility of manipulators

This section considers some issues of the vector and matrix approach to kinematic chains.

The position and orientation of the gripper relative to the manipulator base are determined by the following vector of coordinates:

$$\boldsymbol{w} = [p_1\ p_2\ p_3\ \theta_1\ \theta_2\ \theta_3]^T \tag{2.6}$$

where p_1, p_2, and p_3 are Cartesian coordinates of the origin of the frame attached to the gripper, and θ_1, θ_2, and θ_3 are angular coordinates determining the orientation of the gripper with respect to the manipulator base.

The configuration of the manipulator is determined by the vector of the generalized kinematic pair coordinates (joint coordinates):

$$\boldsymbol{q} = [q_1\ q_2\ \ldots\ldots\ q_n]^T \tag{2.7}$$

where n is the mobility of manipulator, and "T" indicates the transposed matrix.

The relationship between vectors $\boldsymbol{w}$ and $\boldsymbol{q}$:

$$\boldsymbol{w} = F(\boldsymbol{q}) \tag{2.8}$$

can be differentiated with respect to time, obtaining the relation between gripper velocity and the velocities of relative motions in the kinematic pairs

$$\dot{\boldsymbol{w}} = \boldsymbol{J}(\boldsymbol{q})\dot{\boldsymbol{q}} \tag{2.9}$$

where $\boldsymbol{J}(\boldsymbol{q})$ is the Jacobian matrix of function $F(\boldsymbol{q})$.

Replacing the derivatives with the differential increments of the coordinates, we obtain

$$\begin{bmatrix} dp_1 \\ dp_2 \\ dp_3 \\ d\theta_1 \\ d\theta_2 \\ d\theta_3 \end{bmatrix} = \begin{bmatrix} \frac{\partial p_1}{\partial q_1} & \frac{\partial p_1}{\partial q_2} & \frac{\partial p_1}{\partial q_3} & \ldots\ldots & \frac{\partial p_1}{\partial q_n} \\ \frac{\partial p_2}{\partial q_1} & \frac{\partial p_2}{\partial q_2} & \frac{\partial p_2}{\partial q_3} & \ldots\ldots & \frac{\partial p_2}{\partial q_n} \\ \frac{\partial p_3}{\partial q_1} & \frac{\partial p_3}{\partial q_2} & \frac{\partial p_3}{\partial q_3} & \ldots\ldots & \frac{\partial p_3}{\partial q_n} \\ \frac{\partial \theta_1}{\partial q_1} & \frac{\partial \theta_1}{\partial q_2} & \frac{\partial \theta_1}{\partial q_3} & \ldots\ldots & \frac{\partial \theta_1}{\partial q_n} \\ \frac{\partial \theta_2}{\partial q_1} & \frac{\partial \theta_2}{\partial q_2} & \frac{\partial \theta_2}{\partial q_3} & \ldots\ldots & \frac{\partial \theta_2}{\partial q_n} \\ \frac{\partial \theta_3}{\partial q_1} & \frac{\partial \theta_3}{\partial q_2} & \frac{\partial \theta_3}{\partial q_3} & \ldots\ldots & \frac{\partial \theta_3}{\partial q_n} \end{bmatrix} \begin{bmatrix} dq_1 \\ dq_2 \\ \cdot \\ \cdot \\ \cdot \\ dq_n \end{bmatrix} \tag{2.10}$$

The rank of the Jacobian matrix whose elements are the partial derivatives

$$\frac{\partial p_i}{\partial q_j}, \frac{\partial \theta_i}{\partial q_j} \quad (i = 1, 2, 3; \;\; j = 1, 2, ..., \mathrm{n})$$

is determined by the greatest number of linearly independent rows and columns, equal to the number of degrees of freedom of the gripper.

The kinematic chain of the manipulator can be considered a combination of the **regional positioning chain** and the **local orienting chain**. Regional positioning chains in industrial robots usually have three revolute (R) or prismatic (*P*) kinematic pairs, which allow eight combinations: *PPP, PPO, POP, OPP, POO, OPO, OOP* and *OOO* (Fig. 2.10a). The local orienting chain is usually composed of three revolute joints, the axes of which intersect at one point.

The manipulator workspace is a set of geometrical loci of the orienting wrist center, i.e. the point of intersection of the revolute joints of the orientating chain.

Jacobian surfaces are the geometrical loci of those points in the workspace at which the Jacobian matrix is singular. The center of the orienting wrist lying in the Jacobian surface must move in a plane tangent to this surface. Jacobian surfaces may be situated at the boundary of the workspace, or inside it.

If q_1, q_2 and q_3 are generalized coordinates of the first three kinematic pairs of the manipulator, and $\boldsymbol{p}$ is the position vector of the wrist center, then:

$$\begin{bmatrix} dp_{jx} \\ dp_{jy} \\ dp_{jz} \end{bmatrix} = \begin{bmatrix} \frac{\partial p_{jx}}{\partial q_1} & \frac{\partial p_{jx}}{\partial q_2} & \frac{\partial p_{jx}}{\partial q_3} \\ \frac{\partial p_{jx}}{\partial q_1} & \frac{\partial p_{jy}}{\partial q_2} & \frac{\partial p_{jy}}{\partial q_3} \\ \frac{\partial p_{jx}}{\partial q_1} & \frac{\partial p_{jz}}{\partial q_2} & \frac{\partial p_{jz}}{\partial q_3} \end{bmatrix} \begin{bmatrix} dq_1 \\ dq_2 \\ dq_3 \end{bmatrix} \tag{2.11}$$

where $\boldsymbol{p}_j = p_{jx}\,\boldsymbol{i} + p_{jy}\,\boldsymbol{j} + p_{jz}\,\boldsymbol{k}$.

If the joint is revolute, $q_k = \theta_k$; if the joint is prismatic, $q_k = \lambda_k$.

The motion of the wrist center is singular only when the determinant of the Jacobian matrix equals zero. As the determinant of the Jacobian matrix is equal to the determinant of the transposed matrix, the singularity condition can be written as:

$$\begin{bmatrix} \frac{\partial p_{jx}}{\partial q_1} & \frac{\partial p_{jy}}{\partial q_1} & \frac{\partial p_{jz}}{\partial q_1} \\ \frac{\partial p_{jx}}{\partial q_2} & \frac{\partial p_{jy}}{\partial q_2} & \frac{\partial p_{jz}}{\partial q_2} \\ \frac{\partial p_{jx}}{\partial q_3} & \frac{\partial p_{jy}}{\partial q_3} & \frac{\partial p_{jz}}{\partial q_3} \end{bmatrix} = 0 \tag{2.12}$$

Equation (2.12) can also be written using the triple scalar product of the following vectors:

$$\frac{\partial p_j}{\partial q_1} \cdot \left(\frac{\partial p_j}{\partial q_2} \times \frac{\partial p_j}{\partial q_3} \right) = 0 \tag{2.13}$$

where $\frac{\partial p_j}{\partial q_i}$ is the displacement of the end of vector p_j caused by the infinitesimal displacement in the *i-th* kinematic pair;

$$\frac{\partial p_j}{\partial q_i} = e_{ij} \quad \text{in case of prismatic joints; and} \tag{2.14}$$

$$\frac{\partial p_j}{\partial q_i} = e_{ij} \times (p_j - r_{ij}) \quad \text{in case of revolute joints} \tag{2.15}$$

where:

e_{ij} is the unit vector of the *i-th* kinematic pair axis in *j-th* position,

r_{ij} is the position vector of the *i-th* kinematic pair axis in *j-th* position.

For the *OOO*-type regional positioning chain, substituting (2.15) for (2.13), we have

$$\left[e_{1j} \times (p_j - r_{1j})\right] \cdot \left\{\left[e_{2j} \times (p_j - r_{2j})\right] \times \left[e_{3j} \times (p_j - r_{3j})\right]\right\} = 0 \tag{2.16}$$

Equation (2.16) shows that the *OOO*-type kinematic chain is in singular position when unit vector e_2 is parallel to vector $(p_j - r_{2j})$, i.e. when the p_j end lies on the e_2 axis.

Equation (2.13) indicates that three vectors

$$\frac{\partial p_j}{\partial q_i} \quad (i = 1, 2, 3)$$

lie in one plane, and the three axes of vectors e_i intersect the same straight line, which is the line of intersection of three planes containing three pairs of vectors e_i and $(p_j - r_{ij})$ each.

If the number of revolute and prismatic joints in a manipulator is n, there are 2^n different kinematic chains.

To simplify the construction of the manipulator, an arrangement of parallel or perpendicular axes of successive joints is often used. When $n > 3$, the manipulator may to some extent alter the position and orientation of the gripper. Positioning and orienting are often decoupled to simplify control. If the orientation is a function of n parameters ($1 \le n' \le 3$), then n kinematic pairs should be revolute joints with axes intersecting at a point situated at the gripper. When $n = 4$ ($n' = 1, 2, 3$), the manipulator should have at least one revolute joint at the gripper. When $n = 5$ ($n' = 2, 3$), the manipulator should have at least two revolute joints with intersecting axes.

When $n = 6$ ($n' = 3$), the axes of the last three manipulator joints should intersect at one point at the gripper.

The Jacobian surface may be described analytically by assuming known values of two variables q_i, and calculating the value of the third variable from equation (2.15). If the first kinematic pair (at the base) is a revolute joint, then the Jacobian surface is a surface of revolution. Thus it can be assumed that $q_1 = 0$. The situation is similar when the first kinematic pair is a prismatic joint. Consequently, assuming the values of one of the two remaining variables q_2 and q_3, the third can be computed from (2.15). If the first joint is revolute, the Jacobian surface can be obtained by rotating the spatial curve (2.15) around the axis of this joint. If the first joint is prismatic, the Jacobian surface can be obtained by translating the curve along this axis.

2.3. Systematization of manipulators and robots

There are various classifications of manipulators and robots. The systematizations most commonly found in the references are indicated here, together with the author's.

2.3.1. Classification of manipulators

An example of classification based on a control criterion is presented in Table 2.3. Three principal classes can be distinguished: biotechnical manipulators, automatic manipulators, and interactive manipulators. In each group further divisions can be distinguished.

Anthropomorphic manipulators can be divided into classes according to the following criteria:

- similarities of manipulative-reaching operations
- similarities in terms of control and regulation of the motion of the manipulative-reaching system
- applications

Figure 2.12b presents the model of an outrigger manipulator, a partly anthropomorphic mechanism capable of certain manipulative operations. It is compared with the human upper limb in Figure 2.12a.

In the case of a more versatile arm, enabling different kinds of prehension, the similarity of the particular links is anatomic (finger bones, articulations, tendons), geometric (motion ranges in the joints), and kinematic (linear and angular velocities and accelerations). Basically, with the exception of few cases, manipulators combine a manipulative system together with a reaching system to form an artificial limb. It could more appropriately be called an extramitator, since it imitates whole limb functions.

Table 2.3. Systematization of robots

	Class	Variety
I	Biotechnical	– with manual control (non-copying) – copying with unilateral or bilateral operation – semi-automatic (a handle and a controller, or a specialized computer)
II	Automatic	– programmable – adaptive (with sensors and vision) – intelligent (with elements of artificial intelligence) – manipulative robots
III	Interactive	– automated (with a variable working cycle switching to automatic or biotechnical control according to the instructions stored in the memory) – with supervised control (automatic execution of work phases and supervision during transitional periods – conversational (higher form of interaction with the operator – mostly by means of text)

Possible varieties of each of these classes are presented in Table 2.4

Table 2.4. Systematization of anthropomorphic manipulators

	Class	Variety
I	Similar in terms of manipulative-reaching operations	– partly similar to the original in terms of manipulative operations – dissimilar in structure to the original but performing certain manipulative operations – partly similar to the original in terms of manipulative-locomotive operations – dissimilar to the original but enabling the performance of certain manipulative-reaching operations
II	Similar in terms of control and regulation of the motion of the manipulative-reaching system	– a copying manipulator of an open or closed type – a multi-task manipulative-reaching system controlled by a computer – an adaptive manipulator with incipient artificial intelligence – an independent manipulator with artificial intelligence
III	Applications	– processing – transporting – informative-operative

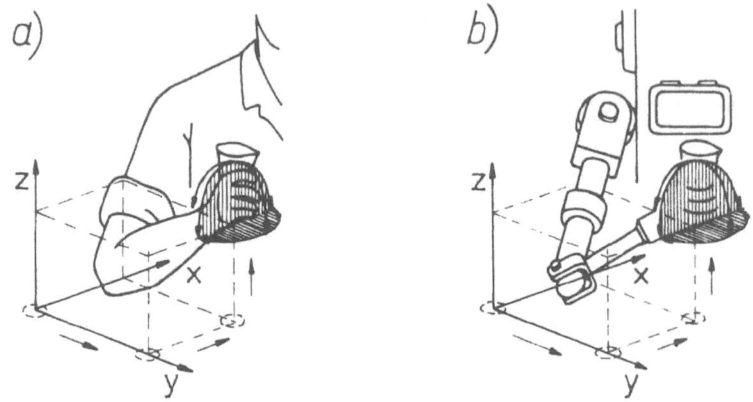

Fig. 2.12. Necessary number of motions of an upper limb manipulator: a) location of the object in space, b) mechanical counterpart of the limb

Similarities of the manipulative-reaching operations require that the proper anatomy, that is links and articulations, be maintained (Fig. 2.12a). Suitable proportions, in terms of link lengths and angles in articulations, are also necessary, involving on one hand, the workspace and on the other, the control and power supplying systems.

A separate class, which in spite of dissimilarities enables the realization of manipulative-reaching functions, allows such deviations in anatomy as the existence of articulations - prismatic joints. Figure 2.13 shows an example of a non-anthropomorphic manipulator of the PUMA type.

Figure 2.14a presents the principle of operation of a copying manipulator, while Figures 2.14b and 2.14c show its kinematics.

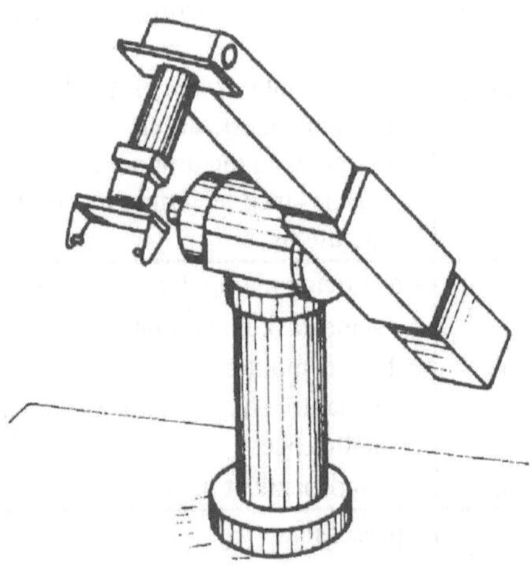

Fig. 2.13. Stanford manipulator

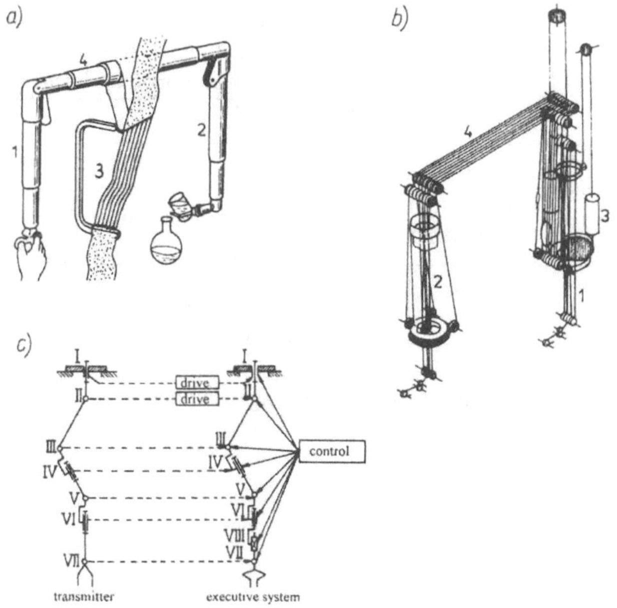

Fig. 2.14. Copying manipulator: a) principle of operation of the copying manipulator, 1 – transmitter, 2 – executive system, 3 – partition, 4 – connection with 1 and 2; b) mechanical copying manipulator developed at Nuclear Centre ANL in USA, 1 – transmitter, 2 – executive system, 3 – counterweight, 4 – connection with 1 and 2; c) manipulator developed for nuclear industry of late USSR, I÷VIII – joints (kinematic pairs)

Typical examples of informative-operative manipulators designed for operation under water, or in space, are shown in Figures 2.15 and 2.17. Their control systems can be of the copying type (Fig. 2.16), or telecopying, or computerized. Figure 2.17 shows a remote control manipulator of the RMS type (Remote Manipulator System), a part of a space vehicle. This is an anthropomorphic manipulator with six degrees of freedom controlled by an astronaut from inside the cabin.

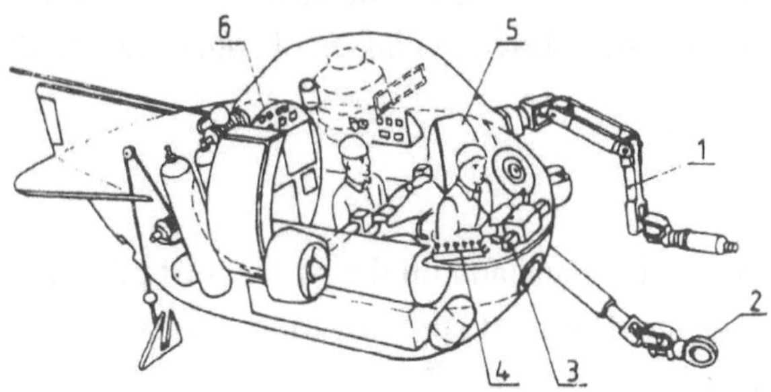

Fig. 2.15. Manipulators installed on a Beaver submarine apparatus built by The Antonetics Corporation: 1 – manipulator, 2 – gripper, 3 – operator, 4 – console, 5 – handle, 6 – controllers

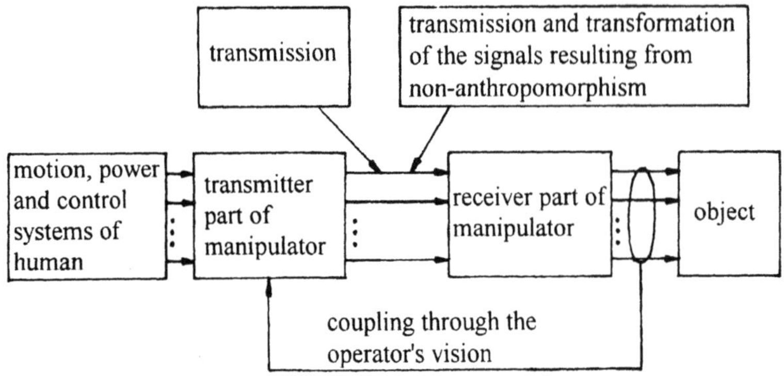

Fig. 2.16. Block diagram of control system of copying manipulator

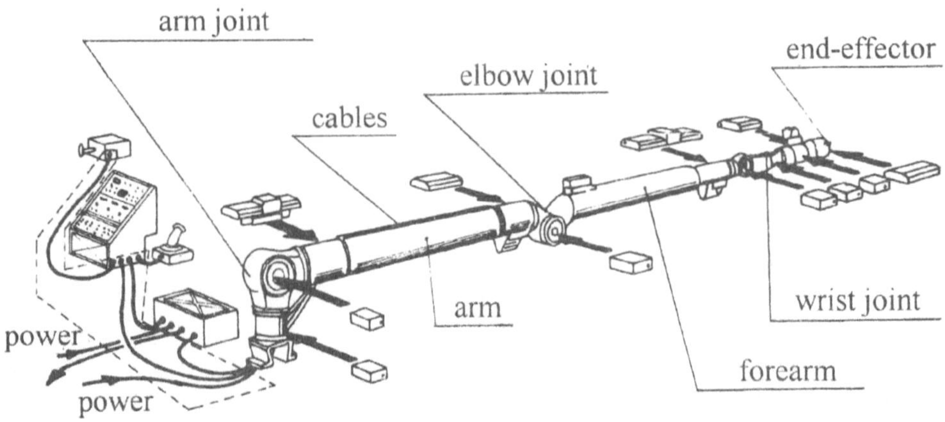

Fig. 2.17. Arm of manipulator attached to space vehicle

2.3.2. Classification of robots

Various classifications can be made on the basis of the criteria assumed. Taking application as the criterion, the following classes of robots can be distinguished:

– for research,

– for training,

– for industry,

– for research under water, in space and on the surfaces of other planets.

Applying the criterion of the degree of specialization, we can distinguish:

– ospecialized robots,

– special robots,

– all-purpose robots.

Robots can also be divided into the following classes according to the type of drive system used:
- with a pneumatic drive system,
- with a hydraulic drive system,
- with an electric drive system,
- with a mixed drive system.

According to the criterion of the geometrical properties, we can distinguish:

a) robots of varying geometric configuration dependent on link dimensions, variable joint ranges, and the linear and angular displacements which determine the motion of the arm, wrist and end-effector;
b) robots operating as coordinated systems (Table 2.5):
   - Cartesian (three mutually perpendicular axes); the end-effector moves in three different directions, for a system which although slower, provides high accuracy;
   - cylindrical (two axes and one angle); the arm moves in horizontal and vertical directions, and rotates about certain point;
   - spherical (one axis and two angles); the arm rotates about a fixed point, and can be raised and lowered from this point;
   - anthropomorphic (three angles); the arm performs bending and abducting motions, just like the human arm, as well as the bending of the elbow joint. This type of system makes it possible to reach positions otherwise unobtainable;
   - Scara type (one angle and two axes); and
   - complex systems (m axes, n angles).

In each of these systems additional motions allow the orientation of the wrist and end-effector in the workspace. The number of axes is often identified with the number of degrees of freedom.

According to the criterion of kinematic properties, robots can be classified by:
- the number of degrees of freedom,
- the number of varieties of operations related to their displacement and velocity.

According to the criterion of dynamic properties, we can distinguish robots with different performances resulting from the minimization of dynamic loading or vibration.

Another criterion can be the accuracy of position and orientation, and the ability to provide the desired precision in certain areas of a workspace. We can distinguish here robots that are superprecise (less then 1μm of error), precise (less than 1mm) and mediocre (from 1mm to 5mm).

Taking mobility as a criterion, we have:
- stationary robots, and
- mobile (ground, gantry) robots.

Table 2.5. Configurations, notation, diagrams and workspaces of robots with different axis systems

Structure of system axes	Configuration	Notation	Diagram	Workspace
Cartesian 3P	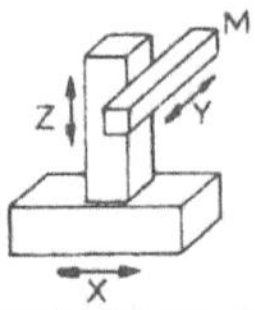	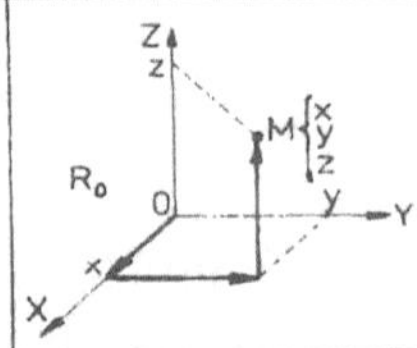	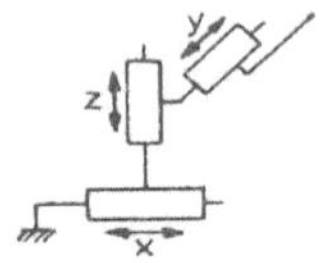	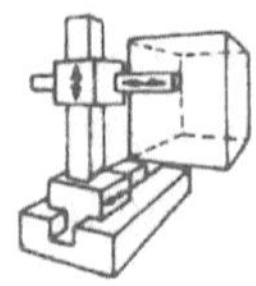
Cylindrical 1O i 2P	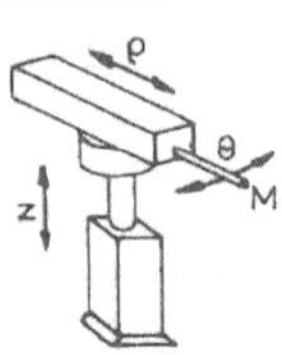	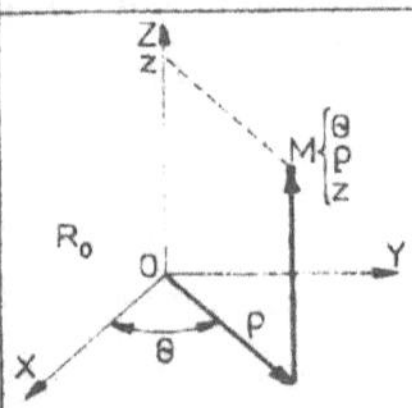	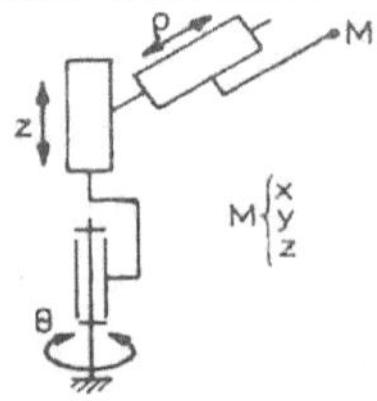	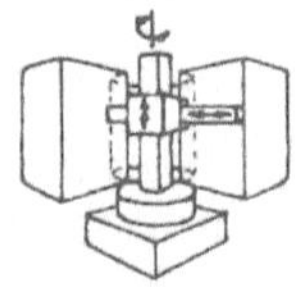
Spherical 2O i 1P	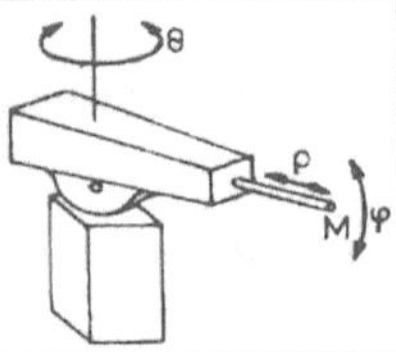	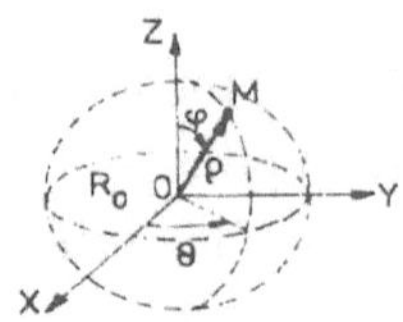	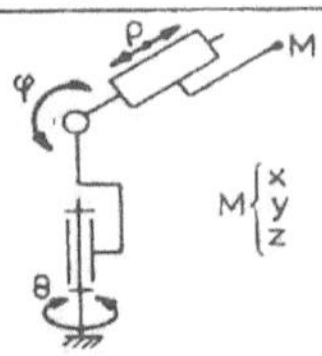	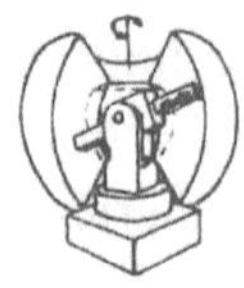
(Anthropomorphic) Polar 3O	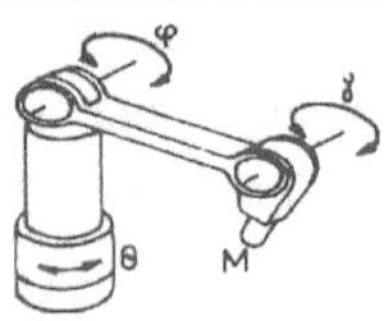	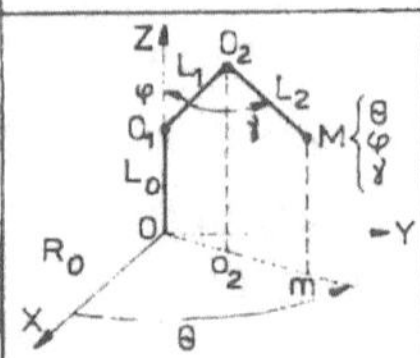	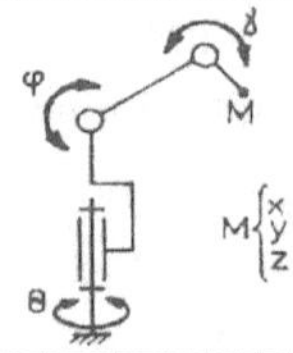	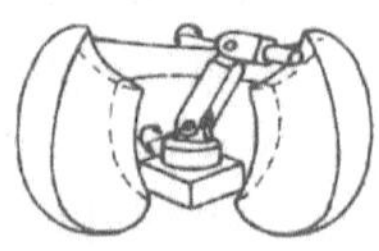
	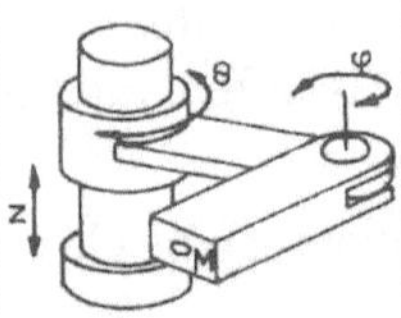	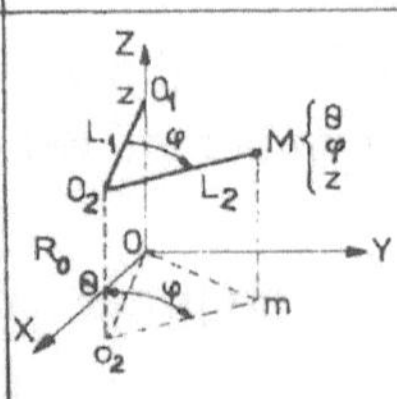	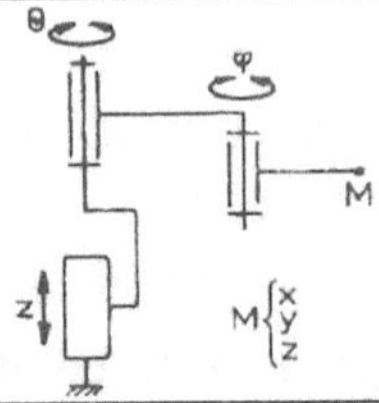	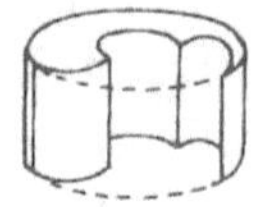
Complex mP nO	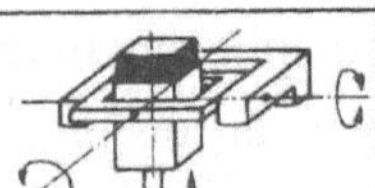 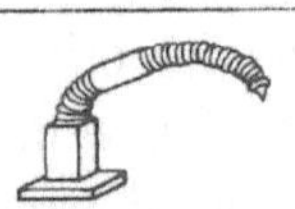			 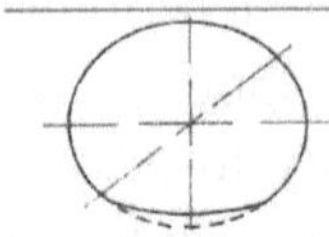

Using the criterion of control, we can distinguish four groups of robots ranging from a simple electromechanical control system fitted with motion limiters and switches, to computer control systems of the on-line type:

– robots with a permanent program (performing successive operations according to a set or changeable sequence - that is the order of the operations can be changed);
– robots with a discrete point-to-point program (PTP): playback robots (with memory to record a set of operations performed manually beforehand);
– robots with continuous control (CP – continuous path control): a robot programmed off-line or on-line and operated in the runtime. The operator teaches the robot by moving its tip along the desired trajectory;
– intelligent robots: using sensor systems (for touch, vision and sound) that enable detection of any change in working conditions and self-adjustment based on decisive-adaptive systems.

Other classifications of robots can be:

– robots for simple operations (without servo type systems)
– programmable robots (with servo type control)
– intelligent robots (with sensor systems – in vision, touch and voice) that modify their performance themselves.

A more common classification is based on the successive generations of robots:

– First generation robots – industrial robots programmed to carry out a certain sequence of operations;
– Second generation robots – robots fitted with a set of sensors that enable interaction with the environment (Ernst's hand 1958).
– Third generation robots – robots of the hand-eye type (recognizing objects).

A structural diagram of the first generation robot is given in Figure 2.18. Position 1 of both comutators represents a learning phase of the program, while position 2 is its playback. During the learning phase the robot receives instructions from a human teacher *IU*; these are subsequently recorded in the memory, after passing through a control block *PC*, a corelation system *RC*, amplifier *Am*, an actuator *At*, and, finally, a mechanical end-effector system. The instructions transferred into the memory through position converters located in the feedback channel, *TP*, coder and interpolator. During the playback phase, when the comutators are in position 2, signals pass through the decoder and are compared with the signals from the position converters (encoders) *TP*. The feedback channel forms a loop within the system, which means that it does not record external information, that is, information resulting from interaction with the environment.

These robots, more amply illustrated in chapter 14, are used to perform different manipulative operations.

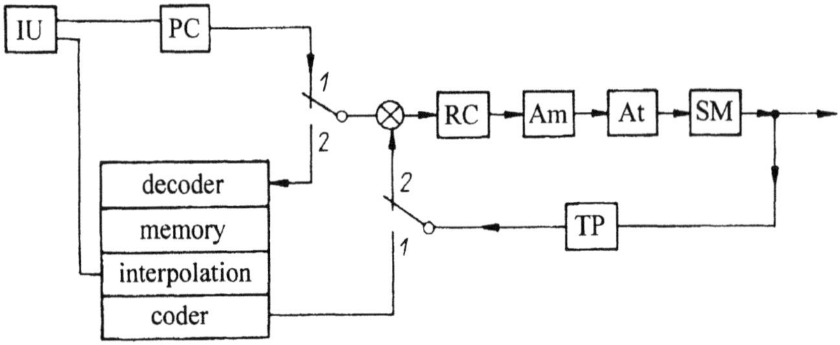

Fig. 2.18. Block (structural) diagram of a first generation robot

Second generation robots are designed to employ those tactics in their contact with environment that will produce an optimal effect. The robots are equipped to recognize a desired object, even when it is mixed in with other objects, and to do so regardless of the position and shape of the object. Those tasks are made possible by a set of sensors. A block diagram of the second generation robot is shown in Figure 2.19. The major blocks represent a work program *PL* and a computational program *PC*. The work program is transferred to manipulators *M* through amplifiers *Am* and *At*, after correction in the *RL* branch. Information about the object is transmitted to a environment investigating system *SIA* (where it is compared with the *PC*), and, at the same time, to two systems called indirect *Ob I* and final *Ob F*, where their comparison takes place and the resulting information is integrated into the program.

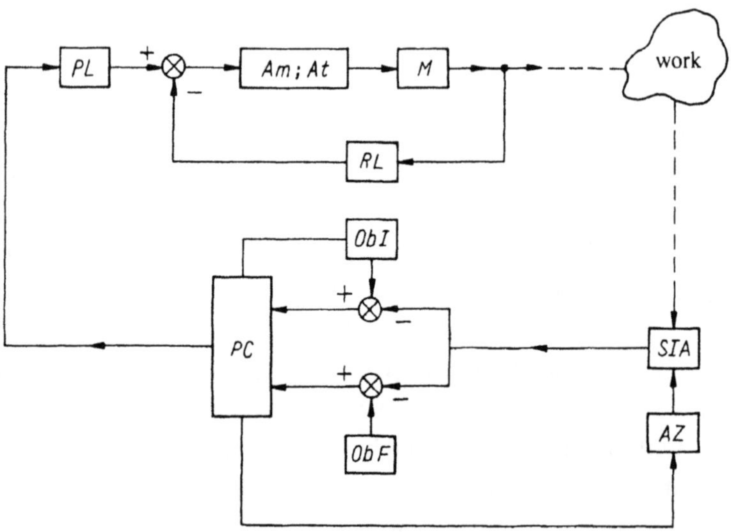

Fig. 2.19. Block (structural) diagram of a second generation robot

An example of the block diagram of the third generation robot is presented in Figure 2.20. In this system information about the environment obtained by means of vision sensors and transferred to the computer allows the performance of a planned task, e.g. a given trajectory, provided that the manipulator kinematics and dynamics, as well as a control criterion, are known. These issues will be discussed in chapter 12.

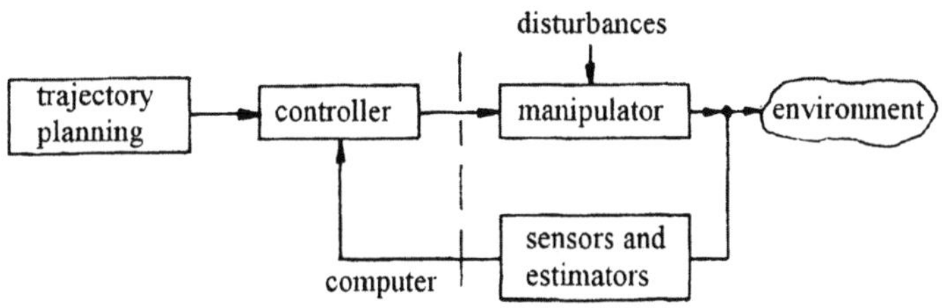

Fig. 2.20. Block (structural) diagram of a third generation robot

The earliest examples of the practical realization of the third generation robots appeared in the seventies:

1. The Shakey robot (mobile), built at The Stanford Research Institute (USA) and capable of single-handedly performing tasks assigned in a simplified natural language (English) in an environment consisting of boxes. To carry out its task it analyzes the image of its surroundings, then plans its actions, and executes them, adjusting itself during operation to unpredictable events. The robot is fitted with a TV camera, optical telemeters, and touch sensors, as well as with a radio transmitter and receiver for communication with the computer controlling the robot's work.
2. An eye-ear-hand type robot, also constructed at Stanford University (USA), which is verbally assigned a task (the manipulation and objects) in a natural language (relatively complex operations require prior planning), and performs this task within the view of a TV camera by means of a mechanical hand with electrical and hydraulic control. Its design includes evaluation analysis and s[illegible]und classification, together with action planning.
3. A series of hand-eye type robots, built in Japan and Scotland, which can car[illegible] various tasks on objects within view of a camera.
4. The HIVID robot, developed at Hitachi Laboratories (Japan), and capable of composing complex objects from simple ones (for example little houses made of building blocks) by means of a manipulator, given a "picture" of the object in three different projections on a plane. The robot observes and analyzes its field of operation, and plans the sequence of motions (trajectories) and objects required to complete the final product (whose representation is obtained through analysis of an image from a second camera).

5. An anthropomorphic robot WABOT, constructed at The University of Waseda (Japan). It has two lower and two upper limbs fitted, among other things, with fingers. The robot can analyze images from two TV cameras, communicate with the operator (bilaterally) in a natural language, and perform given tasks. The robot has twenty-five degrees of freedom in all, and an extensive set of touch receptors, as well as optical and acoustic ones. A great problem, solved by its designers, was the control of such a complex mechanism.

From these examples we see that it is already possible today to built a robot which can single-handedly analyze images of the workspace obtained from vision sensors and perform the required tasks. Basic problems in the practical application of these robots involve computer computational speed and resources for operational memory. The possibility of performing a given task, for example, the assembly of a product from simpler components, depends on the robot's capability of recognizing these elements and determining their position and orientation in the workspace. This information is provided through a visual module as described in chapter 10. (For the sake of simplification, the object reconstructing issues have been omitted there. The reader interested in these problems is referred to [12.29]).

Robot performance is determined by:

- arm geometry,
- the drive system,
- dynamic performance and accuracy,
- reliability and safety.

These factor will be discussed in subsequent chapters

2.4. References

[2.1] *Morecki A.*: Important Definitions. An Outline of Systematic Classification. Basic Problems of Modern Technology. Vol. XXV. Robotics. *By A. Morecki*, Warsaw, PWN, 1987 (*in Polish*).

[2.2] *Morecki A.*: Basic Problems of Modern Robotics. Lecture presented on the session of Committee for Machine Design of Polish Academy of Sciences in April 1992 (*in Polish*).

[2.3] Modelling and Control. (Edited by *P. Coiffet*). Vol. 1. London, Kogan Page, 1981.

[2.4] *Morecki A.*, *Oderfeld J.*: Theory of machines and mechanisms. Warsaw, PWN, 1987 (*in Polish*).

[2.5] *Korendyashev A.I.*: Determining the Number of Degrees of Freedom for the Industrial Robot End-Effector. Masinovedenie, No. 6, 1985 (*in Russian*).

[2.6] Roboter in der Geratetechnik. *Herausgegeben von G. Bogelsack, E. Kallenbach, G. Linnemann*. Berlin, VEB Verlag Technik, 1984.

[2.7] Mechanics of Industrial Robots. 1. Moscow, Vyssaja Skola, 1988 (*in Russian*).

[2.8] *Morecki A., Ekiel J., Fidelus K.*: Cybernetic systems of Limb Movements in Man, Animals and Robots. PWN-Ellis Horwood Limited Publisher, Chichester, 1984.

[2.9] *Morecki A.*: Bionic Manipulators. Warsaw, PWN, 1976 (*in Polish*).

[2.10] Advances in Automation and Robotics. Edited by *G.N.Saridis,* Vol. 1, 1985.

[2.11] *Medvedev W. S., Leskov A. G. Juscenko A. S.*: Control Systems of Manipulators. (edited by *E. Popov*). Moscow, Nauka, 1978 (*in Russian*).

[2.12] *Schraft R. D.* u.a.: Industrieroboter–Technik Einfürung und Anwendung. Export Verlag, 1984.

[2.13] Robot Manipulation Systems. Edited by *A. I. Korendyashev*. Moscow, Masinostroenie, 1989.

[2.14] *Kobrinski A. A., Kobrinski A.E.*: Robot Manipulation Systems. Moscow, Nauka, 1985.

[2.15] *Kourash Etemadi-Zanganeh*: Kinematics of Manipulators with Parallelism, Modularity and Redundancy: Analysis and Design. Ph. D. Thesis, Dept. of Mech. Eng., Montreal, McGill Univ., 1995.

3. MANIPULATOR KINEMATICS

3.1. Introduction

A robotic manipulator is a set of rigid bodies, also called links, performing complex motions in the workspace. In order to describe the displacements, velocities, accelerations, and dynamics of particular links, it is necessary to introduce mathematical tools, that are efficient in terms of computation. The matrix approach (using homogeneous coordinates) describes problems concerning the mechanics of the manipulator in a compact and uniform way, expressing the geometric, kinematic and dynamic relations between manipulator elements and the manipulated objects, and providing control algorithms, along with visual information processing. The use of homogeneous coordinates to describe the complex control system of a robot can considerably simplify the analysis and synthesis of such a system [3.2, 3.13, 3.17].

3.1.1. Homogeneous coordinates

The Cartesian coordinate system OXYZ is introduced in the three-dimensional robot workspace. **Homogeneous coordinates** of a point P in the three-dimensional workspace are any four numbers (x_1, x_2, x_3, x_4), not all equal to zero, related to the Cartesian coordinates (x, y, z) of that point as follows:

$$x = \frac{x_1}{x_4}, \qquad x = \frac{x_2}{x_4}, \qquad x = \frac{x_3}{x_4}$$

The definition of homogeneous coordinates is ambiguous, since if (x_1, x_2, x_3, x_4) are homogeneous coordinates of a certain point from the workspace, the numbers (λx_1, λx_2, λx_3, λx_4), where $\lambda \neq 0$, are homogeneous coordinates of that point too. The coordinate x_4 must be different from zero if the vector $[x_1, x_2, x_3, x_4]^T$ is to describe homogeneous coordinates of a point and not a direction in space.

Example 3.1. Let the numbers (1, 2, 3, 4) be the homogeneous coordinates of a certain point P in the workspace (Fig. 3.1). The coordinates (1·2, 2·2, 3·2, 4·2), (1·1/4, 2·1/4, 3·1/4, 4·1/4), (1·(-8), 2·(-8), 3·(-8), 4 (-8)) etc. are also homogeneous coordinates of the same point.

The definition of homogeneous coordinates implies that for any point in the workspace with Cartesian coordinates (x, y, z), its homogeneous coordinates, for instance (x, y, z, 1), can be defined.

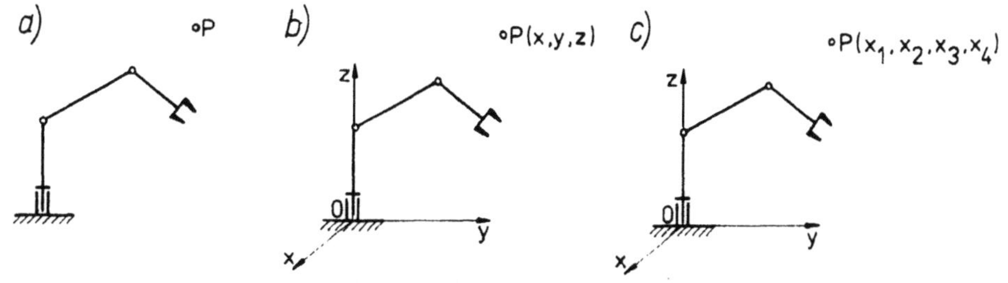

Fig. 3.1. a) A point P in the workspace, b) introduction of the Cartesian coordinate system, c) introduction of homogeneous coordinates

3.1.2. Transformation of homogeneous coordinates

Cartesian coordinates of a point in the workspace in the coordinate system OXYZ can be determined, if we know that point's coordinates in another Cartesian coordinate system, O'X'Y'Z', in accordance with the relations

$$x = c_{11}x' + c_{12}y' + c_{13}z' + r_1$$
$$y = c_{21}x' + c_{22}y' + c_{23}z' + r_2$$
$$z = c_{31}x' + c_{32}y' + c_{33}z' + r_3$$

where r_1, r_2 and r_3 are coordinates of the origin of the O'X'Y'Z' system described in the OXYZ system; and c_{11}, c_{12}, ..., c_{33} are direction cosines of the O'X'Y'Z' system axes relative to the OXYZ system.

If (x_1, x_2, x_3, x_4) represent homogeneous coordinates of a point in the OXYZ system, and the numbers (x_1', x_2', x_3', x_4') are homogeneous coordinates of that point in the O'X'Y'Z' system, then it can always be assumed that $x_4 = x_4'$ (homogeneous coordinates can be multiplied by any number other than zero).

Example 3.2. Let numbers (1, 2, 3, 4) be the homogeneous coordinates of a point in the OXYZ system, while (1/4, –2, 4, 8) are homogeneous coordinates of that point in the O'X'Y'Z' system. A system (1/4·1/2, –2·1/2, 4·1/2, 8·1/2) = = (1/8, –1, 2, 4) represents the homogeneous coordinates of the same point in the workspace in the O'X'Y'Z' system.

Substituting homogeneous coordinates yields the following expressions:

$$x_1 = c_{11}x_1' + c_{12}x_2' + c_{13}x_3' + r_1x_4'$$
$$x_2 = c_{21}x_1' + c_{22}x_2' + c_{23}x_3' + r_2x_4'$$
$$x_3 = c_{31}x_1' + c_{32}x_2' + c_{33}x_3' + r_3x_4'$$
$$x_4 = 0x_1' + 0x_2' + 0x_3' + 1x_4'$$

or, in the matrix:

$$\begin{bmatrix} x_1 \\ x_2 \\ x_3 \\ x_4 \end{bmatrix} = \begin{bmatrix} c_{11} & c_{12} & c_{13} & r_1 \\ c_{21} & c_{22} & c_{23} & r_2 \\ c_{31} & c_{32} & c_{33} & r_3 \\ 0 & 0 & 0 & 1 \end{bmatrix} \begin{bmatrix} x_1' \\ x_2' \\ x_3' \\ x_4' \end{bmatrix}$$

This is how the homogeneous coordinates of a point in different Cartesian coordinate systems are related to each other through linear, non-singular, homogeneous transformations.

When $x_4' = 1$, the first three homogeneous coordinates of a point in the workspace are identical with its Cartesian coordinates, since:

$$x' = \frac{x_1'}{x_4'} = x_1', \qquad y' = \frac{x_2'}{x_4'} = x_2', \qquad z' = \frac{x_3'}{x_4'} = x_3'$$

and $x = \dfrac{x_1}{x_4} = x_1, \qquad y = \dfrac{x_2}{x_4} = x_2, \qquad z = \dfrac{x_3}{x_4} = x_3$

As a result we obtain:

$$\begin{bmatrix} x \\ y \\ z \\ 1 \end{bmatrix} = \begin{bmatrix} c_{11} & c_{12} & c_{13} & r_1 \\ c_{21} & c_{22} & c_{23} & r_2 \\ c_{31} & c_{32} & c_{33} & r_3 \\ 0 & 0 & 0 & 1 \end{bmatrix} \begin{bmatrix} x' \\ y' \\ z' \\ 1 \end{bmatrix}$$

3.2. Rotations and translations of coordinate systems

Homogeneous coordinates of a position vector can be expressed in the form of a column matrix [3.2] $w = [\, w_x \; w_y \; w_z \; 1 \,]^T$, where index T indicates the transposition of a row matrix into a column matrix. The position vector of a point at the origin of the coordinate system is a zero vector, and can be expressed in the form of a column matrix $[0\;\; 0\;\; 0\;\; 1]^T$. The vector $[w_x \; w_y \; w_z \;\; 0]^T$ is a direction vector.

The dot product of the two vectors is determined by a number (scalar):

$$w \cdot v = w_x v_x + w_y v_y + w_z v_z$$

The dot product can also be expressed in the matrix form:

$$w \cdot v = [w]^T[v] = [v]^T[w] = Tr[w][v]^T$$

where Tr indicates the sum of the diagonal elements of the matrix

$$T_r[w][v]^T = T_r \begin{bmatrix} w_x \\ w_y \\ w_z \end{bmatrix} \begin{bmatrix} v_x & v_y & v_z \end{bmatrix} = T_r \begin{bmatrix} w_x v_x & w_x v_y & w_x v_z \\ w_y v_x & w_y v_y & w_y v_z \\ w_z v_x & w_z v_y & w_z v_z \end{bmatrix}$$

The cross product of two vectors is a vector perpendicular to a plane determined by these two vectors, which can be expressed in matrix form:

$$\boldsymbol{w} \times \boldsymbol{v} = \begin{bmatrix} 0 & -w_z & w_y \\ w_z & 0 & -w_x \\ -w_y & w_x & 0 \end{bmatrix} \begin{bmatrix} v_x \\ v_y \\ v_z \end{bmatrix} = \begin{bmatrix} w_y v_z - w_z v_y \\ w_z v_x - w_x v_y \\ w_x v_y - w_y v_x \end{bmatrix}$$

For instance, the velocity vector of a point is defined as a cross product of an angular velocity vector $\dot{\theta}$ and vector $\boldsymbol{p}$ which determines that point's position in a reference system:

$$\boldsymbol{v} = \dot{\theta} \times \boldsymbol{p} = \begin{bmatrix} 0 & -\dot{\theta}_z & \dot{\theta}_y \\ \dot{\theta}_z & 0 & -\dot{\theta}_x \\ -\dot{\theta}_y & \dot{\theta}_x & 0 \end{bmatrix} \begin{bmatrix} p_x \\ p_y \\ p_z \end{bmatrix} = \begin{bmatrix} \dot{\theta}_y p_z - \dot{\theta}_z p_y \\ \dot{\theta}_z p_y - \dot{\theta}_y p_z \\ \dot{\theta}_x p_y - \dot{\theta}_y p_x \end{bmatrix}$$

Homogeneous transformations representing translation by a vector $\boldsymbol{p} = [p_x\, p_y\, p_z\; 1]^T$ rotation about axis x by an angle θ_1, rotation about axis y by an angle θ_2 and rotation about axis z by an angle θ_3, are described by matrices:

$$A_p = \begin{bmatrix} 1 & 0 & 0 & p_x \\ 0 & 1 & 0 & p_y \\ 0 & 0 & 1 & p_z \\ 0 & 0 & 0 & 1 \end{bmatrix}, \qquad A_{z,\theta_3} = \begin{bmatrix} c\theta_3 & -s\theta_3 & 0 & 0 \\ s\theta_3 & c\theta_3 & 0 & 0 \\ 0 & 0 & 1 & 0 \\ 0 & 0 & 0 & 1 \end{bmatrix}$$

$$A_{y,\theta_2} = \begin{bmatrix} c\theta_2 & 0 & s\theta_2 & 0 \\ 0 & 1 & 0 & 0 \\ -s\theta_2 & 0 & c\theta_2 & 0 \\ 0 & 0 & 0 & 1 \end{bmatrix}, \qquad A_{x,\theta_1} = \begin{bmatrix} 1 & 0 & 0 & 0 \\ 0 & c\theta_1 & -s\theta_1 & 0 \\ 0 & s\theta_1 & c\theta_1 & 0 \\ 0 & 0 & 0 & 1 \end{bmatrix}$$

where:

$s\theta_i = \sin\theta_i$, $c\theta_i = \cos\theta_i$, $i = 1, 2, 3$.

The homogeneous transformation of one coordinate system to another is described by the displacement vector $\boldsymbol{p} = [p_x,\ p_y,\ p_z]^T$ and three orientation unit vectors $\boldsymbol{l} = [l_x,\ l_y,\ l_z]^T$, $\boldsymbol{m} = [m_x,\ m_y,\ m_z]^T$, $\boldsymbol{n} = [n_x,\ n_y,\ n_z]^T$ which constitute elements of the transformation matrix from system i to system j:

$$\boldsymbol{T}_{i,j} = \begin{bmatrix} l_x & m_x & n_x & p_x \\ l_y & m_y & n_y & p_y \\ l_z & m_z & n_z & p_z \\ 0 & 0 & 0 & 1 \end{bmatrix} \tag{3.1}$$

Of the nine coordinates of unit vectors ***l***, ***m*** and ***n*** only three are independent, the rest must meet six out of nine relations:

$$\begin{aligned}
&l_x^2 + l_y^2 + l_z^2 = 1 \\
&m_x^2 + m_y^2 + m_z^2 = 1 \\
&n_x^2 + n_y^2 + n_z^2 = 1 \\
&n_x m_x + n_y m_y + n_z m_z = 0 \\
&l_x m_x + l_y m_y + l_z m_z = 0 \\
&l_x n_x + l_y n_y + l_z n_z = 0 \\
&m_y n_z - m_z n_y = l_x \\
&m_z n_x - m_x n_z = l_y \\
&m_x n_y - m_y n_x = l_z
\end{aligned} \tag{3.2}$$

The first three expressions result from the unitary length of the unit vectors ***l***, ***m*** and ***n***, and the rest from the fact that these unit vectors are perpendicular to each other.

If values m_x, m_y and n_x are assumed to be independent, the rest of the values of the $T_{i,j}$ matrix elements can be calculated according to the following formulas [3.7]:

$$\begin{aligned}
&m_z = \pm\sqrt{1 - m_x^2 - m_y^2} \\
&n_y = -m_x\left(m_y n_x \pm m_z\sqrt{1 - m_x^2 - n_x^2}\right)/\left(1 - m_x^2\right) \\
&n_z = \pm\sqrt{1 - n_x^2 - n_y^2} \\
&l_x = m_y n_z - m_z n_y \\
&l_y = m_z n_x - m_x n_z \\
&l_z = m_x n_y - m_y n_x
\end{aligned} \tag{3.2a}$$

The ± signs in the above formulas for m_z, m_y and n_z indicate the possibility of eight different solutions. Those solutions which meet the conditions for unit vectors ***m*** and ***n*** are selected from the possible ones.

In the case of inverse transformation, the matrices are inverted, and observe the following rules:

$$\boldsymbol{T}_{j,i} = \boldsymbol{T}_{i,j}^{-1}, \quad \boldsymbol{T}_{i,j}\boldsymbol{T}_{j,i} = \boldsymbol{E} \tag{3.3}$$

where ***E*** is the unitary matrix whose elements on the diagonal equal one, the rest being zero. Substituting (3.1) into the equation (3.3) yields:

$$\boldsymbol{T}_{i,j}^{-1} = \begin{bmatrix} l_x & l_y & l_z & -\boldsymbol{p}\cdot\boldsymbol{l} \\ m_x & m_y & m_z & -\boldsymbol{p}\cdot\boldsymbol{m} \\ n_x & n_y & n_z & -\boldsymbol{p}\cdot\boldsymbol{n} \\ 0 & 0 & 0 & 1 \end{bmatrix} \tag{3.3a}$$

where:

$$\boldsymbol{p}\cdot\boldsymbol{l} = p_x l_x + p_y l_y + p_z l_z$$

$$\boldsymbol{p}\cdot\boldsymbol{m} = p_x m_x + p_y m_y + p_z m_z$$

$$\boldsymbol{p}\cdot\boldsymbol{n} = p_x n_x + p_y n_y + p_z n_z$$

The rotation of the i system by an angle θ about an arbitrarily oriented axis with a unit vector $\boldsymbol{u}_i$ given in the j system, can be broken down into three successive rotations: 1) The rotation of unit vector $\boldsymbol{u}_i$ until it coincides with the z_j axis, 2) The rotation of angle θ about the z_j axis, and 3) the rotation of system i to the final position. Thus the rotation matrix of system i about the axis of unit vector $\boldsymbol{u}_i$ by the angle θ can be determined from the relation:

$$\boldsymbol{A}_{u,\theta} = \boldsymbol{T}_{i,j}\boldsymbol{A}_{z,\theta}\boldsymbol{T}_{i,j}^{-1} \tag{3.4}$$

Substituting expressions (3.1) and (3.3) in (3.4) yields

$$\boldsymbol{A}_{u,\theta} = \begin{bmatrix} l_x & m_x & n_x & 0 \\ l_y & m_y & n_y & 0 \\ l_z & m_z & n_z & 0 \\ 0 & 0 & 0 & 1 \end{bmatrix} \begin{bmatrix} c\theta & -s\theta & 0 & 0 \\ s\theta & c\theta & 0 & 0 \\ 0 & 0 & 1 & 0 \\ 0 & 0 & 0 & 1 \end{bmatrix} \begin{bmatrix} l_x & l_y & l_z & 0 \\ m_x & m_y & m_z & 0 \\ n_x & n_y & n_z & 0 \\ 0 & 0 & 0 & 1 \end{bmatrix}$$

Multiplying the matrices, starting on the right, in view of the substitution $u_x = n_x$, $u_y = n_y$, $u_z = n_z$, yields:

$$\boldsymbol{A}_{u,\theta} = \begin{bmatrix} u_x^2(1-c\theta)+c\theta & u_x u_y(1-c\theta)-u_z s\theta & u_x u_z(1-c\theta)+u_y s\theta & 0 \\ u_x u_y(1-c\theta)+u_z s\theta & u_y^2(1-c\theta)+c\theta & u_y u_z(1-c\theta)-u_x s\theta & 0 \\ u_x u_z(1-c\theta)-u_y s\theta & u_y u_z(1-c\theta)+u_x s\theta & u_z^2(1-c\theta)+c\theta & 0 \\ 0 & 0 & 0 & 1 \end{bmatrix} \tag{3.5}$$

From the general relation (3.5), we can obtain the matrices of rotation about particular axes. For instance, assuming that $u_x = 1$, $u_y = u_z = 0$, the rotation about axis x by the angle θ is described by the matrix $A_{x,\theta}$. Summing the elements on the diagonal yields:

$$a_{11} + a_{22} + a_{33} = \left(u_x^2 + u_y^2 + u_z^2\right)(1-c\theta) + 3c\theta = 1 + 2c\theta$$

Hence: $c\theta = (a_{11} + a_{22} + a_{33} - 1)/2$

Substracting the elements on the lower side of the matrix (3.5) diagonal across from the respective ones on the upper side and then squaring the differences and performing side summation, we have:

$$(a_{12} - a_{21})^2 + (a_{13} - a_{31})^2 + (a_{23} - a_{32})^2 = 4s^2\theta$$

and hence:

$$2s\theta = \pm\sqrt{(a_{12}-a_{21})^2+(a_{13}-a_{31})^2+(a_{23}-a_{32})^2}$$

with the + sign referring to the angle $0 \le \theta \le \pi$.

Components of the unit vector $\boldsymbol{u}$ are determined as follows:

$$\begin{aligned} u_x &= (a_{32}-a_{23})/2s\theta \\ u_y &= (a_{31}-a_{13})/2s\theta \\ u_x &= (a_{21}-a_{12})/2s\theta \end{aligned} \tag{3.6}$$

When angle θ is small, the numerators and denominators of the expressions (3.6) are small, and the components of the unit vector $\boldsymbol{u}$ can not be accurately determined. To improve the accuracy of the computations the following formulas can be used:

$$\begin{aligned} u_x &= sign(a_{32}-a_{23})\sqrt{(a_{11}-c\theta)/(1-c\theta)} \\ u_y &= sign(a_{31}-a_{13})\sqrt{(a_{22}-c\theta)/(1-c\theta)} \\ u_z &= sign(a_{12}-a_{21})\sqrt{(a_{33}-c\theta)/(1-c\theta)} \end{aligned} \tag{3.7}$$

The greatest component of unit vector $\boldsymbol{u}$, obtained from one of the (3.7) formulas, determines the way the rest of the components are computed. If u_x is the greatest, then the other components are calculated as follows:

$$\begin{aligned} u_y &= (a_{21}+a_{12})/2u_x(1-c\theta) \\ u_z &= (a_{31}+a_{13})/2u_x(1-c\theta) \end{aligned} \tag{3.7a}$$

If u_y is the greatest, then:

$$\begin{aligned} u_x &= (a_{21}+a_{12})/2u_y(1-c\theta) \\ u_z &= (a_{32}+a_{23})/2u_y(1-c\theta) \end{aligned} \tag{3.7b}$$

If u_z is the greatest, then:

$$\begin{aligned} u_x &= (a_{31}+a_{13})/2u_z(1-c\theta) \\ u_y &= (a_{32}+a_{23})/2u_z(1-c\theta) \end{aligned} \tag{3.7c}$$

Example 3.3. Determine an angle and a unit vector of the rotation axis of the transformation

$$A_{u,\theta} = \begin{bmatrix} 0 & -1 & 0 & 0 \\ 0 & 0 & 1 & 0 \\ -1 & 0 & 0 & 0 \\ 0 & 0 & 0 & 1 \end{bmatrix}$$

We first, apply the formulas for $c\theta$ and $s\theta$, obtaining

$$c\theta = (0+0+0-1)/2 = -1/2$$

$$s\theta = \sqrt{(-1+0)^2 + (0+1)^2 + (1-0)^2}/2 = \sqrt{3}/2$$

which yields $\theta = arctg\left(-\sqrt{3}\right) = 2\pi/3$. Then, taking the formulas (3.7) we obtain

$$u_x = 1/\sqrt{3}, \quad u_y = 1/\sqrt{3}, \quad u_z = 1/\sqrt{3}.$$

Euler's angles are often introduced in dynamic equations for a link with a spherical motion which can be described by using three angular displacements: θ_1, precession; θ_2, nutation; and θ_3, rotation. Figure 3.2a shows a base coordinate system xyz attached to the base *0*, a moving coordinate system $x_1y_1z_1$ attached to the link *1* which performs the rotation θ_1 about the axis z_1, and a system $x_2y_2z_2$ attached to the link *2* which performs the rotation θ_2 about the axis x_1, relative to link *1*. The final rotation θ_3 about the axis z_2 is performed by link *3* with respect to link *2*. As in every case of successive rotation, the sequence is important. In the case examined here it can be seen (Fig. 3.2b) that the rotation sequence in the base system will be reversed, that is, the first rotation will be that of θ_3 about axis z, then of θ_2 about axis y and, finally, the rotation of θ_1 about axis z.

The transformation described by Euler's coordinates can be obtained by multiplying the three rotation matrices:

$$\boldsymbol{R}_E = \boldsymbol{A}_{z,\theta_1}\boldsymbol{A}_{y,\theta_2}\boldsymbol{A}_{z,\theta_3} =$$

$$= \begin{bmatrix} c\theta_1 & -s\theta_1 & 0 & 0 \\ s\theta_1 & c\theta_1 & 0 & 0 \\ 0 & 0 & 1 & 0 \\ 0 & 0 & 0 & 1 \end{bmatrix} \begin{bmatrix} c\theta_2 & 0 & s\theta_2 & 0 \\ 0 & 1 & 0 & 0 \\ -s\theta_2 & 0 & c\theta_2 & 0 \\ 0 & 0 & 0 & 1 \end{bmatrix} \begin{bmatrix} c\theta_3 & -s\theta_3 & 0 & 0 \\ s\theta_3 & c\theta_3 & 0 & 0 \\ 0 & 0 & 1 & 0 \\ 0 & 0 & 0 & 1 \end{bmatrix}$$

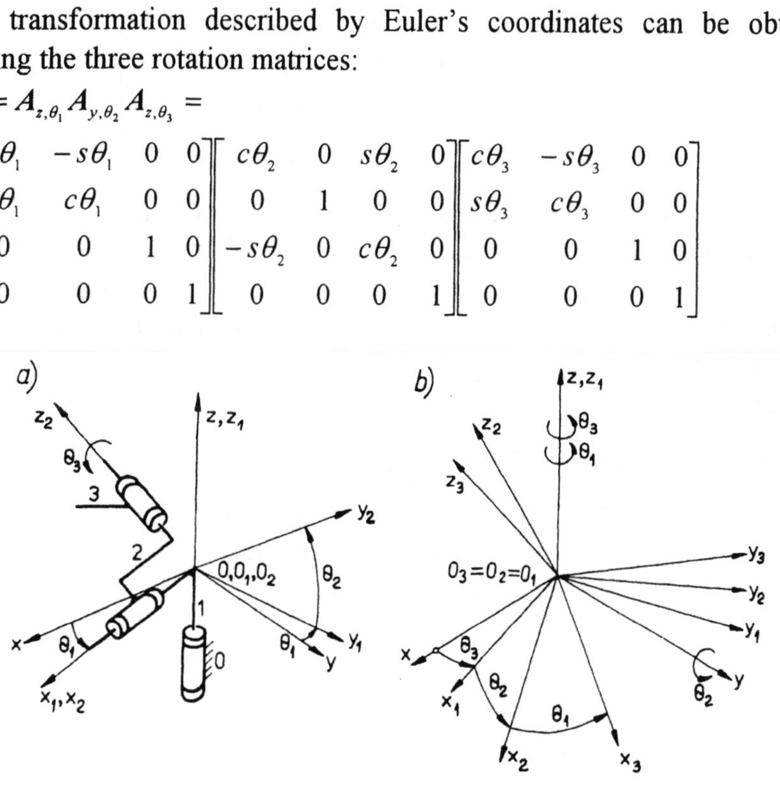

Fig. 3.2. a) An open three-link kinematic chain substituting a spherical joint, dimensioned by means of Euler's angles: θ_1, θ_2, θ_3, b) Coordinate system transformations through rotation by Euler's angles

Hence:

$$R_E = \begin{bmatrix} s\theta_1 c\theta_2 c\theta_3 - s\theta_1 s\theta_3 & -c\theta_1 c\theta_2 s\theta_3 - s\theta_1 s\theta_3 & c\theta_1 s\theta_2 & 0 \\ s\theta_1 c\theta_2 c\theta_3 + c\theta_1 s\theta_3 & -s\theta_1 c\theta_2 s\theta_3 + c\theta_1 c\theta_3 & s\theta_1 s\theta_2 & 0 \\ -s\theta_2 c\theta_3 & s\theta_2 s\theta_3 & c\theta_2 & 0 \\ 0 & 0 & 0 & 1 \end{bmatrix} \tag{3.8}$$

Another sequence of rotation about the x, y, z axes is often used:

$$R_E = A_{z,\theta_1} A_{y,\theta_2} A_{x,\theta_3} =$$

$$= \begin{bmatrix} c\theta_1 & -s\theta_1 & 0 & 0 \\ s\theta_1 & c\theta_1 & 0 & 0 \\ 0 & 0 & 1 & 0 \\ 0 & 0 & 0 & 1 \end{bmatrix} \begin{bmatrix} c\theta_2 & 0 & s\theta_2 & 0 \\ 0 & 1 & 0 & 0 \\ -s\theta_2 & 0 & c\theta_2 & 0 \\ 0 & 0 & 0 & 1 \end{bmatrix} \begin{bmatrix} 1 & 0 & 0 & 0 \\ 0 & c\theta_3 & -s\theta_3 & 0 \\ 0 & s\theta_3 & c\theta_3 & 0 \\ 0 & 0 & 0 & 1 \end{bmatrix}$$

Hence:

$$R_E = \begin{bmatrix} c\theta_1 c\theta_2 & c\theta_1 s\theta_2 s\theta_3 - s\theta_1 c\theta_3 & c\theta_1 s\theta_2 c\theta_3 + s\theta_1 s\theta_3 & 0 \\ s\theta_1 c\theta_2 & s\theta_1 s\theta_2 s\theta_3 + c\theta_1 c\theta_3 & s\theta_1 s\theta_2 c\theta_3 - c\theta_1 s\theta_3 & 0 \\ -s\theta_2 & c\theta_2 s\theta_3 & c\theta_2 c\theta_3 & 0 \\ 0 & 0 & 0 & 1 \end{bmatrix} \tag{3.9}$$

This is a variation of Euler's transformation called Roll, Pitch and Yaw.

Cylindrical coordinates (Fig. 3.3a) are used to determine the position of the body in space by means of three displacements: p_x, the translation along the x axis; θ_z, the rotation about the z axis; and p_z, the translation along the z axis which can be described in the matrix form:

$$C(p_z,\theta,p_x) = A_{(0,0,p_x)} A_{(z,-\theta)} A_{(p_x,0,0)} =$$

$$= \begin{bmatrix} 1 & 0 & 0 & 0 \\ 0 & 1 & 0 & 0 \\ 0 & 0 & 1 & p_z \\ 0 & 0 & 0 & 1 \end{bmatrix} \begin{bmatrix} c\theta & -s\theta & 0 & 0 \\ s\theta & c\theta & 0 & 0 \\ 0 & 0 & 1 & 0 \\ 0 & 0 & 0 & 1 \end{bmatrix} \begin{bmatrix} 1 & 0 & 0 & p_x \\ 0 & 1 & 0 & 0 \\ 0 & 0 & 1 & 0 \\ 0 & 0 & 0 & 1 \end{bmatrix}$$

Hence:

$$C(p_z,\theta,p_x) = \begin{bmatrix} c\theta & -s\theta & 0 & p_x c\theta \\ s\theta & c\theta & 0 & p_x s\theta \\ 0 & 0 & 1 & p_z \\ 0 & 0 & 0 & 1 \end{bmatrix} \tag{3.10}$$

Spherical coordinates are defined by the sequence of three motions: p_z , the translation along the z axis; θ_y , the rotation about the y axis; and θ_z , the rotation around the z axis

$$S(\theta_z,\theta_y,p_z)=A_{z,\theta_z}A_{y,\theta_y}A_{(0,0,p_z)}=$$

$$=\begin{bmatrix} c\theta_z & -s\theta_z & 0 & 0 \\ s\theta_z & c\theta_z & 0 & 0 \\ 0 & 0 & 1 & p_z \\ 0 & 0 & 0 & 1 \end{bmatrix}\begin{bmatrix} c\theta_y & 0 & s\theta_y & 0 \\ 0 & 1 & 0 & 0 \\ -s\theta_y & 0 & c\theta_y & 0 \\ 0 & 0 & 0 & 1 \end{bmatrix}\begin{bmatrix} 1 & 0 & 0 & 0 \\ 0 & 1 & 0 & 0 \\ 0 & 0 & 1 & p_z \\ 0 & 0 & 0 & 1 \end{bmatrix}$$

Hence:

$$S(\theta_z,\theta_y,p_z)=\begin{bmatrix} c\theta_z c\theta_y & -s\theta_z & c\theta_z s\theta_y & p_z c\theta_z s\theta_y \\ s\theta_z c\theta_y & c\theta_z & s\theta_z s\theta_y & p_z s\theta_z s\theta_y \\ -s\theta_y & 0 & c\theta_y & p_z c\theta_y \\ 0 & 0 & 0 & 1 \end{bmatrix} \tag{3.11}$$

Denavit-Hartenberg coordinates [3.2, 3.13] (see Fig. 3.3c) are defined by four parameters: θ_i , the rotation about the z_{i-1} axis; λ_i , the translation along the z_{i-1} axis; l_i , the translation along the x_i axis; α_i , the rotation around the x_i axis. All i links are characterized by two dimensions: l_i , the link length; and α_i , the link twist. All joints are characterized by two dimensions: λ_i , the distance; and θ_i , the angle between two joint links. Homogeneous transformation matrices corresponding to these parameters are expressed in the form

$$A_{z_{i-1},\theta_i}=\begin{bmatrix} c\theta_i & s\theta_i & 0 & 0 \\ -s\theta_i & c\theta_i & 0 & 0 \\ 0 & 0 & 1 & 0 \\ 0 & 0 & 0 & 1 \end{bmatrix}, \qquad A_{z_{i-1},\lambda_i}=\begin{bmatrix} 1 & 0 & 0 & 0 \\ 0 & 1 & 0 & 0 \\ 0 & 0 & 1 & \lambda_i \\ 0 & 0 & 0 & 1 \end{bmatrix}$$

$$A_{x_i,\alpha_i}=\begin{bmatrix} 1 & 0 & 0 & -l_i \\ 0 & 1 & 0 & 0 \\ 0 & 0 & 1 & 0 \\ 0 & 0 & 0 & 1 \end{bmatrix}, \qquad A_{x_i,\alpha_i}=\begin{bmatrix} 1 & 0 & 0 & 0 \\ 0 & c\alpha_i & s\alpha_i & 0 \\ 0 & -s\alpha_i & c\alpha_i & 0 \\ 0 & 0 & 0 & 1 \end{bmatrix}$$

The transformation of the i-1 system into the i system is described by the matrices' product

$$A_{i-1,i}=A_{x_i,\alpha_i}A_{x_i,\alpha_i}A_{z_{i-1},\lambda_i}A_{z_{i-1},\theta_i}$$

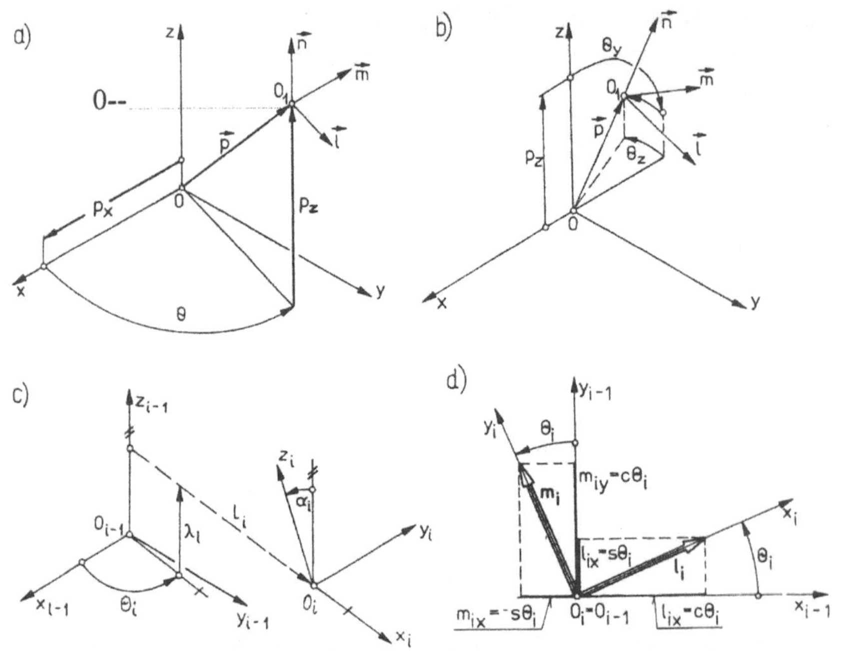

Fig. 3.3. a) Cylindrical coordinates (p_x, θ, p_z), b) Spherical coordinates (p_z, θ_y, θ_z), c) Denavit-Hartenberg coordinates (θ_i, λ_i, l_i, α_i), d) Transformation of unit vectors $\boldsymbol{l}_i$ and $\boldsymbol{m}_i$ by plane rotation θ_i around the $z_{i-1} = z_i$ axis

Hence

$$A_{i-1,i} = \begin{bmatrix} c\theta_i & s\theta_i & 0 & -l_i \\ -s\theta_i c\alpha_i & c\theta_i c\alpha_i & s\alpha_i & -\lambda_i s\alpha_i \\ s\theta_i s\alpha_i & -c\theta_i s\alpha_i & c\alpha_i & -\lambda_i c\alpha_i \\ 0 & 0 & 0 & 1 \end{bmatrix} \tag{3.12}$$

To be more concise, we use the notation: $A_i = A_{i,i-1}$.

In the case of reverse transformation, e.g. the transformation of the i system into the i-1 system, the inverse matrix is used:

$$A_{i-1,i} = A_{i,i-1}^{-1}$$

Hence

$$A_{i,i-1} = \begin{bmatrix} c\theta_i & -s\theta_i c\alpha_i & s\theta_i s\alpha_i & l_i c\theta_i \\ s\theta_i & c\theta_i c\alpha_i & -c\theta_i s\alpha_i & l_i s\theta_i \\ 0 & s\alpha_i & c\alpha_i & \lambda_i \\ 0 & 0 & 0 & 1 \end{bmatrix} \tag{3.13}$$

Note: The sense of the angular displacements is considered positive if the transformation of the i-1 system into the i system is in accordance with the right hand rule, that is the positive sense of angle θ_i is consistent with that of z_{i-1} axis, and that of angle α_i – with the x_i axis.

3.3. Forward kinematics of manipulators

The position of a point P_i belonging to link i is described in the link reference system by the vector $\boldsymbol{r}_{Pi,i} = [\ x_{Pi}\ y_{Pi}\ z_{Pi}\ ]^T$. When point P_i is at the system's origin, e.g. $P_i = O_i$:

$$\boldsymbol{r}_{Pi,i} = \boldsymbol{r}_{Oi,i} = [0\ 0\ 0\ 1]^T$$

The position of the P_i point in the basic reference system 0 (global) can be determined according to the formula

$$\boldsymbol{r}_{Pi,O} = \boldsymbol{r}_{Pi} = \boldsymbol{T}_{i,O}\, \boldsymbol{r}_{Pi,i}$$

where $\boldsymbol{T}_{i,O}$ is the transformation matrix of system i into system 0, obtained as a product of the $\boldsymbol{A}_i$ matrices derived from the (3.13) expression:

$$\boldsymbol{T}_{i,O} = \boldsymbol{A}_1 \boldsymbol{A}_2 \ldots \boldsymbol{A}_i = \boldsymbol{T}_i \tag{3.14a}$$

Matrix $\boldsymbol{T}_i$ can be divided into: $\boldsymbol{B}_i$, an orientation matrix; and $\boldsymbol{p}_i$, a position vector:

$$\boldsymbol{T}_i = \begin{bmatrix} & \boldsymbol{B}_i & & \boldsymbol{p}_i \\ 0 & 0 & 0 & 1 \end{bmatrix} \tag{3.14b}$$

where:

$$\boldsymbol{B}_i = [\boldsymbol{l}_i \quad \boldsymbol{m}_i \quad \boldsymbol{n}_i] = \begin{bmatrix} l_{ix} & m_{ix} & n_{ix} \\ l_{iy} & m_{iy} & n_{iy} \\ l_{iz} & m_{iz} & n_{iz} \end{bmatrix} \tag{3.14c}$$

$$\boldsymbol{p}_i = [p_{ix} \quad p_{iy} \quad p_{iz}]^T \tag{3.14d}$$

Differentiating the (3.14) expression versus time yields the velocity and acceleration of point P_i:

$$\begin{aligned} \boldsymbol{v}_{Pi} &= \sum_{j=1}^{i} \frac{\partial \boldsymbol{T}_i}{\partial q_j} \boldsymbol{r}_{Pi,i} \dot{q}_j \\ \boldsymbol{a}_{Pi} &= \sum_{j=1}^{i} \frac{\partial \boldsymbol{T}_i}{\partial q_j} \boldsymbol{r}_{Pi,i} \ddot{q}_j + \sum_{\substack{j=1 \\ k=1}}^{i} \frac{\partial^2 \boldsymbol{T}_i}{\partial q_j \partial q_k} \boldsymbol{r}_{Pi,i} \dot{q}_j \dot{q}_k \end{aligned} \tag{3.15}$$

where

$$\frac{\partial T_i}{\partial q_j} = A_1 A_2 \ldots A_{j-1} Q_j A_j \ldots A_i, \quad \text{if } j \le i$$

$$\frac{\partial T_i}{\partial q_j} = 0, \quad \text{if } j > i$$

$$\frac{\partial^2 T_i}{\partial q_j \partial q_k} = A_1 A_2 \ldots A_{j-1} Q_j A_j \ldots A_{k-1} Q_k A_k \ldots A_i, \quad \text{if } j < k \le i$$

$$\frac{\partial^2 T_i}{\partial q_j \partial q_k} = A_1 A_2 \ldots A_{j-1} Q_j^2 A_j \ldots A_i, \quad \text{if } j = k \le i$$

$$\frac{\partial^2 T_i}{\partial q_j \partial q_k} = 0, \quad \text{if } j > 1 \text{ and } k > 1 \tag{3.15a}$$

and also noting that:

$$Q_j = \begin{bmatrix} 0 & -1 & 0 & 0 \\ 1 & 0 & 0 & 0 \\ 0 & 0 & 0 & 0 \\ 0 & 0 & 0 & 0 \end{bmatrix}, \qquad \dot{q}_j = \frac{d\theta_j}{dt} = \dot{\theta}_j \qquad \text{for the revolute joint, and}$$

$$Q_j = \begin{bmatrix} 0 & 0 & 0 & 0 \\ 0 & 0 & 0 & 0 \\ 0 & 0 & 0 & 1 \\ 0 & 0 & 0 & 0 \end{bmatrix}, \qquad \dot{q}_j = \frac{d\lambda_j}{dt} = \dot{\lambda}_j \qquad \text{for the prismatic joint.}$$

Considering the relation

$$A_j' = dA_j / dq_j = Q_j A_j$$

we have:

$$dT_{k,0} / dq_j = T_{j-1,0} Q_j T_{k,j-1}^{-1}, \quad \text{for } j \le k \tag{3.16}$$

where

$$T_{k,j-1} = A_j \ldots A_k = (A_1 \ldots A_{j-1})^{-1} (A_j \ldots A_{j-1})(A_j \ldots A_k) = T_{j-1,0}^{-1} T_{k,0}$$

Thus

$$dT_{k,0} / dq_j = T_{j-1,0} Q_j T_{j-1,0}^{-1} T_{k,0} = Q_j^* T_{k,0} \tag{3.17}$$

where

$$Q_j^* = T_{j-1,0} Q_j T_{j-1,0}^{-1}, \quad \text{for } j > 1 \text{ and} \tag{3.17a}$$

$$Q_j^* = Q_j, \quad \text{for } j = 1$$

Substituting (3.16) into (3.15) yields

$$v_{Pk} = \sum_{j=1}^{k} Q_j^* T_{k,0} \dot{q}_j r_{Pk,k} = \sum_{j=1}^{k} Q_j^* \dot{q}_j r_{Pk} \tag{3.18}$$

Hence

$$\boldsymbol{v}_{Pk} = \boldsymbol{D}_k \boldsymbol{r}_{Pk} \tag{3.19}$$

where $\boldsymbol{D}_k$ is an operator matrix of the transformation of the position vector of point P_k belonging to link k into that point's velocity.

The acceleration of point P_k is obtained by differentiation of relation (3.19):

$$\boldsymbol{a}_{pk} = d\boldsymbol{v}_{pk}/dt = \dot{\boldsymbol{D}}_k \boldsymbol{r}_{pk} + \boldsymbol{D}_k \boldsymbol{v}_{pk} \tag{3.20}$$

where

$$\dot{\boldsymbol{D}}_k = \sum_{j=1}^{k} \left(\dot{\boldsymbol{Q}}_j^* \dot{q}_j + \boldsymbol{Q}_j^* \ddot{q}_j\right) = \dot{\boldsymbol{D}}_{k-1} + \dot{\boldsymbol{Q}}_k^* \dot{q}_k + \boldsymbol{Q}_k^* \ddot{q}_k, \quad \text{for } j > 1, \text{ and}$$

$$\dot{\boldsymbol{D}}_1 = \boldsymbol{Q}_1^* \ddot{q}_1$$

The formula for $\dot{\boldsymbol{Q}}_k^*$ has not been introduced before. It is obtained by transforming relation (3.17a) into the form

$$\boldsymbol{Q}_j^* \boldsymbol{T}_{j-1,0} = \boldsymbol{T}_{j-1,0} \boldsymbol{Q}_j$$

and hence

$$\dot{\boldsymbol{Q}}_j^* = \left(\dot{\boldsymbol{T}}_{j-1,0} \boldsymbol{Q}_j - \boldsymbol{Q}_j^* \dot{\boldsymbol{T}}_{j-1,0}\right) \boldsymbol{T}_{j-1,0}^{-1} \tag{3.21}$$

while

$$\dot{\boldsymbol{T}}_{k,0} = \left(\sum_{j=1}^{k} \boldsymbol{Q}_j^* \dot{q}_j\right) \boldsymbol{T}_{k,0} = \boldsymbol{D}_k \boldsymbol{T}_{k,0} \tag{3.22}$$

Substituting (3.22) into (3.21) yields

$$\dot{\boldsymbol{Q}}_j^* = \left(\boldsymbol{D}_{j-1} \boldsymbol{T}_{j-1,0} \boldsymbol{Q}_j - \boldsymbol{Q}_j^* \boldsymbol{D}_{j-1} \boldsymbol{T}_{j-1,0}\right) \boldsymbol{T}_{j-1,0}^{-1}$$

and hence

$$\dot{\boldsymbol{Q}}_j^* = \boldsymbol{D}_{j-1} \boldsymbol{Q}_j^* - \boldsymbol{Q}_j^* \boldsymbol{D}_{j-1} \tag{3.23}$$

Formula (3.20) for the acceleration of point P_k now takes the form of

$$\boldsymbol{a}_{Pk} = \left(\sum_{j=1}^{k} \boldsymbol{Q}_j^* \ddot{q}_j + \dot{\boldsymbol{Q}}_j^* \dot{q}_j\right) \boldsymbol{r}_{Pk} + \boldsymbol{D}_k \boldsymbol{v}_{Pk} \tag{3.24}$$

The angular velocity vector of the k link is determined as follows:

$$\omega_k = \sum_{j=1}^{k} \boldsymbol{B}_{j-1,0} \dot{\theta}_j \boldsymbol{e}_j \tag{3.25}$$

where $\boldsymbol{B}_{j-1,0}$ is the orthogonal matrix of the j-1 system's rotation relative to the 0 base, obtained from the $\boldsymbol{T}_{j-1,0}$ matrix, which is defined by the (3.1) formula, by omitting the last row and column. Thus:

$$\boldsymbol{B}_{j-1,0} = \begin{bmatrix} \boldsymbol{l}_{j-1} & \boldsymbol{m}_{j-1} & \boldsymbol{n}_{j-1} \end{bmatrix} = \begin{bmatrix} l_{j-1,x} & m_{j-1,x} & n_{j-1,x} \\ l_{j-1,y} & m_{j-1,y} & n_{j-1,y} \\ l_{j-1,z} & m_{j-1,z} & n_{j-1,z} \end{bmatrix} \tag{3.25a}$$

where:

$l_{j-1} = [l_{j-1,x}\ l_{j-1,y}\ l_{j-1,z}]^T$, $m_{j-1} = [m_{j-1,x}\ m_{j-1,y}\ m_{j-1,z}]^T$, $n_{j-1} = [n_{j-1,x}\ n_{j-1,y}\ n_{j-1,z}]^T$

are the unit vectors of the j-1 system axes described in system 0.

The projection of a vector ω_k onto the axes of the coordinate system attached to the k link is obtained through transformation:

$$\omega_{k,k} = B_{0,k}\omega_k = \sum_{j=1}^{k} B_{j-1,k}\dot{\theta}_j e_j \tag{3.26}$$

The relation (3.25) may take the form of

$$\omega_k = \omega_{k-1} + B_{k-1,0}\dot{\theta}_k e_k \tag{3.27}$$

Differentiating the relation (5.27) with respect to time yields an angular acceleration of the k link:

$$\dot{\omega}_k = \dot{\omega}_{k-1} + \dot{\omega}_{k,k-1} + \omega_k \times \omega_{k,k-1}$$

where $\omega_{k,k-1} = \dot{\theta}_k e_k$

Example 3.4. A manipulator with six revolute joints, shown in the Figure 3.4, is described by the following parameters:

$\alpha_i = \pi/2,\quad 0,\quad 0, -\pi/2,\quad \pi/2,\quad \pi/2$

$l_i = 0,\quad l_2,\quad l_3,\quad 0,\quad 0,\quad l_6$

$\lambda_i = 0,\quad 0,\quad 0,\quad 0,\quad \lambda_5,\quad \lambda_6$

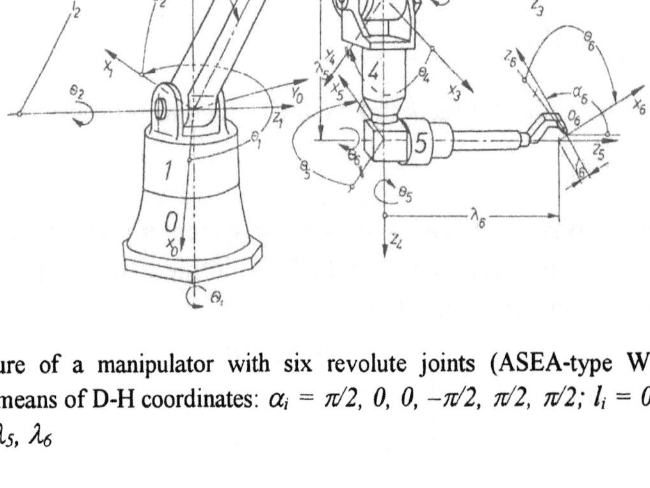

Fig. 3.4. Structure of a manipulator with six revolute joints (ASEA-type WORKMASTER) dimensioned by means of D-H coordinates: $\alpha_i = \pi/2, 0, 0, -\pi/2, \pi/2, \pi/2$; $l_i = 0, l_2, l_3, 0, 0, l_6$; $\lambda_i = 0, 0, 0, 0, \lambda_5, \lambda_6$

The task is to find the coordinates together with the velocity and acceleration of the point $P_6 = O_6$, i.e., where the point P_6 coincides with O_6, the origin of the reference system attached to the 6th link.

The task's solution begins by determining the A_i matrices according to formula (3.13), and applying the given values of the parameters:

$$A_1 = \begin{bmatrix} c\theta_1 & 0 & s\theta_1 & 0 \\ s\theta_1 & 0 & -c\theta_1 & 0 \\ 0 & 1 & 0 & 0 \\ 0 & 0 & 0 & 1 \end{bmatrix}, \quad A_2 = \begin{bmatrix} c\theta_2 & -s\theta_2 & 0 & l_2 c\theta_2 \\ s\theta_2 & c\theta_2 & 0 & l_2 s\theta_2 \\ 0 & 0 & 1 & 0 \\ 0 & 0 & 0 & 1 \end{bmatrix}$$

$$A_3 = \begin{bmatrix} c\theta_3 & -s\theta_3 & 0 & l_3 c\theta_3 \\ s\theta_3 & c\theta_3 & 0 & l_3 s\theta_3 \\ 0 & 0 & 1 & 0 \\ 0 & 0 & 0 & 1 \end{bmatrix}, \quad A_4 = \begin{bmatrix} c\theta_4 & 0 & -s\theta_4 & 0 \\ s\theta_4 & 0 & c\theta_4 & 0 \\ 0 & -1 & 0 & 0 \\ 0 & 0 & 0 & 1 \end{bmatrix} \tag{3.28}$$

$$A_5 = \begin{bmatrix} c\theta_5 & 0 & s\theta_5 & 0 \\ s\theta_5 & 0 & -c\theta_5 & 0 \\ 0 & 1 & 0 & \lambda_5 \\ 0 & 0 & 0 & 1 \end{bmatrix}, \quad A_6 = \begin{bmatrix} c\theta_6 & 0 & s\theta_6 & l_6 c\theta_6 \\ s\theta_6 & 0 & -c\theta_6 & l_6 s\theta_6 \\ 0 & 1 & 0 & \lambda_6 \\ 0 & 0 & 0 & 1 \end{bmatrix}$$

The position of point P_6 in the basic coordinate system is determined according to formula (3.14) which, in this case, takes the form of

$\boldsymbol{r}_{P6,0} = \boldsymbol{T}_6 \boldsymbol{r}_{P6,6}$, where $\boldsymbol{r}_{P6,6} = [0 \; 0 \; 0 \; 1]^T$

$$T_6 = A_1 A_2 A_3 A_4 A_5 A_6 = \begin{bmatrix} l_x & m_x & n_x & p_x \\ l_y & m_y & n_y & p_y \\ l_z & m_z & n_z & p_z \\ 0 & 0 & 0 & 1 \end{bmatrix} \tag{3.29}$$

where

$$l_x = c\theta_1 c(\theta_2+\theta_3+\theta_4)c\theta_5 c\theta_6 - c\theta_1 s(\theta_2+\theta_3+\theta_4)s\theta_6 - s\theta_1 s\theta_5 c\theta_6$$
$$l_y = s\theta_1 c(\theta_2+\theta_3+\theta_4)c\theta_5 c\theta_6 - s\theta_1 s(\theta_2+\theta_3+\theta_4)s\theta_6 + c\theta_1 s\theta_5 c\theta_6$$
$$l_z = s(\theta_2+\theta_3+\theta_4)c\theta_5 c\theta_6 + c(\theta_2+\theta_3+\theta_4)s\theta_6$$
$$m_x = s\theta_1 c\theta_5 + c\theta_1 c(\theta_2+\theta_3+\theta_4)s\theta_5$$
$$m_y = s\theta_1 c(\theta_2+\theta_3+\theta_4)s\theta_5 - c\theta_1 c\theta_5$$
$$m_z = s(\theta_2+\theta_3+\theta_4)s\theta_5$$
$$n_x = c\theta_1 c(\theta_2+\theta_3+\theta_4)c\theta_5 c\theta_6 + c\theta_1 s(\theta_2+\theta_3+\theta_4)c\theta_6 - s\theta_1 s\theta_5 s\theta_6$$
$$n_y = s\theta_1 c(\theta_2+\theta_3+\theta_4)c\theta_5 s\theta_6 + s\theta_1 s(\theta_2+\theta_3+\theta_4)c\theta_6 + c\theta_1 s\theta_5 s\theta_6$$

$$n_z = s(\theta_2+\theta_3+\theta_4)c\theta_5 s\theta_6 - c(\theta_2+\theta_3+\theta_4)c\theta_6$$
$$p_x = l_2 c\theta_1 c\theta_2 + l_3 c\theta_1 c(\theta_2+\theta_3) - \lambda_5 c\theta_1 s(\theta_2+\theta_3+\theta_4) + \lambda_6[s\theta_1 c\theta_5 + c\theta_1 c(\theta_2+\theta_3+\theta_4)s\theta_5] + $$
$$+ l_6[c\theta_1 c(\theta_2+\theta_3+\theta_4)c\theta_5 c\theta_6 - s\theta_1 s\theta_2 c\theta_6 - c\theta_1 s(\theta_2+\theta_3+\theta_4)s\theta_6]$$
$$p_y = l_2 s\theta_1 c\theta_2 + l_3 s\theta_1 c(\theta_2+\theta_3) - \lambda_5 s\theta_1 s(\theta_2+\theta_3+\theta_4) + \lambda_6[-c\theta_1 c\theta_5 + +s\theta_1 c(\theta_2+\theta_3+$$
$$+\theta_4)s\theta_5] + l_6[s\theta_1 c(\theta_2+\theta_3+\theta_4)c\theta_5 c\theta_6 + c\theta_1 s\theta_5 c\theta_6 - s\theta_1 s(\theta_2+\theta_3+\theta_4)s\theta_6]$$
$$p_z = l_2 s\theta_2 + l_3 s(\theta_2+\theta_3) + \lambda_5 c(\theta_2+\theta_3+\theta_4) + \lambda_6 s(\theta_2+\theta_3+\theta_4)s\theta_5$$
$$+ l_6[s(\theta_2+\theta_3+\theta_4)c\theta_5 c\theta_6 + c(\theta_2+\theta_3+\theta_4)s\theta_6] \quad (3.30)$$

The coordinates of point P_6 for the initial manipulator position are obtained by substituting the value $\theta_i = 0$ ($i = 1, 2, 3$) into the relation (3.30). Thus:

$$p_x = l_2 + l_3 + l_6, \qquad p_y = -\lambda_6, \qquad p_z = \lambda_5$$

The velocity and acceleration of point P_6 can be found by differentiating the formulas for this point's coordinates, as is done for compound functions, such as $p_x = F[\theta_i(t)]$.

Example 3.5. Figure 3.5 presents a part of a Stanford-type manipulator, only the first three links are described by the parameters: $\alpha_1 = \alpha_2 = \pi/2$, $l_1 = l_2 = \lambda_1 = 0$, $\lambda_2 = 0.05$ [m].

Given the displacements θ_1, θ_2 and λ_3, as well as the velocities and accelerations in the joints, we must to determine the coordinates of point P_3 of link 3, its velocity and acceleration, and the angular velocities of links 2 and 3. According to the formula (3.13) we find the transformation matrices A_j ($j = 1, 2, 3$), which, in view of the given link parameters, take the form of

$$A_1 = \begin{bmatrix} c\theta_1 & 0 & s\theta_1 & 0 \\ s\theta_1 & 0 & -c\theta_1 & 0 \\ 0 & 1 & 0 & 0 \\ 0 & 0 & 0 & 1 \end{bmatrix}, \qquad A_2 = \begin{bmatrix} c\theta_2 & 0 & s\theta_2 & 0 \\ s\theta_2 & 0 & -c\theta_2 & 0 \\ 0 & 1 & 0 & \lambda_2 \\ 0 & 0 & 0 & 1 \end{bmatrix},$$

$$A_3 = \begin{bmatrix} 1 & 0 & 0 & 0 \\ 0 & 1 & 0 & 0 \\ 0 & 0 & 1 & \lambda_3 \\ 0 & 0 & 0 & 1 \end{bmatrix}$$

The position and orientation matrix in terms of the base are found by applying the formula

$$\boldsymbol{T}_6 = A_1 A_2 A_3 = \begin{bmatrix} c\theta_1 c\theta_2 & s\theta_1 & c\theta_1 s\theta_2 & \lambda_3 c\theta_1 s\theta_2 + \lambda_2 s\theta_1 \\ s\theta_1 c\theta_2 & -c\theta_1 & s\theta_1 s\theta_2 & \lambda_3 s\theta_1 s\theta_2 + \lambda_2 c\theta_1 \\ s\theta_2 & 0 & -c\theta_2 & -\lambda_3 c\theta_2 \\ 0 & 0 & 0 & 1 \end{bmatrix}$$

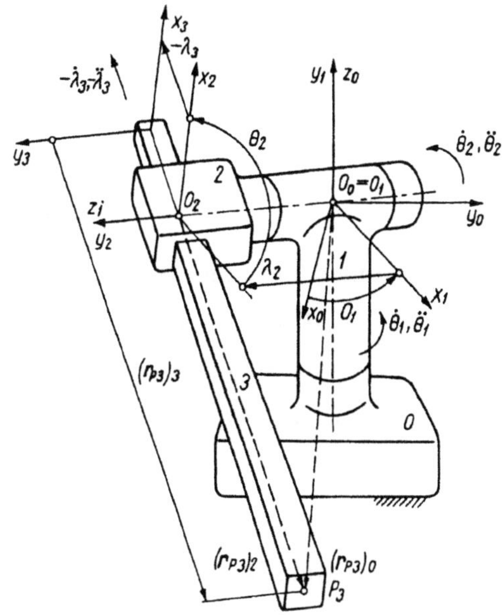

Fig. 3.5. Structure of a Stanford-type manipulator arm dimensioned by means of D-H coordinates: $\alpha_i = 0, 0, 0$; $l_i = 0, 0, 0$; $\lambda_i = 0, \lambda_2, \lambda_3$; θ_i. Notation: $\boldsymbol{r}_{P3,0}$ is a position vector of a point P_3 relative to the base system, $\boldsymbol{r}_{P3,3}$ is a position vector of a point P_3 relative to the link 3 system

The position vector of point P_3 relative to the origin of the 0 system is obtained according to formula (3.14):

$$\boldsymbol{r}_{P3} = \boldsymbol{T}_{3,0}\, \boldsymbol{r}_{P3,3},$$

where $\boldsymbol{r}_{P3,3} = [0 \;\; 0 \;\; z_{p3} \;\; 1\,]^{\mathrm{T}}$ and $x_{P3} = y_{P3} = 0$, because point P_3 lies on the z_3 axis. Hence

$$\boldsymbol{r}_{P3} = \begin{bmatrix} (z_{P3}+\lambda_3)c\theta_1 s\theta_2 + \lambda_2 s\theta_1 \\ (z_{P3}+\lambda_3)s\theta_1 s\theta_2 - \lambda_2 c\theta_1 \\ -(z_{P3}+\lambda_3)c\theta_2 \\ 1 \end{bmatrix}$$

Using formula (3.21) the $\boldsymbol{Q}_j^*$ matrices are obtained:

$$\boldsymbol{Q}_1^* = \boldsymbol{Q}_1$$

$$\boldsymbol{Q}_z^* = \boldsymbol{T}_{1,0}\boldsymbol{Q}_1\boldsymbol{T}_{1,0}^{-1} = \begin{bmatrix} 0 & 0 & -c\theta_1 & 0 \\ 0 & 0 & -s\theta_1 & 0 \\ c\theta_1 & s\theta_1 & 0 & 0 \\ 0 & 0 & 0 & 1 \end{bmatrix}$$

Hence:

$$\boldsymbol{D}_3=\begin{bmatrix} 0 & -\dot\theta_1 & -\dot\theta_2 c\theta_1 & \dot\lambda_3 c\theta_1 s\theta_2 \\ \dot\theta_1 & 0 & -\dot\theta_2 s\theta_1 & \dot\lambda_3 s\theta_1 s\theta_2 \\ \dot\theta_2 c\theta_1 & \dot\theta_2 s\theta_1 & 0 & -\dot\lambda_3 c\theta_2 \\ 0 & 0 & 0 & 1 \end{bmatrix}$$

The velocity of point P_3 is determined by formula (3.19)

$$v_{P3}=\boldsymbol{D}_3\boldsymbol{r}_{P3}=\begin{bmatrix} (z_{P3}+\lambda_3)(-\dot\theta_1 s\theta_1 s\theta_2+\dot\theta_2 c\theta_1 c\theta_2)+\dot\lambda_3 c\theta_1 s\theta_2+\lambda_2\dot\theta_1 c\theta_1 \\ (z_{P3}+\lambda_3)(\dot\theta_1 c\theta_1 s\theta_2+\dot\theta_2 s\theta_1 c\theta_2)+\dot\lambda_3 s\theta_1 s\theta_2+\lambda_2\dot\theta_1 s\theta_1 \\ (z_{P3}+\lambda_3)\dot\theta_2 s\theta_2-\dot\lambda_3 c\theta_2 \\ 1 \end{bmatrix}$$

The acceleration of point P_3 is determined by formula (3.20), with $\dot{\boldsymbol{D}}_3$ determined beforehand by differentiating $\boldsymbol{D}_3$:

$$\dot{\boldsymbol{D}}_3=\begin{bmatrix} 0 & -\ddot\theta_1 & (\dot\theta_1\dot\theta_2 s\theta_1-\ddot\theta_2 c\theta_1) & \ddot\lambda_3 c\theta_1 s\theta_2+\dot\lambda_3(-\dot\theta_1 s\theta_1 s\theta_2+\dot\theta_2 c\theta_1 c\theta_2) \\ \ddot\theta_1 & 0 & (-\dot\theta_1\dot\theta_2 c\theta_1-\ddot\theta_2 s\theta_1) & \ddot\lambda_3 s\theta_1 s\theta_2+\dot\lambda_3(\dot\theta_1 c\theta_1 s\theta_2+\dot\theta_2 s\theta_1 c\theta_2) \\ a_{31} & a_{32} & 0 & -\ddot\lambda_3 c\theta_2+\dot\lambda_3\dot\theta_2 s\theta_2 \\ 0 & 0 & 0 & 1 \end{bmatrix}$$

where $a_{31}=-\dot\theta_1\dot\theta_2 s\theta_1+\ddot\theta_2 c\theta_1$, and $a_{32}=\dot\theta_1\dot\theta_2 c\theta_1+\ddot\theta_2 s\theta_1$

Substituting matrices $\boldsymbol{D}_3$ and $\dot{\boldsymbol{D}}_3$ and vectors $\boldsymbol{r}_{p3}$ and v_{p3} in expression (3.20), we have an acceleration vector of point P_3 in the form of:

$$\boldsymbol{a}_{P3}=[\ a_{P3x}\ \ a_{P3y}\ \ a_{P3z}\ \ 1\ ]^{\mathrm{T}}$$

where

$$a_{P3x}=(z_{P3}+\lambda_3)(-\ddot\theta_1 s\theta_1 s\theta_2+\ddot\theta_2 c\theta_1 c\theta_2-\dot\theta_1^2 c\theta_1 s\theta_2-2\dot\theta_1\dot\theta_2 s\theta_1 c\theta_2-\dot\theta_2^2 c\theta_1 s\theta_2)-$$
$$-2\dot\lambda_3(\dot\theta_1 s\theta_1 s\theta_2-\dot\theta_2 c\theta_1 c\theta_2)+\ddot\lambda_3 c\theta_1 s\theta_2+\lambda_2(\ddot\theta_1 c\theta_1-\dot\theta_1^2 s\theta_1)$$
$$a_{P3y}=(z_{P3}+\lambda_3)(\ddot\theta_1 c\theta_1 s\theta_2+\ddot\theta_2 s\theta_1 c\theta_2-\dot\theta_1^2 s\theta_1 s\theta_2+2\dot\theta_1\dot\theta_2 c\theta_1 c\theta_2-\dot\theta_2^2 s\theta_1 s\theta_2)+$$
$$+2\dot\lambda_3(\dot\theta_1 c\theta_1 s\theta_2+\dot\theta_2 s\theta_1 c\theta_2)+\ddot\lambda_3 s\theta_1 s\theta_2+\lambda_2(\ddot\theta_1 s\theta_1+\dot\theta_1^2 c\theta_1)$$
$$a_{P3z}=(z_{P3}+\lambda_3)(\ddot\theta_2 s\theta_2+\dot\theta_2^2 c\theta_2)+2\dot\lambda_3\dot\theta_2 s\theta_2+\ddot\lambda_3 c\theta_2$$

With a sufficiently dense table of the positions of point P_3 as a function of time and employing a computer to use interpolation formulas, it is possible to determine that point's velocity and acceleration. If an accuracy on the order of a couple of per cent is satisfactory, the simplest formulas will suffice. The more complicated the manipulator, the greater the practical advantage this fact [3.7].

The angular velocities of links *2* and *3* are determined using expressions (3.26) and (3.27):

$$\omega_2 = \dot{\theta}_1 e_1 + B_1 \dot{\theta}_2 e_2 = \begin{bmatrix} 0 \\ 0 \\ \dot{\theta}_1 \end{bmatrix} + \begin{bmatrix} c\theta_1 & 0 & s\theta_1 \\ s\theta_1 & 0 & -c\theta_1 \\ 0 & 1 & 0 \end{bmatrix} \begin{bmatrix} 0 \\ 0 \\ \dot{\theta}_2 \end{bmatrix} = \begin{bmatrix} \dot{\theta}_2 s\theta_1 \\ -\dot{\theta}_2 c\theta_1 \\ \dot{\theta}_1 \end{bmatrix} = \omega_3$$

$$\omega_{2,2} = B_{2,0}^{-1}\omega_2 = \begin{bmatrix} c\theta_1 c\theta_2 & s\theta_1 c\theta_2 & s\theta_2 \\ s\theta_1 & -c\theta_1 & 0 \\ c\theta_1 s\theta_2 & s\theta_1 s\theta_2 & -c\theta_2 \end{bmatrix} \begin{bmatrix} \dot{\theta}_2 s\theta_1 \\ -\dot{\theta}_2 c\theta_1 \\ \dot{\theta}_1 \end{bmatrix} = \begin{bmatrix} \dot{\theta}_1 s\theta_2 \\ \dot{\theta}_2 \\ -\dot{\theta}_1 c\theta_2 \end{bmatrix}$$

where

$$B_{2,0}^{-1} = \begin{bmatrix} l_x & m_x & n_x \\ l_y & m_y & n_y \\ l_z & m_z & n_z \end{bmatrix}$$

It is better to use recurrent formulas in order to facilitate programming:

$$\boldsymbol{r}_{Pi,i-1} = \boldsymbol{A}_i \, \boldsymbol{r}_{Pi,i}$$
$$\boldsymbol{r}_{Pi,i-2} = \boldsymbol{A}_{i-1} \, \boldsymbol{r}_{Pi,i-1}$$

. . . .

$$\boldsymbol{r}_{Pi,0} = \boldsymbol{A}_1 \, \boldsymbol{r}_{Pi,1}$$

Recurrent formulas can also be used instead formulas (3.15)-(3.24) to compute the velocity and acceleration vectors according to:

$$\boldsymbol{v}_{Pi,i-1} = \boldsymbol{Q}_i \, \dot{q}_i \, \boldsymbol{r}_{Pi,i-1}$$
$$\boldsymbol{v}_{Pi,i-2} = \boldsymbol{Q}_{i-1} \, \dot{q}_{i-1} \boldsymbol{r}_{Pi,i-2} + \boldsymbol{A}_{i-1} \, \boldsymbol{v}_{Pi,i-1}$$

. . . .

$$\boldsymbol{v}_{Pi,0} = \boldsymbol{Q}_1 \, \dot{q}_1 \boldsymbol{r}_{Pi,0} + \boldsymbol{A}_1 \, \boldsymbol{v}_{Pi,1}$$

and

$$\boldsymbol{a}_{Pi,i-1} = \boldsymbol{Q}_i \left(\ddot{q}_i + \dot{q}_i^2\right) \boldsymbol{r}_{Pi,i-1}$$

$$\boldsymbol{a}_{Pi,i-2} = \left(\boldsymbol{Q}_{i-1}\ddot{q}_{i-1} + \boldsymbol{Q}_{i-1}^2 \dot{q}_{i-1}^2\right) \boldsymbol{r}_{Pi,i-2} + \boldsymbol{A}_{i-1} \boldsymbol{a}_{Pi,i-1} + 2\boldsymbol{Q}_{i-1}\dot{q}_{i-1}\boldsymbol{A}_{i-1}\boldsymbol{v}_{Pi,i-1}$$

. . . .

$$\boldsymbol{a}_{Pi,0} = \left(\boldsymbol{Q}_1\ddot{q}_1 + \boldsymbol{Q}_1^2 \dot{q}_1^2\right) \boldsymbol{r}_{Pi,0} + \boldsymbol{A}_1 \boldsymbol{a}_{Pi,1} + 2\boldsymbol{Q}_1\dot{q}_1\boldsymbol{A}_1\boldsymbol{v}_{Pi,1}$$

where

$\boldsymbol{Q}_i$, $\dot{q}_i$, $\ddot{q}_i$ are described by the expressions (3.15b).

3.4. References

[3.1] *Angeles J.*: Rational Kinematics. New York, Springer Verlag. 1988.

[3.2] *Craig J. J.*: Introduction to Robotics. Addison-Wesley 1989.

[3.3] *Duffy J.*: Analysis of Mechanisms and Robot Manipulators. London, E.Arnold, 1980.

[3.4] *Frolov K. V., Vorobiev E. J.*: Mechanics of Industrial Robots. Moscow, Vyssaja Skola, 1988 (*in Russian*).

[3.5] *Fu K. S., Gonzales R. C., Lee C. S. G.*: Robotics: Control, Sensing, Vision, and Inteligence. McGraw-Hill Inc. 1987.

[3.6a] *Hunt K. H.*: Robot Kinematics. A Compact Analytic Inverse Solution for Velocities. Trans. ASME, Journal of Mech., Transm. and Autom. in Design, v.109, pp. 42–49, 1987.

[3.6b] *Hunt K. H.*: The particular or the general? (some examples from robot kinematics). Mech. and Mach. Theory, v.21, No 6, pp.481–487, 1989.

[3.7] *Knapczyk J., Lebiediew P.A.*: Theory of Spatial Mechanisms and Manipulators. Warszawa, WNT, 1990 (*in Polish*).

[3.8] *Lee H. Y., Woernle C., Hiller M.*: A Complete Solution for the Inverse Kinematic Problem of the General 6R Robot Manipulator. Trans. ASME, Journal of Mech., Transm. and Autom. in Design, v.113, pp.481–6, 1991.

[3.9] *Lipkin H., Duffy J.*: A Vector Analysis of Robot Manipulators. Recent Advances in Robotics. New York, J.Wiley, 1985.

[3.10] *Litvin F. L., Parenti Castelli V., Phillips R. H.*: Manipulators: Execution of Prescribed Trajectories. Special Link Positions and Versions of Assembly. Mech. and Mach. Theory, v.21, No 2, pp.173–85, 1986.

[3.11] *Lloyd J., Hayward V.*: Kinematics of Common Industrial Robots. Robotics 4 (1988), pp.169–191.

[3.12] *Manseur R., Doty K.L.*: A Fast Algorithm for Inverse Kinematic Analysis of Robot Manipulators. Int. Journal of Robotics Research,.v.7, No 3, pp.52–63, 1988.

[3.13] *Paul R. P.*: Robot Manipulators: Mathematics, Programming and Control. Cambridge. MIT Press, 1981.

[3.14] *Pennock G. R., Yang A. T.*: Application of Dual-Number Matrices to the Inverse Kinematics Problem of Robot Manipulators. Trans. ASME, Journal of Mech. Transm. and Autom. in Design, v.107, No 2, pp.201–208, 1985.

[3.15] *Pieper D. L.*: The Kinematics of Manipulators under Computer Control. Ph. D. Thesis, Stanford Univ. 1968.

[3.16] *Raghavan M., Roth B.*: A General Solution for the Inverse Kinematics of all Series Chains. Proc. of the 8th CISM-IFTOMM Symp. RoManSy'90 Cracow 1990, pp.21–28.

[3.17] *Ranky P. G., Ho C. Y.*: Robot Modelling Control and Applications with Software. Kempston, JFS, 1985.

[3.18] *Tsai L. W., Morgan A.*: Solving the Kinematics of the Most General Six- and Five-Degree of Freedom Mnipulators by Continuation Methods. Trans. ASME, Journal of Mech., Transm. and Autom. in Design, v.107, pp.189–200, 1985.

4. THE INVERSE KINEMATICS OF MANIPULATORS

4.1. Introduction

The objective of inverse kinematics task is to find all the possible sets of angular or linear displacements (configuration coordinates) in the joints that allow of the end-effector (gripper or tool) of the manipulator to assume a certain position and/or orientation. This is a fundamental problem in the programming and control of manipulator motion, when it is necessary to determine how particular configuration coordinates change in time, in order to have the end-effector perform a desired motion of [4.2, 4.6-4.17]. For instance, in the simplest positioning task, "take and place", the initial and final position of the end-effector are given as the time it takes to cover the distance between those two positions. The inverse kinematics solution entails determining the values of the configuration coordinates corresponding to these positions.

For certain kinds of manipulators with five and six degrees of freedom, it is possible to find the inverse kinematics solution in a closed form. Pieper [4.15] has proved that the condition sufficient for the existence of a solution in a closed form is that the axes of the three successive joints intersect in one point.

The problem can be solved in three different ways: matrix, vector, and numerical. In the matrix method, the homogeneous transformation matrices in Denavit-Hartennberg notation are used [3.2, 3.16]; in the vector method [3.7], a three unit vector equation is employed, and in the numerical method, the Newton-Raphson or similar procedure [3.1].

4.2. The matrix method

As we have learned in the subsection 3.3, if the position and orientation of the *i*-th manipulator's link are known, then $T_{i,0}$, the position matrix of that joint relative to the base reference system *0* is also known:

$$T_i = T_{i,0} = \begin{bmatrix} \boldsymbol{l}_i & \boldsymbol{m}_i & \boldsymbol{n}_i & \boldsymbol{p}_i \\ 0 & 0 & 0 & 1 \end{bmatrix} = \begin{bmatrix} & \boldsymbol{B}_i & & \boldsymbol{p}_i \\ 0 & 0 & 0 & 1 \end{bmatrix} \tag{4.1}$$

where

$$B_i = \begin{bmatrix} l_{i,x} & m_{i,x} & n_{i,x} \\ l_{i,y} & m_{i,y} & n_{i,y} \\ l_{i,z} & m_{i,z} & n_{i,z} \end{bmatrix}, \quad p_i = \begin{bmatrix} p_{i,x} \\ p_{i,y} \\ p_{i,z} \end{bmatrix}$$

The configuration variable must satisfy the matrix equation

$$A_1 A_2 \dots A_i = T_i \tag{4.2}$$

The position vector of the origin of the (i-1)-th link's reference system relative to this system is described in the form of a column matrix:

$$p_{i-1,i-1} = [0\ 0\ 0\ 1]^{\mathrm{T}} \tag{4.3a}$$

with respect to the base reference system 0, it is determined according to relation (4.2):

$$p_{i-1} = A_1 A_2 \dots A_i\, p_{i-1,i-1} \tag{4.3b}$$

In view of equation (4.2), relation (4.3b) can be expressed in the form of

$$p_{i-1} = T_i A_i^{-1} p_{i-1,i-1} \tag{4.3c}$$

Hence, considering the expression (4.3), we obtain

$$p_{i-1} = p_i + l_i l_i - (m_i s\alpha_i + n_i c\alpha_i)\lambda_i \tag{4.3d}$$

Relation (4.3d) replaces the left side of equation (4.3b); multiplying the left side of the equation obtained by matrix A_1^{-1} we have:

$$A_1^{-1}\left[p_i + l_i l_i - (m_i s\alpha_i + n_i c\alpha_i)\lambda_i\right] = A_2 A_3 \dots A_i\, p_{i-1,i-1} \tag{4.3e}$$

from which the three configuration variables, giving the position vector p_i and the orientation matrix B_i, can be determined.

The rest of the problem can be solved by applying the orientation equation of the (i-1)-th link relative to the base 0

$$n_{i-1} = A_1 A_2 \dots A_{i-1} n_{i-1,i-1} \tag{4.4a}$$

where the unit vector is described in the base reference system, while

$$n_{i-1,i-1} = [0\ 0\ 1\ 0]^{\mathrm{T}} \tag{4.4b}$$

The unit vector n_{i-1} can be determined from the relation

$$n_{i-1} = T_i A_i^{-1} n_{i-1,i-1} \tag{4.4c}$$

from which we obtain

$$n_{i-1} = m_i s\alpha_i + n_i c\alpha_i \tag{4.4d}$$

Having obtained the unit vectors of orientation of the i-th link, m_i and n_i, by using the relation (4.4d), the unit vector n_{i-1} of the z_{i-1} axis of the (i-1)-th link's reference system can be determined.

Equation (4.4a) can be transformed into

$$A_{i-3}^{-1} \dots A_2^{-1} A_1^{-1} n_{i-1} = A_{i-2} A_{i-1} n_{i-1,i-1}$$

$$A_{i-2}^{-1} \dots A_2^{-1} A_1^{-1} n_{i-1} = A_{i-1} n_{i-1,i-1} \tag{4.4e}$$

From matrix equations (4.4e) adequate equations containing the rest of the unknown configuration variables can be derived. Those equations can be converted into the form of typical trigonometric equations. Six such equations can be found:

(a) $A\,c\theta + B\,s\theta = D$ (4.5a)

where A, B and D are constants.

If $D = 0$, we have two solutions:

$$\theta = Atan\,2(-A,B)$$

or $\theta = \theta + \pi$

If $D \neq 0$ and $A^2 + B^2 > D^2$

we also obtain two solutions:

$$\theta = 2\,Atan\,2\left(B \pm \sqrt{A^2+B^2-D^2}, A+D\right.$$

Atan $2(y, x)$ is a two argument arctangent function [4.2] which yields the value of arctg (y, x), and also verifies whether x and y are positive or negative, in order to determine the quadrant in which the angle lies.

(b) $s\theta = D_1, \quad c\theta = D_2,$ (4.5b)

where D_1 and D_2 are constants; there is only one solution:

$$\theta = Atan\,2(D_1, D_2)$$

(c) $X s\theta = D_1, \quad X c\theta = D_2,$ (4.5c)

where X is variable, D_1 and D_2 are constants; there is only one solution:

$$\theta = Atan\,2(D_1, D_2), \quad \text{if } X > 0$$

$$\theta = \theta + \pi \quad \text{if } X < 0$$

(d) $A\,c\theta_{ij} + B\,c\theta_i = D_1$

$A\,s\theta_{ij} + B\,s\theta_i = D_2$ (4.5d)

where $\theta_{ij} = \theta_i + \theta_j$, while A, B, D_1 and D_2 are constants.

Squaring both equations and adding their respective sides yields two solutions for θ_j:

$$\theta_j = Atan\,2\left(\pm\sqrt{1-K^2}, K\right)$$

where $K = \left(D_1^2 + D_2^2 - A^2 - B^2\right)/2AB$

We then find θ_i according to the formula:

$$\theta_j = Atan\,2(K_1, K_2)$$

where $K_1 = D_2\left(Ac\theta_j + B\right) - D_1\left(As\theta_j\right)$

$$K_2 = D_1\left(Ac\theta_j + B\right) - D_2\left(As\theta_j\right)$$

(e) $A\,s\theta_{ij} + B\,c\theta_{ij} + D\,c\theta_i = F_1$

$A\,c\theta_{ij} + B\,s\theta_{ij} + D\,s\theta_i = F_2$ (4.5e)

where A, B, D, F_1 and F_2 are constants. This system of two equations can be reduced to one equation:

$$F_1 c\theta_i + F_2 s\theta_i = \left(F_1^2 + F_2^2 + D^2 - A^2 - B^2\right)/2D$$

which is solved for θ_i, similarly to the (a) type equation.

Once θ_i has been determined, θ_{ij} can be obtained by means of formula

$$\theta_j = Atan\,2\left(AK_1 + BK_2, BK_1 - AK_2\right)$$

where $K_1 = F_1 - Dc\theta_i$

$$K_2 = F_2 - Ds\theta_i$$

Then the other unknown is computed using:

$$\theta_j = \theta_{ij} - \theta_i$$

(f) $A\, s\theta_{ij} + D\, c\theta_i = F_1$ (4.5f)

$$-A\, c\theta_{ij} + D\, s\theta_i = F_2$$

The (4.5f) system of equations is similar to the system (e), (except that $B = 0$), and is solved in the same way.

Six degrees of freedom suffice to position and orient a link in three dimensional Cartesian space. If $n > 6$, the manipulator has redundant mobility, and the system of equations derived from the matrix equation (4.2) is not determined. If $n < 6$, the system becomes overdeterminate. We shall consider only the case of $n = 6$ here. In the case of $n = 6$, using the equations (4.3) and (4.4), twelve non-trivial scalar equations with six unknown are obtained from the matrix equation (4.2).

Another method for solving the inverse kinematics problem takes advantage of the rotation matrix property. Having written the product of matrices A in the form of

$$A_i A_j = \begin{bmatrix} B_i B_j & B_i p_j + p_i \\ 0 \;\; 0 \;\; 0 & 1 \end{bmatrix} \tag{4.6}$$

equation (4.1) can be broken down into position and orientation equations:

$$\boldsymbol{p} = \boldsymbol{B}_1 (\boldsymbol{B}_2 (\boldsymbol{B}_3 (\boldsymbol{B}_4 (\boldsymbol{B}_5 \boldsymbol{p}_6 + \boldsymbol{p}_5) + \boldsymbol{p}_4) + \boldsymbol{p}_3) + \boldsymbol{p}_2) + \boldsymbol{p}_1 \tag{4.7}$$

$$\boldsymbol{B} = \boldsymbol{B}_1 \boldsymbol{B}_2 \boldsymbol{B}_3 \boldsymbol{B}_4 \boldsymbol{B}_5 \boldsymbol{B}_6$$

where $\boldsymbol{p}_i = [l_i c\theta_i \; l_i s\theta_i \; \lambda_i]^T$ is a position vector of origin 0_i relative to 0_{i-1}, and $\boldsymbol{p}$ is a position vector of the origin of the end-effector reference system relative to the base. When the last joint is the revolute one, then $\boldsymbol{p}_6 = 0$, and the equation (4.7) is simplified to the form of

$$\boldsymbol{p} = \boldsymbol{B}_1 (\boldsymbol{B}_2 (\boldsymbol{B}_3 (\boldsymbol{B}_4 \boldsymbol{p}_5 + \boldsymbol{p}_4) + \boldsymbol{p}_3) + \boldsymbol{p}_2) + \boldsymbol{p}_1 \tag{4.8}$$

Since rotations are orthogonal transformations, they leave dot products unvaried, that is:

$$\boldsymbol{Bu} \cdot \boldsymbol{Bv} = \boldsymbol{u} \cdot \boldsymbol{v} \tag{4.8a}$$

for any rotation matrix $\boldsymbol{B}$ and any vectors $\boldsymbol{u}$ and $\boldsymbol{v}$. A special case of (4.8a), which is often useful, is:

$$\boldsymbol{Bu} \cdot \boldsymbol{v} = \boldsymbol{u} \cdot \boldsymbol{B}^{-1} \boldsymbol{v} \tag{4.8b}$$

That property makes it easier to eliminate algebraic terms and unnecessary variables from the equation (4.8). Thus:

$$\boldsymbol{B}_i^{-1} \boldsymbol{p}_i = \left[l_i \quad \lambda_i s\alpha_i \quad \lambda_i c\alpha_i\right]^T \tag{4.9a}$$

and

$$B_i^{-1}k = [0 \quad s\alpha_i \quad c\alpha_i]^T, \text{ for } k = [0\ 0\ 1]^T \tag{4.9b}$$

are independent of θ_i for i-th joint. Additionally, the end-effector reference system has been chosen so as to have $\alpha_6 = 0$. Consequently,

$$B_6 k = k \tag{4.9c}$$

By the multiple use of the expressions (4.9), four reduced equations are obtained. The first is derived as follows:

$$\begin{aligned} n_z &= \boldsymbol{n} \cdot \boldsymbol{k} = (\boldsymbol{B}\, \boldsymbol{k}) \cdot \boldsymbol{k} \\ n_z &= (\boldsymbol{B}_1\, \boldsymbol{B}_2\, \boldsymbol{B}_3\, \boldsymbol{B}_4\, \boldsymbol{B}_5\, \boldsymbol{B}_6\, \boldsymbol{k}) \cdot \boldsymbol{k} \\ n_z &= (\boldsymbol{B}_1\, \boldsymbol{B}_2\, \boldsymbol{B}_3\, \boldsymbol{B}_4\, \boldsymbol{B}_5\, \boldsymbol{k}) \cdot \boldsymbol{k} \\ n_z &= \boldsymbol{k} \cdot (\boldsymbol{B}_5^{-1}\boldsymbol{B}_4^{-1}\boldsymbol{B}_3^{-1}\boldsymbol{B}_2^{-1}\boldsymbol{B}_1^{-1}\boldsymbol{k}) \end{aligned} \tag{4.10a}$$

The last expression is called equation n_z.

The second reduced equation, called equation p_z, is obtained as follows:

$$\begin{aligned} \boldsymbol{p} &= \boldsymbol{B}_1\, \boldsymbol{B}_2\, \boldsymbol{B}_3\, \boldsymbol{B}_4\, \boldsymbol{q} \\ \boldsymbol{q} &= \boldsymbol{p}_5 + \boldsymbol{B}_4^{-1}\{\boldsymbol{p}_4 + \boldsymbol{B}_3^{-1}[\boldsymbol{p}_3 + \boldsymbol{B}_2^{-1}(\boldsymbol{p}_2 + \boldsymbol{B}_1^{-1}\boldsymbol{p}_1)]\} \\ p_z &= \boldsymbol{p} \cdot \boldsymbol{k} = \boldsymbol{q} \cdot (\boldsymbol{B}_4^{-1}\boldsymbol{B}_3^{-1}\boldsymbol{B}_2^{-1}\boldsymbol{B}_1^{-1})\, \boldsymbol{k} \end{aligned} \tag{4.10b}$$

The third and fourth reduced equations are obtained from dot products:

$$\boldsymbol{p} \cdot \boldsymbol{n} = \boldsymbol{B}_5^{-1}\boldsymbol{q} \cdot \boldsymbol{k} \tag{4.10c}$$

$$\boldsymbol{p} \cdot \boldsymbol{p} = p^2 = \boldsymbol{q} \cdot \boldsymbol{q} = q^2 \tag{4.10d}$$

Since $\boldsymbol{B}_1^{-1}\boldsymbol{p}_1$ and $\boldsymbol{B}_1^{-1}\boldsymbol{k}$ are independent of θ_i, vector $\boldsymbol{q}$ can be considered independent of the first and last joint variables and therefore forms a system of four equations (4.10) with four unknowns ($i = 2, 3, 4, 5$). When the system has been solved, the last two variables can be found by means of the rest of equations (4.2). The simplification which results from the use of the invariants (4.9), is not only economical, but also provides greater insight into the structure and properties of inverse kinematics equations.

Another problem, considering equation (4.10) alone, is the presence of sign ambiguities. In many practical situations one of these equations allows a closed form solution for both $s\theta_i$ and $c\theta_i$. Other functions require the use of Pythagorean identity which yields two values of opposite signs. In some cases the number of sign ambiguities can be reduced by considering other constraints arising from the equations (4.2). Using additional equations can also help eliminate extraneous solutions and simplify the procedure. Equations for the components p_x, p_y, n_x and n_y constitute additional constraints:

$$p_x = (\boldsymbol{B}_1\, (\boldsymbol{B}_2\, (\boldsymbol{B}_3\, (\boldsymbol{B}_4\, \boldsymbol{p}_5 + \boldsymbol{p}_4) + \boldsymbol{p}_3) + \boldsymbol{p}_2) + \boldsymbol{p}_1) \cdot \boldsymbol{i}$$

$$p_y = (\boldsymbol{B}_1\, (\boldsymbol{B}_2\, (\boldsymbol{B}_3\, (\boldsymbol{B}_4\, \boldsymbol{p}_5 + \boldsymbol{p}_4) + \boldsymbol{p}_3) + \boldsymbol{p}_2) + \boldsymbol{p}_1) \cdot \boldsymbol{j}$$

$$n_x = \boldsymbol{B_1 B_2 B_3 B_4 B_5 k \cdot i} \tag{4.11}$$
$$n_y = \boldsymbol{B_1 B_2 B_3 B_4 B_5 k \cdot j}$$

where $\boldsymbol{i} = [1\ 0\ 0]^T$, $\boldsymbol{j} = [0\ 1\ 0]^T$ are the unit vectors of the base reference system.

A serial kinematic chain with revolute and prismatic joints is orthogonal if all twist angles α_i ($i = 1, \ldots, n$) are 0 or $\pi/2$. The orthogonal manipulator with six degrees of freedom can be classified in terms of the values of their twist angles α_i. Since α_6 is always zero, there are only $2^5 = 32$ distinct classes of orthogonal manipulators, eight of which have four or more adjacent parallel joint axes, which reduces the number of degrees of freedom to less than six. As a result, there are only twenty-four orthogonal manipulators with full capability of spatial positioning and orientation. Most industrial robots are orthogonal manipulators.

Matrices $\boldsymbol{A}_i$ for the orthogonal manipulators take one of the following forms:

$$A_i(\alpha = 0) = \begin{bmatrix} c_i & -s_i & 0 & l_i c_i \\ s_i & c_i & 0 & l_i s_i \\ 0 & 0 & 1 & \lambda_i \\ 0 & 0 & 0 & 1 \end{bmatrix} \tag{4.12}$$

$$A_i(\alpha = \pi/2) = \begin{bmatrix} c_i & 0 & s_i & l_i c_i \\ s_i & 0 & -c_i & l_i s_i \\ 0 & 1 & 0 & \lambda_i \\ 0 & 0 & 0 & 1 \end{bmatrix} \tag{4.13}$$

Hence:

$$\boldsymbol{B}_i \boldsymbol{k} = \boldsymbol{B}_i^{-1} \boldsymbol{k} = \boldsymbol{k} \qquad \text{if } \alpha_i = 0$$
$$\boldsymbol{B}_i^{-1} \boldsymbol{k} = \boldsymbol{j} \qquad \text{if } \alpha_i = 0$$

Using this approach, Doty [4.12] has shown that of the twenty-four non-trivial orthogonal manipulators, only those with two non-zero twist angles have a closed form solution. We can specify the following cases:

(a) $\alpha_i = \pi/2, 0, 0, \pi/2, 0$ (b) $\alpha_i = 0, \pi/2, 0, \pi/2, 0$
(c) $\alpha_i = 0, 0, \pi/2, 0, \pi/2$ (d) $\alpha_i = 0, 0, \pi/2, \pi/2, 0$
(e) $\alpha_i = 0, \pi/2, 0, 0, \pi/2$

In the case of the manipulator with six joints with isolated positioning and orienting chains, equation (4.1) may assume the form of:

$$\begin{bmatrix} \boldsymbol{B} & & & \boldsymbol{p} \\ 0 & 0 & 0 & 1 \end{bmatrix} = \begin{bmatrix} \boldsymbol{B}_{3,0} & & & \boldsymbol{p}_r \\ 0 & 0 & 0 & 1 \end{bmatrix} \begin{bmatrix} \boldsymbol{B}_{6,3} & & & \boldsymbol{p}_w \\ 0 & 0 & 0 & 1 \end{bmatrix} \tag{4.14}$$

where

$\boldsymbol{B}_{3,0}$ is an orientation matrix of the third link reference system relative to base,

$\boldsymbol{p}_r$ is a position vector of the origin of the 3rd link system relative to the base,

$\boldsymbol{p}_w$ is a position vector of the origin of the end-effector (6th) reference system relative to the 3rd link system,

$\boldsymbol{B}_{6,3}$ is an orientation matrix of the end-effector reference system relative to the 3rd link system.

On the basis of the equation (4.14) we obtain

$$\boldsymbol{B} = \boldsymbol{B}_{3,0}\,\boldsymbol{B}_{6,3} \qquad (4.15)$$
$$\boldsymbol{p} = \boldsymbol{B}_{3,0}\,\boldsymbol{p}_w + \boldsymbol{p}_r$$

where p_w is a wrist length, that is, the distance between the origin of the end-effector reference system and the origin of the third link system. In most industrial robots the vector $\boldsymbol{p}_w$ is defined by the approach unit vector $\boldsymbol{n}$. In that case the second equation of system (4.15) takes the form

$$\boldsymbol{p}_r = \boldsymbol{p} - p_w\,\boldsymbol{n} \qquad (4.16)$$

So, if the coordinates of the position vector and the orientation unit vector of the end-effector reference system are given, that is, $\boldsymbol{p} = [p_x\ p_y\ p_z]^T$, $\boldsymbol{n} = [n_x\ n_y\ n_z]^T$, together the wrist length p_w, applying formula (4.16) it is possible to determine the coordinates of $\boldsymbol{p}_r$, the position vector of the manipulator arm end. Vector $\boldsymbol{p}_r$ can also be computed as a product of transformation matrices of the first three joints, that is:

$$\boldsymbol{p}_r = \boldsymbol{B}_1\,\boldsymbol{B}_2\,\boldsymbol{p}_3 + \boldsymbol{B}_1\,\boldsymbol{p}_2 + \boldsymbol{p}_1 \qquad (4.17)$$

where

$$\boldsymbol{B}_i = \begin{bmatrix} c\theta_i & -s\theta_i c\alpha_i & s\theta_i s\alpha_i \\ s\theta_i & c\theta_i c\alpha_i & -c\theta_i s\alpha_i \\ 0 & s\alpha_i & c\alpha_i \end{bmatrix}, \quad \boldsymbol{p}_i = \begin{bmatrix} l_i c\theta_i \\ l_i s\theta_i \\ \lambda_1 \end{bmatrix} \qquad (4.17a)$$

When the wrist center is displaced by vector $\boldsymbol{p}_4 = \overline{O_3O_4}$, the following expression is used instead of the formula (4.17):

$$\boldsymbol{p}_r = \boldsymbol{B}_1\,\boldsymbol{B}_2\,\boldsymbol{B}_3\,\boldsymbol{p}_4 + \boldsymbol{B}_1\,\boldsymbol{B}_2\,\boldsymbol{p}_3 + \boldsymbol{B}_1\,\boldsymbol{p}_2 + \boldsymbol{p}_1 \qquad (4.18)$$

Substituting expression (4.17) or (4.18) in equation (4.16) yields a system of three scalar equations with regard to three unknowns θ_1, θ_2 and θ_3.

To find the unknown θ_4, θ_5 and θ_6 , the orientation equation from system (4 is transformed into:

$$\boldsymbol{B}_{6,3} = \boldsymbol{B}_{3,0}^{-1}\boldsymbol{B} \qquad (4.19)$$

noting that:

$$\boldsymbol{B}_{3,0}^{-1}(\boldsymbol{B}_1\boldsymbol{B}_2\boldsymbol{B}_3)^{-1} = \begin{bmatrix} l_{3x} & l_{3y} & l_{3z} \\ m_{3x} & m_{3y} & m_{3z} \\ n_{3x} & n_{3y} & n_{3z} \end{bmatrix} \qquad (4.20)$$

where $\boldsymbol{l}_3$, $\boldsymbol{m}_3$ and $\boldsymbol{n}_3$ are the unit vectors of the reference system attached to the third link.

Having found the values of θ_1, θ_2 and θ_3 from equation (4.20), the coordinates of the unit vectors $\boldsymbol{l}_3$, $\boldsymbol{m}_3$ and $\boldsymbol{n}_3$ can now be determined. From equation (4.19) the elements of the matrix $\boldsymbol{B}_{6,3}$ are calculated as dot products of the particular pairs of unit vectors of the orientation matrices $[\boldsymbol{l}\ \boldsymbol{m}\ \boldsymbol{n}]$ and $[\boldsymbol{l}_3\ \boldsymbol{m}_3\ \boldsymbol{n}_3]$. From equation (4.14) we can derive the relation:

$$\boldsymbol{B}_{6,3} = \boldsymbol{B}_4\boldsymbol{B}_5\boldsymbol{B}_6 \tag{4.21}$$

which can be converted into

$$\boldsymbol{B}_4^{-1}\boldsymbol{B}_{6,3} = \boldsymbol{B}_5\boldsymbol{B}_6 \tag{4.22}$$

In the most prevalent designs, such as the PUMA and ASEA robots, the axes of the wrist joints are mutually perpendicular, that is

$\alpha_4 = \pm\pi/2$, $\alpha_5 = \pm\pi/2$, $\alpha_6 = 0$. Thus, $c\alpha_4 = c\alpha_5 = 0$, $s\alpha_6 = 0$, $c\alpha_6 = 1$.

In view of these relations as well as (4.19) and (4.20), equation (4.22) can be written as

$$\begin{bmatrix} c\theta_4 & s\theta_4 & 0 \\ 0 & 0 & s\alpha_4 \\ s\alpha_4 s\theta_4 & -s\alpha_4 c\theta_4 & 0 \end{bmatrix}\begin{bmatrix} \boldsymbol{l}\cdot\boldsymbol{l}_{3x} & \boldsymbol{m}\cdot\boldsymbol{l}_{3y} & \boldsymbol{n}\cdot\boldsymbol{l}_{3z} \\ \boldsymbol{l}\cdot\boldsymbol{m}_{3x} & \boldsymbol{m}\cdot\boldsymbol{m}_{3y} & \boldsymbol{n}\cdot\boldsymbol{m}_{3z} \\ \boldsymbol{l}\cdot\boldsymbol{n}_{3x} & \boldsymbol{m}\cdot\boldsymbol{n}_{3y} & \boldsymbol{n}\cdot\boldsymbol{n}_{3z} \end{bmatrix} = \begin{bmatrix} c\theta_5 c\theta_6 & -c\theta_5 s\theta_6 & s\alpha_5 s\theta_5 \\ s\theta_5 c\theta_6 & -s\theta_5 s\theta_6 & s\alpha_5 c\theta_5 \\ s\alpha_5 s\theta_6 & s\alpha_5 c\theta_6 & 0 \end{bmatrix} \tag{4.23}$$

where

$$\begin{aligned}
\boldsymbol{l}\cdot\boldsymbol{l}_3 &= l_x c\theta_1 c(\theta_2+\theta_3) + l_y s\theta_1 c(\theta_2+\theta_3) - l_z s(\theta_2+\theta_3) \\
\boldsymbol{l}\cdot\boldsymbol{m}_3 &= -l_x s\theta_1 + l_y c\theta_1 \\
\boldsymbol{l}\cdot\boldsymbol{n}_3 &= l_x c\theta_1 s(\theta_2+\theta_3) + l_y s\theta_1 s(\theta_2+\theta_3) + l_z c(\theta_2+\theta_3) \\
\boldsymbol{m}\cdot\boldsymbol{l}_3 &= m_x c\theta_1 c(\theta_2+\theta_3) + m_y s\theta_1 c(\theta_2+\theta_3) - m_z s(\theta_2+\theta_3) \\
\boldsymbol{m}\cdot\boldsymbol{m}_3 &= -m_x s\theta_1 + m_y c\theta_1 \\
\boldsymbol{m}\cdot\boldsymbol{n}_3 &= m_x c\theta_1 s(\theta_2+\theta_3) + m_y s\theta_1 s(\theta_2+\theta_3) + m_z c(\theta_2+\theta_3) \\
\boldsymbol{n}\cdot\boldsymbol{l}_3 &= n_x c\theta_1 c(\theta_2+\theta_3) + n_y s\theta_1 c(\theta_2+\theta_3) - n_z s(\theta_2+\theta_3) \\
\boldsymbol{n}\cdot\boldsymbol{m}_3 &= -n_x s\theta_1 + n_y c\theta_1 \\
\boldsymbol{n}\cdot\boldsymbol{n}_3 &= n_x c\theta_1 s(\theta_2+\theta_3) + n_y s\theta_1 s(\theta_2+\theta_3) + n_z c(\theta_2+\theta_3)
\end{aligned} \tag{4.23a}$$

Having performed the multiplication on the left side of the equation (4.23), the elements (3,3) on both sides are equated, yielding the formula for θ_4 :

$$(\boldsymbol{n}\cdot\boldsymbol{l}_3)s\theta_4 - (\boldsymbol{n}\cdot\boldsymbol{m}_3)c\theta_4 = 0 \tag{4.24}$$

Equating the respective elements (1,3) and (2,3) on both sides of the equation (4.23) yields the system of equations with regard to $s\theta_5$ and $c\theta_5$:

$$\begin{aligned}
s\theta_5 &= [(\boldsymbol{n}\cdot\boldsymbol{l}_3)c\theta_4 - (\boldsymbol{n}\cdot\boldsymbol{m}_3)s\theta_4]/s\alpha_5 \\
c\theta_5 &= -(\boldsymbol{n}\cdot\boldsymbol{n}_3)s\alpha_4/s\alpha_5
\end{aligned} \tag{4.25}$$

Equating elements (3,1) and (3,2) yields:

$$s\theta_6 = [(\boldsymbol{l}\cdot\boldsymbol{l}_3)s\theta_4 - (\boldsymbol{l}\cdot\boldsymbol{m}_3)c\theta_4]s\alpha_4/s\alpha_5 \qquad (4.26)$$
$$c\theta_6 = [(\boldsymbol{m}\cdot\boldsymbol{l}_3)s\theta_4 - (\boldsymbol{m}\cdot\boldsymbol{m}_3)c\theta_4]s\alpha_4/s\alpha_5$$

The algorithm solving the inverse kinematics problem can be divided into the following blocks:

1) determine the coordinates of the vector $\boldsymbol{p}_6$ (the wrist center), applying formula (4.16)
2) determine the unknowns θ_1, θ_2 and θ_3, using equations (4.16), (4.17) and (4.18)
3) determine the remaining unknowns θ_4, θ_5 and θ_6, applying formulas (4.24)÷(4.26).

Example 4.1. We have a PUMA 560 manipulator with six revolute joints and relative locations (Fig.4.1) described by the following dimensions:

$$\alpha_i = -\pi/2, \quad 0, \quad \pi/2, \quad -\pi/2, \quad \pi/2, \quad 0$$
$$l_i = \quad 0, \quad l_2, \quad 0, \quad 0, \quad 0, \quad 0$$
$$\lambda_i = \quad 0, \quad \lambda_2, \quad 0, \quad \lambda_4, \quad 0, \quad \lambda_6$$

The inverse kinematics problem is to find: the position and orientation of the end-effector, namely $\boldsymbol{p} = [p_x\ p_y\ p_z]^T$ along with unit vectors $\boldsymbol{m} = [m_x\ m_y\ m_z]^T$ and $\boldsymbol{n} = [n_x\ n_y\ n_z]^T$, by finding the values of θ_i ($i = 1, 2, \ldots 6$).

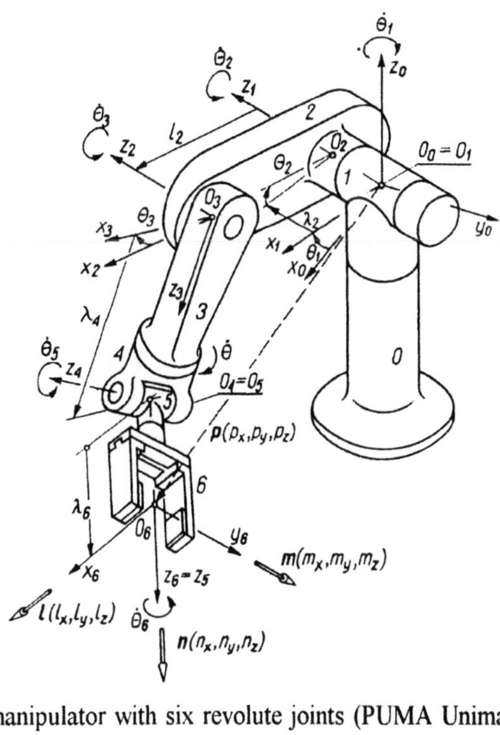

Fig.4.1. Structure of a manipulator with six revolute joints (PUMA Unimation) defined using D-H coordinates: $\alpha_i = -\pi/2, 0, \pi/2, -\pi/2, \pi/2, 0$; $l_i = 0, l_2, 0, 0, 0, 0$; $\lambda_i = 0, \lambda_2, 0, \lambda_4, 0, \lambda_6$

Taking the above given dimensions and using (4.17a) we obtain

$$
\begin{aligned}
&\boldsymbol{B}_1=\begin{bmatrix} c\theta_1 & 0 & -s\theta_1\\ s\theta_1 & 0 & c\theta_1\\ 0 & -1 & 0\end{bmatrix}, && \boldsymbol{p}_1=\begin{bmatrix}0\\0\\0\end{bmatrix}\\
&\boldsymbol{B}_2=\begin{bmatrix} c\theta_2 & -s\theta_2 & 0\\ s\theta_2 & c\theta_2 & 0\\ 0 & 0 & 1\end{bmatrix}, && \boldsymbol{p}_2=\begin{bmatrix}l_2c\theta_2\\l_2s\theta_2\\ \lambda_2\end{bmatrix}\\
&\boldsymbol{B}_3=\begin{bmatrix} c\theta_3 & 0 & s\theta_3\\ s\theta_3 & 0 & -c\theta_3\\ 0 & 1 & 0\end{bmatrix}, && \boldsymbol{p}_3=\begin{bmatrix}0\\0\\0\end{bmatrix}\\
&\boldsymbol{B}_4=\begin{bmatrix} c\theta_4 & 0 & -s\theta_4\\ s\theta_4 & 0 & c\theta_4\\ 0 & -1 & 0\end{bmatrix}, && \boldsymbol{p}_4=\begin{bmatrix}0\\0\\ \lambda_4\end{bmatrix}\\
&\boldsymbol{B}_5=\begin{bmatrix} c\theta_5 & 0 & s\theta_5\\ s\theta_5 & 0 & -c\theta_5\\ 0 & 1 & 0\end{bmatrix}, && \boldsymbol{p}_5=\begin{bmatrix}0\\0\\0\end{bmatrix}\\
&\boldsymbol{B}_6=\begin{bmatrix} c\theta_6 & -s\theta_6 & 0\\ s\theta_6 & c\theta_6 & 0\\ 0 & 0 & 1\end{bmatrix}, && \boldsymbol{p}_6=\begin{bmatrix}0\\0\\ \lambda_6\end{bmatrix}
\end{aligned}
\tag{4.27}
$$

We then determine

$$
\boldsymbol{B}_{3,0}=\boldsymbol{B}_1\boldsymbol{B}_2\boldsymbol{B}_3=\begin{bmatrix} c\theta_1c(\theta_2+\theta_3) & -s\theta_1 & c\theta_1s(\theta_2+\theta_3)\\ s\theta_1c(\theta_2+\theta_3) & c\theta_1 & s\theta_1s(\theta_2+\theta_3)\\ -s(\theta_2+\theta_3) & 0 & c(\theta_2+\theta_3)\end{bmatrix}
\tag{4.28}
$$

From the formula (4.18) we obtain

$$
\boldsymbol{p}_3=\begin{bmatrix} l_2c\theta_1c\theta_2-\lambda_2s\theta_1+\lambda_4c\theta_1s(\theta_2+\theta_3)\\ l_2s\theta_1c\theta_2+\lambda_2c\theta_1+\lambda_4s\theta_1s(\theta_2+\theta_3)\\ -l_2s\theta_2+\lambda_4c(\theta_2+\theta_3)\end{bmatrix}
\tag{4.29}
$$

Since in the manipulator considered $p_w=\lambda_6$, applying formula (4.16) we obtain:

$$
\begin{aligned}
&l_2c\theta_1c\theta_2-\lambda_2s\theta_1+\lambda_4c\theta_1s(\theta_2+\theta_3)=p_x-\lambda_6n_x\\
&l_2s\theta_1c\theta_2+\lambda_2c\theta_1+\lambda_4s\theta_1s(\theta_2+\theta_3)=p_y-\lambda_6n_y\\
&-l_2s\theta_2+\lambda_4c(\theta_2+\theta_3)=p_z-\lambda_6n_z
\end{aligned}
\tag{4.30}
$$

If the first equation of the system (4.30) is multiplied by $s\theta_1$, and the second by $(-c\theta_1)$, summing the sides of these yields

$$(p_x - \lambda_6 n_x)s\theta_1 - (p_y - \lambda_6 n_y)c\theta_1 = -\lambda_2$$

Hence, according to the formula (4.14a) we obtain

$$\theta_1 = Atan2\left(-\lambda_2, \pm\sqrt{p_{rx}^2 + p_{ry}^2 - \lambda_2^2}\right) - Atan2(-p_{ry}, p_{rx}) \tag{4.31}$$

where

$$\begin{aligned} p_{rx} &= p_x - \lambda_6 n_x \\ p_{ry} &= p_y - \lambda_6 n_y \\ p_{rz} &= p_z - \lambda_6 n_z \end{aligned} \tag{4.32}$$

If the first equation of system (4.30) is multiplied by $c\theta_1$, and the second by $s\theta_1$, the equation resulting from the side-summation of the two, together with the third, produces the system of equations

$$\begin{aligned} \lambda_4 s(\theta_2 + \theta_3) &= p_{rx} c\theta_1 + p_{ry} s\theta_1 - l_2 c\theta_2 \\ \lambda_4 c(\theta_2 + \theta_3) &= p_{rz} + l_2 s\theta_2 \end{aligned} \tag{4.33}$$

Squaring both sides of equations (4.33) and adding their respective sides yields the (4.14a) type equation

$$A\, s\theta_2 + B\, c\theta_2 = D$$

where

$$\begin{aligned} A &= -p_{rz} \\ B &= p_{rx}\, c\theta_1 + p_{ry}\, s\theta_1 \\ D &= \left(B^2 + p_{rz}^2 + l_2^2 - \lambda_4^2\right. /2l_2 \end{aligned} \tag{4.34a}$$

Hence, applying the formula (4.14a)

$$\theta_2 = Atan2\left(D, \pm\sqrt{A^2 + B^2 - D^2}\right) - Atan2(B, A) \tag{4.35}$$

After determining θ_2 we can go back to the system (4.33) and solve it with respect to sum $(\theta_2 + \theta_3)$, obtaining:

$$\theta_3 = Atan2(B - l_2 c\theta_2, p_{rz} + l_2 s\theta_2) - \theta_2 \tag{4.36}$$

The remaining unknowns θ_4, θ_5 and θ_6 can be determined by the procedure described above (see equations (4.24)÷(4.26)). From the equation (4.24) we obtain the solution:

$$\theta_4 = Atan2(\mathbf{n} \cdot \mathbf{m}_3, \mathbf{n} \cdot \mathbf{l}_3) \tag{4.37}$$

From expressions (4.35) and (4.35) we obtain

$$\theta_5 = Atan2((\mathbf{n} \cdot \mathbf{l}_3)c\theta_4 + (\mathbf{n} \cdot \mathbf{m}_3)s\theta_4, \mathbf{n} \cdot \mathbf{n}_3) \tag{4.38}$$

$$\theta_6 = Atan2(-(\mathbf{l} \cdot \mathbf{l}_3)s\theta_4 + (\mathbf{l} \cdot \mathbf{m}_3)c\theta_4, -(\mathbf{m} \cdot \mathbf{l}_3)s\theta_4 + (\mathbf{m} \cdot \mathbf{m}_3)c\theta_4) \tag{4.39}$$

This algorithm has been used to develop a computer program for programming a manipulator's motion along a given trajectory, that is the position and orientation of the end-effector as a function of time. It can also be used to determine the boundaries of the workspace and orientation space. For that purpose, the inverse kinematics problem is solved for consecutive points along the trajectory. The lack of solution indicates that the trajectory has gone beyond the boundaries. By using the bisect method for the last two points of trajectory the coordinates of the boundary - points can be determined more accurately.

Example 4.2. For the manipulator PUMA 560 discussed above, we assume the following dimensions: $l_2 = 0.4320;\ \lambda_2 = 0.1495;\ \lambda_4 = 0.4320;\ \lambda_6 = 0.056$, as well as the ranges for admissible values of variables ($\theta_{i,max}, \theta_{i,min}$):

θ_1 (−160°, +160°), θ_2 (−225°, +45°), θ_3 (−45°, +225°),
θ_4 (−110°, +170°), θ_5 (−100°, +100°), θ_6 (−266°, +266°).

The position and orientation of the end effector is described by the matrix

$$T_6 = \begin{bmatrix} -0.0461 & -0.7892 & -0.6124 & 0.3000 \\ 0.6609 & 0.4356 & -0.6112 & 0.2000 \\ 0.7491 & -0.4329 & 0.5015 & 0.1500 \\ 0 & 0 & 0 & 1 \end{bmatrix}$$

First, the coordinates of the arm end are computed, applying the formulas of (4.32):

$$p_{rx} = 0.3346, \qquad p_{ry} = 0.2345, \qquad p_{rz} = 0.1217$$

Then, from formula (4.31), we obtain

$$\theta_1 = Atan2\left(-0.1459,\ \pm\sqrt{0.3346^2 + 0.2345^2 - 0.1495^2}\right) - \\ - Atan2(-0.2345,\ 0.3346) = 35.036° \pm 21.459°$$

For further computations we take $\theta_1 = 13.576°$.
On the grounds of formula (4.35a) we now obtain:

$$A = -0.1217,\ B = 0.3803, \text{ and } D = 0.1846.$$

From formula (4.35b) we obtain:

$$\theta_2 = Atan2\left(0.1846,\ \pm\sqrt{0.1217^2 + 0.3803^2 - 0.1846^2}\right) - \\ - Atan2(-0.3803,\ 0.1217) = 72.249° \pm 27.529°$$

For further computations we take $\theta_2 = 44.720°$.
From formula (4.35b) we obtain:

$$\theta_3 = Atan2(0.3803 - 0.432\cos(44.720°),\ 0.1217 + 0.432\sin(44.720°)) - \\ - Atan2(-0.3803,\ 0.1217) = -34.943°$$

Then formula (4.23a) is used to compute the matrix elements

$$\begin{bmatrix} l\cdot l_3 & m\cdot l_3 & n\cdot l_3 \\ l\cdot m_3 & m\cdot m_3 & n\cdot m_3 \\ l\cdot n_3 & m\cdot n_3 & n\cdot n_3 \end{bmatrix} = \begin{bmatrix} -0.0202 & -0.5817 & -0.8132 \\ 0.6524 & 0.6087 & -0.4516 \\ 0.7576 & -0.5396 & 0.3672 \end{bmatrix}$$

Finally, we use formulas (3.67)÷(3.69) to obtain for one of the configurations:

$$\theta_4 = -Atan2(-0.4516,\ -0.8132) = 209.045°$$

We find the second configuration:

$$\theta_4 = -Atan2(0.4516,\ 0.8132) = 29.045°$$

The first of the values obtained is outside the limits, so we accept $\theta_4 = 29.045°$ and compute:

$$\theta_5 = Atan2(-0.8132\cos(29.045°) - 0.4516\sin(29.045°),\ 0.3672) = -68.458°$$

$$\theta_6 = Atan2(-0.0202\sin(29.045°) + 0.6524\cos(29.045°),\\ 0.5817\sin(29.045°) + 0.6087\cos(29.045°)) = -35.460°$$

Substituting the obtained values of θ_4, θ_5 and θ_6 into equation (4.23) we can verify that they have been computed correctly.

Example 4.3. Solve the inverse kinematics problem for a PUMA-type manipulator with geometric parameters different from those of the previous example:

$$\begin{array}{lcccccc} \alpha_i = & \pi/2, & 0, & \pi/2, & \pi/2, & \pi/2, & 0 \\ l_i = & 0, & l_2, & l_3, & 0, & 0, & 0 \\ \lambda_i = & 0, & 0, & \lambda_3, & \lambda_4, & 0, & 0 \end{array}$$

Equations (4.7)÷(4.11) are used, written in the case considered as:

$$\begin{aligned} n_x &= c\theta_1 c(\theta_2+\theta_3)c\theta_4 s\theta_5 + s\theta_1 s\theta_4 s\theta_5 - c\theta_1 s(\theta_2+\theta_3)c\theta_5 \\ n_y &= s\theta_1 c(\theta_2+\theta_3)c\theta_4 s\theta_5 - c\theta_1 s\theta_4 s\theta_5 - s\theta_1 s(\theta_2+\theta_3)c\theta_5 \\ n_z &= s(\theta_2+\theta_3)c\theta_4 s\theta_5 + c(\theta_2+\theta_3)c\theta_5 \\ p_x &= [\lambda_4 s(\theta_2+\theta_3) + l_3 c(\theta_2+\theta_3) + l_2 c\theta_2]c\theta_1 + \lambda_3 s\theta_1 \\ p_y &= [\lambda_4 s(\theta_2+\theta_3) + l_3 c(\theta_2+\theta_3) + l_2 c\theta_2]s\theta_1 - \lambda_3 c\theta_1 \\ p_z &= l_2 s\theta_2 + l_3 s(\theta_2+\theta_3) - \lambda_4 c(\theta_2+\theta_3) \end{aligned} \tag{4.40}$$

$$\begin{aligned} \boldsymbol{p}\cdot\boldsymbol{n} &= (l_3 + l_2 c\theta_3)c\theta_4 s\theta_5 + \lambda_3 s\theta_4 s\theta_5 - (\lambda_4 + l_2 s\theta_3)c\theta_5 \\ p^2 - l_2^2 - l_3^2 - \lambda_3^2 - \lambda_4^2 &= 2l_2(\lambda_4 s\theta_3 + l_3 c\theta_3) \end{aligned} \tag{4.41}$$

Of the eight equations (4.40) and (4.41), only five are independent; the third and the last two depend on the rest of them. From the equations for p_x and p_y we can now determine θ_1:

$$\theta_1 = Atan2\left(p_x \pm \sqrt{p_x^2 + p_y^2 - \lambda_3^2},\ \lambda_3 - p_y\right)$$

From equation (4.41) we can determine θ_3:

$$\theta_3 = 2Atan2\left(A_3 \pm \sqrt{A_3^2 + B_3^2 - D_3^2},\ B_3 + D_3\right) \tag{4.42}$$

where $A_3 = \lambda_4$, $B_3 = l_3$, $D_3 = \left(p^2 - l_2^2 - l_3^2 - \lambda_3^2 - \lambda_4^2\right)/2l_2$

Then, from the equation for p_z we determine θ_2:

$$\theta_2 = 2Atan2\left(A_2 \pm \sqrt{A_2^2 + B_2^2 - D_2^2},\ B_2 + D_2\right) \tag{4.43}$$

where $A_2 = l_2 + l_3 c\theta_3 + \lambda_4 s\theta_3$, $B_2 = l_3 s\theta_3 - \lambda_4 c\theta_3$, $D_2 = p_z$

In the next step we determine θ_4 and θ_5 from the relations for n_x, n_y and n_z, which are converted into:

$$\begin{aligned} c\theta_1 c\theta_4 s\theta_5 + s\theta_1 c(\theta_2 + \theta_3) s\theta_4 s\theta_5 &= n_x c(\theta_2 + \theta_3) + n_z c\theta_1 s(\theta_2 + \theta_3) \\ s\theta_1 c\theta_4 s\theta_5 - c\theta_1 c(\theta_2 + \theta_3) s\theta_4 s\theta_5 &= n_y c(\theta_2 + \theta_3) + n_z s\theta_1 s(\theta_2 + \theta_3) \end{aligned} \tag{4.45}$$

That system of linear equations can be solved for the products ($c\theta_4 s\theta_5$) and ($s\theta_4 s\theta_5$). When $s\theta_5 \neq 0$, there are two solutions for θ_4:

$$\theta_4 = Atan2(s\theta_4 s\theta_5,\ c\theta_4 s\theta_5) \quad \text{or} \quad \theta_4 = Atan2(-s\theta_4 s\theta_5,\ -c\theta_4 s\theta_5)$$

However, when $s\theta_5 = 0$, the z_3 and z_5 axes are aligned, and the manipulator looses one degree of freedom.

With θ_4 known, the equations for n_x and n_y constitute a linear system of equations with regard to the unknowns $s\theta_5$ and $c\theta_5$. The unknown θ_6 is determined by applying equations (4.23) for l_z and m_z. This procedure will yield eight solutions, which must then be checked for joint variable range limitations.

The invariance of the dot product under rotation of the reference system allows full simplification of the procedure, without multiplying all the transformation matrices. This simplification process also provides good insight into the structure of the inverse kinematic task, and reduces the complex problem to, the solution of a system of four equations with only four unknowns.

Example 4.4. An articulated manipulator with parallel axes of the revolute joints *2*, *3* and *5* is presented in Figure 4.2. Its geometric parameters are:

$$\begin{array}{lllllll} \alpha_i = & \pi/2, & 0, & 0, & -\pi/2, & \pi/2, & 0 \\ l_i = & 0, & l_2, & l_3, & l_4, & 0, & 0 \\ \lambda_i = & 0, & 0, & 0, & 0, & 0, & 0 \end{array}$$

The manipulator's matrices A_i are

$$A_j = \begin{bmatrix} c_j & 0 & s_j & 0 \\ s_j & 0 & -c_j & 0 \\ 0 & 1 & 0 & 0 \\ 0 & 0 & 0 & 1 \end{bmatrix}, \quad \text{for } j = 1, 5$$

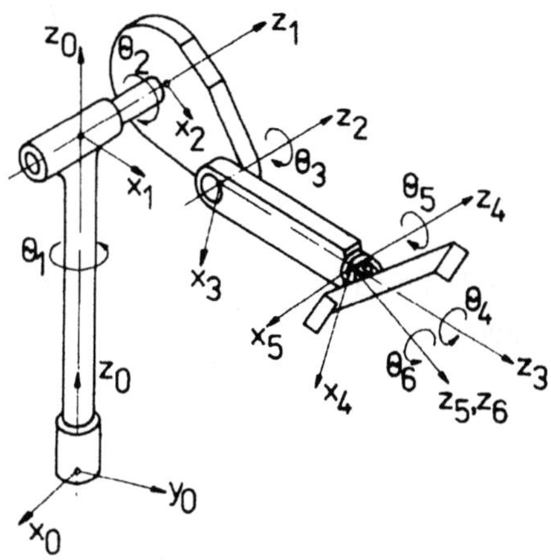

Fig.4.2. Kinematic chain of a PUMA-560 manipulator

$$A_k = \begin{bmatrix} c_k & -s_k & 0 & l_k c_k \\ s_k & c_k & 0 & l_k s_k \\ 0 & 0 & 1 & 0 \\ 0 & 0 & 0 & 1 \end{bmatrix}, \text{ for } k = 2, 3, 4$$

$$A_6 = \begin{bmatrix} c_6 & -s_6 & 0 & 0 \\ s_6 & c_6 & 0 & 0 \\ 0 & 0 & 1 & 0 \\ 0 & 0 & 0 & 1 \end{bmatrix}$$

We first solve the matrix equation

$$A_1^{-1} T_6 = T_{6,1} \tag{4.46}$$

where

$$T_{6,1} = \begin{bmatrix} c_{234}c_5c_6 - s_{234}s_6 & -c_{234}c_5s_6 - s_{234}c_6 & c_{234}s_5 & l_4c_{234} + l_3c_{23} + l_2c_2 \\ s_{234}c_5c_6 + c_{234}s_6 & -s_{234}c_5s_6 + c_{234}c_6 & s_{234}s_5 & l_4s_{234} + l_3s_{23} + l_2s_2 \\ -s_5c_6 & s_5s_6 & c_5 & 0 \\ 0 & 0 & 0 & 1 \end{bmatrix}$$

$$A_1^{-1} T_6 = \begin{bmatrix} l_xc_1 + l_ys_1 & m_xc_1 + m_ys_1 & n_xc_1 + n_ys_1 & p_xc_1 + p_ys_1 \\ -l_z & -m_z & -n_z & -p_z \\ -l_xs_1 + l_yc_1 & -m_xs_1 + m_yc_1 & -n_xs_1 + n_yc_1 & -p_xs_1 + p_yc_1 \\ 0 & 0 & 0 & 1 \end{bmatrix}$$

with $c_{234} = cos(\theta_2 + \theta_3 + \theta_4)$, $c_{23} = cos(\theta_2 + \theta_3)$, $c_2 = cos\theta_2$ and so on.

Equating elements (3,4) on both sides of the equation (5.76) gives us

$$p_x s\theta_1 - p_y c\theta_1 = 0 \tag{4.47}$$

which has the solution for θ_1 in the form of

$$\theta_1 = Atan2\left(p_x \pm \sqrt{p_x^2 + p_y^2},\ -p_y\right) \tag{4.48}$$

Equating elements (3,1), (3,2) and (3,3) on both sides of the equation (4.46) we have the system of equations

$$\begin{aligned} -s_5 c_6 &= l_x s_1 - l_y c_1 \\ s_5 s_6 &= m_x s_1 - m_y c_1 \\ c_5 &= n_x s_1 - n_y c_1 \end{aligned} \tag{4.49}$$

The solution of (4.49) in terms of θ_6 takes the form of

$$\theta_6 = Atan2\left(m_x s_1 - m_y c_1,\ l_y c_1 - l_x s_1\right) \tag{4.50}$$

The solution for θ_5 is obtained as follows:

$$\theta_5 = Atan2\left[s_1\left(l_x c_6 - m_x s_6\right) - c_1\left(l_y c_6 - m_y s_6\right),\ n_x s_1 - n_y c_1\right] \tag{4.51}$$

Equating the first three elements of the second matrix row on both sides of equation (4.46) yields the system of equations for unknowns s_{234} and c_{234}, the solution of which is:

$$\begin{aligned} s_{234} &= -m_z c_5 s_6 + n_z s_5 + l_z c_5 c_6 \\ c_{234} &= l_z s_6 - m_z c_6 \end{aligned} \tag{4.52}$$

The next unknown θ_3 can be determined from equation (4.46) by equating elements (1,4) and (2,4) respectively. As a result we obtain the system of equations:

$$\begin{aligned} p_x c\theta_1 + p_y s\theta_1 &= l_4 c_{234} + l_3 c_{23} + l_2 c_2 \\ p_z &= l_4 s_{234} + l_3 s_{23} + l_2 s_2 \end{aligned} \tag{4.53}$$

Denoting

$$\begin{aligned} p_{cx} &= p_x c\theta_1 + p_y s\theta_1 - l_4 c_{234} \\ p_{cy} &= p_z - l_4 s_{234} \end{aligned} \tag{4.54}$$

the solution can be obtained in the form of

$$\theta_3 = 2Atan2\left(\pm\sqrt{1 + c_3^2},\ 1 + c_3\right) \tag{4.55}$$

where

$$c_3 = \left(l_2^2 + l_3^2 - p_{cx}^2 - p_{cy}^2\right)/2 l_2 l_3 \tag{4.56}$$

and

$$\begin{aligned} \theta_2 &= Atan2\left[l_3 p_{cx} s_3 + \left(l_3 c_3 + l_2\right) p_{cy},\ \left(l_3 c_3 + l_2\right) p_{cx} + l_3 p_{cy} s_3\right] \\ \theta_4 &= Atan2\left(s_{234},\ c_{234}\right) - \theta_3 - \theta_2 \end{aligned} \tag{4.57}$$

Example 4.5. For a Stanford-type manipulator (Figure 4.3) the following values of the kinematic scheme parameters have been assumed:

$$\alpha_i = -\pi/2, \quad \pi/2, \quad 0, \quad -\pi/2, \quad \pi/2, \quad 0$$
$$l_i = \quad 0, \quad 0, \quad 0, \quad 0, \quad 0, \quad 0$$
$$\lambda_i = \quad 0, \quad \lambda_2, \quad \lambda_3, \quad 0, \quad 0, \quad 0$$

where λ_3 is a variable displacement and θ_3 is constant.

Multiplying certain transformation matrices gives us:

$$\mathbf{T}_{6,1} = \begin{bmatrix} c_2(c_4c_5c_6 - s_4s_6) - s_2s_5s_6 & -c_2(c_4c_5s_6 + s_4c_6) + s_2s_5s_6 & c_2c_4s_5 + s_2c_5 & s_2\lambda_3 \\ s_2(c_4c_5c_6 - s_4s_6) + c_2s_5s_6 & -s_2(c_4c_5s_6 + s_4c_6) - c_2s_5s_6 & s_2c_4s_5 + c_2c_5 & -c_2\lambda_3 \\ s_4c_5c_6 + c_4s_6 & -s_4c_5s_6 + c_4c_6 & s_4s_5 & \lambda_2 \\ 0 & 0 & 0 & 1 \end{bmatrix}$$

$$\mathbf{A}_1^{-1}\mathbf{T}_6 = \begin{bmatrix} l_xc_1 + l_ys_1 & m_xc_1 + m_ys_1 & n_xc_1 + n_ys_1 & p_xc_1 + p_ys_1 \\ -l_z & -m_z & -n_z & -p_z \\ -l_xs_1 + l_yc_1 & -m_xs_1 + m_yc_1 & -n_xs_1 + n_yc_1 & -p_xs_1 + p_yc_1 \\ 0 & 0 & 0 & 1 \end{bmatrix} \tag{4.58}$$

Equating elements (3,4) of both matrices we obtain:

$$-p_x s\theta_1 + p_y c\,\theta_1 = \lambda_2 \tag{4.59}$$

Hence

$$\theta_1 = Atan2\left(-p_x \pm \sqrt{p_x^2 + p_y^2 - \lambda_2},\ \lambda_2 + p_y\right)$$

Equating the respective elements (1,4) and (2,4) leads to equations:

$$s_2\lambda_3 = c_1p_x + s_1p_y \tag{4.60}$$
$$-c_2\lambda_3 = -p_z$$

Hence

$$\theta_2 = Atan2\left(p_ys_1 + p_xc_1,\ p_z\right)$$
$$\lambda_3 = \left(p_xc_1 + p_ys_1\right)s_2 + p_zc_2$$

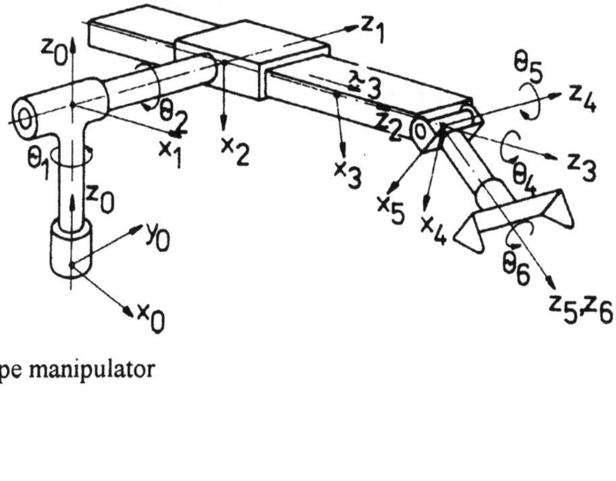

Fig.4.3. Stanford-type manipulator

By equating respective elements of the third column from both matrices we obtain a system of equations the solution of which with respect to θ_4 and θ_5, can be written as follows:

$$\theta_4 = Atan2\left[-n_x s_1 + n_y c_1,\ \left(n_x c_1 + n_y s_1\right)c_2 - n_z s_2\right] \tag{4.61}$$

$$\theta_5 = Atan2\left\{c_4\left[c_2\left(n_x c_1 + n_y s_1\right) - n_z s_2\right] + s_4\left(-n_x s_1 + n_y c_1\right),\ s_2\left(n_x c_1 + n_y s_1\right) + n_z s_2\right\} \tag{4.62}$$

Equating elements (3,1) and (3,2) from both matrices produces a system of linear equations with respect to $s\theta_6$ and $c\theta_6$. Thus

$$\theta_6 = Atan2\left[\left(-l_x s_1 + l_y c_1\right)c_4 - \left(-m_x s_1 + m_y c_1\right)s_4 c_5,\ \left(-l_x s_1 + l_y c_1\right)s_4 c_5 + \left(m_x s_1 + m_y c_1\right)c_4\right] \tag{4.63}$$

Example 4.6. For a SCARA robot, whith three revolute joints and a fourth prismatic joint, all with vertical axes, the following values of the kinematic scheme parameters have been assumed:

$\alpha_i =$	$0,$	$0,$	$0,$	$\pi/2,$	$-\pi/2,$	0
$l_i =$	$l_1,$	$l_2,$	$0,$	$0,$	$0,$	0
$\lambda_i =$	$0,$	$0,$	$\lambda_3,$	$0,$	$0,$	0

The solution of the inverse kinematics using the matrix method for first three unknown θ_1, θ_2 and λ_3 are obtained from the equations

$$\theta_1 = Atan2\left(2l_1 p_y \pm \sqrt{4l_1^2(p_x^2 + p_y^2) - p_{11}^2},\ p_{11} + 2l_1 p_x\right) \tag{4.64}$$

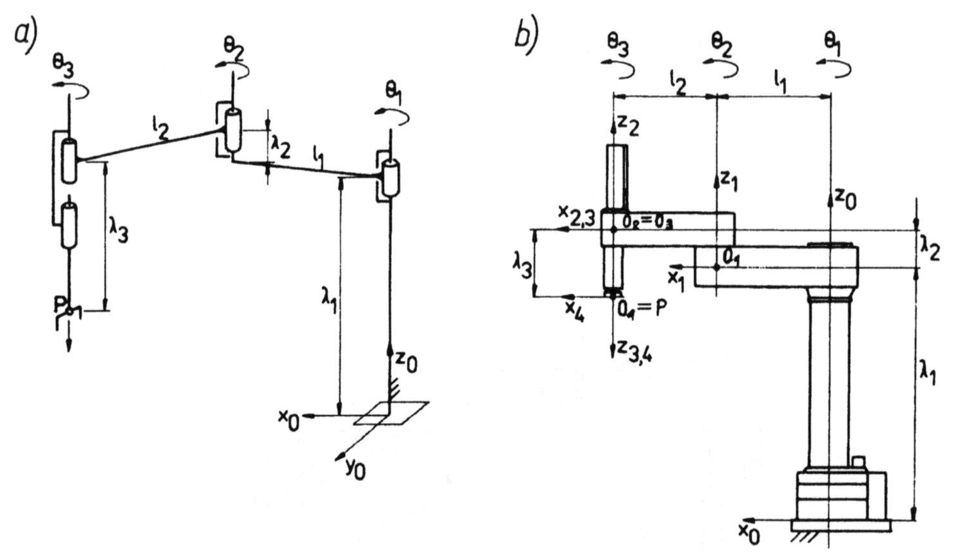

Fig.4.4. SCARA-type manipulator whose arm has three revolute joints ($\theta_1,\theta_2,\theta_3$) with vertical parallel axes and a prismatic joint (λ_3): a) a kinematic diagram, b) the structure of manipulator, defined using D-H coordinates: $\alpha_i = 0, 0, 0$; $l_i = l_1, l_2, 0$; $\lambda_i = 0, 0, \lambda_3$

where

$$p_{11} = p_x^2 + p_y^2 + l_1^2 - l_2^2$$

$$\theta_2 = Atan2(p_y c_1 - p_x s_1, p_x c_1 + p_y s_1 - l_1) \tag{4.65}$$

$$\lambda_3 = p_z \tag{4.66}$$

The remaining unknowns θ_4, θ_5 and θ_6 are found by solving matrix equation

$$A_3^{-1} A_2^{-1} A_1^{-1} T_6 = A_4 A_5 A_6 \tag{4.67}$$

for which the orientation part can be written as:

$$\begin{bmatrix} l_{0x} & m_{0x} & n_{0x} \\ l_{0y} & m_{0y} & n_{0y} \\ l_{0z} & m_{0z} & n_{0z} \end{bmatrix} = \begin{bmatrix} c_4c_5c_6 - s_4s_6 & -c_4c_5s_6 - s_4c_6 & -c_4s_5 \\ s_4c_5c_6 - c_4s_6 & -s_4c_5s_6 + c_4c_6 & -s_4s_5 \\ s_5c_6 & -s_5c_6 & c_5 \end{bmatrix}$$

Thus:

$$\theta_4 = Atan2(-n_{0y}, -n_{0x}) \tag{4.68}$$

$$\theta_5 = Atan2(-n_{0y}s_4 - n_{0x}c_4, n_{0z}) \tag{4.69}$$

$$\theta_6 = Atan2[-m_{0z}s_5 - (m_{0y}s_4 + m_{0x}c_4)c_5, m_{0y}c_4 - m_{0x}s_4] \tag{4.70}$$

Example 4.7. An industrial robot IRb-60 is described by following dimensions (Fig.4.5):

$$\begin{array}{llllll} \alpha_i = & -\pi/2, & 0, & 0, & \pi/2, & 0 \\ l_i = & 0.130, & 0.800, & 1.280, & 0, & 0 \\ \lambda_i = & 0, & 0, & 0, & 0, & 0.400 \end{array}$$

For the value of those parameters we have

$$A_1 = \begin{bmatrix} c\theta_1 & 0 & -s\theta_1 & a_1c\theta_1 \\ s\theta_1 & 0 & c\theta_1 & a_1s\theta_1 \\ 0 & 1 & 0 & 0 \\ 0 & 0 & 0 & 1 \end{bmatrix}, \qquad A_2 = \begin{bmatrix} c\theta_2 & -s\theta_2 & 0 & a_2c\theta_2 \\ s\theta_2 & c\theta_2 & 0 & a_2s\theta_2 \\ 0 & 0 & 1 & 0 \\ 0 & 0 & 0 & 1 \end{bmatrix}$$

$$A_3 = \begin{bmatrix} c\theta_3 & -s\theta_3 & 0 & a_3c\theta_3 \\ s\theta_3 & c\theta_3 & 0 & a_3s\theta_3 \\ 0 & 0 & 1 & 0 \\ 0 & 0 & 0 & 1 \end{bmatrix}, \qquad A_4 = \begin{bmatrix} c\theta_4 & 0 & -s\theta_4 & 0 \\ s\theta_4 & 0 & c\theta_4 & 0 \\ 0 & -1 & 0 & 0 \\ 0 & 0 & 0 & 1 \end{bmatrix}$$

$$A_5 = \begin{bmatrix} c\theta_5 & -s\theta_5 & 0 & 0 \\ s\theta_5 & c\theta_5 & 0 & 0 \\ 0 & 0 & 1 & \lambda_5 \\ 0 & 0 & 0 & 1 \end{bmatrix}$$

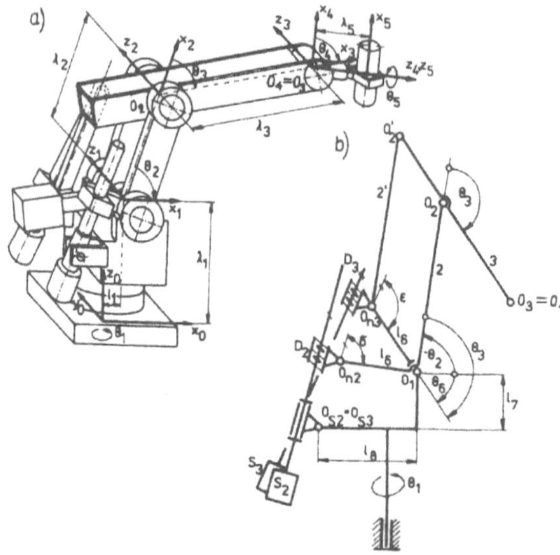

Fig.4.5. IRb-60 robot: a) the manipulator structure; b) the kinematic chain showing screw drive mechanisms: O_{n2} is the pivot of the second link and O_{n3} that of the third

The elements of matrix $T_{6,5}= A_1 A_2 A_3 A_4 A_5$ are

$$l_x = c\theta_1 c\theta_5 c\theta_t + s\theta_1 s\theta_5, \quad m_x = s\theta_1 c\theta_5 - c\theta_1 s\theta_5 c\theta_t, \quad n_x = -c\theta_1 s\theta_t$$
$$l_y = s\theta_1 c\theta_5 c\theta_t - c\theta_1 s\theta_5, \quad m_y = -s\theta_1 s\theta_5 c\theta_t - c\theta_1 c\theta_5, \quad n_y = -s\theta_1 s\theta_t$$
$$l_z = -c\theta_5 s\theta_t, \quad m_z = s\theta_5 s\theta_t, \quad n_z = -c\theta_t$$
$$p_x = c\theta_1[a_1 + a_2 c\theta_2 + a_3 c(\theta_2 + \theta_3) - \lambda_5 s\theta_t]$$
$$p_y = s\theta_1[a_1 + a_2 c\theta_2 + a_3 c(\theta_2 + \theta_3) - \lambda_5 s\theta_t] \tag{4.71}$$
$$p_z = -[a_2 s\theta_2 + a_3 s(\theta_2 + \theta_3) + \lambda_5 c\theta_t]$$

where $\theta_t = \theta_1 + \theta_2 + \theta_3$.

Given the elements of the matrix $T_{5,0}$, we can determine the θ_i angles using the following formulas:

$$\theta_1 = arc\text{tg}(n_y/n_x), \qquad \theta_5 = arc\text{tg}(-m_z/l_z)$$
$$\theta_t = arc\text{tg}\left(\pm\sqrt{l_z^2 + m_z^2}/n_z\right), \qquad \theta_2 = 2arc\text{tg}\frac{A \pm \sqrt{A^2 + B^2 - C^2}}{B + C} \tag{4.72}$$

where

$$A = \lambda_5 n_z - p_z$$
$$B = (p_x - \lambda_5 n_x - a_1 c\theta_1)/c\theta_1,$$
$$C = (A^2 + B^2 + a_2^2 - a_3^2)/2a_2$$
$$\theta_1 + \theta_2 = arc\text{tg}\frac{A - a_2 s\theta_2}{B - a_2 c\theta_2}, \qquad \theta_4 = \theta_t - (\theta_2 + \theta_3)$$

Numerical example. Given a position matrix $T_{5,0}$ for an industrial robot IRb-60:

$$T_{5,0} = \begin{bmatrix} 0.98430 & 0.17357 & 0.03489 & 0.08100 \\ -0.03134 & -0.02324 & 0.99923 & 2.31963 \\ 0.17361 & -0.98464 & -0.01746 & 0.69514 \\ 0 & 0 & 0 & 1 \end{bmatrix}$$

we must find angles θ_i (i = 1, 2, 3, 4, 5).

$$\theta_1 = \operatorname{arctg}(0.99923/0.03489) = 88.000°$$

$$\theta_5 = \operatorname{arctg}(0.98464/0.17361) = 80.000°$$

$$\theta_2 = 2\operatorname{arctg}[(-0.70212 \pm 0.92088)/3.48010] = 7.194° \text{ or } -50.006°$$

with

$$A = -0.70212, \quad B = 1.79106, \quad C = 1.68904.$$

$$\theta_t = \operatorname{arctg}\left(\pm\sqrt{0.17361^2 + 0.98464^2}/0.01746\right) = \pm 89.017°$$

$$\theta_2 + \theta_3 = 2\operatorname{arctg}((-0.70212 - 0.8\sin 50.006°)/(1.79106 - 0.8\cos 50.006°)) = -33.997°$$

$$\theta_3 = -3.997° + 50.006° = 46.009°$$

$$\theta_4 = \pm 89.017° + 3.997° = -85.020° \text{ or } 93.014°$$

Considering the range limitations for joint variable, we can accept only the following solution: θ_i = 88.000°, –50.006°, 46.009°, –85.020°, 80.000°

4.3. Vector method

The orientation of joint axes and manipulator's links relative to each other can be specified in the form of a system of vector relations:

$$\mathbf{e}_i \cdot \mathbf{e}_{i+1} = c\alpha_i, \qquad \mathbf{e}_i \times \mathbf{e}_{i+1} = \mathbf{a}_i s\alpha_i \tag{4.73}$$

$$\mathbf{a}_i \cdot \mathbf{a}_{i+1} = c\theta_i, \qquad \boldsymbol{a}_i \times \boldsymbol{a}_{i+1} = \boldsymbol{e}_{i+1} s\theta_{i+1}, \quad i = 1, 2, \ldots n$$

where $\boldsymbol{e}_i$ is a unit vector of a joint connecting a link i with i-1 one, and $\boldsymbol{a}_i$ is a unit vector perpendicular to the axes of the joints of the link i. From the system (4.73) we can derive the following relations:

$$\mathbf{e}_{i+1} = \mathbf{e}_i c\alpha_i + (\mathbf{a}_i \times \mathbf{e}_i) s\alpha_i, \qquad \mathbf{e}_i = \mathbf{e}_{i+1} c\alpha_i - (\mathbf{a}_{i+1} \times \mathbf{e}_{i+1Á}) s\alpha_{i+1}$$

$$\mathbf{a}_{i+1} = \mathbf{a}_i c\theta_{i+1} + (\mathbf{e}_i \times \mathbf{a}_i) s\theta_{i+1}, \qquad \mathbf{a}_i = \mathbf{a}_{i+1} c\theta_{i+1} - (\mathbf{e}_i \times \mathbf{a}_{i+1}) s\theta_{i+1} \tag{4.73a}$$

with angles α_i and θ_i assumed to be directed. Their sense is established according to the right hand rule. For instance, the sense of angle α_i is said to be positive when the three unit vectors $\boldsymbol{e}_i$, $\boldsymbol{e}_{i+1}$ and $\boldsymbol{a}_i$ constitute a right-handed system.

In a similar way we determine the sense of angle θ_i by means of three unit vectors $\boldsymbol{a}_{i-1}$, $\boldsymbol{a}_i$ and $\boldsymbol{e}_i$ (Figure 4.6).

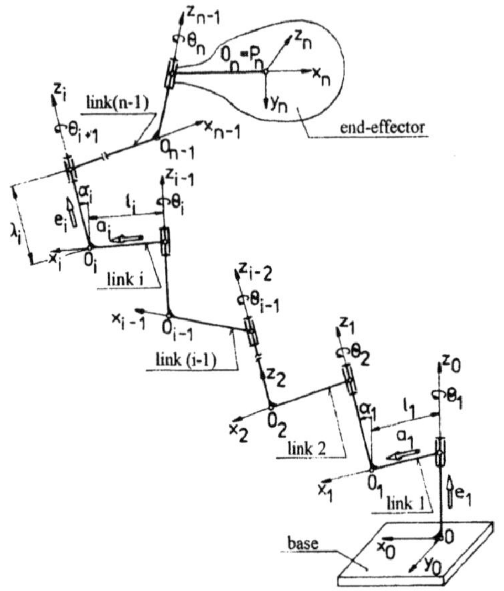

Fig.4.6. Kinematic chain of an n-link general manipulator with D-H coordinate systems: θ_i, λ_i, l_i, α_i; $\boldsymbol{e}_i$ is a unit vector of a revolute joint axis; $\boldsymbol{a}_i$ is a unit vector of a common normal to the joint axes

In the case when two unit vectors $\boldsymbol{e}_i$ and $\boldsymbol{e}_j$ are given as well as two dot products for each of these vectors with the unknown unit vector $\boldsymbol{e}_k$, we can find that vector using the formula

$$\boldsymbol{e}_k = \left[\left(c_{ik} - c_{ij}c_{jk}\right)\boldsymbol{e}_i + \left(c_{jk} - c_{ij}c_{ik}\right)\boldsymbol{e}_j \pm \sqrt{D_k}\left(\boldsymbol{e}_i \times \boldsymbol{e}_j\right)\right] / \left(1 - c_{ij}^2\right) \tag{4.74}$$

where

$$D_k = 1 - c_{ij}^2 - c_{ik}^2 - c_{jk}^2 + 2c_{ij}c_{ik}c_{jk}$$

$$c_{ij} = \boldsymbol{e}_i \cdot \boldsymbol{e}_j, \quad c_{ik} = \boldsymbol{e}_i \cdot \boldsymbol{e}_k, \quad c_{jk} = \boldsymbol{e}_j \cdot \boldsymbol{e}_k$$

The plus sign (+) is put before the root if the three unit vectors $\boldsymbol{e}_i$, $\boldsymbol{e}_j$ and $\boldsymbol{e}_k$ form a right-handed system; otherwise we put the minus sign (–).

A closure equation of a vector polygon takes the form

$$\sum_{i=1}^{n} \lambda_i \boldsymbol{e}_i + \sum_{i=1}^{n-1} l_i \boldsymbol{a}_i = p\boldsymbol{e}_p \tag{4.75}$$

where n is a number of manipulator joints, and $\boldsymbol{e}_p$, a unit vector of a position vector of an origin of the end-effector's reference system (for instance a gripper center).

In the most popular manipulators with six revolute joints, the PUMA or the ASEA, for example, the axes of the last two joints intersect and are mutually perpendicular, that is,

$$\boldsymbol{e}_5 \cdot \boldsymbol{e}_6 = 0, \ \boldsymbol{e}_5 \times \boldsymbol{e}_6 = \boldsymbol{a}_5, \ \boldsymbol{e}_6 = \boldsymbol{l} \times \boldsymbol{n}, \tag{4.76}$$

where $\boldsymbol{l}$ is an orientation vector and $\boldsymbol{n}$, an approach vector of the gripper.

Given the manipulator`s dimensions, as well as the position and orientation of the end-effector, we can use the closure equations (4.75) to determine the position vector of the point of intersection of the last two axes of the joints:

$$p_k \boldsymbol{e}_k = p\boldsymbol{e}_p - l_6 \boldsymbol{m} - \lambda_6 \boldsymbol{e}_6 \tag{4.77}$$

Hence

$$p_k = \sqrt{p^2 \lambda_6^2 + l_6^2 - 2p[\lambda_6(\boldsymbol{e}_p \cdot \boldsymbol{e}_6) + l_6(\boldsymbol{e}_p \cdot \boldsymbol{n})]} \tag{4.78}$$

where

$$\boldsymbol{e}_p \cdot \boldsymbol{e}_6 = \boldsymbol{e}_p \cdot (\boldsymbol{l} \times \boldsymbol{n})$$

$$\boldsymbol{e}_k = (p\boldsymbol{e}_p - l_6 \boldsymbol{m} - \lambda_6 \boldsymbol{e}_6)/p_k \tag{4.79}$$

The further solution of the problem depends on the configuration of the remaining joints of the manipulator. In the case of the ASEA-type manipulator, with the parallel axes of the second, third and fourth joint and perpendicular to these the axes of the first and fifth joint, the following relations occur:

$$\boldsymbol{e}_2 = \boldsymbol{e}_3 = \boldsymbol{e}_4 = \pm(\boldsymbol{e}_k \times \boldsymbol{e}_1) / \sqrt{1 - (\boldsymbol{e}_k \cdot \boldsymbol{e}_1)^2} \tag{4.80}$$

$$\boldsymbol{e}_5 = \pm(\boldsymbol{e}_2 \times \boldsymbol{e}_6) / \sqrt{1 - (\boldsymbol{e}_2 \cdot \boldsymbol{e}_6)^2} \tag{4.81}$$

With the previously determined unit vector $\boldsymbol{e}_5$ from closure equation (4.76) we have

$$p_r \boldsymbol{e}_r = p_k \boldsymbol{e}_k + \lambda_5 \boldsymbol{e}_5 \tag{4.82}$$

Hence

$$p_r = \sqrt{p_k^2 + \lambda_6^2 + l_6^2 - 2p_k \lambda_5 (\boldsymbol{e}_k \cdot \boldsymbol{e}_5)} \tag{4.83}$$

$$\boldsymbol{e}_r = (p_k \boldsymbol{e}_k - \lambda_5 \boldsymbol{e}_5)/p_r \tag{4.84}$$

and from the equation (4.76) we can obtain another closure equation:

$$p_r \boldsymbol{e}_r = l_2 \boldsymbol{a}_2 - l_3 \boldsymbol{a}_3 \tag{4.85}$$

$$c\beta = \boldsymbol{a}_2 \cdot \boldsymbol{e}_r = (l_2^2 + p_r^2 - l_3^2)/2 l_2 p_r \tag{4.86}$$

The expression (4.86) has geometric sense only if its right side is smaller, or equal to, unity. If that condition is not fulfilled, it means that either the origin of the end-effector reference system lies outside the workspace, or the orientation unit vectors of the manipulator lie outside the manipulation space.

If the above condition is fulfilled, we can find

$$\boldsymbol{a}_2 = \boldsymbol{e}_r c\beta \pm (\boldsymbol{e}_r \times \boldsymbol{e}_2) s\beta \tag{4.87}$$

$$\boldsymbol{a}_3 = (p_r \boldsymbol{e}_r - l_2 \boldsymbol{a}_2)/l_3 \tag{4.88}$$

$$\boldsymbol{a}_1 = \boldsymbol{e}_1 \times \boldsymbol{e}_2, \qquad \boldsymbol{a}_4 = \boldsymbol{e}_2 \times \boldsymbol{e}_5, \qquad \boldsymbol{a}_5 = \boldsymbol{e}_5 \times \boldsymbol{e}_6 \tag{4.89}$$

Thus all the unknown unit vectors have been determined as functions of a given $(p\boldsymbol{e}_p)$, the position vector, and $(\boldsymbol{l}, \boldsymbol{n})$, the orientation unit vectors of the end-effector.

The ± sign appearing in formulas (4.81) and (4.87) indicates that four different configurations of the kinematic chain are possible with the same position and orientation of the end-effector.

In particular joints the configuration angles can be calculated using the formulas:

$$c\theta_{i+1} = \boldsymbol{a}_i \cdot \boldsymbol{a}_{i+1},$$
$$s\theta_{i+1} = \boldsymbol{e}_{i+1} \cdot (\boldsymbol{a}_i \times \boldsymbol{a}_{i+1}) \tag{4.90}$$

In the case of the PUMA manipulator where only two successive axes, of those the second and third joint, are parallel, while the rest of the joint axes are perpendicular to the preceding one, the inverse kinematics solving algorithm is as follows:

First, a position vector of the intersection point of the last two joint axes is determined, using formulas (4.78) and (4.79).

Then from the formula (4.74) we obtain

$$\boldsymbol{e}_2 = [(\boldsymbol{e}_2 \cdot \boldsymbol{e}_k)\boldsymbol{e}_k + (\boldsymbol{e}_1 \cdot \boldsymbol{e}_k)(\boldsymbol{e}_2 \cdot \boldsymbol{e}_k)\boldsymbol{e}_1 \pm (\boldsymbol{e}_k \times \boldsymbol{e}_1)\sqrt{D_2}]/[1-(\boldsymbol{e}_1 \cdot \boldsymbol{e}_k)^2] \tag{4.91}$$

where

$$\boldsymbol{e}_2 \cdot \boldsymbol{e}_k = \lambda_2/p_k, \qquad \boldsymbol{e}_1 \cdot \boldsymbol{e}_k = e_{kz}, \qquad D_2 = 1-(\boldsymbol{e}_2 \cdot \boldsymbol{e}_k)^2 - (\boldsymbol{e}_1 \cdot \boldsymbol{e}_k)^2$$

In the next step, from the same formula we obtain:

$$\begin{aligned}\boldsymbol{e}_r = \{&[(\boldsymbol{e}_2 \cdot \boldsymbol{e}_r) - (\boldsymbol{e}_2 \cdot \boldsymbol{e}_k)(\boldsymbol{e}_r \cdot \boldsymbol{e}_k)]\boldsymbol{e}_2 + [(\boldsymbol{e}_r \cdot \boldsymbol{e}_k) - (\boldsymbol{e}_2 \cdot \boldsymbol{e}_k)(\boldsymbol{e}_2 \cdot \boldsymbol{e}_r)]\boldsymbol{e}_k + \\ &\pm (\boldsymbol{e}_2 \times \boldsymbol{e}_k)\sqrt{D_r}\}/[1-(\boldsymbol{e}_2 \cdot \boldsymbol{e}_k)^2]\end{aligned} \tag{4.92}$$

where

$$\boldsymbol{e}_2 \cdot \boldsymbol{e}_r = \lambda_2/p_r, \quad \boldsymbol{e}_r \cdot \boldsymbol{e}_k = (p_k^2 + p_r^2 - \lambda_4^2)/2p_r p_k \quad p_r = \sqrt{\lambda_2^2 + l_2^2}$$

We can then determine

$$\boldsymbol{e}_4 = (p_k \boldsymbol{e}_k - p_r \boldsymbol{e}_r)/\lambda_4 \tag{4.93}$$
$$\boldsymbol{e}_5 = \pm(\boldsymbol{e}_4 \times \boldsymbol{e}_6)/\sqrt{1-(\boldsymbol{e}_4 \cdot \boldsymbol{e}_6)^2} \tag{4.94}$$
$$\boldsymbol{a}_1 = \boldsymbol{e}_1 \times \boldsymbol{e}_2$$
$$\boldsymbol{a}_2 = [\boldsymbol{e}_r(\boldsymbol{a}_2 \cdot \boldsymbol{e}_r) - \boldsymbol{e}_2(\boldsymbol{e}_2 \cdot \boldsymbol{e}_r)]/\sqrt{1-(\boldsymbol{e}_2 \cdot \boldsymbol{e}_r)^2} \tag{4.95}$$

where

$$\boldsymbol{a}_2 \cdot \boldsymbol{e}_r = l_2/p_r, \qquad \boldsymbol{e}_2 \cdot \boldsymbol{e}_r = \lambda_2/p_r$$

According to the relation (4.73) we determine the unit vectors

$$\boldsymbol{a}_i = (\boldsymbol{e}_i \times \boldsymbol{e}_{i+1})/s\alpha_i, \quad i = 3,\ 4,\ 5 \tag{4.96}$$

Finally, applying formulas (4.90) we obtain joint variables θ_j ($j = 1, \ldots 6$). Double ± signs in the expressions (4.91), (4.92) and (4.94) indicate that 2^3 link configurations of the manipulator kinematic chain are possible for the same input data, but only some of these will satisfy the joint variable range limitations.

Of the six coordinates of unit vectors $\boldsymbol{a}_6$ and $\boldsymbol{e}_6$, only three can be freely assumed while the rest of them must satisfy the relations:

$$a_{6x}e_{6x} + a_{6y}e_{6y} + a_{6z}e_{6z} = 0, \qquad a_{6x}^2 + a_{6y}^2 + a_{6z}^2 = 1, \qquad e_{6x}^2 + e_{6y}^2 + e_{6z}^2 = 1,$$

For example, given coordinates a_{6x}, a_{6y} and e_{6x}, the rest of the unit vectors can be calculated using the formulas:

$$\begin{aligned} a_{6z} &= \pm\sqrt{1 - a_{6x}^2 - a_{6y}^2} \\ a_{6y} &= \left(-a_{6x}a_{6y}e_{6x} \pm a_{6z}\sqrt{1 - a_{6x}^2 - e_{6x}^2}\right) / \left(1 - a_{6x}^2\right) \\ e_{6z} &= \pm\sqrt{1 - e_{6x}^2 - e_{6y}^2} \end{aligned} \tag{4.97}$$

The double ± sign allows us to choose one out of the eight possible sign combinations; the sense of unit vector $\boldsymbol{a}_6$ and of unit vector $\boldsymbol{e}_6$ can be considered a aditional criterion.

Example 4.8. Figure 3.4 and Figure 4.7 present the structure and the kinematic chain, respectively, of a manipulator with six revolute joints, the second third and fourth with parallel axes, while the rest of the axes are mutually perpendicular.

The D-H coordinates of the manipulator are:

$\alpha_i =$	90°,	0,	0,	–90°,	90°,	–90°
$l_i =$	0,	0.400,	0.300,	0,	0,	0.050
$\lambda_i =$	0,	0,	0,	0,	0.200,	0.250
$\theta_{i,min} =$	–120°,	–45°,	–120°,	–175°,	–170°,	–170°
$\theta_{i,max} =$	120°,	180°,	120°,	150°,	170°,	170°

The objective is to determine the time profiles of θ_i, the displacement, and $\dot{\theta}_i$, the angular velocities in the particular joints during the motion of the manipulator's P point along a particular line, at a given velocity and orientation.

The coordinates of the two points at the beginning and the end of the line are given: P_0 (P_{0x}, P_{0y}, P_{0z}) and P_k (P_{kx}, P_{ky}, P_{kz}). The approach unit vector $\boldsymbol{n}$ must coincide with the interval $\overline{P_0P_k}$, and the orientation unit vector $\boldsymbol{e}_6 = \boldsymbol{m}$, be perpendicular to the z axis.

The solving algorithm is as follows:

1. The coordinates of unit vectors $\boldsymbol{n}$, $\boldsymbol{m}$ and $\boldsymbol{l}$ are calculated according to the formulas

$$n_x = (p_{kx} - p_{ox})/d, \qquad n_y = (p_{ky} - p_{oy})/d, \qquad n_z = (p_{kz} - p_{oz})/d$$

$$m_x = -(p_{ky} - p_{oy})/d, \qquad m_y = (p_{kx} - p_{ox})/d, \qquad m_z = 0$$

$$l_x = -n_z m_y, \qquad l_y = n_z m_x, \qquad l_z = n_x m_y - n_y m_x$$

where

$$d = \sqrt{(p_{kx} - p_{ox})^2 + (p_{ky} - p_{oy})^2 + (p_{kz} - p_{oz})^2}$$

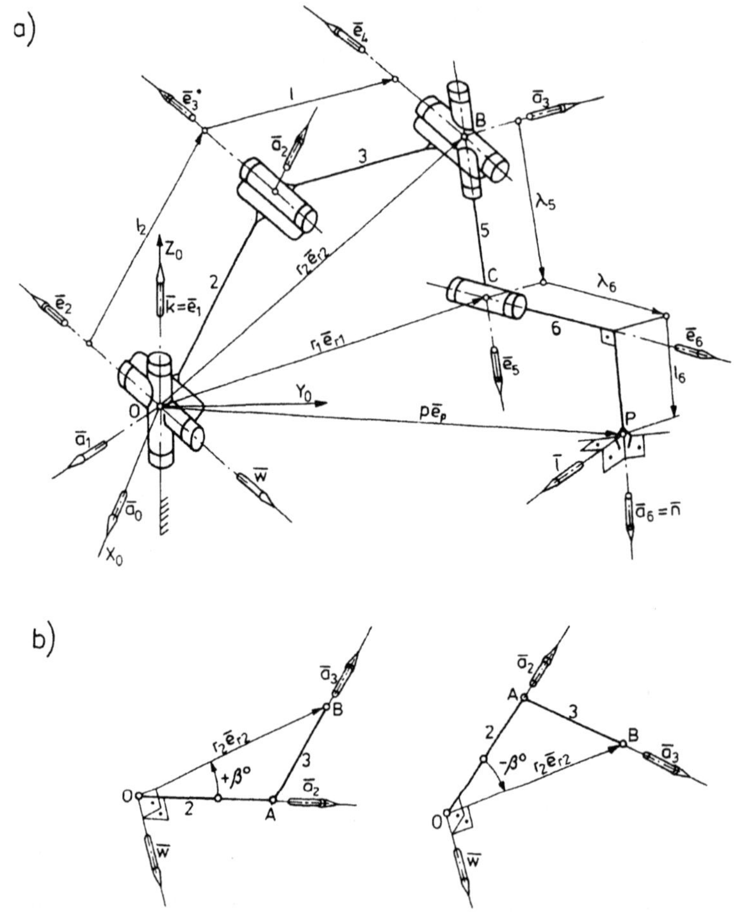

Fig.4.7. a) Kinematic chain of a manipulator with six revolute joints (WORKMASTER) dimensioned in D-H coordinates. Notation: $r\boldsymbol{e}_r$ is a gripper centre position vector, and $r\boldsymbol{e}_{r2}$ is a wrist center position vector (point B). b) Two possible arm configurations: $+\beta$ in the lower, $-\beta$ in the upper with the same wrist center position

2. A unit vector $\boldsymbol{e}_p$ of a P_0 point position vector is calculated:

$$e_{px} = p_{ox}/p, \qquad e_{py} = p_{oy}/p, \qquad e_{pz} = p_{oz}/p$$

where

$$p = \sqrt{(p_{kx} - p_{ox})^2 + (p_{ky} - p_{oy})^2 + (p_{kz} - p_{oz})^2}$$

3. The dot products formed by unit vector $\boldsymbol{e}_p$ and unit vectors $\boldsymbol{m}$ and $\boldsymbol{n}$ are calculated:

$$\boldsymbol{e}_p \cdot \boldsymbol{m} = e_{px}m_x + e_{py}m_y + e_{pz}m_z$$

$$\boldsymbol{e}_p \cdot \boldsymbol{n} = e_{px}n_x + e_{py}n_y + e_{pz}n_z$$

4. Unit vector e_k is calculated as well as the distance $p_k = OC$, that is the distance from the point of intersection of the axes of the fifth and sixth revolute joints to the origin of the base reference system:

$$p_k = \sqrt{p^2 + \lambda_6^2 + l_6^2 - 2p\lambda_6\left[(e_p \cdot m) + l_6(e_p \cdot n)\right]}$$
$$e_{kx} = (pe_{px} - l_6 n_x - \lambda_6 m_x)/p_k$$
$$e_{ky} = (pe_{py} - l_6 n_y - \lambda_6 m_y)/p_k$$
$$e_{kz} = (pe_{pz} - l_6 n_z - \lambda_6 m_z)/p_k$$

5. Formulas (4.80) and (4.81) are used to calculate the coordinates of the unit vectors e_2 and e_5:

$$e_{2x} = \pm(e_{ky}e_{1z} - e_{kz}e_{1y})/\sqrt{1-(e_k \cdot e_1)^2}$$
$$e_{2y} = \pm(e_{kz}e_{1x} - e_{kx}e_{1z})/\sqrt{1-(e_k \cdot e_1)^2}$$
$$e_{2z} = \pm(e_{kx}e_{1y} - e_{ky}e_{1x})/\sqrt{1-(e_k \cdot e_1)^2}$$
$$e_{5x} = \pm(e_{2y}m_z - e_{2z}m_y)/\sqrt{1-(e_2 \cdot m)^2}$$
$$e_{5y} = \pm(e_{2z}m_x - e_{2x}m_z)/\sqrt{1-(e_2 \cdot m)^2}$$
$$e_{5z} = \pm(e_{2x}m_y - e_{2y}m_z)/\sqrt{1-(e_2 \cdot m)^2}$$

where

$$e_k \cdot e_1 = e_{kx}e_{1x} + e_{ky}e_{1y} + e_{kz}e_{1z}, \qquad e_2 \cdot m = e_{2x}m_x + e_{2y}m_y + e_{2z}m_z$$

6. With formulas (4.83) and (4.84) the distance $p_r = OB$ from the point of intersection of the axes of the fourth and the fifth revolute joints to the origin of the base reference system is calculated, together with the coordinates of the unit vector e_r:

$$p_r = \sqrt{p_k^2 + \lambda_5^2 - 2p_k\lambda_5(e_k \cdot e_5)}$$

where

$$e_k \cdot e_5 = e_{kx}e_{5x} + e_{ky}e_{5y} + e_{kz}e_{5z}$$
$$e_{rx} = (p_k e_{kx} - \lambda_5 e_{5x})/p_r, \quad e_{ry} = (p_k e_{ky} - \lambda_5 e_{5y})/p_r, \quad e_{rz} = (p_k e_{kz} - \lambda_5 e_{5z})/p_r$$

7. According to formula (4.86), the value of $c\beta = a_2 \cdot e_r$ is calculated and the condition $c\beta \leq 1$ is verified. If the condition is fulfilled, then the coordinates of unit vectors a_2, a_3, a_1, a_4 and a_5 are calculated according to (4.87)÷(4.89) using the formulas

$$c\beta = (l_2^2 + p_r^2 - l_3^2)/2l_2 p_r$$
$$a_{2x} = e_{rx}c\beta \pm (e_{ry}e_{2z} - e_{rz}e_{2y})s\beta$$
$$a_{2y} = e_{ry}c\beta \pm (e_{rz}e_{2x} - e_{rx}e_{2z})s\beta$$
$$a_{2z} = e_{rz}c\beta \pm (e_{rx}e_{2y} - e_{ry}e_{2x})s\beta$$

$$a_{3x}=\left(p_r e_{rx}-l_2 a_{2x}\right)/l_3$$
$$a_{3y}=\left(p_r e_{ry}-l_2 a_{2y}\right)/l_3$$
$$a_{3z}=\left(p_r e_{rz}-l_2 a_{2z}\right)/l_3$$
$$a_{4x}=\left(e_{2y}e_{5z}-e_{2z}e_{5y}\right)$$
$$a_{4y}=\left(e_{2z}e_{5x}-e_{2x}e_{5z}\right)$$
$$a_{4z}=\left(e_{2x}e_{5y}-e_{2y}e_{5x}\right)$$
$$a_{5x}=\left(e_{5y}e_{6z}-e_{5z}e_{6y}\right)$$
$$a_{5y}=\left(e_{5z}e_{6x}-e_{5x}e_{6z}\right)$$
$$a_{5z}=\left(e_{5x}e_{6y}-e_{5y}e_{6x}\right)$$

8. Joint variables θ_i ($i = 1, \ldots 6$) are calculated with formulas (4.90):

$$c\theta_i = a_{i,x}a_{i+1,x} + a_{i,y}a_{i+1,y} + a_{i,z}a_{i+1,z}$$
$$s\theta_i = e_{i+1,x}\left(a_{i,y}a_{i+1,z} - a_{i,z}a_{i+1,y}\right) + e_{i+1,y}\left(a_{i,z}a_{i+1,x} - a_{i,x}a_{i+1,z}\right) + \\ + e_{i+1,z}\left(a_{i,x}a_{i+1,y} - a_{i,y}a_{i+1,x}\right)$$

Using this algorithm the task has been solved for the given initial point P_0 (0.050; 0.740; 0.070) we have:

$\boldsymbol{n} = [0.4767; -0.5721; 0.7682]^{\mathrm{T}}$, $\boldsymbol{e}_6 = \boldsymbol{m} = [0.7682; 0.6402; 0]^{\mathrm{T}}$
$\boldsymbol{l} = [-0.4273; 0.5123; 0.7447]^{\mathrm{T}}$, $\boldsymbol{e}_p = [0.0671; 0.9933; 0.0940]^{\mathrm{T}}$
$p = 0.745$, $\boldsymbol{e}_p \cdot \boldsymbol{m} = 0.6875$, $\boldsymbol{e}_p \cdot \boldsymbol{n} = -0.4736$
$p_k = 0.6313$, $\boldsymbol{e}_k = [-0.2626; 0.9632; 0.0580]^{\mathrm{T}}$
$\boldsymbol{e}_2 = [0.9648; 0.2630; 0]^{\mathrm{T}}$, $\boldsymbol{e}_2 \cdot \boldsymbol{m} = 0.9095$
$\boldsymbol{e}_5 = [0; 0; 1]^{\mathrm{T}}$, $\boldsymbol{e}_k \cdot \boldsymbol{e}_5 = 0.0580$
$p_r = 0.6516$, $\boldsymbol{e}_r = [-0.2546; 0.9340; -0.2507]^{\mathrm{T}}$
$c\beta = 0.9488$, $\boldsymbol{a}_2 = [-0.2207; 0.8097; 0.5438]^{\mathrm{T}}$
$\boldsymbol{a}_3 = [-0.2587; 0.9490; 0.1805]^{\mathrm{T}}$, $\boldsymbol{a}_4 = [0.2630; -0.9648; 0]^{\mathrm{T}}$
$\boldsymbol{a}_5 = [-0.6402; 0.7682; 0]^{\mathrm{T}}$
$\theta_1 = 105.2°$, $\theta_2 = 32.9°; -3.9°$, $\theta_3 = -43.3°; 43.3°$
$\theta_4 = -169.6°; 140.6°$, $\theta_5 = -155.4°; -155.4°$, $\theta_2 = 138.1°$

A computer program developed for the algorithm described above is used to calculate the values of the joint variables θ_i ($i = 1, \ldots, 6$) for the given straight line trajectory between points

P_0 (0.050; 0.740; 0.070) and P_k (0.500; 0.200; 0.700)

with fixed orientation unit vectors of the end-effector:

$\boldsymbol{n} = [0.4767; -0.5721; 0.7682]^{\mathrm{T}}$ and $\boldsymbol{m} = [0.7682; 0.6402; 0]^{\mathrm{T}}$.

The results of these calculations for two configurations are shown in the Table here below.

i	0	1	. . .	41	42	. . .	88	89	90
P_{ix}	0.050	0.055		0.300	0.305		0.490	0.495	0.500
P_{iy}	0.740	0.734		0.440	0.434		0.212	0.206	0.200
P_{iz}	0.070	0.077		0.420	0.427		0.686	0.693	0.700
θ_1	105.2	105.0		74.8	73.6		16.4	15.0	13.6
θ_2	−3.9	−5.9		−75.9	−77.2		−91.4	−90.7	−90.0
θ_3	43.3	47.1		118.1	118.1		81.3	79.3	77.2
θ_4	140.6	138.8		137.9	139.1		169.9	168.5	−167.2
θ_5	−155.4	−155.1		−124.9	−123.8		−66.6	−65.2	−63.8
θ_6	138.1	138.1		138.1	138.1		138.1	138.1	138.1
θ_1	105.2	105.0		74.8	73.6		16.4	15.0	13.6
θ_2	32.9	34.1		15.4	14.1		−24.1	−25	−25.8
θ_3	−43.3	−47.1		−118.1	−118.1		−81.3	−79.3	−77.2
θ_4	−169.6	−167.0		−77.3	−76.0		−74.6	−75.8	−77.0
θ_5	−155.4	−155.1		−124.9	−123.8		−66.6	−65.2	−63.8
θ_6	138.1	138.1		138.1	138.1		138.1	138.1	138.1

4.4. Iterative method

In the case of manipulators for which an closed-form solution can not be found, numerical methods of solving inverse kinematics equations are used. Most of these methods are based on an inverse Jacobian and use of the Newton-Raphson method.

Here we present a method introduced by R. Manseur and K. Dotty [4.12] which can be applied to any 6 DOF manipulator for which all joint variables can be found in a closed form when one joint variable is known. For manipulators satisfying this condition, the inverse kinematic problem can be reduced to a one-dimensional root-finding technique, such as the one-dimensional Newton-Raphson, or the scant method.

The equation (4.2) can be expressed in the form

$$A_2 A_3 A_4 A_5 A_6 = A_1^{-1} T_6 = T_{6,1} \tag{4.98}$$

where the right side is a function of θ_1 only. Assuming that θ_1 is known, the problem is reduced to five unknowns, and can, for many manipulators, be solved in a closed form by the method presented here.

From equation (4.98) we obtain a position vector of the end-effector reference system origin relative to the reference system of the first link:

$$p' = B_2 (B_3 (B_4 p_5 + p_4) + p_3) + p_2 = B_2 B_3 B_4 q' \tag{4.99}$$

where

$$T_{6,1} = \begin{bmatrix} l' & m' & n' & p' \\ 0 & 0 & 0 & 1 \end{bmatrix}$$

$$q' = p_5 + B_4^{-1}\left(p_4 + B_3^{-1}\left(p_3 + B_2^{-1} p_2\right)\right) \tag{4.100}$$

With repetitive use of rotation orthogonality, we have

$$n'_z = n' \cdot k = B_2 B_3 B_4 B_5 k \cdot k = k \cdot \left(B_5^{-1} B_4^{-1} B_3^{-1} B_2^{-1}\right) k \tag{4.101}$$

$$p'_z = p' \cdot k = q' \cdot \left(B_4^{-1} B_3^{-1} B_2^{-1}\right) k = B_1^{-1}(p - p_1) \cdot k \tag{4.102}$$

$$p' \cdot n = B_5^{-1} q \cdot k = p \cdot n \qquad p' \cdot p' = q' \cdot q' \tag{4.103}$$

From equations (4.101) and (4.103), we obtain four equations independent of θ_2 and θ_6. These equations are the basis for the iterative solving method. To avoid extraneous solutions and sign ambiguities, additional equations containing θ_2 are introduced:

$$n'_x = k \cdot \left(B_5^{-1} B_4^{-1} B_3^{-1} B_2^{-1}\right) i = B_1^{-1} n \cdot i \tag{4.104}$$

$$n'_y = k \cdot \left(B_5^{-1} B_4^{-1} B_3^{-1} B_2^{-1}\right) j = B_1^{-1} n \cdot j \tag{4.105}$$

$$p'_x = p' \cdot i = q' \cdot B_4^{-1} B_3^{-1} B_2^{-1} i = B_1^{-1}(p - p_1) \cdot i \tag{4.106}$$

$$p'_y = p' \cdot j = q' \cdot B_4^{-1} B_3^{-1} B_2^{-1} j = B_1^{-1}(p - p_1) \cdot j \tag{4.107}$$

The method consists of the following steps:

(1) Derive equations (4.101)÷(4.107) in detail.

(2) Given θ_1, verify that the remaining variables can be obtained in a closed form on the basis of equations (4.101)÷(4.104).

(3) Find a real value of θ_1 by means of one of the equations that can be computed for any value of θ_1. For example, given θ_1 and using the remaining equations, the equation (4.103) can be evaluated, and, consequently, a good candidate function obtained:

$$f(\theta_1) = B_5^{-1}\left(p_5 + B_4^{-1}\left(p_4 + B_3^{-1}\left(p_3 + B_2^{-1} p_2\right)\right)\right) \cdot k - p \cdot n \tag{4.108}$$

(4) Determine a root of the f function from the initial supposed value of θ_1, using the Newton-Raphson, or another numerical method.

Having determined the value of the root of f, the values of the remaining joint variables can be obtained in a closed form.

Example 4.9. For a GP66 (Stanford-type) manipulator with five revolute joints and one prismatic joint, the following values of parameters are assumed:

$$\begin{array}{lcccccc} \alpha_i = & \pi/2, & \pi/2, & 0, & \pi/2, & \pi/2, & 0 \\ l_i = & 0, & l_2, & 0, & 0, & 0, & 0 \\ \lambda_i = & 0, & 0, & \lambda_3, & 0, & \lambda_3, & 0 \end{array}$$

This manipulator belongs to the class of manipulators for which the inverse kinematics solution can not be obtained in explicit form.

Assuming a value of θ_1, the coordinates x and z of vector $\boldsymbol{p}'$ and unit vector $\boldsymbol{m}'$ can be computed:

$$p'_x = p_x c_1 + p_y s_1 \qquad n'_x = n_x c_1 + n_y s_1$$
$$p'_z = p_x s_1 - p_y c_1 \qquad n'_z = n_x s_1 - n_y c_1 \tag{4.109}$$

For the GP66 manipulator:

$$p_1 = p_4 = p_6 = 0, \qquad \boldsymbol{B}_2^{-1}\boldsymbol{p}_2 = l_2\boldsymbol{i}$$
$$\boldsymbol{p}_3 = \lambda_3\boldsymbol{k}, \qquad \boldsymbol{p}_5 = \lambda_5\boldsymbol{k}, \qquad \boldsymbol{B}_3 = \boldsymbol{I}$$

Thus:

$$\boldsymbol{p} = \boldsymbol{B}_1\boldsymbol{B}_2(\lambda_5\boldsymbol{B}_4\boldsymbol{k} + \lambda_3\boldsymbol{k} + l_2\boldsymbol{i})$$
$$\boldsymbol{p}' = \boldsymbol{B}_2(\lambda_5\boldsymbol{B}_4\boldsymbol{k} + \lambda_3\boldsymbol{k} + l_2\boldsymbol{i}) \tag{4.110a}$$
$$\boldsymbol{n}' = \boldsymbol{B}_1^{-1}\boldsymbol{n} = \boldsymbol{B}_2\boldsymbol{B}_4\boldsymbol{B}_5\boldsymbol{k}$$

Hence

$$n'_z = \boldsymbol{n}'\cdot\boldsymbol{k} = \boldsymbol{B}_5\boldsymbol{k}\cdot\boldsymbol{B}_4^{-1}\boldsymbol{j} \tag{4.110b}$$

And since

$$\boldsymbol{B}_5\boldsymbol{k} = [s_5, -c_5, 0]^{\mathrm{T}} \quad \text{and} \quad \boldsymbol{B}_4^{-1}\boldsymbol{j} = [s_4, 0, -c_4]^{\mathrm{T}} \tag{4.111a}$$

we have

$$n'_z = s_4 s_5 \tag{4.111b}$$

Since

$$p'_z = \boldsymbol{p}'\cdot\boldsymbol{k} \text{ and } \boldsymbol{B}_2^{-1}\boldsymbol{k} = \boldsymbol{j} \tag{4.112a}$$

hence

$$p'_z = (\lambda_5\boldsymbol{B}_4\boldsymbol{k} + \lambda_3\boldsymbol{k} + l_2\boldsymbol{i})\cdot\boldsymbol{j}$$
$$p'_z = -\lambda_5 c_4 \tag{4.112b}$$

Equations (4.110a) enable us to determine:

$$\boldsymbol{p}'\cdot\boldsymbol{p}' = \boldsymbol{n}\cdot\boldsymbol{p} = \boldsymbol{B}_5\boldsymbol{k}\cdot\left(\lambda_5\boldsymbol{k} + \lambda_3\boldsymbol{B}_4^{-1}\boldsymbol{k} + l_2\boldsymbol{B}_4^{-1}\boldsymbol{i}\right)$$

Since $\boldsymbol{B}_2^{-1}\boldsymbol{k} = \boldsymbol{j}$ and $\boldsymbol{B}_5\boldsymbol{k}\cdot\lambda_5\boldsymbol{k} = 0$, then

$$\boldsymbol{n}\cdot\boldsymbol{p} = -\lambda_3 c_5 + l_2 c_4 s_5 \tag{4.113a}$$

Equations (4.113a) can be used to determine the real function of θ_1

$$f(\theta_1) = -\lambda_3 c_5 + l_2 c_4 s_5 - \boldsymbol{n}\cdot\boldsymbol{p} \tag{4.113b}$$

Values of θ_1 obtained by solving the inverse kinematics must be roots of the f function. For the known θ_1 we can use (4.109) to calculate p'_x, p'_z, n'_x and n'_z.

From equation (4.112b) we obtain

$$c_4 = -p'_z/\lambda_5 \tag{4.114}$$

and $s_4 = \pm\sqrt{1-c_4^2}$

where ± expresses a sign ambiguity.

The variable λ_3 can be determined according to the formula

$$\lambda_3 = \sqrt{p^2 - l_2^2 - \lambda_5^2 - 2l_2\lambda_5 s_4} \tag{4.115}$$

Using the equation (4.111b) the value of s_5 can be computed if s_4 is non-zero

$$s_5 = -m_z'/s_4$$

and $c_5 = \pm\sqrt{1-s_5^2}$ (4.116)

These additional ambiguities can be avoided if we take into consideration equations (4.116) and (4.117), the solution of which yields:

$$s_2 = (\lambda_3 p_x' + k_0 p_z)/(\lambda_3^2 + k_0^2)$$

$$c_2 = (k_0 p_x' - \lambda_3 p_z)/(\lambda_3^2 + k_0^2 \tag{4.117}$$

where

$$k_0 = l_2 + \lambda_5 s_4$$

From equations (4.104) and (4.105) we obtain

$$c_5 = (n_z' - s_2 c_4 s_5)/c_2 \quad \text{or} \quad c_5 = (c_2 c_4 s_5 - n_x')/s_2 \tag{4.118}$$

When the values of λ_3, c_4 and c_5 are calculated, then the value of $f(\theta_l)$ is fully determined with formula (4.113b). The derivative of $f(\theta_l)$ can also be found. By differentiating the relation (4.112b) with respect to θ_l we obtain

$$dc_4/d\theta_1 = -p_x'/\lambda_5 \tag{4.119}$$

By differentiating identity

$$s_4^2 + c_4^2 = 1$$

we obtain

$$\frac{ds_4}{d\theta_1} = -c_4\left(\frac{dc_4}{d\theta_1}\right)\bigg/ s_4 = \frac{c_4 p_x'}{s_4\lambda_5} \tag{4.120}$$

where $dp_z'/d\theta_l = p_x'$

We then determine

$$\frac{d\lambda_3}{d\theta_1} = -l_2\lambda_5\left(\frac{dc_4}{d\theta_1}\right)\bigg/\lambda_3 = \frac{-l_2\lambda_5 c_4 p_x'}{s_4\lambda_3\lambda_5}$$

$$\frac{ds_4}{d\theta_1} = \left[n_x' - s_5\left(\frac{ds_4}{d\theta_1}\right)\right]\bigg/ s_4 \tag{4.121}$$

$$\frac{dc_5}{d\theta_1} = -s_5\left(\frac{ds_5}{d\theta_1}\right)\bigg/ c_5$$

Finally, (4.113b) is differentiated to obtain

$$\frac{df}{d\theta_1} = l_2\left(c_4\frac{ds_5}{d\theta_1} + s_5\frac{dc_4}{d\theta_1}\right) - \left(\lambda_3\frac{dc_5}{d\theta_1} + c_5\frac{d\lambda_3}{d\theta_1}\right) \tag{4.122}$$

New estimates of θ_1 can be obtained with the one-dimensional Newton-Raphson iterative method:

$$\theta_{1,new} = \theta_1 - \frac{f(\theta_1)}{df/d\theta_1} \tag{4.123}$$

When θ_1 has been determined with given accuracy, the remaining joint variables θ_3, θ_4, θ_5, and λ_3, are calculated from the last iteration. If θ_1 is close enough to exact solution, then complicated calculations of $df/d\theta_1$ can be reduced by numerical differentiation:

$$\left(\frac{df}{d\theta_1}\right)_i = \frac{f_{i-1} - f_i}{\theta_{1,i-1} - \theta_{1,i}} \tag{4.124}$$

where i denotes the number of iterations in which the variables are being computed. This saves the computational cost of equations (4.119)-(4.123) and avoids the problem of singularities that occur when division by a number close to zero is needed.

Numerical example. The dimensions: $l_2 = 0.3600$ and $\lambda_5 = 0.1900$ are given, as well as the position and orientation matrix in the initial position:

$$T_6 = \begin{bmatrix} \sqrt{2}/2 & 0 & \sqrt{2}/2 & 1 \\ \sqrt{2}/2 & 0 & -\sqrt{2}/2 & -1/2 \\ 0 & 1 & 0 & -1/2 \\ 0 & 0 & 0 & 1 \end{bmatrix}$$

The computations begin with $\theta_1 = -20°$. The results of successive calculations are assumed as the initial values for the next iteration. The number of iterations not exceed a maximum of six. The results of the calculations are as follows:

$\theta_1 = -19.072°$; $\theta_2 = 54.427°$; $\theta_4 = -140.114°$; $\theta_5 = -137.013°$; $\theta_6 = -121.439°$
$\lambda_3 = 1.192$.

Example 4.10. For a SMART-3S (COMAU) manipulator with six revolute joints, the following values of parameters are assumed:

$$\begin{array}{lllllll} \alpha_i = & \pi/2, & 0, & \pi/2, & -\pi/2, & \pi/2, & 0 \\ l_i = & l_1, & l_2, & l_3, & 0, & 0, & 0 \\ \lambda_i = & 0, & 0, & 0, & \lambda_4, & \lambda_5, & \lambda_6 \end{array}$$

This manipulator belongs to the class of manipulators for which the inverse kinematics solution can not be obtained in explicit form.

Assuming a value of θ_1, the coordinates of vector $\boldsymbol{p}'$ and unit vectors $\boldsymbol{l}'$, $\boldsymbol{m}'$ and $\boldsymbol{n}'$ can be computed. From equation (4.98) we obtain a position vector and orientation matrix of the end-effector reference system origin relative to the reference system of the first link:

$$\boldsymbol{T}_{6,1} = A_2 A_3 A_4 A_5 A_6 = \begin{bmatrix} l' & m' & n' & p' \\ 0 & 0 & 0 & 1 \end{bmatrix} \tag{4.125}$$

where

$$\begin{aligned}
p'_x &= \lambda_6(c_{23}c_4s_5 + s_{23}c_5) - \lambda_5 c_{23}s_4 + \lambda_4 s_{23} + l_3 c_{23} + l_2 c_2 \\
p'_y &= \lambda_6(s_{23}c_4s_5 - c_{23}c_5) - \lambda_5 s_{23}s_4 - \lambda_4 c_{23} + l_3 s_{23} + l_2 s_2 \\
p'_z &= \lambda_6 s_4 s_5 + \lambda_5 c_4 \\
l'_x &= c_{23}(c_4c_5c_6 - s_4s_6) - s_{23}s_5c_6 \\
l'_y &= s_{23}(c_4c_5c_6 - s_4s_6) + c_{23}s_5c_6 \\
l'_z &= s_4c_5c_6 + c_4s_6 \\
m'_x &= -c_{23}(c_4c_5s_6 + s_4c_6) + s_{23}s_5s_6 \\
m'_y &= -s_{23}(c_4c_5s_6 + s_4c_6) - c_{23}s_5s_6 \\
m'_z &= -s_4c_5s_6 + c_4c_6 \\
n'_x &= c_{23}c_4s_5 + s_{23}c_5 \\
n'_y &= s_{23}c_4s_5 - c_{23}c_5 \\
n'_z &= s_4s_5
\end{aligned} \tag{4.125a}$$

and

$$\boldsymbol{T}_{6,1} = A_1^{-1} T_6 \tag{4.126}$$

$$\begin{aligned}
p'_x &= p_x c_1 + p_y s_1 - l_1 & m'_x &= m_x c_1 + m_y s_1 \\
p'_y &= p_z & m'_y &= m_z \\
p'_z &= p_x s_1 - p_y c_1 & m'_z &= m_x s_1 - m_y c_1 \\
l'_x &= l_x c_1 + l_y s_1 & n'_x &= n_x c_1 + n_y s_1 \\
l'_y &= l_z & n'_y &= n_z \\
l'_z &= l_x s_1 - l_y c_1 & n'_z &= n_x s_1 - n_y c_1
\end{aligned} \tag{4.126a}$$

Using equation (4.125a) we obtain

$$\begin{aligned}
& c_4 = (p'_z - \lambda_6 n'_z)/\lambda_5 \quad \text{and} \quad s_4 = \pm\sqrt{1 - c_4^2} \\
& \theta_6 = Atan2\left[c_4, \pm\sqrt{l'^2_z + m'^2_z - c_4^2}\right] - Atan2(m'_z, l'_z) \\
& s_{23} = -(m'_y c_6 + l'_y s_6)/s_4 \\
& c_{23} = -(m'_x c_6 + l'_y s_6)/s_4 \\
& s_5 = n'_z / s_4 \qquad c_5 = n'_x s_{23} - n'_y c_{23}
\end{aligned} \tag{4.127}$$

Equations that can be used to define a real function of θ_1

$$f(\theta_1) = l_2(s_3c_5 + c_3c_4s_5) + l_3c_4s_5 + \lambda_4c_5 + \lambda_6 - p'_xn'_x - p'_yn'_y - p'_zn'_z \tag{4.128}$$

or

$$\begin{aligned} f(\theta_1) = {} & 2l_2(\lambda_6s_5c_4 - \lambda_5s_4 + l_3)c_3 + 2l_2(\lambda_6c_5 + \lambda_4)s_3 + 2l_3(\lambda_6s_5c_4 - \lambda_5s_4) + \\ & +2\lambda_6\lambda_4c_5 + \lambda_6^2 + \lambda_5^2 + \lambda_4^2 + l_3^2 + l_2^2 - p'_xp'_x - p'_yp'_y - p'_zp'_z \end{aligned} \tag{4.129}$$

Numerical example. The dimensions of the SMART-3S (COMAU) manipulator:

$l_1 = 0.160$, $l_2 = 0.610$, $l_3 = 0.110$, $\lambda_4 = 0.610$, $\lambda_5 = 0.113$, λ_6 0.078,

are given, as well as the position and orientation matrix of the end-effector:

$$T_6 = \begin{bmatrix} 0{,}7120 & 0{,}0580 & 0{,}7000 & 0{,}200 \\ -0{,}2668 & -0{,}8995 & 0{,}3460 & -0{,}500 \\ 0{,}6495 & -0{,}4330 & 0{,}6250 & 0{,}050 \\ 0 & 0 & 0 & 1 \end{bmatrix}$$

The computations begin with $\theta_1 = -70°$. The results of successive calculations are assumed as the initial values for the next iteration. The assumed ranges of variables are:

$\theta_{1min} = -165°$; $\theta_{2min} = -45°$; $\theta_{3min} = -75°$; $\theta_{4min} = -175°$; $\theta_{5min} = -140°$; $\theta_{6min} = -175°$
$\theta_{1max} = 165°$; $\theta_{2max} = 170°$; $\theta_{3max} = 90°$; $\theta_{4max} = 175°$; $\theta_{5max} = 140°$; $\theta_{6max} = 175°$

The results of the calculations describe only two configurations with variables:

1) $\theta_1 = -69.16°$; $\theta_2 = 85{,}45°$; $\theta_3 = -52.94°$; $\theta_4 = 62.84°$; $\theta_5 = -60.85°$; $\theta_6 = -108.53°$
2) $\theta_1 = -60.29°$; $\theta_2 = 90{,}85°$; $\theta_3 = -72.21°$; $\theta_4 = -118.79°$; $\theta_5 = 58.60°$; $\theta_6 = 54.72°$

A fast iterative inverse kinematic method based on a one-dimensional Newton-Raphson technique was presented. This method neither requires computation of the Jacobian nor the inverse Jacobian of the manipulator. It can be applied to any manipulator that does not allow closed-form solutions, but for which knowledge of one of the joint variables allows closed-form solutions for the remaining joint variables.

4.5. References

[4.1] *Angeles J.*: Fundamentals of Robotic Mechanical Systems: Theory, Methods, and Algorithms. New York, Springer Verlag. 1997.

[4.2] *Craig J.J.*: Introduction to Robotics. Addison-Wesley 1989

[4.3] *Duffy J.*: Analysis of Mechanisms and Robot Manipulators. London, E. Arnold, 1980.

[4.4] *Frolov K. V., Vorobiev E. J.*: Mechanics of Industrial Robots. Vyssaja Skola, Moscow 1988.

[4.5] *Fu K. S., Gonzales R. C., Lee C. S. G.*: Robotics: Control, Sensing, Vision, and Inteligence. McGraw-Hill 1987.

[4.6a] *Hunt K. H.*: Robot Kinematics. A Compact Analytic Inverse Solution for Velocities. Trans. ASME, Mech., Transm. and Autom. in Design, vol. 109, s. 42–49, 1987.

[4.6b] *Hunt K. H.*: The particular or the general? (some examples from robot kinematics). Mech. and Mach. Theory, vol. 21, no 6, s.481–487, 1989.

[4.7] *Knapczyk J., Lebiediew P. A.*: Theory of Spatial Mechanisms and Manipulators. Warszawa, WNT, 1990 (*in Polish*).

[4.8] *Lee H. Y., Woernle C., Hiller M.*: A Complete Solution for the Inverse Kinematic Problem of the General 6R Robot Manipulator. Trans. ASME, Mech., Transm. and Autom. in Design, vol. 113, s. 481–486, 1991.

[4.9] *Lipkin H., Duffy J.*: A Vector Analysis of Robot Manipulators. Recent Advances in Robotics. New York, Wiley, 1985.

[4.10] *Litvin F. L., Parenti Castelli V., Phillips R. H.*: Manipulators: Execution of Prescribed Trajectories. Special Link Positions and Versions of Assembly. Mech. and Mach. Theory, vol. 21, no 2, s. 173–185, 1986.

[4.11] *Lloyd J., Hayward V.*: Kinematics of Common Industrial Robots. Robotics 4 (1988), s. 169–191.

[4.12] *Manseur R., Doty K. L.*: A Fast Algorithm for Inverse Kinematic Analysis of Robot Manipulators. Int. Robotics Research, vol. 7, no 3, s. 52–63, 1988.

[4.13] *Paul R. P.*: Robot Manipulators: Mathematics, Programming and Control. Cambridge, MIT Press, 1981.

[4.14] *Pennock G. R., Yang A. T.*: Application of Dual-Number Matrices to the Inverse Kinematics Problem of Robot Manipulators. Trans. ASME, Mech. Transm. and Autom. in Design, vol. 107, no 2, s. 201–208, 1985.

[4.15] *Pieper D.L.*: The Kinematics of Manipulators under Computer Control. Ph. D.Thesis, Stanford Univ. 1968.

[4.16] *Raghavan M., Roth B.*: A General Solution for the Inverse Kinematics of all Series Chains. Proc. of the 8th CISM-IFToMM Symp. RoManSy'90 Cracow 1990, s.21–28.

[4.17] *Ranky P. G., Ho C. Y.*: Robot Modelling Control and Applications with Software. Kempston, JFS 1985.

[4.18] *Tsai L. W., Morgan A.*: Solving the Kinematics of the Most General Six- and Five-Degree-of-Freedom Mnipulators by Continuation Methods. Trans. ASME, Mech., Transm. and Autom. in Design, vol. 107, s.189–200, 1985.

[4.19] *Angeles J., Zanganeh K. E.*: The semigraphical Determination of All Real Solutions of General Six-Revolute Manipulators. Proc. of the 9th CISM-IFToMM Symp. RoManSy'92, Udine. Springer Verlag 1993, s. 23–32.

5. STATICS AND DYNAMICS OF MANIPULATORS

5.1. Statics of manipulators

Each of the manipulator joints usually has a separate drive. Drive forces, as well as torques acting in joints, are distributed through the manipulator links to the end-effector (gripper) on which the environment also exerts a force and torque. The relationship between input (drive) and output (acting on the end-effector) forces and torques constitutes the basis of a control system.

Forces and torques are vector quantities described in a certain coordinate system. Force vectors are denoted as $\boldsymbol{F}$, and torque vectors as $\boldsymbol{M}$. Force vectors are assumed to be attached to the coordinate system origin.

If we have a torque and a force applied at the origin of a certain coordinate system attached to a link, it is possible to find an equivalent force and torque described in a different coordinate system, attached to this link as well. To solve this kind of problem we use the principle of virtual work. We consider the force and torque applied to a link that cause a differential, so-called virtual displacement, and performing virtual work. Since the displacement is infinitesimal and does not change the total energy of the mechanical system, the sum of the virtual work of all the forces acting on the link is equal to zero.

The virtual work, done by force $\boldsymbol{F}$, is defined as

$$\delta W = \boldsymbol{F}^T \boldsymbol{D} \tag{5.1}$$

where

$$\boldsymbol{F}^T = [F_x \ F_y \ F_z \ M_x \ M_y \ M_z] \tag{5.1a}$$

$$\boldsymbol{D} = [d_x \ d_y \ d_z \ \delta_x \ \delta_y \ \delta_z]^T$$

and $\boldsymbol{D}$ is a differential displacement vector, consisting of a translation vector $\boldsymbol{d}$ and a rotation vector δ.

The same displacement $\boldsymbol{D}_j$ of the i link may result from load $\boldsymbol{F}_j$, that is a torque and a force exerted at a different point of this link and described in the j link system, performing the same virtual work

$$\boldsymbol{F}_i^T \boldsymbol{D}_i = \boldsymbol{F}_j^T \boldsymbol{D}_j \tag{5.2}$$

As the virtual displacement $\boldsymbol{D}_j$ in j system is equivalent to the virtual displacement $\boldsymbol{D}_i$ in i system, we have

$$\boldsymbol{D}_j = \boldsymbol{J}\,\boldsymbol{D}_i \tag{5.3}$$

where J is the Jacobian describing the transformation of an infinitesimal displacement in the i system into the j system. Substituting (5.3) in (5.2) and ordering, we obtain:

$$F_i = J^T F_j \quad \text{or} \quad F_i^T = F_j^T J \tag{5.4}$$

If the relative orientation and position of the i and j systems is described by means of a matrix:

$$T_{i,j} = \begin{bmatrix} l & m & n & p \\ 0 & 0 & 0 & 1 \end{bmatrix} \tag{5.5}$$

then, considering (5.1a), the relationship (5.4) can be converted into

$$\begin{aligned} M_{j,x} &= l \cdot [(F_i \times p) + M_i] \\ M_{j,y} &= m \cdot [(F_i \times p) + M_i] \\ M_{j,z} &= n \cdot [(F_i \times p) + M_i] \end{aligned} \tag{5.5a}$$

$$\begin{aligned} F_{j,x} &= l \cdot F_i \\ F_{j,y} &= m \cdot F_i \\ F_{j,z} &= n \cdot F_i \end{aligned} \tag{5.5b}$$

Example 5.1. A force and torque are given in the base system *0*

$$F = [10 \ \ 0 \ \ 0]^T$$
$$M = [0 \ \ 100 \ \ 0]^T$$

An equivalent force and torque in the six- link system are to be found, knowing that

$$T_6 = \begin{bmatrix} 0 & 0 & 1 & 10 \\ 1 & 0 & 0 & 5 \\ 0 & 1 & 0 & 0 \\ 0 & 0 & 0 & 1 \end{bmatrix}$$

where

$$l = [0 \ \ 1 \ \ 0]^T, \ m = [0 \ \ 0 \ \ 1]^T, \ n = [1 \ \ 0 \ \ 0]^T, \ p = [10 \ \ 5 \ \ 0]^T$$

Substituting the coordinates of vectors F and M, as well as the elements of the matrix $T_{6,0}$, in formula (5.5) gives us

$$F \times p = \begin{bmatrix} i & j & k \\ 10 & 0 & 0 \\ 10 & 5 & 0 \end{bmatrix} = 0i + 0j + 50k$$

$$F \times p + M = [0 \ \ 100 \ \ 50]^T$$

Finally: $M_6 = [100 \ \ 50 \ \ 0]^T$, $F_6 = [0 \ \ 0 \ \ 10]^T$

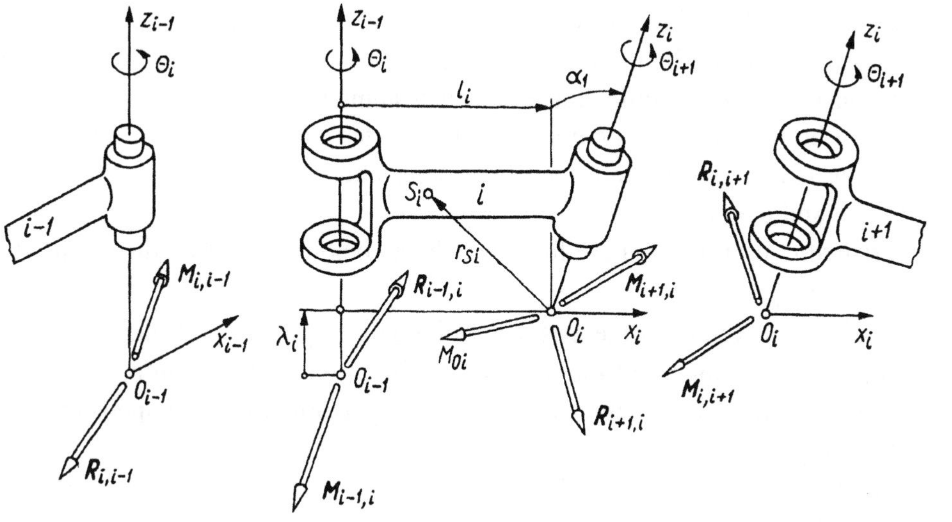

Fig. 5.1. System of forces and torques acting on the i link of the kinematic chain with revolute joints. Notation: Ri-1,i, Mi-1,i – vectors of the force and torque exerted on the i link by the i-1 link

Now let us consider a force system acting on a single link i of a manipulator's kinematic chain. Figure 5.1 shows the forces and torques exerted on the i link, which is connected with the links i–1 and i+1 through revolute joints. A force imposed at a point O_{i-1} is denoted as a vector $\boldsymbol{R}_{i-1,i}$, signifying that it is exerted by the i–1 link on the i link. A vector $\boldsymbol{R}_{i,i+1}$ denotes a force exerted on the i+1 link by the i link. The force of gravity is applied at the mass center S_i and equals $m_i\boldsymbol{g}$, where m_i is the mass of the i link and $\boldsymbol{g}$ is the gravity acceleration vector. The equilibrium of forces acting on the i link is expressed by the equation:

$$\boldsymbol{R}_{i-1,i} - \boldsymbol{R}_{i,i+1} + m_i\boldsymbol{g} = 0 \quad (i = 1, \ldots, n) \tag{5.6}$$

noting that all the vectors are defined in the base coordinate system. In a similar way we can express an equation for torques relative to the S_i point:

$$\boldsymbol{M}_{i-1,i} - \boldsymbol{M}_{i,i+1} - (\boldsymbol{p}_{i-1,i} + \boldsymbol{r}_{Si})\times\boldsymbol{R}_{i-1,i} - (-\boldsymbol{r}_{Si})\times\boldsymbol{R}_{i,i+1} = 0 \quad (i = 1, \ldots, n) \tag{5.7}$$

where

$\boldsymbol{p}_{i-1,i}$ is a position vector of point O_i relative to O_{i-1} (that is the vector $\overline{O_{i-1}O_i}$ expressed in the base system; and

$\boldsymbol{r}_{Si}$ is a vector $\overline{O_iS_i}$ defining the position of the point S_i relative to O_i, also expressed in the base system.

The external forces and torques caused by the environment, written as vectors: $\boldsymbol{F}_{n,n+1}$ and $\boldsymbol{M}_{n,n+1}$, and an additional link n+1 act on the end-effector, interacting with environment.

Equations (5.6) and (5.7) can be written for all links, thus the number of vector equations is $2n$, while the number of introduced forces and torques is $2(n+1)$. Two vectors must be given in order for equation systems (5.6) and (5.7) to be solvable. These may be the external force and torque exerted on the end-effector. Assuming that friction is negligibly small, the drive forces and torques in the joints have been determined and denoted as f_i.

For a prismatic joint, f_i is a drive force which can be obtained from the dot product

$$f_i = \boldsymbol{e}_i^T \boldsymbol{R}_{i-1,i} \tag{5.8}$$

where $\boldsymbol{e}_i$ is a unit vector of the prismatic joint axis. Equation (5.8) indicates that the drive force is a component of the force $\boldsymbol{R}_{i-1,i}$ of the joint axis direction, with the rest of the components, internal forces which do not perform work, compensated by the joint housing.

For a revolute joint f_i is a drive torque

$$f_i = \boldsymbol{e}_i^T \boldsymbol{M}_{i-1,i} \tag{5.9}$$

The rest of the $\boldsymbol{M}_{i-1,i}$ components are borne by the joint housing.

All the driven forces and torques can be put together and defined as a n-dimensional vector in the form of

$$\boldsymbol{f} = [f_1\ f_2\ \dots\ f_n]^T \tag{5.10}$$

The relationship between drive forces and torques and an external force and torque applied to the end-effector is described by the equation

$$\boldsymbol{f} = \boldsymbol{J}^T \boldsymbol{F} \tag{5.11}$$

where

$$\boldsymbol{F} = [\boldsymbol{R}_{n,n+1}\quad \boldsymbol{M}_{n,n+1}]^T$$

$\boldsymbol{J}$ denotes a 6×n matrix, called the Jacobian matrix, which describes the relationship between differential displacements in the joints and those of the end-effector. It is important to note that the system of forces does not include gravity, or other forces except for those applied on the end-effector.

Example 5.2. Figure 5.2 shows a two-link planar manipulator exerting a force $\boldsymbol{F} = [F_x\ \ F_y]^T$ on a surface. The task is to find equivalent drive torques $\boldsymbol{f} = [f_1\ f_2]^T$, assuming that friction in the joints is negligibly small.

The Jacobian of the examined manipulator is defined as follows

$$\boldsymbol{J} = \begin{bmatrix} -l_1 s\theta_1 - l_2 s(\theta_1+\theta_2) & -l_2 s(\theta_1+\theta_2) \\ l_1 c\theta_1 + l_2 c(\theta_1+\theta_2) & l_2 c(\theta_1+\theta_2) \end{bmatrix} \tag{5.12}$$

Substituting (5.12) into equation (5.11) we have the equivalent drive torques

$$\begin{bmatrix} f_1 \\ f_2 \end{bmatrix} = \begin{bmatrix} -l_1 s\theta_1 - l_2 s(\theta_1+\theta_2) & l_1 c\theta_1 + l_2 c(\theta_1+\theta_2) \\ -l_2 s(\theta_1+\theta_2) & l_2 c(\theta_1+\theta_2) \end{bmatrix} \begin{bmatrix} F_x \\ F_y \end{bmatrix} \tag{5.13}$$

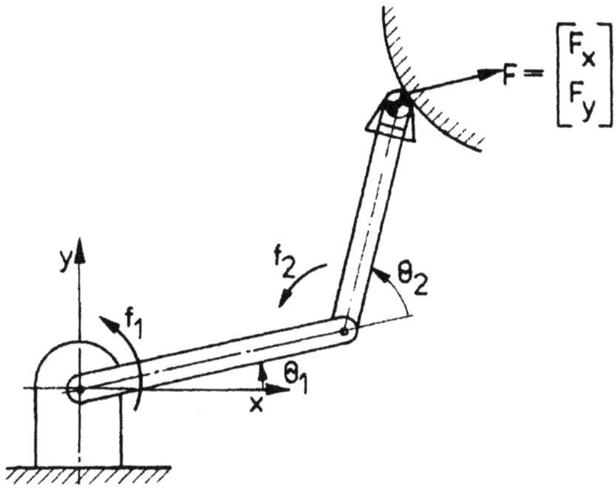

Fig. 5.2. Planar two-link manipulator with revolute joints. Force $\boldsymbol{F} = [F_x\ F_y]^T$ is applied on a surface, with drive torques f_1 and f_2 acting in the joints

From equation (5.13) we may draw the conclusion that the drive forces and torques are related through the Jacobian matrix to the external forces and torques applied on the end-effector. Thus, a static load is closely related to the instantaneous motion of the manipulator. The order of the Jacobian matrix is defined by the number of all the possible velocities generated by movements in the joints. When the matrix is degenerate, certain manipulator positions are singular, and there are directions along which the end-effector can not move.

Drive forces and torques in the joints can be determined only for a particular end-effector's load. However, in certain positions of the manipulator (or with a force system acting on the end-effector), all loads can be compensated by the manipulator structure. If the external force is applied to the end-effector in the direction along which the link can move, then that force must be balanced by drive forces and torques. If, however, the Jacobian matrix is degenerate (or the manipulator is in the singular configuration), the external force is partly compensated for by the manipulator's structure.

Example 5.3. The position and orientation of the Stanford-type manipulator of example 5.8 is defined by the matrix

$$\boldsymbol{T}_6 = \begin{bmatrix} 0 & 1 & 0 & 20 \\ 1 & 0 & 0 & 6 \\ 0 & 0 & -1 & 0 \\ 0 & 0 & 0 & 1 \end{bmatrix}$$

with the configuration coordinates

$$\theta_1 = 0, \quad \theta_2 = \pi/2, \quad \lambda_3 = 20, \quad \theta_4 = 0, \quad \theta_5 = \pi/2, \quad \theta_6 = \pi/2$$

and the Jacobian matrix

$$\frac{\partial T_6}{\partial q_i} = \begin{bmatrix} 20 & 0 & 0 & 0 & 0 & 0 \\ -6 & 0 & 1 & 0 & 0 & 0 \\ 0 & 20 & 0 & 0 & 0 & 0 \\ 0 & 1 & 0 & 0 & 1 & 0 \\ 0 & 0 & 0 & 1 & 0 & 0 \\ -1 & 0 & 0 & 0 & 0 & 1 \end{bmatrix}$$

The problem: calculate the drive forces and torques that balance the force and torque exerted on the end-effector, given in its coordinate system

$$F = [0\ \ 0\ \ 10\ \ 0\ \ -200\ \ 1000]^T$$

Solution: the equation (5.11) which in this particular case takes the form of:

$$\begin{bmatrix} M_1 \\ M_2 \\ M_3 \\ M_4 \\ M_5 \\ M_6 \end{bmatrix} = \begin{bmatrix} 20 & -6 & 0 & 0 & 0 & -1 \\ 0 & 0 & 20 & 1 & 0 & 0 \\ 0 & 1 & 0 & 0 & 0 & 0 \\ 0 & 0 & 0 & 0 & 1 & 0 \\ 0 & 0 & 0 & 1 & 0 & 0 \\ 0 & 0 & 0 & 0 & 0 & 1 \end{bmatrix} \begin{bmatrix} 0 \\ 0 \\ 100 \\ 0 \\ -200 \\ 1000 \end{bmatrix} = \begin{bmatrix} -1000 \\ 2000 \\ 0 \\ -200 \\ 0 \\ 1000 \end{bmatrix}$$

5.2. Dynamics of manipulators

Two types of problems are found here. In the first, which arises in manipulator motion control, given a trajectory point q_i, $\dot{q}_i$ and $\ddot{q}_i$, we must find the drive force and torque vectors *f*. In the second we must determine the manipulator motion caused by drive forces and torques, that is, given the *f* vector, q_i, $\dot{q}_i$ and $\ddot{q}_i$ must to be found.

The dynamic properties of a manipulator are expressed as time profiles of its position, subject to drive forces and torques. Those relationships can be described through the system of motion differential equations. Two methods are used: Newton-Euler and the Lagrange. Newton-Euler equations are directly derived from the Newton`s second principle of dynamics, which describes the behavior of a system to which forces and torques are applied. Lagrange equations describe the dynamic properties of the system in joint space in terms of the manipulator's kinetic and potential energy.

5.2.1. Rigid body mass distribution

For a rigid body performing a rotational motion the notion of moment of inertia is used. However, if a rigid body performs a spatial motion, we introduce the notion of the tensor of inertia, which is a generalization of the scalar moment of inertia. The tensor of inertia is usually defined relative to the coordinate system attached to the body:

$$\mathbf{I} = \begin{bmatrix} I_{xx} & -I_{xy} & -I_{xz} \\ -I_{xy} & I_{yy} & -I_{yz} \\ -I_{xz} & -I_{yz} & I_{zz} \end{bmatrix} \tag{5.14}$$

where

$$I_{xx} = \iiint_V \left(y^2 + z^2\right)\rho dv, \qquad I_{yy} = \iiint_V \left(x^2 + z^2\right)\rho dv, \qquad I_{zz} = \iiint_V \left(x^2 + y^2\right)\rho dv$$

$$I_{xy} = \iiint_V xy\rho dv, \qquad I_{xz} = \iiint_V xz\rho dv, \qquad I_{yz} = \iiint_V yz\rho dv$$

assuming that the rigid body consists of differential elements of volume dv and density ρ. The position of each element is described by the vector $\boldsymbol{p} = [x \;\; y \;\; z]^T$.

Example 5.4. The problem: find an inertia tensor for a cube-shaped body of uniform density ρ, relative to the system whose axes x, y and z coincide with the edges of the cube; w, l and h are the edge's lengths, respectively. To solve this problem we first compute I_{xx}:

$$I_{xx} = \int_0^h \int_0^l \int_0^w \left(y^2 + z^2\right)\rho dx dy dz = \int_0^h \int_0^l \left(y^2 + z^2\right) w\rho dy dz = \int_0^h \left(\frac{l^3}{3} + z^2 l\right) w\rho dz = \frac{m}{3}\left(l^2 + h^2\right)$$

where m is the total mass of the body. Permuting terms we have

$$I_{yy} = \frac{m}{3}\left(w^2 + h^2\right) \qquad I_{zz} = \frac{m}{3}\left(w^2 + h^2\right)$$

We next compute

$$I_{xy} = \int_0^h \int_0^l \int_0^w xy\rho dx dy dz = \int_0^h \int_0^l \frac{w^2}{2} y\rho dy dz = \int_0^h \frac{w^2 l^2}{4}\rho dz = \frac{m}{4} wl$$

Permuting the terms gives us

$$I_{xz} = \frac{m}{4} hw \qquad I_{yz} = \frac{m}{4} hl$$

Thus, the inertia tensor matrix for that object takes the form of

$$\boldsymbol{I} = \begin{bmatrix} \frac{m}{3}\left(l^2 + h^2\right) & -\frac{m}{4} wl & -\frac{m}{4} hw \\ -\frac{m}{4} wl & \frac{m}{3}\left(w^2 + h^2\right) & -\frac{m}{4} hl \\ -\frac{m}{4} hw & -\frac{m}{4} hl & \frac{m}{3}\left(l^2 + w^2\right) \end{bmatrix}$$

The inertia tensor is a function of the position and orientation of the reference system. To compute the changes in the inertia tensor under translations of the reference coordinate system, we can use the parallel axis theorem, which relates the inertia tensor in the coordinate system which has its origin in the center of mass, to another reference system. If a system {S} is attached at the center of mass and a system {O} is translated by a vector $r_S = \overline{OS} = [x_S\ y_S\ z_S]^T$, this theorem can be expressed by the formulas

$$I_{Ozz} = I_{Szz} + m\left(x_S^2 + y_S^2\right) \qquad (5.15)$$
$$I_{Oxy} = I_{Sxy} - m x_S y_S$$

The remaining moments and products of inertia are determined by permutation of the indices x, y and z in (5.15). This theorem can be stated in a vector-matrix form

$$\mathbf{I}_O = \mathbf{I}_S + m\left[p_S^T \mathbf{p}_S \mathbf{I}_3 - \mathbf{p}_S \mathbf{p}_S^T\right] \qquad (5.16)$$

where $\boldsymbol{I}_3$ denotes the 3×3 identity matrix

Example 5.5. The problem: find the inertia tensor for the solid body described in example 5.4, when it is described in the coordinate system with its origin at this body's center of mass, that is:

$$\begin{bmatrix} x_S \\ y_S \\ z_S \end{bmatrix} = \frac{1}{2}\begin{bmatrix} w \\ l \\ h \end{bmatrix}$$

Applying the (5.15) theorem, we have

$$I_{Ozz} = \frac{m}{12}\left(w^2 + l^2\right)$$
$$I_{Oxy} = 0$$

The other elements are found by symmetry. The resulting inertia tensor, written in the reference system at the center of mass, is

$$\mathbf{I}_S = \begin{bmatrix} \frac{m}{12}\left(h^2 + l^2\right) & 0 & 0 \\ 0 & \frac{m}{12}\left(w^2 + h^2\right) & 0 \\ 0 & 0 & \frac{m}{12}\left(l^2 + w^2\right) \end{bmatrix}$$

Since the resultant matrix is diagonal, system {O} represents the principal axes of this body.

Some additional facts about inertia tensors:

1. If two axes of a reference system form a plane of symmetry, then products of inertia having as their index the coordinate normal to that plane will be zero.
2. Moments of inertia must be positive; products of inertia may have either sign.

3. The sum of the three moments of inertia does not vary with changes in the reference system.
4. The eigenvalues of a inertia tensor are the principal moments for the body; the associated eigenvectors are the principal axes.

The geometry of the links of most manipulators is quite complicated, and this makes the application of the formulas (5.14)÷(5.16) difficult. Consequently, it is easier to measure, rather than calculate, the moments of inertia of each link by means of some experimental procedure.

5.2.2. Newton-Euler equations

We shall consider the links of a manipulator as rigid bodies. If we know the position of the center of mass and the inertia tensor of the link, then its distribution of mass is completely characterized. A force $\boldsymbol{F}$ applied at the body's center of mass and making it move with acceleration $\dot{\boldsymbol{v}}_S$ is defined by Newton's equation:

Dynamics of manipulators The torque $\boldsymbol{M}$ which must be exerted on the body to cause rotational motion with angular acceleration $\dot{\boldsymbol{\omega}}$ is given by Euler's equation:

$$\boldsymbol{F} = m\dot{\boldsymbol{v}}_S \tag{5.17}$$

$$\boldsymbol{M} = \boldsymbol{I}_S\dot{\boldsymbol{\omega}} + \boldsymbol{\omega} \times \boldsymbol{I}_S\boldsymbol{\omega} \tag{5.18}$$

where $\boldsymbol{I}_S$ is the body's inertia tensor written in the reference system {S} whose origin is at the center of mass.

Having computed the acceleration of the center of mass as well as the angular velocity and acceleration of each link in its reference system, we can compute the inertial force and torque with equations (5.17) and (5.18). The joint forces and torques must still be found. This can be done by writing force and moment balance equations for each separate link, in the form of (5.6) and (5.7), together with the inertial forces and torques. As the inertial forces and torques are defined in the reference system attached to the link, the forces and torques with which the neighbouring links act on that link should be defined in the same reference system.

Gravitational force can be taken into account by including in the kinematics equations $\dot{\boldsymbol{v}}_0 = \boldsymbol{g}$, where $\boldsymbol{g}$ is the gravity vector and $\dot{\boldsymbol{v}}_O$ is a ficticious acceleration of the base of the manipulator in the vertical upward direction, causing the same result, in terms of link loadings, that the force of gravity would. In this way, we can account for the effect of the gravity forces in the inertial forces of the links, without additional computational effort.

The equilibrium equations for the forces and torques exerted on link i, written in that link's reference system are:

$$\boldsymbol{F}_i = \boldsymbol{R}_{i-1,i} - \boldsymbol{B}_{i+1,i}\boldsymbol{R}_{i+1,i} \tag{5.19}$$

$$M_i = M_{i-1,i} - B_{i+1,1}M_{i+1,i} - r_{Si} \times F_i - p_{i+1,i} \times B_{i+1,i}R_{i+1,i} \tag{5.20}$$

where

$B_{i+1,}$ is the rotation matrix of system i+1 relative to system i ;

r_{Si} is the position vector of the center of mass S_i of link i relative to the system i origin, that is $r_{Si} = \overline{O_iS_i}$;

$p_{i+1,i}$ is the position vector of point O_{i+1} of system i+1 origin relative to the point O_i – the system i origin;

$R_{i+1,i}$ and $M_{i+1,i}$ are the force and torque with which link i+1 acts on link i .

Equations (5.19) and (5.20) can be converted into a form which allows us to determine forces and torques in sequence, starting from link $i = n$:

$$R_{i-1,i} = B_{i+1,1}R_{i+1,i} + F_i \tag{5.21}$$

$$M_{i-1,i} = B_{i+1,1}M_{i+1,i} + r_{Si} \times F_i + p_{i+1,i} \times B_{i+1,i}R_{i+1,i} + M_i \tag{5.22}$$

If no non-potential external forces are exerted on the end-effector n, then $R_{n+1,n} = 0$ and $M_{n+1,n} = 0$. However, if the manipulator is in contact with the environment, the force $R_{n+1,n}$ and the torque $M_{n+1,n}$ are nonzero. The relationships (5.8) and (5.9) are used to compute drive forces and torques,.

The complete algorithm for calculating drive forces and torques in the joints consists of two parts. First, link velocities and accelerations are computed iteratively from link 1 out to link n, and the inertial forces and torques are determined by the Newton-Euler equations of (5.17) and (5.18). Then, the forces and torques acting between links are computed recursively, using (5.21) and (5.22), and starting from link n. In the second phase of the algorithm, the drive forces and torques are also determined. For a manipulator with revolute joints the procedure as follows:

$$\begin{aligned}
\omega_{i+1} &= B_{i,i+1}\omega_i + \dot{\theta}_{i+1}e_{i+1} \\
\dot{\omega}_{i+1} &= B_{i,i+1}\dot{\omega}_i + B_{i,i+1}\omega_i \times \dot{\theta}_{i+1}e_{i+1} + \ddot{\theta}_{i+1}e_{i+1} \\
\dot{v}_{i+1} &= B_{i,i+1}\left[\dot{\omega}_i \times p_{i+1,i} + \omega_i \times \left(\omega_i \times p_{i+1,i}\right) + \dot{v}_i\right] \\
\dot{v}_{Si} &= \dot{\omega}_{i+1} \times r_{Si+1} + \omega_{i+1} \times \left(\omega_{i+1} \times r_{Si+1}\right) + \dot{v}_{i+1} \\
F_{i+1} &= m_{i+1}\dot{v}_{Si+1} \\
M_{i+1} &= I_{Si+1}\dot{\omega}_{i+1} + \omega_{i+1} \times I_{Si+1}\omega_{i+1} \qquad i = 0, 1 \ldots 5
\end{aligned} \tag{5.23}$$

$$f_i = M_{i-1,i} \cdot e_i \tag{5.24}$$

where $e_i = [0 \;\; 0 \;\; 1]^T$.

In the beginning of the computational process, the kinematic quantities, described in reference systems attached to the moving links, are determined. If they are determined for link i , and we want them to be included in the kinematic quantities of link i+1, they must be multiplied by a corresponding rotation matrix $B_{i,i+1}$. When links i and i+1 are connected by a prismatic joint, the angular velocities and accelerations of both links are the same.

Example 5.6. The problem: a two-link manipulator with revolute joints is assumed to have its links' masses concentrated at the end-points (that is $S_1 = O_1$, $S_2 = O_2$). No force is exerted on the end-effector, that is:

$\boldsymbol{R}_{3,2} = 0, \quad \boldsymbol{M}_{3,2} = 0$

The solution: applying formulas (5.23) we have $\omega_0 = 0, \quad \dot{\omega}_0 = 0, \quad \dot{v}_0 = 0$

$$\boldsymbol{B}_{i+1,i} = \begin{bmatrix} c\theta_{i+1} & -s\theta_{i+1} & 0 \\ s\theta_{i+1} & c\theta_{i+1} & 0 \\ 0 & 0 & 1 \end{bmatrix}, \qquad \boldsymbol{B}_{i,i+1} = \begin{bmatrix} c\theta_{i+1} & s\theta_{i+1} & 0 \\ -s\theta_{i+1} & c\theta_{i+1} & 0 \\ 0 & 0 & 1 \end{bmatrix},$$

$$\omega_1 = \dot{\theta}_1 \boldsymbol{e}_1 = \begin{bmatrix} 0 \\ 0 \\ \dot{\theta}_1 \end{bmatrix}, \qquad \dot{\omega}_1 = \ddot{\theta}_1 \boldsymbol{e}_1 = \begin{bmatrix} 0 \\ 0 \\ \ddot{\theta}_1 \end{bmatrix},$$

$$\dot{v}_1 = \begin{bmatrix} c\theta_1 & s\theta_1 & 0 \\ -s\theta_1 & c\theta_1 & 0 \\ 0 & 0 & 1 \end{bmatrix} \begin{bmatrix} 0 \\ g \\ 0 \end{bmatrix} = \begin{bmatrix} gs\theta_1 \\ gc\theta_1 \\ 0 \end{bmatrix},$$

$$\dot{v}_{S1} = \begin{bmatrix} 0 \\ l_1\ddot{\theta}_1 \\ 0 \end{bmatrix} + \begin{bmatrix} -l_1\dot{\theta}_1^2 \\ 0 \\ 0 \end{bmatrix} + \begin{bmatrix} gs\theta_1 \\ gc\theta_1 \\ 0 \end{bmatrix} = \begin{bmatrix} -l_1\dot{\theta}_1^2 + gs\theta_1 \\ l_1\ddot{\theta}_1 + gc\theta_1 \\ 0 \end{bmatrix},$$

$$\boldsymbol{F}_1 = \begin{bmatrix} -m_1 l_1 \dot{\theta}_1^2 + m_1 gs\theta_1 \\ m_1 l_1 \ddot{\theta}_1 + m_1 gc\theta_1 \\ 0 \end{bmatrix}, \qquad \boldsymbol{M}_1 = \begin{bmatrix} 0 \\ 0 \\ 0 \end{bmatrix},$$

$$\omega_2 = \begin{bmatrix} 0 \\ 0 \\ \dot{\theta}_1 + \dot{\theta}_2 \end{bmatrix}, \qquad \dot{\omega}_2 = \begin{bmatrix} 0 \\ 0 \\ \ddot{\theta}_1 + \ddot{\theta}_2 \end{bmatrix},$$

$$\dot{v}_2 = \begin{bmatrix} c\theta_2 & s\theta_2 & 0 \\ -s\theta_2 & c\theta_2 & 0 \\ 0 & 0 & 1 \end{bmatrix} \begin{bmatrix} -l_1\dot{\theta}_1^2 + gs\theta_1 \\ l_1\ddot{\theta}_1 + gc\theta_1 \\ 0 \end{bmatrix} = \begin{bmatrix} l_1\ddot{\theta}_1 s\theta_2 - l_1\dot{\theta}_1^2 c\theta_2 + gs(\theta_1 + \theta_2) \\ l_1\ddot{\theta}_1 c\theta_2 + l_1\dot{\theta}_1^2 s\theta_2 + gc(\theta_1 + \theta_2) \\ 0 \end{bmatrix},$$

$$\dot{v}_{S2} = \begin{bmatrix} 0 \\ l_2(\ddot{\theta}_1 + \ddot{\theta}_2) \\ 0 \end{bmatrix} + \begin{bmatrix} -l_2(\dot{\theta}_1 + \dot{\theta}_2)^2 \\ 0 \\ 0 \end{bmatrix} + \begin{bmatrix} l_1\ddot{\theta}_1 s\theta_2 - l_1\dot{\theta}_1^2 c\theta_2 + gs(\theta_1 + \theta_2) \\ l_1\ddot{\theta}_1 c\theta_2 + l_1\dot{\theta}_1^2 s\theta_2 + gc(\theta_1 + \theta_2) \\ 0 \end{bmatrix},$$

$$F_2 = m_2 \begin{bmatrix} l_1\ddot{\theta}_1 s\theta_2 - l_1\dot{\theta}_1^2 c\theta_2 + gs(\theta_1+\theta_2) - l_2(\dot{\theta}_1+\dot{\theta}_2)^2 \\ l_1\ddot{\theta}_1 c\theta_2 + l_1\dot{\theta}_1^2 s\theta_2 + gc(\theta_1+\theta_2) + l_2(\ddot{\theta}_1+\ddot{\theta}_2) \\ 0 \end{bmatrix}, \quad M_2 = \begin{bmatrix} 0 \\ 0 \\ 0 \end{bmatrix}$$

We then apply formulas (5.24) to determine the following quantities:

$R_{1,2} = F_2$

$$M_{1,2} = m_2 \begin{bmatrix} 0 \\ 0 \\ l_1 l_2 \ddot{\theta}_1 c\theta_2 + l_1 l_2 \dot{\theta}_1^2 s\theta_2 + l_2 gc(\theta_1+\theta_2) + l_2^2(\ddot{\theta}_1+\ddot{\theta}_2) \end{bmatrix}$$

$$R_{0,1} = m_2 \begin{bmatrix} c\theta_2 & -s\theta_2 & 0 \\ s\theta_2 & c\theta_2 & 0 \\ 0 & 0 & 1 \end{bmatrix} \begin{bmatrix} l_1\ddot{\theta}_1 s\theta_2 - l_1\dot{\theta}_1^2 c\theta_2 + gs(\theta_1+\theta_2) - l_2(\dot{\theta}_1+\dot{\theta}_2)^2 \\ l_1\ddot{\theta}_1 c\theta_2 + l_1\dot{\theta}_1^2 s\theta_2 + gc(\theta_1+\theta_2) + l_2(\ddot{\theta}_1+\ddot{\theta}_2) \\ 0 \end{bmatrix} +$$

$$+ m_1 \begin{bmatrix} -l_1\dot{\theta}_1^2 + gs\theta_1 \\ l_1\ddot{\theta}_1 + gc\theta_1 \\ 0 \end{bmatrix}$$

$$M_{1,0} = m_2 \begin{bmatrix} 0 \\ 0 \\ l_1 l_2 \ddot{\theta}_1 c\theta_2 + l_1 l_2 \dot{\theta}_1^2 s\theta_2 + l_2 gc(\theta_1+\theta_2) + l_2^2(\ddot{\theta}_1+\ddot{\theta}_2) \end{bmatrix} + m_1 \begin{bmatrix} 0 \\ 0 \\ l_1^2\ddot{\theta}_1 + l_1 gc\theta_1 \end{bmatrix} +$$

$$+ m_2 \begin{bmatrix} 0 \\ 0 \\ l_1^2\ddot{\theta}_1 - l_1 l_2(\dot{\theta}_1+\dot{\theta}_2)^2 s\theta_2 + l_1 gs\theta_2 s(\theta_1+\theta_2) + l_1 l_2 c\theta_2(\ddot{\theta}_1+\ddot{\theta}_2) + l_1 gc\theta_2 c(\theta_1+\theta_2) \end{bmatrix}$$

Extracting the z components from the expressions for $M_{i,i-1}$ we obtain the actuator torques:

$$\begin{aligned} f_1 &= m_2 l_2^2(\ddot{\theta}_1+\ddot{\theta}_2) + m_2 l_1 l_2(2\ddot{\theta}_1+\ddot{\theta}_2)c\theta_2 + (m_1+m_2)l_1^2\ddot{\theta}_1 - m_2 l_1 l_2 \dot{\theta}_2^2 s\theta_2 - \\ &- 2m_2 l_1 l_2 \dot{\theta}_1\dot{\theta}_2 s\theta_2 + m_2 l_2 gc(\theta_1+\theta_2) + (m_1+m_2)l_1 gc\theta_1 \\ f_2 &= m_2 l_1 l_2 \ddot{\theta}_1 c\theta_2 + m_2 l_1 l_2 \dot{\theta}_1^2 s\theta_2 + m_2 l_2 gc(\theta_1+\theta_2) + m_2 l_2^2(\ddot{\theta}_1+\ddot{\theta}_2) \end{aligned} \quad (5.25)$$

The above relationships present drive torques in the joints as functions of joint space variables, velocities and accelerations.

It is often convenient to express dynamic equations for a manipulator in the form of a one-matrix equation with a structure, such as:

$$f = M(\Theta)\ddot{\Theta} + V(\Theta,\dot{\Theta}) + G(\Theta) \quad (5.26)$$

where: $M(\Theta)$ is the $n\times n$ mass matrix of the manipulator, $V\left(\Theta,\,\dot{\Theta}\right)$ is an $n\times 1$ vector of centrifugal and Coriolis forces, and $G(\Theta)$ is an $n\times 1$ vector of gravitational forces. Each element of $M(\Theta)$ and $G(\Theta)$ is a complex function of the joint variables defining the manipulator's position, and each element of $V\left(\Theta,\,\dot{\Theta}\right)$ is a complex function of Θ as well as of $\dot{\Theta}$.

The relationships (5.25) which are the solution to the problem of the example 5.6 can be written in the form of (5.26), the mass matrix being composed of all the terms that include $\ddot{\Theta}$ and are functions of Θ:

$$M(\Theta)=\begin{bmatrix} l_2^2 m_2 + 2l_1 l_2 m_2 c\theta_2 + l_1^2(m_1+m_2) & l_2^2 m_2 + l_1 l_2 m_2 c\theta_2 \\ l_2^2 m_2 + l_1 l_2 m_2 c\theta_2 & l_2^2 m_2 \end{bmatrix}$$

The manipulator's inertia matrix is symmetrical and positive definite, and is, therefore, always invertible.

Vector $V\left(\Theta,\dot{\Theta}\right)$ contains all those terms that depend on joint velocity:

$$V\left(\Theta,\dot{\Theta}\right)=\begin{bmatrix} -m_2 l_1 l_2 \dot{\theta}_2^2 s\theta_2 - 2m_2 l_1 l_2 \dot{\theta}_1 \dot{\theta}_2 s\theta_2 \\ m_2 l_1 l_2 \dot{\theta}_1^2 s\theta_2 \end{bmatrix}$$

A term dependent on $\dot{\theta}_i^2$ represents a centrifugal force, whereas a term containing the product $\dot{\theta}_i\dot{\theta}_j$ is a Coriolis force.

The gravitational force vector, containing the gravitational acceleration, takes the following form here:

$$G(\Theta)=\begin{bmatrix} m_2 l_2 g c(\theta_1+\theta_2) + (m_1+m_2) l_1 c\theta_1 \\ m_2 l_2 g c(\theta_1+\theta_2) \end{bmatrix}$$

Note that the gravitational term depends only on the manipulator's position, defined by θ_i, and not on the velocity.

The equation (5.26) can also be written in a slightly different form:

$$f = M(\Theta)\ddot{\Theta} + B(\Theta)\left[\dot{\Theta}\dot{\Theta}\right] + C(\Theta)\left[\dot{\Theta}^2\right] + G(\Theta) \tag{5.27}$$

where:

$B(\Theta)$ is an $n\times(n-1)/2$ matrix of the Coriolis force coefficients,

$[\dot{\Theta}\dot{\Theta}]$ is an $n(n-1)/2\times 1$ vector of the velocity products

$$\left[\dot{\Theta}\dot{\Theta}\right] = \left[\dot{\theta}_1\dot{\theta}_2\ \dot{\theta}_1\dot{\theta}_3 \ldots \dot{\theta}_{n-1}\dot{\theta}_n\right]^T \tag{5.27a}$$

$C(\Theta)$ is an $n\times n$ matrix of the centrifugal coefficients,

$[\dot{\Theta}^2]$ is an $n\times 1$ vector of the squared velocities

$$\left[\dot{\Theta}^2\right] = \left[\dot{\theta}_1^2\ \dot{\theta}_2^2 \ldots \dot{\theta}_n^2\right]^T \tag{5.27b}$$

Equation (5.27) is called the **configuration space equation,** as the above matrices are functions only of the manipulator's position.

For the manipulator being examined here, we have:

$$[\dot{\Theta}\dot{\Theta}] = [\dot{\theta}_1\dot{\theta}_2] \qquad [\dot{\Theta}^2] = [\dot{\theta}_1^2\ \dot{\theta}_2^2]^T$$

$$B(\Theta) = \begin{bmatrix} -2m_2 l_1 l_2 s\theta_2 \\ 0 \end{bmatrix} \quad C(\Theta) = \begin{bmatrix} 0 & m_2 l_1 l_2 s\theta_2 \\ m_2 l_1 l_2 s\theta_2 & 0 \end{bmatrix}$$

Example 5.7. The problem: the Stanford-type manipulator shown in Figure 5.3, has three joints connected by two revolute joints and a prismatic joint. Link dimensions are those given in example 3.4. Since the coordinate system`s axes coincide with the links` symmetry axes, we have

$$r_{S1} = [0\ \ -y_{S1}\ \ 0]^T, \qquad r_{S2} = [0\ \ -y_{S2}\ \ 0]^T, \qquad r_{S3} = [0\ \ 0\ \ z_{S3}]^T,$$

$$I_{0i} = m_i \begin{bmatrix} k_{ix}^2 & 0 & 0 \\ 0 & k_{iy}^2 & 0 \\ 0 & 0 & k_{iz}^2 \end{bmatrix}$$

where $k_{ij} = I_{ij}/m_i$, $j = x, y, z$, and k_{ij} is an inertia radius of link i.

The inertial forces and torques acting on link 3 (see Fig. 5.3) are obtained, with the use of the formulas (5.24), in the form:

$$F_3 = m_3 \begin{bmatrix} (\lambda_3 + z_{S3})(\ddot{\theta}_2 - \dot{\theta}_1^2 s\theta_2 c\theta_2) + \lambda_2\ddot{\theta}_1 c\theta_2 + 2\dot{\lambda}_3\dot{\theta}_2 \\ -(\lambda_3 + z_{S3})(\ddot{\theta}_1 s\theta_2 + 2\dot{\theta}_1\dot{\theta}_2 c\theta_2) - \lambda_2\dot{\theta}_1^2 + 2\dot{\lambda}_3\dot{\theta}_1 s\theta_2 \\ -(\lambda_3 + z_{S3})(\dot{\theta}_2^2 + \dot{\theta}_1^2 s\theta_2) + \lambda_2\dot{\theta}_1^2 s\theta_2 - 2\ddot{\lambda}_3 \end{bmatrix}$$

$$M_3 = m_3 \begin{bmatrix} (k_{3x}^2 + \lambda_3 z_{S3})(\ddot{\theta}_1 s\theta_2 - 2\dot{\theta}_1\dot{\theta}_2 c\theta_2) - k_{3z}^2\dot{\theta}_1\dot{\theta}_2 c\theta_2 + (\lambda_3 + z_{S3})\lambda_2\dot{\theta}_1^2 + \\ + 2z_{S3}\dot{\lambda}_3\dot{\theta}_1 s\theta_2 \\ (k_{3x}^2 + \lambda_3 z_{S3})(\ddot{\theta}_2 - \dot{\theta}_1^2 s\theta_2 c\theta_2) + k_{3z}^2\dot{\theta}_1^2 s\theta_2 c\theta_2 + 2(\lambda_3 + z_{S3})\dot{\lambda}_3\dot{\theta}_2 + \\ + 2\lambda_2 z_{S3}\dot{\theta}_1\dot{\theta}_2 s\theta_2 + \lambda_2 z_{S3}\ddot{\theta}_1 c\theta_2 \\ -k_{3z}^2(\ddot{\theta}_1 - \dot{\theta}_1\dot{\theta}_2 s\theta_2) \end{bmatrix}$$

where component F_{3z} is the actuator force acting on the link with the prismatic joint, that is $R_{2,3z} = F_{3z}$.

The drive torque of the second revolute joint equals the z_1 axis component of the torque exerted by link 1 on link 2:

$$f_2 = M_{1,2z} = m_2 k_{2z}^2\ddot{\theta}_2 + m_3\{(k_{3x}^2 + 2\lambda_3 z_{S3})(\ddot{\theta}_2 - \dot{\theta}_1^2 s\theta_2 c\theta_2) + k_{3z}^2\dot{\theta}_1^2 s\theta_2 c\theta_2 +$$
$$+ 2(\lambda_3 + z_{S3})\dot{\lambda}_3\dot{\theta}_2 + (\lambda_2 z_{S3})\lambda_2\ddot{\theta}_1 c\theta_2 + 2\lambda_2\lambda_3\dot{\theta}_1\dot{\theta}_2 s\theta_2 + \lambda_3^2(\ddot{\theta}_2 - \ddot{\theta}_1 s\theta_2 c\theta_2)\}$$

The drive torque of the first revolute joint is:

$$
\begin{aligned}
f_1 = M_{0,1z} &= m_1 k_{1z}^2 \ddot{\theta}_1 + m_2\left(k_{2x}^2 + 2\lambda_2 y_{S2} + \lambda_2^2\right)\ddot{\theta}_1 + \\
&+ m_3\Big\{\left(k_{3x}^2 + \lambda_3 z_{S3}\right)\left(\ddot{\theta}_1 s\theta_2 + 2\dot{\theta}_1\dot{\theta}_2 c\theta_2\right)s\theta_2 + k_{3z}^2\left(\ddot{\theta}_1 - 2\dot{\theta}_1\dot{\theta}_2 s\theta_2\right)c\theta_2 + \\
&+ \left(\lambda_3 + z_{S3}\right)\lambda_2\dot{\theta}_1^2 s\theta_2 + \lambda_3\left(\lambda_3 + z_{S3}\right)\left(\dot{\theta}_2^2 + \dot{\theta}_1^2 s^2\theta_2\right) + 2\lambda_3 z_{S3}\ddot{\theta}_1 s^2\theta_2 - \\
&- \lambda_2\lambda_3\ddot{\theta}_1 s^2\theta_2 - \lambda_3\ddot{\lambda}_3 s\theta_2\Big\}
\end{aligned}
$$

The detailed derivation of these relationships can be found in [5.2, 5.5].

Example 5.8. For a PUMA manipulator, the diagram and dimensions are those presented in example 3.6, and the local coordinate systems are assumed to coincide with the symmetry axes of the links. Mass distribution is described by the center of mass coordinates:

$$
\boldsymbol{r}_{S1} = [0 \;\; -y_{S1} \;\; 0]^T, \qquad \boldsymbol{r}_{S2} = [x_{S2} \;\; 0 \;\; 0]^T, \qquad \boldsymbol{r}_{S3} = [0 \;\; 0 \;\; z_{S3}]^T,
$$

$$
\boldsymbol{I}_{0i} = m_i \begin{bmatrix} k_{ix}^2 & 0 & 0 \\ 0 & k_{iy}^2 & 0 \\ 0 & 0 & k_{iz}^2 \end{bmatrix}, \quad i = 1, 2, 3
$$

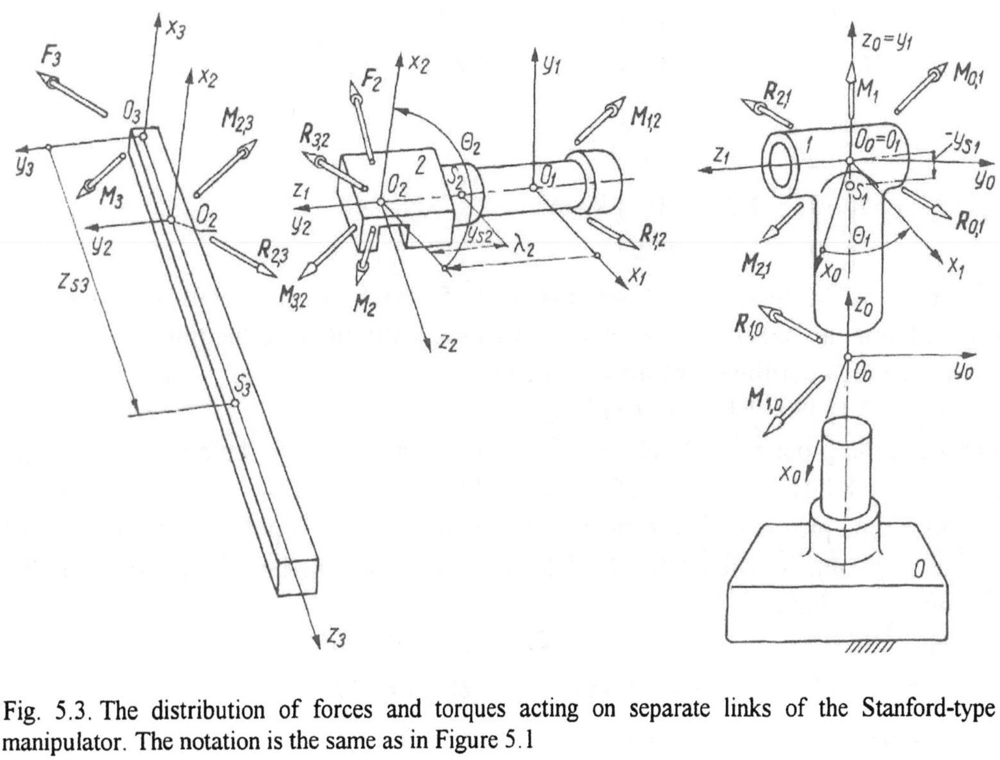

Fig. 5.3. The distribution of forces and torques acting on separate links of the Stanford-type manipulator. The notation is the same as in Figure 5.1

The actuators' torques in this particular joint are

$$f_3 = M_{2,3z} = m_3 \left\{ k_{3z}^2 \left(\ddot{\theta}_2 + \ddot{\theta}_3 + z_{S3} l_2 \left[\ddot{\theta}_2 c\theta_3 + \dot{\theta}_2^2 s\theta_{21} + \dot{\theta}_2 \dot{\theta}_3 (s\theta_2 - s\theta_3) + \right. \right. \right.$$
$$\left. \left. \left. + \dot{\theta}_1^2 c\theta_2 s(\theta_2 + \theta_3) \right] \right) \right\}$$

$$f_2 = M_{1,2z} = m_2 k_{2z}^2 \ddot{\theta}_2 + M_{2,3z} + m_3 \left\{ l_2^2 \left(\ddot{\theta}_2 + \dot{\theta}_1^2 s\theta_2 c\theta_2 \right) + \right.$$
$$\left. + z_{S3} l_2 \left[\dot{\theta}_1^2 s\theta_2 c(\theta_2 + \theta_3) - \left(\dot{\theta}_2 + \dot{\theta}_3 \right)^2 s\theta_3 + \left(\ddot{\theta}_2 + \ddot{\theta}_3 \right) c\theta_3 \right] \right.$$

$$f_1 = M_{0,1z} = m_1 k_{1y}^2 \ddot{\theta}_1 + m_2 k_{2x}^2 \ddot{\theta}_1 + m_3 \left\{ k_{3x}^2 \left[\ddot{\theta}_1 - \dot{\theta}_1 \left(\dot{\theta}_2 + \dot{\theta}_3 \right) s(\theta_2 + \theta_3) c(\theta_2 + \theta_3) \right] + \right.$$
$$+ k_{3z}^2 \left(\dot{\theta}_2 + \dot{\theta}_3 \right)^2 c(\theta_2 + \theta_3) + z_{S3} l_2 \left[2\ddot{\theta}_1 c\theta_2 - 2\dot{\theta}_1 \dot{\theta}_2 s\theta_2 + \dot{\theta}_2 \left(\dot{\theta}_2 + \dot{\theta}_3 \right) c\theta_3 \right] c(\theta_2 + \theta_3) -$$
$$\left. - l_2^2 \dot{\theta}_1 \dot{\theta}_2 s\theta_2 c\theta_2 \right\}$$

Example 5.9. For a PUMA manipulator we consider, for the sake of simplification, only the first three links (counting from the base). The loading caused by the wrist and gripper's links have been omitted since it is small compared with the load of the first three links. For dynamic computations, the following mass distribution has been assumed:

$m_1 = 50\ kg$, $m_2 = 109\ kg$, $m_3 = 57\ kg$,

$r_{S1} = [0\ \ 0\ \ 0]^T\ m$, $r_{S2} = [-0.200\ \ 0\ \ 0]^T\ m$, $r_{S3} = [0.060\ \ 0\ \ 0]^T\ m$,

$$I_{01} = 50 \begin{bmatrix} 3.37 & 0 & 0 \\ 0 & 0.15 & 0 \\ 0 & 0 & 3.37 \end{bmatrix}, \quad I_{01} = 109 \begin{bmatrix} 15.1 & 0 & 0 \\ 0 & 15.1 & 0 \\ 0 & 0 & 10.8 \end{bmatrix}$$

$$I_{01} = 57 \begin{bmatrix} 4.20 & 0 & 0 \\ 0 & 4.20 & 0 \\ 0 & 0 & 0.75 \end{bmatrix} \text{kg·m}^2$$

The mass of the rest of the links, that is 4, 5 and 6, together with the object of manipulation, are taken into account in the mass distribution of the link *3*.

Given the coordinates of the arm's end:

$p_r = [-0.13390\ \ 0.14360\ \ 0.31192]^T$,

and using the procedure of solution of inverse kinematics we obtain the following values: $\theta_1 = 83.41°$, $\theta_2 = -0.75°$, $\theta_3 = -44.10°$

As a time profile of θ_i for the motion duration $t = 4\ s$, a polynomial of the third degree is assumed, with certain variables` values obtained half way through the motion range, that is for $t = 2\ s$:

$\theta_1 = 68.61°$, $\theta_2 = -6.59°$, $\theta_3 = -22.90°$

$\dot{\theta}_1 = -13.87\,s^{-1}$, $\dot{\theta}_2 = -5.48\,s^{-1}$, $\dot{\theta}_3 = 19.87\,s^{-1}$

$\ddot{\theta}_1 = 0\,s^{-2}$, $\ddot{\theta}_2 = 0\,s^{-2}$, $\ddot{\theta}_3 = 0\,s^{-2}$

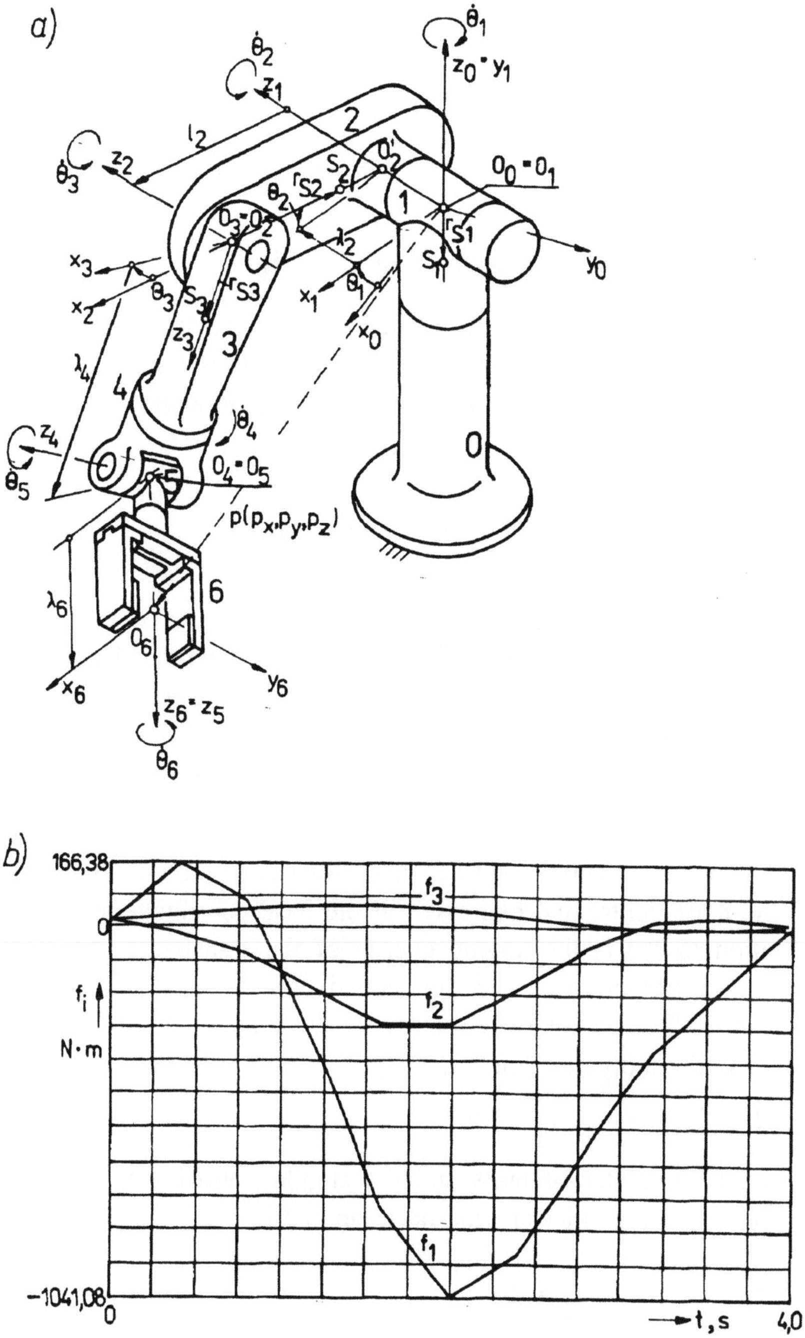

Fig. 5.4. a) A manipulator of PUMA 560 manipulator with the geometrical dimensions and mass distribution indicated in the figure, b) time profiles of drive torques in the first three revolute joints of the manipulator, computed for configuration coordinate functions of the polynomial type

Using the above formulas, the drive torques are calculated:

$f_3 = M_{2,3z} = -284.45\ Nm,$

$f_2 = M_{1,2z} = 39.14\ Nm,$

$f_1 = M_{0,1x} = -1041\ Nm.$

The time profiles of these torques are shown in Figure 5.4b.

5.2.3. Lagrangian equations of the II order

If we let $q_1, \ldots, q_n$ be the generalized coordinates that fully define a system's position, and E_k and E_p be the total kinetic energy and the potential energy of the system, respectively, introducing the notion of the Lagrangian function (kinetic potential)

$$L(q_i, \dot{q}_i) = E_k - E_p$$

we have the motion equations of the system in the form:

$$\frac{d}{dt}\frac{\partial L}{\partial \dot{q}_i} - \frac{\partial L}{\partial q_i} = Q_i, \quad i = 1, \ldots, n \tag{5.28}$$

where Q_i is a generalized force corresponding to the generalized displacement q_i. That force can be found by determining the virtual work performed by the active forces (non-conservative) exerted on the system.

The kinetic energy of link i is described by the expression

$$E_{ki} = \tfrac{1}{2} m_i \boldsymbol{v}_{Si}^T \cdot \boldsymbol{v}_{Si} + \tfrac{1}{2} \boldsymbol{\omega}_i^T \boldsymbol{I}_{Si} \cdot \boldsymbol{\omega}_i \tag{5.29}$$

where m_i is the mass of the link, and I_{Si} – a 3×3 matrix of the inertia tensor, defined relative to the mass center and expressed in the base coordinate system. The first term of relation (5.29) is the kinetic energy of the translatory motion with the velocity of the center of mass and the second term – the kinetic energy of the rotational motion. The total kinetic energy of the manipulator is the sum of the kinetic energy of the individual link

$$E_k = \sum_{i=1}^{n} E_{ki} \tag{5.29a}$$

The velocity of the center of mass and the angular velocity of the link can be expressed using joint velocities and the Jacobian matrix

$$\begin{bmatrix} \boldsymbol{v}_{Si} \\ \boldsymbol{\omega}_{Si} \end{bmatrix} = \boldsymbol{J} \begin{bmatrix} \dot{q}_1 \\ \vdots \\ \dot{q}_n \end{bmatrix} = \boldsymbol{J} [\dot{q}_1, \ldots, \dot{q}_n]^T \tag{5.30}$$

where $\boldsymbol{J}$ is the Jacobian matrix, the first three rows of which refer to the linear velocity and the last three, to the angular velocity.

If we denote by J_{vi} and $J_{\omega i}$ the 3×1 column vectors of the Jacobian matrix corresponding to the linear and angular velocity, then:

$$v_{Si} = J_{v1}\dot{q}_1 + J_{v2}\dot{q}_2 + \ldots + J_{vn}\dot{q}_n \tag{5.30a}$$

$$\omega_i = J_{\omega 1}\dot{q}_1 + J_{\omega 2}\dot{q}_2 + \ldots + J_{\omega n}\dot{q}_n$$

noting that for a prismatic joint

$$J_{vi} = e_{i-1}, \qquad J_{\omega i} = 0$$

and for a revolute one

$$J_{vi} = e_{i-1} \times p_{Si}, \qquad J_{\omega i} = e_{i-1}$$

where p_{si} is a position vector of the center of mass S_i relative to the base system, and e_{i-1} is a unit vector of the $i-1$ axis.

Substituting the relationships (5.30a) in (5.29) give us

$$E_k = \tfrac{1}{2}\sum_{i=1}^{n}\left(m_i\dot{q}^T J_{vi}^T J_{vi}\dot{q} + \dot{q}^T J_{\omega i}^T I_{Si} J_{\omega i}\dot{q}\right) = \frac{1}{2}\dot{q}^T M \dot{q} \tag{5.31}$$

$$M = \sum_{i=1}^{n}\left(m_i J_{vi}^T J_{vi} + J_{\omega i}^T I_{Si} J_{\omega i}\right) \tag{5.31a}$$

where M is an $n \times n$ matrix taking into account all the mass properties of the manipulator related to the joints, and thus dependent on the position. Denoting by d_{ij} the element $[ij]$ of the matrix M, equation (5.31) can be written in the form

$$E_k = \tfrac{1}{2}\sum_{i=1}^{n}\sum_{i=1}^{n} d_{ij}\dot{q}_i\dot{q}_j \tag{5.32}$$

with d_{ij} being a function of $q_1, q_2, \ldots, q_n$.

The potential energy of a manipulator can be described by the formula

$$E_p = \sum_{i=1}^{n} m_i g^T p_{Si} \tag{5.33}$$

where p_{si} is a position vector of the center of mass of the link i, dependent on the manipulator's position.

The generalized forces Q_i include all the forces and torques acting on the links, except for gravitational forces and inertial forces. It is assumed that drive forces or torques $f = [f_1 f_2 \ldots f_n]^T$ are exerted in the joints and the external active forces F applied on the end-effector. The generalised force can be found by computing the virtual work performed by those forces. Thus we have

$$Q = f + J^T F_z \tag{5.34}$$

The kinetic energy derivatives occurring in the Lagrangian equation (5.28), considering (5.32), are

$$\frac{d}{dt}\left(\frac{dE_k}{d\dot{q}_i}\right) = \frac{d}{dt}\left(\sum_{i=1}^{n} d_{ij}\dot{q}_j\right) = \sum_{i=1}^{n} d_{ij}\ddot{q}_j + \frac{dd_{ij}}{dt}\sum_{i=1}^{n}\dot{q}_j \tag{5.35}$$

where

$$\frac{d}{dt}\left(d_{ij}\right)=\sum_{k=1}^{n}\frac{\partial d_{ij}}{\partial q_k}\dot{q}_k \tag{5.35a}$$

$$\frac{\partial E_k}{\partial q_i}=\frac{1}{2}\sum_{j=1}^{n}\sum_{k=1}^{n}\frac{\partial d_{ij}}{\partial q_i}\dot{q}_j\dot{q}_k \tag{5.36}$$

The derivative of potential energy (5.33) is obtained in the form of

$$G_i=\frac{\partial E_p}{\partial q_i}=\sum_{i=1}^{n}m_j g^T J_{vj} \tag{5.37}$$

as a partial derivative of the p_{Si} vector with respect to q_i is the same as j-th column vector of the J_{vi} matrix.

Substituting the relationships (5.37), (5.36) and (5.28) yields

$$\sum_{j=1}^{n}d_{ij}\ddot{q}_j+\sum_{j=1}^{n}\sum_{k=1}^{n}c_{ijk}\dot{q}_j\dot{q}_k+G_i=Q_i,\quad i=1,\ldots,n \tag{5.38}$$

where

$$c_{ijk}=\frac{\partial d_{ij}}{\partial q_k}-\frac{1}{2}\frac{\partial d_{jk}}{\partial q_i} \tag{5.39}$$

The first term in equation (5.38) represents the inertial torques. The second is due to the Coriolis and centrifugal forces, and the third to gravitational forces. The torques of interaction between joints, represented by the terms $d_{ij}\ddot{q}_j$ (j≠1), are described by elements lying outside the diagonal of the inertia matrix, whereas the Coriolis and centrifugal forces result from the dependence of the inertia matrix of the manipulator on its position.

Example 5.9. The problem: for the two-link manipulator of example 5.6, derive the Lagrangian dynamic equations. To solve this problem we begin by determining the manipulator's inertia matrix $[d_{ij}]$. Using formula (5.30a) the linear velocities of the links' centers of mass are computed

$$v_{S1}=\begin{bmatrix}-r_{S1}s\theta_1 & 0\\ r_{S1}c\theta_1 & 0\end{bmatrix}\begin{bmatrix}\dot{\theta}_1\\ \dot{\theta}_2\end{bmatrix}$$

$$v_{S2}=\begin{bmatrix}-l_1 s\theta_1-r_{S2}s(\theta_1+\theta_2) & -r_{S2}s(\theta_1+\theta_2)\\ l_1 c\theta_1+r_{S2}c(\theta_1+\theta_2) & r_{S2}c(\theta_1+\theta_2)\end{bmatrix}\begin{bmatrix}\dot{\theta}_1\\ \dot{\theta}_2\end{bmatrix}$$

where the 2×2 matrices are the Jacobian matrices J_{vi} of the system (5.30a). The angular velocities are defined by the formulas

$$\omega_1=\dot{\theta}_1=[1\quad 0]^T\dot{q}$$

$$\omega_2=\dot{\theta}_1+\dot{\omega}_2=[1\quad 1]^T\dot{q}$$

Substituting these relationships into the formula (5.35) gives us

$$M = \begin{bmatrix} m_1 r_{S1}^2 + J_1 + m_2\left(l_1^2 + r_{S2}^2 + 2l_1 r_{S2} c\theta_2\right) + J_2 & m_2 l_1 r_{S2} c\theta_2 + m_2 r_{S2}^2 + J_2 \\ m_2 l_1 r_{S2} c\theta_2 + m_2 r_{S2}^2 + J_2 & m_2 r_{S1}^2 + J_2 \end{bmatrix}$$

The elements of the above matrix of the manipulator's inertia tensor are functions of position.

The coefficients of the Coriolis and inertia forces, computed using (5.39), for the manipulator in question,

$$c_{111} = 0, \qquad c_{122} = -m_2 l_1 r_{S2} s\theta_2, \qquad c_{112} + c_{121} = -2m_2 l_1 r_{S2} s\theta_2$$

$$c_{211} = m_2 l_1 r_{S2} s\theta_2, \qquad c_{222} = 0, \qquad c_{212} + c_{221} = 0,$$

The coefficients of the gravitational forces, determined according to the formula (5.37), are

$$G_1 = \mathbf{g}^T \left[m_1 J_{v1,1} + m_2 J_{v1,2}\right]$$

$$G_2 = \mathbf{g}^T \left[m_1 J_{v2,1} + m_2 J_{v2,2}\right]$$

Finally, we can write equations (6.38) in the form

$$d_{11}\ddot{\theta}_1 + d_{12}\ddot{\theta}_2 + c_{122}\dot{\theta}_2^2 + \left(c_{112} + c_{121}\right)\dot{\theta}_1\dot{\theta}_2 + G_1 = f_1$$

$$d_{22}\ddot{\theta}_2 + d_{12}\ddot{\theta}_1 + c_{211}\dot{\theta}_1^2 + G_2 = f_2$$

As we have assumed that no external force is exerted on the end-effector, the generalized forces f_i are drive torques in the joints:

Example 5.10. The problem: for a planar two-link manipulator with a revolute joint at the base, and a prismatic joint connecting the moving links (see Figure 5.5), the following mass distribution is assumed:

$$\mathbf{r}_{S1} = [l_1\ 0\ 0], \qquad \mathbf{r}_{S2} = [\lambda_2\ 0\ 0]$$

$$\mathbf{I}_{S1} = \begin{bmatrix} k_{x1}^2 & 0 & 0 \\ 0 & k_{y1}^2 & 0 \\ 0 & 0 & k_{z1}^2 \end{bmatrix}, \qquad \mathbf{I}_{S2} = \begin{bmatrix} k_{x2}^2 & 0 & 0 \\ 0 & k_{y2}^2 & 0 \\ 0 & 0 & k_{z2}^2 \end{bmatrix}$$

The solution: we use formula (5.29) to find the kinetic energy of the links *1* and *2* :

$$E_{k1} = \tfrac{1}{2} m_1 \left(l_1^2 \dot{\theta}_1^2 + k_{z1}^2 \dot{\theta}_1^2\right), \qquad E_{k2} = \tfrac{1}{2} m_2 \left(\lambda_2^2 \dot{\theta}_1^2 + \dot{\lambda}_2^2 + k_{z2}^2 \dot{\theta}_1^2\right)$$

In the same way, using formula (5.29), we compute the potential energy of those links:

$$E_{p1} = m_1 l_1 g\left(s\theta_1 + 1\right)$$

$$E_{p2} = m_2 g\left(\lambda_2 s\theta_1 + \lambda_{2\max}\right)$$

where λ_{2max} is the maximum extension of the prismatic joint.

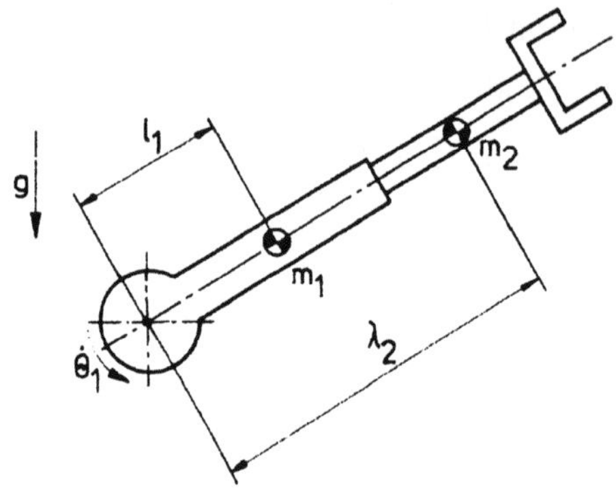

Fig. 5.5. A planar two-link manipulator with a revolute joint at the base, and a prismatic one, showing the mass distribution of the links.

Now we compute the partial derivatives of the kinetic and potential energy:

$$\frac{\partial E_k}{\partial \dot{\Theta}} = \begin{bmatrix} m_1\left(l_1^2 + k_{z1}^2\right)\dot{\theta}_1 + m_2\left(\lambda_2^2 + k_{z2}^2\right)\dot{\theta}_1 \\ m_2\dot{\lambda}_2 \end{bmatrix}$$

$$\frac{\partial E_k}{\partial \Theta} = \begin{bmatrix} 0 \\ m_2\lambda_2\dot{\theta}_1^2 \end{bmatrix}$$

$$\frac{\partial E_p}{\partial \Theta} = \begin{bmatrix} (m_1 l_1 + m_2\lambda_2)gc\theta_1 \\ m_2 gs\theta_1 \end{bmatrix}$$

Hence, according to the formula (5.26) we have

$$M(\Theta) = \left[d_{ij}\right] = \begin{bmatrix} m_1\left(l_1^2 + k_{z1}^2\right) + m_2\left(\lambda_2^2 + k_{z2}^2\right) & 0 \\ 0 & m_2 \end{bmatrix}$$

$$V(\Theta, \dot{\Theta}) = \begin{bmatrix} 2m_2\lambda_2\dot{\theta}_1\dot{\lambda}_2 \\ -m_2\lambda_2\dot{\theta}_1^2 \end{bmatrix}$$

$$G(\Theta) = \begin{bmatrix} (m_1 l_1 + m_2\lambda_2)gc\theta_1 \\ m_2 gs\theta_1 \end{bmatrix}$$

$$f_1 = \left[m_1\left(l_1^2 + k_{z1}^2\right) + m_2\left(\lambda_2^2 + k_{z2}^2\right)\right]\ddot{\theta}_1 + 2m_2\lambda_2\dot{\theta}_1\dot{\lambda}_2 + (m_1 l_1 + m_2\lambda_2)gc\theta_1$$

$$f_2 = m_2\ddot{\lambda}_2 - m_2\lambda_2\dot{\theta}_1^2 + m_2 gs\theta_1$$

5.2.4. Simulation of manipulator motion

For computer simulation we can use the previously derived dynamic equations, written in the form (6.26), and solved for acceleration

$$\ddot{\Theta} = M^{-1}(\Theta)\left[f - V(\Theta,\dot{\Theta}) - G(\Theta)\right] \qquad (5.40)$$

Any numerical integration method can be employed to compute the joint coordinates and velocities. The initial conditions for the manipulator motion are usually stated in the form of

$$\Theta(0)=\Theta_0, \qquad \dot{\Theta}(0)=0$$

and the equation (5.40) is numerically integrated forward in time, with an increment Δt. In the simplest method of integration, called Euler integration, the iteration begins with $t = 0$:

$$\dot{\Theta}(t+\Delta t)=\dot{\Theta}(t)+\ddot{\Theta}(t)\Delta t$$

$$\Theta(t+\Delta t)=\Theta(t)+\dot{\Theta}(t)\Delta t+\tfrac{1}{2}\ddot{\Theta}(t)\Delta t^2$$

noting that for each iteration, $\ddot{\Theta}$ is computed using (5.40). In this way we can determine the position, velocity and acceleration of a manipulator caused by given drive forces and torques.

5.3. Selected theoretical problems*)

5.3.1. Computer-aided analysis of the kinematics of multi-link systems

Kinematic analysis of mechanical systems often requires that kinematic equations be formulated and solved numerically in a systematic way. In this manner we can solve kinematics problems on the basis of general data read in a problem-oriented language. These data contain information about the structure of the system (the number of links, type of joint between links, and so on) as well as about the velocity, position and acceleration of certain links of the system that are entered during kinematic analysis. In this section the links of the mechanical system are assumed to be connected in a way that can be described by holonomic constraint equations. The non-linear task of position finding is solved by means of Newton-Raphson and Fletcher-Powell algorithms [5.24]. Velocities and accelerations are determined using classical methods for solving linear equations. The procedure presented is based on the results of [5.30, 5.31] and is suitable for non-holonomic constraints as well.

5.3.1.1. The generalized coordinates of a free system of rigid links

The position and orientation of a link k, not subject to the influence of constraints (see Figure 5.6), is described by a Cartesian coordinate vector $\boldsymbol{r}_k = [x\ y\ z]^T$ of the mass

*) Based on [5.24 and 5.25]

center of the sm_k link in the inertial system, as well as by a vector $\boldsymbol{p}_k = [\theta_0\ \theta_1\ \theta_2\ \theta_3]^T$ of four Euler parameters [5.24].

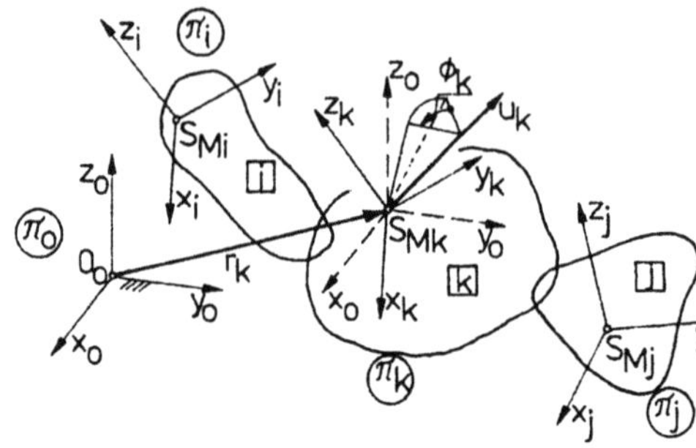

Fig. 5.6. A description of the position and orientation of a link

The resultant vector of the generalized coordinates $\boldsymbol{q}_k$ of link k is seven-dimensional:

$$\boldsymbol{q}_k = \left[\boldsymbol{r}^T \quad \boldsymbol{p}^T\right]_k^T = \left[x\ y\ z\ \theta_0\ \theta_1\ \theta_2\ \theta_3\right]_k^T \tag{5.41}$$

A generalized coordinate vector of a complete spatial system of c free rigid links has n components ($n = 7c$)

$$\boldsymbol{q} = \left[q_1^T\, q_2^T \ldots q_c^T\right]^T \tag{5.42}$$

5.3.1.2. Constraint equations

Constraint equations result from mutual geometric relationships between generalized coordinates describing neighboring links. For example, spherical joints (Fig. 5.7) are described by the vector equation

$$\boldsymbol{r}_r + A_k \overline{O_k P_k}' - \left(\boldsymbol{r}_j + A_j \overline{O_j P_j}'\right) = 0 \tag{5.43}$$

where O_j is the origin of the coordinate system j, r_j and r_k are radius vectors of the origins of systems j and k, (') denotes local coordinates of the vectors, and A_k is the cosine matrix expressed in terms of Euler parameters.

Most kinematic pairs classified in the theory of machines and manipulators can be described analytically [5.15].

The equation of holonomic constraints imposed on a system of c rigid links can be written in vector form as

$$\boldsymbol{F}^P(q) = \left[\left(\boldsymbol{F}_1^P(q)\right)^T \left(\boldsymbol{F}_2^P(q)\right)^T \ldots \left(\boldsymbol{F}_p^P(q)\right)^T\right]^T = 0 \tag{5.44}$$

If the number of independent equations is L, then the number of degrees of freedom is n-L. Kinematic analysis requires the definition of n-L additional equations, called guiding equations

$$\boldsymbol{F}^D(q,t) = \left[\left(\boldsymbol{F}_1^D(q,t)\right)^T \left(\boldsymbol{F}_2^D(q,t)\right)^T \ldots \left(\boldsymbol{F}_d^D(q,t)\right)^T\right]^T = 0 \tag{5.45}$$

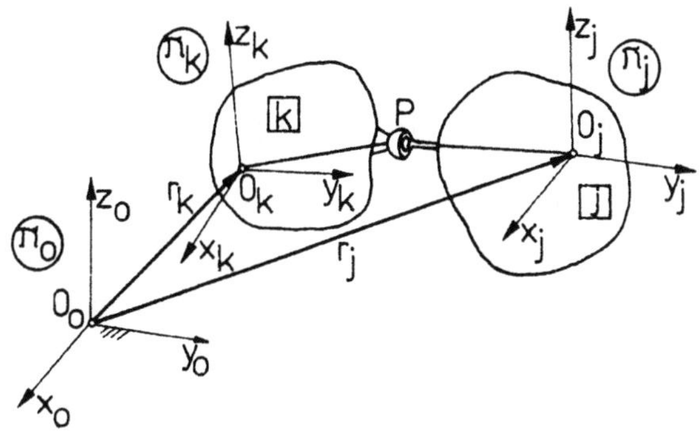

Fig. 5.7. Spherical joint

Those equations result from the relationships describing a particular kinematics problem (direct or inverse). Equations (5.44) and (5.45) constitute a complete system of constraints.

5.3.1.3. Kinematic analysis

The solution of the position finding task is obtained by the numerical solution of the system of non-linear algebraic equations in the form:

$$\begin{cases} \boldsymbol{F}^D = \boldsymbol{F}^D(q) = 0 \\ \boldsymbol{F}^D = \boldsymbol{F}^D(q,t) = 0 \end{cases} \tag{5.46}$$

One and two-fold differentiation of equation (5.46) with respect to time give us the equations for velocities and accelerations:

$$\begin{bmatrix} \boldsymbol{F}_q^P \\ \boldsymbol{F}_q^D \end{bmatrix} \dot{q} = \begin{bmatrix} 0 \\ -\boldsymbol{F}_t^D \end{bmatrix} \tag{5.47}$$

$$\begin{bmatrix} \boldsymbol{F}_q^P \\ \boldsymbol{F}_q^D \end{bmatrix} \ddot{q} = \begin{bmatrix} -\left(\boldsymbol{F}_q^P \dot{q}\right)_q \dot{q} \\ -\left(\boldsymbol{F}_q^D \dot{q}\right)_q \dot{q} - 2\boldsymbol{F}_{qt}^D \dot{q} - \boldsymbol{F}_{tt}^D \end{bmatrix} \tag{5.48}$$

System (5.47) is solved using the classical Newton algorithm. Systems (5.47) and (5.48), are both systems of linear equations (with respect to $\dot{q}$ and $\ddot{q}$). The numerical algorithm for solving these systems of equations exploit the fact that the Jacobian matrix of constraint equations is a sparse matrix enabling us to use more efficient numerical procedures [5.30, 5.31].

5.3.1.4. Choosing initial approximation

When solving the task for positions a good aproximate initial position q_0 must be chosen. This is often quite difficult. In a computer implementation of the algorithm, presented here, aproximate initial position is selected using Fletcher-Powell's optimization algorithm. The objective function is defined in the form of [5.15]:

$$\Psi(q,t_0,\mu)=[q-q_0]^T[q-q_0]+\mu F^T(q,t_0)F(q,t_0) \tag{5.49}$$

where μ is the weight multiplier.

The selection of μ is performed by steps. First the minimum of the objective function for a certain assumed value of μ is sought, then the value of μ is increased, for instance by one order, and the minimum is sought again. The process ends when the constraint equations are satisfied with the desired accuracy.

5.3.1.5. The case of the singular Jacobian matrix

There are at least four qualitatively different cases in which the Jacobian matrix is singular.

1. In system (5.48) there are dependent equations due to overconstraints (redundant constraints) in the mechanism [5.15, 5.18].
2. In system (5.48) there are dependent equations due to guiding overconstraints.
3. In system (5.48) there are dependent equations resulting from the fact that during kinematic analysis the mechanism has reached the position beyond which it can not move without "tearing" the constraint.
4. During kinematic analysis the mechanism reached singularity (the so-called point of bifurcation of system (5.48) solution) [5.30].

The methodology for each of the selected cases is presented in [5.24]. Redundant constraints are analyzed by means of the Gauss algorithm with partial and free selection of the initial element. Constraints' "tearing" position of the multi-link system is analyzed in terms of the mechanism's velocities and accelerations. An optimization algorithm is used, instead of the Newton-Raphson for singularities.

5.3.1.6. Computer program

A computer program, designed for an IBM PC-type computer, has been developed for the automatic analysis of multi-link systems using the above relationships. This program provides a kinematic analysis of large multi-link systems on the basis of general data expressed in a problem-oriented language. Constraint equations are formed within the program automatically, depending upon the types of joints selected by the user, who also defines the guiding constraints, together with errors with which the kinematic analysis is carried out. If there are redundant constraints, the program sends a warning. In the current version of the program the results can be edited in the form of charts, or numerical results.

5.3.2. Modeling and dynamic analysis of multi-link systems

Dynamic analysis concerns those mechanical systems for which a dynamic model can be built under the assumption that the analyzed structure constitutes a system of rigid bodies with constraints that are holonomic, dependent on time, bilateral, independent and ideal . In particular, the kinematic chain of the links must be closed.

By dynamic analysis we mean here the numerical solution of the dynamic task. Friction forces and clearances in the joints are neglected. The computer program presented also allows for friction forces [5.24], but this issue is not considered here [5.30].

5.3.2.1. Generalized coordinates

The motion of the system of c links is described with respect to the stationary, global, and orthonormal coordinate system Π_0, attached rigidly to the base. To each link k ($k = 1, 2, ..., c$) is fixed the orthonormal system Π_k with origin in the link's center of mass sm_k (Fig. 5.6). In order to transform the coordinates from system Π_k to system Π_0, we convert the latter into a system Π_k through translation by a vector $\boldsymbol{r}_k^T = [x\ y\ z]$ and rotation about the unitary vector of a direction cosine matrix $\boldsymbol{A}_k$.

A system of seven parameters has been assumed as the generalized coordinates of a free link $\boldsymbol{q}_k$:

$$\boldsymbol{q}_k^T = [x\ y\ z\ \theta_0\ \theta_1\ \theta_2\ \theta_3]_k = [\boldsymbol{r}_k^T\ \boldsymbol{p}_k^T] \tag{5.50}$$

where

$$\boldsymbol{r}_k^T = [x\ y\ z]_k, \qquad \boldsymbol{p}_k^T = [\theta_0\ \theta_1\ \theta_2\ \theta_3]_k \tag{5.51}$$

and

$$\boldsymbol{u}_k^T = [u_1\ u_2\ u_3]_k, \qquad (\theta_0)_k = \cos(\phi_k/2)$$

$(\theta_0)_k$ parameters are defined by the formula

$$(\theta_0)_k = (u_0)_k \sin(\phi_k/2), \quad i = 1, 2, 3 \tag{5.52}$$

For each k the Euclid's norm of the vector $\boldsymbol{p}_k$ must be equal to 1, that is

$$\| \boldsymbol{p}_k \| = 1, \quad (k = 1, 2, ..., c) \tag{5.53}$$

The vector $\boldsymbol{p}_k$ coordinates are called Euler parameters [5.24] and in analytical studies are treated as normalized elements $\boldsymbol{r}$ of the algebra with division quaternions [5.24].

Parametrization of the three-dimensional group of rotations in Euclidian space $\boldsymbol{R}^3$ with the use of quaternions has two advantages which are decisive in the context of computer applications:

1. It is global and non-singular.
2. The matrix A_k of direction cosines, defined for Euler parameters by formula [5.24]

$$A_k = \begin{bmatrix} \theta_0^2+\theta_1^2-0.5 & \theta_1\theta_2-\theta_0\theta_3 & \theta_1\theta_3-\theta_0\theta_2 \\ \theta_1\theta_2-\theta_0\theta_3 & \theta_0^2+\theta_2^2-0.5 & \theta_2\theta_3-\theta_0\theta_1 \\ \theta_1\theta_3-\theta_0\theta_2 & \theta_2\theta_3-\theta_0\theta_1 & \theta_0^2+\theta_3^2-0.5 \end{bmatrix} \quad (5.54)$$

does not contain trigonometric functions, time-consuming in terms of computer computation.

A drawback to its use is, instead, the necessity dealing with the additional constraints set by the relationship (5.54). The vector of generalized coordinates of the spatial system of c free rigid links has n components ($n = 7 \cdot c$):

$$\boldsymbol{q} = \left[q_1^T \; q_2^T \ldots q_k^T \ldots q_c^T\right]^T \quad (5.55)$$

5.3.2.2. Motion equations of the rigid link system

In deriving motion equations we assume that the constraints imposed on the system of c rigid links are described by the equation of holonomic, time-dependent, bilateral, and ideal constraints

$$F(q, t) = 0 \quad (5.56)$$

We assume, for all purposes, that the representation $\boldsymbol{F}$, being a function of the $\boldsymbol{q}$ variables, is a non-singular representation for any $\boldsymbol{t}$. The initial state of the system at time t_0 is described by the initial conditions set by the equations

$$\boldsymbol{q}(t_0) = \boldsymbol{q}_0, \quad \dot{\boldsymbol{q}}(t_0) = \dot{\boldsymbol{q}}_0 \quad (5.57)$$

The initial coordinates $\boldsymbol{q}_0$ and velocities $\dot{\boldsymbol{q}}_0$ must satisfy the system of constraint equations.

Motion equations of the system of rigid links, subject to the action of constraints of the form (6.46), can be written using first order Langrangian equations [5.24, 5.25]. Those equations can be reduced to the form [5.24]

$$\begin{cases} \boldsymbol{M}\ddot{\boldsymbol{q}} = \boldsymbol{g} - \boldsymbol{F}_q^T \boldsymbol{A} \\ \boldsymbol{F}(q,t) = 0 \end{cases} \quad (5.58)$$

where

$\boldsymbol{M}$ is a diagonal, positive definite mass matrix,

$\boldsymbol{g}$ is a forces vector,

$\boldsymbol{F}_q^T$ is the Jacobian matrix,

$\boldsymbol{A}$ is a vector of Lagrangian multipliers.

Equation (5.58) is a differential-algebraic equation with an index of *3* [5.24]. As a result, methods for the numerical integration of ordinary differential equations cannot usually be apply directly to equation (5.58). Consequently, the index reduction method [5.30] is used.

5.3.2.3. Numerical solution of the motion equation system

For a clear geometric interpretation, the constraints have been assumed to be independent of time. This assumption does not declass the objects being analyzed (see [5.24]).

The configuration manifold of a system with constraints can be immersed in space $\boldsymbol{R}^n$. The tangent space $\boldsymbol{TM}_q$ of *n-l* dimension (*l* is a number of scalar equations of constraints) exists in each point of this manifold. If we denote as $\boldsymbol{D}$ the matrix of column base vectors of the space $\boldsymbol{Tm}_q$, then

$$\dot{q} = \boldsymbol{D}\dot{w}, \tag{5.59}$$

where $\dot{w}$ is the independent, generalized velocity.

Taking (5.58), system (5.59) can be converted into the form:

$$\begin{cases} \boldsymbol{D}^T \boldsymbol{M}\boldsymbol{D}\ddot{w} = \boldsymbol{D}^T g - \boldsymbol{D}^T \boldsymbol{M}\dot{\boldsymbol{D}}\dot{w} \\ F(q,t) = 0 \end{cases} \tag{5.60}$$

Since the matrix $\boldsymbol{D}^T\boldsymbol{M}\boldsymbol{D}$ is positive definite, the system of equations (5.60) can be solved by applying classical procedures for integrating ordinary differential equations. The dependent velocities are determined using relationship (5.59), and the coordinates $\boldsymbol{q}$ – by solving the non-linear task of finding the positions (5.58) in each integration step.

From the point of view of the numerical efficiency of this method, the key issue is the selection of the matrix $\boldsymbol{D}$. This matric is computed on the basis of the implicit function theorem. The independent coordinates are selected every for certain sequence of integration steps, using the Gauss-Jordan algorithm. The efficiency of the algorithm is enhanced by the fact that the Jacobian matrix f_q is a sparse matrix [5.24, 5.31]. This matrix is automatically computed for each group of joints. The product $\dot{\boldsymbol{D}}\dot{w}$ is computed in a similar manner.

5.3.2.4. Computer program and numerical examples

The above methods have been used to develop a computer program, designed for an IBM PC. In the current version it enables numerical solution of the forward dynamic task for a rigid links system with a number of links and degrees of freedom. The structure and mass characteristics of the multi-link system are expressed in a problem-oriented language [5.24]. The joints connecting the links can be defined individually in the form of computer subroutines, or taken from what is called the standard joints library, a compendium of descriptions of joints commonly used in practice (revolute, prismatic, cylindrical and so on). If the user chooses the second option, the desired data can be imported by giving the characteristic dimensions of the joint (a rotation axis, displacement direction and so on).

Another computer program has been developed to compute the dynamics of an elastic manipulator built at the Department of Robotics and Technical Biomechanics the Warsaw University of Technology. A segment of the manipulator and its kinematic diagram are shown in Figure 5.8. The segment has ten links and a mobility equal to 2 [5.15]. We can distinguish two kinematic pairs of the second class and six pairs of the fifth class. The segment constitutes a closed kinematic chain (see chapter 16).

Solving the problem of motion equations with the aid of a computer called for reading data describing masses and inertia tensors of the segment links, types of joints from the standard joints library, coordinates of points determining the rotation axes of fifth class joints, as well as the linear displacement axes of ball pivots in the second class joints. The hydraulic servo-motor forces driving the segment were entered in the form of subroutines written in FORTRAN.

The results obtained have been presented in detail in [5.25]. There is also a program for animation, showing the segment's motion under the influence of forces.

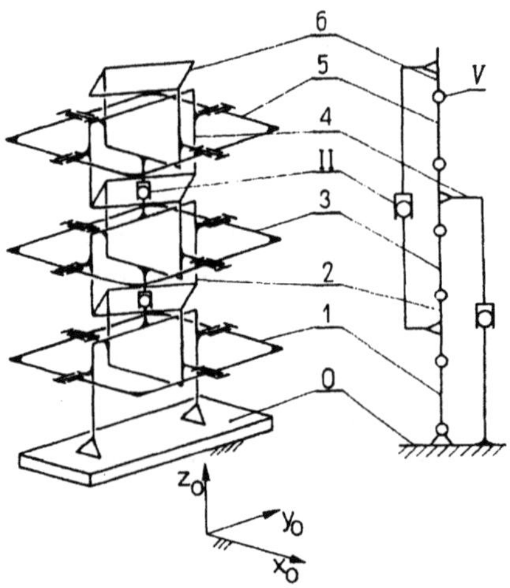

Fig. 5.8. Kinematic chain of a manipulator segment. Notation: 0 is a base, 1, 2, ..., 6 –movable links, II a kinematic pair of the second class, V a kinematic pair of the fifth class

5.3.3. Computer program

A computer program for the automatic analysis of multi-link systems using the above given relationships has been developed for the IBM PC. The program has a three-part structure.

A **preliminary program**, called the **preprocessor**, provides the input for the main program on the basis of data prepared by the user in a problem-oriented language. A simplified diagram of data entry is presented here below.

A **final program**, called the **postprocessor**, is used to present the calculated results in the form of charts or rough data (Figure 5.9).

The **main program**, shown the computation diagram (Figure 5.9) as MAIN (for systems without friction) or MAINT (for systems with friction), performs the major computations. On the grounds of the data obtained from the preprocessor, it automatically generates constraint equations, the elements of the Jacobian matrices for particular joints, and the right sides of velocity and acceleration equations. These operations draw on a joints library created specific for that purpose. Further computations are performed in three basic blocks of all the program: an optimization block, a factoring block, and a Newton-Raphson iteration block.

In the optimization block an initial position, that satisfies the constraint equations with an error not greater than that assumed, is determined. In that way we can verify, on the basis of the input data, whether such a mechanism can actually be built. We also obtain a good starting point, needed in the Newton-Raphson algorithm. Since in the initial step the constraints in the system of constraint equations (which entails singularity of the Jacobian matrix) may be mutually dependent, it is not possible to employ the classical Newton-Raphson method to solve the system of non-linear equations. The solution of the system is possible, however, using optimization methods. Among these we have chosen the **method of the greatest gradient descent**, developed by R. Fletcher and C. M. Reeves.

Well assembling a multi-link system with the use of the standard elements library in the constraint equation system the order of the Jacobian matrix may turn out to be lower than the number of equations, due to the singularity, or redundancy of the constraints.

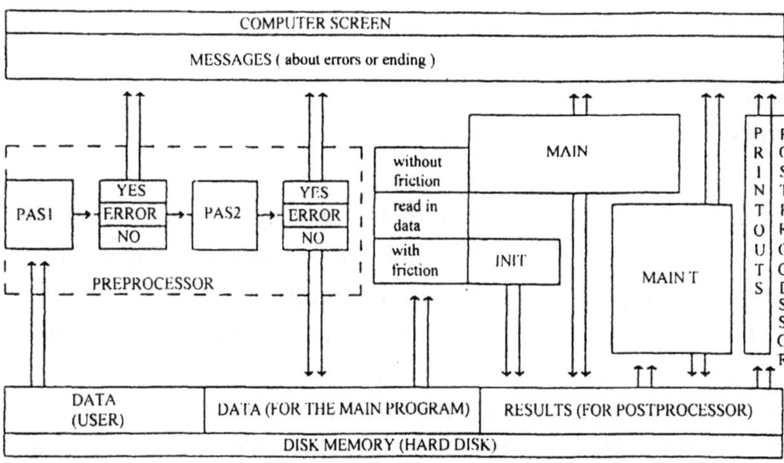

Fig. 5.9. Simplified diagram of the DAMS program package

Redundant constraints are identified and eliminated in the factoring block by examining the order of the Jacobian matrix of joint equations in the initial position and arbitrary selected neighboring position. The LU factoring method with a double selection of the basic element has been chosen to determine the matrix order, because of its useful property of numerical stability. Having found and rejected the redundant constraints, the mechanism's number of degrees of freedom is calculated. Guiding constraints are then added to the system of constraint equations. As before, these must be independent of one another, as well as of the joints defining the constraints. The factoring algorithm is used again for the Jacobian matrix of the full system of constraint equations to verify that independence.

Positions, velocities and accelerations are calculated in the Newton-Raphson iteration block. These calculations take advantage of the fact that the Jacobian matrix of constraint equations is a sparse matrix. The matrix is being factored symbolically before iteration begins. The solution of the system of linear equations is found using the factoring formula derived by the preliminary procedures.

Data is entered into the preprocessor in a problem-oriented language, using the following key-words:

- read links;
- read topology;
- read types of joints;
- read specification of joints;
- read additional constraints;
- read additional data;
- read Bryant;
- read initial positions;
- read constant coordinates;
- read durations;
- read limits;
- read errors;
- end.

Using the above key words, the following data are entered: the number of links, the number of joints, link indices of the link joints, types of joints and the number of joints of particular types, the quantities defining guiding constraints, the initial position vector, coordinates which can not be changed during optimization, the time of the beginning and the end of the simulation, the time delay in writing the intermediate results on the hard disk, the maximum number of iterations, admissible errors in the Newton-Raphson procedure, the admissible error in satisfying the right sides of the equations. The instruction *read Bryant* means that the orientation of particular links will be given in Bryant angles. Omission of the instruction means that the orientation will be defined by Euler parameters. The key-word *end* signifies that data-entry has terminated. After a key-word, a description sequence concerning this word appears. Each sequence ends with a comma; instruction descriptions end with a semicolon. A comment, denoted by the quotation "comment", may be placed in any given point of the list.

This program presented was used for computations regarding a spine-type manipulator. These computations have proved very helpful in the design of the control system. Part of the manipulator's animation is shown in Figure 5.10.

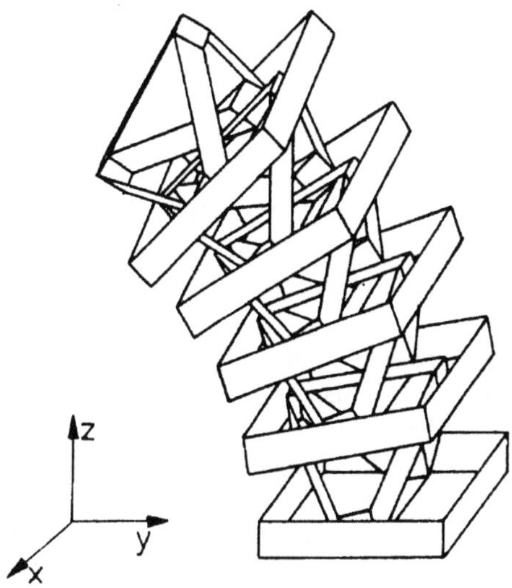

Fig. 5.10. Part of the manipulator's animation

5.4. References

[5.1] *Armstrong B., Khatib O., Burdick J.*: The Explicit Dynamic Model and Inertial Parameters of the PUMA 560 Arm: Proc. of the IEEE, Int. Conf. on Robotics and Automation, San Francisco 1986, s. 510–518.

[5.2] *Asada H., Slotine J. J. E.*: Robot Analysis and Control. New York, Wiley 1986.

[5.3] *Brady M.* at all: Robot Motion. MIT Press 1983.

[5.4] *Burdick J. W.*: An Algorithm for Generation of Efficient Manipulator Dynamic Equations. Proc. of the IEEE, Int. Conf. on Robotics and Automation, San Francisco 1986, s. 212–218.

[5.5] *Craig J. J.*: Introduction to Robotics. Mechanics and Control. Addison-Wesley 1989

[5.6] *Frolov K. V., Vorobiev E. J.* Mechanics of Industrial Robots. Moscow, Vyssaja Skola, 1988 (*in Russian*).

[5.7] *Gupta S., Townsend M. A.*: On the Equations of Motion for Robot Arms and open Kinematic Chains. Trans. ASME, J. Mech., Transm. Autom. in Design, vol. 110, sept.1988, s. 287–294.

[5.8] *Haug E. J.*: Computer Aided Kinematics and Dynamics of Mechanical Systems. Vol. 1 Basic Methods. Allyn and Bacon, Boston 1989.

[5.9] *Knapczyk J., Lebiediew P. A.*: Theory of Spatial Mechanisms and Manipulators. Warsaw, WNT, 1990 (*in Polish*).

[5.10] *Kolovskij M. E., Sluosz A. W.*: Bases of Dynamics of Industrial Robots. Moscow, Nauka 1988 (*in Russian*).

[5.11] *Olędzki A.*: Bases of Theory of Machines and Mechanismes. Warsaw, WNT 1987 (*in Polish*).

[5.12] *Paul R. P.*: Robot Manipulators: Mathematics, Programming and Control. Cambridge, MIT Press 1981.

[5.13] *Pennock G. R., Yang A. T.*: Dynamic Analysis of Multi-Rigid Body Open Chain Systems. Trans.ASME, J. Mech., Transm. and Autom. Design, vol. 105, s. 28–34, 1983.

[5.14] *Vukobratovic M., Potkonjak V.*: Dynamics of Manipulation Robots. Berlin, Springer, 1983.

[5.15] *Buśko Z., Frączek J., Kamiński D., Morecki A.*: Designing and Modeling of Elastic Manipulators. Proc. of XII Conf. TMM, Bielsko-Biała 1989.

[5.16] *Roth B., Rastegor J., Scheinman V.*: On the design of Computer Controlled Manipulators. First CISM–IFtoMM Symposium on Theory and Practice of Robots and Manipulators, Vol. 1, Udine, Springer-Verlag 1974.

[5.17] *Thring M. W.*: Robots and Telechirs, Ellis Horwood Limited, Publishers. Chichester, 1983.

[5.18] *Umetani Y, Hirose S.*: Biomechanical Study of Serpentine Locomotion, First CISM-IFtoMM Symposium on Theory and Practice of Robots and Manipulators, Vol. 1, Udine, Springer-Verlag 1974.

[5.19] *Hirose S., Ikuta K., Umetani Y.*: A New Design Method of Servoactuators Based on the Shape Memory Effect, Proc. Ro.Man.Sy'84; The Fifth CISM-IFtoMM Symposium. London, Kogan Page,. Hermes Publishing 1985.

[5.20] *Malczyk G., Morecki A.*: Elastic Manipulator of the Elephant Trunk Type, Biocybernetics and Biomedical Engineering, 1987, Vol. 7, No 1–4, s. 155–168.

[5.21] *Morecki A., Ekiel J., Fidelus K.*: Bionics of Motion. Warsaw, PWN, 1971 (*in Polish*).

[5.22] *Wilson J. F.* at all: A continuum model of Elephant trunks, Manuscript (presented of Piza Conference, 1978)

[5.23] *Cieslak R.*: Elastic Manipulator, Construction and Static Analysis. Master Thesis. Warsaw Univ. of Technology, Warsaw 1990.

[5.24] *Frączek J.*: Computer Method of Dynamic Analysis of Kinematic Chain of Manipulators. Doct. Dissertation, Warsaw Univ. of Technology, Warsaw 1990.

[5.25] *Morecki A., Busko Z., Czerwinski M., Fraczek J.*: Elaboration of Robot Model With Elastic Manipulation Chain. Proc. of MERA–PIAP, CPBR 7.1, Warsaw1989. (*in Polish*)

[5.26] Spine-Robotics, Sweden 1983.

[5.27] Spine News, Sweden 1984.

[5.28] *Lhote F.*, at all: Robots Components and Systems. Robot Technology. Vol. 4, Kogan Page, London, 1984.

[5.29] *Wilson J. F., Mahajan V.*: The Mechanics and Positioning of Highly Flexible Manipulator Limbs. Jnl of Mechanisms. Transm. and Autom. in Design., Trans. ASME, June 1989.

[5.30] *Frączek J.*: Dynamics of Mechanical Systems with Coulomb Friction. Archive of Mechanical Engineering, Vol 40, No 1, Warsaw 1993.

[5.31] *Frączek J.*: Dynamic Analysis of Elastic Manipulators. Archive of Mechanical Engineering, Vol 38, No 3, Warsaw 1991.

6. GEOMETRICAL AND FUNCTIONAL CHARACTERISTICS AND MANIPULATOR MOTION PLANNING

6.1. Introduction

Industrial robots are widely used to perform such tasks as bonding, welding, painting and assembling. In spite of this variety of industrial applications, robots have many geometrical properties in common. The geometrical and design parameters of an industrial robot are usually the result of an optimal compromise between compactness and rigidity, maintaining a maximum volume of the reachable workspace.

Serial manipulators feature a number of kinematic pairs equal to the number of degrees of freedom. Most manipulators are designed so that the last $n-1$ revolute joints of the kinematic chain, which orient the end-effector, have axes intersecting at a point called a **wrist center**. The first three kinematic pairs determine the position of the wrist center. This is the reason why two manipulator chains are distinguished: a regional **positioning chain** (the arm) and a local **orientation chain** (the wrist).

The positioning chain is almost always very simple. Twist angles of the joint axes are 0 or $\pm\pi/2$ while many link lengths or offsets equal 0. Manipulator joint variables are usually limited, as it is impossible to realize unlimited motion in prismatic or screw pairs. The ranges of joint variables affect the shape and size of the workspace, and the number of possible configurations.

A robot's **positioning chain** determines the shape and volume of the workspace. It can be proved that for a manipulator with six revolute joints, the workspace volume is at its maximum when the axes of the first two joints intersect with each other at a right angle, and the axis of the third joint is parallel to the axis of the second joint.

An **orientation chain** defines the ability to achieve, or change the orientation of the end-effector. As has been proved, optimal orientation ability is obtained when the axes of the last three revolute joints intersect at a right angle with one another successively, with joints 4 and 6 that can rotate through at least 2π and joint 5, through π. When the three joint axes are orthogonal and intersect at one point, they correspond to a spherical joint. Thus the ideal structure of a manipulator may be of the 3R-S type (R being a revolute joint, and S, a spherical joint) or the 3P-S type (P being a prismatic joint).

The **reachable workspace** is a geometrical locus of particular end-effector's points (for instance, thse center of a gripper, the tip of a tool, or the wrist center) attainable with the use of the full range of all joint variables. The choice of a reference point, or a system affects the size of the workspace. Since not all the points of the workspace can be reached from any direction, the part of the workspace where the end-effector can assume an arbitrary orientation is distinguished as the **dextrous workspace**. In the remaining workspace, the orientation of the end-effector is limited.

The tasks that a manipulator can perform depend on its design and on general indicators such as: lifting capacity, high-speed capacity, workspace dimensions, operational accuracy and repeatability. In some applications the manipulator's dimensions are also considered as well as its power consumption and cost. The lifting capacity depends on link dimensions, and the power developed by the drive system. It also depends on the end-effector's position in the workspace, the manipulator's configuration and the inertial forces [6.1, 6.4].

We can also distinguish the maximum velocity of the end-effector, and the time cycle for a particular task. Quite often the acceleration and deceleration phases account for most of the cycle time; consequently, the end-effector's acceleration capacity may be of greater importance than its maximum velocity.

6.2. Analysis of manipulator workspace

The manipulator workspace can be divided into separate sub-spaces in which the end-effector can reach a desired position and orientation in a certain configuration. On the boundaries of these sub-spaces, the configuration may change. The workspace boundaries of an n-link robot are difficult to describe using general displacement equations. It is easier to take plane sections of the workspace and determine the contour of the section in the plane. Rotating or translating the plane yields a three-dimensional workspace.

We shall present an analytical method of determining workspace boundaries, taking the general case of a six revolute joint manipulator which can be divided into two separate chains: positioning and orientation (It is admissible to separate these two chains only if the axes of the last three revolute joints intersect at the point called the wrist center).

Determining the workspace for any given manipulator with six revolute joints in its Cartesian space is a very complex and time consuming task. As a result, this task is usually limited to determining the workspace of a manipulator with three revolute joints, assuming that the positioning structure has a fully general geometry. We shall also show that the closure equation for the kinematic chain of this manipulator can be written in the form of a four order polynomial of one variable, for example $t = tg\theta_1$. The polynomial coefficients are functions of the D-H parameters, as well as Cartesian coordinates of the end-effector.

As is shown in Figure 6.1a, the position of the wrist center W with respect to the base system can be determined by means of a vector $\boldsymbol{p}_r = [p_x \;\; p_y \;\; p_z]^T$. Substituting relationships (5.49) into (5.48) yields three equations with three variables θ_i [6.3]:

$$
\begin{aligned}
&Ac\theta_2 + Bs\theta_2 = l_2 + l_3c\theta_3 + \lambda_4s\alpha_4s\theta_3 \\
&-(As\theta_2 - Bc\theta_2)c\alpha_2 + D\,s\alpha_2 = \lambda_2s\alpha_2 + l_3s\theta_3 - \lambda_4s\alpha_3c\theta_3 \\
&(As\theta_2 - Bc\theta_2)s\alpha_2 + D\,c\alpha_2 = \lambda_2s\alpha_2 + \lambda_3 - \lambda_4c\alpha_3
\end{aligned}
\tag{6.1}
$$

where

$$
\begin{aligned}
&A = p_xc\theta_1 + p_ys\theta_1 - l_1 \\
&B = (p_xs\theta_1 - p_yc\theta_1)c\alpha_1 + p_3s\alpha_1 \\
&D = -(p_xs\theta_1 - p_yc\theta_1)s\alpha_1 + p_3c\alpha_1 \\
&p_3 = p_2 - \lambda_1
\end{aligned}
\tag{6.2}
$$

λ_4 is a displacement of the wrist's center relative to the origin of system "3".

Given the position of the wrist center, that is the coordinates p_x, p_y and p_z, the solution of system (6.1) for θ_1, θ_2 and θ_3 represents the geometrical configuration of the positioning chain. In order to obtain one equation for θ_1, variables θ_2 and θ_3 must be eliminated. To eliminate θ_3, the first two equations of system (6.1) are transformed so that only functions of θ_3 remain on their right sides. The equations are then squared and added together, to yield:

$$
(Ac\theta_2 + Bs\theta_2)l_2 = Q - (p_xc\theta_1 + p_ys\theta_1)l_1 - \lambda_2D \tag{6.3}
$$

where

$$
2Q = p_x^2 + p_y^2 + p_3^2 + l_1^2 + l_2^2 - l_3^2 + \lambda_2^2 - \lambda_3^2 - \lambda_4^2 - 2\lambda_3\lambda_4c\alpha_3
$$

The variable θ_2 can be eliminated from the third equation of system (6.1), and from equation (6.3), converting them into the form

$$
\begin{aligned}
&(As\theta_2 - Bc\theta_2)l_2s\alpha_2 = E \\
&(Ac\theta_2 + Bs\theta_2)l_2s\alpha_2 = F
\end{aligned}
$$

where

$$
\begin{aligned}
&E = (R - Dc\alpha_2)l_2 \\
&F = [Q - (p_xc\theta_1 + p_ys\theta_1)l_1 - \lambda_2D]\,s\alpha_2 \\
&R = \lambda_2c\alpha_2 + \lambda_3 + \lambda_4c\alpha_3
\end{aligned}
\tag{6.5}
$$

Squaring equations (6.4) and adding their respective sides, we obtain a second order trigonometric polynomial of one variable θ_1

$$
K_1c^2\theta_1 + K_2s^2\theta_1 + K_3c\theta_1s\theta_1 + K_4c\theta_1 + K_5s\theta_1 + K_6 = 0 \tag{6.6}
$$

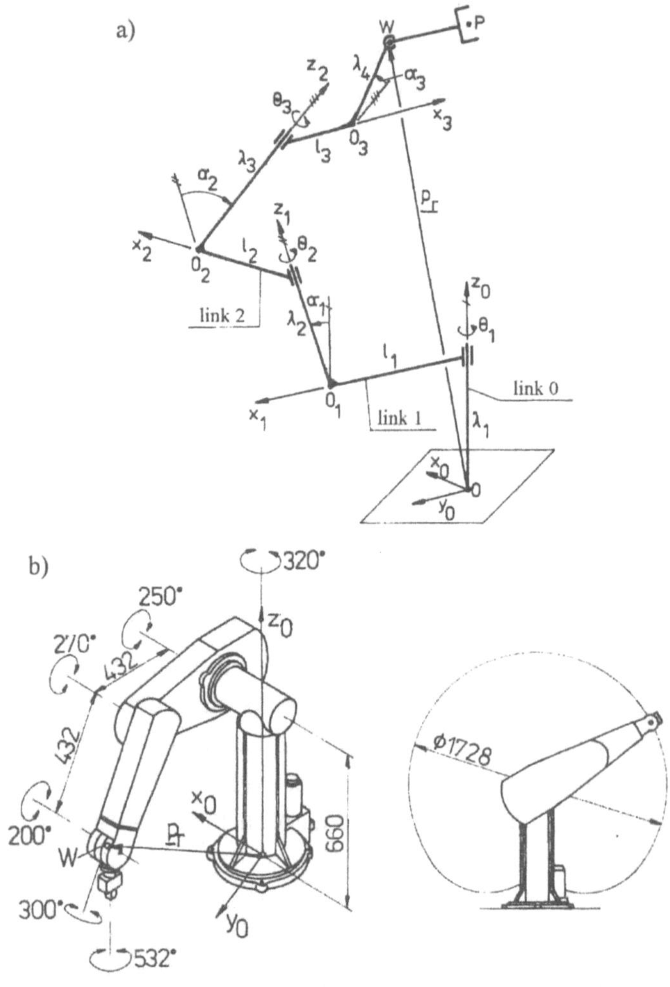

Fig. 6.1. a) General manipulator with 6 revolute joints, the axes of its last three joints intersecting at W. Notations l_i, α_i, λ_i are geometrical dimensions of the link i, while pr is a position vector of the wrist center (point W) relative to the base, b) Structure of the PUMA 560 manipulator and its workspace

where

$$K_1 = -\left(p_x^2 + p_y^2\right) l_2^2 s^2\alpha_2 + \left(p_x l_1 - p_y \lambda_2 s\alpha_1\right)^2 s^2\alpha_2 + p_y^2 l_2^2 s^2\alpha_1$$

$$K_2 = -\left(p_x^2 + p_y^2\right) l_2^2 s^2\alpha_2 + \left(p_y l_1 + p_x \lambda_2 s\alpha_1\right)^2 s^2\alpha_2 + p_x^2 l_2^2 s^2\alpha_1$$

$$K_3 = 2\left[p_x p_y \left(l_1^2 - \lambda_2 s^2\alpha_1\right) + \left(p_x^2 - p_y^2\right) l_1 \lambda_2 s\alpha_1\right] s^2\alpha_2 - 2 p_x p_y l_2^2 s^2\alpha_1$$

$$K_4 = 2\left[\left(p_x l_1 - p_y p_3 s\alpha_1 c\alpha_1\right) l_2^2 s^2\alpha_2 - \left(p_x l_1 + p_y \lambda_2 s\alpha_1\right)\left(Q - p_3 \lambda_2 c\alpha_1\right) s^2\alpha_2 + \right.$$

$$\left. + p_x s\alpha_1 \left(R - p_3 c\alpha_1 c\alpha_2\right) l_2^2\right]$$

$$K_5 = 2\left[\left(p_y l_1 - p_x p_3 s\alpha_1 c\alpha_1\right)l_2^2 s^2\alpha_2 - \left(p_y l_1 + p_x \lambda_2 s\alpha_1\right)\left(Q - p_3\lambda_2 c\alpha_1\right)s^2\alpha_2 + \right.$$
$$\left. + p_x s\alpha_1\left(R - p_3 c\alpha_1 c\alpha_2\right)l_2^2\right]$$
$$K_6 = -\left(l_1^2 + p_3^2\right)l_2^2 s^2\alpha_2 + \left(Q - p_3\lambda_2 c\alpha_1\right)^2 s^2\alpha_2 + \left(R - p_3 c\alpha_1 c\alpha_2\right)^2$$

Finally, substituting

$$t = tg\theta_1/2, \qquad s\theta_1 = 2t/(1+t^2), \qquad c\theta_1 = (1-t^2)/(1+t^2)$$

we have

$$t^4 + at^3 + bt^2 + ct + d = 0 \tag{6.7}$$

Two special cases often occur in industrial robots:

1. When $\alpha_2 = 0$, equation (6.7) takes the form

$$(L_1 t^2 + 2L_2 t + L_3)^2 = 0 \tag{6.8}$$

where

$$L_1 = -R + p_y s\alpha_1 + (p_z - \lambda_1)c\alpha_1$$
$$L_2 = p_y s\alpha_1$$
$$L_3 = -R - p_y s\alpha_1 + (p_z - \lambda_1)c\alpha_1$$
$$R = \lambda_2 + \lambda_3 + \lambda_4 c\alpha_3$$

The solution of equation (6.7) is

$$\theta_1 = 2ar\operatorname{ctg}\left(-L_2 \pm \sqrt{L_2^2 - L_1 L_3}\right)/L_1 \tag{6.9}$$

2. When $l_2 = 0$, equation (6.7) takes the form

$$N_1 t^2 + 2N_2 t + N_3 = 0 \tag{6.10}$$

where

$$N_1 = l_1 p_x - \lambda_2 p_y s\alpha_1 - \lambda_2(p_z - \lambda_1)c\alpha_1 + Q$$
$$N_2 = -(l_1 p_y + \lambda_2 p_x s\alpha_1)$$
$$N_1 = -l_1 p_x + \lambda_2 p_y s\alpha_1 - \lambda_2(p_z - \lambda_1)c\alpha_1 + Q$$
$$Q = \tfrac{1}{2}\left[\left(p_x^2 + p_y^2 + p_3^2\right) + \left(l_1^2 - l_3^2\right) + \left(\lambda_2^2 - \lambda_3^2 - \lambda_4^2\right) - 2\lambda_3\lambda_4 c\alpha_3\right.$$

For any given point of the positioning workspace of a robot there are four different solutions of equation (6.7), corresponding to four different configurations of the manipulator. If the given point lies outside the workspace, the solutions will be imaginary. However, for a point lying on the workspace boundary, there will be two identical real solutions, and two imaginary ones.

If we take a planar section of the workspace made by a plane parallel to the axis x_0y_0 of the base coordinate system, then this boundary is a curve drawn by the wrist center when the manipulator rotates about the first joint axis and the positioning chain links are in a fully extended configuration. The position of the section is defined by p_2 and, consequently, we need only two equations containing p_x and p_y as unknown coordinates of the wrist center on the workspace boundary. Since the value of θ_1 is arbitrary here, $\theta_1 = 0$ is assumed and with $d = 0$ obtained from equation (6.8), we have

$$K_1 + K_4 + K_6 = 0 \tag{6.11}$$

where K_1, K_4 and K_6 are coefficients of equation (6.6) and depend on p_x and p_y. The second equation for p_x and p_y can be obtained on the grounds of the fact that equation (6.7) has two identical real solutions for $\theta_1 = 0$. Thus, by differentiating (6.7) with regard to t, and substituting $\theta_1 = 0$ in the resulting equation, we have $c = 0$, and hence

$$K_3 + K_5 = 0 \tag{6.12}$$

Substituting $c = 0$ and $d = 0$ into equation (6.7) we obtain

$$t^2(t^2 + at + b) = 0 \tag{6.13}$$

A diagram drawn for the left side of equation (6.13) is a symmetrical function of θ_1 in the neighbourhood of $\theta_1 = 0$, and periodic in the interval $-\pi/2 < \theta_1 < \pi/2$. Ignoring the trivial case of all the roots being zero, we have two complex non-zero roots lying on the asymptote line of $\theta_1 = \pm p/2$. Equations (6.11) and (6.12) constitute a non-linear system of equations with two unknowns, p_x and p_y. The inner and outer boundary of the workspace section on the plane parallel to x_0y_0 can be determined from the relationship

$$R_{1,2} = \sqrt{p_x^2 + p_y^2} \tag{6.14}$$

With the known value of p_3, the new values of p_x and p_y can be found for the inner and outer workspace boundary. Repeating this procedure, and then rotating about the axis z_0, we can produce a three-dimensional representation of the workspace boundaries (Figure 6.1b).

To minimize energy consumption during operation, the displacements in kinematic pairs of the positioning (regional) chain should be as small as possible. So, it is preferable that the positioning structure has large dimensions, and the orientation structure, small dimensions. In that case the workspace is dominated by the positioning structure.

Manipulators are usually built to have motion limitations in the joints. These result from the design of the mechanical, hydraulic or electric drive systems. In some cases the required workspace and number of configurations can be achieved without full rotation in each revolute joint. Motion limitations are a requisite in the prismatic and screw joints. The influence of these limitations on the shape and shape of the workspace, as well as on the number of possible configurations, should be known when designing a manipulator and programming its motion.

A manipulator workspace can be divided into two separate subspaces. In each of these the end-effector can attain a desired position and orientation, and the number of possible configurations in both subspaces being the same. On the boundary separating these subspaces, the corresponding pairs of configurations of both subspaces coincide, meaning that the configuration can change only on certain boundaries. This defines the reachable space and the workspace for each configuration. On the boundary of this workspace a given configuration may coincide with another one.

Example 6.1. A planar, two link manipulator with revolute joints can reach any point in the workspace with one of two possible configurations (Figure 6.2). Taking the end-point of the end-effector $P(p_x, p_y)$, we obtain the boundary curve separating the workspaces for both configurations, setting $\theta_2 = 0$ and $\theta_2 = \pi$ (Figure 6.2b and c). The workspace for the first configuration corresponds with the range of the value of θ_1 within the interval $(0, 2\pi)$, while the value of θ_2 lies within the interval $(0, \pi)$.

When the joint variables are limited to the intervals $-75° \le \theta_1 \le 45°$ and $-\pi/2 \le \theta_2 \le \pi/2$, the constraints imposed on θ_1 affect the workspaces of both configurations symmetrically. The range of the value of θ_2 can be divided into two separate intervals. The first interval, $\pi \le \theta_2 \le (3/2)\pi$, refers to the workspace of the first configuration only.

The second interval, $\pi/2 \le \theta_2 \le \pi$, influences only the second configuration of the workspace. The workspaces of both configurations with their limitations are shown in Figure 6.2b (the hatched area). The manipulator workspace is obtained by combining the workspaces of both configurations (Figure 6.2d) [6.14].

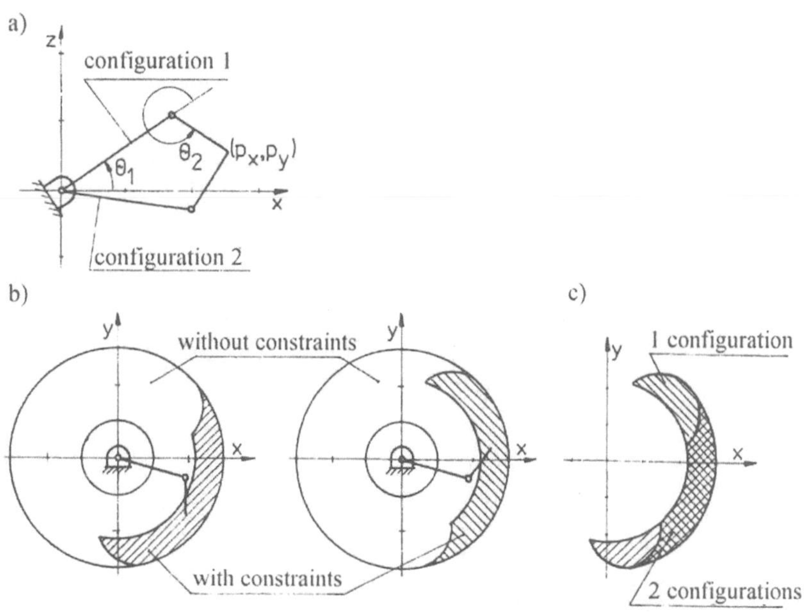

Fig. 6.2. a) Planar two-link manipulator with revolute joints, reaching point (p_x, p_y) with two configurations, b) the workspace of the first configuration without and with limitations of motion in the joints (the hatched area), c) the workspace of the second configuration, d) the manipulator's workspace

Example 6.2. To analyze the workspace of a manipulator with three revolute joints (Figure 6.3) whose dimensions are: $\alpha_1 = 0$, $\alpha_2 = \pi/2$, $l_1 = 3.5$, $l_2 = 1.5$, $l_3 = 1$, $\lambda_1 = \lambda_2 = 0$, we consider the influence of the second and third joint as variable limitations. Due to symmetry relative to the axis z_0, the section made by a plane containing this axis fully defines the shape and reach of the workspace.

The manipulator considered has four distinct configurations. The boundaries refer to $\theta_2 = 0$, π and $\theta_3 = \pm\pi/2$. The variable intervals of admissible displacements in the joints are listed in Table 6.1.

Table 6.1. Intervals of admissible displacements in the revolute joints of the manipulator of example 6.2

configuration	1	2	3	4
$\theta_{1min}, \theta_{1max}$	$0, 2\pi$	$0, 2\pi$	$0, 2\pi$	$0, 2\pi$
$\theta_{2min}, \theta_{2max}$	$\pi, 2\pi$	$0, \pi$	$\pi, 2\pi$	$0, \pi$
$\theta_{3min}, \theta_{3max}$	$-\pi/2, \pi/2$	$-\pi/2, \pi/2$	$\pi/2, (3/2)\pi$	$\pi/2, (3/2)\pi$

The influence of the range of the third joint variable on the workspace is shown in Figure 6.4. The shape of the workspace shown in the bottom row is established by superimposing the workspaces obtained for the different configurations. The numbers of configurations in each sub-space are also given.

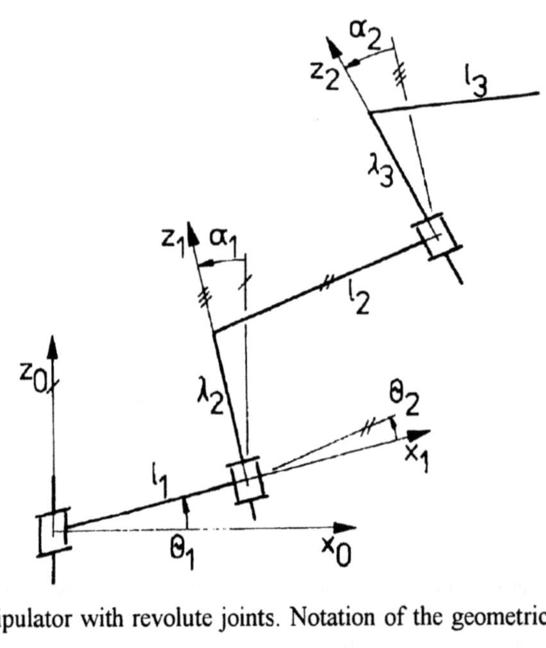

Fig. 6.3. Spatial manipulator with revolute joints. Notation of the geometrical dimensions: l_i, α_i, λ_i ($i = 1, 2, 3$)

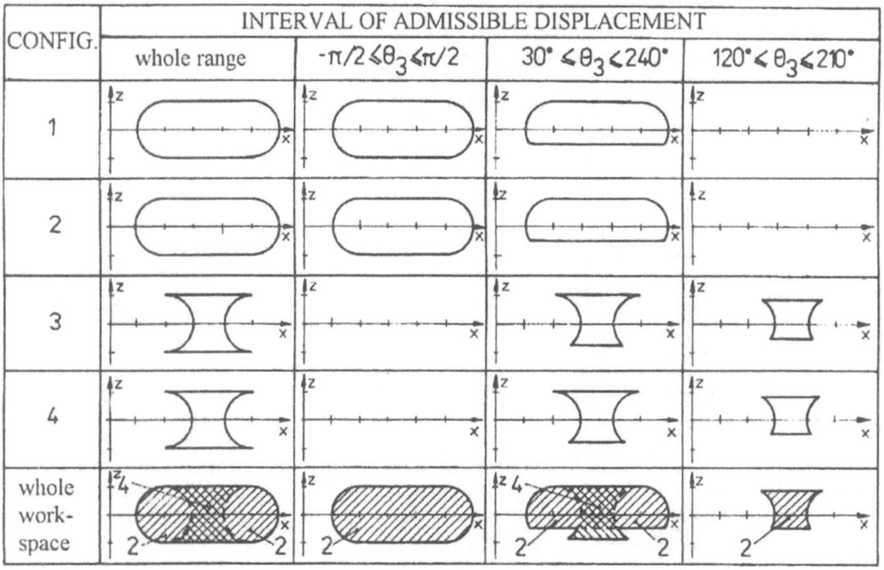

Fig. 6.4. Influence of motion limitations in the third revolute joint on the shape of the manipulator workspace

Since the third joint variable is limited to the interval $-\pi/2 \leq \theta_3 \leq \pi/2$ in the first two configurations, the workspaces do not change (see second column). However, there are no workspaces for the last two configurations. The operating range of the manipulator remains unchanged because the manipulator's reach in the first two configurations covers its whole workspace.

The result of the motion limitations in the second and third revolute joint is presented in Figure 6.5.

The influence of variable limitations of the joint on the workspace should be examined for each configuration. The manipulator workspace and its subspaces are determined by summing the workspaces obtained for all configurations. If there are no limitations, the pairs of configurations give rise to similar reach covering the whole manipulator workspace.

The change of a configuration is possible only on the boundaries where the pairs of configurations coincide. At the contact points of the different boundaries several configurations may coincide.

Manipulators with one fixed revolute axis have symmetrical workspace for each configuration. In this case any limitation of rotation in the joint decreases the manipulator's reach. In practice, prismatic, cylindrical, and screw joints always have motion limitations which affect the workspace. In the case of revolute joints, motion limitations may not decrease the workspace.

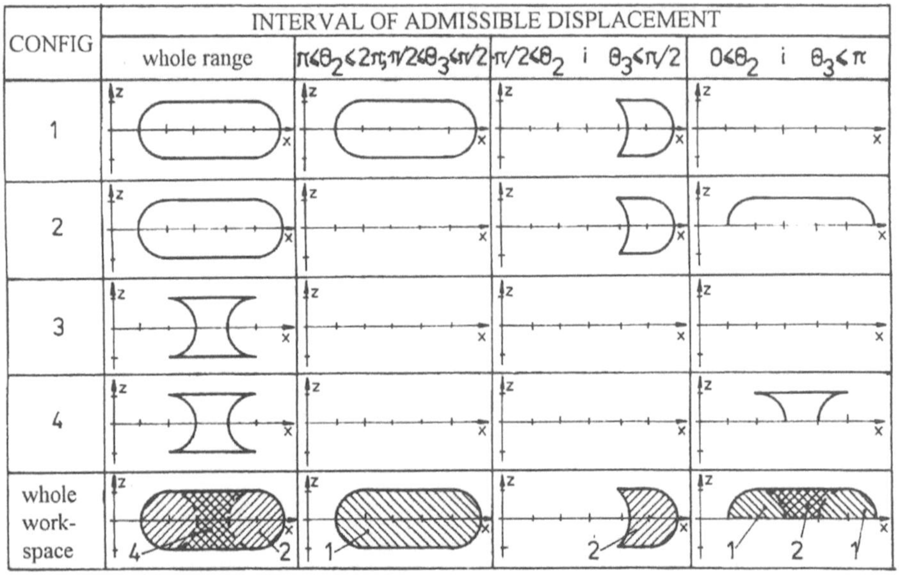

Fig. 6.5. Shapes of the workspace of the manipulator of Figure 6.3 resulting from different motion limitations in the second and third revolute joint

For a quantitative evaluation of the manipulator workspace, the **coefficient of link length efficiency** is introduced. Based on a dimensional analysis of the workspace volume, it can be defined by a general formula

$$V = \sum_{ijk}^{2n} \mu_{ijk} l_i l_j l_k \tag{6.15}$$

where: i, j, and k are integers from the range $(1, 2n)$ (n being the number of movable links); and μ_{ijk} is a coefficient of term $l_i\, l_j\, l_k$ whose value depends on the structure of the manipulator.

There is a definite relationship between the workspace volume and the total length of the manipulator links. It has been established [6.10] that for a given manipulator (that is, for a given set of data: twist angles, motion limitations on the joints, and link length ratios), we obtain a constant value for the ratio between the workspace volume and the volume of a cube the edge of which equals the total length of links, that is:

$$k_V = \frac{V}{\left[\sum_{i=1}^{n}(l_i + \lambda_i)\right]^3} = \frac{V}{L^3} = const \tag{6.16}$$

where l_i and λ_i are link lengths and offsets in the revolute joints of the manipulator with n links. The volume index k_V defines the efficiency of the link lengths in terms of the reachable workspace. This index can be normalized dividing it by its maximum possible value. If the total length of the links is L, the maximum workspace of the manipulator is a sphere with the radius of L. Consequently, the maximum value of the volume index is

$$k_{V,max} = \frac{4\pi L^3/3}{L^3} = \frac{4\pi}{3} = 4.1888$$

For example, for the PUMA 600 manipulator, these indices are: $k_V = 1.39$, and $k_{V,znorm} = 0.331$; for the Cincinnati Milacron T3 manipulator, they are: $k_V = 2.23$, and $k_{V,znorm} = 0.532$. Comparing these indices, we may conclude that the second manipulator utilizes the length of its links better. In the case of a Cartesian manipulator, the value of the index k_V is maximum when all three prismatic joints have the same ranges of variable λ_i and equal indices $k_V = 1/9 = 0.1111$.

Example 6.3. For the Scara-type manipulator we assume the following linear dimensions: $l_1 = l_2 = l/2$ and λ_3. We also assume, for the sake of simplification, that there are no variable limitations of the joints. The total length of the links is

$$L = l_1 + l_2 + \lambda_3 = l + \lambda_3$$

Since the workspace is a cylinder with the radius of l and height of λ_3,

$$V = \pi l^2 \lambda_3$$

The maximum value of the volume index k_V, calculated according to formula (6.16), is obtained when $\lambda_3 = l/2$ and equals [6.4]

$$k_V = \frac{\pi l^3}{2\left(\frac{3}{2}l\right)^3} = \frac{2\pi}{9} = 0.6981$$

At singular points, a manipulator loses one or more degrees of freedom. This may mean that certain tasks can not be performed at these points, and in the proximity of these points manipulator actions can not be well conditioned. In a sense, the further away the manipulator is from the singular position, the greater its capacity for achieving uniform motion and exerting steady forces in all directions. To obtain a manipulator with a maximum and well conditioned workspace, many criteria have been suggested for the qualitative estimation of this influence.

As singular configurations are determined by the solution of

$$det(J(\Theta)) = 0 \tag{6.17}$$

it is natural to use the value of the Jacobian determinant to evaluate a manipulator's mobility. A **mobility measure** in the form

$$w = \sqrt{\det\left(J(\Theta)\right)J^T(\Theta)} \tag{6.18}$$

is suggested. For a non-redundant manipulator this takes the form

$$w = |\, det(J(\Theta))\,| \tag{6.18a}$$

A well designed manipulator has a workspace with a large value of w in most of it.

Other mobility measures, based on analysis of the acceleration, or the end-effector's ability to exert forces in all directions [6.4], have also been suggested.

6.3. Analysis of manipulator accuracy

Many industrial applications, such as the assembly of electronic, or optical units, require high accuracy. For the industrial robot, to achieve such accuracy, certain conditions must be satisfied: proper arm rigidity, lack of backlashes in drive systems, and adequate power-assist and control systems.

Repeatability is the notion most often used in the qualitative evaluation of the manipulator's ability to place the end-effector in the same position at successive trials.

Accuracy is a measure of the manipulator's ability to attain the programmed position of the end-effector.

Most of the industrial robots used today are much better at repeatability, than at accuracy. Repeatability is important when "teaching" a manipulator to take a desired position by means of a portable programming unit (keyboard). At present, the tendency is to use the "off-line" method of programming manipulator motion. In this case manipulator accuracy becomes more important and, consequently, both positioning and orientation errors must be analyzed.

Errors in the positioning and orientation of a link as well as in manipulator control are functions of the manipulator's position and configuration and are caused by many factors, such as dimensional errors in manufacture, elastic strains in the links, backlashes in drive systems, and settings errors of the joints variables.

In the case of a joint whose elements are ideally shaped (for example, a revolute joint formed by ideal cylinders), the positioning error can be determined on the basis of the conditions of the contact in the joint, which depend on the forces and torques exerted in that joint. When the joint elements are not ideally shaped, the contact conditions also depend on the value of the joint variable.

The real values of Denavit-Hartenberg parameters do not usually equal the ideal ones. Consequently, we increase these values by the values of error, that is, $l_i + \Delta l_i$, $\alpha_i + \Delta\alpha_i$, $\theta_i + \Delta\theta_i$, $\lambda_i + \Delta\lambda_i$. The errors of positioning: $\Delta\theta_i$ in revolute joints and $\Delta\lambda_i$ in prismatic ones are caused by inaccurate position sensors, while dimensional errors $\Delta\alpha_i$ and Δl_i are usually caused by manufacturing errors.

If we assume that all the errors are of the same order, the orientation error of reference system i relative to system $i\text{-}1$ can be determined by the following formula:

$$\Delta\delta_{i,i-1} = \Delta\alpha_i \boldsymbol{l}_i + \Delta\theta_i \boldsymbol{n}_{i-1} \tag{6.19a}$$

where $\boldsymbol{l}_i$, $\boldsymbol{m}_i$, and $\boldsymbol{n}_i$ denote the unit vectors of the axes of the reference system i.

The error of the position of point O_i (the origin of reference system i) relative to point O_{i-1} (the origin of reference system $i-1$) depends on the errors of all the parameters. Thus we have:

$$\Delta \boldsymbol{p}_{i,i-1} = \Delta l_i \boldsymbol{l}_i + \Delta\lambda_i \boldsymbol{n}_i + \Delta\alpha_i \lambda_i (\boldsymbol{l}_i \times \boldsymbol{n}_{i-1}) + \Delta\theta_i l_i (\boldsymbol{n}_{i-1} \times \boldsymbol{l}_i)$$

Hence:

$$\Delta \boldsymbol{p}_{i,i-1} = (\Delta l_i + \Delta\theta_i \lambda_i s\alpha_i)\, \boldsymbol{l}_i + (\Delta\theta_i l_i - \Delta\alpha_i \lambda_i c\alpha_i)\, \boldsymbol{m}_i + (\Delta\alpha_i \lambda_i s\alpha_i)\, \boldsymbol{n}_i + \Delta\lambda_i \boldsymbol{n}_{i-1} \tag{6.19b}$$

Assuming that the magnitudes of the joint variable setting errors and the tolerances of dimensions are small in comparison with the values of the joint variables as well as of the nominal dimensions, the errors of positioning and orientation of the end-effector can be written as total differentials [6.7, 6.9, 6.16] in the form

$$\Delta \boldsymbol{p} = \frac{\partial \boldsymbol{p}}{\partial \boldsymbol{l}} \Delta \boldsymbol{l} + \frac{\partial \boldsymbol{p}}{\partial \boldsymbol{\lambda}} \Delta \boldsymbol{\lambda} + \frac{\partial \boldsymbol{p}}{\partial \boldsymbol{\alpha}} \Delta \boldsymbol{\alpha} + \frac{\partial \boldsymbol{p}}{\partial \boldsymbol{\theta}} \Delta \boldsymbol{\theta} \tag{6.19c}$$

$$\Delta \boldsymbol{\delta} = \frac{\partial \boldsymbol{\delta}}{\partial \boldsymbol{\alpha}} \Delta \boldsymbol{\alpha} + \frac{\partial \boldsymbol{\delta}}{\partial \boldsymbol{\theta}} \Delta \boldsymbol{\theta} \tag{6.19d}$$

where

$\boldsymbol{p} = [p_x\ p_y\ p_z]^T$ is a position vector of the origin of the end-effector reference system relative to the base system;

$\boldsymbol{\delta} = [\delta_x\ \delta_y\ \delta_z]^T$ is an orientation vector of the end-effector;

$\boldsymbol{l} = [l_1\ l_2\ \ldots\ l_n]^T$, $\boldsymbol{\lambda} = [\lambda_1\ \lambda_2\ \ldots\ \lambda_n]^T$, and $\boldsymbol{\alpha} = [\alpha_1\ \alpha_2\ \ldots\ \alpha_n]^T$ are vectors of linear and angular dimensions of the manipulator links;

Δl_i and $\Delta\alpha_i$ are errors (tolerances) of linear and angular dimensions;

$\Delta\theta_i$ is a setting error of a joint variable in the revolute pair, or an error (tolerance) of the angular dimension in the prismatic pair;

$\Delta\lambda_i$ – a setting error of a joint variable in the prismatic pair, or an error (tolerance) of the linear dimension in the revolute pair.

Since the errors of geometric dimensions can be identified during calibration and properly compensated [6.2, 6.5, 6.6], then only joint variable setting errors ($\Delta\theta_i$ and/or $\Delta\lambda_i$) are considered here.

A matrix of partial derivatives of the position vector $\boldsymbol{p}$ for a manipulator with six revolute joints is of $3\times n$ dimension:

$$\frac{\partial \boldsymbol{p}}{\partial \boldsymbol{\theta}} = \begin{bmatrix} \frac{\partial p_x}{\partial \theta_1} & \frac{\partial p_x}{\partial \theta_2} & \cdots & \frac{\partial p_x}{\partial \theta_n} \\ \frac{\partial p_y}{\partial \theta_1} & \frac{\partial p_y}{\partial \theta_2} & \cdots & \frac{\partial p_y}{\partial \theta_n} \\ \frac{\partial p_z}{\partial \theta_1} & \frac{\partial p_z}{\partial \theta_2} & \cdots & \frac{\partial p_z}{\partial \theta_n} \end{bmatrix} \tag{6.20}$$

Small increments in position and orientation vectors can be expressed by the Jacobian matrix (6.3)

$$\begin{bmatrix} \Delta\delta \\ \hline \Delta p \end{bmatrix} = J[\Delta\Theta] \tag{6.21}$$

where

$$J = \begin{bmatrix} \dfrac{\partial\delta}{\partial\theta} \\ \dfrac{\partial p}{\partial\theta} \end{bmatrix} = \begin{bmatrix} \gamma_{1x} & \gamma_{2x} & \cdots & \gamma_{nx} \\ \gamma_{1y} & \gamma_{2y} & \cdots & \gamma_{ny} \\ \gamma_{1z} & \gamma_{2z} & \cdots & \gamma_{nz} \\ \beta_{1x} & \beta_{2x} & \cdots & \beta_{nx} \\ \beta_{1y} & \beta_{2y} & \cdots & \beta_{ny} \\ \beta_{1z} & \beta_{2z} & \cdots & \beta_{nz} \end{bmatrix}$$

and J is a $(6\times n)$ matrix whose elements can be expressed in the end-effector reference system. For a revolute joint

$$\gamma_i = [l_z \quad m_z \quad n_z]_{i-1}^T \tag{6.22a}$$

$$\beta_i = [(l_y p_x - l_x p_y) \quad (m_y p_x - m_x p_y) \quad (n_y p_x - n_x p_y)]_{i-1}^T$$

and for a prismatic joint

$$\gamma_i = [0 \quad 0 \quad 0]_{i-1}^T \tag{6.22b}$$

$$\beta_i = [l_z \quad m_z \quad n_z]_{i-1}^T$$

where the index "*i-1*" in vectors γ_i and β_i means that the unit vectors $\boldsymbol{l}$, $\boldsymbol{m}$, and $\boldsymbol{n}$, as well as vector $\boldsymbol{p}$, are referred to system "*i-1*". The vectors of differential displacement and rotation of the end-effector relative to its reference system are:

$$\Delta\delta = \sum_{i=1}^{6} \gamma_i \Delta\theta_i, \qquad \Delta p = \sum_{i=1}^{6} \beta_i \Delta\theta_i \tag{6.23}$$

where $\Delta\delta = [\Delta\delta_x \ \Delta\delta_y \ \Delta\delta_z]^T$ and $\Delta p = [\Delta p_x \ \Delta p_y \ \Delta p_z]^T$ are the vectors of positioning and orientation errors. Multiplying vectors $\Delta\delta$ and Δp by matrix $T_{6,0}$ we obtain the positioning and orientation errors expressed in the base reference system.

Vectors (6.23) can be put together in the form of the error matrix in the end-effector reference system:

$$\Delta^n = \begin{bmatrix} 0 & \Delta\delta_z & \Delta\delta_y & \Delta p_x \\ \Delta\delta_z & 0 & -\Delta\delta_x & \Delta p_y \\ -\Delta\delta_y & \Delta\delta_x & 0 & \Delta p_z \\ 0 & 0 & 0 & 1 \end{bmatrix}$$

This matrix can be transformed into the base reference system, using the formula

$$\Delta = T_n \Delta^n$$

If the joint variable errors are definite, then the control system according to relation (6.23) can compensate the positioning and orientation errors. But in practice these errors depend on position and configuration, so it is difficult to foresee all possible errors. The vectors of the joint variable errors are usually defined by their lower and upper limits

$$-\Delta\Theta^d \leq \Delta\Theta \leq \Delta\Theta^g$$

The positioning and orientation errors (6.23), caused by these errors (6.24), are, therefore, incorporated in what are called **tolerance solids** [6.16]. (The analysis of errors is relatively easy, if errors $\Delta\Theta$ are regarded as determined variables. However, that approach gives us the worst results.)

Tolerance solids are obtained by assuming the absolute values of the elements of the Jacobian matrix, as well as the maximum absolute values of the joint variable errors in equation (6.21). Hence, positioning and orientation errors are determined by the inequality

$$-\Delta p^* \leq \Delta p \leq \Delta p^*$$

$$-\Delta\delta^* \leq \Delta\delta \leq \Delta\delta^*$$

where

$$\Delta p^* = \sum_{i=1}^{6} |\beta_i| \Delta\theta_i^*, \quad \Delta\delta^* = \sum_{i=1}^{6} |\gamma_i| \Delta\theta_i^*$$

and $\Delta\theta_i^* = max\left(\Delta\theta_i^d, \Delta\theta_i^g\right)$

Example 6.4. For a planar manipulator with two revolute joints, the Jacobian matrix (see formula (6.12)) takes the form

$$J = \begin{bmatrix} -l_1 s\theta_1 - l_2 s(\theta_1 + \theta_2) & -l_2 s(\theta_1 + \theta_2) \\ l_1 c\theta_1 + l_2 c(\theta_1 + \theta_2) & l_2 c(\theta_1 + \theta_2) \end{bmatrix}$$

The following dimensions have been assumed: $l_2 = 1.4142$, $l_1 = 1.000$, and the values of the joint variables are $\theta_1 = 0°$, $\theta_2 = 135°$. Thus

$$J = \begin{bmatrix} -0.7071 & -0.7071 \\ 0.7071 & -0.7071 \end{bmatrix}$$

According to relationship (6.21) the Jacobian matrix obtained refers to the anti-clockwise, 135° rotation of vector $\Delta\Theta$.

The tolerance area V_T^o (Figure 6.6b), resulting from the deterministic approach, completely covers area V_T, regarded as a confidence interval in the probabilistic approach. Assuming that the non-overlapping fraction is small, 0.3, for example, we can obtain a tolerance area V_T^* much smaller than V_T^o, which is shown in Figure 6.6c.

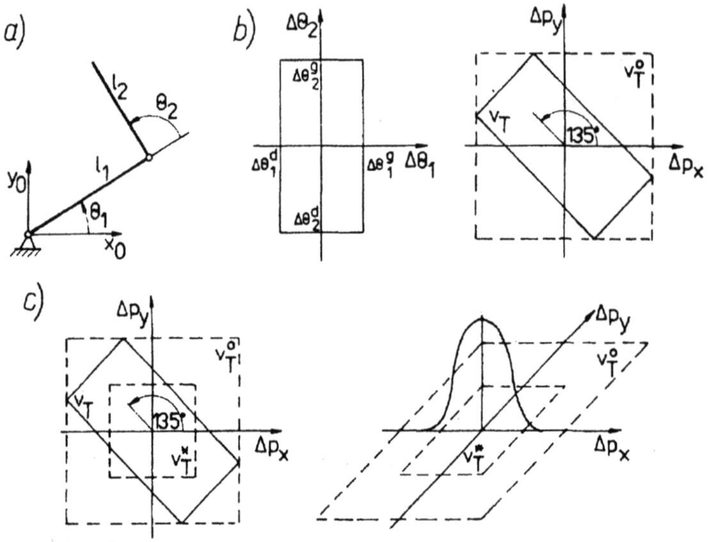

Fig. 6.6. a) Planar manipulator with two revolute joints, b) Area of the joint variable intervals converted into the area of the positioning tolerances V_T, c) Area of the positioning tolerances V_T diminished to V_T* with normal variable distribution

Example 6.5. For the industrial robot RPA-80 (Figure 6.7a) whose kinematic chain is the same as that of the PUMA 560 manipulator, the geometrical dimensions shown in Table 6.2 are

Table 6.2

i	1	2	3	4	5	6
α_i	$-\pi/2$	0	$\pi/2$	$-\pi/2$	$\pi/2$	0
l_i	0	1.000	0	0	0	0
λ_i	0	0	0	0.300	0	0.450
$\theta_{i,min}$	-135°	-55°	-90°	-90°	-90°	-90°
$\theta_{i,max}$	135°	55°	90°	90°	90°	90°

We assume that the manipulator position is determined by the following values of the joint variables: $\theta_1 = 135°$, $\theta_2 = 0°$, $\theta_3 = -90°$, $\theta_4 = \theta_5 = \theta_6 = 0°$, while the values of the joint variable setting errors are the same in all joints and equal to $\Delta\theta_l = 0.2\cdot 10^{-3}$ [*rad*].

For the manipulator considered, the elements of the Jacobian matrix are:

$$\gamma_{1x} = -s_{23}(c_4c_5c_6 - s_4s_6) - c_{23}s_5c_6$$

$$\gamma_{1y} = s_{23}(c_4c_5s_6 + s_4c_6) + c_{23}s_5s_6$$

$$\gamma_{1z} = -s_{23}c_4s_5 + c_{23}c_5$$

$$\beta_{1x} = [s_{23}(c_4c_5s_6 + s_4c_6) + c_{23}s_5s_6]\lambda_6 + s_{23}(s_4c_5c_6 + c_4s_6)\lambda_4 -$$
$$- [c_{23}(c_4c_5c_6 - s_4s_6) - s_{23}s_5c_6]\lambda_2 + c_2(s_4c_5c_6 + c_4s_6)l_2$$
$$\beta_{1y} = [s_{23}(c_4c_5c_6 - s_4s_6) + c_{23}s_5c_6]\lambda_6 + s_{23}(-s_4c_5s_6 + c_4c_6)\lambda_4 -$$
$$- [-c_{23}(c_4c_5s_6 + s_4c_6) + s_{23}s_5s_6]\lambda_2 + c_2(-s_4c_5s_6 + c_4c_6)l_2$$
$$\beta_{1z} = s_{23}s_4s_5\lambda_4 + c_2s_4s_5l_2 - (c_{23}c_4s_5 + s_{23}c_5)\lambda_2$$

where

$$s_{23} = sin(\theta_2 + \theta_3)$$

$$\beta_2 = \begin{bmatrix} (c_4c_6 - s_4c_5s_6)\lambda_6 + (c_4c_5c_6 - s_4s_6)\lambda_4 + s_3(c_4c_5c_6 - s_4s_6) + c_3s_5c_6l_2 \\ -(c_4s_6 - s_4c_5c_6)\lambda_6 - (c_4c_5s_6 - s_4c_6)\lambda_4 - s_3(c_4c_5c_6 - s_4s_6) + c_3s_5c_6l_2 \\ c_4s_5\lambda_4 + (s_3c_4s_5 - c_3c_5)l_2 \end{bmatrix}$$

$$\beta_3 = \begin{bmatrix} (c_4c_6 - s_4c_5s_6)\lambda_6 + (c_4c_5c_6 - s_4s_6)\lambda_4 \\ -(c_4s_6 - s_4c_5c_6)\lambda_6 - (c_4c_5s_6 - s_4c_6)\lambda_4 \\ c_4s_5\lambda_4 \end{bmatrix}$$

$$\gamma_2 = \begin{bmatrix} s_4c_5c_6 + c_4s_6 \\ -s_4c_5s_6 + c_4c_6 \\ s_4s_5 \end{bmatrix}, \qquad \gamma_3 = \begin{bmatrix} s_4c_5c_6 + c_4s_6 \\ -s_4c_5s_6 + c_4c_6 \\ s_4s_5 \end{bmatrix}$$

$$\gamma_4 = \begin{bmatrix} -s_5c_6 \\ s_5s_6 \\ c_5 \end{bmatrix}, \qquad \gamma_5 = \begin{bmatrix} s_6 \\ c_6 \\ 0 \end{bmatrix}, \qquad \gamma_6 = \begin{bmatrix} 0 \\ 0 \\ 1 \end{bmatrix}$$

$$\beta_4 = \begin{bmatrix} s_5s_6\lambda_6 \\ s_5c_6\lambda_6 \\ 0 \end{bmatrix}, \qquad \beta_5 = \begin{bmatrix} c_6\lambda_6 \\ -s_6\lambda_6 \\ 0 \end{bmatrix}, \qquad \beta_6 = \begin{bmatrix} 0 \\ 0 \\ 0 \end{bmatrix}$$

Substituting the numerical data in the above formulas gives us the Jacobian matrix in the form

$$J = \begin{bmatrix} 1 & 0 & 0 & 0 & 0 & 0 \\ 0 & 1 & 1 & 0 & 1 & 0 \\ 0 & 0 & 0 & 1 & 0 & 1 \\ 0 & 0.350 & 1.350 & 0 & 0.450 & 0 \\ -0.350 & 1.000 & 0 & 0 & 0 & 0 \\ 0 & 0 & 0 & 0 & 0 & 0 \end{bmatrix}$$

which converts the given joint variable errors into tolerances of positioning and orientation of the end-effector in Cartesian coordinates.

The results of calculations according to the above algorithm are shown in Figure 6.7b, in the form of the section of the RPA-80 manipulator workspace, created by the plane *Oxz*, and showing the lines of constant accuracy. These results indicate the differences in tolerances within an interval of ($0.20 \div 0.88$), and the values are close to those given by the manufacturer [6.7].

6.4. Manipulator trajectory planning in joint space

The sequence of desired positions, velocities and accelerations of some point such as the end-effector, or the manipulator's tip is called the trajectory of a robot manipulator. The process of computing this sequence is called trajectory planning. It converts the description of a desired motion into a trajectory, defining the time sequence of intermediate configurations of the manipulator. As the trajectory is executed, the tip of the end-effector traces a curve in space (the trajectory path), and changes its orientation accordingly.

Planning a manipulator trajectory in joint space requires finding time profiles of the relative displacements, velocities and accelerations for each joint, covering the distance between the initial and final point of the trajectory. This requires solving the inverse kinematics task for the initial and final configurations, that is, we must determine formulas for joint coordinates $q_i(t)$ which ensure that the end-effector will perform the required motion. The solution of this task is usually nonunique and requires a choice among the possible motions so as to lead to an optimal solution. The way we formulate an optimal solution depends on the conditions and constraints. One criteria for optimalization can be to determine algorithms which lead to the simplest formulas for $q_i(t)$. Other criteria may be the minimum period of the time it takes to perform a given task, minimum energy consumption, minimum reactions of the manipulator masses, and so on. Only after deciding on one or more of these criteria (multi-criterion synthesis) can the algorithms be developed.

In the simplest task of manipulator motion planning, encountered in pick-and-place operations, we have two end-effector positions, the initial and final one, together with the time required to complete the trajectory. If the manipulator has n degrees of freedom, then we can determine the values of the coordinates for the given positions q_{ip}, q_{ik}, for $i = 1, ..., n$, with the indices "*p*" and "*k*" meaning the initial and final configurations respectively. However, more than one solution can be obtained because for each position of the end-effector, several configurations of the manipulator kinematic chain are possible.

If there were no additional limitations for admissible values of q_i, we could use the task solving algorithm based on the priority of the generalized coordinate of the joint closest to the base, because a change there entails the motion of the whole manipulator. Clearly, changing any given coordinate entails a motion of greater mass than the change of the following one. Thus, priority is given to the solution in which the absolute value of a difference $(q_{1k} - q_{1p})$ is minimum, that is the motion relative to the base is as little as possible. If those values are equal for different configurations, we choose the one for which $| q_{2k} - q_{2p} |$, and so on.

In real manipulators there are design limitations to admissible values of q_i coordinates. If the initial and final position and configuration of a manipulator are assumed considering limitations of the type: $q_{i\,min} \leq q_i \leq q_{i\,max}$, there are at least four constraints, called the initial and final constraint equations:

$$q_i(t_{ip}) = q_{ip}, \qquad q_i(t_{ik}) = q_{ik}, \qquad \dot{q}_i(t_{ip}) = \dot{q}_i(t_{ik}) = 0 \tag{6.24}$$

These four conditions can be met by a polynomial of at least the third degree

$$q(\tau) = a_0 + a_1\tau + a_2\tau^2 + a_3\tau^3 \tag{6.25}$$

where $\tau = t/t_k$, $\tau \in [0, 1]$ is a normalized time variable.

To find four unknown coefficients a_i on the basis of the conditions (6.24), relationship (6.25) must be differentiated with respect to time. We obtain

$$\dot{q}(\tau) = (a_1 + 2a_2\tau + 3a_3\tau^2)/t_k \tag{6.26}$$

Taking into consideration relations (6.25) and (6.26), and equation (6.24), we have:

$$\begin{aligned} a_o &= q_p \\ a_0 + a_1 + a_2 + a_3 &= q_k \\ a_1 &= 0 \\ a_1 + 2a_2 + 3a_3 &= 0 \end{aligned} \tag{6.27}$$

The solution of equation (6.27) for the unknown a_i is

$$\begin{aligned} a_0 &= q_p \\ a_1 &= 0 \\ a_2 &= 3(q_k - q_p) \\ a_3 &= -2(q_k - q_p) \end{aligned} \tag{6.28}$$

Thus, the trajectory planned in the configuration coordinates takes the form

$$q(\tau) = q_p + (3\tau^2 - 2\tau^3)(q_k - q_p). \tag{6.29}$$

If the trajectory has to be continuous up to the second derivative, the following conditions must be met:

$$\begin{aligned} q(0) &= q_p, & q(t_k) &= q_k \\ \dot{q}(0) &= 0, & \dot{q}(t_k) &= 0 \\ \ddot{q}(0) &= 0, & \ddot{q}(t_k) &= 0 \end{aligned} \tag{6.30}$$

Since there are six conditions to be satisfied, we use a fifth degree polynomial, the coefficients of which must be adjusted. This polynomial is written as

$$q(\tau) = a_0 + a_1\tau + a_2\tau^2 + a_3\tau^3 + a_4\tau^4 + a_5\tau^5 \tag{6.31}$$

Differentiating relation (6.31) twice gives us

$$\dot{q}(\tau) = (a_1 + 2a_2\tau + 3a_3\tau^2 + 4a_4\tau^3 + 5a_5\tau^4)/t_k$$
$$\ddot{q}(\tau) = (2a_2 + 6a_3\tau + 12a_4\tau^2 + 20a_5\tau^3)/t_k^2 \qquad (6.32)$$

Considering relations (6.31) and (6.32) with reference to equations (6.30) we obtain

$$a_o = q_p$$
$$a_1 = a_2 = 0$$
$$a_3 + a_4 + a_5 = q_k \qquad (6.33)$$
$$3a_3 + 4a_4 + 5a_5 = 0$$
$$6a_3 + 12a_4 + 20a_5 = 0$$

The solutions the last three equations (6.33) are:

$$a_3 = 10(q_k - q_p), \quad a_4 = -15 \quad (q_k - q_p), \quad a_5 = 6(q_k - q_p)$$

Thus, polynomial (5.160) can be written in the form

$$q(\tau) = q_p + (10\tau^3 - 15\tau^4 + 6\tau^5)(q_k - q_p) \qquad (6.34)$$

However, in real applications it is often desirable to use lift-off and set-down configurations to ensure acceptable departure and arrival motion. In this case, if we maintain the requirement of continuity up to the second derivative, the following conditions must be added to those above:

$$q(t_r) = q_r$$
$$q(t_w) = q_w \qquad (6.35)$$

The values of q_r and q_w are determined as inverse kinematics solutions, or are assumed values that fulfil the task conditions. In this case a 7-th degree polynomial can be used:

$$q(\tau) = a_0 + a_1\tau + a_2\tau^2 + \ldots + a_7\tau^6 \qquad (6.36)$$

The polynomial coefficients are subject to the coordinates of the intermediate points. However, using a polynomial of the seventh degree, or higher has certain drawbacks. In particular, it is difficult to determine its extreme values; there are unexpected "pulses" in the displacement function which cause unnecessary movements, and so on [6.21].

Another approach to the problem divides the trajectory into several segments, each one of which is interpolated by a polynomial of the lowest possible degree. A solution consisting of three polynomials of 4, 3, and 4 degrees is often used. In this type of approach, the continuity between the different portions of the trajectory is imposed. Sometimes polynomials of 5, 3, and 5 degrees are used, as are polynomials of 3, 3, 3, 3, and 3 degrees [6.5].

To compute the "4-3-4"-type trajectory, a time variable is introduced, normalized according to the formula

$$\tau = (t - t_{i-1})/(t_i - t_{i-1}), \quad t \in [t_{i-1}, t_i] \qquad (6.37)$$

where t_i is the duration of the i-th segment of the trajectory; and $(t_i - t_{i-1})$, an interval of the real time it takes to cover the i-th segment of the trajectory. The successive trajectory segments are described by the polynomials

$$\begin{aligned} q_1 &= a_{10} + a_{11}\tau + a_{12}\tau^2 + a_{13}\tau^3 + a_{14}\tau^4 \\ q_2 &= a_{20} + a_{21}\tau + a_{22}\tau^2 + a_{23}\tau^3 \\ q_3 &= a_{30} + a_{31}\tau + a_{32}\tau^2 + a_{33}\tau^3 + a_{34}\tau^4 \end{aligned} \tag{6.38}$$

where the index "i" at the configuration variable q_i denotes the number of the trajectory segment, and the indices at the unknown coefficients a_{ij} are: i, the segment number, and j, the coefficient number.

To determine the fourteen unknown coefficients of polynomials (6.38), a constraint equation must be applied. The first and second derivatives of these polynomial functions with respect to real time, velocities and accelerations respectively, are determined according to the following formulas, taking into account expressions (6.37):

$$\begin{aligned} \dot{q}_j &= \frac{dq_j}{d\tau}\left(t_j - t_{j-1}\right)^{-1} \\ \ddot{q}_j &= \frac{d^2 q_j}{d\tau^2}\left(t_j - t_{j-1}\right)^{-2} \end{aligned} \tag{6.39}$$

noting that for $j = 1$, $t_j - t_{j-1} = t_r$, for $j = 2$, $t_j - t_{j-1} = t_b$; and for $j = 3$, $t_j - t_{j-1} = t_w$. Thus velocities and accelerations on the successive trajectory segments can be written in the form of polynomial functions $\tau \in [0, 1]$:

$$\begin{aligned} \dot{q}_1(\tau) &= \left(4a_{14}\tau^3 + 3a_{13}\tau^2 + 2a_{12}\tau + a_{11}\right)/t_r \\ \ddot{q}_1(\tau) &= \left(12a_{14}\tau^2 + 6a_{13}\tau + 2a_{12}\right)/t_r^2 \\ \dot{q}_2(\tau) &= \left(3a_{23}\tau^2 + 2a_{22}\tau + a_{21}\right)/t_b \\ \ddot{q}_2(\tau) &= \left(6a_{23}\tau + 2a_{22}\right)/t_b^2 \\ \dot{q}_3(\tau) &= \left(4a_{34}\tau^3 + 3a_{33}\tau^2 + 2a_{32}\tau + a_{31}\right)/t_w \\ \ddot{q}_3(\tau) &= \left(12a_{34}\tau^2 + 6a_{33}\tau + 2a_{32}\right)/t_w^2 \end{aligned} \tag{6.40}$$

The boundary conditions imposed on the displacement functions (6.38) as well as on velocities and accelerations (6.40) at the initial point ($t = t_p$) and final point ($t = t_k$) provide six equations, from which we can determine:

$$\begin{aligned} a_{10} &= q_p, \quad a_{11} = \dot{q}_p t_r, \quad a_{12} = \tfrac{1}{2}\ddot{q}_p t_r^2 \\ a_{30} &= q_k, \quad a_{31} = \dot{q}_k t_w, \quad a_{32} = \tfrac{1}{2}\ddot{q}_k t_w^2 \end{aligned} \tag{6.41}$$

The continuity requirement for two intermediate points (1, the end of acceleration, and 2, the beginning of deceleration) allows us to formulate the next six equations which, considering (6.41), take the form:

$$a_{14} + a_{13} - a_{20} = -\left(q_p + \dot{q}_p t_r + \tfrac{1}{2}\ddot{q}_p t_r^2\right)$$

$$
\begin{aligned}
&(4a_{14}+3a_{13})/t_r - a_{21}/t_b = -(\dot{q}_p + \ddot{q}_p t_r)\\
&(12a_{14}+6a_{13})/t_r^2 - 2a_{22}/t_b^2 = -\ddot{q}_p\\
&a_{23}+a_{22}+a_{21}+a_{20}-a_{34}-a_{33} = q_k - \dot{q}_k t_w + \tfrac{1}{2}\ddot{q}_k t_w^2\\
&(3a_{23}+2a_{22}+a_{21})/t_b - (4a_{34}+3a_{33})/t_w = -\dot{q}_k + \ddot{q}_k t_w\\
&(6a_{23}+2a_{22})/t_b^2 - (12a_{34}+6a_{33})/t_w^2 = -\ddot{q}_k
\end{aligned}
\tag{6.42}
$$

Increments in the value of the variable q on each trajectory can be determined from the following formulas:

$$
\begin{aligned}
&\Delta q_1 = q_1 - q_p = a_{14} + a_{13} + \tfrac{1}{2}\ddot{q}_p t_r^2 + \dot{q}_p t_r\\
&\Delta q_2 = q_2 - q_1 = a_{23} + a_{22} + a_{21}\\
&\Delta q_3 = q_k - q_2 = -a_{34} + a_{33} - \tfrac{1}{2}\ddot{q}_k t_w^2 + \dot{q}_k t_w
\end{aligned}
\tag{6.43}
$$

All the unknown coefficient values occurring in equations (6.42) and (6.43) can be determined from the system they constitute, written in the matrix form

$$\boldsymbol{y} = \boldsymbol{A}\,\boldsymbol{x} \tag{6.44}$$

$$
\begin{bmatrix}
\Delta q_1 - \frac{1}{2}\ddot{q}_p t_r^2 - \dot{q}_p t_r\\
-\frac{1}{2}\ddot{q}_p t_r - \dot{q}_p\\
-\frac{1}{2}\ddot{q}_p\\
\Delta q_2\\
-\frac{1}{2}\ddot{q}_k t_w + \dot{q}_k\\
-\ddot{q}_k\\
\Delta q_3 + \frac{1}{2}\ddot{q}_k t_w^2 - \dot{q}_k t_w
\end{bmatrix}
=
\begin{bmatrix}
1 & 1 & 0 & 0 & 0 & 0 & 0\\
3/t_r & 4/t_r & -1/t_b & 0 & 0 & 0 & 0\\
6/t_r^2 & 12/t_r^2 & 0 & -2/t_b^2 & 0 & 0 & 0\\
0 & 0 & 1 & 1 & 1 & 0 & 0\\
0 & 0 & 1/t_b & 2/t_b & 3/t_b & -3/t_w & 4/t_w\\
0 & 0 & 0 & 2/t_b^2 & 6/t_b^2 & 6/t_w^2 & -12/t_w^2\\
0 & 0 & 0 & 0 & 1 & 1 & -1
\end{bmatrix}
\begin{bmatrix}
a_{13}\\ a_{14}\\ a_{21}\\ a_{22}\\ a_{23}\\ a_{33}\\ a_{34}
\end{bmatrix}
$$

Thus, the trajectory planning task is completed by solving the equation $\boldsymbol{x} = \boldsymbol{A}^{-1}\boldsymbol{y}$ The matrix $\boldsymbol{A}$ structure allows us to easily determine unknown coefficient values as an inverse matrix always exists, provided the lengths of the trajectory segments, that is t_r, t_b and t_w, do not equal zero.

The solution of the "4-3-4" trajectory planning task can be obtained according to the following formulas:

$$
\begin{aligned}
&q_1(\tau) = \left(\Delta q_1 - \dot{q}_p t_r - \tfrac{1}{2}\ddot{q}_p t_r^2 - \sigma\right)\tau^4 + \sigma\tau^3 + \left(\tfrac{1}{2}\ddot{q}_p t_r^2\right)\tau^2 + \left(\dot{q}_p t_r\right)\tau + q_p\\
&\dot{q}_1 = \dot{q}_1(1) = 4\Delta q_1/t_r - 3\dot{q}_p - \ddot{q}_p t_r - \sigma/t_r\\
&\ddot{q}_1 = \ddot{q}_1(1) = 12\Delta q_1/t_r^2 - 12\dot{q}_p/t_r - 5\ddot{q}_p - \sigma/t_r^2\\
&q_2(\tau) = \left(\Delta q_2 - \dot{q}_1 t_b - \tfrac{1}{2}\ddot{q}_1 t_b^2\right)\tau^3 + \left(\tfrac{1}{2}\ddot{q}_1 t_b^2\right)\tau^2 + \left(\dot{q}_1 t_b\right)\tau + q_1\\
&\dot{q}_2 = \dot{q}_2(1) = 3\Delta q_2/t_b - 2\dot{q}_1 - \tfrac{1}{2}\ddot{q}_1 t_b\\
&\ddot{q}_2 = \ddot{q}_2(1) = 6\Delta q_2/t_b^2 - 6\dot{q}_1/t_b - 2\ddot{q}_1
\end{aligned}
\tag{6.45}
$$

$$q_3(\tau)=\left(9\Delta q_3-4\dot{q}_2t_w-\tfrac{1}{2}\ddot{q}_2t_w^2-5\dot{q}_kt_w+\tfrac{1}{2}\ddot{q}_kt_w^2\right)\tau^4+$$
$$+\left(-8\Delta q_3+5\dot{q}_kt_w-\tfrac{1}{2}\ddot{q}_kt_w^2+3\dot{q}_2t_w\right)\tau^3+\left(\tfrac{1}{2}\ddot{q}_2t_w^2\right)\tau^2+\left(\dot{q}_2t_w\right)\tau+q_2$$

where

$$\sigma=f/g$$
$$f=2\Delta q_1(4+2t_w/t_b+2t_w/t_r+3t_b/t_r)-\Delta q_2\,t_r/t_b(3+t_w/t_b)+$$
$$+2\Delta q_3\,t_r/t_w-\dot{q}_0t_r(6+6t_b/t_r+4t_w/t_r+3t_b/t_r)-\dot{q}_kt_r-$$
$$-\ddot{q}_pt_rt_w(\tfrac{5}{3}+t_r/t_b+2t_rt_w+\tfrac{5}{3}t_b/t_w)+\ddot{q}_kt_rt_w$$
$$g=t_w/t_b+2\,t_w/t_r+2+3\,t_b/t_r$$

The following are assumed to be the input values: Δq_1, Δq_2, Δq_3, $\dot{q}_p$, $\ddot{q}_p$, $\dot{q}_k$, $\ddot{q}_k$, t_r, t_b, t_w.

In practice, velocity and acceleration values at the beginning and at the end of the trajectory are assumed to be zero, that is $\dot{q}_p=\ddot{q}_p=\dot{q}_k=\ddot{q}_k=0$, and the acceleration and deceleration are assumed to be symmetrical: $t_r=t_w$, $\Delta q_1=\Delta q_3$. In that case a quite complicated algorithm is simplified to take the form of:

$$q_1(\tau)=(\Delta q_1-\sigma)\tau^4+\sigma\tau^3+q_p$$
$$\dot{q}_1=\dot{q}_1(1)=4\Delta q_1/t_r-\sigma/t_r$$
$$\ddot{q}_1=\ddot{q}_1(1)=12\Delta q_1/t_r^2-\sigma/t_r^2$$
$$q_2(\tau)=\left(\Delta q_2-\dot{q}_1t_b-\tfrac{1}{2}\ddot{q}_1t_b^2\right)\tau^3+\left(\tfrac{1}{2}\ddot{q}_1t_b^2\right)\tau^2+(\dot{q}_1t_b)\tau+q_1$$
$$\dot{q}_2=\dot{q}_2(1)=3\Delta q_2/t_b-2\dot{q}_1-\tfrac{1}{2}\ddot{q}_1t_b$$
$$\ddot{q}_2=\ddot{q}_2(1)=6\Delta q_2/t_b^2-6\dot{q}_1/t_b-2\ddot{q}_1$$
$$q_3(\tau)=\left(9\Delta q_1-4\dot{q}_2t_r-\tfrac{1}{2}\ddot{q}_2t_r^2\right)\tau^4+(-8\Delta q_1+3\dot{q}_2t_r)\tau^3+(\dot{q}_2t_r)\tau+q_2$$

where

$$\sigma=f/g$$
$$f=2\Delta q_1(6+2t_r/t_b+3t_b/t_r)-\Delta q_2\,t_r/t_b(3+t_r/t_b)+2\Delta q_1$$
$$g=4+t_r/t_b+3t_b/t_r$$

Numerical example. The geometrical parameters and ranges of angular displacement in the joints of a Puma-type manipulator are assumed to be those shown in Table 3.

The initial and final positions of the end-effector are defined by matrices:

$$T_{6,0p}=\begin{bmatrix}-0.0474 & -0.979 & -0.196 & 0.566\\ 0.789 & -0.157 & 0.594 & 0.225\\ -0.612 & -0.127 & 0.780 & 0.668\\ 0 & 0 & 0 & 1\end{bmatrix}$$

$$T_{6,0k} = \begin{bmatrix} -0.0474 & -0.979 & -0.196 & 0.2 \\ 0.789 & -0.157 & 0.594 & 0.3 \\ -0.612 & -0.127 & 0.780 & 0.4 \\ 0 & 0 & 0 & 1 \end{bmatrix}$$

The procedure used to solve the inverse kinematics task gives the values of the configuration angles for the initial and final manipulator position in a particular configuration. The results computed are shown in Table 4, where the values of extreme velocities and accelerations in the joints are also given.

Table 3

	l_i [m]	λ_i [m]	α_i [rad]	$(\theta_{i\,min}, \theta_{i\,max})$ [rad]
1	0	0	90	(–160°, 160°)
2	0.432	0.15	0	(–125°, 125°)
3	0	0	–90	(–135°, 135°)
4	0	0.432	90	(–140°, 140°)
5	0	0	-90	(–100°, 100°)
6	0	0.1	0	(–260°, 260°)

Table 4

	1	2	3	4	5	6
θ_{ip} [°]	30	60	–120	–45	60	90
θ_{ik} [°]	75.025	108.008	-209.880	-21.496	69.443	23.288
$\dot{\theta}_{iextr}$ [rad/s]	0.393	0.419	–0.784	0.205	0.0824	-0.582
$\ddot{\theta}_{iextr}$ [rad/s^2]	0.589	0.628	1.177	0.308	0.124	0.873

Motion duration is assumed to be t = 3s. The displacements, velocities and accelerations profiles are computed assuming that the periods of acceleration, transition and deceleration all have the same duration, while angular displacements in particular joints for the successive periods are assumed to have the ratio 1:2:1, that is:

$t_r = t_b = t_w = 1s,\ \dot{q}_p = \ddot{q}_p = \dot{q}_k = \ddot{q}_k = 0,\ \Delta q_1/\Delta q_2/\Delta q_3 = 1/2/1$

Under these assumptions, for consecutive joints (i = $1,..6$) we have obtained the $q_{ij}(\tau)$ relationships for particular time intervals (j = $1,..3$), as well as the derivative values on the interval boundaries.

For $i = 1$ (first joint):

$$q_{11}(\tau) = -11.256\tau^4 + 22.513\tau^3 + 30$$
$$\dot{q}_{11} = \dot{q}_{11}(1) = 22.513$$
$$\ddot{q}_{11} = \ddot{q}_{11}(1) = 0$$
$$q_{12}(\tau) = 22.513\tau + 41.256$$
$$\dot{q}_{12} = \dot{q}_{12}(1) = 22.513$$
$$\ddot{q}_{12} = \ddot{q}_{12}(1) = 0$$
$$q_{13}(\tau) = 11.256\tau^4 - 22.513\tau^3 + 22.513\tau + 63.769$$
$$f = 180.100, \quad g = 8.000, \quad \sigma = 22.513$$

For $i = 2$ (second joint):

$$q_{21}(\tau) = -12.002\tau^4 + 24.004\tau^3 + 60$$
$$\dot{q}_{21} = \dot{q}_{21}(1) = 24.004$$
$$\ddot{q}_{21} = \ddot{q}_{21}(1) = 0$$
$$q_{22}(\tau) = 24.004\tau + 72.002$$
$$\dot{q}_{22} = \dot{q}_{22}(1) = 24.004$$
$$\ddot{q}_{22} = \ddot{q}_{22}(1) = 0$$
$$q_{23}(\tau) = 12.002\tau^4 - 24.004\tau^3 + 24.004\tau + 96.006$$
$$f = 192.033, \quad g = 8.000, \quad \sigma = 24.004$$

For $i = 3$ (third joint):

$$q_{31}(\tau) = 22.470\tau^4 - 44.940\tau^3 - 120$$
$$\dot{q}_{31} = \dot{q}_{31}(1) = -44.940$$
$$\ddot{q}_{31} = \ddot{q}_{31}(1) = 0$$
$$q_{32}(\tau) = -44.940\tau - 142.470$$
$$\dot{q}_{32} = \dot{q}_{32}(1) = -44.940$$
$$\ddot{q}_{32} = \ddot{q}_{32}(1) = 0$$
$$q_{33}(\tau) = -22.470\tau^4 + 44.940\tau^3 - 44.940\tau - 187.410$$
$$f = -359.519, \quad g = 8.000, \quad \sigma = -44.940$$

For $i = 4$ (fourth joint):

$$q_{41}(\tau) = -5.876\tau^4 + 11.752\tau^3 - 45$$
$$\dot{q}_{41} = \dot{q}_{41}(1) = 11.752$$
$$\ddot{q}_{41} = \ddot{q}_{41}(1) = 0$$
$$q_{42}(\tau) = 11.752\tau - 39.124$$
$$\dot{q}_{42} = \dot{q}_{42}(1) = 11.752$$

$$\ddot{q}_{42} = \ddot{q}_{42}(1) = 0$$

$$q_{43}(\tau) = 5.876\tau^4 - 11.752\tau^3 + 11.752\tau - 27.372$$

$f = 94.016,\quad g = 8.000,\quad \sigma = 11.752$

For $i = 5$ (fifth joint):

$$q_{51}(\tau) = -2.361\tau^4 + 4.722\tau^3 + 60$$

$$\dot{q}_{51} = \dot{q}_{51}(1) = 4.722$$

$$\ddot{q}_{51} = \ddot{q}_{51}(1) = 0$$

$$q_{52}(\tau) = 4.722\tau + 62.361$$

$$\dot{q}_{52} = \dot{q}_{52}(1) = 4.722$$

$$\ddot{q}_{52} = \ddot{q}_{52}(1) = 0$$

$$q_{53}(\tau) = 2.361\tau^4 - 4.722\tau^3 + 4.722\tau + 67.083$$

$f = 36.774,\quad g = 8.000,\quad \sigma = 4.722$

For $i = 6$ (sixth joint):

$$q_{61}(\tau) = 16.678\tau^4 - 33.356\tau^3 + 90$$

$$\dot{q}_{61} = \dot{q}_{61}(1) = -33.356$$

$$\ddot{q}_{61} = \ddot{q}_{61}(1) = 0$$

$$q_{62}(\tau) = -33.356\tau + 73.322$$

$$\dot{q}_{62} = \dot{q}_{62}(1) = -33.356$$

$$\ddot{q}_{62} = \ddot{q}_{62}(1) = 0$$

$$q_{63}(\tau) = -16.678\tau^4 + 33.356\tau^3 - 33.356\tau + 39.966$$

$f = -266.848,\quad g = 8.000,\quad \sigma = -33.356$

Note that $\dot{q}_{ij}$ and $\ddot{q}_{ij}$ have units [°/s] and [°/s^2] respectively.

The profiles of displacement, velocity and acceleration are presented in Figure 6.8

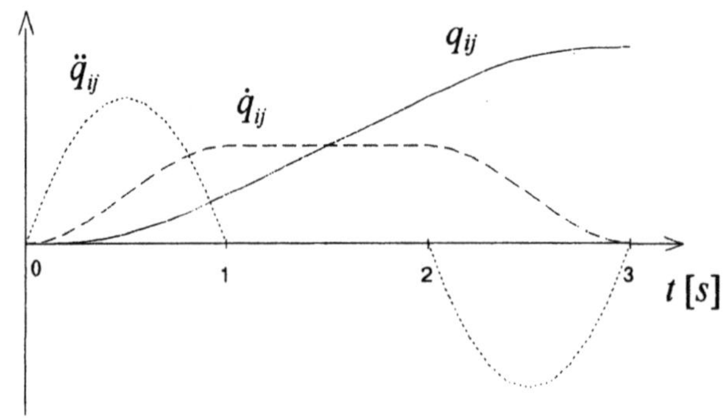

Fig. 6.8. Trajectory time profiles of the displacement (continuous line), velocity (dotted line) and acceleration (dashed line)

6.5. References

[6.1] *Asada H., Slotine J. E.*:Analysis and Control. New York, Wiley 1986.

[6.2] *Broderick P. L.,Cipra R. J.*: A Method for Determining and Correcting Robot Position and Orientation Errors Due to Manufacturing. Trans. ASME, J. Mech., Transm. and Autom. in Design, 86–DET–87.

[6.3] *Cimino W. W., Pennock G. R.*: Workspace of a Six-Revolute Decoupled Robot Manipulator. Proc. of the IEEE Int. Conf. on Robotics, San Francisco 1986, vol. 1, s. 1848–1952.

[6.4] *Craig J. J.*: Introduction to Robotics. Mechanics and Control. Addison-Wesley 1989

[6.5] *Fu K. S., Gonzales R. C., Lee C. S. G.*: Robotics: Control, Sensing, Vision and Inteligence. New York, McGraw-Hill 1987.

[6.6] *Kazerounian K., Qian G. Z.*: Kinematic Callibration of Robotic Manipulators. Trans. ASME, J. Mech, Transm., Autom. in Design, vol. 111, Dec.1989, s. 482–487.

[6.7] *Knapczyk J., Morecki A.*: Analysis of the Positioning and Orientation Accuracy in 6R Manipulators (Direct task). Proc. of the 6th CISM IFToMM Symp. on Theory and Practice of Robots and Manipulators. Edited by A. Morecki, G. Bianchi and K. Kędzior, The MIT Press, Cambridge 1987.

[6.8] *Knapczyk J., Lebiediew P. A.*: Theory Spatial Mechanisms and Manipulators. Warszawa, WNT, 1990 (*in Polish*).

[6.9] *Kumar A.*: Characteristics of Manipulator Geometry. Ph. D.Thesis, Univ. of Houston 1980.

[6.10] *Lee T. W., Yang D. C. H.*: On the Evaluation of Manipulator Workspace Trans. ASME, J. Mech., Transm., Aut. Design, vol. 105, 1983, s. 70.

[6.11] *Meng C. H., Born A. H.*: Statistical Characterization of Position Errors of an Assemble of Robots and Its Applications. Trans. ASME, J. Mech. Transm. Autom. in Design, vol. 11, June 1989, s. 215–222.

[6.12] *Ming-Shu Hsu, Kohli D.*: Boundary Surfaces and Accesibility Regions for Regional Structures of Manipulators. Mechanism and Machine Theory. vol. 22, No. 3, s. 277–289, 1987.

[6.13] *Ranky P. G., Ho C. Y.*: Robot Modelling, Control and Applications with Software. IFS LTD, Kempston 1985.

[6.14] *Rastegar J., Deravi P.*: The Effect of Jonit Motion Constraints on the Workspace and Number of Configurations of Manipulators. Mech. and Mach. Theory, vol. 22, No. 5, s. 401–409, 1987.

[6.15] *Tsai M. J., Waldron K. J.*: Geometric Characteristics of Industrial Robots. Trans. ASME, J. Mech., Transm., Aut.in Design, 86–DET–141.

[6.16] *Woo-Jong Lee, Woo T. C.*: Tolerance Volume Due to Joint Variable Errors in Robots. Trans. ASME, J. Mech., Transm., Autom. in Design, vol. 111, Dec.1989, s. 597–604.

[6.17] *Ranjbaran F., Angeles J.*: The Determination of the Cartesian Workspace of General Three–Axes Manipulators. 3rd Int. Workshops on Advances in Robot Kinematics, Ferrara 92, Proc., s. 318–323.

[6.18] *Wang H. S., Roth B.*: Position Errors of Manipulators. Proc. of the 7th CISM-IFtoMM Symp. RoManSy'88, Udine. Hermes 1990, s. 310–317.

[6.19] *Everett L., Hsu T.*: The theory of Kinematic Parameter Identification for Industrial Robots. Trans. ASME, J. Dyn. Syst., Measurements and Control, vol. 110, 1988, s. 96–99.

[6.20] *Szczepinski W.*: Theory of Polyhedrons of Positioning Accuracy of Manipulators. Mech. Mach. Theory, v.26, No 7, s. 697–709, 1991.

[6.21] *Gosselin C.M,* Hadj-Messaoud A.: Automatic Planning of Smooth Trajectories for Pick-and-Place Operations. Trans. ASME, Jnl of Mech. Design v.115, Sept 1993.

7. PLATFORM PARALLEL MANIPULATORS

7.1. Introduction

Parallel manipulators are closed chains with one or more loops where only some kinematic pairs are actively controlled. The most common arrangement comprises a base connected via serial, parallel, or hybrid (serial-parallel) chains to the output link (platform). Actuated pairs provide the platform with the desired degrees of freedom (DOF) relative to the base. Fully parallel manipulators are characterized by six binary legs, each attached to the base and the platform through spherical joints. There may be up to six legs of controllable, variable length, providing the platform with a corresponding number of dofs. Each leg can be regarded as a serial manipulator with six revolutes, or five revolutes and one prismatic pair. Usually, three of the revolutes with intersecting axes in one point constitute the spherical joint at the platform, while two more constitute a universal joint at the base, i.e. the concatenation of two revolutes with intersecting axes. Of these six joints, only one is actuated.

Various architectures have been developed, or proposed for platform parallel manipulators. The most general provides six legs, each with an actuated prismatic joint, connected to the base through six distinct universal joints, and to the mobile platform through six distinct spherical joints. This is the 6(S-S) manipulator, indicating the numbers of joints on the mobile platform and the base. When the six joints are replaced by three concentric spherical joints, the manipulator is referred to as the 3-6. Other variations are obtained through a different arrangement of the joints on base and mobile platform.

Mechanical systems, such as the one seen in Figure 7.1, were first introduced to test tires (Gough 1948); later Stewart (1965) suggested they be used as flight simulators. Potential applications for the parallel manipulator arise whenever there is a need for high structural stiffness, or high performance dynamics, and when it is desirable to have the actuators as close as possible to the base, for example, in mining machines, and vehicle, or airplane simulators. Many industrial applications demand powerful and compact multiple DOF systems to support and position universal fixtures and tool-holders. Fully parallel manipulators can maneuver heavy loads for complex tasks in machining and assembly processes. For example the NIST ROBOCRANE [19.23] can manipulate tools and devices to perform a variety of tasks, such as cutting, shaping and finishing, excavating and grading; it also acts as flexible fixturing. The ARTISAN [19.37] is a ten DOF manipulation system designed to support research in dynamical control and dexterous manipulation. The system includes a three DOF micro-manipulator, and a parallel actuated system placed between the wrist and the hand. This wrist/micro-manipulator combination is a hybrid series/ parallel manipulator, which forms a six DOF, high bandwidth and a high resolution manipulation system.

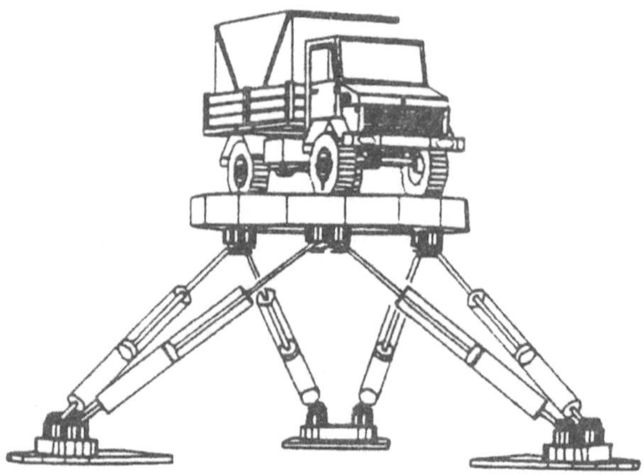

Fig.7.1. The 6 DOF parallel manipulator used as multi-axis test facility driven by six actuators

7.2. Planar platform manipulators

Potential applications for planar parallel manipulators include metal cutting, deburring, pick-and-place operations over a plane surface; they also serve as mobile bases for serial manipulators. A planar parallel manipulator consists of three base joints and three joints on the platform, are connected by three actuated legs, so that by changing the distances between the corresponding base and platform joints the platform has three DOF and, therefore, within a specified workspace, can take any position with any desired orientation.

There are two types of 3 DOF planar parallel manipulators, the 3(RRR) manipulator shown in Fig.7.2, and the 3(RPR) manipulator shown in Fig.7.5. The former has a greater load capacity, the latter, a greater workspace. As only the distances between the axes of the rotary joints at the leg ends feature in the equations, both types of mechanisms can be treated in the same way mathematically. The planar motion (translation and rotation) of the platform is determined by the time-dependent lengths of the legs in the case of the 3(RPR) mechanism, and the time-dependent configuration angles of the legs, in the case of the 3(RRR) mechanism.

7.2.1. Direct kinematic analysis of the 3(RRR) planar parallel manipulator

The direct position analysis (DPA) of the 3(RRR) manipulator can be stated in the following way: given the three base points O_1, O_2, O_3 in an arbitrary fixed base coordinate system, the three platform points O_7, O_8, O_9 in an arbitrary platform coordinate system, and the distances O_1O_7, O_2O_8 and O_3O_9 (the leg lengths), find the position(s) of platform.

The fixed Cartesian frame of reference is denoted by (xy); the origin is coincident with O_1 (base joint) and oriented so that the x axis is coincident with the line $O_1\, O_2$. The input angular displacements are denoted by θ_i (i = 1, 2, 3), and the output angular displacements, by θ_j (j = 4, 5, 6, 7). All displacement variables are measured from the x - axis and are positive in the counterclockwise direction.

To obtain the platform angular displacement variable θ_7 for a given set of input displacements, we must consider two of the three loop-closure equations. For example, for two loops: $O_1\, O_4\, O_7\, O_8\, O_5\, O_2\, O_1$ and $O_2\, O_5\, O_8\, O_9\, O_6\, O_3\, O_2$, we can write the following four scalar eqs:

$$
\begin{aligned}
&a_{58}\, c\theta_5 - a_{69}\, c\theta_6 + a_{89}\, c(\theta_7 + \gamma_7 + \gamma_9) = a_{23}\, c\gamma_2 - a_{25}\, c\theta_2 + a_{36}\, c\theta_3 \\
&a_{58}\, s\theta_5 - a_{69}\, s\theta_6 + a_{89}\, s(\theta_7 + \gamma_7 + \gamma_9) = a_{23}\, s\gamma_2 - a_{25}\, s\theta_3 + a_{36}\, s\theta_3 \\
&- a_{47}\, c\theta_4 + a_{58}\, c\theta_5 - a_{78}\, c\theta_7 = - a_{12} + a_{14}\, c\theta_1 - a_{25}\, c\theta_2 \\
&- a_{47}\, s\theta_4 + a_{58}\, c\theta_5 + a_{78}\, s\theta_7 = a_{14}\, s\theta_1 - a_{25}\, s\theta_2
\end{aligned}
\tag{7.1}
$$

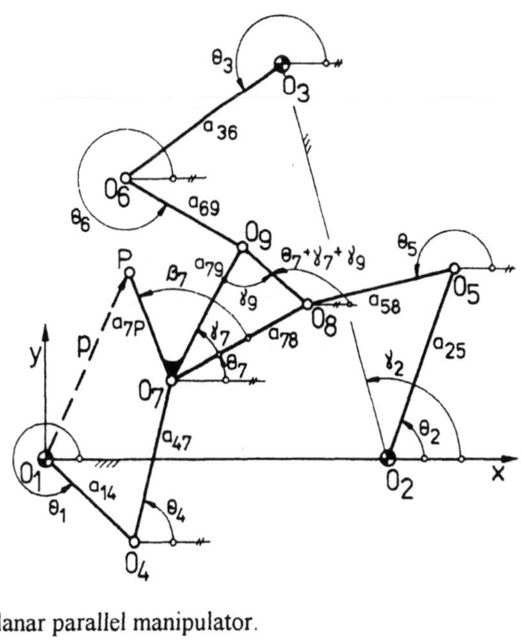

Fig.7.2. The 3(RRR) planar parallel manipulator.

where:

$c\theta_i = cos\theta_i$, $s\theta_i = sin\theta_i$

Now we eliminate θ_4, θ_5 and θ_6, using the following procedure:
The equations (7.1) are rearranged in the form:

$$\begin{aligned} c\theta_6 &= (a_{58}c\theta_5 + u_1)/a_{69} \\ s\theta_6 &= (a_{58}s\theta_5 + u_2)/a_{69} \\ c\theta_4 &= (a_{58}c\theta_5 + u_3)/a_{47} \\ s\theta_4 &= (a_{58}c\theta_5 + u_4)/a_{47} \end{aligned} \tag{7.2}$$

where

$$\begin{aligned} u_1 &= a_{89}c(\theta_7 + \gamma_7 + \gamma_9) - f_1, & f_1 &= a_{23}c\gamma_2 - a_{25}c\theta_2 + a_{36}c\theta_3 \\ u_2 &= a_{89}s(\theta_7 + \gamma_7 + \gamma_9) - f_2, & f_2 &= a_{23}s\gamma_2 - a_{25}s\theta_2 + a_{36}s\theta_3 \\ u_3 &= - a_{78}c\theta_7 - f_3, & f_3 &= - a_{12} + a_{14}c\theta_1 - a_{25}c\theta_2 \\ u_4 &= - a_{78}s\theta_7 - f_4, & f_4 &= a_{14}s\theta_1 - a_{25}s\theta_2 \end{aligned} \tag{7.2a, b}$$

with f_k $(k = 1, 2, 3, 4)$ as the functions of input variables θ_i $(i = 1, 2, 3)$, and u_k $(k = 1, 2, 3, 4)$ as the functions of the output variable θ_7.

Squaring and adding the corresponding equations (7.2), and applying the trigonometric identities $s^2\theta_i + c^2\theta_i = 1$ $(i = 4, 6)$, yields

$$\begin{aligned} 2a_{58}(u_1c\theta_5 + u_2s\theta_5) + u_5 &= 0 \\ 2a_{58}(u_3c\theta_5 + u_4s\theta_5) + u_6 &= 0 \end{aligned} \tag{7.3}$$

where

$$\begin{aligned} u_5 &= u_1^2 + u_2^2 + a_{58}^2 - a_{69}^2 \\ u_6 &= u_3^2 + u_4^2 + a_{58}^2 - a_{47}^2 \end{aligned} \tag{7.3a}$$

Writing equations (7.3) in matrix form and using Cramer`s rule gives

$$c\theta_5 = \Delta_1/\Delta, \qquad s\theta_5 = \Delta_2/\Delta \tag{7.4}$$

where

$$\begin{aligned} \Delta &= 2a_{58}(u_1u_4 - u_2u_3) \\ \Delta_1 &= u_2u_6 - u_4u_5 \\ \Delta_2 &= u_3u_5 - u_1u_6 \end{aligned} \tag{7.4a}$$

Squaring and adding the corresponding equations (7.4), and applying the trigonometric identity $s^2\theta_5 + c^2\theta_5 = 1$, yields

$$F(\theta_7) \equiv \Delta_1^2 + \Delta_2^2 - \Delta^2 = 0 \tag{7.5}$$

Substituting the relations (7.4a) and (7.2a) into equation (7.5) gives cubic equation in $s\theta_7$ and $c\theta_7$

$$F(\theta_7) = A_1s^3\theta_7 + A_2s^2\theta_7c\theta_7 + A_3s^2\theta_7 + A_4s\theta_7c\theta_7 + A_5s\theta_7 + A_6c\theta_7 + A_7 \tag{7.6}$$

where A_i $(i = 1, 2, ..., 7)$ are expressed in terms of the input variables θ_i $(i = 1, 2, 3)$. Their analytical forms are very complicated and are not shown here.

The values of A_i can be determined by interpolation in the following way: for the given values of θ_i ($i = 1, 2, 3$) the corresponding values of f_k ($k=1, 2, 3, 4$) are calculated using the formula (7.2b); assuming seven values of θ_7, the corresponding values of $F(\theta_7)$ are calculated using equation (7.5), applying (7.2a) for u_k ($k=1, 2, 3, 4$), and (7.4a) for Δ, Δ_1 and Δ_2 ; we can now solve the system of seven linear equations with seven unknowns A_i $i=1, 2, ..., 7$, by substituting the assumed values of θ_7 and calculated values of $F(\theta_7)$ into equation (7.6).

The calculation procedure can be simplified by taking the values of θ_7 that give the specified values of $s\theta_7$ and $c\theta_7$. For example, 0 and 1; 0 and -1; -1 and 0; 1 and 0; 0,6 and 0,8; 0,6 and -0,8; -0,6 and 0,8; -0,6 and -0,8. Substituting the specified values for $s\theta_7$ and $c\theta_7$ in equation (7.6), we obtain the system of seven linear equations in terms of A_i ($i=1, 2, ...7$), which gives the following solution:

$$A_1 = \frac{5}{96}\left[16(F_1 + F_2) + 24F_3 - 6F_4 - 25(F_5 + F_6)\right]$$

$$A_2 = \frac{5}{72}\left[-36F_1 + 4F_2 - 9(F_3 + F_4) + 25(F_5 + F_7)\right]$$

$$A_3 = \frac{1}{2}(-F_1 - F_2 + F_3 + F_4) \tag{7.7}$$

$$A_4 = \frac{1}{24}\left[16(F_1 + F_2) + 9(F_3 + F_4) - 25(F_6 + F_7)\right]$$

$$A_5 = \frac{1}{2}(F_3 - F_4) - A_1, \quad A_6 = \frac{1}{2}(F_1 - F_2), \quad A_7 = \frac{1}{2}(F_1 + F_2)$$

Then, using the trigonometric identity $s^2\theta_7 + c^2\theta_7 = 1$, equation (7.6) yields a sixth order polynomial in

$$z = s\theta_7$$

$$F(z) = B_1 z^6 + B_2 z^5 + B_3 z^4 + B_4 z^3 + B_5 z^2 + B_6 z + B_7 = 0 \tag{7.8}$$

where:

$$B_1 = A_1^2 + A_2^2$$

$$B_2 = 2(A_1A_3 + A_2A_4)$$

$$B_3 = A_3^2 + A_4^2 - A_2^2 + 2A_1A_5 + 2A_2A_6$$

$$B_4 = 2(A_3A_5 + A_4A_6 + A_1A_7 - A_2A_4) \tag{7.8a}$$

$$B_5 = A_5^2 + A_6^2 - A_4^2 + 2A_3A_7 - 2A_2A_6$$

$$B_6 = 2(A_5A_7 - A_4A_6)$$

$$B_7 = A_7^2 - A_6^2$$

There are six possible values of θ_7 that can satisfy equation (7.8); i.e. six assembly configurations are possible for a given set of input variables. This may be interpreted geometrically as the prescribed position of the input links fixes the positions of points O_4, O_5 and O_6. The remainder of the mechanism can then be regarded as a four-bar linkage $O_4O_7O_8O_5$ and a binary link O_6O_9. Since a circle can intersect the sixth-order coupler curve at up to six real points, six assembly configurations are possible. The point of tangency corresponds to the real repeated roots. The stationary configurations are important in determining of the workspace.

The configuration variables of the legs can be calculated for any given value of θ_7 by using equations (7.4) to find the corresponding value θ_5, and equation (7.2) to find θ_4 and θ_6.

If the position of the 3(RRR) manipulator is known, and the input variables are given as a function of time, then the angular velocity of the output links $\dot{\theta}_j$ ($j = 4, 5, 6, 7$) can be derived by differentiating equations (7.1) with respect to time:

$$\sum_{j=4}^{7} \frac{\partial l_k}{\partial \theta_j} \dot{\theta}_j = \sum_{i=1}^{3} \frac{\partial r_k}{\partial \theta_i} \dot{\theta}_i \tag{7.9}$$

where l_k and r_k are the left and right side of k-th equation (7.1)

Introducing the vectors of the input and output variables, we can write equation (7.9) as matrix equation

$$\boldsymbol{A}\,[\dot{\theta}_4 \;\; \dot{\theta}_5 \;\; \dot{\theta}_6 \;\; \dot{\theta}_7]^{\mathrm{T}} = \boldsymbol{B}\,[\dot{\theta}_1 \;\; \dot{\theta}_2 \;\; \dot{\theta}_3]^{\mathrm{T}} \tag{7.10}$$

where

$$\mathrm{A} = \begin{bmatrix} 0 & -a_{58}s\theta_5 & a_{69}s\theta_6 & -a_{89}s(\theta_7+\gamma_7+\gamma_9) \\ 0 & a_{58}c\theta_5 & -a_{69}c\theta_6 & a_{89}c(\theta_7+\gamma_7+\gamma_9) \\ a_{47}s\theta_4 & -a_{58}s\theta_5 & 0 & a_{78}s\theta_7 \\ -a_{47}c\theta_4 & a_{58}c\theta_5 & 0 & -a_{78}c\theta_7 \end{bmatrix}$$

$$\mathrm{B} = \begin{bmatrix} 0 & a_{25}s\theta_2 & -a_{36}s\theta_3 \\ 0 & -a_{25}c\theta_2 & a_{36}c\theta_3 \\ -a_{14}s\theta_1 & a_{25}s\theta_2 & 0 \\ a_{14}c\theta_1 & -a_{25}c\theta_2 & 0 \end{bmatrix}$$

We then solve the resultant system of linear equation with respect to $\dot{\theta}_j$:

$$[\dot{\theta}_4 \;\; \dot{\theta}_5 \;\; \dot{\theta}_6 \;\; \dot{\theta}_7]^{\mathrm{T}} = (\boldsymbol{A}^{-1}\boldsymbol{B})\,[\dot{\theta}_1 \;\; \dot{\theta}_2 \;\; \dot{\theta}_3]^{\mathrm{T}} = \boldsymbol{J}\,[\dot{\theta}_1 \;\; \dot{\theta}_2 \;\; \dot{\theta}_3]^{\mathrm{T}} \tag{7.11}$$

where $\boldsymbol{J} = \boldsymbol{A}^{-1}\boldsymbol{B}$ is the Jacobian matrix representing the transformation mapping the input angular velocities into output angular velocities.

The singular configuration of the 3(RRR) manipulator occurs when $\det[A] = 0$, then

$$a_{78}\, s(\theta_6 - \theta_5)\, s(\theta_7 + \gamma_7 - \theta_4) - a_{89}\, s(\theta_4 - \theta_5)\, s(\theta_7 + \gamma_7 + \gamma_9 - \theta_6) = 0 \tag{7.12}$$

The conclusion is that, for a singular configuration, lines O_4O_7 , O_5O_8 and O_6O_9 must either intersect at a common point, or be parallel, i.e. intersect at infinity.

The position of platform point P with respect to the origin of the base coordinate system is described by vector $\boldsymbol{p} = [x_P\ y_P]^T$, where

$$x_P = a_{14}c\theta_1 + a_{47}c\theta_4 + a_{7P}c(\theta_7+\beta_7)$$
$$y_P = a_{14}s\theta_1 + a_{47}s\theta_4 + a_{7P}s(\theta_7+\beta_7) \qquad (7.13)$$

The velocity and acceleration of platform point P can be obtained by differentiating equation (7.13) with respect to time.

Numerical example 7.1. The link lengths of the 3(RRR) manipulator are specified [7.34]: $a_{12}= 256$, $a_{23}= 308$, $a_{14}= 184$, $a_{25}= 160$, $a_{36}= 144$, $a_{47}= 220$
$a_{58}= 116$, $a_{69}= 212$, $a_{78}= 224$, $a_{89}= 204$, $a_{79}= 413$, $a_{7P}= 224$
$a_{8P}= 156$, $a_{9P}= 236$, $\gamma_2= 10°$, $\gamma_7+\gamma_9= 30°$

The input variables are given as follows:

$\theta_1 = 5°$, $\theta_2 = 20°$, $\theta_3 = -10°$, and $\dot{\theta}_1 = 1$ rad/s, $\dot{\theta}_2 = 5$ rad/s, $\dot{\theta}_3 = -2$ rad/s .

Using the above procedure, we obtain six corresponding angular displacements of the platform: θ_7 = -64,29°; 36,78°; -27,54°; 40,58°; -44,54°; -16,50°.

The configuration angles of the legs are calculated, and the values obtained corresponding to θ_7 = -64,29°, are: θ_4 = 37,82°, θ_5 = -66,36°, θ_6 =-112,32°.

The six solutions are shown in Figure 7.3 as the intersections of the coupler curves and the corresponding circular loci.

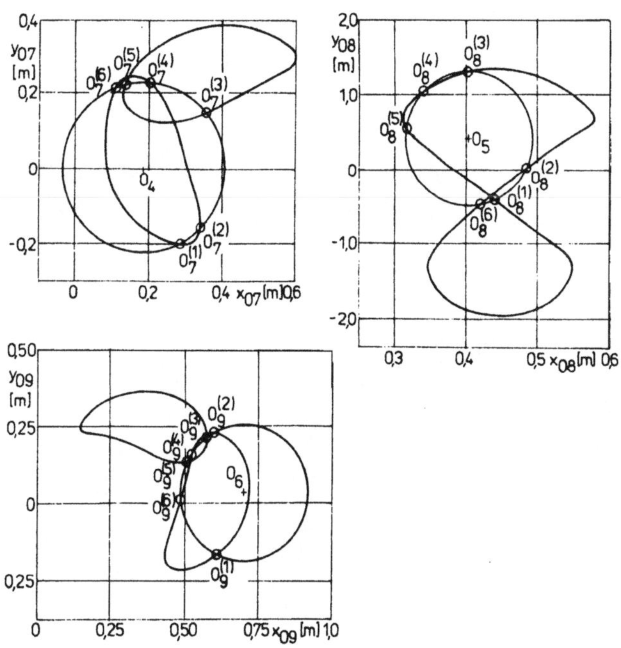

Fig.7.3. The coupler curves of platform points O_7, O_8 and O_9 of the 3(RRR) manipulator and their intersection points $O_i^{(j)}$ (i = 7, 8, 9; j = 1, 2, ... 6) by the circles with radius $O_i\,O_{i+3}$ (i = 4, 5, 6)

The angular velocity of the platform is calculated using equation (7.11) and taking the first configuration with $\theta_7 = -64.29°$, $\theta_4 = 37.82°$, $\theta_5 = -66.36°$, $\theta_6 = 112.32°$

$$\dot{\theta}_7 = -0.20461\dot{\theta}_1 - 0.22685\dot{\theta}_2 + 0.38925\dot{\theta}_3 = -2.12 \text{ rad/s}$$

The position and velocity of platform point P are calculated using equation (7.12) and its time derivative

$$x_P = 561.6, \qquad y_P = 60.6 \text{ [mm]}$$

$$\dot{x}_p = 74.90\dot{\theta}_1 - 122.44\dot{\theta}_2 + 36.79\dot{\theta}_3 = -610{,}9 \text{ [mm/s]}$$

$$\dot{y}_p = 0.490\dot{\theta}_1 + 84.91\dot{\theta}_2 + 77.53\dot{\theta}_3 = 270 \text{ [mm/s]}$$

Numerical example 7.2. The link lengths of the 3(RRR) manipulator are given: $a_{47} = 50$; $a_{58} = 78$; $a_{69} = 70$; $a_{78} = 135$; $a_{89} = 70$; $\gamma_7 = -15°$;
the coordinates of leg joints are calculated for the given input variables:

$$x_{O4} = a_{14}\,c\theta_1 = 91.5, \qquad y_{O4} = a_{14}\,s\theta_1 = -122$$

$$x_{O5} = a_{25}\,c\theta_2 = -10, \qquad y_{O5} = a_{25}\,s\theta_2 = 0$$

$$x_{O6} = a_{36}c\theta_3 = 19.5, \qquad y_{O6} = a_{36}\,s\theta_3 = -122$$

Using the procedure presented above, we obtain six corresponding angular displacements of the platform: θ_7 = 12.64°; 13.23°; 347.74°; 275.87°; 292.49°; 334.73° The six respective configurations of the platform are presented in Figure 7.4.

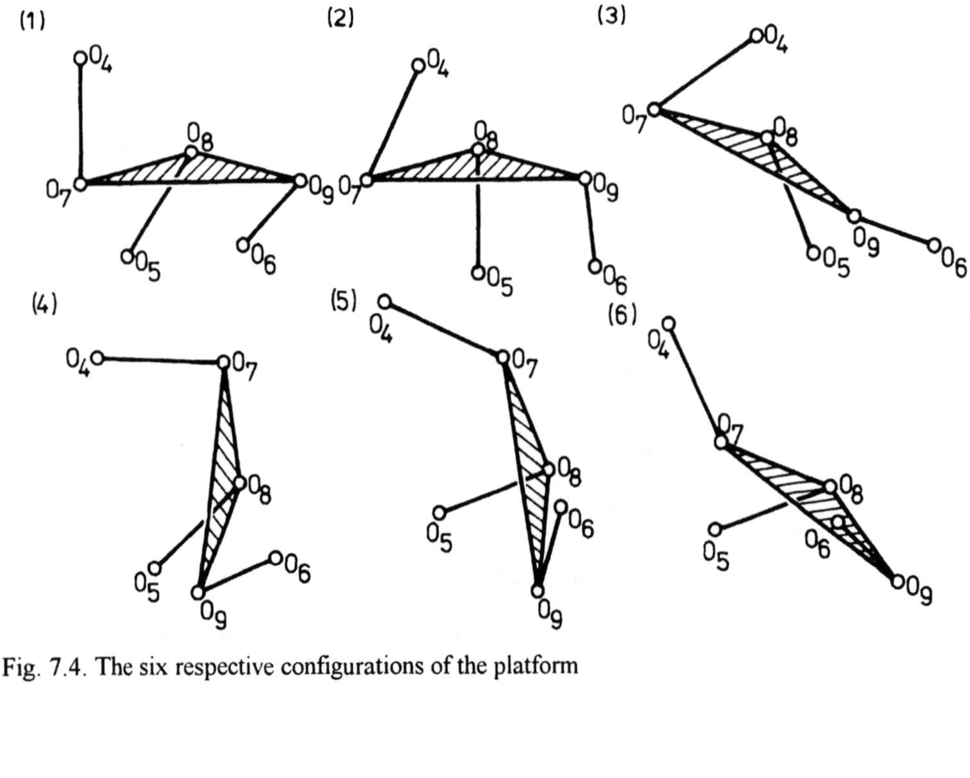

Fig. 7.4. The six respective configurations of the platform

7.2.2. Direct kinematic analysis of the 3(RPR) manipulator [7.28]

The motion (translation and rotation) of the platform is determined by the time-dependent lengths $d_i(t) = O_iO_{i+6}$ ($i=1, 2, 3$) of the legs. We assume that: x_i , y_i are the coordinates of three base joints O_i described in the base reference system, and x_{i+6}^*, y_{i+6}^*, the coordinates of three platform joints O_{i+6} described in the platform reference system Px^*y^* (see Figure 7.5).

The position of the platform relative to the base can be expressed by x_P, y_P, the coordinates of platform point P, and ϕ, the rotation angle. The values x_P, y_P and ϕ can serve therefore as the output variables to be determined for a given set of input variables d_i (i=1, 2, 3).

The input variables and output variables are interconnected by the following conditions:

$$[(x_i - x_P) - (x_{i+6}^* c\phi - y_{i+6}^* s\phi)]^2 + [(y_i - y_P) - (x_{i+6}^* s\phi + y_{i+6}^* c\phi)]^2 - d_i^2 = = G_i = 0, \quad (i=1, 2, 3) \tag{7.14}$$

Equation (7.14) can be rearranged in the form

$$(x_{i+6}^* c\phi - y_{i+6}^* s\phi - x_i)\, x_P + (x_{i+6}^* s\phi + y_{i+6}^* c\phi - y_i) y_P + 0.5(x_{i+6}^*)^2 + +0.5\,(y_{i+6}^*)^2 + A_{i+6}\, c\phi + B_{i+6}\, s\phi + C_{i+6} = 0, \quad i=1, 2, 3 \tag{7.14a}$$

where

$$A_{i+6} = -(x_{i+6}^* x_i + y_{i+6}^* y_i)$$
$$B_{i+6} = -(x_{i+6}^* y_i - y_{i+6}^* x_i)$$
$$C_{i+6} = [(x_{i+6}^*)^2 + (y_{i+6}^*)^2 + x_i^2 + y_i^2 - d_i^2]/2 \tag{7.14b}$$

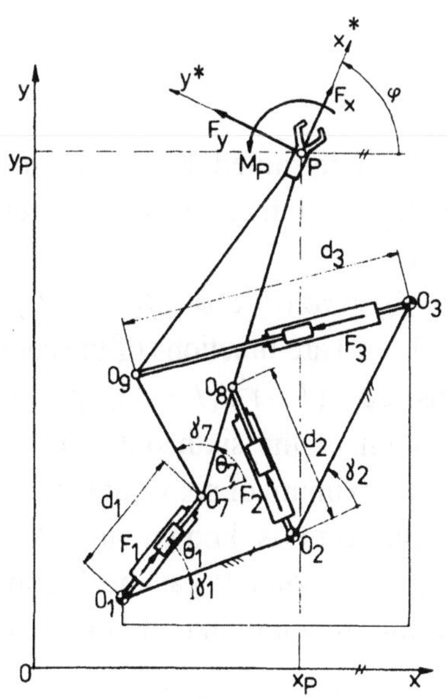

Fig. 7.5. The 3(RPR) planar parallel manipulator

Two equations that are linear in the output variables x_P and y_P and independent from each other can be derived from the set of equations (7.14) by subtracting the second and the third equation from the first:

$$D_{11} x_P + D_{12} y_P = -D_{13}$$
$$D_{21} x_P + D_{22} y_P = -D_{23} \qquad (7.15)$$

where

$$D_{11} = x_{78}^{*} c\phi - y_{78}^{*} s\phi - x_{12}$$
$$D_{12} = x_{78}^{*} s\phi + y_{78}^{*} c\phi - y_{12}$$
$$D_{13} = A_{78} c\phi + B_{78} s\phi + C_{78} \qquad (7.15a)$$
$$D_{21} = x_{89}^{*} c\phi - y_{89}^{*} s\phi - x_{23}$$
$$D_{22} = x_{89}^{*} s\phi + y_{89}^{*} c\phi - y_{23}$$
$$D_{23} = A_{89} c\phi + B_{89} s\phi + C_{89}$$

and

$$x_{78}^{*} = x_7^{*} - x_8^{*}, \qquad y_{78}^{*} = y_7^{*} - y_8^{*}$$
$$x_{89}^{*} = x_8^{*} - x_9^{*}, \qquad y_{89}^{*} = y_8^{*} - y_9^{*}$$
$$x_{12} = x_1 - x_2, \qquad y_{12} = y_1 - y_2 \qquad (7.15b)$$
$$x_{23} = x_2 - x_3, \qquad y_{23} = y_2 - y_3$$
$$A_{78} = A_7 - A_8, \qquad B_{78} = B_7 - B_8, \qquad C_{78} = C_7 - C_8$$
$$A_{89} = A_8 - A_9, \qquad B_{89} = B_8 - B_9, \qquad C_{89} = C_8 - C_9$$

The solution of (7.15) is

$$x_P = \Delta_1/\Delta, \qquad y_P = \Delta_2/\Delta \qquad (7.16)$$

where

$$\Delta = D_{11}D_{22} - D_{21}D_{12}$$
$$\Delta_1 = -D_{13}D_{22} + D_{12}D_{23} \qquad (7.17)$$
$$\Delta_2 = -D_{11}D_{23} + D_{13}D_{21}$$

Substituting x_P and y_P as determined by (7.16) into the first (i=1) equation (7.14) leads to the final equation which allows us to determine the rotation angle ϕ. This equation has the form of

$$(f_1 c^3\phi + f_2 c^2\phi + f_3 c\phi + f_4) + s\phi (f_5 c^2\phi + f_6 c\phi + f_7) = \mathrm{F}(\phi) = 0 \qquad (7.18)$$

The coefficients f_k (k=1, 2, ...7) are functions of the coordinates x_i, y_i and x_{i+6}^{*}, y_{i+6}^{*} (i = 1, 2, 3). The substitutions: $c\phi = (1 - t^2)/(1 + t^2)$, $s\phi = 2t/(1 + t^2)$ transform equation (7.18) into an algebraic equation of the sixth order in $t = tg(\phi/2)$, which allows the numerically determination of a maximum of six values for ϕ, the platform rotation angle.

If the position of the platform is known and the leg lengths are given as functions of time, the velocity of point P can be found by solving a set of three linear equations. Differentiation of equation (7.13) with respect to time yields

$$\frac{\partial G_i}{\partial x_p}\dot{x}_p + \frac{\partial G_i}{\partial y_p}\dot{y}_p + \frac{\partial G_i}{\partial \varphi}\dot{\varphi} = 2d_i\dot{d}_i, \quad i = 1, 2, 3 \qquad (7.19)$$

The equation (7.19) can be written in matrix from

$$A \cdot \dot{p} = B \cdot \dot{d} \tag{7.20}$$

where

$$A = \begin{bmatrix} \frac{\partial G_i}{\partial x_p} & \frac{\partial G_i}{\partial y_p} & \frac{\partial G_i}{\partial \varphi} \end{bmatrix}, \; i = 1, 2, 3$$

$$\dot{p} = \begin{bmatrix} \dot{x}_p & \dot{y}_p & \dot{\varphi} \end{bmatrix}^T$$

$$B = \begin{bmatrix} 2d_1 & 0 & 0 \\ 0 & 2d_2 & 0 \\ 0 & 0 & 2d_3 \end{bmatrix}$$

$$\dot{d} = \begin{bmatrix} \dot{d}_1 & \dot{d}_2 & \dot{d}_3 \end{bmatrix}^T \tag{7.21}$$

The solution of equation (7.20) yields

$$\dot{\mathrm{p}} = (\mathrm{A}^{-1}\mathrm{B})\dot{\mathrm{d}} = \mathrm{J}\begin{bmatrix} \dot{d}_1 & \dot{d}_2 & \dot{d}_3 \end{bmatrix}^T \tag{7.22}$$

where $J = A^{-1}B$ is the Jacobian matrix of the 3(RPR) manipulator.

Differentiation with respect to time of equation (7.20) gives

$$\left(\frac{\partial A}{\partial x_p}\dot{x}_p + \frac{\partial A}{\partial y_p}\dot{y}_p + \frac{\partial A}{\partial \varphi}\dot{\varphi} \right)\dot{p} + A\ddot{p} = \dot{B}\dot{d} + B\ddot{d} \tag{7.23}$$

Equation (7.23) can be solved with respect to the acceleration of point P.:

$$\ddot{p} = A^{-1}\left[\dot{B}\dot{d} - \left(\frac{\partial A}{\partial x_p}\dot{x}_p + \frac{\partial A}{\partial y_p}\dot{y}_p + \frac{\partial A}{\partial \varphi}\dot{\varphi} \right)\dot{p} \right] + A^{-1}B\ddot{d} \tag{7.24}$$

where

$$\ddot{p} = \begin{bmatrix} \ddot{x}_p & \ddot{y}_p & \ddot{\varphi} \end{bmatrix}^T$$

$$\dot{B} = \begin{bmatrix} 2\dot{d}_1 & 0 & 0 \\ 0 & 2\dot{d}_2 & 0 \\ 0 & 0 & 2\dot{d}_3 \end{bmatrix}$$

$$\ddot{d} = \begin{bmatrix} \ddot{d}_1 & \ddot{d}_2 & \ddot{d}_3 \end{bmatrix}^T \tag{7.25}$$

Numerical example 7.3. The coordinates of points O_i (i=1, 2, 3) are given in the base reference system Oxy, and the coordinates of points O_{i+6} are given in the platform coordinate system Px^*y^*:

$x_1 = 83$, $y_1 = 56$, $x_2 = 243$, $y_2 = 108$, $x_3 = 350$, $y_3 = 310$;
$x_7^* = -310$, $y_7^* = -430$, $x_8^* = -209$, $y_8^* = -290$, $x_9^* = -238$, $y_9^* = 60$ mm

The leg lengths are given as input variables with the values:

$d_1 = 112$, $d_2 = 142$, $d_3 = 264$ mm

The equation (7.17) takes the form

$$1.1524\, c^3\phi - 2.9832\, c^2\phi + 2.3457\, c\phi - 0.5242 + s\phi\,(\,-1.2497\, c^2\phi + $$
$$+ 1.7391\, c\phi - 0.5274) = 0$$

This equation has six real solutions:

$$\phi = 63.43°, \quad 18.62°, \quad -23.63°, \quad -90.30°, \quad -50.03°, \quad -14.73°$$

The corresponding coordinates of point P are then:

$$x_P = 246.2, \quad 331.5, \quad 303.8, \quad 134.8, \quad 425.4, \quad 475.8$$
$$y_P = 440.4, \quad 303.2, \quad 49.1, \quad -142.7, \quad -173.5, \quad -57.5$$

Considering the first configuration with the following input and output variables:

$$x_P = 246.2, \quad y_P = 440.4; \quad \phi = 63.43°, \quad d_1 = 112, \quad d_2 = 142, \quad d_3 = 264$$

from equation (7.22) we obtain

$$\dot{x}_p = -2{,}4248\, \dot{d}_1 + 1{,}2040\, \dot{d}_2 - 3{,}1591\, \dot{d}_3$$
$$\dot{y}_p = 1{,}0341\, \dot{d}_1 + 0{,}3756\, \dot{d}_2 + 0{,}5411\, \dot{d}_3$$
$$\dot{\varphi} = 0{,}1406\, \dot{d}_1 - 0{,}0834\, \dot{d}_2 + 0{,}1292\, d_3$$

and from equation (7.24) with $\dot{d}_1 = -0{,}12;\ \dot{d}_2 = -0{,}09;\ \dot{d}_3 = 0{,}14$ m/s

$$\ddot{x}_p = -0{,}0198 - 2{,}4248\, \ddot{d}_1 + 1{,}2040\, \ddot{d}_2 - 3{,}1591\, \ddot{d}_3$$
$$\ddot{y}_p = -2{,}1739 + 1{,}0341\, \ddot{d}_1 + 0{,}3756\, \ddot{d}_2 + 0{,}5411\, \ddot{d}_3$$
$$\ddot{\varphi} = -0{,}05572 + 0{,}1406\, \ddot{d}_1 - 0{,}0834\, \ddot{d}_2 + 0{,}1292\, \ddot{d}_3$$

7.2.3. Inverse kinematic analysis of the 3(RRR) manipulator

It is now well known that the inverse kinematic problem of the parallel manipulators is, in general, much easier to solve than the direct one. In the inverse position problem of the 3(RRR) manipulator the input variables $\boldsymbol{p} = [x_P\ y_P]^T$, the position of point P, and $\phi = \theta_7$, the orientation angle of the platform with respect to the base coordinate system, are given; the corresponding configuration angles θ_i (i = 1, 2, 3) must be solved. The solution of this problem gives a maximum of eight configurations.

The coordinates of the platform joints O_{i+6} (i = 1, 2, 3) are readily computed as functions of the input variables

$$\begin{aligned}
x_7 &= x_P - a_{7P}\, c(\theta_7 - \beta_7) \\
y_7 &= y_P - a_{7P}\, s(\theta_7 + \beta_7) \\
x_8 &= x_P - a_{7P}\, c(\theta_7 + \beta_7) + a_{78}\, c\theta_7 \\
y_8 &= y_P - a_{7P}\, s(\theta_7 + \beta_7) + a_{78}\, s\theta_7 \\
x_9 &= x_P - a_{7P}\, c(\theta_7 + \beta_7) + a_{79}\, c(\theta_7 + \gamma_7) \\
y_9 &= y_P - a_{7P}\, s(\theta_7 + \beta_7) + a_{79}\, s(\theta_7 + \gamma_7)
\end{aligned} \tag{7.26}$$

These coordinates can be written as functions of the output variables, i.e. the leg configuration angles

$$x_{i+6} = a_{i,i+3}\, c\theta_i + a_{i+3,i+6}\, c\theta_{i+3}$$
$$y_{i+6} = a_{i,i+3}\, s\theta_i + a_{i+3,i+6}\, s\theta_{i+3},\ i=1, 2, 3 \quad (7.27)$$

Using the identities $s^2\theta_{i+3} + c^2\theta_{i+3} = 1$, the angles θ_{i+3} ($i = 1, 2, 3$) can be eliminated from (7.27) and trigonometric equations for θ_i written in the form

$$A_i\, c\theta_i + B_i\, s\theta_i = D_i,\ i = 1, 2, 3 \quad (7.28)$$

where

$$A_1 = x_7 = x_P - a_{7P}\, c(\theta_7 + \beta_7)$$
$$B_1 = y_7 = y_P - a_{7P}\, s(\theta_7 + \beta_7)$$
$$D_1 = (A_1^2 + B_1^2 + a_{14}^2 - a_{47}^2)/2a_{14}$$
$$A_2 = x_P - a_{7P}\, c(\theta_7 + \beta_7) + a_{78}\, c\theta_7 - a_{12} \quad (7.28a)$$
$$B_2 = y_P - a_{7P}\, s(\theta_7 + \beta_7) + a_{78}\, s\theta_7$$
$$D_2 = (A_2^2 + B_2^2 + a_{25}^2 - a_{58}^2)/2a_{25}$$
$$A_3 = x_P - a_{7P}\, c(\theta_7 + \beta_7) + a_{79}\, c(\theta_7 + \gamma_7) - a_{12} - a_{23} c\gamma_2$$
$$B_3 = y_P - a_{7P}\, s(\theta_7 + \beta_7) + a_{79}\, s(\theta_7 + \gamma_7) - a_{23}\, s\gamma_2$$
$$D_3 = (A_3^2 + B_3^2 + a_{36}^2 - a_{69}^2)/2a_{36}$$

The solutions of equations (7.28) yield

$$\theta_i = \mathrm{Atan2}(A_i, -B_i) - \mathrm{Atan2}(D_i, \pm\sqrt{A_i^2 + B_i^2 - D_i^2}),\ (i= 1, 2, 3) \quad (7.29)$$

7.2.4. Workspace analysis of the 3(RRR) manipulator

The total (or reachable) workspace of a planar manipulator is the area that comprises all points which a fixed point in the end-effector can reach. The region of the workspace where the end-effector can assume any arbitrary orientation is referred to as the primary (or dextrous) workspace. The remainder of the workspace, where platform orientation is restricted, is called the secondary workspace. The dexterity of the manipulator is limited near the boundaries of the workspace, and at the boundary the manipulator looses one, or more, DOF.

The mobility region can be found for each leg. This is the region delimited by the singularity curve, which is the closed curve separating the region where the leg has mobility from that in which it does not. For a given leg to have mobility at the prescribed Cartesian coordinates $[x_P\ y_P\ \phi]^T$ of the platform, the expression under square root in (7.29) must be positive. Thus the area of the workspace, for the i-th leg is obtained by writing

$$A_i^2 + B_i^2 - D_i^2 = 0,\ i=1, 2, 3 \quad (7.30)$$

where A_i, B_i and D_i are given by equations (7.28a)

The workspace of the manipulator considered, dashed in Fig.7.6a, is then obtained by the intersection of the above three mobility regions. Each of these regions is bounded by two concentric circles with centers, C_i $(x_P - A_i\ ,\ y_P - B_i\ )$ described by the equation

$$(x - x_P + A_i)^2 + (y - y_P + B_i)^2 = (a_{i,i+3} \pm a_{i+3,i+6})^2 \qquad (7.31)$$

The concentric circles are obtained by choosing alternatively the plus and minus sign in equation (7.31), which gives rise to the radii of the outer and inner circles $(a_{i,i+3} + a_{i+3,i+6})$ and det $(a_{i,i+3} - a_{i+3,i+6})$, respectively. The centers, C_i , are located on the circumference of three other circles, centered at the base joints O_i . Their location on this circumference depends on the end-effector orientation given by the angle $\phi = \theta_7$.

Example 7.4. The 3(RRR) manipulator is symmetric, i.e., the base joints O_i and the platform joints O_{i+6} (i=1, 2, 3) are located on the vertices of equilateral triangles, and all the legs have the same link length [7.2].

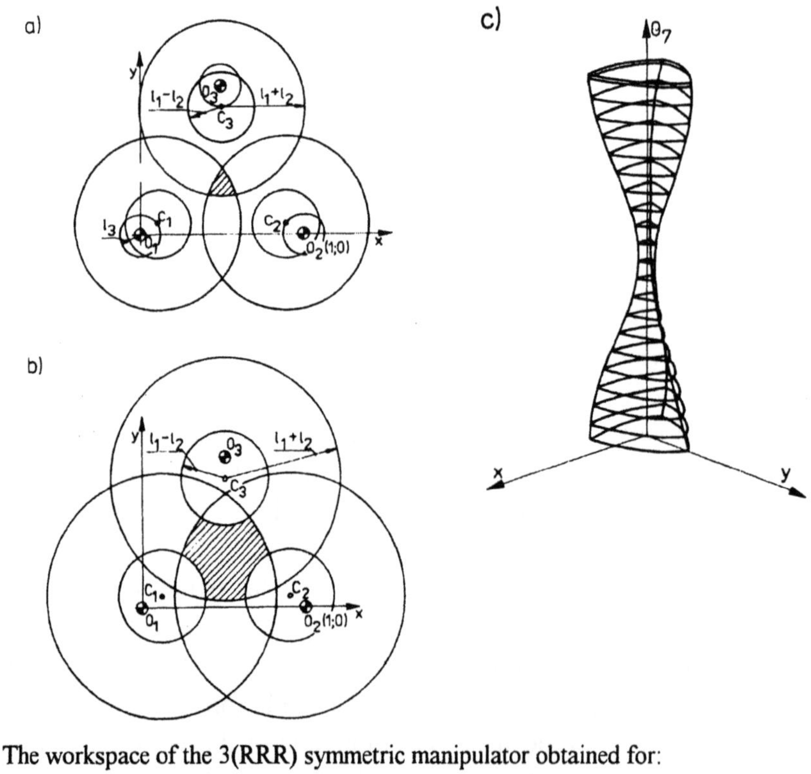

Fig.7.6. The workspace of the 3(RRR) symmetric manipulator obtained for:
a) $a_{i,i+3} = 0.355$, $a_{i+3,i+6} = 0.150$, $a_{7P} = a_{8P} = a_{9P} = 0.125$, $\phi = 0$,
b) $a_{i,i+3} = 0.486$, $a_{i+3,i+6} = 0.221$, $a_{7P} = a_{8P} = a_{9P} = 0{,}125$, $\phi = 0$,
c) $a_{i,i+3} = a_{i+3,i+6} = 0.355$, $a_{7P} = a_{8P} = a_{9P} = 0.1$

$$a_{12} = a_{23} = a_{31} = 1$$
$$a_{i,i+3} = l_1 \qquad a_{i+3,i+6} = l_2 \qquad i = 1, 2, 3$$
$$a_{7P} = a_{8P} = a_{9P} = l_3$$
$$\gamma_2 = \gamma_3 = 2\pi/3 \qquad \gamma_7 = 2\beta_7 = \pi/3$$

The singularity curve, i.e. the limits of the workspace, is obtained for the i-th leg by using equation (7.31) which, in this case takes the form of

$$(x - x_{Ci})^2 + (y - y_{Ci})^2 = (l_1 \pm l_2)^2 \quad i = 1, 2, 3 \tag{7.32}$$

where

$$\begin{aligned} x_{C1} &= l_3\, c(\phi + \pi/6) & y_{C1} &= l_3\, s(\phi + \pi/6) \\ x_{C2} &= 1 - l_2\, c(\phi - \pi/6) & y_{C2} &= -\, l_3\, s(\phi - \pi/6) \\ x_{C3} &= s(\pi/6) - l_3\, s\phi & y_{C3} &= c(\pi/6) - l_3\, c\phi \end{aligned} \tag{7.32a}$$

From the particular form of the expressions of the coordinates of the centers, C_i, given in equations (7.32a), these may be located on the circumference of three other circles of radii l_3, centered at the base joints O_i. Their location on this circumference depends on the end-effector orientation given by angle ϕ. For example, if $\phi = 0$, the centers C_i are located as shown in Fig.7.6a, i.e., bringing the annular regions as close as possible to each other. For $\phi = \pi$, they reach the configuration shown in Figure 7.6b. The area of the workspace will be minimum when angle $\phi = \pi$, since in this case the distance between the regions, the intersection of which defines the workspace, is maximum.

One important criterion for the usefulness of the manipulator is that we have7 a nonvanishing workspace for every value of orientation angle ϕ. This can be represented as a volume in the (x, y, ϕ) space achieved by imposing both a nonvanishing workspace for $\phi=\pi$, and the condition that the intersection of two of the circles defining the outer boundary of the annular regions be inside the third, and that the three circles defining the inner boundary do not intersect. This leads to

$$3(l_1 - l_2)^2 \le (\sqrt{3}\, l_3 + 1)^2 \le 3(l_1 + l_2)^2 \tag{7.33}$$

The manipulator verifying these conditions will have a nonvanishing workspace for every angle ϕ.

7.2.5. Jacobian matrix of the 3(RRR) manipulator

The Jacobian matrix of a manipulator is generally defined as the matrix representing the transformation mapping the joint rates into Cartesian velocities. However, in the case of the parallel manipulator, since the role of the direct and inverse kinematic problems is exchanged, the Jacobian matrix will be defined as the representation of the inverse transformation, i.e.

$$J\begin{bmatrix}\dot{x} & \dot{y} & \dot{\theta}_7\end{bmatrix}^T = \begin{bmatrix}\dot{\theta}_1 & \dot{\theta}_2 & \dot{\theta}_3\end{bmatrix}^T \tag{7.34}$$

where $[\dot{x} \quad \dot{y} \quad \dot{\theta}_7]^T$ is the vector of Cartesian velocities and $[\dot{\theta}_1 \quad \dot{\theta}_2 \quad \dot{\theta}_3]^T$, the vector of joint rates.

The accuracy of manipulator control depends on the condition number of the Jacobian matrix. This number can be thought of as a factor amplifying the errors of the actuators, and hence directly affecting the precision of end-effector motion. The condition number must be kept as small as possible, the smallest value that can be attained being 1. One simple way of doing so is to render this matrix proportional to an orthogonal matrix, which is known to have a condition number 1. This, however, is possible only in specific configurations, since the Jacobian matrix is configuration dependent.

7.3. Spherical platform manipulators

Spherical parallel manipulators can be used to perform a pure orientation of a body in three dimensional space (with a fixed point). Solar panels, telescopes, radar antennas, mirrors for laser beams, spherical wrists for manipulators, and suspension systems for vehicle wheels are some of the major applications. Another application, in bio-mechanical engineering, to replace currently used artificial hips, is being studied.

The spherical parallel manipulator consists of three kinematic subchains (RRR) or (SPS) connecting the end-effector to the base. On each chain there is one actuated revolute or prismatic joint. The characteristic feature of the spherical 3(RRR) manipulator with actuated revolute joints is that the axes of rotation of all joints intersect at a common point called the center of the manipulator. The base and the end-effector can be thought of as two pyramidal modules having one vertex in common. The axes of revolute joints are located on the edges of the pyramids.

A spherical 3DOF manipulator can also be constructed using prismatic actuators. In this case, the end-effector is mounted on a fixed spherical joint, the center of spherical motion, and connected to the base by means of three (SPS) legs. Each leg is attached to the end-effector by a spherical joint, and to the base by a universal joint. Prismatic actuators are used to control the lengths of the legs.

The inverse kinematic problem yields to a single solution in the case of prismatic actuators, whereas it leads to eight solutions in the case of revolute actuators. Theoretical and practical considerations lead to the conclusion that both arrangements offer potential advantages, depending on design criteria. For example, the manipulator with prismatic actuators allows simpler kinematic equations, while the manipulator with revolute actuators will, in general, have a larger workspace. The working space is also greatly affected by the interferences among links.

7.3.1. Direct kinematic analysis of the spherical 3(SPS) manipulator

Direct position analysis is used to find the orientation of the platform of the spherical parallel manipulator when a set of actuator displacements is given. The platform (see Fig.7.7) is connected to the base through a spherical joint centered at point Q, while the three legs meet both the base and platform at three distinct points A_i and B_i (i=1, 2, 3), the centers of universal and spherical joints, respectively.

The leg lengths $d_i = A_iB_i$ ($i=1, 2, 3$) are known, and $\boldsymbol{a}_i = \overline{QA_i}$, the position vectors of points A_i, are given in the base reference system, $x_b y_b z_b$, taken with the origin coincident with point Q and the z_b axis directed from point Q to point A_1. Moreover, $\boldsymbol{b}_i = =\overline{QB_i}$, the position vectors of points B_i are given in the platform reference system, $x_p y_p z_p$, chosen with origin coincident with Q and z_p axis directed from point Q toward point B_1.

Echelon form solution [7.22]

The orientation of the platform with respect to the base can be parametrized as follows. When the leg lengths are given, the triangle A_1QB_1 remains defined, and can be considered a binary link connected to the base and platform by two revolute joints with axes z_b and z_p respectively. Indeed, if legs 2 and 3 are momentarily removed from the platform and base, the 2DOF auxiliary open chain formed by the base, link 1 and the platform can be considered. Link 1 can rotate about z_p , showing that the platform orientation can be parametrized by using only two angles θ_1 and θ_2.

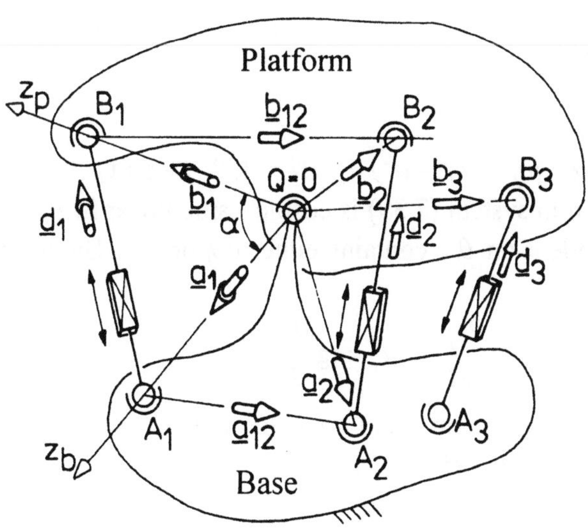

Fig.7.7. The 3(SPS) spherical parallel manipulator

Let us consider the 3 x 3 rotation matrix $\boldsymbol{R}$ for the coordinate transformation from system $x_p y_p z_p$ to system $x_b y_b z_b$

$$\boldsymbol{R}=\begin{bmatrix} c\theta_1 c\theta_2 - s\theta_1 s\theta_2 c\alpha & -c\theta_1 s\theta_2 - s\theta_1 c\theta_2 c\alpha & s\theta_1 s\alpha \\ s\theta_1 c\theta_2 + c\theta_1 s\theta_2 c\alpha & -s\theta_1 s\theta_2 + c\theta_1 c\theta_2 c\alpha & -c\theta_1 s\alpha \\ s\theta_2 s\alpha & c\theta_2 s\alpha & c\alpha \end{bmatrix} \tag{7.35}$$

where α is the angle between the axes z_b and z_p .

With reference to Fig.7.7 we may written

$$c\alpha=\frac{a_1^2+b_1^2-d_1^2}{2a_1b_1}, \quad s\alpha=\sqrt{1-c^2\alpha} \tag{7.36}$$

The closure equations for the manipulator in question are:

$$(\boldsymbol{R}\,\boldsymbol{b}_j-\boldsymbol{a}_j)^2=d_j^2, \quad (j=2, 3) \tag{7.37}$$

which can written as

$$\boldsymbol{a}_j^T\,\boldsymbol{R}\,\boldsymbol{b}_j=(a_j^2+b_j^2-d_j^2)/2 \tag{7.38}$$

Expansion of equation (7.38) in terms of coordinates of the vectors $\boldsymbol{a}_j=[a_{jx}\ a_{jy}\ a_{jz}]^T$ and $\boldsymbol{b}_j=[b_{jx}\ b_{jy}\ b_{jz}]^T$ yields

$$D_{1i}\,c\theta_1 c\theta_2+D_{2i}\,c\theta_1 s\theta_2+D_{3i}\,s\theta_1 c\theta_2+D_{4i}\,s\theta_1 s\theta_2+D_{5i}\,c\theta_1+D_{6i}\,s\theta_1+$$
$$+D_{7i}\,c\theta_2+D_{8i}\,s\theta_2+D_{9i}=0, \quad (i=1, 2) \tag{7.39}$$

where

$$D_{1i}=a_{jx}b_{jx}+a_{jy}b_{jy}c\alpha$$
$$D_{2i}=-a_{jx}b_{jy}+a_{jy}b_{jx}c\alpha$$
$$D_{3i}=a_{jy}b_{jx}-a_{jx}b_{jy}c\alpha$$
$$D_{4i}=-a_{jy}b_{jy}-a_{jx}b_{jx}c\alpha$$
$$D_{5i}=-a_{jy}\,b_{jz}\,s\alpha$$
$$D_{6i}=a_{jx}\,b_{jz}\,s\alpha$$
$$D_{7i}=a_{jz}\,b_{jy}\,s\alpha$$
$$D_{8i}=a_{jz}\,b_{jx}\,s\alpha$$
$$D_{9i}=a_{jz}\,b_{jz}\,c\alpha-(a_j^2+b_j^2-d_j^2)/2, \quad (i=1, 2; j=i+1)$$

Every equation in system (7.39) is cosine - sine linear in θ_1 and θ_2 ; i.e. in each, any term, if dependent on θ_j , contains either $c\theta_j$ or $s\theta_j$. By substituting in system (7.39) the identities:

$$c\theta_j=\frac{1-t_j^2}{1+t_j^2} \quad s\theta_j=\frac{2t_j}{1+t_j^2} \quad t_j=\mathrm{tg}\frac{\theta_j}{2}$$

the system of two equations on t_1 and t_2 can be written

$$E_i\,t_2^2+F_i\,t_2+G_i=0 \quad (i=1, 2) \tag{7.40}$$

where

$$E_i=a_{22i}\,t_1^2+a_{12i}\,t_1+a_{02i}$$

$$F_i = a_{21i}t_1^2 + a_{11i}t_1 + a_{01i}$$
$$G_i = a_{20i}t_1^2 + a_{10i}t_1 + a_{00i} \qquad (7.40a)$$

and

$$a_{00i} = D_{1i} + D_{5i} + D_{7i} + D_{9i}$$
$$a_{01i} = 2(D_{2i} + D_{8i})$$
$$a_{02i} = -D_{1i} + D_{5i} - D_{7i} + D_{9i}$$
$$a_{10i} = 2(D_{3i} + D_{6i})$$
$$a_{11i} = 4D_{4i} \qquad (7.40b)$$
$$a_{12i} = 2(-D_{3i} + D_{6i})$$
$$a_{20i} = -D_{1i} - D_{5i} + D_{7i} + D_{9i}$$
$$a_{21i} = 2(-D_{2i} + D_{8i})$$
$$a_{22i} = D_{1i} - D_{5i} - D_{7i} + D_{9i}$$

The elimination of the unknown t_2 from the system (7.40) can be worked out by imposing the condition

$$\begin{vmatrix} 0 & E_1 & F_1 & G_1 \\ E_1 & F_1 & G_1 & 0 \\ 0 & E_2 & F_2 & G_2 \\ E_2 & F_2 & G_2 & 0 \end{vmatrix} = 0 \qquad (19.41)$$

where only the unknown t_1 appears. The left side of equation (7.41) is the eliminant of equations (7.40), and equation (7.41) itself is the condition under which the closure equations (7.40) have the same solutions for t_2.

By developing equation (7.41) it follows that

$$(E_1G_2 - G_1E_2)^2 + (E_1F_2 - F_1E_2)(G_1F_2 - F_1G_2) = 0 \qquad (7.42)$$

represents an algebraic equation of the eight order in t_1. Hence

$$w_8 t_1^8 + w_7 t_1^7 + \ldots + w_1 t_1 + w_0 = 0 \qquad (7.43)$$

Equation (7.43) has eight solutions in the complex field. Let t_{1k} (k = 1, 2, ..., 8) be a generic root of equation (7.43); the polynomials E_i , F_i , G_i assume definite values, and system (7.40) can be linearly solved in the unknowns $(t_2)^2$ and t_2.

Vector form solution [7.16]

The subject of our investigation is the spherical 3-DOF parallel manipulator with three linear actuators, each of them attached to the base with a universal joint and to the platform with a spherical joint (see Fig.7.8). Fourth spherical joint between the platform and the base permits the spherical motion of the platform.

The vectors listed below are described in the base coordinate system $Ox_by_bz_b$:

$\boldsymbol{a}_i$ = OA$_i$ = const., $\quad$ $\boldsymbol{b}_i$ = OB$_i$ = var.,

$\boldsymbol{a}_{ij} = \boldsymbol{a}_j - \boldsymbol{a}_i$, $\quad$ $\boldsymbol{b}_{ij} = \boldsymbol{b}_j - \boldsymbol{b}_i$, $\quad$ $\boldsymbol{d}_i = \boldsymbol{a}_i - \boldsymbol{b}_i$, $\quad$ $\boldsymbol{a}_i^o = \boldsymbol{a}_i/a_i$, $\quad$ $\boldsymbol{b}_i^o = \boldsymbol{b}_i/b_i$

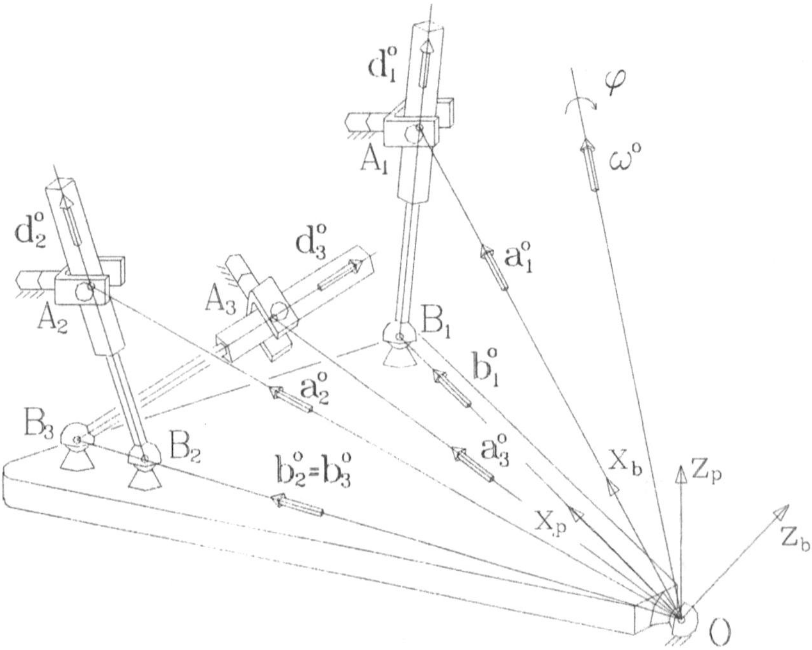

Fig.7.8. The 3-DOF spherical parallel manipulator

The geometry of the manipulator considered is given. In particular, the positions of points A_i (i=1, 2, 3) are given in an arbitrary reference system fixed to the base; the positions of points B_i are given in an arbitrary reference system fixed to the platform; the leg lengths $A_iB_i = d_i$ are known. Without loss of generality, reference systems are chosen with origins coincident with point O , the x_b axis directed from point O toward point A_1 , the x_p axis directed toward point B_1, the y_b axis in the plane OA_1A_2 and the y_p axis in the plane OB_1B_2.

Direct position analysis for the general fully-parallel 3-DOF spherical manipulator results in a two-equation system in echelon form; the first equation is of the eighth order, the other is linear. Consequently, when a set of actuator displacements is given, eight configurations of manipulator are possible. The forward position of the general spherical manipulator can be solved in compact closed-form when the three platform points O, B_2 and B_3 are collinear so the unit orientation vectors of points B_2 and B_3 are equal ($\boldsymbol{b}_2^{\circ} = \boldsymbol{b}_3^{\circ}$).

Two unknown unit vectors $\boldsymbol{b}_1^{\circ}$ and $\boldsymbol{b}_2^{\circ}$ can be found directly using twice the general formula to find one of the three unit vectors [7.25] in the following way:

$$b_2^{\circ} = \frac{\left[\left(a_2^{\circ} \cdot b_2^{\circ}\right) - \left(a_2^{\circ} \cdot a_3^{\circ}\right)\left(a_3^{\circ} \cdot b_2^{\circ}\right)\right] a_2^{\circ} + \left[\left(a_3^{\circ} \cdot b_2^{\circ}\right) - \left(a_2^{\circ} \cdot a_3^{\circ}\right)\left(a_2^{\circ} \cdot b_2^{\circ}\right)\right] a_3^{\circ} \pm \sqrt{D_2}\left(a_2^{\circ} \times a_3^{\circ}\right)}{1 - \left(a_2^{\circ} \cdot a_3^{\circ}\right)^2} \quad (7.44)$$

where

$$D_2 = 1-\left(a_2^o \cdot b_2^o\right)^2 - \left(a_2^o \cdot a_3^o\right)^2 - \left(a_3^o \cdot b_2^o\right)^2 + 2\left(a_2^o \cdot b_2^o\right)\left(a_2^o \cdot a_3^o\right)\left(a_3^o \cdot b_3^o\right) \tag{7.44a}$$

$$b_1^o = \frac{\left[\left(a_1^o \cdot b_1^o\right)-\left(a_1^o \cdot b_2^o\right)\left(b_1^o \cdot b_2^o\right)\right]a_1^o + \left[\left(b_1^o \cdot b_2^o\right)-\left(a_1^o \cdot b_2^o\right)\left(a_1^o \cdot b_1^o\right)\right]b_2^o \pm \sqrt{D_1}\left(a_1^o \times b_2^o\right)}{1-\left(a_1^o \cdot b_2^o\right)^2} \tag{7.45}$$

where

$$D_1 = 1-\left(a_1^o \cdot b_1^o\right)^2 - \left(a_1^o \cdot b_2^o\right)^2 - \left(b_1^o \cdot b_2^o\right)^2 + 2\left(a_1^o \cdot b_1^o\right)\left(a_1^o \cdot b_2^o\right)\left(b_1^o \cdot b_2^o\right) \tag{7.45a}$$

The dot products of the corresponding unit vectors can be obtained as follows:

$$a_i^o \cdot a_j^o = \frac{a_i^2 + a_j^2 - a_{ij}^2}{2a_i a_j}, \quad b_i^o \cdot b_j^o = \frac{b_i^2 + b_j^2 - b_{ij}^2}{2b_i b_j}, \quad a_i^o \cdot b_i^o = \frac{a_i^2 + b_i^2 - d_i^2}{2a_i b_i}, \tag{7.46}$$

In order to solve the direct velocity problem, the vector closure equation for the ith leg:

$$\boldsymbol{b}_i + \boldsymbol{d}_i = \boldsymbol{a}_i$$

is differentiated with respect to time. Taking the dot product of both sides of the resulting equation by $\boldsymbol{d}_i$ we obtain:

$$\dot{\boldsymbol{b}}_i \cdot \boldsymbol{a}_i = -d_i \dot{d}_i \tag{7.47}$$

On the other hand:

$$\dot{\boldsymbol{b}}_i = \omega \times \boldsymbol{b}_i \tag{7.48}$$

where ω is the platform angular velocity.
Substituting (7.47) into (7.48) we can obtain

$$\boldsymbol{a}_i \times \boldsymbol{b}_i \cdot \omega = d_i \dot{d}_i \tag{7.49}$$

The vector equation (7.49) can be expressed in matrix form:

$$\boldsymbol{G}\,\omega = \boldsymbol{D}\,\dot{\boldsymbol{d}} \tag{7.50}$$

or $$\omega = \boldsymbol{J}_\omega\,[\dot{d}_1 \;\; \dot{d}_2 \;\; \dot{d}_3]^T \tag{7.50a}$$

where:

$$\boldsymbol{J}_\omega = \boldsymbol{G}^{-1}\boldsymbol{D} \tag{7.51}$$

$\boldsymbol{G} = \left[\boldsymbol{a}_1 \times \boldsymbol{b}_1 \quad \boldsymbol{a}_2 \times \boldsymbol{b}_2 \quad \boldsymbol{a}_3 \times \boldsymbol{b}_3\right]^T$, $\boldsymbol{D} = \boldsymbol{I}d$, $\boldsymbol{I}$=3×3 - identity matrix,

$d = \left[d_1 \quad d_2 \quad d_3\right]^T$, $\dot{d} = \left[\dot{d}_1 \quad \dot{d}_2 \quad \dot{d}_3\right]^T$

Numerical example 7.5. The coordinates of the joints of the spherical 3(SPS) manipulator in question were determined by measurements of a road building machine and described in the base reference system (for the base joints) and in the platform frame (for the platform joints). These coordinates are given as follows:

$\boldsymbol{a}_1 = [2{,}47\ 0\ 0]^T$, $\boldsymbol{a}_2 = [2{,}25\ 1{,}02\ 0]^T$, $\boldsymbol{a}_3 = [2{,}72\ 0{,}13\ -0{,}46]^T$
$\boldsymbol{b}_{1p} = [2{,}40\ 0\ 0]^T$, $\boldsymbol{b}_{2p} = [2{,}18\ 1{,}01\ 0]^T$, $\boldsymbol{b}_{3p} = [2{,}41\ 1{,}07\ -0{,}03]^T$

Using the formulas (7.44) - (7.46) and taking the leg lengths

$\boldsymbol{d} = [0{,}60\ 0{,}65\ 1{,}33]^T$

the calculation results are presented as coordinates of the platform joints in the base reference system:

$\boldsymbol{b}_1 = [2{,}328\ 0{,}286\ -0{,}506]^T$, $\boldsymbol{b}_2 = [1{,}975\ 1{,}253\ -0{,}539]^T$, $\boldsymbol{b}_3 = [2{,}181\ 1{,}340\ -0{,}625]^T$

The Jacobian matrix is calculated by using (7.51)

$$J_\omega = \begin{bmatrix} 6.518 & -1.945 & 1.441 \\ -1.945 & -1.029 & 0.523 \\ 1.441 & 0.523 & -0.379 \end{bmatrix}$$

7.3.2. Inverse kinematic analysis of the spherical 3(RRR) manipulator [7.3]

The spherical 3(RRR) parallel manipulator shown in Fig.7.9 consists in a platform connected to a fixed base by means three kinematic chains. Each chain is composed of two intermediate links and three revolute joints. The axes of all nine revolute joints intersect at one common point, which is referred to as the center of the mechanism. The three motors are fixed to the base and consequently, only the revolute joints connecting each of the three chains to the base are actuated.

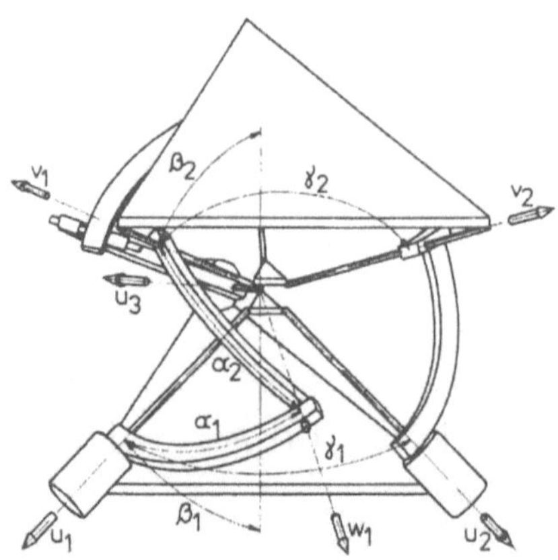

Fig.7.9. The 3(RRR) spherical parallel manipulator

The unit vectors along the axes of the actuators (fixed axes) are indicated as $\boldsymbol{u}_i$ (i=1, 2, 3) while the unit vectors along the axes attached to the platform are indicated as $\boldsymbol{v}_i$ (i=1, 2, 3). Finally, the unit vectors defined along the axes of intermediate joints are indicated as $\boldsymbol{w}_i$. The link angles on each of the kinematic chains are assumed to be identical and are called α_1 and α_2. The platform and the base are assumed to be symmetric, and the angle between the fixed axes (base) is γ_1 while the angle between the axes on platform is γ_2. The actuator angles are θ_i (i =1, 2, 3).

The inverse kinematic problem for this manipulator consists in finding the motor angles corresponding to a given orientation of the platform. The solution of this problem contains eight branches, two branches per leg, since the solution for the input motor angles θ_i, are completely decoupled.

For the symmetric manipulator we can write

$$\boldsymbol{u}_1 \cdot \boldsymbol{u}_2 = \boldsymbol{u}_2 \cdot \boldsymbol{u}_3 = \boldsymbol{u}_1 \cdot \boldsymbol{u}_3 = c\,\gamma_1$$

$$\boldsymbol{v}_1 \cdot \boldsymbol{v}_2 = \boldsymbol{v}_2 \cdot \boldsymbol{v}_3 = \boldsymbol{v}_1 \cdot \boldsymbol{v}_3 = c\,\gamma_2$$

$$\boldsymbol{u}_i \cdot \boldsymbol{w}_i = c\,\alpha_1 \qquad \boldsymbol{w}_i \cdot \boldsymbol{v}_i = c\,\alpha_2$$

In the case of a manipulator with coplanar three base axes, $\gamma_1 = 2\pi/3$ and the reference configuration be choosen as the one in which $\boldsymbol{u}_i = \boldsymbol{v}_i$. We can define

$$\boldsymbol{w}_i = [c\eta_i\, s\alpha_1\, c\theta_i + s\,\eta_i\, c\alpha_1 s\eta_i\, s\alpha_1\, c\theta_i - c\eta_i\, c\alpha_1\, s\alpha_1 s\theta_i\,]^T \tag{7.52a}$$

$$\boldsymbol{u}_i = [s\eta_i - c\eta_i\, 0]^T \tag{7.52b}$$

where η_i (i = 1, 2, 3) represent the angles between the projections of the axes of the actuated revolutes on the base plane and a given reference in that plane.

We have chosen: $\eta_1 = \pi/2$, $\eta_2 = -5\pi/6$, $\eta_3 = -\pi/6$

or $\boldsymbol{u}_1 = [1\;0\;0]^T$, $\boldsymbol{u}_2 = [-1/2 \ldots 0]^T$, $\boldsymbol{u}_3 = [-1/2 \ldots 0]^T$

The solution of the inverse kinematic problem is then obtained by writing the closure equation as

$$\boldsymbol{w}_i \cdot \boldsymbol{v}_i = c\,\alpha_2 \tag{7.53}$$

which, for each leg, leads to a quadratic equation in terms of $t_i = tg\;\theta_i/2$

$$A_i\, t_i^2 + 2B_i\, t_i + C_i = 0, \qquad (i = 1, 2, 3) \tag{7.54}$$

where

$$A_i = (v_{ix}\, s\eta_i - v_{iy}\, c\eta_i)\, c\alpha_1 - (v_{ix}\, c\eta_i + v_{iy}\, s\eta_i)\, s\alpha_1 - c\alpha_2$$

$$B_i = v_{iz}\, s\alpha_1$$

$$C_i = (v_{ix}\, s\eta_i - v_{iy}\, c\eta_i)\, c\alpha_1 + (v_{ix}\, c\eta_i + v_{iy}\, s\eta_i)\, s\alpha_1 - c\alpha_2 \tag{7.54a}$$

The coordinates of the unit vectors $v_i = [v_{ix}\; v_{iy}\; v_{iz}]^T$ are given by the coordinates of platform orientation.

The mobility region for each individual leg is given by the set of possible orientations that platform can attain, given the link dimensions of the leg. This region is bounded by the singularity surface, which is for each leg the closed surface separating the region where the leg has mobility from that in which it does not. The global mobility region will then be the intersection of all these regions. The singularity surface can be obtained using equation (7.54) and set the discriminant at zero

$$z_i = B_i^2 - A_i C_i = 0$$

From this we can determine that the conditions under which the manipulator is capable of producing all possible rotations are: $\alpha_1 = \alpha_2$ and $\alpha_1+\alpha_2 = \pi$, which means $\alpha_1=\alpha_2= \pi/2$. This result is similar to that obtained for an open-loop three axes spherical wrist, for which this condition must be met if we want the wrist to be able to reach all 0possible orientations.

Furthermore, we note that, when the manipulator has $\alpha_1=\alpha_2= \pi/2$, the solution to the inverse kinematic problem is simplified to

$$\theta_i = arctg\,[(v_{ix}\, c\eta_i + v_{iy}\, s\eta_i)/(-v_{iz})] \tag{7.55}$$

The Jacobian matrix of the 3(RRR) spherical parallel manipulator is similar to that used for planar manipulator, i.e.

$$\boldsymbol{J}\,\omega = [\dot{\theta}_1 \quad \dot{\theta}_2 \quad \dot{\theta}_3]^{\mathrm{T}} \tag{7.56}$$

where ω is the angular velocity of the platform. The Jacobian matrix can be found by the differentiation of both sides of equationn (7.53), which leads to

$$\dot{\boldsymbol{w}}_i \cdot \boldsymbol{v}_i + \boldsymbol{w}_i \cdot \dot{\boldsymbol{v}}_i = 0 \tag{7.57}$$

where

$$\dot{\boldsymbol{w}}_i = \dot{\theta}_i\,(\boldsymbol{u}_i \times \boldsymbol{w}_i)$$

$$\boldsymbol{w}_i \cdot \dot{\boldsymbol{v}}_i = -\,(\boldsymbol{w}_i \times \boldsymbol{v}_i) \cdot \omega \tag{7.57a}$$

then

$$\dot{\theta}_i = \frac{(\boldsymbol{w}_i \times \boldsymbol{v}_i) \cdot \omega}{(\boldsymbol{u}_i \times \boldsymbol{w}_i) \cdot \boldsymbol{v}_i} \tag{7.58}$$

The i-th row of the Jacobian matrix can then be written as

$$\boldsymbol{J}_i = \frac{(\boldsymbol{w}_i \times \boldsymbol{v}_i)^T}{(\boldsymbol{u}_i \times \boldsymbol{w}_i) \cdot \boldsymbol{v}_i} \tag{7.59}$$

where

$$\boldsymbol{u}_i \times \boldsymbol{w}_i = \begin{bmatrix} -c\eta_i s\theta_i s\alpha_1 \\ -s\eta_i s\theta_i s\alpha_1 \\ c\theta_i s\alpha_1 \end{bmatrix} \tag{7.59a}$$

We observe that the i-th row of the Jacobian matrix is, in fact, a vector orthogonal to the plane defined by vectors v_i and w_i, i.e. a vector orthogonal to the plane defined by the i-th secondary link.

As pointed out above, parallel manipulators may present two kinds of singularities. The first, i.e. the one that lies on the boudary of the workspace, which gives det $[J] \Rightarrow \infty$, has been investigated and used to analyse the workspace. The condition under which this type of singularity arises can also be obtained from the expression of the Jacobian, which produces

$$(\boldsymbol{u}_i \times \boldsymbol{w}_i) \cdot \boldsymbol{v}_i = 0 \tag{7.60}$$

and states that vectors $\boldsymbol{u}_i$, $\boldsymbol{v}_i$ and $\boldsymbol{w}_i$ are coplanar. In this type of singularity, there is a set of velocities of the platform that cannot be produced.

However, there is another kind of singularity, when det $[J] = 0$, for which the different motor rates are not independent. When the manipulator is in that configuration, there is a set of nonzero velocities of the end-effector that lead to a velocity of zero at each of the motors. This type of singularity corresponds to a gain of (unwanted) freedom by the manipulator. This configuration should be avoided, as the manipulator is not controllable in that situation.

Example 7.6. We consider the case for which $\gamma_1 = \gamma_2 = 2\pi/3$, i.e. a spherical manipulator with coplanar actuators. Let $\alpha_1 = 2\pi/3$, $\alpha_2 = \pi/2$. The corresponding manipulator is isotropic. The Jacobian matrix is written as

$$J = \begin{bmatrix} 1 & -1 & 0 \\ \sqrt{3}/3 & \sqrt{3}/3 & -2\sqrt{3}/3 \\ -\sqrt{6}/3 & -\sqrt{6}/3 & -\sqrt{6}/3 \end{bmatrix}$$

Figure 7.10 shows an isotropic spherical parallel manipulator with coplanar actuators. In order to avoid mechanical interference between the legs, each of the links rotates on the surface of a different concentric sphere. This manipulator is used to orient a camera with high speed and precision.

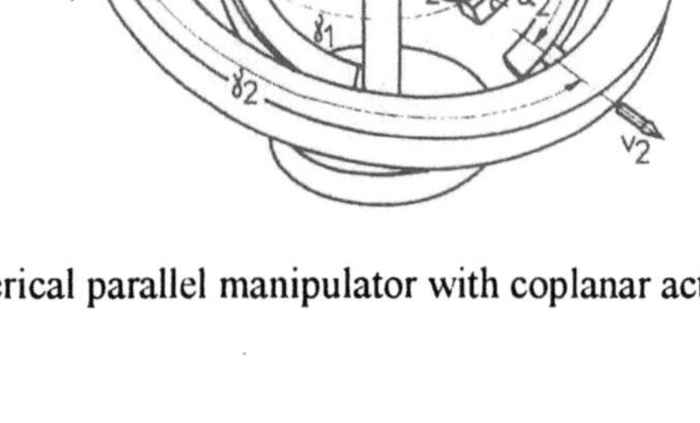

Fig.7.10. The isotropic spherical parallel manipulator with coplanar actuators

7.4. The parallel manipulator of the DELTA robot

The manipulation of lightweight objects is a task often encontered in the microengineering, electronic, food and pharmaceutical sectors. Due to the required standards of hygiene and reliability of the products, the presence of man operated workstations is undesirable.

The geometry of robot DELTA 4 (see Figure 7.11) is such that it can displace a mobile travelling plate in a position of 3-dimensional space corresponding to 3 DOF (x, y, z); the travelling plate stays constantly parallel to a reference plane and cannot rotate about the axis perpendicular to this plane; this configuration of the robot with three dof thus produces a parallelogram in space, while the fourth DOF is the aforementioned rotation obtained by the serialization of an axis at the travelling plate. This rotation is effected by a motor fixed to the robot base, connected in turn to the end-effector by a telescopic arm with two universal joints.

The displacement of the travelling plate is the result of the movement of the three articulated arms mounted on the base, each of which is connected to a pair of parallel rods. The three parallelograms ensure the stability of the whole (three orientations are eliminated by joining the arms in a common termination). The articulations that join the three pairs of parallel rods to the travelling plate are situated on the three sides of an equilateral triangle. For simplicity, the reference plane is horizontal, and the planes in which the arms move, are vertical.

Figure 7.11a gives a simplified view of DELTA 4. The base coordinate frame has its axis x directed toward one of the actuators, indicated as $i = 1$. The z axis is orthogonal to the base plate. The end-effector is connected to the plate by three identical extensions. Each extension is made of two parts: the arm which rotates with the actuator in a vertical plane, and two ball-socket joints connecting the arm to the lower parallelogram, that drives the travelling plate. To simplify the drawings, the parallelogram is represented as one rod.

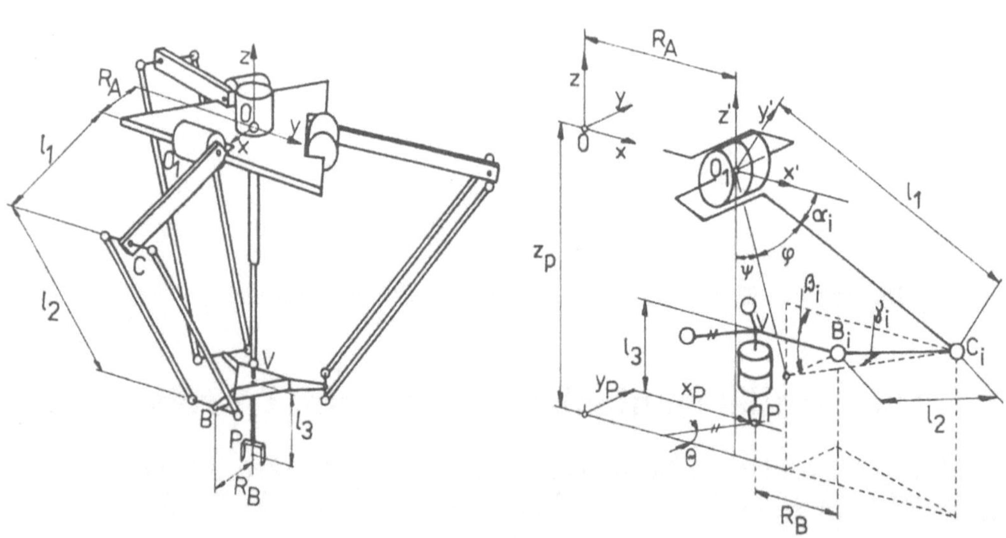

Fig. 7.11. The DELTA 4 robot

Notations: R_A is the distance between the origin O of the base frame and the actuator axis; l_1 , the length of an arm; l_2 , the length of a parallelogram rod; l_3 ,the distance between the center of the end-effector and the center of the travelling plate; R_B , the distance between the end-effector axis and the nearer side of parallelogram; α_i, the joint variable of the i-th arm, referred to the base plate plane; β_i, the angle between the i-th parallelogram plane and the base plane, measured in the vertical plane which contains the i-th arm; γ_i, the angle between the i-th vertical arm plane and a rod of parallelogram. The cartesian coordinates of the end-effector are: x_P, y_P, z_P, the center coordinates; and θ, angular displacement of the gripper referred to the x axis (roll angle).

7.4.1. Direct position analysis of the DELTA robot [7.26]

Direct position analysis expresses position and orientation of the end-effector x_P, y_P, z_P and θ as functions of four joint variables α_i (i = 1, 2, 3, 4). The coordinate z_{Ci} of point C_i can be derived easily from α_i, i.e. $z_{Ci} = l_1\, c\alpha_i$:

Let Q_i be the translation of C_i parallel to vector B_iV, and $VQ_i = l_2$. We can calculate the coordinates of V as follows. Given M_1, the midpoint of segment Q_1Q_2, and M_2, the midpoint of Q_1Q_3 , we call π_1 and π_2 the planes containing M_1 or M_2 respectively, and orthogonal to Q_1Q_2 or Q_1Q_3 respectively; π_3 is the plane of Q_i. The equations of planes π_1 , π_2 and π_3 are

$$\begin{aligned} A_1 x + B_1 y + C_1 z + D_1 &= 0 \\ A_2 x + B_2 y + C_2 z + D_2 &= 0 \\ A_3 x + B_3 y + C_3 z + D_3 &= 0 \end{aligned} \tag{7.61}$$

Let $U\,(x_u , y_u , z_u)$ be the intersection of three planes π_1 , π_2 and π_3 . The coordinates of point U can be obtained as the solution of system (7.61):

$$\begin{aligned} z_u &= [(B_3 s - D_3 q)A_1 - (B_1D_2 - B_2D_1)A_3]/[(-B_3 r + C_3 q)A_1 - (C_1B_2 - C_2B_1)A_3] \\ y_u &= -(r z_u + s)/q \\ x_u &= -(B_1y_u + C_1z_u + D_1)/A_1 \end{aligned} \tag{7.62}$$

where

$$\begin{aligned} q &= B_2 - B_1A_2/A_1 \\ r &= C_2 - C_1A_2/A_1 \\ s &= D_2 - D_1A_2/A_1 \end{aligned} \tag{7.62a}$$

Point V belongs to the straight line orthogonal to π_3 and passing though point U. The distance $Q_2V = l_2$. Since Q_2UV is a right-angled triangle, the translation of U can be calculated as:

$$d = UV = [l_2{}^2 - (Q_2U)^2]^{1/2} \tag{7.63}$$

Subsequently we may find V.

7.4.2. Inverse position analysis of the DELTA robot

Inverse position analysis expresses the joint variables α_i (i = 1, 2, 3) as functions of the position and orientation of the end-effector: Let us define a new frame $O_1x`y`z`$ with the $y`$ axis lying on the actuator axis and the $z`$ axis being parallel to the z axis. In this new frame we may pass from $W(x_P, y_P, z_P)$ to $B_1(x_{B1}, y_{B1}, z_{B1})$ as follows

$$x_{B1} = x_P + R_B - R_A$$
$$y_{B1} = y_P$$
$$z_{B1} = z_P - l_3 \tag{7.64}$$

The coordinates of point B_1 can be expressed in terms of the angles α_1, β_1, γ_1

$$x_{B1} = l_1\, c\alpha_1 - l_2\, c\gamma_1\, c\beta_1$$
$$y_{B1} = l_2\, s\gamma_1 \tag{7.65}$$

From equations (7.65) we derive

$$\gamma_1 = \arcsin(y_{B1}/l_2)$$
$$\beta_1 = \arccos[(-x_{B1} + l_1\, c\alpha_1)/l_2\, c\gamma_1] \tag{7.66}$$

Knowing that θ is complementary angle to $(\alpha_1 + \phi)$ in the plane of the arm, the joint variable α_1 is determined:

$$\psi = \text{arctg}(-x_{B1} / z_{B1}) \tag{7.67}$$

By the cosine theorem:

$$l_2^2\, c^2\gamma_1 = l_1^2 + (x_{B1}/s\psi)^2 - 2l_1\,(x_{B1}/s\psi)\, c\phi$$

or

$$\phi = \arccos\{[l_1^2 + (x_{B1}/s\psi)^2 - l_2^2\, c^2\gamma_1]/\, 2l_1\,(x_{B1}/s\psi)\} \tag{7.68}$$

Finally, we may express the required joint variable as

$$\alpha_1 = \pi/2 - (\psi + \phi)$$

The other joint variables, α_2 and α_3, can be derived in the same way,once the position of V has been expressed in terms of the corresponding new reference frame. The joint variable a_4, as already seen, corresponds to ROLL.

7.4.3. Dynamic analysis of the DELTA robot

Dynamic analysis of robot manipulators is required when: (a) The maximum drive torques to be developed by joint motors must be determined, knowing the velocities, accelerations and payload characteristics. (b) High-speed and/or high acceleration continuous motion must be programmed, since in such cases a feed-forward signal for motor control might be required. Expressions for motor torques required to execute a certain programmed motion contain many coupling, motion-related nonlinear terms, the importance of which increases to the velocity and range of motion. In the context of control, the objective of the dynamic model is to determine the torques which must be exerted by the actuators in order to produce a given trajectory of the manipulator. The model can be obtained using two main approaches, namely, the Newton-Euler formulation or Lagrangian formulation.

The Cartesian coordinates (x_P, y_P, z_P) of the end-effector of the DELTA robot are expressed as the complicated functions of the joint coordinates (α_i), and it is rather difficult to use these functions to derive Lagrange equations. In this case it is much easier to use both the dependent and independent coordinates as the generalized co-ordinates, i.e.

$$\{q_j\} = \{x_P, y_P, z_P, \alpha_1, \alpha_2, \alpha_3\}, \quad (j = 1, 2, ...6) \tag{7.69}$$

The geometrical constraints can be written as

$$f_i = [x_P - (R_A - R_B + l_1 c\alpha_i)c\delta_i]^2 + [y_P - (R_A - R_B + l_1 c\alpha_i)s\delta_i]^2 + (z_P + l_1 s\alpha_i)^2 - l_2^2 = 0, \quad i = 1, 2, 3 \tag{7.70}$$

where

$$\delta_1 = 0,\ \delta_2 = 2\pi/3\ ,\ \delta_3 = 4\pi/3$$

The Lagrange equations with multipliers

$$\frac{d}{dt}\cdot\frac{\partial L}{d\dot{q}_j} - \frac{\partial L}{dq_j} = Q_j + \sum_{i=1}^{3}\lambda_i A_{ij}\,, \quad (j = 1, 2, ...6) \tag{7.71}$$

consisting six generalized coordinates satisfying three constraint equations (7.70). Then it is possible to determine three Lagrange`s multipliers $\lambda_i\,(i = 1, 2, 3)$ and three drive torques Q_i .

For simplification in the dynamic model the inertial moments of parallelogram rods are assumed to be equal to zero, and the masses of the rods reduced partially to the arms and to the platform. This simplification makes it possible to introduce an assumption, that reaction forces exerted by rods on the platform act on rod directions.

The potential energy of the manipulator is described by the equation

$$E_p = \left(m_p + \frac{3}{2}m_2\right)gz_p - \frac{1}{2}(m_1 + m_2 + 2m_c)gl_1(s\alpha_1 + s\alpha_2 + s\alpha_3) \tag{7.72}$$

where:

m_p is the mass of the platform with end effector,

m_1 – the mass of the arm reduced to its center,

m_2 – the mass of the parallelogram,

m_C – the mass of joint C .

The kinetic energy can be expressed as

$$E_k = \frac{1}{2}(J_1 + m_c l_1^2)(\dot{\alpha}_1^2 + \dot{\alpha}_2^2 + \dot{\alpha}_3^2) + \frac{1}{2}(m_p + m_2)(\dot{x}_p^2 + \dot{y}_p^2 + \dot{z}_p^2) \tag{7.73}$$

where J_1 is the inertial moment of the arm versus its rotation axis.

The Lagrangian $L = E_k - E_p$ can be expressed as a function of the six coordinates and their derivatives. The partial derivatives of L are:

$$\frac{\partial L}{dx} = \frac{\partial L}{dy} = 0$$
$$\frac{\partial L}{dz} = -\left(m_p + \frac{3}{2}m_2\right)g$$
$$\frac{\partial L}{d\alpha_i} = \left(J_1 + m_c l_1^2\right)\ddot{\alpha}_i, \qquad (i = 1, 2, 3)$$
$$\frac{\partial L}{d\dot{q}_i} = \left(m_p + m_2\right)\dot{q}_i,$$
$$\dot{q}_1 = \dot{x}, \quad \dot{q}_2 = \dot{y}, \quad \dot{q}_3 = \dot{z} \tag{7.74}$$
$$\frac{\partial L}{d\dot{\alpha}_i} = \left(J_1 + m_c l_1^2\right)\dot{\alpha}_i$$
$$\frac{d}{dt}\left(\frac{\partial L}{d\dot{q}_i}\right) = \left(m_p + m_2\right)\ddot{q}_i$$
$$\ddot{q}_1 = \ddot{x}, \quad \ddot{q}_2 = \ddot{y}, \quad \ddot{q}_3 = \ddot{z}$$
$$\frac{d}{dt}\left(\frac{\partial L}{d\dot{\alpha}_i}\right) = \left(J_1 + m_c l_1^2\right)\ddot{\alpha}_i$$

The partial derivatives of the constraint equations (7.70) are obtained in the form

$$A_{ij} = \frac{\partial f_i}{dq_j}, \qquad (i = 1, 2, 3, \quad j = 1, 2, \ldots 6)$$
$$A_{i1} = 2[(R_A - R_B + l_1\, c\alpha_i)c\delta_i - x_P]$$
$$A_{i2} = 2[(R_A - R_B + l_1\, c\alpha_i)s\delta_i - y_P] \tag{7.75}$$
$$A_{i3} = 2(l_1\, s\alpha_i + z_P]$$
$$A_{i,i+3} = 2l_1\, z_P\, c\alpha_i - 2l_1\, [(R_A - R_B) + x_P\, c\delta_i + y_P\, s\delta_{ij}\, s\alpha_i, \quad (i = 1, 2, 3)$$
$$A_{24} = A_{26} = A_{15} = A_{16} = A_{34} = A_{35} = 0$$

The first three Lagrange eqs (7.71), taking into account (7.74), can be written in the form

$$\left(m_p + m_2\right)\ddot{x} = \sum_{i=1} \lambda_i A_{i1}$$
$$\left(m_p + m_2\right)\ddot{y} = \sum_{i=1} \lambda_i A_{i2} \tag{7.76}$$
$$\left(m_p + m_2\right)\ddot{z} = \sum_{i=1} \lambda_i A_{i3}, \quad (i = 1, 2, 3)$$

The Lagrange multipliers λ_i $(i = 1, 2, 3)$ can be solved by (7.76) for the given coordinates and their derivatives.

The last three Lagrange equations (7.71) can be used to determine the drive torques:

$$M_1 = Q_4 = \frac{d}{dt}\cdot\left(\frac{\partial L}{d\dot{\alpha}_1}\right) - \frac{\partial L}{d\alpha_1} - x_1 A_{14}$$
$$M_2 = Q_5 = \frac{d}{dt}\cdot\left(\frac{\partial L}{d\dot{\alpha}_2}\right) - \frac{\partial L}{d\alpha_2} - x_2 A_{25} \qquad (7.77)$$
$$M_3 = Q_6 = \frac{d}{dt}\cdot\left(\frac{\partial L}{d\dot{\alpha}_3}\right) - \frac{\partial L}{d\alpha_3} - x_3 A_{36}$$

Numerical example 7.7. The given parameters of robot DELTA-4 are:

$l_1 = 0.2$ m, $l_2 = 0.4$ m,
$m_1 = 0.1$ kg, $m_2 = 0.045$ kg,
$m_p = 0.1$ kg, $m_C = 0.018$ kg,
$R_A = 0.151$ m, $R_B = 0.025$ m, $J_1 = 0.002$ kgm^2

Point P must to move along the elliptic trajectory in the xz plane, passing through the initial, midle, and final position with given coordinates:

$P_i = [-0.200\ 0\ -0.348]^T$, $P_m = [0\ 0\ -0.248]^T$ and $P_f = [0.200\ 0\ -0.348]^T$.

The point acceleration is assumed as a sine function with maximum of acceleration of 100 m/s^2 and a maximum velocity of 5 m/s.

The calculated results give the extreme values of drive torques:

$M_{1min} = -10.25$ Nm, $M_{1max} = 12.95$ Nm
$M_{2min} = -6.21$ Nm, $M_{2max} = 8.49$ Nm

The DELTA robot has the following advantages:

The motors and gearing are fixed, which results in a low mobile mass; the choice of motor is simple, as the stator mass has no influence; the motors lie outside of the working volume, which limits collision possibilities; the link between motors and sensors and the control are simple and reliable (no moving cable).

The mechanical construction is also simple and cheap; certain components appeare six and twelve times in identical form in the same robot; there is no load on prismatic joints. The natural frequency is high, which results in minimal error between the motor and end-effector position. The working volume is large without central forbidden zones. Two of the three orientations of the end-effector are mechanically defined.

The disadvantages are: all rectilinear displacements of the end-effector require a contribution from all three motors, which might prove to be a difficulty in component insertion. The displacement increment is variable, and is a function of the position of the travelling plate as well as of the direction of movement.

7.5. Stewart-Gough platform manipulators

Several 6-6 leg arrangements with special geometry, that is with either/both base and/or platform plane, or symmetrically shaped have been studied to find a closed form solution of the direct position analysis (DPA). In particular the 6-6 parallel manipulator with either base or platform planar has been solved in closed form [7.1]. While for the case with base and platform planar an univariate polynomial equation of degree twenty in the square of variable has been found. An analytical solution of the general Stewart-Gough platform manipulator (SPM) has been proposed by Husty [7.6] and the univariate equation of degree forty has been found. Numerical procedures have also been proposed for the solution of the DPA of the 6-6 general SPM. Most of them are based on the solution of a system of six nonlinear equations in six unknowns.

In recent years some new methods for the (DPA) of the generalized SPM have been developed, especially numerical iterative methods in the case when only some solutions of DPA are concerned. All the possible solutions can be found if the problem can be solved in analytical form. Direct displacement analysis for parallel mechanisms, that is, to find the position and orientation of the output link when the input variables are given, is a difficult analytical problem.

Two main steps are required to obtain the analytical solution: (1) a suitable set of nonlinear equations, such as the closure equations or the constraint conditions must be found; (2) non-desired unknowns from some equations must be eliminated in order to obtain a sequentially solvable set of equations without introducing extraneous solutions. Many researchers have considered the problem of forward kinematics solution for SPM and computer simulation of platform motion in real time. However, a number of problems remain: a) matching the roots generated by an algorithm to the appropriate configuration of SPM; b) tracking SPM singularities by used algorithm, c) detection of representational singularities which do not necessarily coincide with the SPM singularities, and d) prediction of convergence rates and ultimate real time performance.

7.5.1. Direct kinematics of the 6-5 spatial parallel manipulator

The 6-5 spatial parallel manipulator (SPM) is one of more general cases of the SPM with 6-DOF, in which two of the six spherical joints at the platform have a common center. The platform position and orientation can be changed by six actuator displacements as independent variables. The vector method [7.15] was developed for kinematic analysis of 6-5 SPM is shown in Figure 7.12. The vector approach, based on the tetrahedron solution, has been extended for multi-loop spatial chains, such as, the five-rod wheel guiding mechanism applied for to rear suspension of the Mercedes 190.

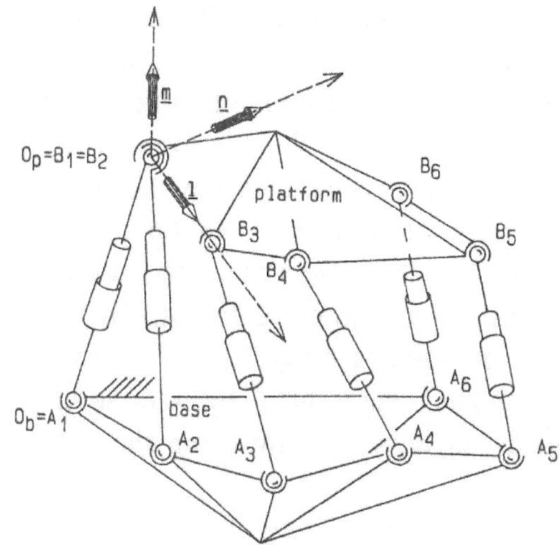

Fig.7.12. Kinematic scheme of the 6-5 SPM.

Here we consider the 6-5 SPM which, although a special geometry, is very close to the 6-6 general manipulator. This manipulator is called 6-5 because in the base six spherical joints can be chosen completely arbitrarily, while on the moving platform they must be chosen so that two of them coincide; concequently there are actually only five spherical joints.

The general formula (7.75) for finding one of three unit vectors can be treated as a subroutine used to calculate the unknown unit vector e_k, when two unit vectors e_i and e_j and the dot products of each of these vectors with the unknown unit vector e_k, i.e. $e_i^T e_k$ and $e_j^T e_k$, are known. Using the formula (7.75) the following unit vectors can be determined in the specified order.

The following system of coordinates has been applied to describe the position of a SPM mechanism,. The centers of the base joints, where the legs are joined to the base, are denoted by A_i , (i= 1, 2, ...6). $A_1 = O_b$ can be taken as the origin of the base coordinate frame. If A_2 is located on the x_b axis and A_3 on the plane $z_b = 0$, their coordinates satisfy the condition $x_{A1} = y_{A1} = z_{A1} = y_{A2} = z_{A2} = z_{A3} = 0$. Thus, 12 coordinates remain to defining the base joints for a non-planar configuration.

The platform joints of the 6-5 SPM are denoted by B_i (i = 1, 2,. ..., 5), where $B_1 = B_2$, and the geometry of the non-planar platform is defined by 9 parameters. This means, a total of 21 constant parameters and 6 input variables, namely the 6 lengths of the legs, $d_{ii} = A_iB_i$, for a total of 27 parameters in the general case of the 6-5 SPM mechanism. For some practical applications the number of input variables can be reduced; for example, only one or two input variable parameters are used for five-rod wheel guiding mechanism, considered a special case of the 6-5 SPM.

For the kinematic chain shown in Figure 7.12, the leg axes are directed by unit vectors $\boldsymbol{d}_{ii}^{o}$ (i = 1, 2,...6). The adjacent joint axes, which are chosen along the common center line between the adjacent joint centers, are directed by the unit vectors $\boldsymbol{a}_{ij}^{o}$, for the base and $\boldsymbol{b}_{ij}^{o}$, for the platform. The direction of the line connecting the joint centers A_i and B_j is described by the unit vector $\boldsymbol{d}_{ij}^{o}$.

The geometrical structure of the SPM can be considered a set of vector tetrahedrons described by three joint centers of the base (A_i , A_j , A_k) and one (B_i) of the platform (see Fig.7.13a) or one joint center (A_i) of the base and three joint centers (B_i , B_j , B_k) of the platform (see Fig.7.13b). In this way the direct position analysis of the SPM can be represented by the solutions of these vector tetrahedrons. The problem can be solved by using the formula of the three unit vectors.

$$\boldsymbol{e}_k = [(c_{ik} - c_{ij}\, c_{jk})\boldsymbol{e}_i + (c_{jk} - c_{ij}\, c_{ik})\boldsymbol{e}_j + \sigma D_{ij}^{\;1/2}\, (\boldsymbol{e}_i \times \boldsymbol{e}_j)]\, /(1\text{-}c_{ij}^{\;2}), \tag{7.78}$$

where:

$\sigma = 1$ or $\sigma = -1$

$c_{ij} = \boldsymbol{e}_i^T \boldsymbol{e}_j, \qquad c_{ik} = \boldsymbol{e}_i^T \boldsymbol{e}_k, \qquad c_{jk} = \boldsymbol{e}_j^T \boldsymbol{e}_k$

$D_{ij} = 1\text{-}\, c_{ij}^{\;2} - c_{ik}^{\;2} - c_{jk}^{\;2} + 2c_{ij}\, c_{ik}\, c_{jk}$

The first type of vector tetrahedron ($A_i\ A_j\ A_k\ B_i$, see Fig.7.13a) is completely defined by three non-coplanar vectors, for example $\boldsymbol{a}_{ij}$, $\boldsymbol{a}_{ik}$ and $\boldsymbol{d}_{ii}$. The length d_{ii} of leg i is given for the considered tetrahedron but the unit vector $\boldsymbol{d}_{ii}^{o}$ remains unknown and can be determined as a function of the additionaly assumed input variable $z = \boldsymbol{a}_{ik}^{oT}\boldsymbol{d}_{ii}^{o} = \cos\,\delta_i$. The unknown unit vector $\boldsymbol{d}_{ii}^{o}$ can be computed in terms of two known unit vectors, $\boldsymbol{a}_{ij}^{o}$ and $\boldsymbol{a}_{jk}^{o}$, and two known dot products ($\boldsymbol{a}_{ij}^{oT}\boldsymbol{a}_{ik}^{o}$) and ($\boldsymbol{a}_{ij}^{oT}\boldsymbol{d}_{ii}^{o}$), using formula (7.78)

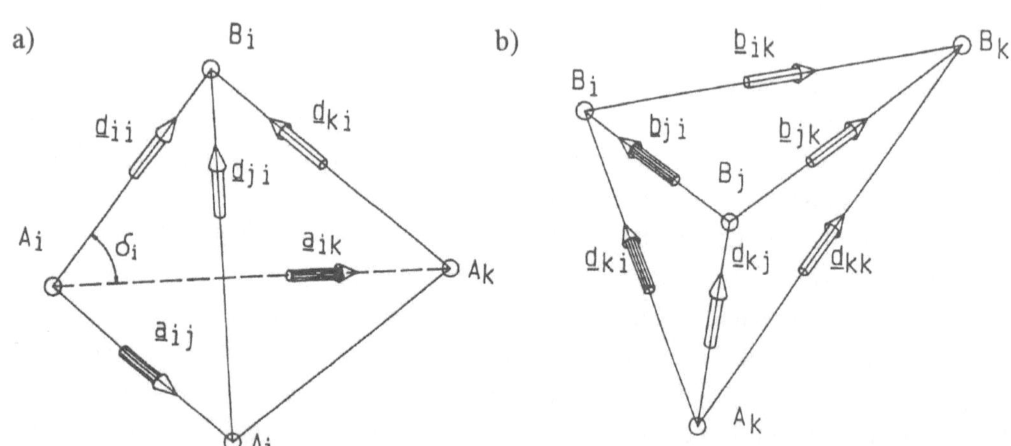

Fig.7.13. Two types of tetrahedron used to solve the direct position analysis of the SPM.

$$\boldsymbol{d}_{ii}^{o} = \{[(\boldsymbol{a}_{ij}^{oT}\, \boldsymbol{d}_{ii}) - z\,(\boldsymbol{a}_{ij}^{oT}\boldsymbol{a}_{ik}^{o})]\, \boldsymbol{a}_{ij}^{o} + [z - (\boldsymbol{a}_{ij}^{oT}\boldsymbol{d}_{ii}^{o})(\boldsymbol{a}_{ij}^{oT}\boldsymbol{a}_{ik}^{o})]\, \boldsymbol{a}_{ik}^{o} + \\ + \sigma D_{ii}^{1/2}\,(\boldsymbol{a}_{ij}^{o} \times \boldsymbol{a}_{ik}^{o})\} \,/\, [1 - (\boldsymbol{a}_{ij}^{oT}\boldsymbol{a}_{ik}^{o})]^2 \tag{7.79}$$

where: $\sigma = 1$ or $\sigma = -1$

$$D_{ii} = 1 - (\boldsymbol{a}_{ij}^{oT}\boldsymbol{d}_{ii}^{o})^2 - z^2 - (\boldsymbol{a}_{ij}^{oT}\boldsymbol{a}_{ik}^{o})^2 + 2z\,(\boldsymbol{a}_{ij}^{oT}\boldsymbol{d}_{ii}^{o})\,(\boldsymbol{a}_{ij}^{oT}\boldsymbol{a}_{ik}^{o})$$

$$\boldsymbol{a}_{ij}^{oT}\boldsymbol{d}_{ii}^{o} = (d_{11}^{\;2} + a_{ij}^{\;2} - d_{ii}^{\;2})/2d_{ii}a_{ij}$$

The unknown vector $\boldsymbol{d}_{ik}$ can be determined from the vector triangle $A_i\, A_k\, B_i$:

$$\boldsymbol{d}_{ik} = \boldsymbol{d}_{ii} - \boldsymbol{a}_{ik} \tag{7.80}$$

where

$$\boldsymbol{d}_{ii} = \boldsymbol{d}_{ii}^{o}\, d_{ii}\,,\ \boldsymbol{a}_{ik} = \boldsymbol{a}_{ik}^{o}\, a_{ik}$$

The second type of vector tetrahedron ($B_i\, B_j\, B_k\, A_k$, see Fig.7.13b) is described by three vectors $\boldsymbol{d}_{ki}$, $\boldsymbol{b}_{ij}$ and $\boldsymbol{b}_{jk}$. If the unit vector $\boldsymbol{b}_{jk}^{o}$ is unknown, the formula (7.78) can be used to describe it:

$$\boldsymbol{b}_{jk}^{o} = \{[(\boldsymbol{d}_{ki}^{oT}\boldsymbol{b}_{jk}^{o}) - (\boldsymbol{d}_{ki}^{oT}\boldsymbol{b}_{ji}^{o})(\boldsymbol{b}_{ji}^{oT}\boldsymbol{b}_{jk}^{o})]\, \boldsymbol{d}_{ki}^{o} + [(\boldsymbol{b}_{ji}^{oT}\boldsymbol{b}_{jk}^{o}) - (\boldsymbol{d}_{ki}^{oT}\boldsymbol{b}_{ji}^{o})(\boldsymbol{d}_{ki}^{oT}\boldsymbol{b}_{jk}^{o})]\boldsymbol{b}_{ji}^{o} + \\ + \sigma D_{ik}^{1/2}\,(\boldsymbol{d}_{ki}^{o} \times \boldsymbol{b}_{ji}^{o})\} \,/\, [1 - (\boldsymbol{d}_{ki}^{oT}\, \boldsymbol{b}_{ji}^{o})^2], \tag{7.81}$$

where $\sigma = 1$ or $\sigma = -1$

$$D_{ik} = 1 - (\boldsymbol{d}_{ki}^{oT}\boldsymbol{b}_{jk}^{o})^2 - (\boldsymbol{d}_{ki}^{oT}\boldsymbol{b}_{ji}^{o})^2 - (\boldsymbol{b}_{ji}^{oT}\, \boldsymbol{b}_{jk}^{o})^2 + 2(\boldsymbol{d}_{ki}^{oT}\boldsymbol{b}_{jk}^{o})(\boldsymbol{d}_{ki}^{oT}\boldsymbol{b}_{ji}^{o})(\boldsymbol{b}_{ji}^{oT}\boldsymbol{b}_{jk}^{o})$$

$$\boldsymbol{d}_{ki}^{oT}\boldsymbol{b}_{ji}^{o} = -\,(d_{ki}^{\;2} + b_{ji}^{\;2} - d_{kj}^{\;2})/2d_{ki}\, b_{ji}$$

The unknown vector $\boldsymbol{b}_{ki}$ can be determined from the vector triangle $A_k\, B_i\, B_k$:

$$\boldsymbol{b}_{ki} = \boldsymbol{d}_{kk} - \boldsymbol{d}_{ki}\,, \tag{7.82}$$

Using formulas (7.79) to (7.82), any of the unknown vectors $\boldsymbol{d}_{ii}$ and $\boldsymbol{b}_{ki}$can be obtained as the vector functions of the angles δ_i and used to described the position vectors of the joint centers B_i . The governing direct kinematics problem is now determined by imposing the constraints that points B_i must satisfy. Since the distance between B_i and A_i is fixed as is the given actuator length d_{ii} , the position vectors $\boldsymbol{r}_{Bi} = O_bB_i$ and $\boldsymbol{r}_{Ai} = O_bA_i$ must satisfy the conditions:

$$\boldsymbol{r}_{Bk} - \boldsymbol{r}_{Ak} = \boldsymbol{d}_{kk} \tag{7.83}$$

where:

$$\boldsymbol{r}_{Bk} = \boldsymbol{d}_{ii} + \boldsymbol{b}_{ik}\,,$$

$$\boldsymbol{r}_{Ak} = \boldsymbol{a}_{ik}\,,\quad k = i + 2$$

Substituting the expressions for $\boldsymbol{d}_{kk}$ and $\boldsymbol{b}_{ik}$ into the equations (7.83) we derive the displacement equations of the SPM concerned. The system of nonlinear equations (7.83) can be solved numerically.

This vector method can be used to solve the direct kinematics problem of the generalized SPM, and of the simplified type, such as the 5-5. The vector method is efficient, accurate and easy to apply for a SPM mechanism with a limited range of motion. By using this method one can determine the SPM displacement functions for any configuration, including singular positions.

Position analysis of the 6-5 SPM

The position analysis of the 6-5 SPM can be accomplished in the following way. Legs 4 and 5 can be removed, and the joint centers B_4 and B_5 treated as the coupler points of the transformed mechanism. The lengths of the other legs are treated as constants. The position of this mechanism (now with two degrees of freedom) can be described by the two angles, δ_1 and δ_3, shown in Figure 7.12, the cosines of which $\cos\delta_i = z_i$ (i =1, 3) can be assumed as two independent input variables. The positions of the coupler points, B_4 and B_5, can be found using the vector method, and the distances $d_5 = A_5B_5$ and $d_6 = A_6B_6$ can be determined as functions of z_1 and z_3, in the way described below.

The following unit vectors can be determined using formula (7.78) in the following order (the vectors $\boldsymbol{a}_{ij}$ with the unit vectors $\boldsymbol{a}_{ij}^o$ are assumed as known). In the first step we describe the unit vector $\boldsymbol{d}_{11}^o$ as the function of z_1 (see Figure 7.14a)

$$\boldsymbol{d}_{11}^o = \{[(\boldsymbol{a}_{12}^{oT}\boldsymbol{d}_{11}^o) - z_1(\boldsymbol{a}_{12}^{oT}\boldsymbol{a}_{16}^o)]\,\boldsymbol{a}_{12}^o + [z_1 - (\boldsymbol{a}_{12}^{oT}\boldsymbol{d}_{11}^o)(\boldsymbol{a}_{12}^{oT}\boldsymbol{a}_{16}^o)]\,\boldsymbol{a}_{16}^o + \sigma D_{11}^{1/2}(\boldsymbol{a}_{12}^o \times \boldsymbol{a}_{16}^o)\} / [1 - (\boldsymbol{a}_{12}^{oT}\boldsymbol{a}_{16}^o)^2] \tag{7.84}$$

where $\sigma = 1$ or $\sigma = -1$

$$D_{11} = 1 - (\boldsymbol{a}_{12}^{oT}\boldsymbol{d}_{11}^o)^2 - z_1^{\,2} - (\boldsymbol{a}_{12}^{oT}\boldsymbol{a}_{16}^o)^2 + 2z_1(\boldsymbol{a}_{12}^{oT}\boldsymbol{d}_{11}^o)(\boldsymbol{a}_{12}^{oT}\boldsymbol{a}_{16}^o)$$

$$\boldsymbol{a}_{12}^{oT}\boldsymbol{d}_{11}^o = (d_{11}^{\,2} + a_{12}^{\,2} - d_{21}^{\,2}) / 2d_{11}a_{12}$$

$$\boldsymbol{a}_{16}^{oT}\boldsymbol{d}_{11}^o = z_1 = \cos\delta_1 \quad \text{(see Figure 7.14a)}$$

$$\boldsymbol{d}_{ii} = \boldsymbol{d}_{ii}^o\, d_{ii}, \qquad \boldsymbol{d}_{ij}^o = \boldsymbol{d}_{ij} / d_i$$

$$\boldsymbol{d}_{31} = \boldsymbol{d}_{11} - \boldsymbol{a}_{13}\,,\; \boldsymbol{d}_{31} = \boldsymbol{d}_{11} - \boldsymbol{a}_{16}$$

In the second step we describe the unit vector $\boldsymbol{d}_{31}^o$ as a function of z_3 (see Figure 7.14b)

$$\boldsymbol{d}_{33}^o = \{[z_3 - (\boldsymbol{d}_{31}^{oT}\boldsymbol{d}_{33}^o)(\boldsymbol{a}_{36}^{oT}\boldsymbol{d}_{31}^o)]\,\boldsymbol{a}_{36}^o + [(\boldsymbol{a}_{16}^{oT}\boldsymbol{d}_{11}^o) - z_3(\boldsymbol{a}_{36}^{oT}\boldsymbol{d}_{31}^o)]\,\boldsymbol{d}_{31}^o + \sigma D_{33}^{1/2}(\boldsymbol{a}_{36}^o \times \boldsymbol{d}_{31}^o)\} / [1 - (\boldsymbol{a}_{36}^{oT}\boldsymbol{d}_{31}^o)^2] \tag{7.85}$$

where $\sigma = 1$ olr $\sigma = -1$

$$D_{33} = 1 - z_3^{\,2} - (\boldsymbol{d}_{31}^{oT}\boldsymbol{d}_{33}^o)^2 - (\boldsymbol{a}_{36}^{oT}\boldsymbol{d}_{31}^o)^2 + 2z_3(\boldsymbol{d}_{31}^{oT}\boldsymbol{d}_{33}^o)(\boldsymbol{a}_{36}^{oT}\boldsymbol{d}_{31}^o)$$

$$\boldsymbol{d}_{31}^{oT}\boldsymbol{d}_{33}^o = (d_{31}^{\,2} + d_{33}^{\,2} - b_{13}^{\,2}) / 2d_{31}d_{33}$$

$$\boldsymbol{a}_{36}^{oT}\boldsymbol{d}_{31}^o = (d_{31}^{\,2} + a_{36}^{\,2} - d_{61}^{\,2}) / 2\,d_{31}a_{36}$$

In the next step we describe the unit vector $\boldsymbol{b}_{34}^o$ (see Figure 7.14c):

$$\boldsymbol{b}_{34}^o = \{[(\boldsymbol{b}_{34}^{oT}\boldsymbol{b}_{13}^o) - (\boldsymbol{b}_{34}^{oT}\boldsymbol{d}_{43}^o)(\boldsymbol{b}_{13}^{oT}\boldsymbol{d}_{43}^o)]\boldsymbol{b}_{13}^o + - [-(\boldsymbol{b}_{34}^{oT}\boldsymbol{d}_{43}^o) + (\boldsymbol{b}_{34}^{oT}\boldsymbol{b}_{13}^o)(\boldsymbol{b}_{13}^{oT}\boldsymbol{d}_{43}^o)]\boldsymbol{d}_{43}^o + \sigma D_{34}^{1/2}(\boldsymbol{b}_{13}^o \times \boldsymbol{d}_{43}^o)\} / [1 - (\boldsymbol{b}_{13}^{oT}\boldsymbol{d}_{43}^o)^2], \tag{7.86}$$

where $\sigma = 1$ or $\sigma = -1$

$$D_{34} = 1 - (\boldsymbol{b}_{34}^{oT}\boldsymbol{b}_{13}^o)^2 - (\boldsymbol{b}_{34}^{oT}\boldsymbol{d}_{43}^o)^2 - (\boldsymbol{b}_{13}^{oT}\boldsymbol{d}_{43}^o)^2 + 2(\boldsymbol{b}_{34}^{oT}\boldsymbol{b}_{13}^o)(\boldsymbol{b}_{34}^{oT}\boldsymbol{d}_{43}^o)(\boldsymbol{b}_{13}^{oT}\boldsymbol{d}_{43}^o)$$

$$\boldsymbol{b}_{34}^{oT}\boldsymbol{b}_{13}^o = -(b_{13}^{\,2} + b_{34}^{\,2} - b_{14}^{\,2}) / 2b_{13}b_{34}$$

$$\boldsymbol{b}_{34}^{oT}\boldsymbol{d}_{43}^o = (b_{34}^{\,2} + d_{43}^{\,2} - d_{44}^{\,2}) / 2b_{34}d_{43}$$

$$\boldsymbol{b}_{13}^{oT}\boldsymbol{d}_{43}^o = (b_{13}^{\,2} + d_{43}^{\,2} - d_{14}^{\,2}) / 2b_{13}d_{43}$$

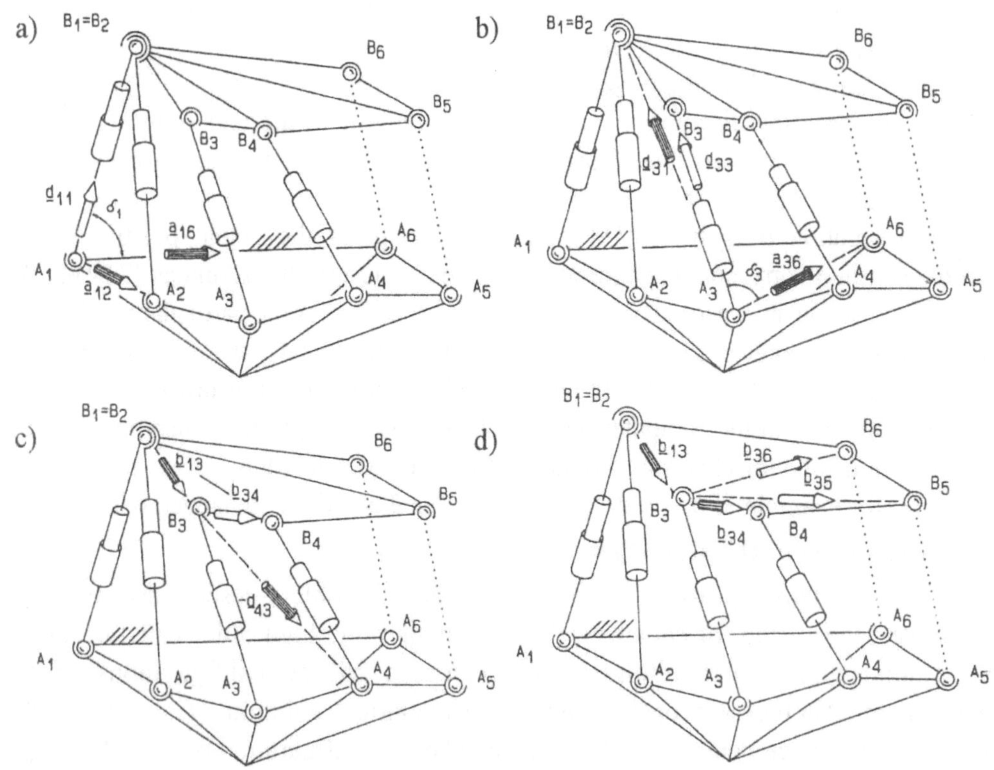

Fig.7.14a-d. The successive stages for solution of the DPA problem for the 6-5 SPM.

In the next two steps we describe the unit vectors $\boldsymbol{b}_{35}^o$ and $\boldsymbol{b}_{36}^o$ (see Figure 7.14d):

$$\boldsymbol{b}_{35}^o = \{[(\boldsymbol{b}_{35}^{oT}\boldsymbol{b}_{13}^o) - (\boldsymbol{b}_{34}^{oT}\boldsymbol{b}_{35}^o)(\boldsymbol{b}_{13}^{oT}\boldsymbol{b}_{34}^o)]\boldsymbol{b}_{13}^o + [(\boldsymbol{b}_{34}^{oT}\boldsymbol{b}_{35}^o) - (\boldsymbol{b}_{35}^{oT}\boldsymbol{b}_{13}^o)(\boldsymbol{b}_{13}^{oT}\boldsymbol{b}_{34}^o)]\boldsymbol{b}_{34}^o + \sigma D_{35}^{1/2}(\boldsymbol{b}_{13}^o \times \boldsymbol{b}_{34}^o)\} / [1 - (\boldsymbol{b}_{13}^{oT}\boldsymbol{b}_{34}^o)^2] \quad (7.87)$$

where $\sigma = 1$ or $\sigma = -1$

$$D_{35} = 1 - (\boldsymbol{b}_{34}^{oT}\boldsymbol{b}_{13}^o)^2 - (\boldsymbol{b}_{35}^{oT}\boldsymbol{b}_{13}^o)^2 - (\boldsymbol{b}_{34}^{oT}\boldsymbol{b}_{35}^o)^2 + 2(\boldsymbol{b}_{34}^{oT}\boldsymbol{b}_{13}^o)(\boldsymbol{b}_{34}^{oT}\boldsymbol{b}_{35}^o)(\boldsymbol{b}_{13}^{oT}\boldsymbol{b}_{35}^o)$$

$$\boldsymbol{b}_{34}^{oT}\boldsymbol{b}_{35}^o = (b_{34}^2 + b_{35}^2 - b_{45}^2)/ 2b_{34}b_{35}$$

$$\boldsymbol{b}_{13}^{oT}\boldsymbol{b}_{35}^o = (b_{13}^2 + b_{35}^2 - b_{15}^2)/ 2b_{13}b_{35}$$

$$\boldsymbol{b}_{36}^o = \{[(\boldsymbol{b}_{36}^{oT}\boldsymbol{b}_{13}^o) - (\boldsymbol{b}_{34}^{oT}\boldsymbol{b}_{36}^o)(\boldsymbol{b}_{13}^{oT}\boldsymbol{b}_{34}^o)]\boldsymbol{b}_{13}^o + [(\boldsymbol{b}_{35}^{oT}\boldsymbol{b}_{36}^o) - (\boldsymbol{b}_{36}^{oT}\boldsymbol{b}_{13}^o)(\boldsymbol{b}_{13}^{oT}\boldsymbol{b}_{34}^o)]\boldsymbol{b}_{34}^o + \sigma D_{36}^{1/2}(\boldsymbol{b}_{13}^o \times \boldsymbol{b}_{34}^o)\} / [1 - (\boldsymbol{b}_{13}^{oT}\boldsymbol{b}_{34}^o)^2] \quad (7.88)$$

where $\sigma = 1$ or $\sigma = -1$

$$D_{36} = 1 - (\boldsymbol{b}_{34}^{oT}\boldsymbol{b}_{13}^o)^2 - (\boldsymbol{b}_{36}^{oT}\boldsymbol{b}_{13}^o)^2 - (\boldsymbol{b}_{34}^{oT}\boldsymbol{b}_{36}^o)^2 + 2(\boldsymbol{b}_{34}^{oT}\boldsymbol{b}_{13}^o)(\boldsymbol{b}_{34}^{oT}\boldsymbol{b}_{36}^o)(\boldsymbol{b}_{13}^{oT}\boldsymbol{b}_{36}^o)$$

$$\boldsymbol{b}_{34}^{oT}\boldsymbol{b}_{36}^o = (b_{34}^2 + b_{36}^2 - b_{46}^2)/ 2b_{34}b_{36}$$

$$\boldsymbol{b}_{13}^{oT}\boldsymbol{b}_{36}^o = (b_{13}^2 + b_{36}^2 - b_{16}^2)/ 2b_{13}b_{36}$$

The closure equations of the vector loops $A_5A_3B_3B_5$ and $A_6A_3B_3B_6$ can be written as

$$\boldsymbol{r}_{B5} - \boldsymbol{r}_{A5} = \boldsymbol{d}_{33} + \boldsymbol{b}_{35} - \boldsymbol{a}_{35} = \boldsymbol{r}_{55}$$

$$\boldsymbol{r}_{B6} - \boldsymbol{r}_{A6} = \boldsymbol{d}_{33} + \boldsymbol{b}_{36} - \boldsymbol{a}_{36} = \boldsymbol{r}_{66} \tag{7.89}$$

By equating the distances r_{55} and r_{66} to the lengths of the legs 5 and 6, d_{55} and d_{66}, respectively, one gets the following system of two constraint equations:

$$r_{55}(z_1, z_3) = d_{55} \quad r_{66}(z_1, z_3) = d_{66} \tag{7.90}$$

Any solution of system (7.90) for z_1 and z_3 gives the actual position of the mechanism in a corresponding configuration. The signs σ in the formulas (7.84)-(7.88) must be chosen in any case according to the joints location for the particular mechanism. As some of the angles are determined using the cosines law, the sides of triangles including legs 1, 2, 3 and 6 should have reasonable proportions in order to ensure the good performance of the computer program. If there is any significant disproportion, the description of the mechanism should be transformed according to its symmetrical structure (for example, if $b_{13} \ll d_{11}$ or $b_{13} \ll d_{33}$, leg 3 could be exchanged with leg 6).

Description of the algorithm

There are two approaches to the solution of the problem described above. One transforms the system of equations (7.90) to obtain one high order equation of the polynomial type, which allows us to determine all the configurations of the mechanism. Another approach assumes the numerical solution of system (7.90), finding the displacements of the links in one chosen configuration of the SPM.

In order to solve the system of nonlinear equations (7.90), some numerical methods can be used, for example, the method based on minimization. On the basis of equations (7.90) the following objective function has been formulated:

$$F(z_1, z_3) = (d_{66} - r_{66})^2 + (d_{55} - r_{55})^2 \tag{7.91}$$

where z_1 and z_3 are the decisive variables. The objective function has been minimized toward zero (with the assumed accuracy Δ).

After minimization, the procedure has been repeated for the values of d_{ii} changed by $\Delta\, d_{ii}$ as many times as needed to cover the entire range of the analyzed motion of the platform, described by the minimum and maximum lengths of the legs. The estimated increments of the decisive variables between two positions of the mechanism, facilitating the calculations in each successive step, were evaluated on the basis of the two previous steps.

The input data for the minimization algorithm were the following:

- the distances between the platform joints' centers;
- the coordinates of the centers of the base spherical joints, given in the base frame;
- an approximate initial position of the mechanism (for the minimization at the first position in the range of the allowed motion) and a rough estimate of the increments of the decisive variables between the first and the second position.

The dimensional data can also be entered by giving the coordinates of all the joints' centers for the initial position of the mechanism. The flowchart of the displacement analysis algorithm is shown in Figure7.15.

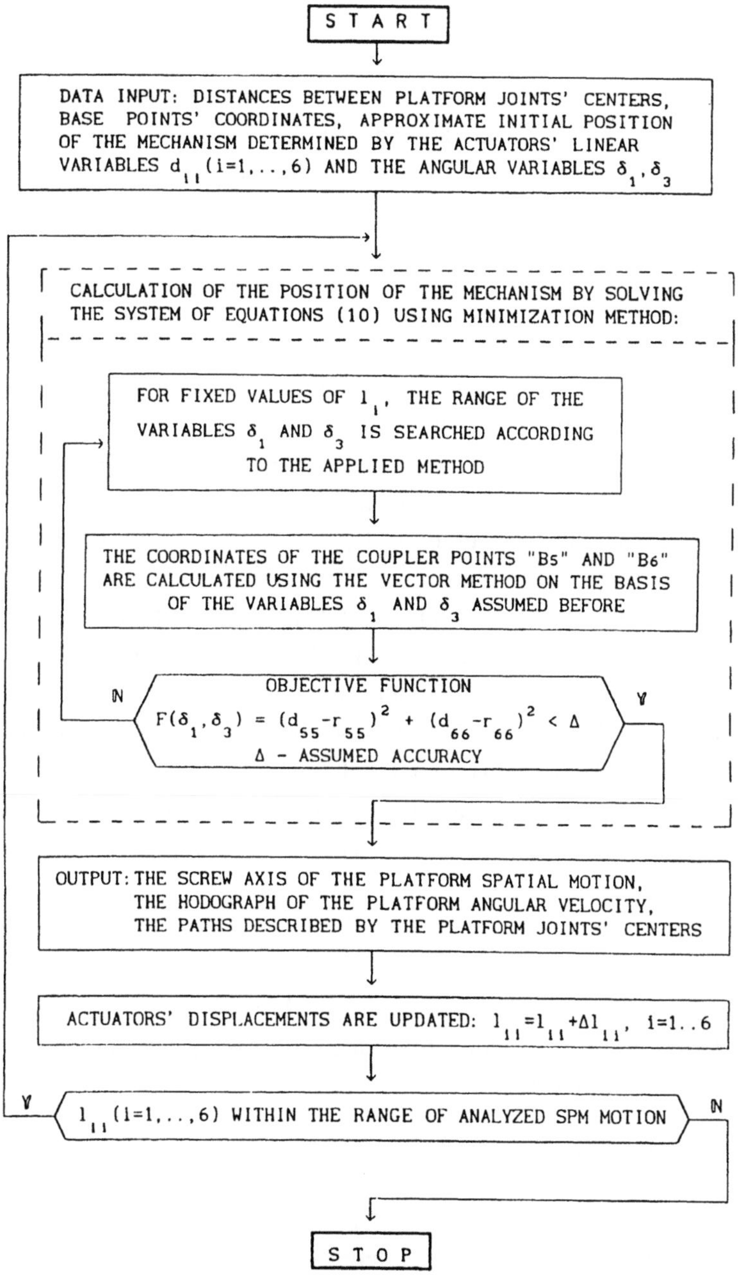

Fig.7.15. Flowchart of the algorithm for displacement analysis of the 6S-5S SPM.

A computer program for PC computers has been written in Turbo Pascal 6.0 on the basis of the algorithm described above. The evaluation of the objective function (calculated using the vector method) requires about 500 elementary operations (such as addition, multiplication, and root extraction). In order to minimize the objective function a sequential optimization method with variable step has been applied. The range of the displacements allowed in each leg actuator is divided into 100 discrete positions. The objective function is calculated about 100 times for any mechanism position (this number depends on the accuracy required by the user). Thus position analysis in the whole range requires about $100 \cdot 100 \cdot 500 = 5.000.000$ elementary operations. This requires about 20 seconds of calculation time for a PC/486 computer.

Platform motion analysis

The spatial motion of the platform can be described as a screw motion, which is a combination of rotation about a screw axis and a translation along the same axis. Thus the displacement of the platform can be completely specified by a set of screw parameters: the screw axis, and the linear and angular displacements relative to this axis. Screw parameters can be determined if two successive positions (corresponding to assumed increments in leg lengths) of three non-collinear points are known. As the range of the platform motion is divided into a number of finitely separated positions, the screw parameters of motion between any two positions can be found by using one of the methods described by Fenton and Shi (1990).

The unit vectors $\boldsymbol{l}$, $\boldsymbol{m}$, $\boldsymbol{n}$ of the frame attached to the platform are calculated using the coordinates of three platform joint centers B_1, B_2, B_3 (see Figure 7.12):

$$\boldsymbol{l} = \boldsymbol{b}_{12}\,,\ \boldsymbol{m} = (\boldsymbol{b}_{12} \times \boldsymbol{b}_{13})/\sin(\boldsymbol{b}_{12}, \boldsymbol{b}_{13}),\ \boldsymbol{n} = \boldsymbol{l} \times \boldsymbol{m}$$

These vectors are applied to form the platform orientation matrix $\boldsymbol{R}$ for the i- th position. The transformation from i-th to (i+1)-th orientation is described by the rotation matrix:

$$\boldsymbol{R}_{i,i+1} = \boldsymbol{R}_{i+1}\, \boldsymbol{R}_i^{-1}$$

where

$$\boldsymbol{R} = [\boldsymbol{l}\ \boldsymbol{m}\ \boldsymbol{n}] \text{ and } \boldsymbol{R}_{i,i+1} = [r_{ij}],\ (i = 1, 2, 3;\ \ j = 1, 2, 3)$$

The screw rotation angle $\theta_{i,i+1}$ and screw unit vector $\boldsymbol{u}_{i,i+1}$ can be determined by the elements of the rotation matrix:

$$\theta_{i,i+1} = \operatorname{atan2}\left(\pm\frac{1}{2}\sqrt{(r_{12}-r_{21})^2 + (r_{13}-r_{31})^2 + (r_{23}-r_{32})^2} \right)$$

$$2\,\boldsymbol{u}_{i,i+1} \sin\theta = [r_{32} - r_{23}\ \ \ r_{31} - r_{13}\ \ \ r_{21} - r_{12}]^{\mathrm{T}}$$

These parameters can be applied to determine a hodograph of angular velocity vector (unified in time domain).

Numerical example. The input data concerned the Mercedes 190 rear suspension, which can be considered a kind of the 6-5 type SPM, and have been determined on the basis of technical drawings using identification procedure. The initial position of the mechanism has been described using Cartesian coordinates of the following points:

$A_1(0, 0, 0)$,	$B_1(59, 457, -23)$
$A_2(-7, 223, 180)$,	$B_2 = B_1$
$A_3(423, 291, 20)$,	$B_3(146, 427, -68)$
$A_4(301, 223, 54)$,	$B_4(243, 461, 38)$
$A_5(311, 232, 171)$,	$B_5(180, 457, 166)$
$A_6(97, 151, 185)$,	$B_6(97, 461, 196)$

The analysis has been performed for the variable length of leg 2 , changed in the range Δd_2 [-80.0, 80.0] with step 1.6 assuming unit velocity ratio). Some numerical results are shown in Figure 7.16 as the hodograph of the platform angular velocity and the trajectory of joint center B.

a)

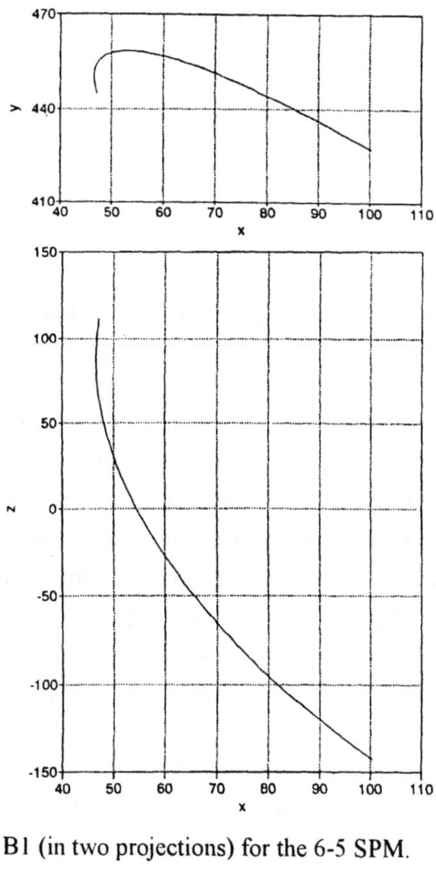

Fig.7.16a. Trajectory of point B1 (in two projections) for the 6-5 SPM.

b)

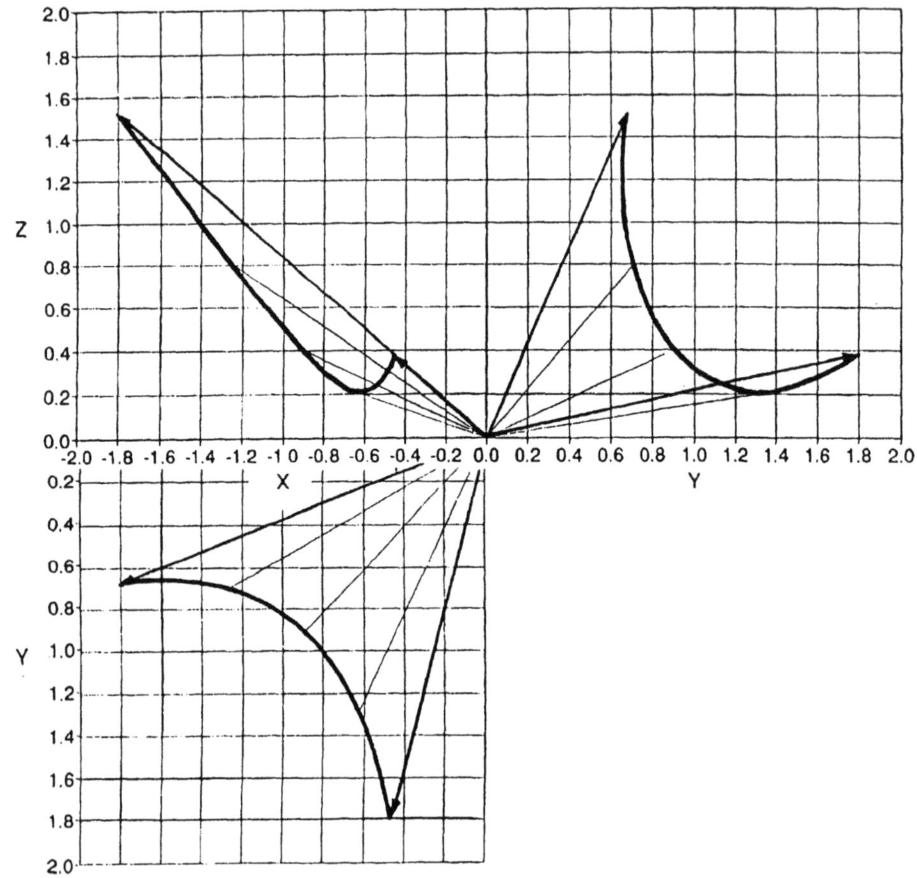

Fig.7.16b. Hodograph of the platform angular velocity obtained for the 6S-5S SPM from example 1.

The vector method presented above is efficient, accurate, and easy to apply for the SPM mechanism with a limited range of motion. By using this method we can determine the SPM displacement functions for any configuration, including singular positions. The functions mentioned above can be used as an input for velocity and acceleration analysis, which can be performed by numerical differentiation.

The numerical method applied in the algorithm presented allows us to reach the required accuracy very quickly. The algorithm can be used for the computer simulation of the platform motion, resulting from the given motion of one or more leg actuators simultaneously.

Some additional information about the relative motion parameters (such as angular displacements in the joints) can be easily obtained as intermediate results. The geometrical data obtained from technical drawings or measurements can be easily transformed into input data for the computer program, while output results can be presented as arrays, or graphs, and can be used for the identification procedure, if the number or accuracy of the given geometrical parameters is insufficient.

7.6. References

[7.1] *Dasgupta B., Mruthyunjaya T.S.*: A canonical formulation of the direct position kinematics problem for a general 6-6 Stewart Platform. Mech.Mach.Theory, Vol.29, No 6, pp.819-27, 1994.

[7.2] *Gosselin C., Angeles J.*: The Optimum Kinematic Design of a Planar 3-DOF Parallel Manipulator. Trans. ASME, Jnl of Mech. Transm. Autom. Design, Vol.110, pp.35-41, March 1988.

[7.3] *Gosselin C., Angeles J.*: The Optimum Kinematic Design of a Spherical 3-DOF Parallel Manipulator. Trans. ASME, Jnl of Mech. Transm. Autom. Design, Vol.111, No 2, pp.202-207, 1989.

[7.4] *Gosselin C.M.*: Parallel Computational Algorithms for the Kinematics and Dynamics of Spatial Parallel Manipulators. Trans.ASME, Jnl Dyn.Syst. Measur. Control, Vol,118, pp.22-28, March 1996.

[7.5] *Hunt K.H., Primrose E.J.*: Assembly Configurations of Some In-Parallel-Actuated Manipulators. Mech. Mach. Theory, Vol.28, No 1, pp.31-42, 1993.

[7.6] *Husty M.L.*: An Algorithm for solving the direct kinematics of general Stewart-Gough Platforms. Mech. Mach. Theory, Vol.31, No 4, pp.365-380, 1996.

[7.7] *Husain M., Waldron K.J.*: Direct Position Kinematics of the 3-1-1-1 Stewart Platform. Trans.ASME, Jnl of Mech.Design, Vol.116, Dec.1994, pp.1102-7.

[7.8] *Innocenti C.*: Algorithm for Kinematic Calibration of Fully Parallel Manipulators. Computational Kinematics. Kluwer Acad.Publ. 1995, pp.241-250.

[7.9] *Innocenti C.*: Direct Kinematics in Analytical Form of the 6-4 Fully-Prallel Mechanism. Trans.ASME, Jnl of Mech.Design, Vol.117, March 1995, pp.89-95.

[7.10] *Innocenti C., Parenti-Castelli V.*: Symbolic-Form Forward Kinematics of a 5-4 Fully-Parallel Manipulator. Advances in Robot Kinematics and Comp. Geometry, Kluwer Acad.Publ. 1994, pp.429-438.

[7.11] *Innocenti C., Parenti-Castelli V.*: Closed-Form Direct Positions Analysis of 5-5 Parallel Mechanism. Trans.ASME, Jnl of Mech.Design, Vol.115, Sept. 1993, pp.515-521.

[7.12] *Innocenti C., Parenti-Castelli V.*: Echelon Form Solution of Direct Kinematics for the General Fully-Parallel Spherical Wrist. Mech. Mach. Theory, Vol.28, No 4, pp.553-561, 1993.

[7.13] *Ji Z.*: Analysis of Design Parameters in Platform Manipulators. Trans.ASME, Jnl of Mech.Design, Vol.118, Dec.1996, pp.526-531.

[7.14] *Knapczyk J., Dzierżek S.*: Displacement and Force Analysisi of Five-Rod Suspension with Flexible Joints. Trans.ASME, Jnl of Mech.Design, Vol.117, Dec.1995, pp.532-538.

[7.15] *Knapczyk J., Dzierżek S.*: Kinematic Analysis of 6S-5S Stewart Platform by Using Vector Method. Prc. of the 3rd Int. Workshop on Advances in Robot Kinematics, Ferrara 1992, pp.123-128.

[7.16] *Knapczyk J., Tora G.*: An Inverse Force Analysis of Spherical 3 DOF Parallel Manipulator with Three Linear Actuators Considered as Spring System. Proc. the 11th CISM-IFToMM Symp. Romansy'11, Springer Wien New York 1997, pp.53-62.

[7.17] *Lin W., Griffis M., Duffy J.*: Forward Displacement Analysis of the 4-4 Stewart Platform. Trans.ASME, Jnl of Mech.Design, Vol.114, Sept.1992, pp.444-450.

[7.18] *Luh C.M. at all*: Working Capability Analysis of Stewart Platforms. Trans.ASME, Jnl of Mech.Design, Vol.118, June.1996, pp.220-227.

[7.19] *Merlet J.P.*: Les robots paralles. Hermes, Paris 1990.

[7.20] *Murthy V., Waldron K.J.*: Position Kinematics of the Generalized Lobster Arm and Its Series-Parallel Dual. Trans.ASME, Jnl of Mech.Design, Vol.114, Sept.1992, pp.406-413.

[7.21] *Nielsen J., Roth B.*: The direct kinematics of the general 6-5 Stewart-Gough mechanism. Recent Advances in Robot Kinematics. Kluwer Acad.Publ. 1996, pp.7-16.

[7.22] *Parenti-Castelli V.*: Recent techniques for direct position analysis of the generalized Stewart platform mechanism. Proc. of the 3rd Int. Workshop on Advances in Robot Kinematics, Ferrara 1992, pp.129-135.

[7.23] *Raghavan M.*: The Stewart Platform of General Geometry Has 40 Configurations. Trans.ASME, Jnl of Mech.Design, Vol.115, June 1993, pp.277-282.

[7.24] *Pennock G.R., Kassner D.J.*: Kinematic Analysis of a Planar Eight-Bar Linakage: Application to Platform Type Robot. Trans.ASME, Jnl of Mech.Design, Vol.114, March 1992, pp.87-95.

[7.25] *Shi X., Fenton R.G.*: Solution to the Forward Instantaneous Kinematics for a General 6-DOF Stewart Platform. Mech. Mach. Theory, Vol.27, No 3, pp.251-259, 1992.

[7.26] *Sternheim F.*: Computation of the direct and inverse geometric models of the Delta 4 parallel robot. Robotersysteme 3, pp.199-203, 1987.

[7.27] *Waldron K.J., Raghavan M., Roth B.*: Kinematics of a Hybrid Series-Parallel Manipulation System. Trans.ASME, Jnl of Dyn. Syst. Meas. Control, Vol.3, pp.211-221, 1989.

[7.28] *Wohlhart K.*: Direct Kinematic Solution of the General Planar Stewart Platform. Proc. of Int. Conf. on Computer Integrated Manufacturing, pp.401-411, Zakopane 1992.

[7.29] *Wohlhart K*: A Parallel Redundant Manipulator Based on the Assur Group (3,4). Proc. of Int. Conf. on Computer Integrated Manufacturing, pp.371-8, Zakopane 1994.

[7.30] *Wohlhart K*: Position Analysis of the Rhombic Assur Group. Proc. of the 10th CISM-IFToMM Symp. Romansy`10, Gdańsk 1994.

[7.31] *Yang P.H., Waldron K.J.*: Kinematics of a 3-DOF Motion Platform for a Low-Cost Driving Simulator. Recent Advances in Robot Kinematics, pp.89-98, Kluwer Acad. Publ. 1996.

Part Two

GRIPPERS, DRIVES AND SENSORS OF MANIPULATORS AND ROBOTS

8. MANIPULATOR AND ROBOT GRIPPERS

The main function of a gripper is to grasp the object of manipulation, hold it during manipulation, and release it at its destination. Various factors influence gripper design and construction: these are mainly object-related features, such as its shape, mass, dimensions, mass distribution, material, surface, temperature, etc. Various systems have been proposed for the classification of the many different grippers that can be found in practice.

8.1. Gripper functions

Prehension is one of the basic actions of men and animals. The most highly developed natural gripper is the human hand. Consisting of a palm with five fingers (Figure 8.1a), it contains 18 links and 17 revolute (1 and 2 degree-of-freedom) kinematic pairs. In the context of the theory of mechanisms, the hand is a mechanism described by the following parameters:

$n = 18, \quad p_5 = 13, \quad p_4 = 5,$

where:

n is the number of movable links, and

p_5 and p_4 are kinematic pairs of the fifth and fourth class respectively.

Thus the mobility (w) calculated with respect to the radiocarpal joint, treated as the base, is (Figure 8.1b):

$$w = 6n - 5p_5 - 4p_4 = 6 \cdot 18 - 5 \cdot 13 - 4 \cdot 5 = 23 \qquad (8.1)$$

Six basic types of prehension can be identified: cylindrical (Figure 8.2a), pinch (Figure 8.2b), hook (Figure 8.2c), palm (Figure 8.2a), spherical (Figure 8.2e) and lateral (Figure 8.2f). Other, different classifications of prehensile motion can be found, where distinguishing even nine various types of prehension [8.3].

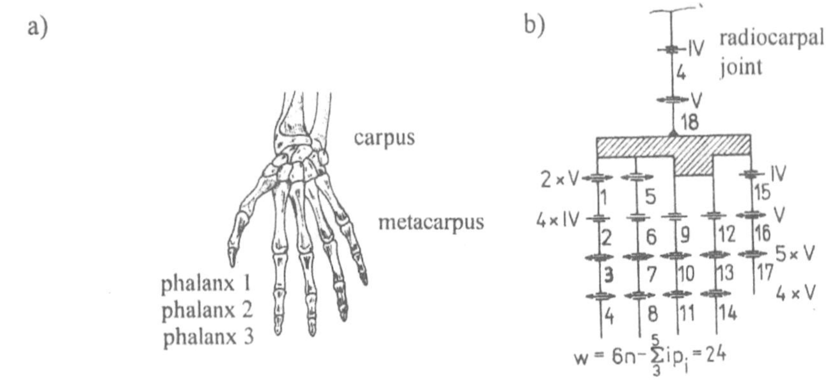

Fig. 8.1. The human hand: a) palm with fingers, b) structural diagram

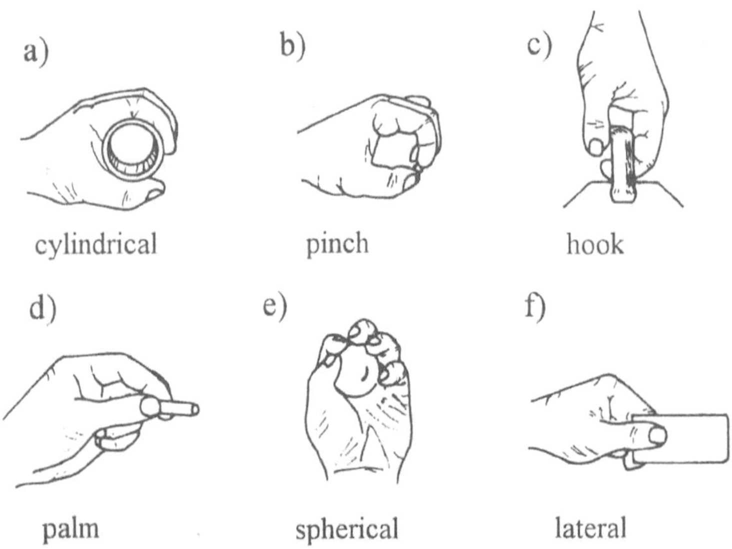

Fig. 8.2. Types of prehension

8.2. Systematic classification of grippers

Various criteria lead to various classification schemes. These criteria for gripper classification may be the number of fingers, joints, degrees of freedom, or the type of drive. Different combinations of these factors lead to different gripper models. Assuming a classification according to the number of fingers, typical designs with two, three and five fingers can be distinguished. In industrial applications two-finger grippers (Figure 8.3a) are the most common. Three- and five-fingered grippers are usually employed in prosthetic manipulators (figs 8.3b, c, d).

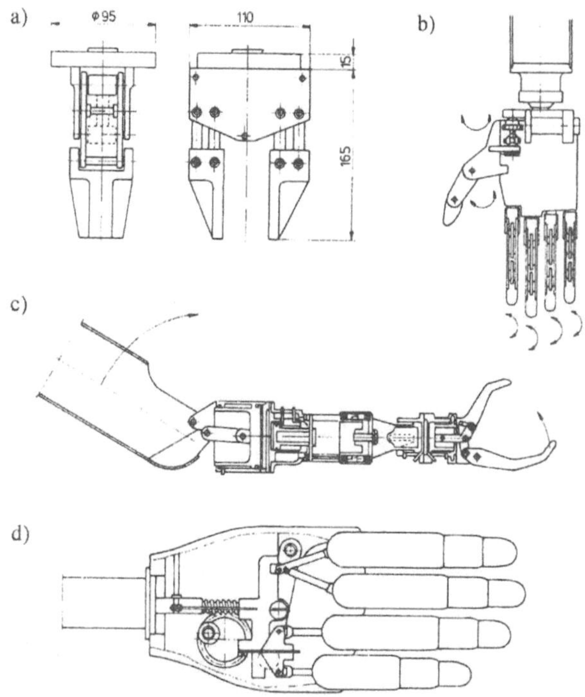

Fig. 8.3. Types of grippers: a) gripper with translating jaws, b) artificial hand with five fingers, c) Warsaw hand, d) Belgrade hand

Another classification divides grippers into single or multiple gripper structures situated on the radiocarpal joint of the manipulator's wrist. The wrist allows the gripper to perform several functions simultaneously (Figure 8.4). External (Figure 8.5a) and internal (Figure 8.5b) grippers can also be defined. Figure 8.6 shows a multiphalangeal, "soft" gripper, designed for enfolding objects of different shapes. Figure 8.7 shows a vacuum gripper with vacuum cups.

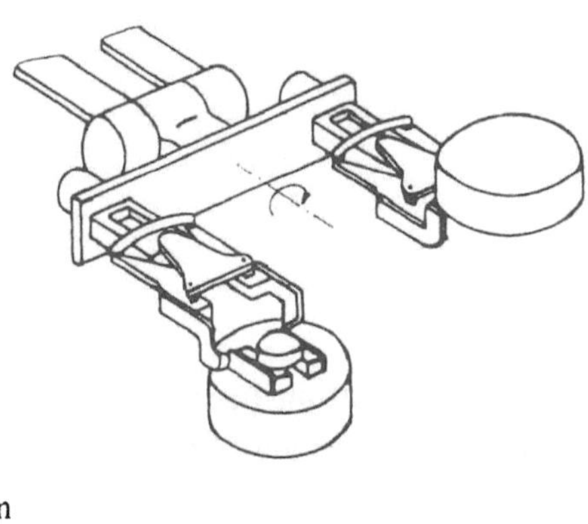

Fig. 8.4. Multi-gripper system

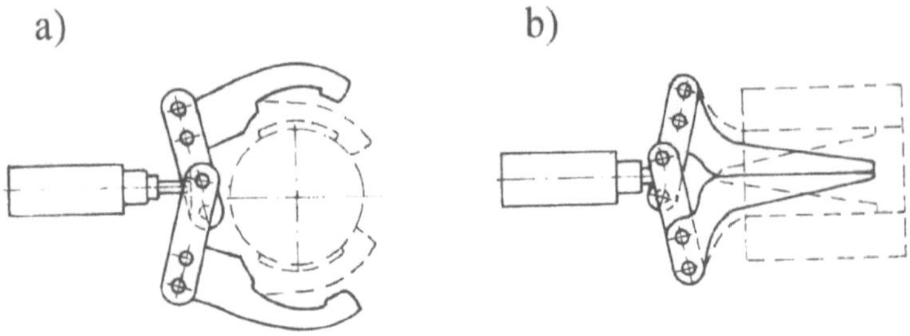

Fig. 8.5. Types of grippers: a) external gripper, b) internal gripper

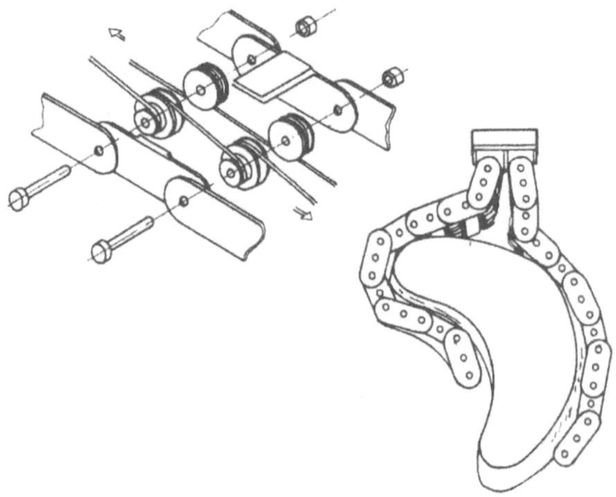

Fig. 8.6. Two-fingered gripper which can adapt to different object shapes

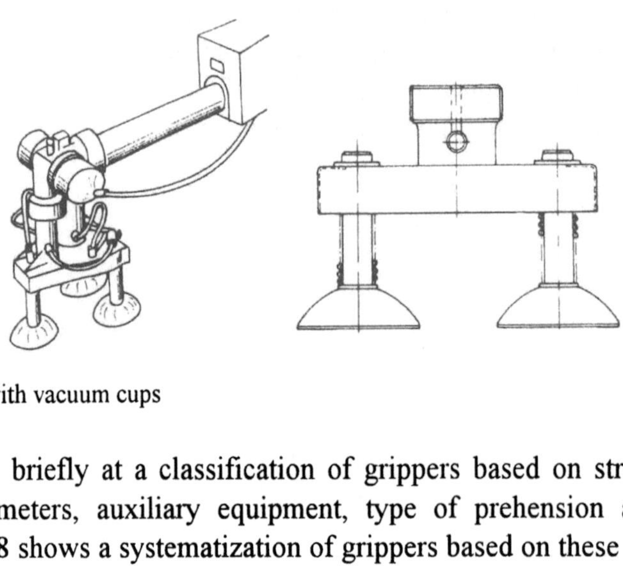

Fig. 8.7. Gripper with vacuum cups

Let us look briefly at a classification of grippers based on structural criteria, functional parameters, auxiliary equipment, type of prehension and attachment mode. Figure 8.8 shows a systematization of grippers based on these criteria.

Industrial robot grippers

- Construction
  - Drive system
    - Pneumatic
    - Hydraulic
    - Electric
    - Electromagnetic
      - with pneumatic release
      - without pneumatic release
    - Adhesive
  - Transmission of driving motion
    - Linkage
      - vice-type
      - shear-type
      - finger-type
      - clamp-type
    - Yoke
    - Wedge
    - Gear
    - Rope (string)
  - End-effector
    - Rigid fingers
      - two-fingered
      - three-fingered
      - multi-fingered
    - Elastic fingers
    - Flexible fingers
- Operating parameters
  - Grasping force
    - Constant
    - Adjustable
    - Controlled automatically
  - Object dimensions
    - Maximum
    - Minimum
  - Grasping time
- Auxiliary equipment
  - Sensors
    - Touch sensors
    - Proximity detectors
    - Force sensors
    - Temperature sensors
  - Auxiliary mechanisms
    - Flexible clamps
    - Rotation
    - Linear displacement
  - Processing equipment
    - Cutter
    - Air blower
    - Spray nozzle
- Prehension type
  - Force prehension
    - Pressure
    - Adhesion
  - Shape prehension
- Fixing system
  - Manual
  - Automatic

Fig. 8.8. Systematic classification of industrial robot grippers

Most grippers are fitted with hydraulic or pneumatic drives, linkage transmissions and rigid fingers to grasp the object with a calculated (and adjustable) force. Sensor systems are still no widely used in industrial applications. Figure 8.9 shows the flow chart of a gripper. Considering the wide variety of gripper designs, a very convenient criterion of systematization is the classification by the number, class and sequence of kinematic pairs [8.1].

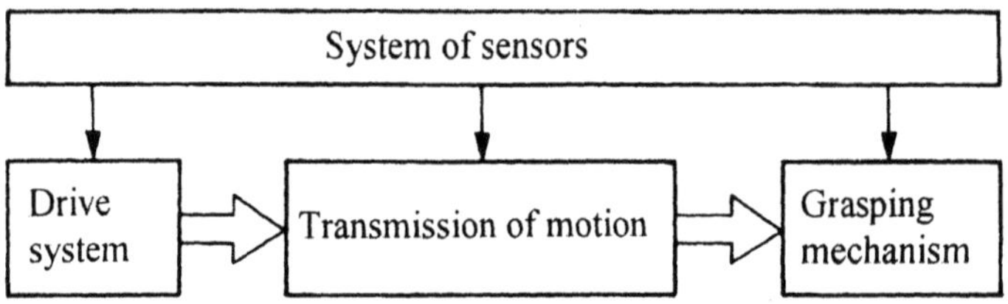

Fig. 8.9. Flow chart of a gripper

The most common drive transmission systems are shown in Figure 8.10. The driving motion can be transmitted by a linkage (Figure 8.10a), wedge (Figure 8.10b), yoke (Figure 8.10c), gears (Figure 8.10d), or chain (Figure 8.10e). The mobility w of such mechanisms, computed from the formulas given below, equals 1.

$w = 3n - 2p_5 = 3 \cdot 5 - 2 \cdot 7 = 1$ (Figure 8.10a)

$w = 3 \cdot 3 - 2 \cdot 3 - 2 = 1$ (Figure 8.10b)

$w = 3 \cdot 3 - 2 \cdot 3 - 2 = 1$ (Figure 8.10c)

$w = 3 \cdot 3 - 2 \cdot 4 = 1$ (Figure 8.10d)

$w = 3 \cdot 3 - 2 \cdot 4 = 1$ (Figure 8.10e)

Let us analyze structural variants of grasping mechanisms featuring a mobility of one, and containing only fifth-class kinematic pairs. The relationship

$$3n - 2p_5 = 1 \tag{8.2}$$

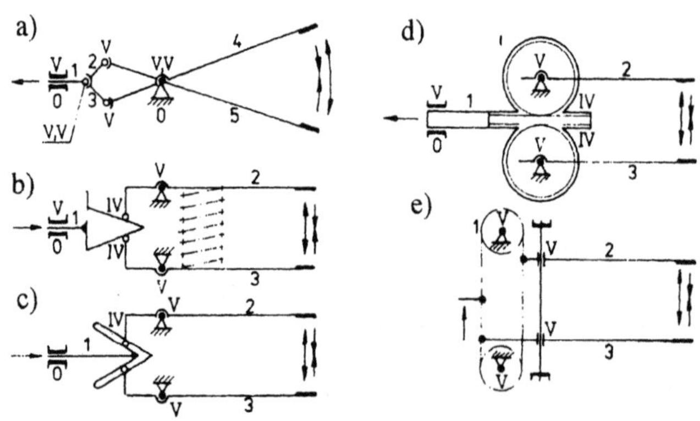

Fig. 8.10. Transmission of driving motion: a) linkage, b) wedge, c) yoke, d) gear, e) chain

yields

$$n = \frac{1+2p_5}{3} \tag{8.3}$$

The pairs of numbers (n, p_5) satisfying equation (8.3) are specified in Table 8.1.

Table 8.1. Corresponding numbers of movable links (n) and fifth-class kinematic pairs (p_5) contained in gripping mechanisms with $w = 1$

N	1	3	5	7	9	11	13	....
p_5	1	4	7	10	13	16	19	....

Typical grasping mechanisms using different (n, p_5) combinations are illustrated in Table 8.2. The simplest grippers have one movable link (n) and one fifth-class kinematic pair (p_5). These are pneumatic or hydraulic actuators. One jaw of these actuators is connected to a fixed cylinder, the other one to a movable piston. Grippers of this kind present major drawbacks, such as the limited of changing the direction of motion.

A commonly used group of mechanisms are those featuring five moving links and seven fifth-class kinematic pairs. The advantages of this type of grippers are their compact construction and structural symmetry. Grippers with a greater number of links and joints, namely (n = 7, p_5 = 10) and (n = 9, p_5= 13) are also found. Figure 8.11 shows an example of this. As the number of joints increases, the gripper's construction becomes more complex and difficult to engineer. Planar mechanisms may also have fourth-class kinematic pairs. In this case the relationship

$$3n - 2p_5 - p_4 = 1 \tag{8.4}$$

yields

$$n = \frac{1+p_4+2p_5}{3} \tag{8.5}$$

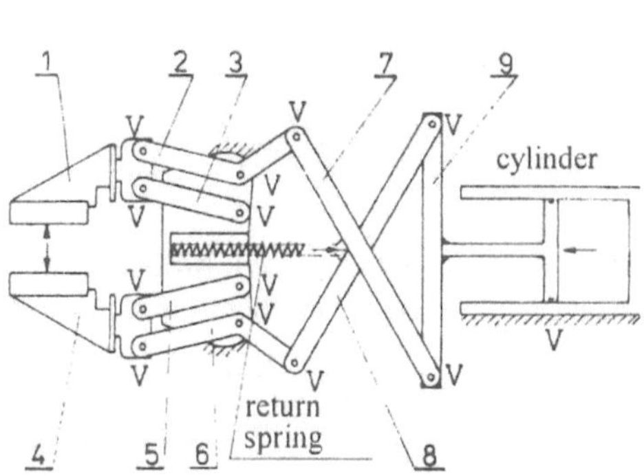

Fig. 8.11. Linkage-type gripper

Table 8.3 specifies selected combinations of movable links and fourth and fifth-class kinematic pairs.

Table 8.2. Grasping mechanisms with $w = 1$, various numbers of links, and fifth-class kinematic pairs.

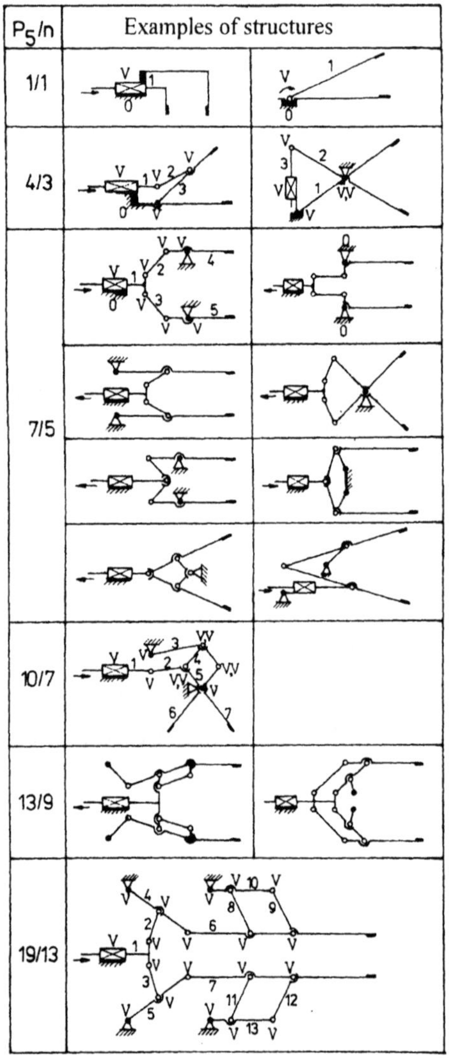

Table 8.3. Corresponding numbers of movable links (n), fifth-class kinematic pairs (p_5) and fourth-class kinematic pairs (p_4) contained in prehension mechanisms with $w = 1$.

n	2	4	6	8	....	3	5	7	9	....	4	6	....	5	7	9	....	6	....	7	....
p_5	2	5	8	11	....	3	6	9	12	....	4	7	....	5	8	11	....	6	....	7	....
p_4	1	1	1	1	....	2	2	2	2	....	3	3	....	4	4	4	....	5	....	6	....

Table 8.4. Prehension mechanisms containing fourth- and fifth-class kinematic pairs.

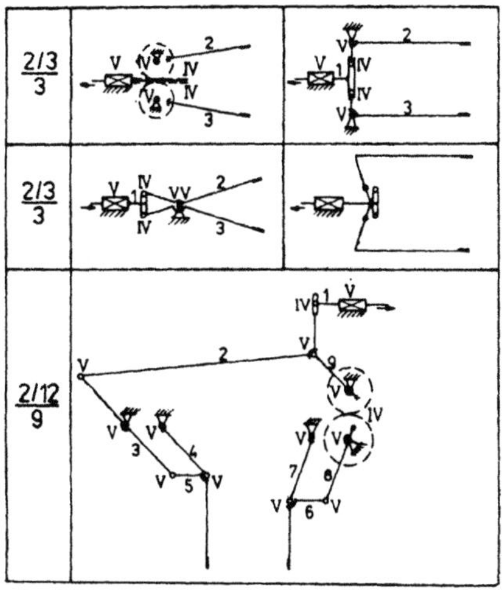

One advantage of grasping mechanisms containing fourth- and fifth-class kinematic pairs is the lower number of intermediate links transmitting the motion. Their drawback is the generation of considerable friction forces in the fourth-class kinematic pair. Up-to-date solutions of industrial robot grippers usually contain structures with two fourth-class kinematic pairs and three fifth-class kinematic pairs (Table 8.3). Table 8.4 shows a more developed structure consisting of two fourth-class kinematic pairs, twelve fifth-class kinematic pairs and nine movable links.

Figure 8.12 shows a gripper composed of four movable links and three fourth-class kinematic pairs and four fifth-class kinematic pairs, with a mobility equal to one. Figure 8.13 presents a typical solution for an industrial robot gripper ("Universal 5") with a lifting capacity of 5 kg. The gripper is fitted with a pneumatic actuator and develops a gripping force of 800 N. Figure 8.14 shows an example of a two-fingered gripper. Figures 8.15 and 8.16 show examples of self-centering grippers providing passive correction of position. The three-fingered gripper illustrated in Figure 8.17 has a lifting capacity of 2 kg (minimum mass: 100 g) and is designed to grasp cylindrical objects and rings from both the inside and the outside. The diameter of an object grasped from the outside may range from 2 to 100 mm, while the diameter of an object grasped from the inside may range from 25 to 100 mm. The prehensile motion is driven by a double-action pneumatic actuator. The gripping mechanism consists of a gear drive (two sector gears, 2) and a lever (1). The gripper is connected to an arm. A double-jawed gripper (Figure 8.18) can be used for the simultaneous manipulation of two objects. Another interesting construction (Figure 8.19) permits the gripping of small balls [8.2, 8.4].

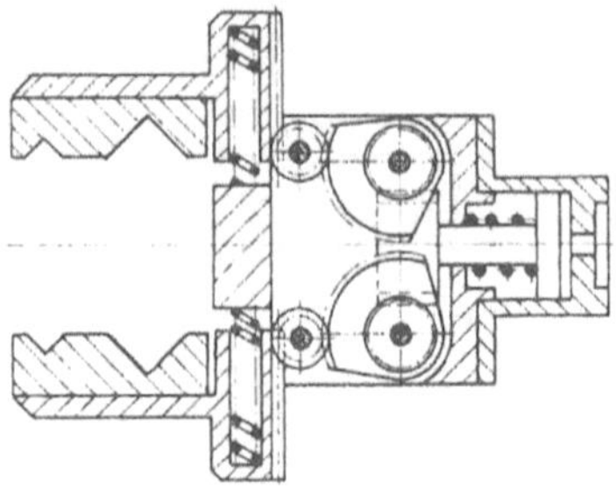

Fig. 8.12. Gripper with gear drive

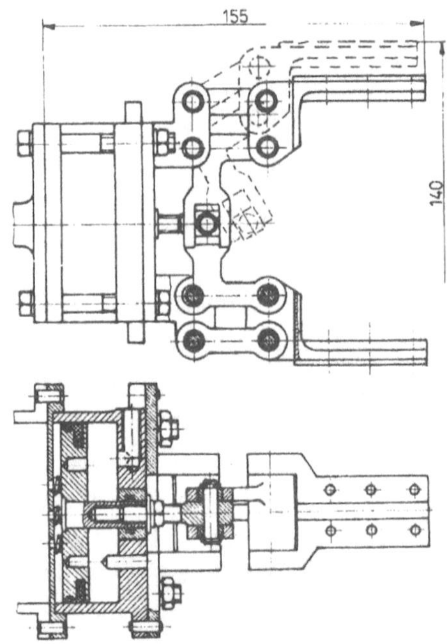

Fig. 8.13. Air-driven gripper of "Universal 5" robot

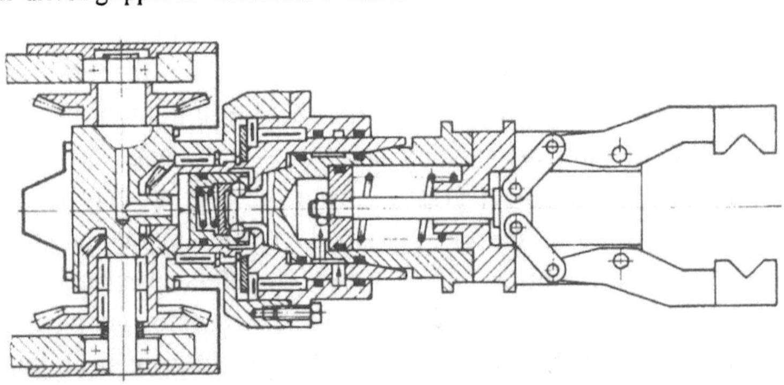

Fig. 8.14. Gripper for handling heavy objects

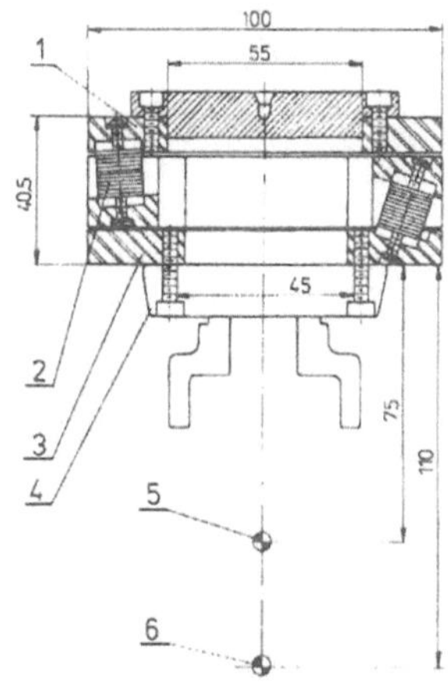

Fig. 8.15. Self-adjustable gripper, 1 - surface for internal connection 55 mm, 2 - compliant plates, 3 - surface for internal connection 45 mm, 4 - gripper face, 5, 6 -correction centre AST-100-110

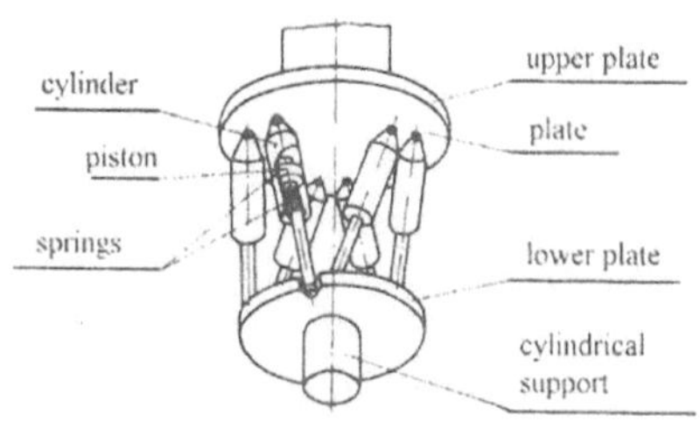

Fig. 8.16. Gripper with passive self-correction of position

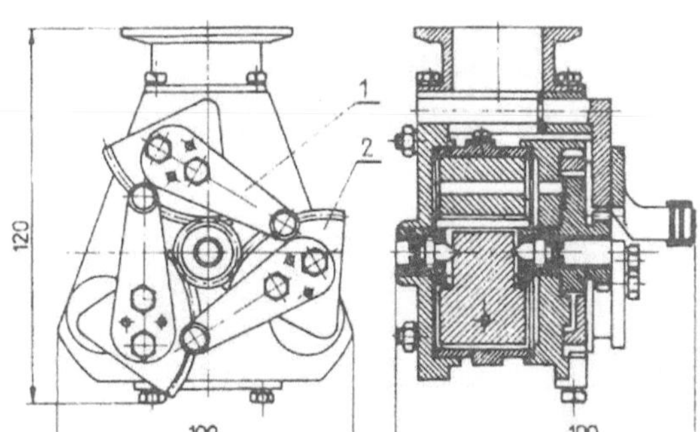

Fig. 8.17 Gripper with three fingers and a gear drive, capable of lifting up to 2 kg

Figure 8.20 shows the design of the Stanford hand (IPL) installed on the Unimation 600 industrial robot. A characteristic feature of this solution is the application of a pulley-and-rope drive. The arrangement of the ropes is shown in Figure 8.21, and the resistance sensors for measuring the force developed are illustrated in Figure 8.22 (56).

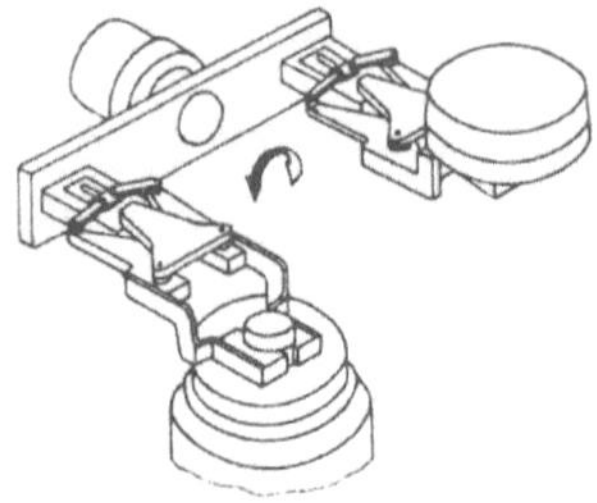

Fig. 8.18. Gripper with doubled jaws

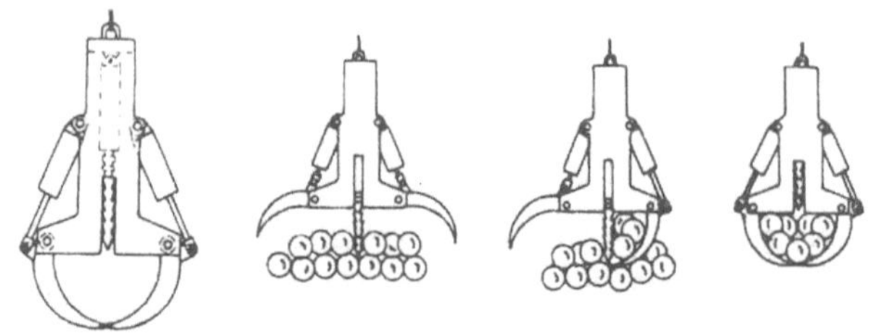

Fig. 8.19. Gripper for bulk materials

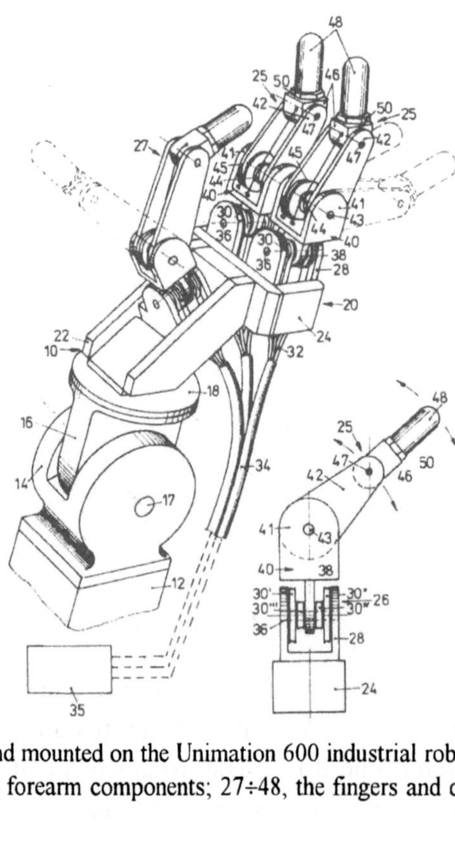

Fig. 8.20. Stanford IPL hand mounted on the Unimation 600 industrial robot; 10, 12, 14, 16, 17, 18, 20, 22, 24 are the arm and forearm components; 27÷48, the fingers and drive links; 35, the pulley-and-rope control.

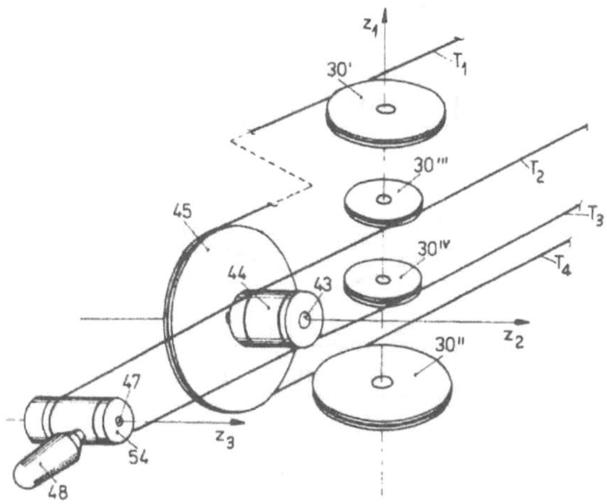

Fig. 8.21. Arrangement of the ropes (T_1 - T_4) providing the development of torques in the three finger joints.

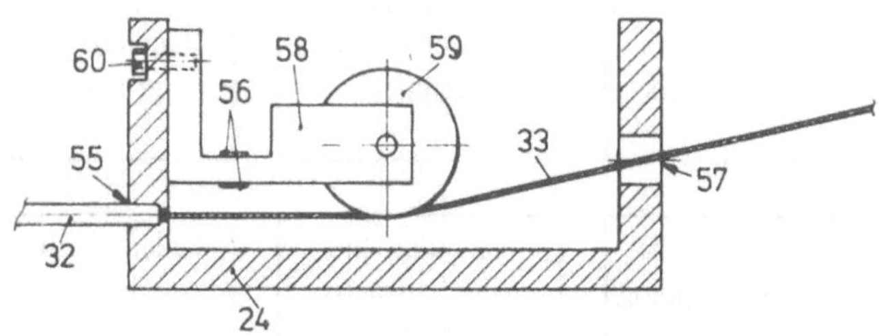

Fig. 8.22. Rope tension sensor. The tension can be measured by means of a resistance sensor 56 (Mason and Salisbury, Robot Hands and Mechanics of Manipulation): 24 ÷ 60 are the rope guiding and sensor mounting elements.

8.3. Methodology of design and selection of industrial robots grippers

The suitable design (or selection) of the gripper is a decisive factor in the manipulation process. The methodology of gripper selection includes the following steps:

1) selection of the type of prehension,
2) selection of the type of gripper (and principle of operation),
3) selection of gripper design parameters,
4) adaptation of the gripper jaws to the shape of the object's surface.

The flow chart for the selection of the industrial robot gripper is shown in Figure 8.23.

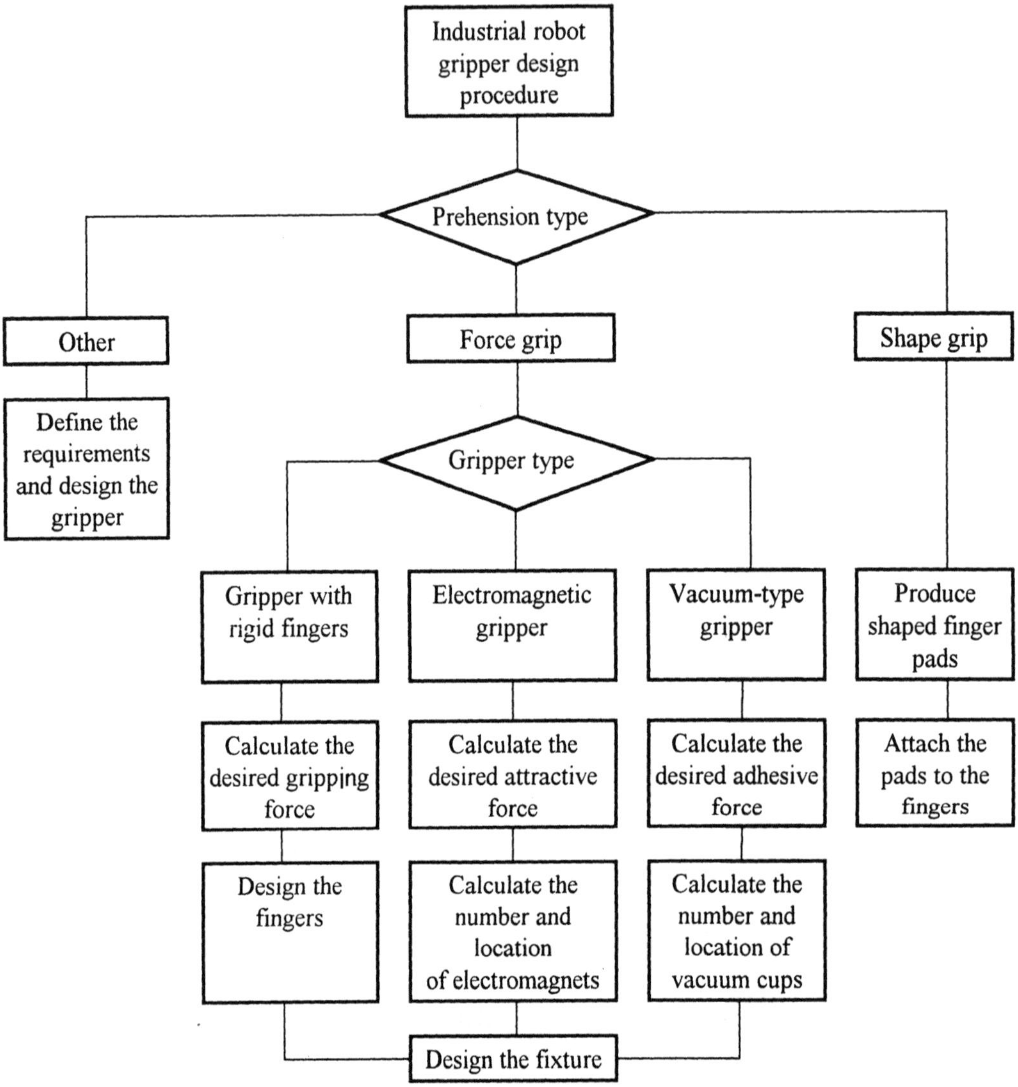

Fig. 8.23. Industrial robot gripper design procedure

8.3.1. Selection of type of grasping

One of the first things we must decide in selecting a gripper is how it should grasp the object. Proper grasping depends on:

- the method of fixing the object in the gripper,
- the parameters of the object of manipulation,
- the initial position of the object with reference to the gripper,
- the dynamic conditions of the manipulation process.

The grasping mechanism is a specific assembly of an industrial robot; it interacts directly with the manipulated object and fixes it in the gripper's jaws. To perform the desired manipulation properly and accurately there must be no possibility of displacement of the object in the gripper. The gripper holds the object from six directions to prevent this: along three axes of the coordinate system and around these axes. Fixing can be provided in two ways:

1) by exerting forces on the object of manipulation (force grip),
2) by depriving the object of all six degrees of freedom (shape grip).

This action must be able to prevent the object of manipulation from changing its position relative to the gripper during the desired manipulation, i.e. it must neither slide out of the gripper's jaws, nor change its orientation.

In the case of force prehension the force should prevent any displacement of the object during manipulation. In the case of shape prehension the force makes the fingers close, but it is not applied directly on the object of manipulation situated between the gripper pads. It is the pads on the gripper fingers, shaped to provide a seat for the object (or for some of its parts), that deprive the object of all six degrees of freedom. Shape prehension is used for fragile objects, flexible parts, and the like, where no contact forces should be applied to their surfaces.

Table 8.5. Examples of manipulated object shapes and recommended finger pad types

Gripper type / Object of manipulation	With rigid fingers	Vacuum-type	Electromagnetic	Other
Shafts, sleeves	For a wide range of diameters and lengths	Not recommended	Only small lengths and large diameters	Pneumatic grippers with elastic diaphragms
Plates	Not recommended	Recommended	Recommended	For small parts - grippers with elastic fingers
Sheets, plates	Not recommended	Recommended, with several vacuum cups	Recommended, with several electromagnets	
Boxes	Only special designs	Recommended, with numerous spatially arranged vacuum cups	Recommended, with numerous spatially arranged electromagnets	
Complex shapes	Not recommended	Recommended, with appropriately arranged numerous vacuum cups	Recommended, with appropriately arranged numerous electromagnets	Special designs of grippers

In industrial practice combined force and shape prehension is commonly used; the fingers partially limit the freedom of the objects by their shape, while simultaneously applying suitable forces on the object, preventing its displacement.

If the shape prehension option is selected, shaped finger pads must be manufactured (in most cases by preparing an exact negative of the object's surface shape) and these pads attached to the fingers of a standard gripper.

If the force prehension is selected, we must determine the most reliable way of grasping the given object, considering the shape of the object as the most important factor. Examples of manipulated object shapes are shown in Table 8.5, together with the possibilities of application of different types of grippers.

8.3.2. Selection of the type of gripper

As the objects of manipulation feature a wide variety of shapes, sizes, materials, surfaces, and weights, a very great number of gripper designs have been developed to date. But no catalogues of industrial robot grippers are available on the market, and although in possession of various technical solutions, manufacturers they usually supply a manipulator with one of three types of grippers: with rigid fingers, with vacuum cups, or electromagnetic. The user must choose the grippers suitable for the objects manipulated.

The main **technical parameters** of industrial robot grippers include:

- for grippers with rigid fingers:
- rated lifting capacity [kg],
- grasping force [N],
- grasping time [s],
- release time [s],
- maximum and minimum dimensions of the object [mm],
- motion of fingers,
- output force of the drive system [N],
- actuator travel [mm],
- hydraulic fluid pressure [Pa].
- for vacuum-type grippers:
- vacuum cup area [mm^2],
- rated vacuum pressure in the cup [Pa].
- for electromagnetic grippers:
- number of turns of the winding,
- dimensions of the contact area between the gripper face and the object [mm^2],

- supply voltage.

Because of their simple construction, vacuum-type and electromagnetic grippers are commonly used whenever possible.

The main components of a **vacuum-type gripper** are vacuum cups and a vacuum generator. The cups are made of rubber, or flexible plastics. In most cases, the vacuum is generated by an ejector pump using compressed air power. Vacuum grippers are also mounted on joints which provide for the proper positioning and orientation of the gripper, allowing energy savings of vacuum or compressed air (e.g. switching the vacuum on only when the cup touches the surface of the object).

When deciding to use a vacuum-type gripper one must remember that:

- those grippers may be used with flat and even surfaces of various materials,
- those grippers provide limited adhesive force for a given surface,
- positioning accuracy is reduced due to cup flexibility,
- there must be no dust or dirt between the cup contact surface and the object of manipulation,
- grasping time is longer as it takes time to generate the desired vacuum,
- connections must be air-tight,
- gripper durability is reduced by the wear and tear on vacuum cups.

An **electromagnetic gripper** consists of a magnet core and a winding. Permanent magnets are sometimes used, but this requires an additional device for releasing the object. Magnetic grippers are usually employed to carry ferromagnetic parts with shaped surfaces that cannot be grasped by other types of grippers.

When using an electromagnetic gripper one must remember that:

- these grippers can be used only for ferromagnetic materials,
- these grippers can generate a high force of attraction per area unit,
- a sudden displacement during grasping may impair positioning accuracy; in some cases the desired accuracy can be obtained by means of a stiff bolt.
- the presence of residual magnetism creates the risk of attracting dirt onto the surface of the gripper and the object of manipulation,
- prehension times are very short,
- the winding generates heat when the gripper is in operation,
- the core and windings may be made by the user.

8.3.3. Design parameters of the gripper

Having selected the type of gripper its parameters must be established; in particular the desired gripping force must be calculated.

The force of attraction of an electromagnet can be determined as:

$$F = \frac{(In)^2}{25P(R_p + R_m)} \tag{8.6}$$

where:

I is the current in the winding,
n, the number of turns,
P, the contact area between the object and the electromagnet, and
R_p, R_m are the magnetic reluctance of air and metal respectively.

The adhesive force of a vacuum-type gripper can be determined as

$$F = k_p P(p_a - p_p) \tag{8.7}$$

where

P is the working area of the vacuum cup,
p_a, the atmospheric pressure,
p_a, the pressure inside the vacuum cup,
$k_p = 0.85$ is a coefficient allowing for changes in the atmospheric pressure and in the vacuum cup working area.

In both cases the values of the adhesive force of force of attraction computed as above, can be increased by the application of several electromagnets or vacuum cups.

The ability of a vacuum-type, or electromagnetic gripper to hold the object of manipulation can be determined by the following relationship:

$$\mu\left(\frac{F}{mk_1k_2} - g\cos\theta_0 + a\cos\theta\right) \geq g\sin\varphi_0\sin\theta_0 + a\sin\varphi\sin\theta \tag{8.8}$$

where:

F is the adhesive force [N],
m, the object mass,
g, the gravitational acceleration [m/s^2],
a, the acceleration of the gripper [m/s^2],
θ, the angle between the acceleration vector and the gripper axis,
θ_0, the angle between the vector of gravitational acceleration and the gripper axis,
φ, the angle between the projection of the acceleration vector on the plane perpendicular to the Z-axis and the Y-axis, and
φ_0, the angle between the projection of the vector of gravitational acceleration on the plane perpendicular to the Z-axis and the Y-axis (Figure 8.24).

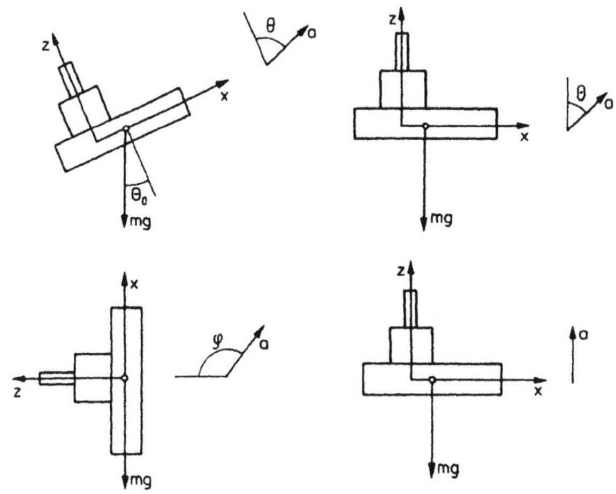

Fig. 8.24. Special cases of manipulation when the object is held by a vacuum-type, or electromagnetic gripper.

$$
\begin{cases}
\varphi = \varphi_0 = \pi/2 \\
\mu\left(\dfrac{F}{mk_1k_2} - g\cos\theta_0 + a\cos\theta\right) \geq g\sin\theta_0 - a\sin\theta \\
\varphi = \varphi_0 = \pi/2, \quad \theta_0 = 0 \\
\mu\left(\dfrac{F}{mk_1k_2} - g + a\cos\theta\right) \geq a\sin\theta \\
\varphi = \varphi_0 = \pi/2, \quad \theta_0 = \pi/2 \\
\mu\left(\dfrac{F}{mk_1k_2} - a\cos\theta\right) \geq g + a\sin\theta \\
\varphi = \varphi_0 = \pi/2, \quad \theta_0 = \theta = \pi/2 \\
\mu\dfrac{F}{mk_1k_2} \geq g + a
\end{cases}
\tag{8.9}
$$

where:

μ is the coefficient of friction between the object and the gripper (between steel and rubber $\mu = 0.3$; between steel and other metals, $\mu = 0.17$),

k_1 is the safety factor (surplus of gripping force) depending on gripper operating conditions,

k_2 is a factor allowing for the shift between the point at which the adhesive force is applied and the center of gravity of the object, depending on the gripper construction:

$$k_2 = \frac{r}{r + \Delta x} \tag{8.10}$$

r is the mean radius of the gripper surface,

Δx is the shift between the gripper axis and the center of gravity of the object.

In the case of grippers with rigid fingers, the most significant parameters in obtaining the proper prehension for a given object are: finger displacement ranges, gripping force, gripper dimensions, type of prehension, prehension depth and other design parameters of the gripper. The user, who as a rule possesses only one solution for a gripper with rigid fingers, has a very narrow margin for changing the design parameters of the gripper (limited to changing the fingers, or the power supply). Lower pressure supplied to the gripper leads to lower gripping forces; the gripping force changes usually with any change in finger length.

If the grippers delivered by the gripper manufacturer allow no possibility of grasping the object with the desired parameters, one should attempt to design his own gripper, following the recommendations specified in 8.4.

8.3.4. Adaptation of the gripper fingers to the shape of the object of manipulation

Most gripper designs contain two opposite bolts on which rigid fingers are mounted. By changing the size and shape of the fingers, the gripper may be adapted for a given object of manipulation. Such adaptation may include:

- a change in the shape of the fingers (adaptation to the object's shape),
- a change in the fingers' dimensions (adaptation to the object size),
- a change in the contact surface of the fingers (touching the object),
- the application of special materials on the contact surface of the fingers (pads with increased friction, elastic pads, etc.)

8.4. Industrial robot grippers - principles of design

The main stages of the gripper design process are as follows:

1) location of the grip,
2) calculation of the forces and moments acting on the object in dynamic conditions,
3) determination of the number of fingers, their type of motion, and range of displacement,
4) determination of the structure of the drive transmission system,
5) calculation of the drive system,
6) design of the gripper's construction, including connection with the manipulator.

Fig. 8.25 shows the sequence of individual stages of industrial robot gripper design.

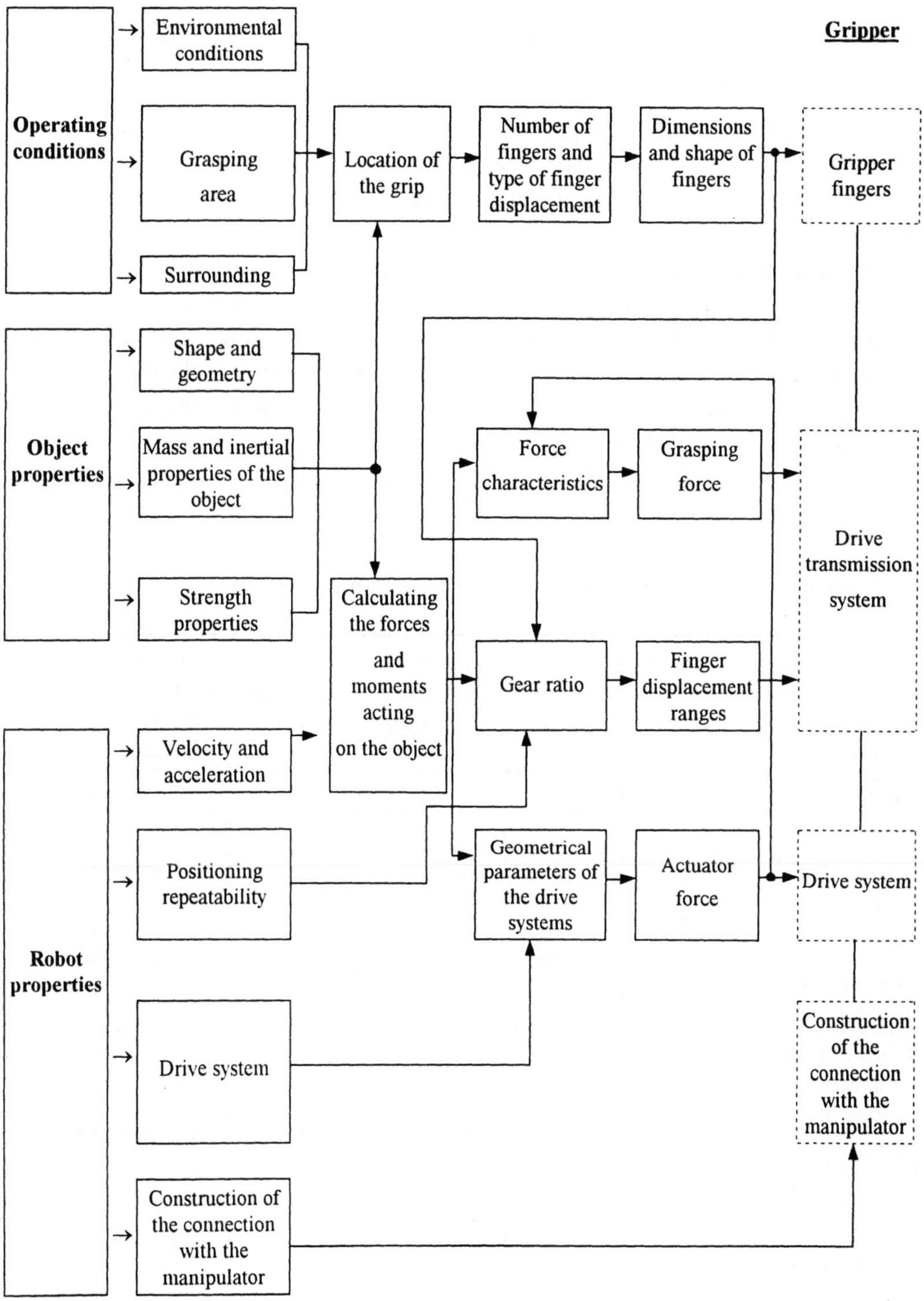

Fig. 8.25. Gripper design stages.

8.4.1. Defining the input parameters for the gripper design procedure

The input data for the gripper design procedure are defined on the basis of the operating conditions of the gripper, the geometrical characteristics of the robotized workstation, the features of the object of manipulation, and the characteristics of the industrial robot.

Operating conditions and geometrical characteristics of the workstation:

1) Operating conditions that determine the construction of the gripper (standard, dust-proof, spark-proof, etc.), presented as the permissible limits of ambient conditions that will allow the desired technical functionality of the gripper.
2) The dimensions of the gripping area, the way of locating the object, orientation of the object during gripping.
3) The dimensions of the releasing area, the way of locating the object, orientation of the object during releasing.

Table 8.6. The influence of characteristics of the object on gripper functionality.

Gripper / Object	Type of prehension	Shape of gripping elements	Geometrical parameters of gripping elements	Surface of gripping elements	Force developed by the drive system	Sensors
Mass					+	+
Position of the center of gravity	+				+	
Moments of inertia	+				+	
Shape	+	+	+	+		+
Geometrical parameters	+	+	+			
Manufacturing tolerances of the contact area	+	+	+			+
Surface quality of the contact area	+	+		+	+	
Stiffness	+				+	
Resistance to external stress	+				+	+

Characteristics of the object:

1) The shape and geometrical parameters of the object determine the shape, number, and other properties of the gripper fingers, the gripper configuration and the opening of the fingers.

2) The mass and inertial characteristics of the object permit the determination of the inertial forces acting on the object during manipulation.
3) The strength of the object sets the maximum limit of the gripping force.

Table 8.6 lists the properties of the object of manipulation and their influence on gripper functionality.

Characteristics of the robot:

1) The velocities and accelerations of particular links and assemblies determine the inertial forces acting on the object.
2) The positioning repeatability of the robot is taken into consideration when determining the opening of the fingers.
3) The existing robot drive systems may have an influence on the gripper drive system.
4) The end of the robot arm should be considered the point of gripper connection.

8.4.2. Determining the location of the gripper relative to the object

The location of the gripper relative to the object is determined by considering its properties (shape, dimensions, mass, strength, surface quality), as well as operating conditions (environment, situation of the object in space, means of placing the object in the gripper, reference points or surfaces).

Having determined the location of the gripper, and knowing the shape and geometry of the object, as well as the gripping and releasing area, we can define the number of fingers, their shape and size, and the kind of motion to employ. Information about the motion of the fingers is also used to calculate the parameters of the drive transmission system. The number of fingers may vary: a standard gripper has two opposite fingers. These fingers may be rigid, not flexing when the object of manipulation is grasped. However, undesired dynamic phenomena that occur when a rigid finger comes in contact with the hard surface of an object, and often specific requirements set often by the manufacturing process, or by the object itself may force the designer to use elastic gripper fingers.

Standard equipment in industrial robots usually includes a gripper with incomplete fingers (mounting bolts only). The user selects suitable finger pads depending on the shape of the object of manipulation. This kind of gripper design allows the user to the obtain required gripping conditions without modifying the gripper drive system, and the drive transmission mechanism, but simply by changing the length, shape and material of the finger pads.

Finger pads may perform their functions in various manenrs, for example:

- enclose the whole object,
- correct the position of the object (slight corrections for cylindrical objects, for instance),
- provide the desired parameters in the contact between the object and the finger,
- ensure a soft grip,
- provide thermal insulation of the gripper.

8.4.3. Calculating forces and moments acting on the object

Forces and moments acting on the object can be found by various methods. The most common are **analytical methods.** The calculation procedure requires a knowledge of the inertial characteristics of the object; then the forces and moments acting on the grasped object can be computed, and the permissible ranges of changes in force and moment determined. The calculations are performed for the most unfavorable cases of prehension and the most extreme operating conditions of the manipulator (e.g. rotations on the maximum radius, at maximum acceleration, etc.).

The sum of the static and dynamic forces and moments acting on the gripper carrying the object of manipulation are

$$\boldsymbol{F} = -\boldsymbol{T} + \boldsymbol{P} \tag{8.11}$$

$$\boldsymbol{M} = -\boldsymbol{L}_c + \boldsymbol{r}_c \times \boldsymbol{F} \tag{8.12}$$

the dynamic forces T determine the effects of gripper and object motion:

$$T = \frac{d\theta}{dt} + \omega \times \theta$$

$$\theta = m v_c \tag{8.13}$$

where ω is the angular velocity vector, v_c is the linear velocity vector, and m is the mass of the object.

The angular momentum of the object K_c is given by

$$\boldsymbol{K}_c = \boldsymbol{I}_c \omega + m \boldsymbol{r}_c \times v_c \tag{8.14}$$

$$\boldsymbol{L}_c = \frac{d\boldsymbol{K}_c}{dt} + \omega \times \boldsymbol{K}_c \tag{8.15}$$

where

$\boldsymbol{I}_c$ is the moment of inertia of the object, and

$\boldsymbol{r}_c$ is the vector describing the location of the center of gravity.

To develop these equations the user must have an extensive knowledge of the dynamics of the particular industrial robot.

Another problem is the procedure transforming the forces and moments determined for the object of manipulation to the coordinate system attached to any of the manipulator links or any other reference system.

Having determined the components *Fx, Fy,* and *Fz* of the vector of force ***F*** and *Mx, My,* and *Mz* of the vector of moment ***M***, it is possible to analyze the reliability of the prehension in various load conditions.

Usually it is reasonable to use the maximum values of the forces and moments acting at each manipulation stage. The detailed analysis of the envisaged manipulation process makes it possible to orient of the object so that the most undesired cases of loading (from the dynamic point of view) can be eliminated.

Figure 8.26 presents an example of calculation of the grasping force $\boldsymbol{F_G}$ for a given actuator force $\boldsymbol{F_A}$ and displacement s_A. The quotient $\boldsymbol{F_G}/\boldsymbol{F_A}$ for $v_G = \omega_{41} \cdot l_5$ has the same form as in [8.5].

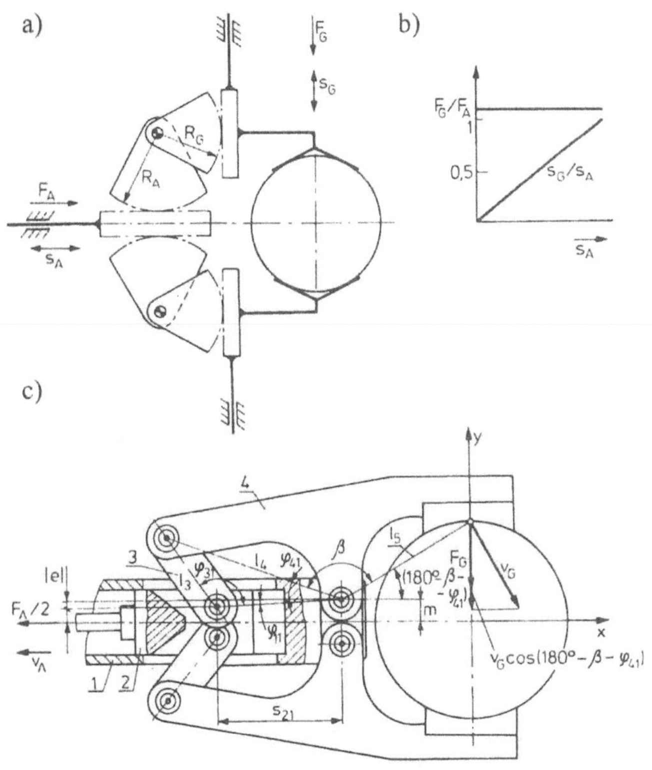

Fig. 8.26. Gripper with constant grasping force: a) grasping mechanism, b) graph of forces, c) calculation scheme

8.4 References

[8.1] *Barczyk J.*: Grippers of Industrial Robots. Basic Problems of Contemporary Technics. Vol. 25. Robotics. Edited by A. Morecki. Polish Scientific Publ. Warsaw 1987. (*in Polish*)

[8.2] Mechanics of Industrial Robots, 3, Moscow, Wyssaja Skola 1989.

[8.3] *Kato I:*. Mechanical Hands, Tokyo, 1992.

[8.4] *Czelpanov I. B., Kolpasnikow S. N.*: Grippers of Industrial Robots. Leningrad, Masinostroenie 1989.

[8.5] Volmer J. (Hrsg.) : Industrieroboter - Entwicklung, Berlin, VEB Verlag Technik 1983

[8.6] *Barczyk J.*: Method of Selecting and Principles of Design of Grippers of Industrial Robots. Warsaw 1991. (manuscript *in Polish*)

9. DRIVES AND MECHANISMS USED IN ROBOTS

9.1. Robot drives

Three main types of actuators are currently used in robots: pneumatic, hydraulic and electric, as well as different combinations of these three. Every robot is fitted with a system of actuators on the arms or in the joints, constituting a drive system.

Figure 9.1 shows the percentages of different types of actuators used in industrial robots. The data presented in Figure 9.1a were collected in 1977 on a sample of 118 industrial robots. Figure 9.1b shows data gathered recently. With the rapid development of new types of electric motors or actuators, such as stepping motors, linear induction motors, printed circuit disc motors, the share of electric drives has increased from 12.7% in 1977 to about 50% in 1990. The share of pneumatic drives fell in the same period from 45% to 10%. Hydraulic drives are still important, being the main power source for robots featuring high lifting capacity. However, in the last few years new solutions of pneumatic **servomotors** have appeared, which may contribute to a revival of this kind of drive. Each drive type is described in detail here below.

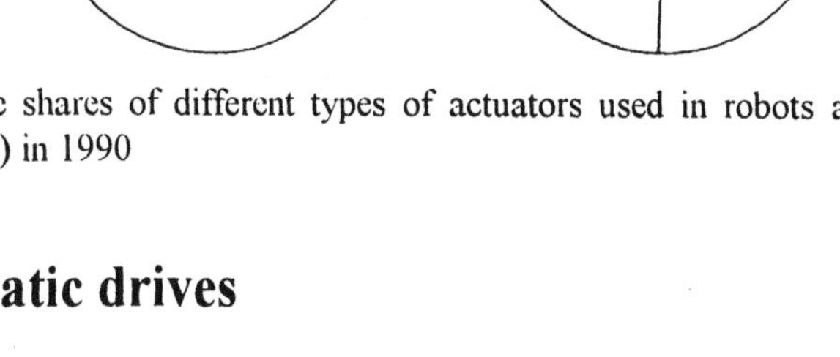

Fig. 9.1. Percentage shares of different types of actuators used in robots according to [9.9] and [9.12]: a) in 1977, b) in 1990

9.1.1. Pneumatic drives

Pneumatic drives feature the use of compressible fluids, usually compressed air. The advantages of these drives are the an abundantly available and easily accessed power medium (air) that can be released when the task is finished. Their pressure, compared with hydraulic drives, also makes these systems very safe to operate. In addition, unlike all liquids, air has good dynamic properties, no viscosity and low stiffness (high compliance). Pneumatic drives are used mainly in small, simple robots performing "pick-and place" tasks.

The system shown in Figure 9.2 generally operates with compressed air at a pressure of 0.4÷0.8 MPa. A linear (or sometimes rotary) actuator (1) is supplied with air through a two-position, four-way valve (2) controlled by an electric or pneumatic logic system (5). Limit switches (3) transmit a logic signal that arrests the piston at end stops (4). The system usually has only two positions (3). If a pneumatic manipulator with several two-position joints is equipped with these drives, the duration of the cycle can be programmed, but the range of displacement must be changed manually. The system described above is relatively inexpensive, but also much less flexible than a proportional one [9.2].

A simple proportional system is shown in Figure 9.3. The heart of this system is a two-stage amplifier. The first part is an impact piston amplifier. The second is an air transmitter that provide a large volume flow. The error signal e is used to provide feedback. The balance between the pistons' positions and the feedback is ensured by two springs (of k_z and k_f stiffness coefficients) and the suitable ratio of the areas A_1 and A_2.

Another example of a pneumatic control system is the controller shown in Figure 9.4. It is essentially an enhancement of the system shown in Figure 9.3. A differentiating, or integrating option can be obtained by modifying the bellows arrangement. Such modifications can lead to a pneumatic PID controller. Each controller can be used to actuate a prismatic joint of a robot. A four-way valve and a cylinder with a piston act as a hydraulic system. It is also possible to actuate a rotational joint by means of a turbine motor. The motor generates a torque proportional to the output pressure p_o of the controllers described above and independent of the velocity of the lever. High pressure compressed air ensures quick and accurate movements with the use of mechanical end stops at individual joints.

When handling large loads that require high lifting capacities and maintaining the desired trajectory, hydraulic actuators must be used.

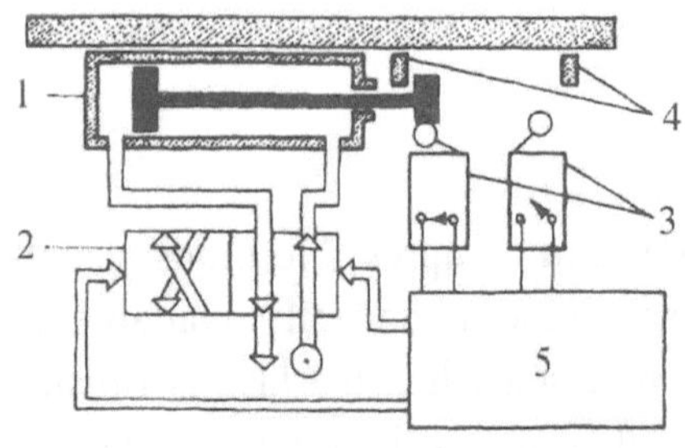

Fig. 9.2. U-1 pneumatic drive system - principle of operation [according to 9.2]

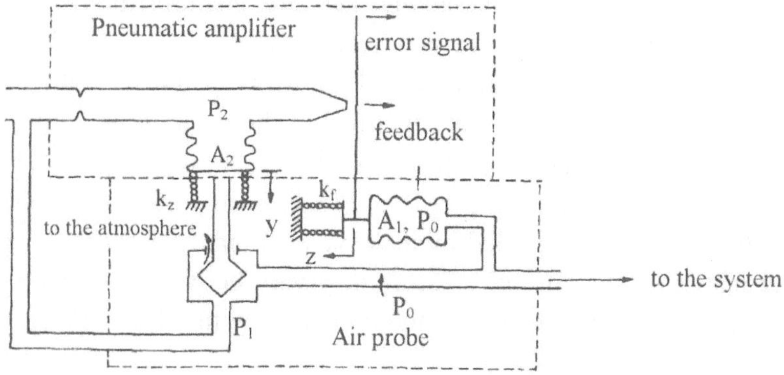

Fig. 9.3. Proportional pneumatic control system - principle of operation, according to [9.2]

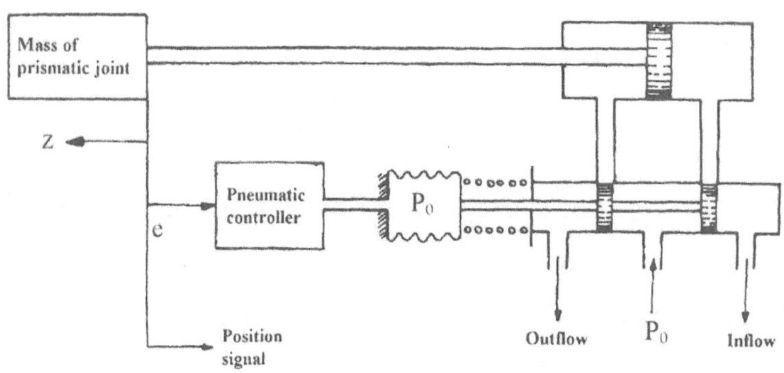

Fig. 9.4. Enhanced pneumatic control system

9.1.1.1. Linear pneumatic actuators

Figures 9.5 through 9.7 present various designs of linear pneumatic actuators. The first actuator (Figure 9.5) is a binary-type actuator of uncontrolled speed [9.3].

Larger actuators (cylinder diameters of over 25 mm) feature a more complex construction. They are usually double-acting cylinders with adjustable dampers (cushions) on both ends of the cylinder.

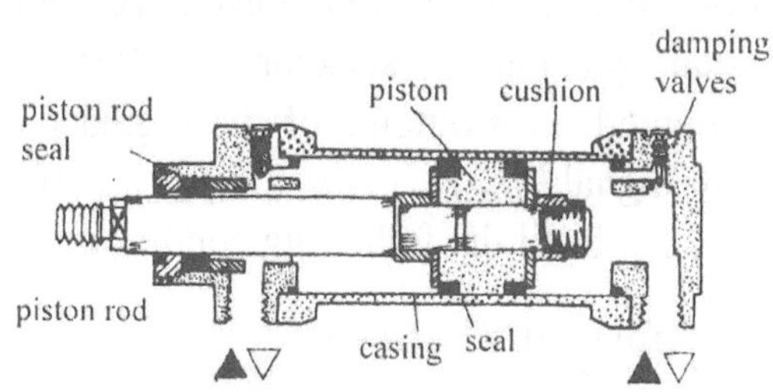

Fig. 9.5. Linear pneumatic actuator (by Rexroth Sigma) according to [9.3]

Figure 9.6 shows a double-acting actuator manufactured by the Herion company. Piston (*1*) is fitted with two sealing rings (*3* and *4*). When the piston approaches one of the end positions, the sleeves (or spears) (*5* and *6*) enter sealing rings *7* and *8* and compress the air contained in the cushion chamber, slowing the piston. Valves *9* and *10* are used to adjust the cushioning action. The body (*11*) contains an additional scraper ring (*12*) and sealing ring (*13*), which seal the piston rod (*14*). The piston rod is guided by a sleeve (*15*) made of bronze or another self-lubricating material. The control pressure may vary between 0.6 and 1.6 MPa. The air supplied to the actuator is filtered and lubricated with oil mist [9.3].

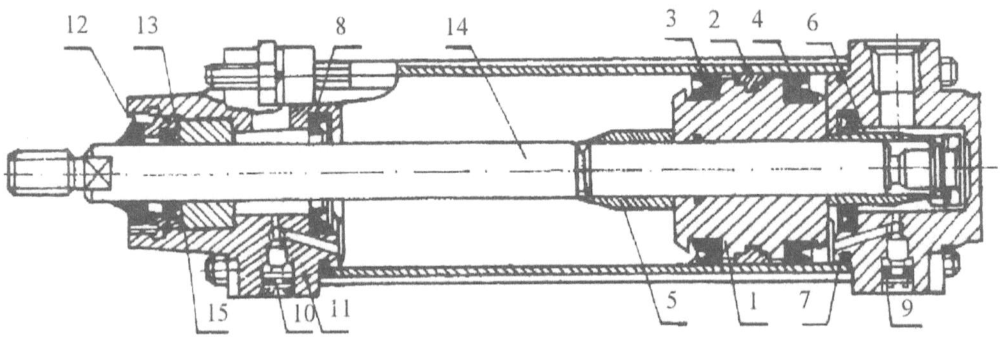

Fig. 9.6. Double-acting actuator with adjustable damping at both ends according to [9.3]

Another group of pneumatic actuators are multi-position actuators. **Multi-position pneumatic actuators** convert digital pneumatic signals into quasi-analogue displacement of the actuator piston rod. These actuators can be divided into:

- **two-position pneumatic actuators with digital position adjusters,**
- **systems of two-position pneumatic actuators with mechanical summation systems adding their piston rod displacements,**
- **multi-chamber pneumatic actuators with bumper systems.**

Digital pneumatic actuators may be of the piston, ball, or diaphragm type.

A two-position pneumatic actuator with a position adjuster (Fig. 9.7) is composed of a double-acting, two-position, piston-type pneumatic actuator (*1*), controlled by a digital pneumatic selector *2*. Pneumatic shut-off valves $Z_0, Z_1, \ldots, Z_i, \ldots, Z_{n-1}$ are mounted on n outputs of the digital pneumatic selector. The valves are controlled by signals $p_{x0}, p_{x\,1}, \ldots, p_{x\,i}, \ldots, p_{x\,(n-1)}$. If the binary pneumatic signals $p_{x0}, p_{x\,1}, \ldots, p_{x\,i}, \ldots, p_{x\,(n-1)}$ fulfil the following condition:

$$\sum_{\substack{k=0 \\ k \neq i}}^{n-1} p_{sk} = 0, \quad p_{xi} = 1, \quad \text{for } i = 0, 1, \ldots, (n-1) \tag{9.1}$$

then the digital pneumatic signal,

$$p_x = a_{n-1} p_{x\,(n-1)} + \ldots + a_i p_{x\,i} + \ldots + a_1 p_{x1} + a_0 p_{x0} \tag{9.2}$$

is converted into a displacement l_y of the actuator piston rod according to the relationship:

$$l_y = a_i \Delta l + l_0 \text{ for } i = 0, 1, \ldots, (n-1) \tag{9.3}$$

where

a_i is the weight coefficient for the digital pneumatic signal p_x,

l_y is the displacement.

The chambers are supplied through pneumatic resistors *3* and *4*. Simple construction, high force output and relatively high positioning accuracy in a predetermined position are the main advantages of the digital pneumatic actuator described above. In the design presented in Figure 9.7, the piston rod may assume the function of the selector.

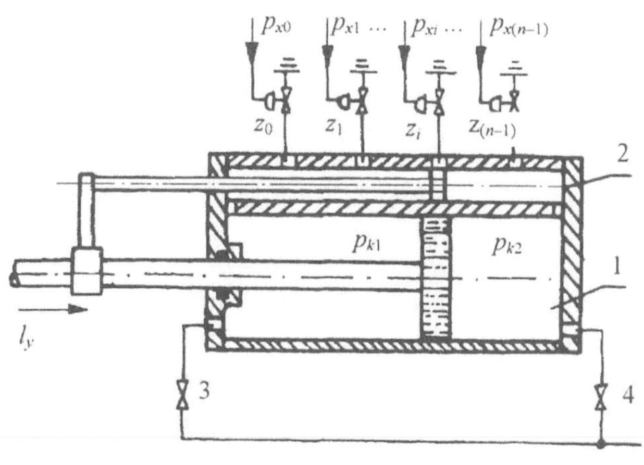

Fig. 9.7. Diagram of a two-position, piston-type actuator with a digital position adjuster according to [9.3]

9.1.1.2. Angular pneumatic actuators

Angular pneumatic actuators are used to control the angular position of the actuator shaft. Such actuators can be divided into:

- **analogue pneumatic actuators,**
- **digital pneumatic actuators.**

When a high torque is required on the output shaft over the whole range of the angular displacement φ_y, double acting piston-type or diaphragm-type actuators are used as the main components. In the solution presented in Figure 9.8 the double-acting cylinder *1* actuates the output shaft *2* by means of a rack-and-pinion transmission mechanism.

The signal p_x is delivered to the input bellows of a precision adjuster (*3*). A movable lever of the adjuster controls two pneumatic vector cascades operating in a differential arrangement. The cascade pressures p_{k1} and p_{k2} are delivered to the input chambers of the actuator *1*. Rotation of the output shaft (*2*), causes (through string 4) the tensioning of a measuring spring in the feedback loop (*1-3*). In spite of resistance forces, in the steady state there is a linear relationship between φ_y and p_x.

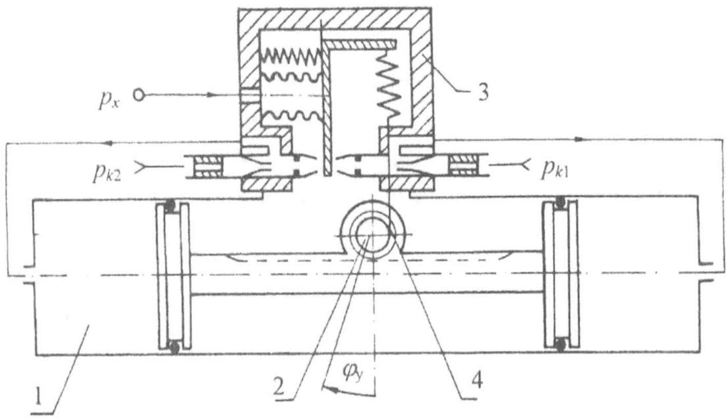

Fig. 9.8. Diagram of an pneumatic actuator with angular displacement of the output shaft according to [9.3]

The use of pneumatic cascades provides high accuracy in the processing of p_x/φ_y as well as an increase in the output force and torque. The input chambers of the actuator (*1*) must be air-tihght. Another solution of rack-and-pinion actuator is shown in Figure 9.9. This actuator is used when the range of rotation is limited (for example, in the radiocarpal joint of a manipulator).

Other mechanisms (screws, cams) are used as well as pneumatic diaphragm actuators, i.e. artificial pneumatic muscles [9.3], to convert the pressure in the cylinder into rotational motion.

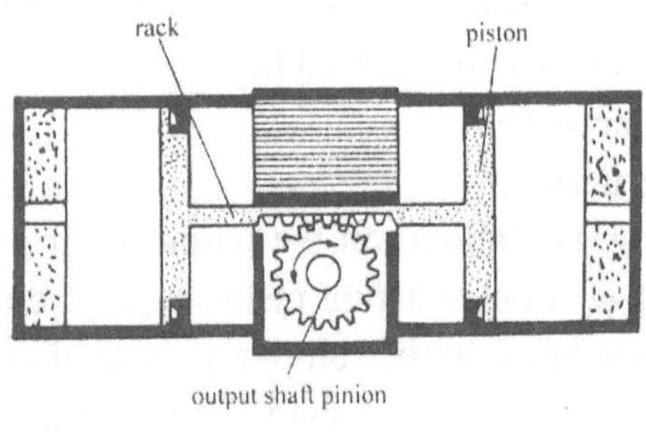

Fig. 9.9. Pneumatic actuator with rack-and-pinion gear.

9.1.1.3. Mathematical modeling

In the following pages we describe of a linear mathematical model of an electropneumatic servodrive for the case in which the cylinders are controlled separately, and that in which both cylinders are controlled together.

9.1.1.3a. Linearized model of an electropneumatic servodrive with separately controlled cylinders

The mathematical model presented below describes the dynamics of a servodrive by means of partially simplified, nonlinear differential equations [9.17].

The following basic simplifications have been made in order to linearize the equations and decrease the order of the nonlinear differential equations:

1. The dynamics of the valve itself have been neglected. The relationship between the valve controller voltage and the opening of the valve is always linear.
2. The description of pressure changes in the cylinder, and of flow rate dynamics in the valve has been limited to linear relationships. High order components (the influence of friction) have been neglected in linearizing the differential equations around the working point
3. The friction force in the cylinder has been assumed as proportional only to the velocity.

Applying these simplifications and the description of the state space the following state variables and resistance and output signals can be specified:

$$x_1(t) = s(t), \quad x_2(t) = v(t), \quad x_3(t) = P_1(t), \quad x_4(t) = P_2(t)$$
$$u_1(t) = U_1(t), \quad u_2(t) = U_2(t), \quad z(t) = F_z(t), \quad y(t) = x_1(t)$$

where

$U_1(t)$, $U_2(t)$ are the control voltages of the control valves of both cylinders,
$s(t)$ is the displacement of the pistons,
$v(t)$ is the velocity of the pistons,
$P_1(t)$, $P_2(t)$ are the pressures in the cylinders (referring to the valves), and
$F_z(t)$ is the resistance force of the drive system.

The state and output equations for the servodrive presented in Figure 9.10a are:

$$\left.\begin{aligned}
\frac{dx_1(t)}{dt} &= x_2(t) \\
\frac{dx_2(t)}{dt} &= -\frac{k_R}{m}x_2(t) + \frac{A_1}{m}x_3(t) - \frac{A_2}{m}x_4(t) + \frac{1}{m}z(t) \\
\frac{dx_3(t)}{dt} &= -\frac{n_1 P_{10} A_1}{V_{10}}x_2(t) + \frac{n_1 R \vartheta_0 k_{MU1}}{V_{10}}u_1(t) \\
\frac{dx_4(t)}{dt} &= -\frac{n_2 P_{20} A_2}{V_{20}}x_2(t) + \frac{n_2 R \vartheta_0 k_{MU2}}{V_{20}}u_2(t)
\end{aligned}\right\} \qquad (9.4)$$

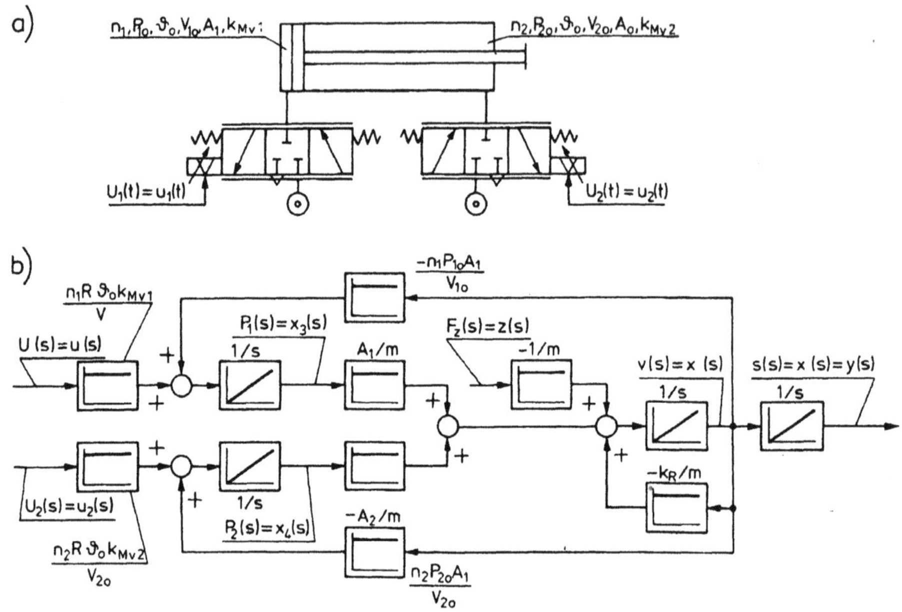

Fig. 9.10 Linearized dynamic model of an electropneumatic servodrive with separately controlled cylinders: a) schematic diagram of the drive, b) flow chart of the linearized drive.

$$y(t)=x_1(t) \tag{9.5}$$

where:

the k_R are the constant characteristics of friction in the cylinder, $f_R(v)$;

$$k_R=\frac{df_r}{dv}\bigg|_{v=v_0} \tag{9.6}$$

m is the total mass loading the drive system,

A_1, A_2 are the areas of the pistons on the input and output port sides respectively,

n_1, n_2 is the polytropic exponents for the air flow in the cylinders,

ϑ_0 is the temperature of air in the working point,

P_{10}, P_{20} are the absolute pressures in the cylinders in the working point,

V_{10}, V_{20} are the total volumes of the cylinders in the working point,

$$\left.\begin{aligned} V_{10} &= V_{tot1}+A_1 s_0 \\ V_{20} &= V_{tot2}+A_2\,(s_{max}\;\; s_0) \end{aligned}\right\} \tag{9.7}$$

V_{tot1}, V_{tot2} are the total volumes of the joined cylinders,

s_{max} is the length of the cylinders,

k_{MU1}, k_{MU2} are the constant characteristics of mass flow of the control valve for the input and output port respectively

$$\left.\begin{aligned} k_{MU1} &= \frac{df_m(u)}{du}\bigg|_{u_1=u_{10}} \\ k_{MU2} &= \frac{df_m(u)}{du}\bigg|_{u_2=u_{20}} \end{aligned}\right\} \tag{9.8}$$

Hence

for $u_1 > 0$ and $u_2 > 0$

$$k_{MUi} = C_{imax}\frac{k_{Vi}}{s_{Vimax}}\rho_0 P_V \psi(P_{i0}/P_V),\quad i=1,2 \tag{9.9}$$

and for $u_1 < 0$ and $u_2 < 0$

$$k_{MUi} = C_{imax}\frac{k_{Vi}}{s_{Vimax}}\rho_0 P_{i0} \psi(P_{ath}/P_{i0}),\quad i=1,2 \tag{9.10}$$

where

C_1, C_2 are the main values of mass volume flow through the corresponding control ports of the control valves,

k_{V1}, k_{V2} are the constants of the voltage conversion for the corresponding valves during displacement of the pistons,

$$k_{Vi} = \frac{s_{imax}}{U_{imax}},\quad i=1,2 \tag{9.11}$$

P_V, P_{ath} are the supply pressure and atmospheric pressure respectively, and ρ_0 is the air density (as defined in the standards).

The flow function is defined if P' and P'' are the primary and secondary pressures, and b, the critical pressure:

$$\left.\begin{aligned} (P''/P) &= \sqrt{1-\left(\frac{P''/P'-b}{1-b}\right)^2} && \text{when}\ \frac{P''}{P'} < b \\ (P''/P) &= 1 && \text{when}\ \frac{P''}{P'} > b \end{aligned}\right\} \tag{9.12}$$

The matrices of a linearized representation of the state space for the electropneumatic servodrive are

$$\mathbf{A} = \begin{bmatrix} 0 & a_{12} & 0 & 0 \\ 0 & a_{22} & a_{23} & a_{24} \\ 0 & a_{32} & 0 & 0 \\ 0 & a_{42} & 0 & 0 \end{bmatrix} \tag{9.13}$$

$$\mathbf{B} = \begin{bmatrix} 0 & 0 \\ 0 & 0 \\ b_{31} & 0 \\ 0 & b_{42} \end{bmatrix} \tag{9.14}$$

$$\mathbf{C} = \begin{bmatrix} c_{11} & 0 & 0 & 0 \end{bmatrix} \tag{9.15}$$

$$\mathbf{D} = \begin{bmatrix} 0 \\ d_{21} \\ 0 \\ 0 \end{bmatrix} \tag{9.16}$$

where

$$a_{12} = 1, \qquad a_{22} = -\frac{k_R}{m}, \quad a_{32} = -\frac{n_1 P_{10} A_1}{V_{10}}, \qquad a_{42} = -\frac{n_2 P_{20} A_2}{V_{20}}$$

$$a_{23} = \frac{A_1}{m}, \qquad a_{24} = \frac{A_2}{m}$$

$$b_{31} = \frac{n_1 R \vartheta_0 k_{MU1}}{V_{10}}, \qquad b_{42} = \frac{n_2 R \vartheta_0 k_{MU2}}{V_{20}}$$

$$c_{11} = 1, \qquad d_{21} = -\frac{1}{m}$$

The flow chart of the linearized servodrive is shown in Figure 9.10b. For a model having the same structure and the same parameters as the linearized drive, the following matrices are obtained for $\alpha = \alpha_0$ (working point):

$$\boldsymbol{A}_{\mathrm{M}}(\alpha_0) = \boldsymbol{A}(\alpha_0) = [a^{@}_{ij}] \tag{9.17}$$

$$\boldsymbol{B}_{\mathrm{M}}(\alpha_0) = \boldsymbol{B}(\alpha_0) = [b^{@}_{il}] \tag{9.18}$$

for $i = 1, 2, \ldots n; j = 1, 2, \ldots n; l = 1, 2, \ldots r;$ when $n = 4$ and $r = 4$.

In order to study the controllability and observability of the models the controllability matrix:

$$\boldsymbol{S}_S = [\boldsymbol{B}_M \mid \boldsymbol{A}_M \boldsymbol{B}_M \mid \boldsymbol{A}_M^2 \boldsymbol{B}_M \mid \boldsymbol{A}_M^3 \boldsymbol{B}_M] \tag{9.19}$$

and the observability matrix:

$$\boldsymbol{S}_B^T = [\boldsymbol{C}_M^T \mid (\boldsymbol{C}_M \boldsymbol{A}_M)^T \mid (\boldsymbol{C}_M \boldsymbol{A}_M^2)^T \mid \boldsymbol{C}_M^3 (\boldsymbol{C}_M \boldsymbol{A}_M^3)^T] \tag{9.20}$$

must be formulated. The rank of these matrices is $n = 4$, and they have non-zero determinants det $\boldsymbol{S}_S$ and det $\boldsymbol{S}_B^T$. The model is thus fully controllable and observable.

9.1.1.3b. Linearized model of a drive system with joint cylinder control

The order of the model of an electropneumatic servodrive with connected cylinders (illustrated in Figure 9.11a) can be reduced to $n = 3$ by introducing only one control signal

$$u(t) = u_1(t) = -u_2(t) \tag{9.21}$$

and one new state variable

$$\Delta P(t) = P_1(t) - P_2(t) \tag{9.22}$$

The signals valid for this model are

$$x_1(t) = s(t), \qquad x_2(t) = v(t), \qquad x_3(t) = \Delta P\,(t)$$
$$u(t) = U\,(t), \qquad z(t) = F_z(t), \qquad y(t) = x_1(t)$$

where

$U(t)$ is the control voltage from the 5/3 control valve of both cylinders,
$s(t)$ is the displacement of the cylinders, or pistons,
$v(t)$ is the velocity of the cylinders, or pistons,
$\Delta P(t)$ is the pressure difference, or the loading pressure of both cylinders,
$F_z(t)$ is the resistance force of the drive system.

The state and output equations for the servodrive presented in Figure 9.11a can be obtained from equations (9.4) - (9.12), adding the equation describing the relationship between the surfaces of both cylinder's (provided that the cylinder is of the double-end type) $A_1 = A_2 = A$:

$$\left.\begin{aligned} \frac{dx_1(t)}{dt} &= x_2(t) \\ \frac{dx_2(t)}{dt} &= -\frac{k_R}{m}x_2(t) + \frac{A}{m}x_3(t) - \frac{1}{m}z(t) \\ \frac{dx_3(t)}{dt} &= -A\left(\frac{n_1 P_{10}}{V_{10}} + \frac{n_2 P_{20}}{V_{20}}\right)x_2(t) + R\vartheta_0\left(\frac{n_1 k_{MU1}}{V_{10}} + \frac{n_2 k_{MU2}}{V_{20}}\right)u(t) \end{aligned}\right\} \tag{9.23}$$

$$y(t) = x_1(t) \tag{9.24}$$

The state space of the drive system is represented by the following matrices:

$$\mathbf{A} = \begin{bmatrix} 0 & a_{12} & 0 & 0 \\ 0 & a_{22} & a_{23} & a_{24} \\ 0 & a_{32} & 0 & 0 \end{bmatrix} \tag{9.25}$$

$$\mathbf{B} = \begin{bmatrix} 0 \\ 0 \\ b_{31} \end{bmatrix} \tag{9.26}$$

$$\mathbf{C} = \begin{bmatrix} c_{11} & 0 & 0 \end{bmatrix} \tag{9.27}$$

$$\mathbf{D} = \begin{bmatrix} 0 \\ b_{21} \\ 0 \end{bmatrix} \tag{9.28}$$

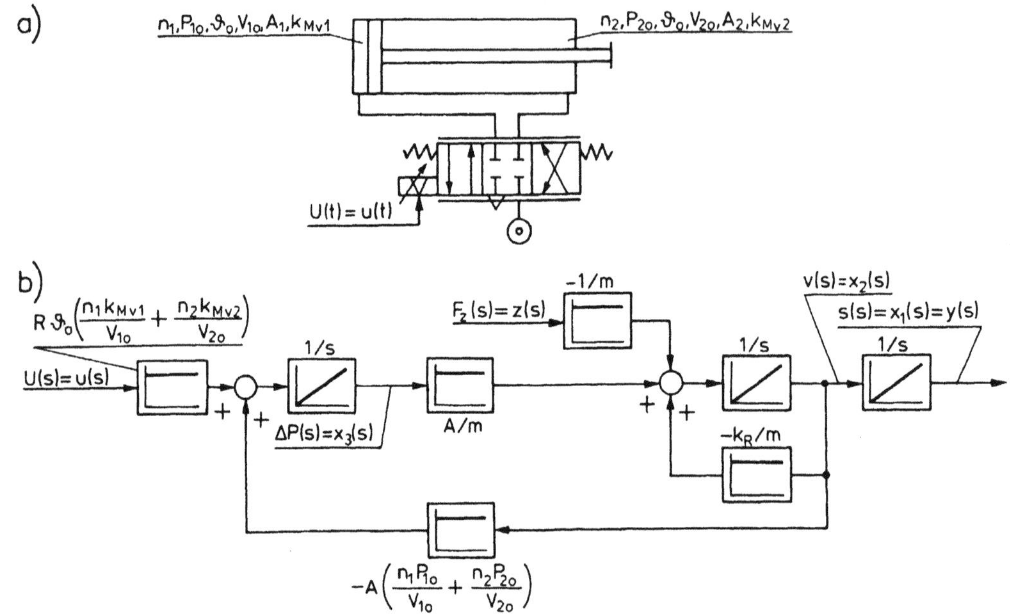

Fig. 9.11. Linearized dynamic model of electropneumatic servodrive with joined cylinder control: a) schematic diagram of the drive, b) flow chart of the linearized drive.

where

$$a_{12}=1, \qquad a_{22}=-\frac{k_R}{m}, \qquad a_{32}=-A\left(\frac{n_1P_{10}}{V_{10}}+\frac{n_2P_{20}}{V_{20}}\right)$$

$$a_{23}=\frac{A}{m}$$

$$b_{31}=R\vartheta_0\left(\frac{n_1k_{MU1}}{V_{10}}+\frac{n_2k_{MU2}}{V_{20}}\right)$$

$$c_{11}=1$$

$$d_{21}=-\frac{1}{m}$$

The flow chart of the linearized servodrive is shown in Figure 9.11b. For a model having the same structure and the same parameters as the linearized drive, the following matrices $A_M(\alpha_0)$ and $B(\alpha_0)$ are obtained for $n = 3$ and $r = 1$. The model considered here is also fully controllable and observable as the determinants of both matrices:

$$\det S_S = \det [B_M \mid A_M B_M \mid A_M^2 B_M] \tag{9.29}$$

$$\det S_B^T = \det [C_M^T \mid (C_M A_M)^T \mid (C_M A_M^2)^T] \tag{9.30}$$

are different from zero, and both matrices S_S and S_B^T are thus of rank 3.

9.1.2. Hydraulic drives

The hydraulic drive, in spite of the ever more numerous application of the electric drive, is still one of the most common drive systems, especially when rapid movement with significant load capacity is necessary.

Let us consider the example of an hydraulic actuator with a cross-section of 50 cm^2 supplied with hydraulic fluid from a source providing 21 MPa fluid pressure. When 14.0 MPa acts on the piston of such an actuator, 70 000 N of force are developed.

A very short start-up time (from several miliseconds to one second) is another advantage of the hydraulic drive.

9.1.2.1. Linear hydraulic actuators

Actuators of this type are similar to pneumatic ones, and can develop the same power although they are smaller and lighter. There are three types of such actuators: **double-acting, differential,** and **single-acting.**

Figure 9.12a shows a linear, double-acting hydraulic actuator. The pressures on the sides of the piston are p_1 and p_2 respectively. As the piston rod is attached to one side of the piston only, the working areas of the piston are different on each side.
Thus the output force is not a linear function of pressure and can be described as:

$$F = (S + s)\, p_1 - S\, p_2 \tag{9.31}$$

This drawback can be eliminated by using a piston with a rod extending from each side of the piston. In this case the force is

$$F = S\,(p_1 - p_2) \tag{9.32}$$

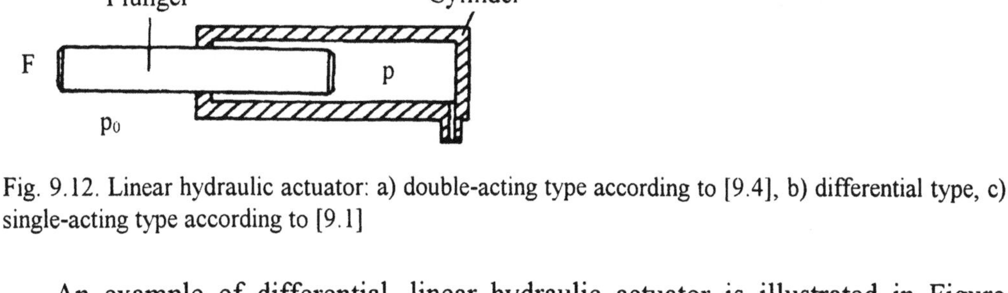

Fig. 9.12. Linear hydraulic actuator: a) double-acting type according to [9.4], b) differential type, c) single-acting type according to [9.1]

An example of differential, linear hydraulic actuator is illustrated in Figure 9.12b. In this case the cross-sectional area of the piston rod equals the remaining piston area (excluding the rod): $s = S$. On the rod side there is a high pressure p_a and on the other side, a variable pressure p. The force generated by the actuator is:

$$F = s\,(2p - p_a) \tag{9.33}$$

Fig. 9.12c shows an example of linear, ram-type, single-acting hydraulic actuator. In this design a variable pressure p is exerted on one side of the ram, while the other side is subjected to a low pressure p_0.

The force is developed in one direction only:

$$0 \le F = S\,(p - p_b) \le S\,(p_a - p_0)\,. \tag{9.34}$$

The ram is retracted by an external force. The linear hydraulic actuators described here provide high force outputs and long ranges of displacement [9.1].

9.1.2.2. Rotary hydraulic actuators

Two types of rotary hydraulic actuators are used to obtain a limited angular displacement of the actuator output link: **hydraulic actuators transforming back-and-forth piston movement into output shaft rotation**, and **hydraulic actuators with rotating pistons**.

The first type of actuator is similar in design to the angular pneumatic actuator with a rack-and-pinion mechanism; it is shown in Figure 9.13. Another solution is an actuator with two parallel linear cylinders, illustrated in Figure 9.14. The range of angular displacement in this case is limited to $2\pi/3$. The rotary hydraulic actuator with rack-and-pinion gear features low angular velocity and constant torque developed under constant pressure.

Figures 9.15a and 9.15b show a simple design of a vane hydraulic actuator with stationary barriers and rotating vanes mounted on the output shaft. The solution with one vane (Fig. 9.15a) provides less than 2π of angular displacement (about $11\pi/6$). The displacement range for an actuator with two vanes is twice as small (about $5\pi/6$), but the output torque can be doubled. The rotary piston has less inertia and friction, compared with the linear one. These pistons should be used when the power to weight ratio is high, and when the actuator acts directly on the load. The most suitable application is the actuation of the joint between two adjacent robot links.

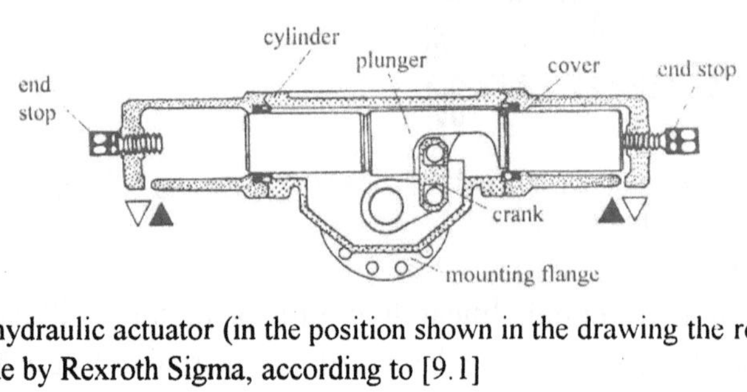

Fig. 9.13. Angular hydraulic actuator (in the position shown in the drawing the rotation of the output shaft is locked) made by Rexroth Sigma, according to [9.1]

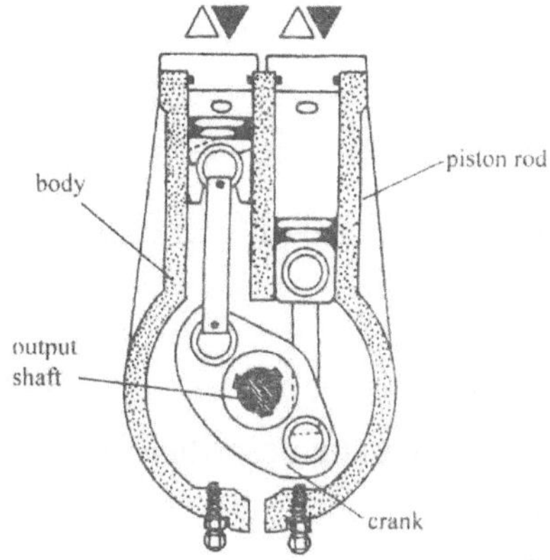

Fig. 9.14. Actuator with parallel cylinders (Rexroth Sigma)

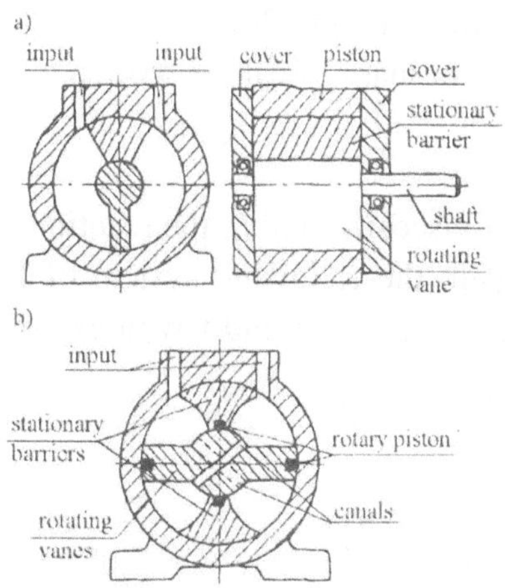

Fig. 9.15. Rotary vane actuator according to [9.1]: a) single-vane model, b) double-vane model.

9.1.2.3. Mathematical modeling

In the case of a symmetrical, double-acting hydraulic actuator the equations are [9.12]

$$\begin{aligned} \frac{S(h+x)}{B}\frac{dp_1}{dt} &= q_1 - q_d - q_f \\ \frac{S(h-x)}{B}\frac{dp_2}{dt} &= q_d + q_f - q_2 \end{aligned} \qquad (9.35)$$

where

x is piston position [m],

v, the velocity [m/s],

$2h$, the full stroke [m],
S, the channel cross-section [m^2],
p, the pressure [Pa],
q, the flow rate [m^3/s],
B, the volumetric modulus of elasticity of hydraulic fluid [Pa],
$q_d = \frac{dV}{dt} = S\frac{dx}{dt}$, V - volume [m^3],
$V = s(h + x)$ or $V = C(\theta_0 + \theta)$.

Pressure losses resulting from leaks can be described as

$p_1 - p_2 = Rq_f$ (laminar flow)
$p_1 - p_2 = rq_f^2$ (turbulent flow) (9.36)

The equation in hydromechanical coupling (piston-liquid) is

$$q_d = Sv \tag{9.37}$$

where q_d is the deformation of hydraulic fluid.

The equation of motion for mass M can be written as

$$M\frac{dV}{dt} = f - cv - f_r \tag{9.38}$$

where c is the coefficient of viscosity, and f_r, the reaction force.

By elimination we obtain the following system of four first-order differential equations linking state variables p_1, p_2, v and x with external variables q_1, q_2 and f_r

$$\begin{aligned}
&\frac{S(h+x)}{B}\frac{dp_1}{dt} = -\frac{p_1}{R} + \frac{p_2}{R} - s_u + q_1 \\
&\frac{S(h-x)}{B}\frac{dp_2}{dt} = \frac{p_1}{R} - \frac{p_2}{R} - s_v + q_2 \\
&M\frac{dv}{dt} = Sp_1 - Sp_2 - cv - f_r \\
&\frac{dx}{dt} = v
\end{aligned} \tag{9.39}$$

In the case of the rotary hydraulic actuator the mathematical modeling is similar to that shown in 9.1.2.3. The following relationships can be obtained:

$$\begin{aligned}
&c = C(p_1 - p_2) \\
&q_d = C\frac{d\theta}{dt} \\
&\frac{C(\theta_0+\theta)}{B}\frac{dp_1}{dt} = -\frac{p_1}{R} + \frac{p_2}{R} - c\omega + q_1 \\
&\frac{S(\theta_0-x)}{B}\frac{dp_2}{dt} = \frac{p_1}{R} - \frac{p_2}{R} + c\omega + q_2
\end{aligned} \tag{9.40}$$

$$J\frac{d\omega}{dt} = Cp_1 - Cp_2 - \varphi\omega - c_r$$

$$\frac{d\theta}{dt} = \omega$$

where

C is the coefficient of hydraulic coupling; equal to the volume of transferred fluid [m^3/rad] [9.1].

9.1.2.4. Hydraulic motors

Various types of hydraulic motors are used to obtain unlimited rotary motion. They are mainly reversible piston machines, that also act as pumps. The basic and most commons designs are: the axial-piston motor with an angled cam plate, the axial-piston motor with an angled cylinder, the radial-piston motor, and the gear motor.

The axial-piston motor with an angled cylinder is shown in Figure 9.16. The principle of operation is similar to that of a motor with an angled plate. The ball joints connect the pistons to the plate, which is attached to and rotates together with the output shaft. The barrel, rotating with the pistons, bears against the spherical surface of the direction-control valve. Cylinder capacity can be changed by adjusting the angle between the drum axis and the output shaft axis.

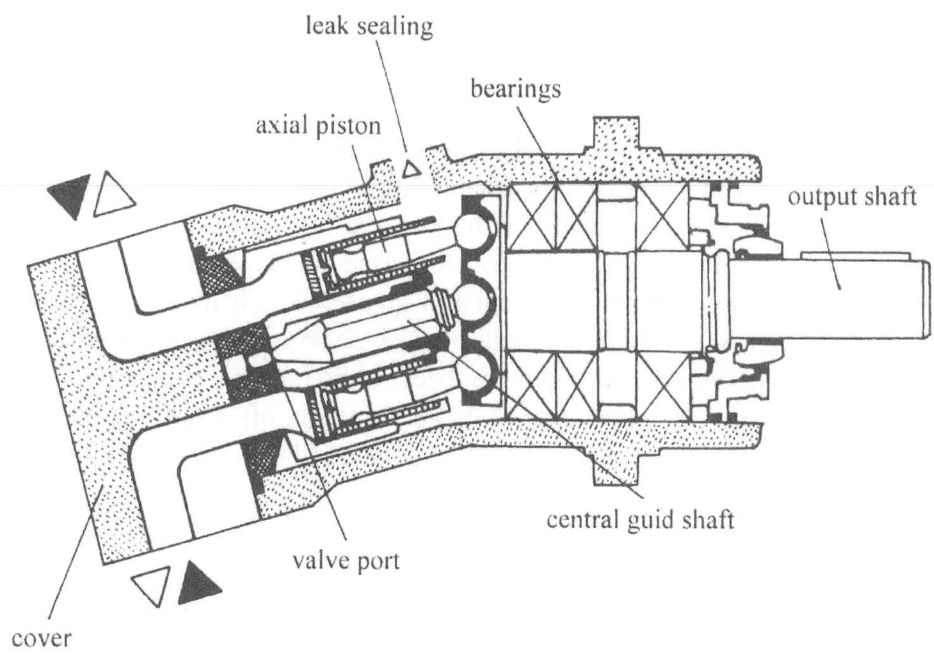

Fig. 9.16. Axial-piston motor with angled cylinder according to [9.1]

The operating pressure and the maximum permissible pressure are the main parameters specified for hydraulic actuators. Other characteristics depend on the type of actuator. In the case of piston actuators, the displacement, volume, velocity and mass of the moving parts are important factors. Important characteristics for hydraulic motors are the ratio of volume to angular displacement in m^3/rad, the maximum angular velocity, the rated torque, and the characteristics shown in Figure 9.17 [9.1].

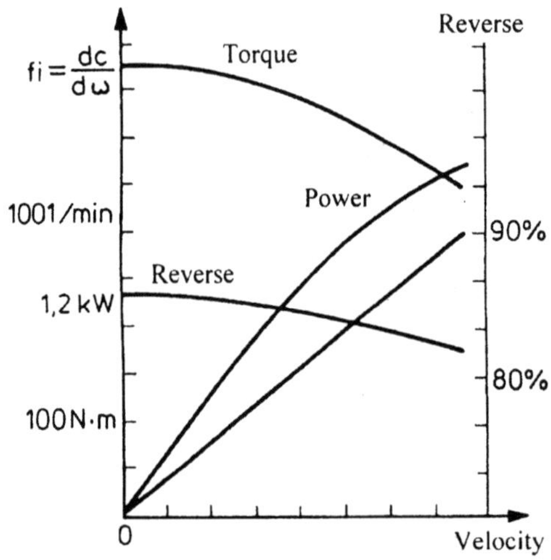

Fig. 9.17. Characteristics of the blade motor according to [9.1]

9.1.2.5. Direction-control valves, amplifiers and hydraulic servovalves

The main task of a direction-control valve is to direct the stream of hydraulic fluid to the desired place within the hydraulic circuit (this is usually one of the chambers of a hydraulic actuator or hydraulic motor). With regard to construction, **direction-control valves** can be divided into two groups: **spool valves** and **poppet valves**.

Figure 9.18a is the schematic diagram of a spool-type direction-control valve (a three-position, four-way valve) and Figure 9.18b shows its graphic symbol. The spool (*2*) moving inside the body (*1*) may occupy three different positions. In the central position both chambers of the motor are cut off from the pump, or accumulator, and from the reservoir: the hydraulic fluid can not flow through the valve. In the extreme right position of the spool (*2*), the pump (*P*) is connected to chamber *A* of the motor, while the chamber *B* of the motor is connected to the reservoir: the hydraulic fluid is directed to chamber *A* of the hydraulic motor. In the extreme left position of the spool, pump *P* is connected to chamber *B* of the motor, while chamber *A* of the motor is connected to the reservoir: the hydraulic fluid is directed to chamber *B* of the hydraulic motor [9.16].

With respect to the means of control, hydraulic direction-control valves can be divided into **single-stage direction-control valves, and two-stage direction-control valves**. Single-stage direction-control valves can be positioned mechanically, hydraulically, pneumatically, or electrically; two-stage direction-control valves are electrically actuated.

Figure 9.18c presents the simplified longitudinal section of an electrically actuated single-stage, two-position, four-way spool valve produced by Rexroth; Figure 9.18d shows its graphic symbol. In order to compare the different electromagnet types, on the left we have drawn an AC electromagnet *1*, and on the right, a DC electromagnet *2*. As both operate in air, production valves are fitted with two identical electromagnets. An axial force is generated by applying voltage to the solenoid; this moves the solenoid plunger, which acts directly on the valve spool, moving it to the extreme position. The position of the valve shown in the figure is obtained by applying voltage to solenoid *1*. The inside and outside sealing rings mounted in the sleeves (*3*) fixed by means of springs (*4*) prevent fluid from entering the solenoids. Electrically actuated direction-control valves are built to cope with flows reaching 1.66 dm^3/s (100 dm^3/min) and pressures up to 32 MPa.

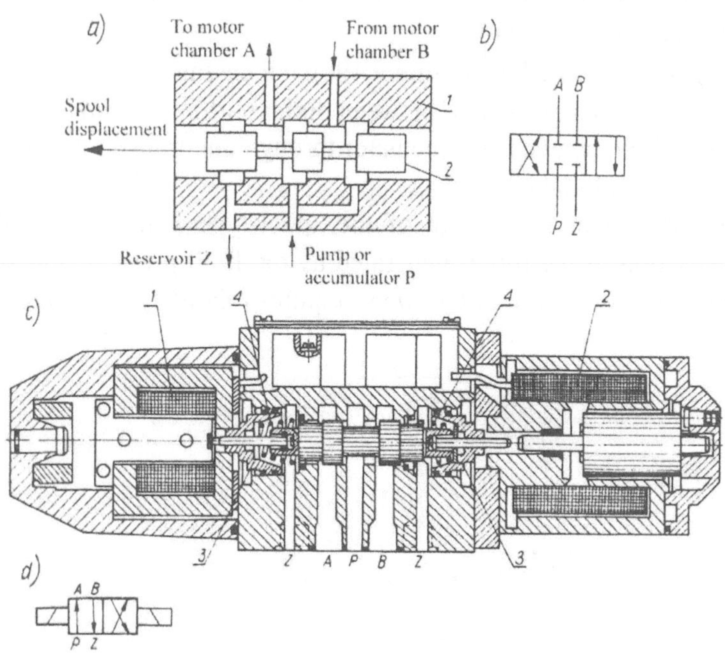

Fig. 9.18. Three-position, four-way spool valve: a) schematic diagram according to [9.3], b) graphical symbol, c) Electrically controlled single-stage, two-position, four-way spool valve made by Rexroth - longitudinal section, d) graphical symbol.

Hydraulic servovalves can be divided into flow-control and pressure-control valves. **Flow-control servovalves**, also called **hydraulic amplifiers,** intensify and continuously transform a small input signal into a large output signal in the form of the hydraulic fluid flow rate. **Pressure-control servovalves** convert a small input signal into a large output pressure signal at low flow rates.

Electrically controlled hydraulic servovalves are known as **electrohydraulic amplifiers**. Figure 9.19 illustrates the principle of operation of a spool-type hydraulic amplifier.

The spool-type hydraulic amplifier consists of a spool moving within a sleeve with five ports. The inlet port (*4*) delivers the hydraulic fluid from the pump; the transfer ports (*1* and *2*) supply the hydraulic fluid to, and discharge it from, the actuator chambers; the outlet ports direct the fluid to the hydraulic fluid reservoir. The position of the spool is the input signal of the spool-type amplifier:

- when the spool is in the center (Figure 9.19a), the inlet port (*4*) and the outlet ports (*3* and *5*) are closed and the hydraulic fluid cannot flow through the amplifier;
- when the spool is on the right (Figure 9.19b), the passage between inlet port *4* and transfer port *1* is open, as is that between outlet port *5* and transfer port *2*; thus the hydraulic fluid can enter the actuator through transfer port *1* and exit the actuator through transfer port *2*;
- when the spool is on the left (Figure 9.19c), the passage between inlet port *4* and transfer port *2* is open, as is that between outlet port *5* and transfer port *1*. Thus the hydraulic fluid can enter the actuator through one transfer port (*2*) and exit the actuator through the other (*1*).

An important feature in the design of the spool-type hydraulic amplifier is the partial compensation of the forces generated by the fluid pressure and acting on the pistons; as a result, the sliding of the spool requires relatively low forces.

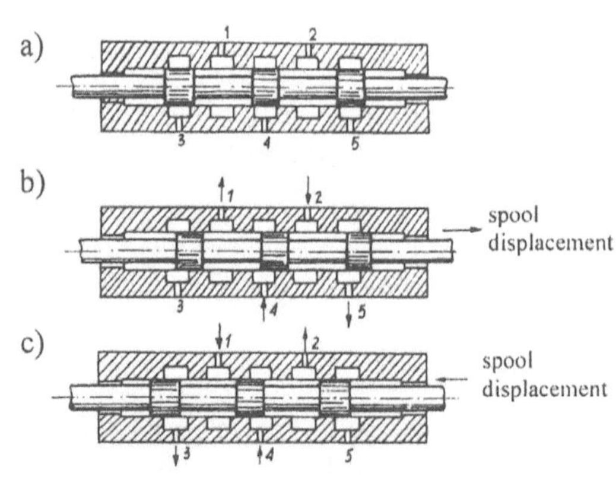

Fig. 9.19 Spool-type hydraulic amplifier

Figure 9.20 shows the hydraulic fluid circuit of a linear actuator controlled by means of a spool amplifier. When the spool is centered, the inlet port (*4*) and the outlet ports (*3* and *5*) are covered by the spool pistons. These pistons must be well sealed from the sleeve, as it influences spool stability, despite the forces acting on it. Sealing has significantly improved in quality in the last ten years as a result of developments in machining technology and fluid filtration.

Shifting the spool to the left, or to the right from the position shown in Figure 9.20 causes the hydraulic fluid to flow into the right, or left actuator chamber respectively. The fluid from the chamber, the volume of which is being reduced as a result of the spool's movement is discharged into the hydraulic fluid reservoir through the amplifier. The fluid is then delivered from the reservoir to the hydraulic amplifier by a pump driven by an electric induction motor. Before entering the amplifier the fluid flows through a filter fitted with a pressure gauge. This filter is designed to remove from the fluid mechanical impurities, which would be dangerous for the sealings of the spool's pistons. If the pressure of the hydraulic fluid in the filter exceeds the value set in the overflow valve, this valve discharges a part of the fluid contained in the filter to the fluid reservoir, limiting in this way the hydraulic fluid pressure and the maximum force of the actuator, protecting the actuator from overloading. When overload conditions no longer exist, the overflow valve is closed, and the system resumes normal operation.

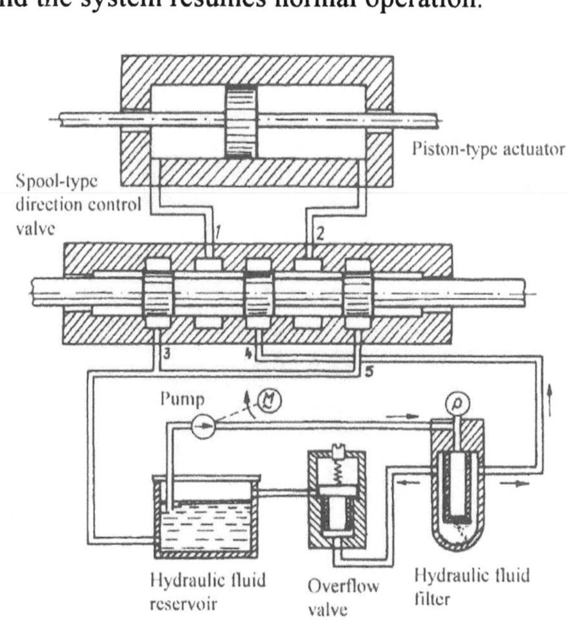

Fig. 9.20. Hydraulic circuit of a linear actuator controlled by an amplifier. *1* - *5*: ports of the spool-type amplifier [9.4]

Servomechanisms used in robotics are usually multi-stage devices. Their function is to move the spool of the valve in such a way that the port is uncovered in proportion to the voltage or current control signal.

There are two types of hydraulic servovalves; they differ in the way the spool is controlled. One is an **electrohydromechanic valve** (shown in Figure 9.21); the other is a **hydraulic servovalve with direct electric control**. Figure 9.21 shows a hydraulic valve produced by D.B.A. Air Equipment. The first stage is a two-position hydraulic motor with a diaphragm mounted in its rotor. The diaphragm is fitted with a return spring, which brings it back to its neutral position. The edges of the diaphragm lie in front of two calibrated high-pressure nozzles. With the diaphragm in neutral position, the flow rates through the nozzles are equal, and the pressures in the control ducts of the spool are equal. As a result, the spool takes a position midway between two calibrated countersprings.

The current flowing through the coil causes the diaphragm to turn, disturbing the balance of flow rates in the nozzles. A pressure difference is generated in the control ducts of the spool, and the spool moves to a distance at which the return spring compensates the disturbance. Proper selection of the components ensures that the displacement of the spool is proportional to the current flowing through the coil.

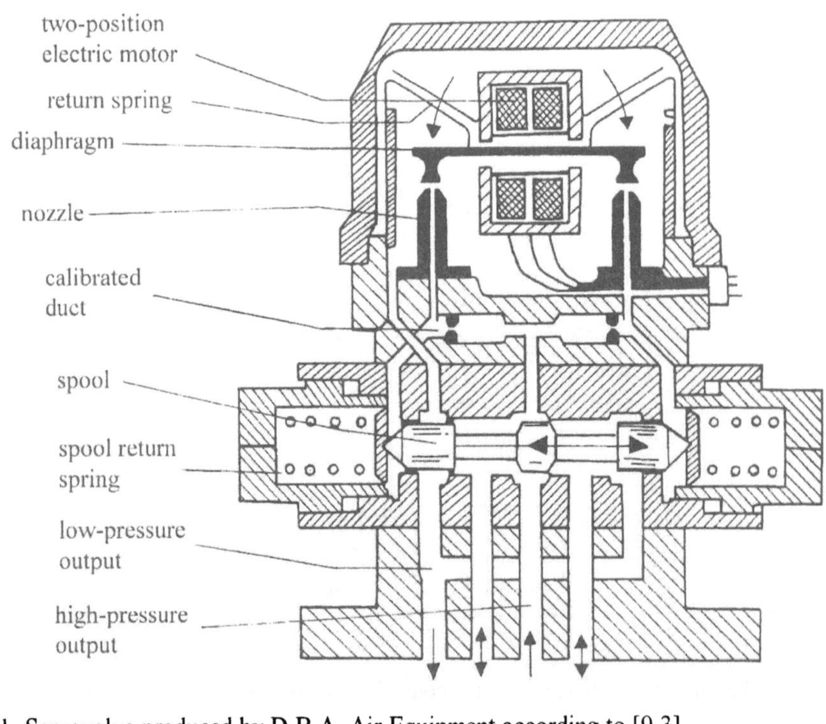

Fig. 9.21. Servovalve produced by D.B.A. Air Equipment according to [9.3]

Figure 9.22a shows a diagram of the WP3-10 two-stage electrohydraulic amplifier, currently produced in Poland by PZL-Hydral Wrocław; Figure 9.22b shows its characteristics. The first stage consists of a nozzle-diaphragm assembly; the second is a four-position, spool control valve. The diaphragm (*1*) is connected to the core (*2*) of the electromechanical torque converter *3* by means of a torsional tube *4*. The hydraulic fluid is delivered from the pump of the accumulator to the nozzles *5* and *6* through two fixed reducing pipes *7* and *8* and a very fine filter *9* built into the electrohydraulic amplifier (Fig. 9.22c).

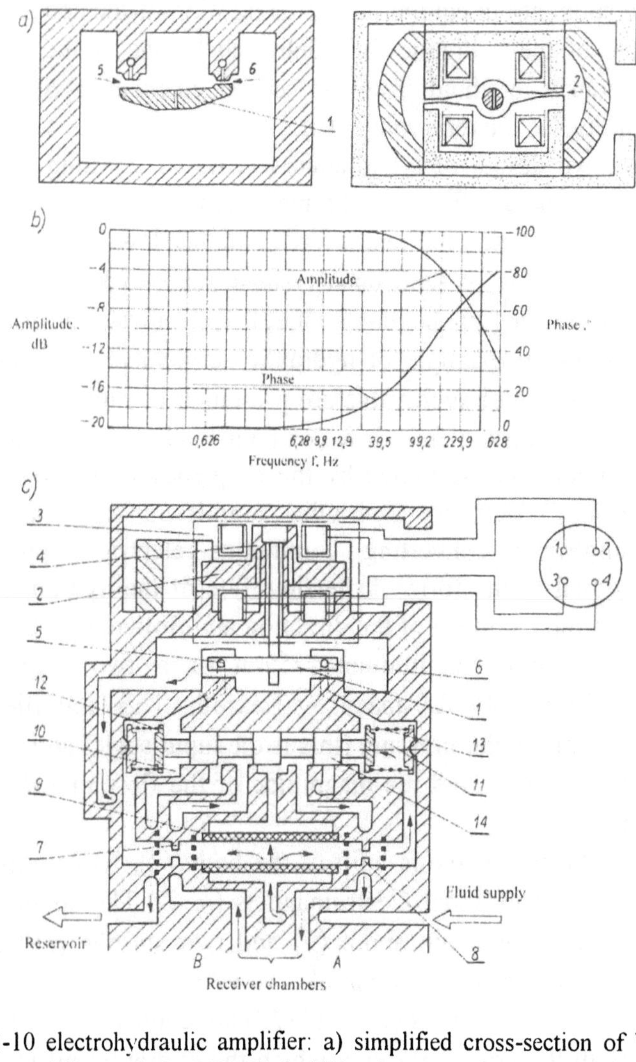

Fig. 9.22. WP3-10 electrohydraulic amplifier: a) simplified cross-section of WP3-10 - two-stage electrohydraulic amplifier made by PZL-Hydral Wrocław according to [9.3], b) phase and amplitude characteristics, c) construction of the amplifier.

If equal currents are flowing in opposite directions through the windings of both coils of the torque converter *3*, the electromagnetically generated torques are balanced, the diaphragm *1* rests at the same distance from the nozzles *5* and *6*, equal pressures exists in the chambers *10* and *11* and the control spool *14* is moved by the springs *12* and *13* to the position in which both chambers of the hydraulic motor are disconnected from the pump and from the reservoir. If the currents flowing through the windings of the coils are different a torque is generated turning the diaphragm *1* and causing a pressure difference between the chambers *10* and *11*. This difference acts on the spool pistons *14* producing a resultant force which moves the control spool *14* to the position in which all forces acting on the spool are in equilibrium. This position of the spool corresponds to a particular flow rate of the hydraulic fluid delivered to one of the hydraulic motor chambers. Simultaneously the fluid from the second chamber is discharged by the electrohydraulic amplifier to the reservoir [9.16].

Another well-known design of a hydraulic servovalve is Moog valve. The flow chart of this valve is presented in Figure 11.11 and its principle of operation is illustrated in Figure 16.27b. In this solution the return force is rather low and the valve contains only one thin, flat spring. At present, mainly single-stage hydraulic servovalves are used, applying a high-power hydraulic motor controlled electronically [9.16].

Figure 9.23a shows an example of hydraulic servovalve with rotating piston. The dynamic characteristics of such valve are presented in Figure 9.23b. These are transfer functions describing the free flow (without load) for a given pressure and flow rate. The bandwidth is determined by the frequency (from -6 dB) and phase ($-\pi/2$). An important group of hydraulic amplifiers contains proportional electrohydraulic amplifiers converting continuous electric input signal into continuous output signal in the form of the flow rate of the hydraulic fluid. These devices are used for remote control of the direction and flow rate of the fluid.

Figure 9.24 shows the longitudinal section of a two-stage, proportional electrohydraulic amplifier - 4WRZ made by Rexroth. This amplifier consists of a hydraulic valve controlling the pressure (this is the input stage) and a spool control valve controlling the direction and the magnitude of the flow rate (being the main stage of the amplifier).

The pressure controlling valve is built of the body (*1*), two spools (*2, 3*), two springs (*4, 5*) and two proportional electromagnets (*6, 7*). The spool control valve consists of the body (*8*), the main spool (*9*) and two springs (*10*) and (*11*). Its principle of operation of this electrohydraulic valve is following: when the input electric signal is zero, both chambers (*12*) and (*13*) are connected by the ports in the spools (*2*) and (*3*) with the port *Y*, through which the hydraulic oil is discharged to the reservoir.

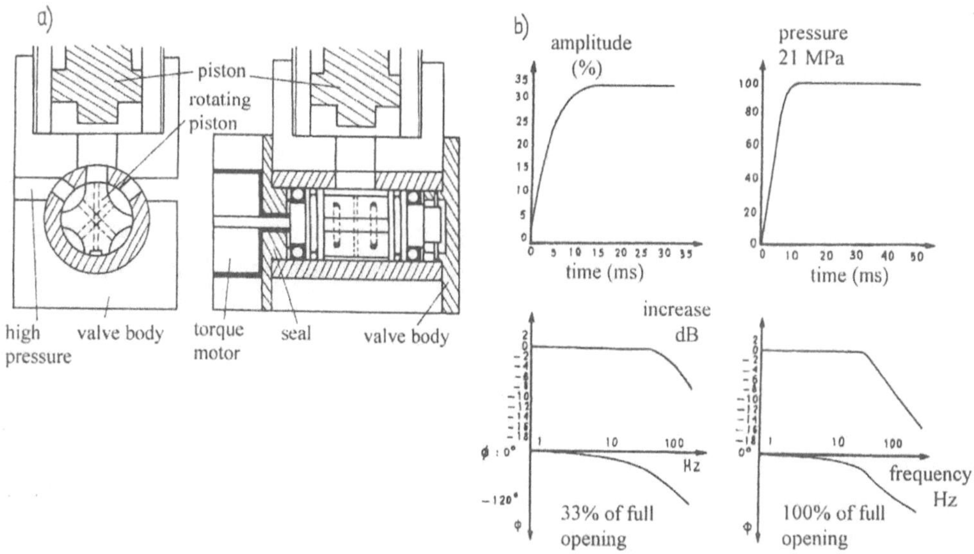

Fig. 9.23. Servovalve with rotating piston: a) servovalve with rotating piston controlled by a differential spool, b) dynamic characteristics of the valve, according to [9.12].

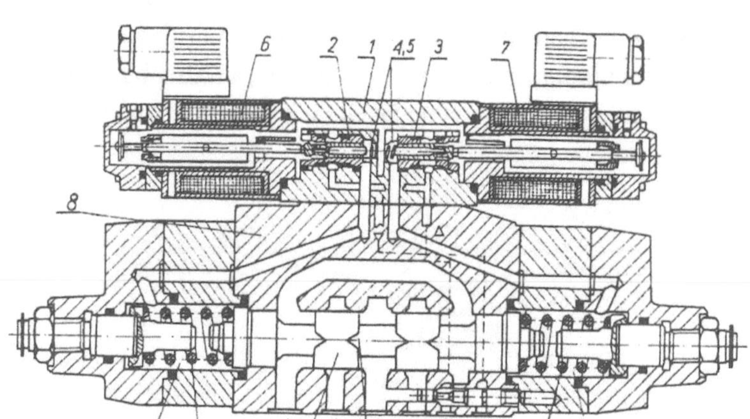

Fig. 9.24. Longitudinal section of Rexroth 4WRZ - a two-stage, proportional electrohydraulic amplifier [according to 9.16].

Thus the pressures is zero in both chambers (*12*, *13*) and the spool (*9*) is maintained by the springs (*10*) and (*11*) in the mid-position, in which the receiver ports *A* and *B* are disconnected from the source port *P* and the reservoir port *Z*. When the input current is applied, for instance, to the coil of the proportional electromagnet *6*, the spool *2* overcomes the resistance of the spring *4* and moves to the right, connecting the chamber

12 through the longitudinal and lateral ducts in the spool *2* with the internal supply port *P* or external supply port *X*. Simultaneously the spool disconnects the chamber *12* from the discharge port *Y*. Now in the chamber 12 a pressure proportional to the input current signal applied to the coil of the proportional electromagnet *6* is generated. This pressure acts on the front face of the main spool *9* causes it to move the spool to the right, to the position in which all forces acting on the spool are in equilibrium. As the result, the main spool *9* is now in the position proportional to the input current signal, in which the supply port *P* is connected with port *B* of the receiver and port *A* of the receiver is connected with port *Z* of the reservoir.

In case when the input current signal is applied to the coil of the proportional electromagnet *7*, the main spool *9* moves to the left, to the position in which the supply port *P* is connected with port *A* of the receiver and port *B* of the receiver is connected with port *Z* of the reservoir. Symmetric cut-out around the circumference of the main spool *9* cause that the area of the spool valve ports increases together with the displacement of the spool [9.16].

9.1.2.6. Electrohydraulic servomechanisms

Figure 9.25 shows a schematic diagram and Figure 9.26 shows a flow chart of an electrohydraulic servomechanism. In this solution the servovalve of the actuator is used to control the relative motion in a prismatic joint. It can also be used for a rotational joint, if the linear hydraulic actuator is replaced by an angular (rotary) one.

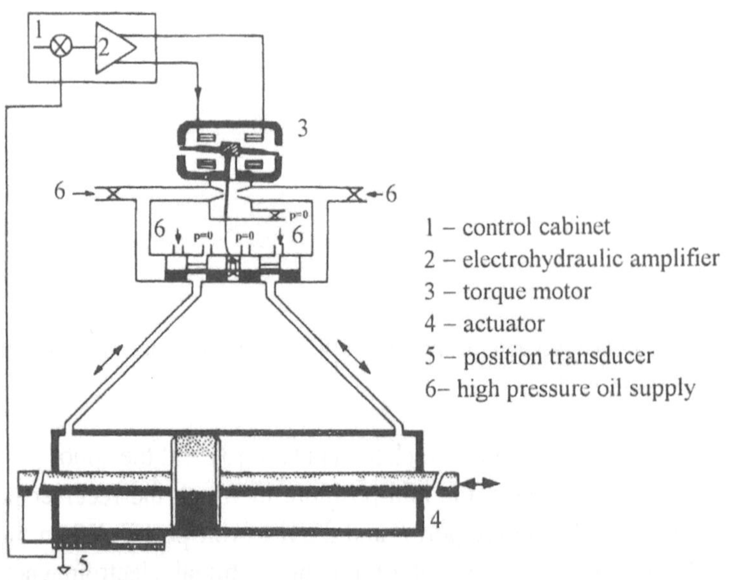

Fig. 9.25. Electrohydraulic servomechanism according to [9.2]

In Figure 9.26b Q and Q_0 are respectively the input and output of the hydraulic servovalve, A is the cross-sectional area of the piston, K_f is the signal of the position transducer, K_{pq} and K_q are coefficients allowing for the fluid leaks, K_a and K_s are stiffnesses of the fluid in the actuator.

Figure 9.27 shows typical static characteristics of a hydraulic valve with controlled flow, where: q is the flow rate in [m^3/s], p_s is the supply pressure in [Pa], Δp is the pressure difference across the piston in [Pa], i is the current of the electric torque motor, i_r is the current used to control the hydraulic valve. Various models of hydraulic valves are available for the flow rate ranging from $0.15 \cdot 10^{-3}$ to $37.8 \cdot 10^{-3}$ [m^3/s], what corresponds to the power output ranging from 2.2 to 52 kW [9.2].

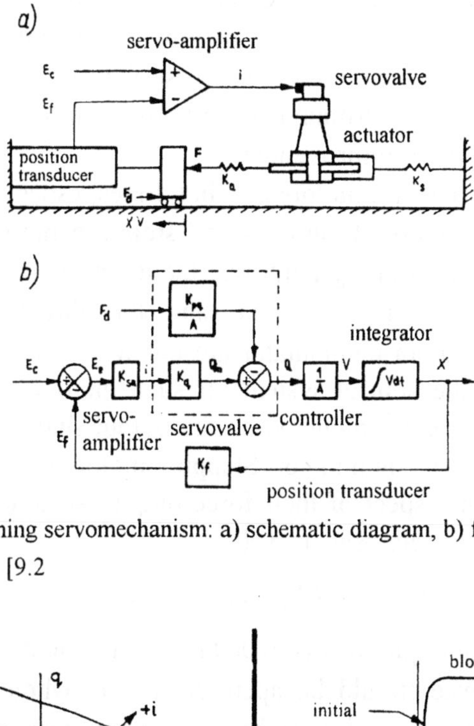

Fig. 9.26. Typical positioning servomechanism: a) schematic diagram, b) flow chart (Moog Inc. East Aurora, NY) according to [9.2

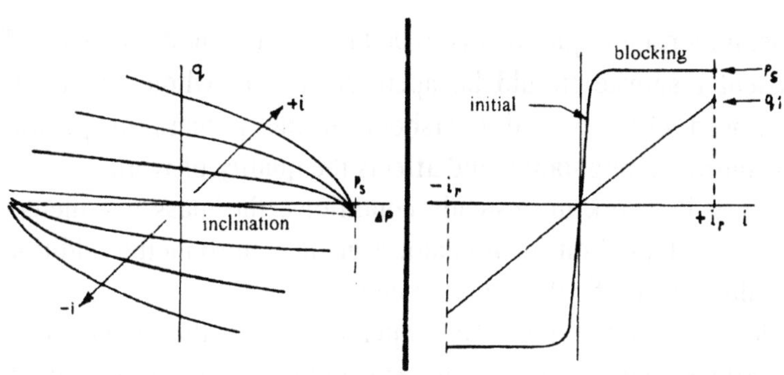

Fig. 9.27. Static characteristics curves of a typival hydraulic valve with controlled flow (Moog Inc. East Aurora, NY) according to [9.2]

9.1.3. Electric drives

9.1.3.1. Importance of electric drives

During the first development stage of industrial robots mainly pneumatic and hydraulic drives were used. Higher requirements set by the users of second and later generations of robots resulted in a fast development of electric drives. It is estimated that 50% of currently used robots are fitted with electric drives. Electric drives have numerous advantages, such as simplicity of power supply, availability of electric motors (the most modern models do not require gear reducers and can be mounted directly in the joints), accuracy and easy control, low energy losses and low environment pollution. The most significant drawback of electric drives is the issue of power. The ratio of power to weight or torque to weight for electric motors is worse than in hydraulic ones, which is caused by the weight of electromagnetic circuits. An application of permanent magnets can bring about an improvement.

The most common electric motors in robot drives, especially in so-called direct drives are **commutator-type DC motors** and **stepping motors**. The latter type is especially suitable for positioning purposes. Induction motors without commutator are not advantageous when the r.p.m. and the direction of rotation vary. Synchronous motors without commutator are not widely used, although they have a series of virtues. In the control system for such motors a magnetoindtuction or optoelectronic pulse generator mounted on the output shaft, which is also required in case of DC motors with commutator. Stepping motors needs further development of their construction in respect of their force output, accuracy and control. Linear electric motors are not widely used as well.

9.1.3.2. Requirements to be met by electric drives

The most important requirements to be met by electric drives are as follows:

- the transient response should be aperiodic in the whole range of parameters change. Speed changes and overshoot in the process of positioning cause problems during manipulation and affects the quality of work;
- as robots usually work in unsteady conditions, the mass parameters of electric motors (often situated on manipulator arms and introducing additional dynamic load) should fulfil particular requirements;
- electric drives should feature stable operation when the moments of inertia and loading torques change. Electric drives should also protect the robot against the effects of link and joint compliances (should provide additional damping);

- electric drives should be safe (explosion-proof), free from sparking, reliable and economical (drive economy is described by power consumption per each operating cycle);
- electric drives should maintain desired parameters of motion, namely positioning accuracy, continuous and smooth adjustment of current in a wide range, possibility of maintaining one position for prolonged period of time, even in case of power supply failure;
- requirements concerning work safety should be also met (self-locking of the drive in case of failure);
- the system controlling the angular position of the electric motor output shaft should permit teaching the manipulator (storage of motion program) in manual and automatic teaching mode.

Additional advantages of electric drives, such as low noise level, no lubricating oil, immediate start-up and independence of basic characteristic from temperature are also important factors.

Current development stage of microelectronics technology leads to microminiaturization of control system for electric drives. Electric drives of modular design consisting of an electric motor, harmonic drive and feedback sensors and transducers are the next development stage. Even today electric motors with high dynamic properties are constructed, providing the r.p.m. span reaching 100000:1, up to 100 Hz frequency range and overload factor of 10.

9.1.3.3. DC motors with commutator

DC motors with commutator (working as servodrives) have found the widest scope of applications in robotics. This results mainly from simple adjustment of speed and torque. As such motors have been used for a long time in processing machines, the control systems for such motors are well-known and many solutions have been tested in industrial practice.

As robot drives usually work in difficult dynamic conditions, it is desirable to increase peak accelerations reached in transient states and defined as

$$\varepsilon_{\max} = \frac{M_{dop}}{I_s + I_{red}} \tag{9.41}$$

where:

M_{dop} - the maximum permissible torque developed by the motor,

I_s - the moment of inertia of the rotor of electric motor,

I_{red} - the reduced moment of inertia of the driven mechanism.

For a given electric motor, its acceleration can be increased only by increasing the torque developed by the motor or by decreasing its moment of inertia.

Figure 9.28 shows the schematic diagram of a position control system. This system has two feedback loops: the position feedback and the velocity feedback, thus controlling the motion of an electric motor with variable direction of rotation [9.12].

The equation of motion for the electrical system shown in Figure 9.29a in the case when the current flowing through the excitation winding is constant can be written as

$$u(t) = Ri(t) + L\frac{di}{dt} + K\omega(t) \tag{9.42}$$

where:

R - the resistance of the armature winding,

L - the self-inductance of the armature winding.

The equation of motion for the mechanical system shown in Figure 9.29a in the case when the current flowing through the armature winding is constant can be written as

$$M_m = J\frac{d\omega}{dt} + f\omega + M_f + M_r \tag{9.43}$$

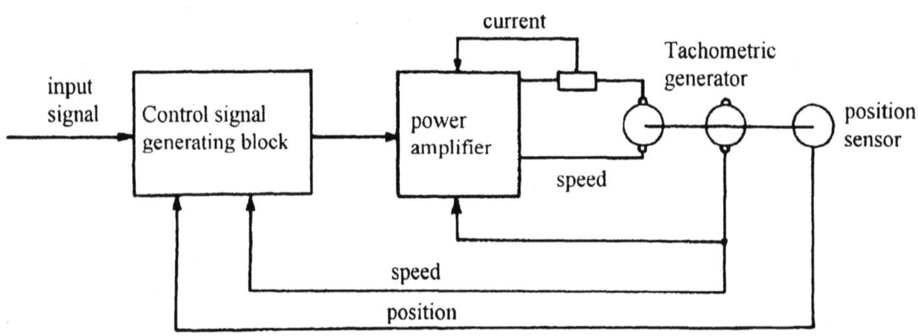

Fig. 9.28. Position control system - principle of operation according to [9.16]

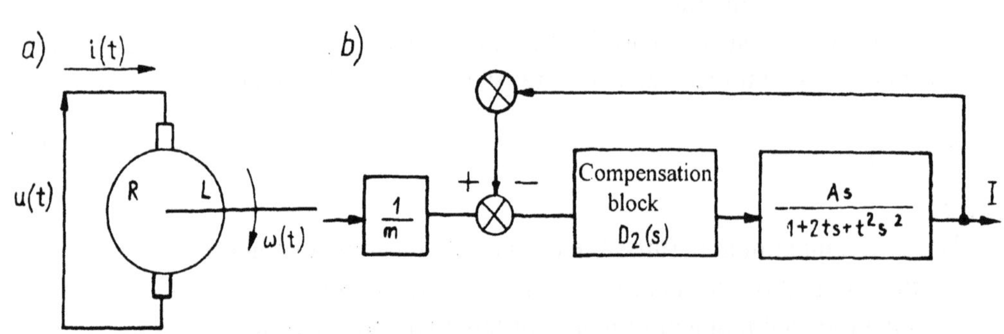

Fig. 9.29. Schematic diagram of a motor: a) momentary position of the motor, b) current loop for very small or zero torque according to [9.16]

where:

J - the moment of inertia of rotating parts,

f - the viscous friction coefficient,

M_r - the torque resulting from dry friction,

M_f - the loading torque,

ω - the angular velocity.

The electromagnetic torque M_m is proportional to the armature current $i(t)$:

$M_m = K\, i(t)$, (K is a constant)

Having performed Laplace transformation for both sides of the equation at zero initial conditions, equations (9.42) and (9.43) yield:

$$U(s) = (R + L_s)\, I(s) + k\omega(s) \tag{9.44}$$

$$K\, I(s) = (J_s + f)\, \Omega(s) + C_f(s) + C_r(s) \tag{9.45}$$

The transfer function linking the velocity and the voltage can be written as

$$G(s) = \frac{K}{I_s^2 + (RJ + Lf)s + Rf + k^2} \tag{9.46}$$

$$G(s) = A_0 \frac{1}{1 + 2\xi rs + r^2 s^2} \tag{9.47}$$

where $\xi >> 1$.

If f is small, the following two time constants can be introduced:

$$\tau_e = \frac{L}{R}, \qquad \tau_m = \frac{RI}{k^2} \tag{9.48}$$

where τ_e is the electric constant and τ_m is the electromechanical constant in the following equation:

$$1 + 2\xi rs + r^2 s^2 = (1 + \tau_e s)(1 + \tau_m s) \tag{9.49}$$

The electrical system for $f = 0$ can be described as shown in Figure 9.29b. If $D_2(s)$ provides proportional and integrating action, the transfer function refers only to the open system constituted by the motor and the current circuit. In such cas the motor has the only constant τ_m as τ_e has a very low value. In general, electric m have very low constant τ_m which allows the moments of inertia to be reduced internal friction to be increased.

9.1.3.4. Electric motors used in robots

Electric motors used in robots have long rotors with relatively small diameter. This is aimed at the reduction of time constant τ_m, which is less than 10 or 20 ms. Such motors are durable and feature high thermal capacity (time constant > 1 h), good mechanical properties and small self-inductance of the armature winding (about 100 μH).

Let us give several examples of technical data taken from AXEM catalogues (table 9.1). Motor no. 2 features efficient ventilation (air flow rate of $80 \cdot 10^{-3}$ m^3/s) which explains its especially high power rating.

Figure 9.30a presents a printed circuit disc rotor of an electric motor, Figure 9.30b shows the general view of the motor, Figure 9.30c shows its schematic diagram and Figure 9.30d illustrates its magnetic circuit. The shapes of a common type motor and a printed circuit disc motor [9.6] are compared in Figure 9.31a while some properties of various drives are compared in Figure 9.31b.

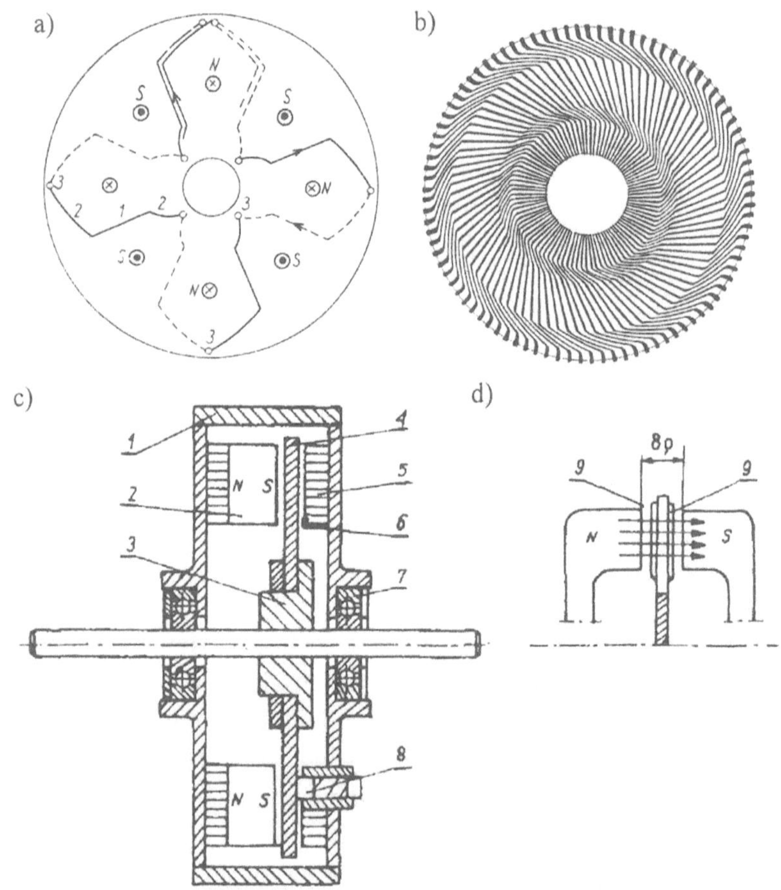

Fig. 9.30. Printed circuit disc motor according to [9.16] a) printed winding layout: continuous lines - winding on one side, dashed lines - winding on the other side of the isolating disc. 1 - active part of the wire, 2 - joints of wires on one side,3 - joints of wires printed on different sides of the disc; N,S - polarity of the magnetic flux. b) general view of the rotor, c) schematic diagram of the motor, d) magnetic circuit: 1 - housing, 2 - pole, 3 - hub, 4 - disc, 5 - coil, 6 - cover, 7 - bearings, 8 - sleeve, 9 - disc.

Table 9.1. Example of technical data from AXEM catalogue

Parameters	**Electric motor**		
	1	**2**	**3**
Rated output torque [N·m]	3.2	28.5	1.2
Rated armature current [A]	7.2	37.5	6.0
Rated power output [kW]	1.0	9.0	0.37
Rated armature voltage [V]	164	272	83
Rated speed [r.p.m.]	3000	3000	3000
Moment of inertia of the rotor [10^{-3} kg·m]	1	7.3	0.79
Time constant [ms]	6.5	5.5	30
Maximum theoretical angular acceleration [rad/s^2]	24400	22500	14500
Mass [kg]	9.7	58	6.5

The typical electric motor may compete with the printed circuit disc motor in terms of the torque to weight ratio. On the other hand, the printed circuit motor is quick and responsive, but it has lower starting torque. The printed circuit disc motor has been designed to reduce the time constant and the inertia of the rotor, which are significant factors for the control system designer.

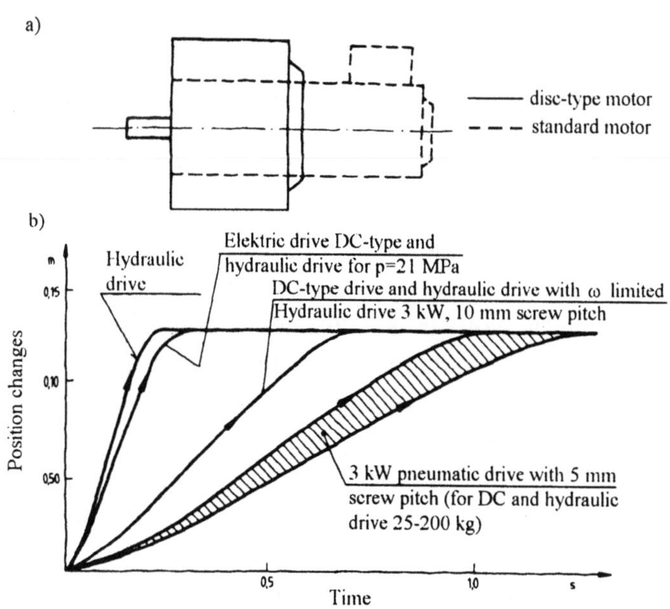

Fig. 9.31. Comparison of a) shapes of a standard and a disc-type motors, b) properties ofpneumatic, hydraulic and electric drive.

9.1.3.5. Electric stepping motors

The principle of operation of all stepping motors is based on discrete changes of the electromagnetic field in the motor gap. The changes in the situation of the excitation field axis move the rotor which cycles through predetermined positions. The number of these positions is always higher than two per one full revolution of the rotor [9.19].

Stepping motors transforms electric signal changing in a discrete manner into increments of angular displacement. They are of synchronous type, which ensures correlation between the number of input pulses and the final position of the rotor [9.1], [9.6].

The main vices of stepping motors are: discontinuous acceleration profiles, limited performance and variable torque depending on the rotor position. Three types of stepping motors can be listed here:

- **magnetoelectric steppers** (with permanent magnet),
- **variable reluctance steppers,**
- **hybrid steppers** (combining the features of both former types).

Figure 9.32 illustrates the principle of operation of a stepping motor with a double-pole permanent magnet rotor. The figure shows four consequent phases of the commutation cycle. When a voltage is supplied to the coil of pole *1* (phase *1*), a magnetic flux $\phi_1 = \phi^s$ directed as shown in the drawing will be generated. The rotor, which is now polarized and has two poles, will move to the position in which the direction of its magnetic flux is the same as the magnetic flux of the stator. If the coil of pole *2* is powered simultaneously (phase *2*), the magnetic flux of the stator ϕ^s, being the resultant magnetic flux of ϕ_1 and ϕ_2, will change its spatial position. As a consequence of the interaction between the magnetic fluxes of the stator and rotor a torque will start to act on the rotor, causing it to turn until its magnetic axis aligns with the magnetic flux of the stator. Another movement of the rotor will take place after disconnection the coil of pole *1* from the voltage source (phase 3) and the fourth phase - when the voltage is connected additionally to the coil of pole *3*. After eight phases of such commutation the vector of the magnetic flux of the stator and the rotor itself make one full revolution. A disadvantage of such control of the stepping motor is the necessity of applying pulses of different polarity.

Figure 9.33 shows a diagram of a stepping motor control system. Energy losses in stepping motors are similar to the losses in commutator-type DC motors. Parameters of a four-phase stepping motors with a step of $\pi/12$ (24 steps per one full revolution) are given in table 9.2. These parameters refer to a particular power supply system using an electronic commutator based on semiconductor-type, MOSFET bipolar electric valves. Four-phase stepping motors are most common in robotic drives.

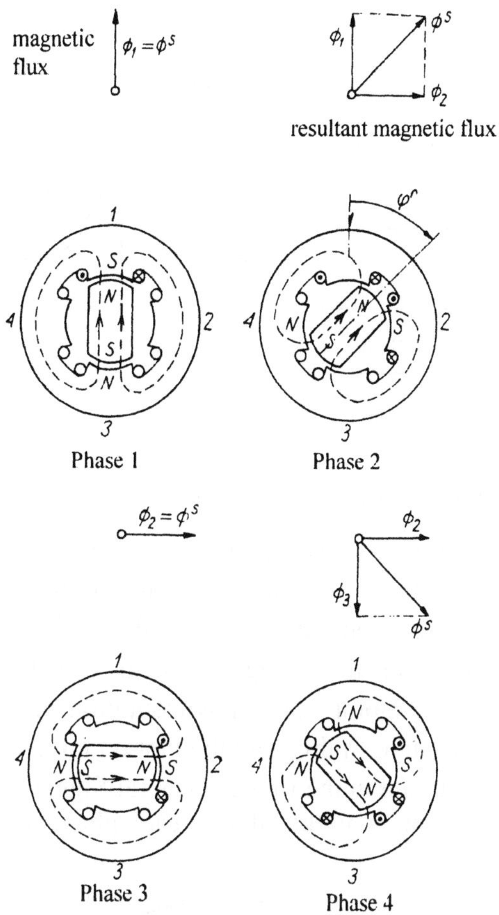

Fig. 9.32. The principle of operation of a stepping motor with a double-pole permanent magnet rotor according to [9.6].

Table 9.2. Example of stepping motors technical data

Maximum static torque [g·cm]	2300
Maximum frequency [Hz]	400
Maximum dynamic torque [g·cm]	1400
Weight [g]	600
Current [A]	2.8
Voltage [V]	28
Mechanical time constant τ [ms]	0.67
Pulse	116
Real frequency [Hz]	474

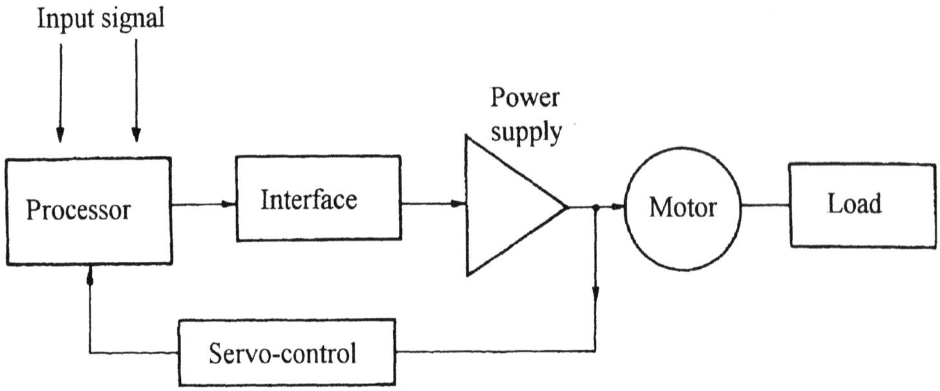

Fig. 9.33. Stepping motor control system.

9.1.3.6. Linear electric motors

Linear induction motors will be given only a brief description. Figure 9.34 shows the main components of such motor. This motor is built of a permanent magnet moving in a housing filled with epoxide filling. The moving link slides on a pneumatic bearing 0.0132 mm thick. Figs 9.35a, b, c and d show the phases of motion of the linear electric motor. Each new excitation phase causes the slider to move by 1/4 of the slider pitch. In this way the motion covering 3/4 of the pitch can be realized, which is illustrated in the figure (General Signal Corp. Santa Clara, CA) [9.5]. The direction of current is marked during each phase (Fig. 9.35a, b, c, d).

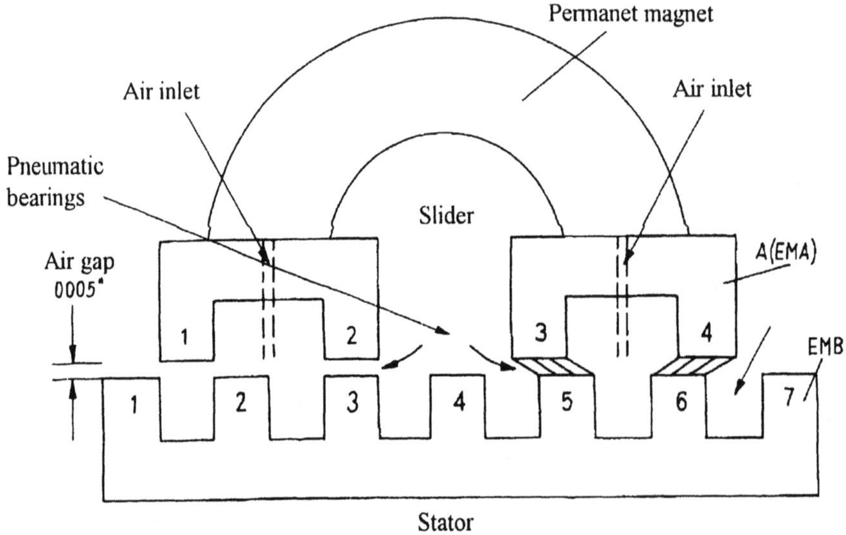

Fig. 9.34. Main components of two-phase linear electric motor (General Signal Corp. Santa Clara, CA)

If electromagnet A (EMA) is activated, maximum magnetic flux density is generated in pole *2* and the motor rests in the position shown in Figure 9.35a. If EMA is not activated and EMB is activated, maximum magnetic flux density is generated in pole *3* and minimum magnetic flux density is generated in pole *4*. As a consequence, the slider moves to the right.

Sometimes a linear induction motor may be used in a direct drive system. Usually the position of the slider relative to the housing is discretized. The pitch is determined by the area of the teeth (poles). A typical pitch equals 10.16 mm. For the sequence shown in Figure 9.35 the resolution equals 2.6 mm (a quarter of the full pitch). This resolution is also called "the basic step".

The linear electric motor may also house two orthogonal sliders moving in one housing.

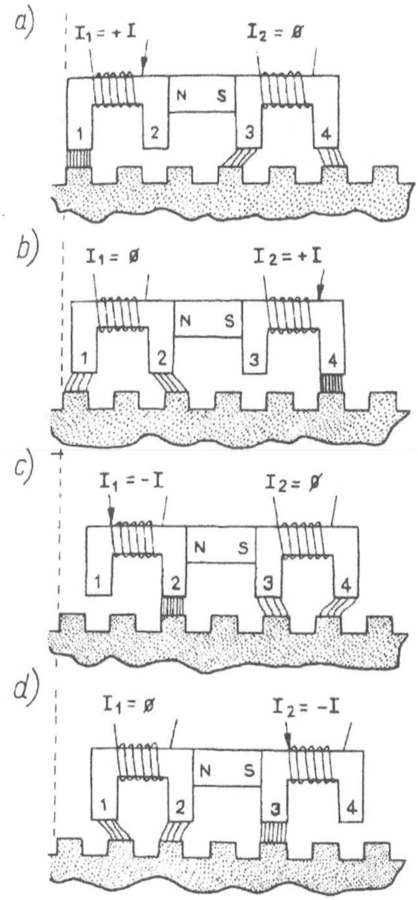

Fig. 9.35. Two-phase linear electric motor - phases of motion

9.1.3.7. Direct drive electric motors

A new type of electric motor was designed in the 80's. Such motor permits an application of a direct drive (without any gear). This new electromechanical converter was produced by Motornetics Corporation and called a "Megatorque motor system". The drive generates high output torque (507 ÷ 14480 Nm) at very low angular velocities (30 r.p.m.) without the need to use gear-type speed reducers. Additionally an integral part of the system is a position sensor with a resolution comparable to that of electronic position transducers used nowadays in manipulators.

"Megatorque" electric motor is a three-phase AC motor fitted with an electronic commutator, similarly to a DC motor with mechanical commutator. "Megatorque" electric motor is a variable-reluctance machine and contains no permanent magnets (fig. 9.36).

The heart of the motor is a set of plates joining the rotor and the stator. Fig. 9.37 shows an example of such plate. A thin, ring-shaped rotor is mounted between two concentric stators. Both stators interact with the rotor. A high number of magnetic teeth of the rotor together with the effect of two stators allows the motor to produce an amplified, high torque.

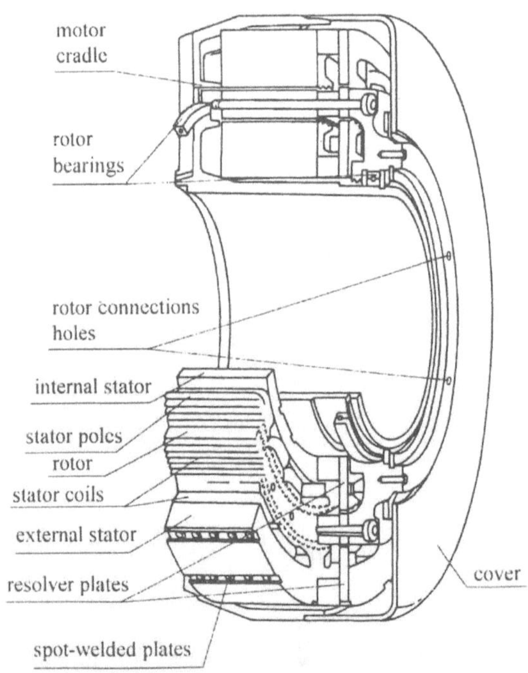

Fig. 9.36. Cross-section of a "Megatorque" electric motor - direct drive system (B.Powell & Corp., Santa Rosa, CA). Model 1140, ultra-high torque according to [9.5].

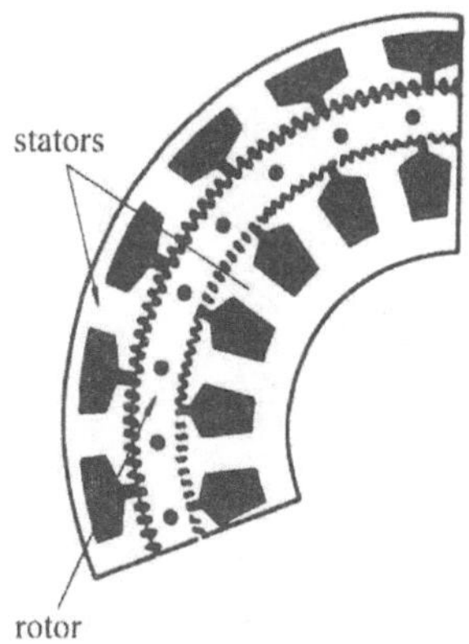

Fig. 9.37. Quarter of "Megatorque" plate section with two stators placed around the rotor according to [9.5]

Three-phase magnetic field is generated by 36 coils of two stator windings (18 coils per each stator). Each of the stators has 150 teeth; the rotor acts as a pole of the electric motor. The torque is produced by sequential excitation of these 12 poles. For each full revolution of the rotor 150 variable cycles take place giving a gear ratio of 150:1, which corresponds to the amplification of the electromechanical torque.

It must be noted that by the situation of the rotor between two stators (as in the design described above) one can obtain the same rated output parameters as in the case a conventional electric motor with 300 poles or 900 coils in the stator winding. Shortening the magnetic flux path is another advantage of such situation of the rotor [9.5], [9.8].

During the last few years a quick development of similar electric reluctance motors and robots equipped with such drives (e.g. SCARA industrial robot) can be noticed.

9.2 Transmission of motion in robots

9.2.1. Arrangement of actuators in robots

Motion transmission mechanisms transmit the motion of the motor (or motors) into the motion of the links of the manipulator's open kinematic chain containing rotational or prismatic joints.

In some cases, when the kinematic chains are of closed type (for example in the pantograph mechanism), the relative displacements of the links are not independent and the motion transmission mechanisms must be situated in other ways.

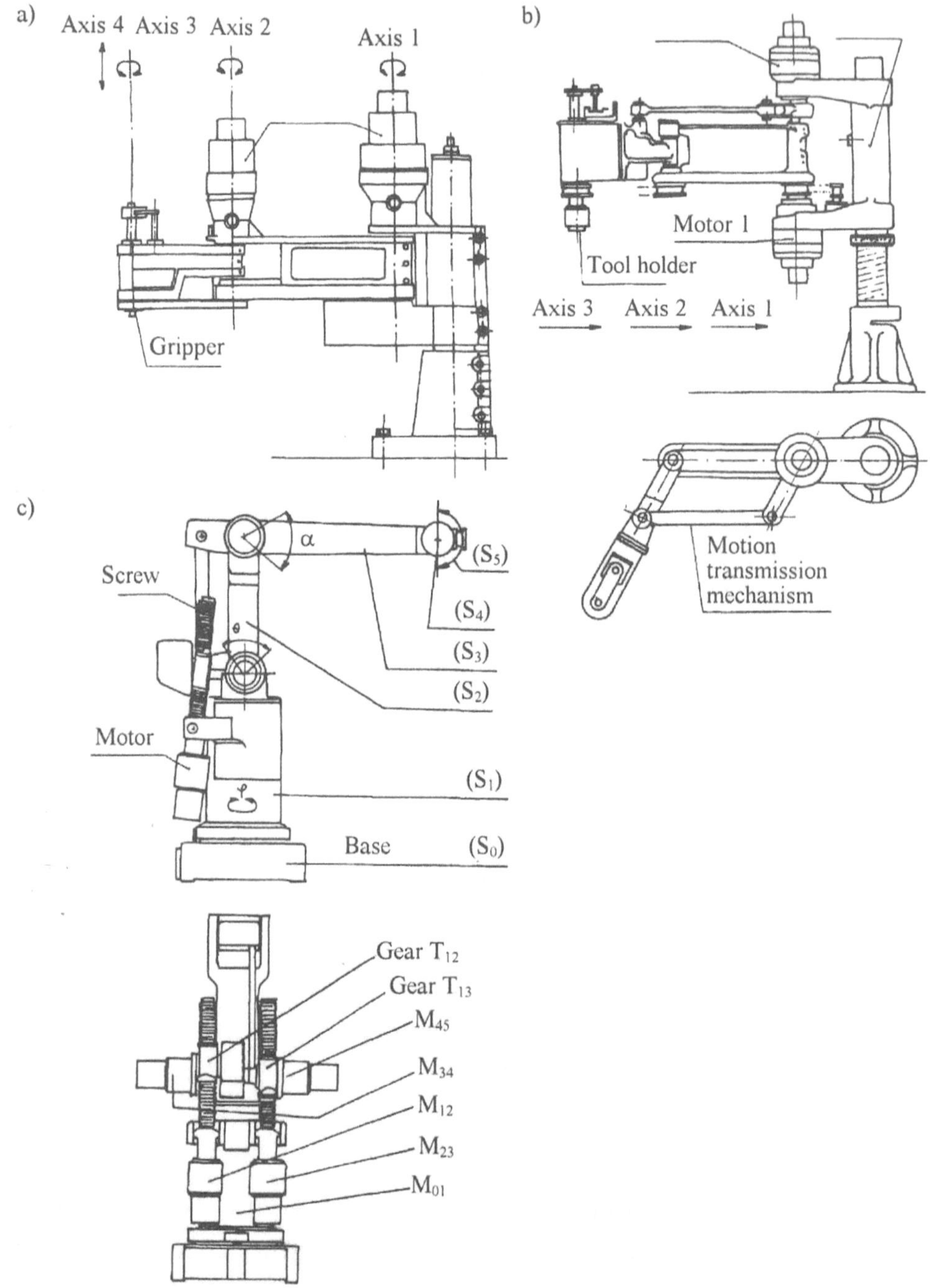

Fig. 9.38. Examples of motion transmission mechanisms used in robots: a) SKILAM type assembly robot, b) PUMA-2 type assembly robot, c) MOTOMAN or ASEA robot; motion is transmitted by the actuators mounted on link S_l according to [9.1], [9.10]

One should be aware that during the transmission of motion, as the result of play in joints, backlash in gears, friction, compliance, loading, etc. unfavourable dynamic phenomena (like vibration) occur affecting the positioning accuracy of the end-effector. Thus the arrangement of actuators and gears should help to reduce this negative influence of dynamic and static phenomena. In figs 9.38a, b and c three examples are given to illustrate these issues.

Figure 9.38a shows an open kinematic chain of SKILAM type assembly robot with four controlled axes. Motors *1* and *2* are situated near joints *1* and *2* which are driven by these motors. Axis *3* is driven by a stepping motor and axis 4 is driven by a pneumatic actuator. Fig. 9.38b shows a closed kinematic chain of PUMA-2 type assembly robot fitted with a 4-DOF gripper. The mechanism of a four-stage reduction gear is mounted on the base of the robot in order to reduce the influence of dynamic phenomena on the gripper motion. Fig. 9.38c shows a 5-DOF robot. The base is denoted by S_0, the links of the robots are S_1, S_2, S_3, S_4 and S_5. The motor driving the joint connecting S_4 and S_5 (M_{45}) is installed near S_1. Similar arrangement applies also to the motors marked as M_{34}, M_{23} and M_{12} [9.1] with the aim of reducing the effects of inertia in the relative motion of S_0 and S_1 (MOTOMAN 104 robot).

Let us now give a brief description of various motion transmission mechanisms used in robot drives.

9.2.2. Overview of various motion transmission mechanisms

9.2.2.1. Belt and chain drives

This type of mechanism can transmit rotary motion between two parallel shafts as well as to convert rotary motion into translatory motion and vice-versa.

Figure 9.39 shows an example of the belt drive [9.5]. The primary drawback of such drive is a slip between the belt and the pulley. In order to eliminate the potential for slip, transmission mechanisms with synchronous (timing) belts (having a toothed profile that mates with corresponding grooves in the pulleys) have been applied recently (fig. 9.40).

Fig. 9.41 illustrates a mechanisms of two-legged walking robot driven by means of timing belts. Motion transmission mechanisms shown in this figure have found a wide application owing to its quiet operation and high efficiency.

Another motion transmission mechanism is a chain drive shown in Figure 9.42a. It works properly at low speeds and is usually mounted between the links. Figure 9.42b shows an example of chain drive application in MOTOMAN robot (made by YASUKAWA) [9.1].

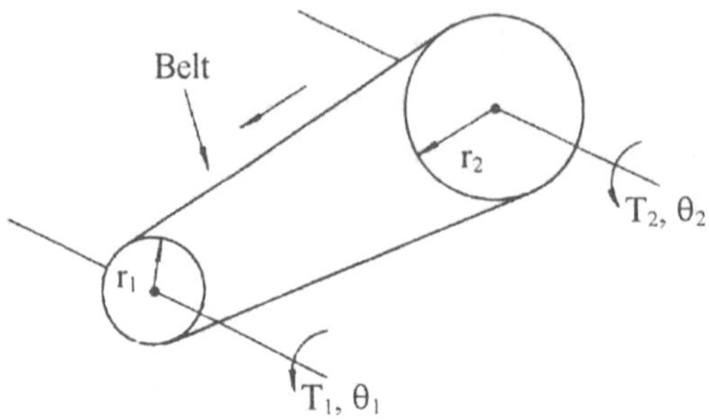

Fig. 9.39. Belt drive; r_i, T_i and θ_i are the parameters according to [9.5]

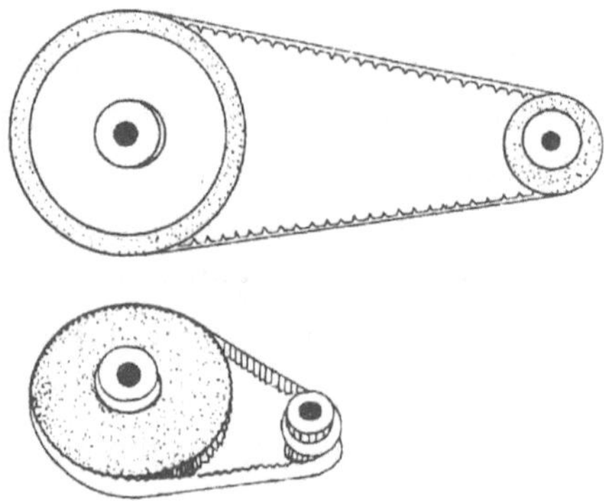

Fig. 9.40. Timing belt drive according to [9.5]

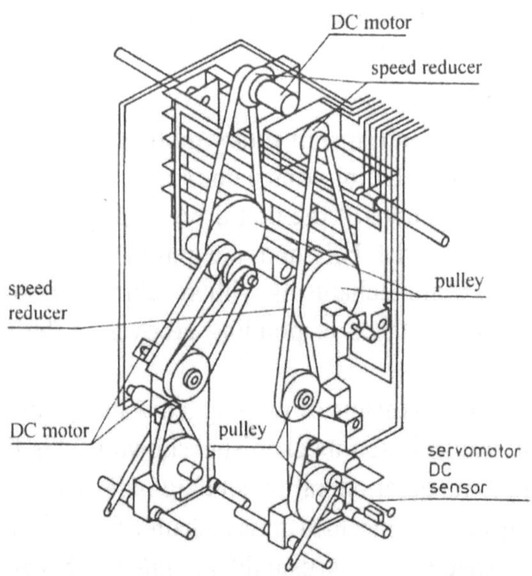

Fig. 9.41. Example of application of a motion transmission mechanisms with timing belts in a two-legged walking machine according to [9.9]

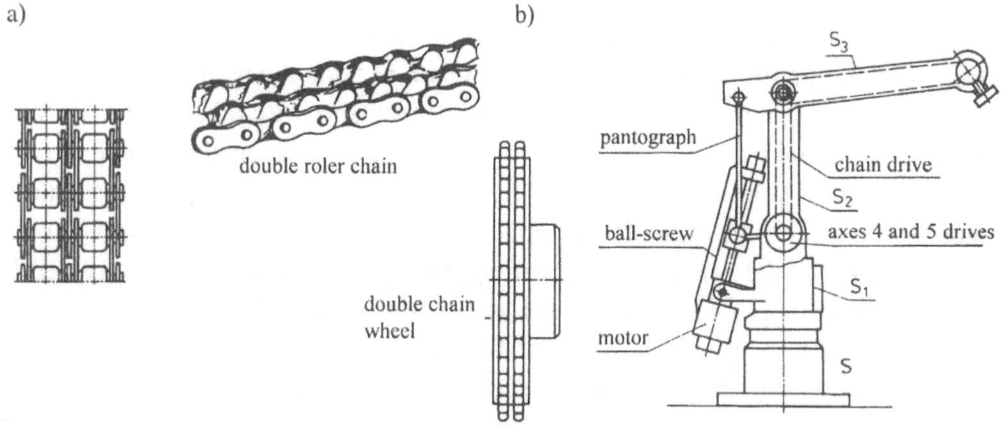

Fig. 9.42. Chain drive and its applications: a) chain with rollers, b) MOTOMAN robot drive according to [9.1].

9.2.2.2. Screw-type and rack-and-pinion transmission mechanisms

Other motion transmission mechanisms commonly used in robots are screw-type and rack-and-pinion mechanisms. Figs 9.43a and b show an example of the ball screw drive system. Fig. 9.43c shows the nut and the screw of such drive.

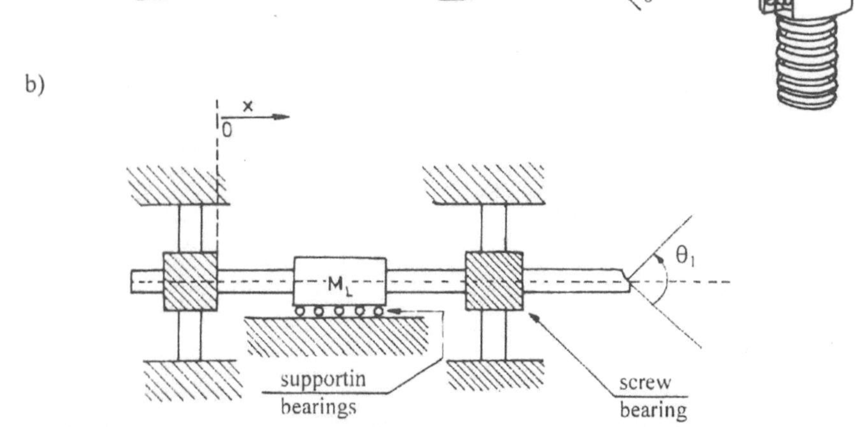

Fig. 9.43. Ball screw drive: a) general view, b) model, c) ball screw (TRANSROLL) according to [9.5].

From the kinematic viewpoint both these types of mechanisms (mechanisms applying screws and rack-and-pinion mechanisms) are equivalent. Both convert the rotary motion into the translatory motion and both provide the reduction of speed. It must be, however, noted that the screw-type mechanisms with a long screw is very sensitive to lateral vibration. Suitable location of the mechanisms relative to the robot links may help to increase the gear ratio and to achieve higher stiffness. Figure 9.44a shows an example of application of such screw mechanisms in the drive of axis *4* of the robot. The screw-type transmission mechanisms presented in Figure 9.44b is supported on both sides by two self-aligning bearings. Such design is favourable in respect of its stiffness, stability, resistance to bending and vibration.

The solution shown in fig. 9.44a uses screw-type kinematic pairs to transmit the motion in axes *2* and *4*. The drive system of axis *2* (the motor, screw and nut) changes its moment of inertia during the motion of axis 1 insignificantly. Fig. 9.44b shows an example of an in-parallel manipulator with screw-type drives [9.11]. By applying ball screws or roller screws the performance of such transmission mechanisms may be enhanced (fig. 9.43c).

Fig. 9.45 shows a schematic diagram of the rack-and-pinion drive used to transfer the load (M). The transmission ratio in such system is determined as:

$$x = 2\pi r\theta \tag{9.50}$$

Equation (9.50) shows that the output linear displacement x is proportional to the number of revolutions of the input shaft θ. The value of the constant of proportionality depends on the characteristics of the pinion. The moment of inertia of mass M reduced on the pinion axis can be written as:

$$I = Mr^2 \tag{9.51}$$

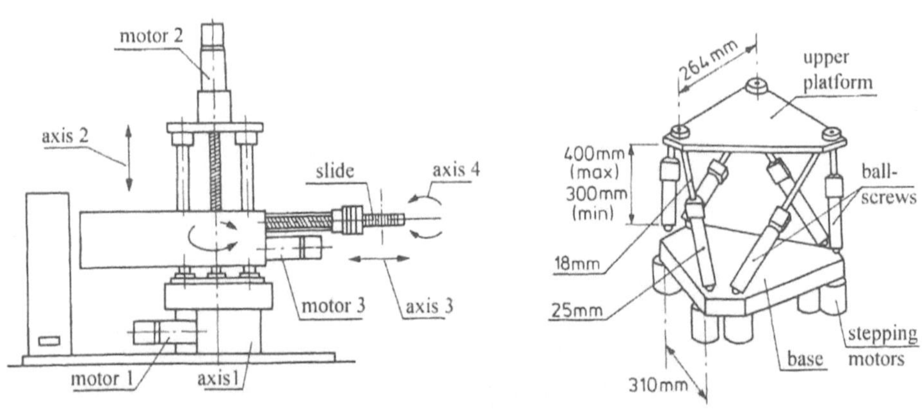

Fig. 9.44: a) SIROBOT 1 by Siemens - design scheme, b) in-parallel manipulator - Stewart platform according to [9.19].

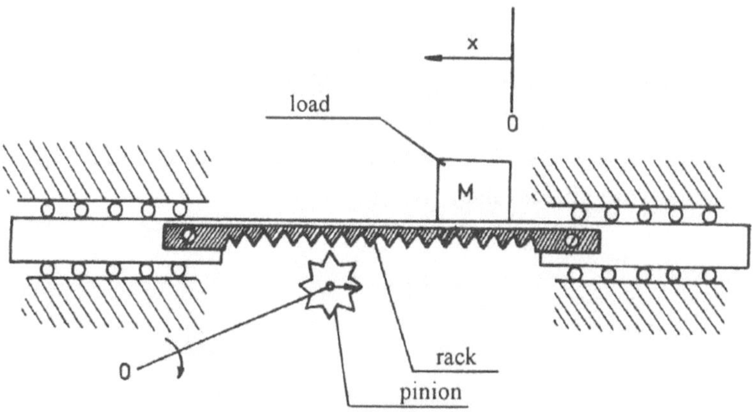

Fig. 9.45. Rack and pinion mechanism used to transfer the load.

9.2.2.3. Linkage-type transmission mechanisms

This chapter is limited to some examples only. Fig. 9.46 shows the kinematic diagram of a mechanism applied in KUKA robot. The mechanism is a parallelogram in which the axes of rotation are perpendicular to the plane of motion. The motion of one of the links can be controlled only in that plane and causes identical motion of the opposite link of the parallelogram. The orientation of the other link is maintained invariant with respect to the first link of the parallelogram. Such configuration features high stiffness.

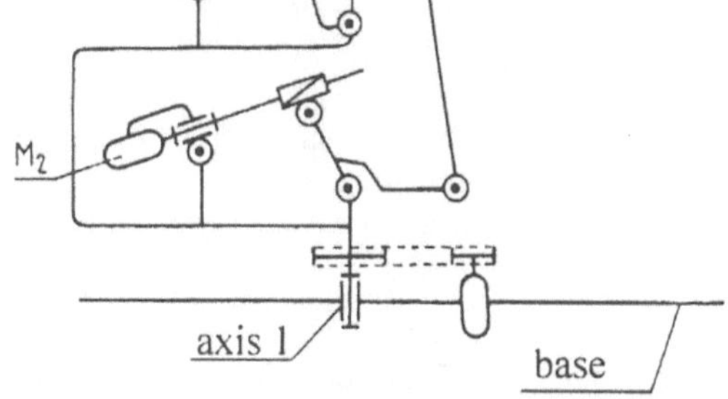

Fig. 9.46. Linkage transmitting the motion of motors and reduction gears M_2 and M_3 to robot axes *2* and *3*, according to [9.1].

Figure 9.47 shows a MELKONG manipulator used for rehabilitation purposes. The motion of two plates (*A* and *B*) constituting a platform supporting the disabled person is controlled by five kinematic chains *1-5*, each consisting of a motor, a reducing gear, a screw and a nut. The motion of the plates is a copy of the motion of links *1* to *5*. As the amplitude of motion is less than $\pi/2$, the linkage-type mechanism can compete with the chain drive (angular velocity less than $10\pi/18$ rad/s).

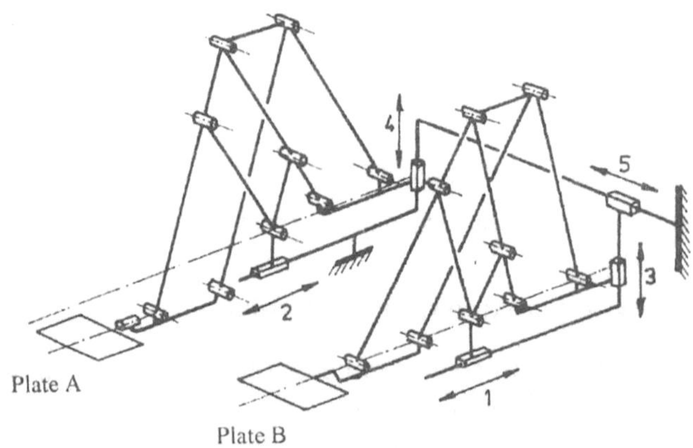

Fig. 9.47. Kinematic diagram of MELKONG robot according to [9.1]

Figure 9.48 shows the transmission of motion in M7 master-slave manipulator made in USA. The motion is transmitted from the motion transmitter *1* (master) to the motion receiver *2* (slave) by means of a combined linkage and string drive [9.12].

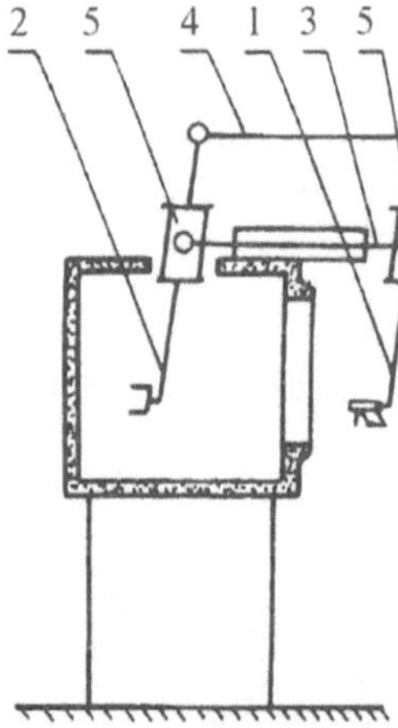

Fig. 9.48. Transmission of motion by means of a parallelogram (M7 manipulator, USA, 1956, still in operation) [according to 9.10]: *1* - master, *2* - slave, *3*, *4*, and *5* - combined linkage and string drive

A mechanism transmitting motion between axes *2* and *3* of ASEA IRb-6 robot in order to actuate the manipulator's wrist is illustrated in Fig. 9.49 [9.1].

In DELTA-4 manipulator, shown in Figure 9.50, the grippers (the upper platform) is driven by means of a system of linkages [9.9].

Figure 9.51 shows an in-parallel mechanism transmitting the motion of three actuators to the manipulator's arm. The mechanism is a linkage combined ball screws [9.14].

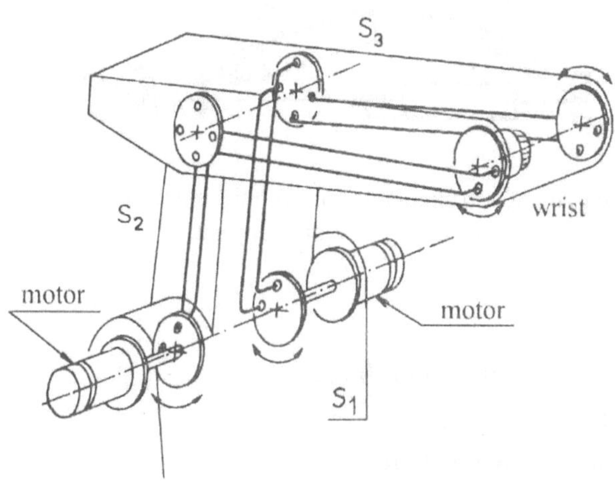

Fig. 9.49. Transmission of motion between axes *2* and *3* in ASEA IRb-6 robot according to [9.5].

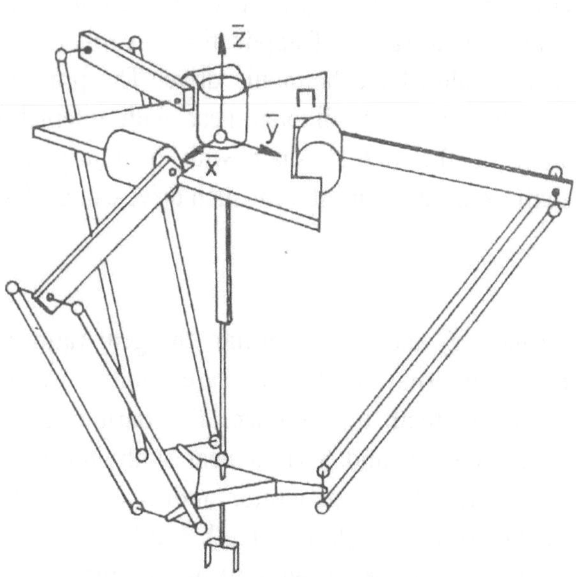

Fig. 9.50. Schematic diagram of the DELTA-4 robot according to [9.5].

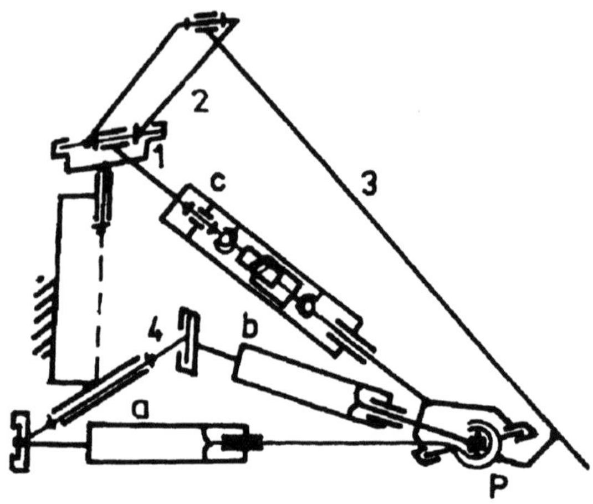

Fig. 9.51. Kinematic diagram of a manipulator arm with three-actuator in-parallel drive according to [9.14]; *a, b, c* - actuators; *1, 2, 3, 4* - linkage, *P* - gripper joint.

9.2.2.4. Gears

Two types of gears are used in robotics:

- gears with fixed shafts,
- epicyclic gears with moving shafts.

Figure 9.52 shows an example of planetary gear with all shafts fixed. Numerous modifications of such gears can be found in industrial practice.

Let us now describe another type of gear drive, namely the harmonic drive. This type of gear have found a wide range of applications in robots with electric drives. Figure 9.53 shows an example of the harmonic drive. The principle of operation is illustrated in Figure 9.53a using a traditional gear with internal mesh. Such gear produces rotary motion of the output shaft with a minor, cyclic displacement perpendicular to the shaft axis. The gear ratio for such drive can be determined as:

$$\frac{\omega_{20}}{\omega_{30}} = \frac{z_3}{z_3 - z_0} \tag{9.52}$$

If the numbers of teeth z_3 and z_0 are similar, the gear ratio can be quite high. The ratios of traditional, internal mesh gears are limited by the condition that $z_0 - z_3 > 8$ in order to avoid teeth interference. The harmonic drive consists of a flexible gear with external teeth and a rigid gear with internal teeth. A usually elliptical ball-bearing assembly (wave generator) induces a travelling wave of mesh between the gears, By the proper selection of teeth numbers, module pitch and other parameters the gear may work properly, without interference, for a relatively small difference $z_0 - z_3 \geq 2$ (Fig. 9.53b). The gear ratio may reach even 320.

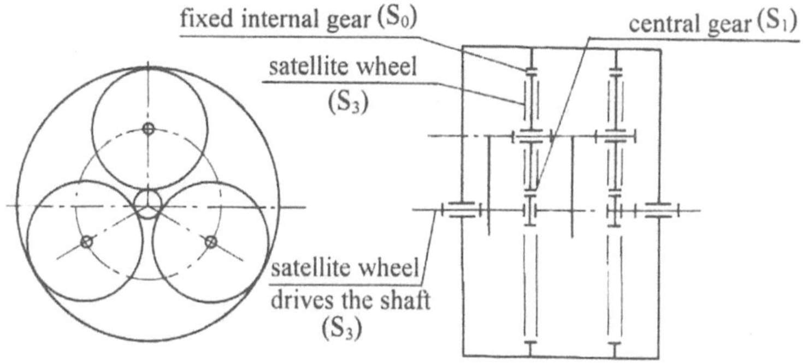

Fig. 9.52. Two-stage planetary gear

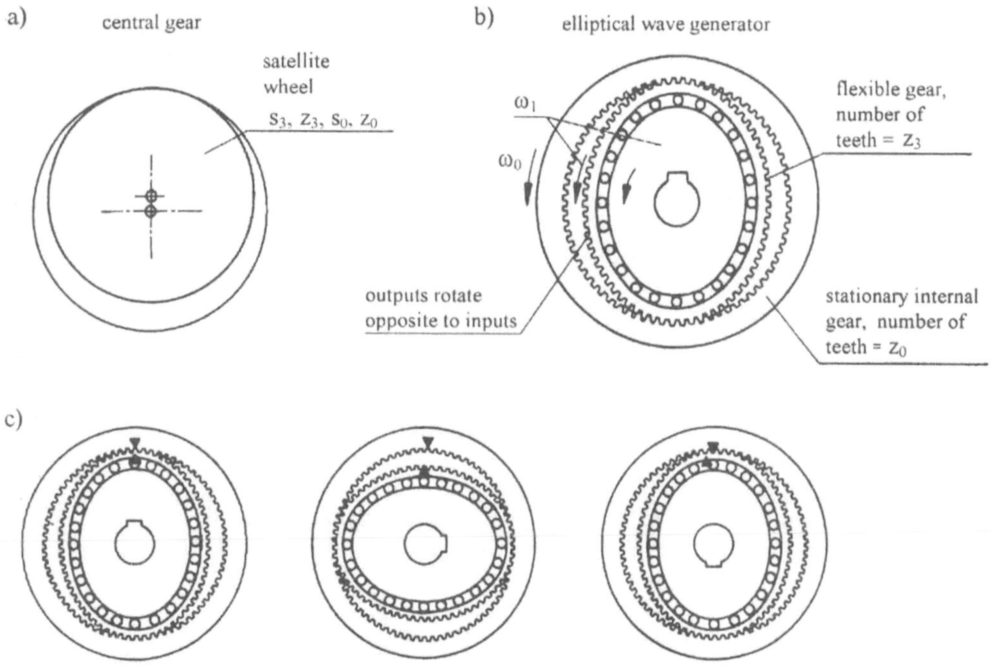

Fig. 9.53. Harmonic drive according to [9.1].

Fig. 9.53c shows subsequent stages of harmonic drive operation and fig. 9.54 is a drawing showing the construction of a harmonic drive. This drive can perform several functions: it can act as a speed reducer (the wave generator is the input link, the flexible gear is the output link and they rotate in opposite directions) or a speed multiplier (the flexible gear is the input link, the wave generator is the output link and they both rotate in the same direction). Other arrangements are also possible: the wave generator can be the input link, the rigid gear the output link and both links rotate in the same direction. The harmonic drive may also work as a differential gear. If this is the case, the wave generator remains fixed, the flexible gear is the input link and the rigid gear is the output link, rotating in the same direction.

The wave generator is usually a two-phase, ball bearing unit and causes oval (almost elliptical) deformation of the flexible gear. The mesh is obtained at both ends of the long axis of the ellipse while on both ends of the short axis the teeth maintain necessary tip clearance.

Let us conclude this section with some examples.

The single-stage harmonic drive with involute teeth and the pressure angle of $\pi/6$ rad the numbers of teeth are: $z_2 = 202$, $z_1 = 200$; $m = 0.8$, gear ratio $i = 100$, efficiency equals 0.85 and the output power reaches 2.35 kW.

In case of a two-stage harmonic drive with involute teeth and the pressure angle of $\pi/6$ rad the numbers of teeth of the first stage are 200/202 and of the second stage are 198/200. Each stage provides gear ratio $i = 100$ and the total reduction ratio $i = 10000$ [9.1, 9.7].

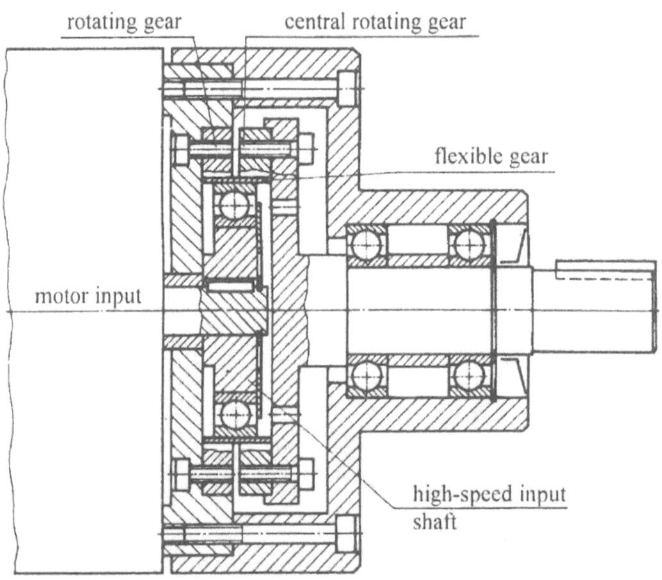

Fig. 9.54 The construction of a harmonic drive.

9.2.2.5. Transmission of motion by means of hydraulic actuators

This solution is similar to the mechanisms using screws. In this case the amplitude of the angular motion does not exceed 40° relative to the mid-position. Such mechanisms are used to transmit motion in painting robots.

Fig. 9.55 shows an example of a 3-DOF ACMA-Renault painting robot [9.1].

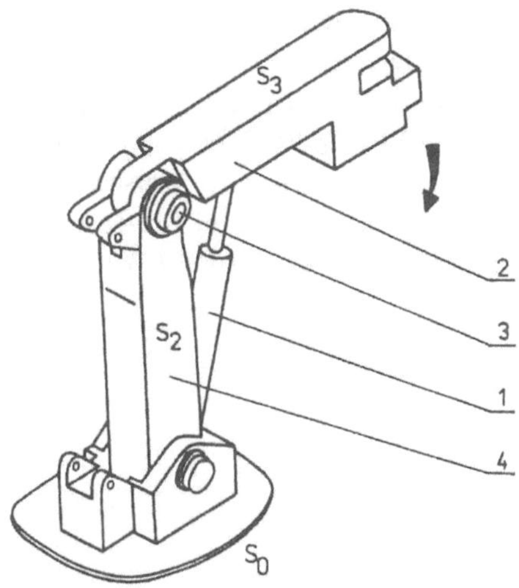

Fig. 9.55 Robot with hydraulic drives - painting robot produced by ACMA-Renault. *1* - hydraulic cylinder, *2* - link S_3, *3* - axis, *4* - link S_2, S_0 is the base according to [9.5]

9.3 References

[9.1] *Lhote F., Kauffman J. M., Andre, P. and Taillard J. P.*: Robot Components and System. Vol. 4, Robot technology. London, Edited by P. Coiffet. Kogan Page, 1983.

[9.2] *Jutord A., B, Betemps M.*: Les Robots. 11 Paris, Hermes 1988.

[9.3] *Lammel L., Osiadacz A.*: Pneumatic Transducers of Automatics Scientific. Techn. Publ., Warsaw 1978 (*in Polish*).

[9.4] *Niederlinski A.*: Industrial Robots. School and Pedagog. Publ., Warsaw 1981 (*in Polish*).

[9.5] Inter. Encyclopedia of Robotics: Applications and Automation. Vol. 1. Edited by R. C. Dorf, Willey 1988.

[9.6] Electrical and Mechanical Elements of Automatics. Edited by J. Owczarek. Sc. Tech. Pub., Warsaw 1983.

[9.7] *Müller L.*: Planetary Gears. Polish Sc. Pub., Warsaw 1983.

[9.8] Control of Actuators in Robotic Mechanisms, Chap. 4. Vol. 1. Edited by R. C. Dorf, Willey 1988.

[9.9] Proceed. of the 18th Inter-Symp. on Industrial Robots. Edited by C. W. Burckhardt. Lausanne, IFS, U. K, 1988.

[9.10] *Vertut J. Coiffet P.*: Teleoperation and Robotics Applications and Technology, Robot Technology. Vol. 38. London – Paris – Lausanne, Hermes Publishing 1985.

[9.11] *Merlet J. P.*: Les robots parallelés. Paris, Hermes 1990.

[9.12] *Coiffet P.*: La Robotique, principles et applications. Paris – London – Lausanne, Hermes 1986.

[9.13] The IMP robots and their Applications. Robotics 2. Edited by J. Buć, Sc. Tech. Pub. Warsaw 1986.

[9.14] *Mianowski K,, Nazarczuk K.*: Parallel Drive Manipulator Arm. Proceed. of the 8th CISM--IFToMM Symp. on the Theory and Practice of Robots and Manipulators. Ro.Man.Sy'90, Cracow 1990 Pub. of Warsaw Univ. of Poland, 1992.

[9.15] *Afonow Ju. M.*: Elektroprivody Promyslennych Robotov. Moskva, Energoatomizdat, 1990.
[9.16] *Pizoń A.*: Hydraulic and Electrohydraulic Control Systems. Sc. Tech. Pub., Warsaw 1987.
[9.17] *Olszewski M.*: Modelladoptive Laneregelung von elektropneumatischen Servantrieben. FESTO Forschungsbericht Nr EP1176/0149, Esslingen/Berkheim 1991.
[9.18] Electrical Drive System. Edited by Z. Grunwald, Warsaw 1977.
[9.19] *McCallion H., Truong P. D.*: The Analysis of Six–Degree–of Freedom Work Station for Mechanical Assembly. Proceedings of the IFToMM World Congress. Montreal 1979.

10. SENSORS AND TRANSDUCERS USED IN ROBOTS

10.1. Functions of sensors in robots

The interaction between the robot and the environment must be informed by data on the state of the environment, and the features of the individual objects contained in the robot's workspace, and information on the status of the robot itself and its mechanisms. We must know something about the mechanical and physical properties of the environment and particular objects, their position and orientation in space, the position and orientation of the robot links and the end-effector, the forces and velocities reached by these links, etc. These data are needed for the proper control of the robot using feedback loops (Fig. 10.1), and can be provided by means of sensors and transducers, either intelligent (integrated with computer systems, such as vision systems), or non-intelligent (providing input data for the computer).

The sensors used in robotics can be divided into two groups: sensors delivering information about parameters describing the state of robot, and sensors delivering information about the state of the environment. All these data are present in the outputs of the sensors as signals representing corresponding parameter values. **The parameters describing the state of therobot** include the *position* and *velocity* of its particular links and the *forces* exerted by those links.

The parameters describing the state of the environment include the *shapes*, *position* and *orientation* of the objects grasped by the robot, the colors of these objects, *parameters of the disturbances* acting on the robot during operation, and other *specific environmental data* which must be known in order to perform the given operation.

Robot sensors (for example vision sensors) are usually located within the robot links, or their close vicinity. The gripper usually features the highest concentration of sensors.

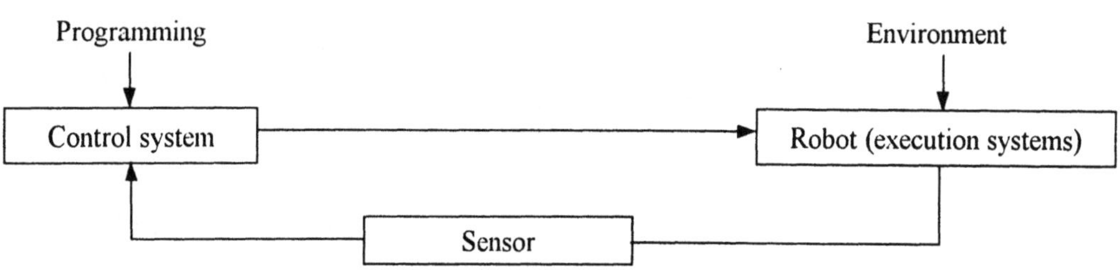

Fig. 10.1. Flow chart illustrating the robot control principle using feedback

Table 10.1. Classification of sensors used in robotics according to measured properties

No.	Properties measured	Sensor type and construction	Notes
1	2	3	4
1	Angular displacement	Potentiometric sensors	Sensitivity: 0.01÷0.15%
		Inductive sensors: - synchros - resolvers	Rotation angle measurement accuracy: 0.1÷0.2%
		Pulse counters	
2	Linear displacement	Potentiometric sensors	Sensitivity 0.01% in the range up to several tens of centimeters
		Inductive sensors	Measuring range 80÷100 mm
		Differential sensors	
		Pulse counters	
	Including slip	Optomechanical sensors: - roller with photosensor - ball with photosensor - needle with photosensor - fibre optics sensors [1)]	Displacement value Displacement direction and value
3	Velocity and acceleration	Tachometric generators Accelerometers	
4	Force	Strain gauges	
		Pressure sensors using semiconductors, piezoelectric elements, metallic polimers, fiber optics [1)], carbon plates	Combination of several individual sensors or a sensor matrix, including artificial skin
5	Touch sensors	Contact sensors: - microswitches - fibre optics	
		Inductive sensors	To be used with metal workpieces
		Pneumatic sensors	Accuracy 0.5%
		Pressure sensors	See item 4
6	Proximity detectors, Location and orientation sensors	Ultrasound sensors	Also in less transparent media
		Photometric sensors: - measuring the intensity of reflected luminous flux, - interrupting the luminous flux	Sensing range: 50÷70 mm Flat and long surfaces, long sensing ranges
		Laser sensors	> 3 mm
		Inductive sensors	Metal pieces between 5÷40 mm
		Optical sensors	

Table 10.1. ( )

No.	Properties measured	Sensor type and construction	Notes
	Proximity detectors, Location and orientation sensors	Television vidicon sensors	
		Pneumatic sensors	Several milimeters, accuracy 0.5%
		Collision detector - "filler" type	Several centimeters
		Fiber optics sensors[1)]	Several centimeters
7	Object identification	Television vidicon sensors	1 or 2 TV cameras 2 or 3 dimensions
		Filters	Color
		Holographic sensors	3 dimensions, color
		Integrating optical sensors	
		Fibre optics sensors[1)]	Shape, contour, color
		Inductive sensors	Identification of metal pieces
		Photometric sensors	
		Utrasound sensors	Also in less transparent media
		Laser sensors	Object contour
		Pressure sensors	See item 4
8	Temperature	Resistive temperature detectors (RTD)	Accuracy: 0.1÷1 K Measuring range: 15÷1273 K
		Bimetallic sensors	Accuracy: 0.5÷3 K Measuring range: 93÷223 K
		Thermistors and semiconductor sensors	Accuracy: 0.1÷0.5 K Measuring range: 73÷1673 K
		Pyrometers	Accuracy: 2% Measuring range: 673÷3773 K
		Fibre optics sensors[1)]	Used with pyrometers
9	Distinction between gases and steam	Sensors measuring: - gas resistance changes - thermal conductivity changes - absorption of infrared radiation - colour changes of chemically active substances	
		Oxygen (using its paramagnetic properties)	
		Steam hydrometers	

[1)] These sensors, still under development [10.8] have been included in Tables 1 and 2 in order to indicate their place in the classification of other sensors.

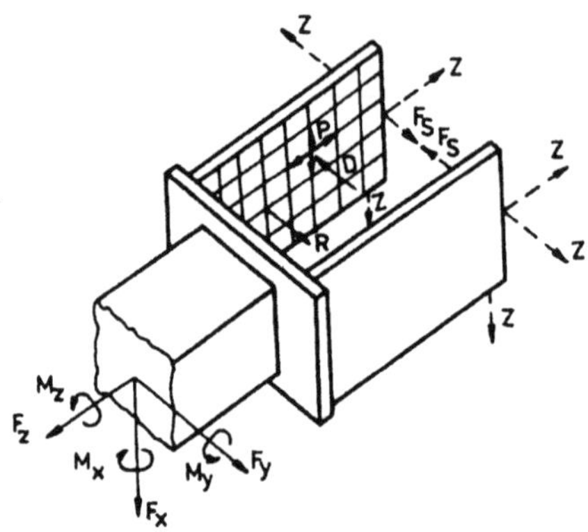

Fig. 10.2. Simplified model of a two-finger robot gripper. Notation: Z - directions of measurements, D - touch, F_s - gripping forces, M - torque, M_x, M_y, M_z - torque components, F_x, F_y, F_z - force components.

Figure 10.2 presents a simplified drawing of a two-finger gripper. The arrows indicate the measuring directions of the proximity detectors Z, touch sensors D, grip force sensors F_S, slip sensors P, force sensor F and torque sensor M, of the installed on the manipultor's arm. The gripper may be also equipped with a recognition system R implemented by touch sensors, proximity detectors, and vision sensors suitably located on the inside surfaces of the gripper. The drawing does not show the displacement and velocity sensors, usually located on the robot's arm.

Table 10.1 lists the sensors used in robotics, grouped according to the properties measured. A detailed description of the most important sensors is given here below.

10.2. Mathematical description and methods of dynamic analysis of transducers

The mathematical description of transducer properties involves the definition of relationships between input and output signals. In the case of analogue transducers with a single input and a single output this is usually an ordinary differential equation. This equation is derived from our knowledge of the physical phenomena taking place during measurement. The real device is commonly replaced with a so-called dynamical model, describing only those properties which are requisite in the analysis performed. In many real cases it is not possible to define the relationship between the transducer's input and output signals strictly by means of a linear differential equation. In these cases the equations are linearized around a point of static equilibrium [10.2].

In dynamic analysis performed either in the time domain, or in the frequency domain, operational calculus is usually applied. The basis of this calculus is Laplace transform, which simplifies the task of determining the transducer's response, $y(t)$, to the given input signal $x(t)$.

A linearized differential equation describing the relationship between the input and the output signals usually takes the form of

$$a_n y^{(n)} + a_{n-1} y^{(n-1)} + \ldots + a_2 \ddot{y} + a_1 \frac{dy}{dt} + a_0 y += b_0 x + b_1 \frac{dx}{dt} + \ldots + b_m x^{(m)} \qquad (10.1)$$

In these equations x and y are increments of the input and output signals from the values corresponding to the state of static equilibrium. They are often presented in the normalized form

$$T_n^n y^{(n)} + \ldots + T_2^2 \ddot{y} + T_1 \dot{y} + y = r_0 x + \ldots + r_m x^{(m)} \qquad (10.2)$$

where

$$T_1 = \frac{a_1}{a_0}, \qquad T_2^2 = \frac{a_2}{a_0}, \ldots \quad T_n^n = \frac{a_n}{a_0}$$

$$r_0 = \frac{b_0}{a_0}, \qquad r_1 = \frac{b_1}{a_0}, \ldots \quad r_m = \frac{b_m}{a_0}$$

Constants T_1, T_2, ..., T_n are **time constants.**

Coefficient

$$r_0 = \frac{b_0}{a_0} = k$$

is called **the static gain coefficient**.

Transforming both sides of equation (10.2) by Laplace transform at zero initial conditions, we obtain

$$G(s) = \frac{Y(s)}{X(s)} = \frac{r_m s^m + r_{m-1} s^{m-1} + \ldots + r_1 s + r_0}{T_n^n s^n + T_{n-1} s^{n-1} + \ldots + a_0} = \frac{M(s)}{N(s)}$$

Formula (10.3) shows us that the transmittance is a rational function of variable s, that is, the quotient of two polynomials $M(s)$ and $N(s)$. For a physical system the order of polynomial $M(s)$ cannot be higher than the order of polynomial $N(s)$: $m \leq n$.

Equation $N(s) = 0$ is the typical form of the differential equation (10.3), and its roots are called **operator transmitance poles**. The roots of equation $M(s) = 0$ are called **operator transmitance zeros**.

Equation (10.3) yields

$$y(t) = L^{-1}\,[X(s) \cdot G(s)] \qquad (10.4)$$

When measuring transducers, we usually search for responses to some standard signals, such as:

a) Dirac delta,
b) step function,
c) velocity step function.

In the case of periodic input, what is most interesting is usually not the whole output signal profile, but the steady-state response, determined on the basis of the output after the transient processes have disappeared.
When the input signal has the form of

$$x = x_m \cos(\omega t + \phi) \tag{10.5}$$

where x_m is the amplitude, and ϕ is the phase angle, after necessary transformations [10.2] we obtain a **spectral transmittance** transmittance in the form

$$G(s) = \frac{b_m(j\omega)^m + \ldots + b_1(j\omega) + b_0}{a_n(j\omega)^n + \ldots + a_1(j\omega) + a_0} = G(s)\Big|_{s=j\omega} \tag{10.6}$$

The spectral transmittance is usually written as

$$G(j\omega) = M(\omega)e^{j\varphi(\omega)} = P(\omega) + jQ(\omega) \tag{10.7}$$

where

$M(\omega) = |G(\omega)| = \sqrt{P^2(\omega) + Q^2(\omega)} = \frac{y_m}{x_m}$ - is the spectral transmittance module,

and

$\varphi(\omega) = \arg G(j\omega) = arctan\frac{Q(\omega)}{P(\omega)}$ - is the spectral transmittance argument

The graph of spectral transmittance $G(j\omega)$ on the complex plane is called the amplitude-phase respronse. Two other graphs may be used instead of the $G(j\omega)$ relationship to describe the system: $M(\omega)$ (the **amplitude-frequency response**) and $\varphi(\omega)$ (the **phase-frequency response**).

Basic linear elements used in dynamic measurements are: non-inertial (proportional) elements, inertial elements of the first order, oscillating elements, derivative elements, integral elements, and delay elements.

Table 10.2 shows the transfer functions of transducers used in dynamic measurements, and their amplitude and phase responses, as well as step and pulse responses. The table 10.2 also shows the loci of the ends of the transmittance vector, together with the transmittance poles and zeros.

Table 10.3 lists the typical signal distortions, with their most probable causes. A series of square pulses is assumed, as the input function [10.6].

Table 10.2. Transfer functions of transducers used in dynamic measurement, their amplitudes, phases, steps and responses

Mathematical expresion	Locus of transmitance vector ends	Phase and amplitude graphs	Pole (x) and (0) graphs	Time function: Transfer function $x_{e/1}=h(t)$	Time function: Weigh function $a_{e/1}=g(t)$	Example of system described by these characteristics
k	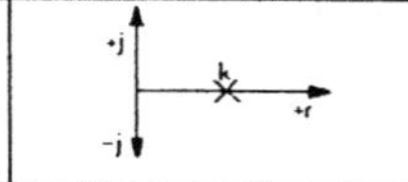	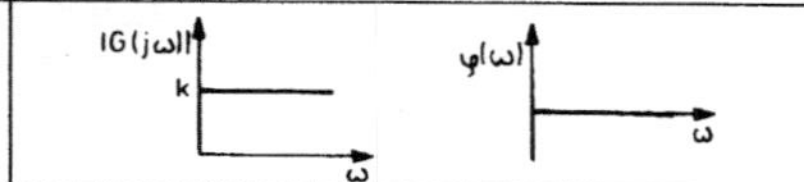	Lacking	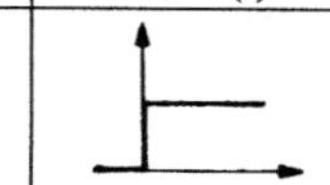	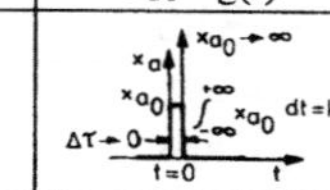	Idealized systems with proportional characteristics
$\frac{k}{1+T_1 s}$	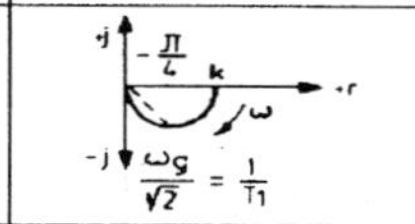	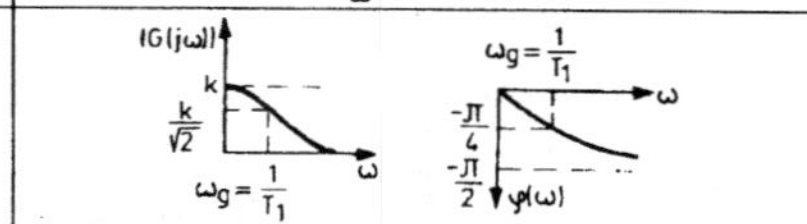	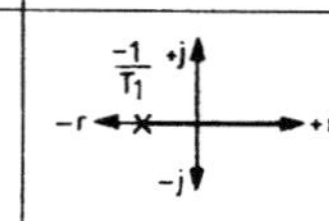	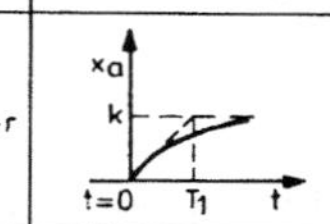	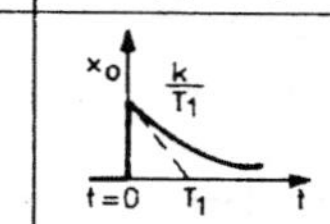	Idealized damping systems; systems with first-order delay and compensation
$\frac{k}{1+T_1 s+T_2^2 s^2}$	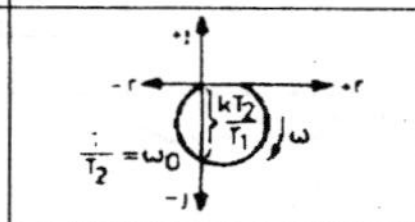	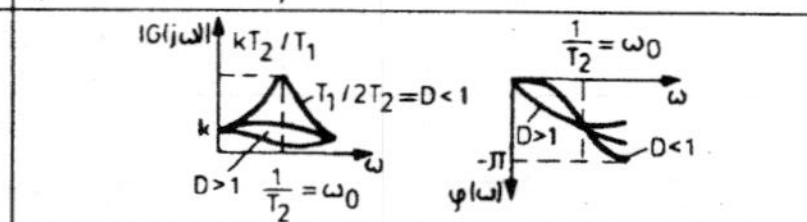	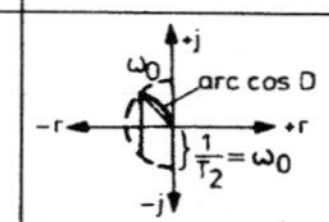	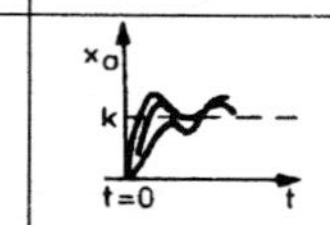	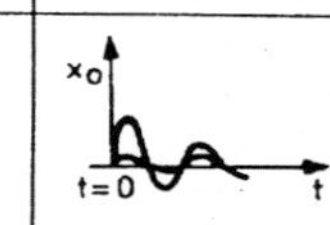	Mass, spring and damper systems; D>1 – damping sysytems; D<1 – vibrating systems
$\frac{-ks^2}{1+T_1 s+T_2^2 s^2}$	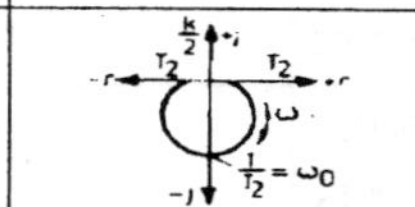	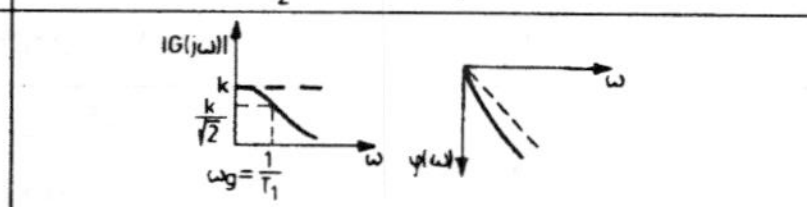	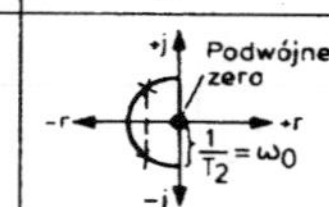	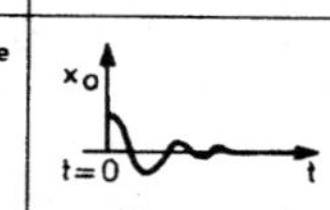	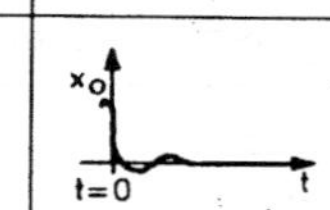	Mass, spring and damper systems; without fixed point D<1
$e^{-t_t s}$	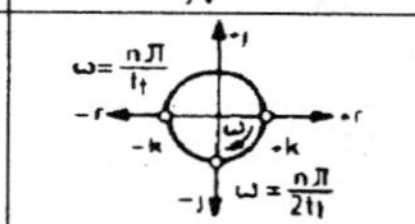	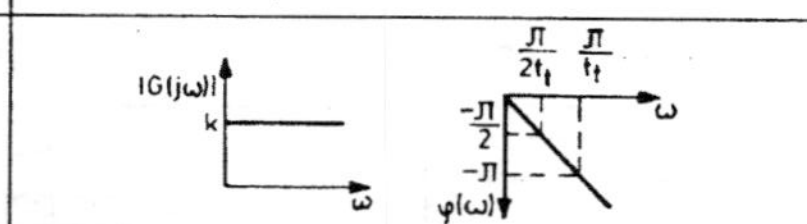	Cannot be precisely illustrated	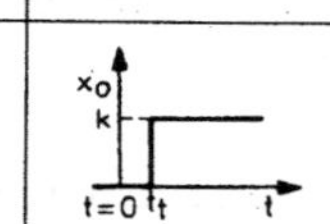	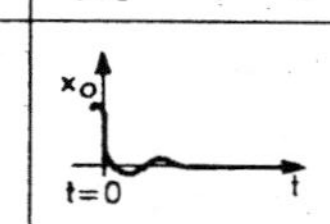	Ideal system with dead-time resulting from an existence of pipeline, transport over some distance
$\frac{ke^{-t_1 s}}{1+T_1 s}$	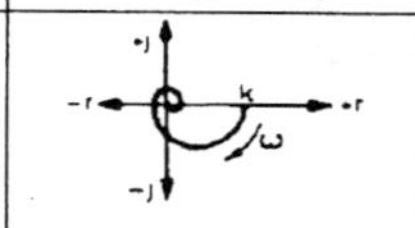	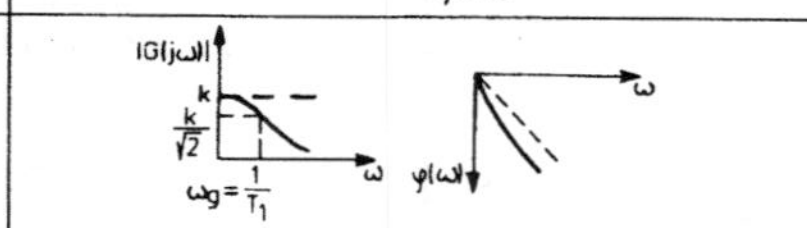	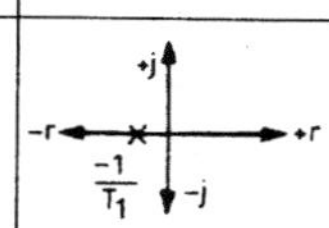	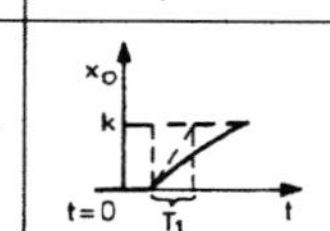	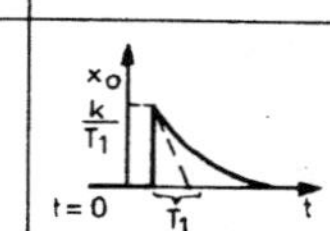	Real dead-time systems
$\frac{k}{s}\,\frac{k}{1+T_1 s}$	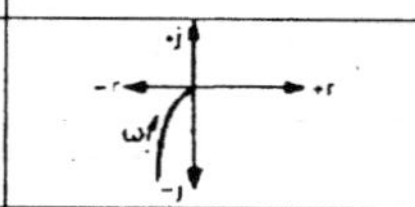	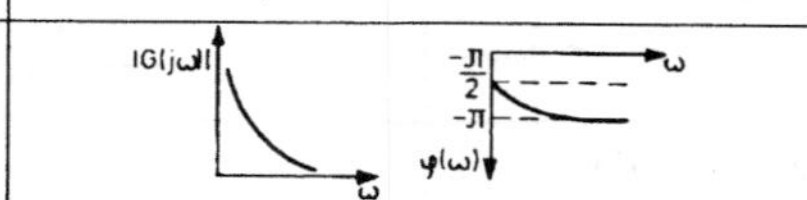	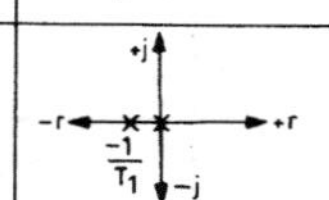	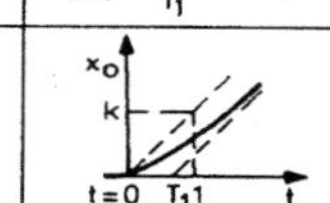	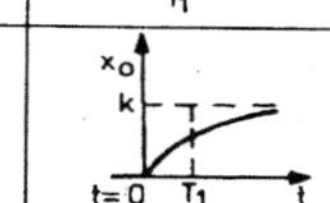	Real integrator
$\frac{ks}{1+T_1 s}$	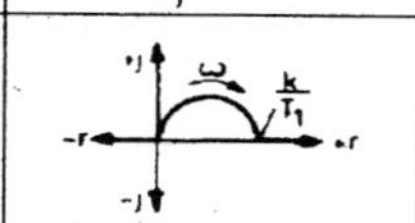	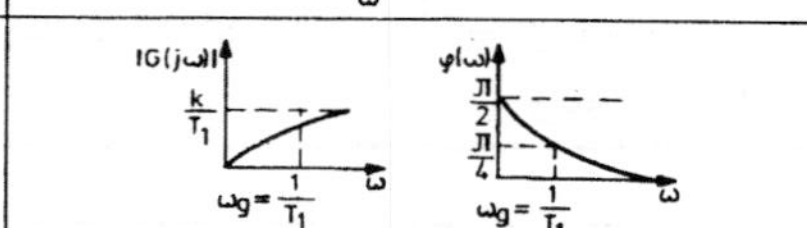	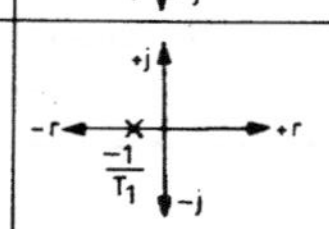	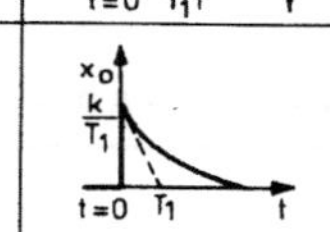	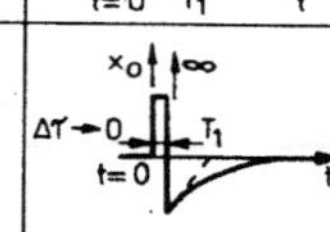	Real differentiator

Table 10.3. Typical signal distortions and most probable causes

Signal profile: —— input signal $x_e(t)$ - - - output signal $x_e(t)$	Typical properties	Causes, phase and amplitude characteristics $\lvert G(j\omega)\rvert$; $\varphi(\omega)$	Typical properties	Notes
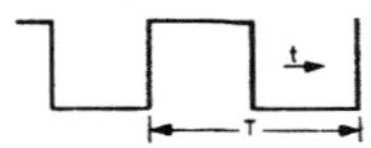	Sharp corners Steep edges Flat top	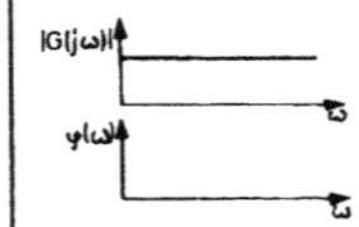	Ideal system.	Cannot be implemented in practice; implemented only approximately for very slow processes (very low frequencies).
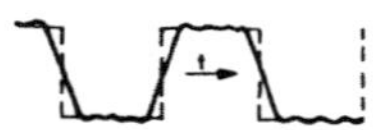	Rounded corners Inclined edges Flat top	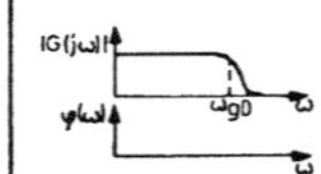	Lower limit frequency = 0; upper limit frequency is too low.	Short input pulses cannot be reproduced or are reproduced but with improper amplitude.
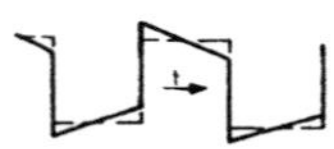	Sharp corners Steep edges Sloped top	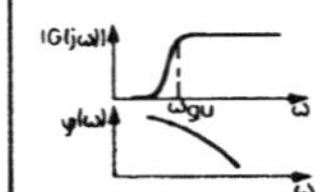	Lower limit frequency is too high; upper limit frequency = ∞.	Short pulses are reproduced properly. Errors are bigger in slower processes.
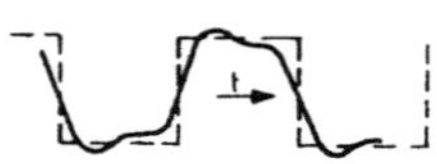	Rounded corners Inclined edges Sloped top	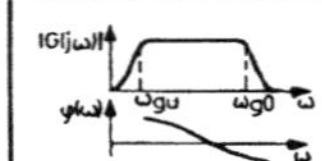	Improper upper and lower limit frequencies.	Signal distortion similar as in 2 and 3; short pulses reproduced incorretly.
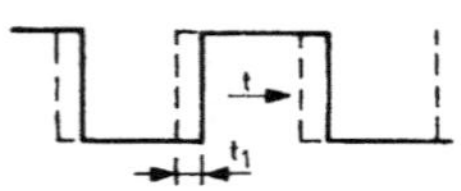	Profile reproduced accurately but shifted by f_1	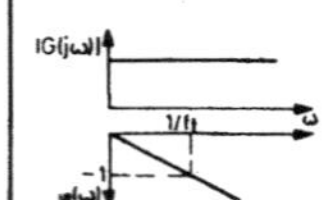	Ideal amplitude characteristic, linear phase characteristic.	No changes in the output signal except for phase shift; this is usually meaningless in measurements, but unfavourable in control systems; ideal case: when $G(j\omega)$ = const additional distortion as in 2, 3, 4.
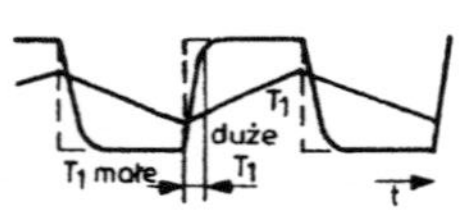	Profile containing pure "e"-functions with time constant T_1 Inclined edges Flat top	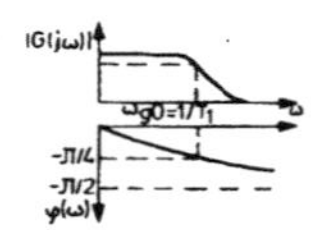	Upper limit frequency is too low.	Short pulses are reproduced incorrectly; idealized proportional systems with strong damping properties.

Table 10.3. ( )

Signal profile: —— input signal $x_e(t)$ - - - output signal $x_e(t)$	Typical properties	Causes, phase and amplitude characteristics $\lvert G(j\omega) \rvert$; $\varphi(\omega)$	Typical properties	Notes
	D>1: Rounded corners Inclined edges D<1: Vibration	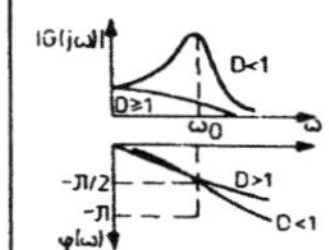	Upper limit frequency is too low; additionally resonance for D<1.	Short pulses are reproduced incorrectly. D≥1 - mass-spring-damper systems, e.g. vibration testers.
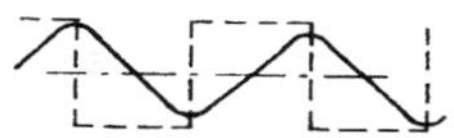	The curve after integration does not have sharp corners at points of x_e step change, absolutely correct only for low frequencies.	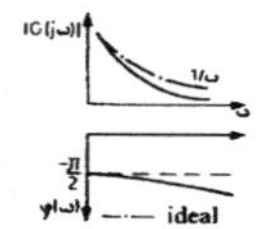	Integrator introduces unstable phase shift by $\pi/2$; Amplitude profile falls faster than $1/\omega$.	In non-electric integrating systems this effect is caused by masses; in electric systems - too low upper limit frequency.
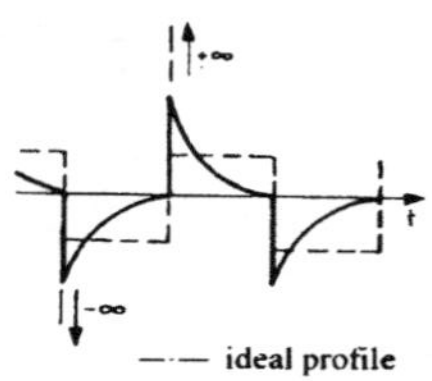	The curve after differentiation is rounded. Correct only for low frequencies	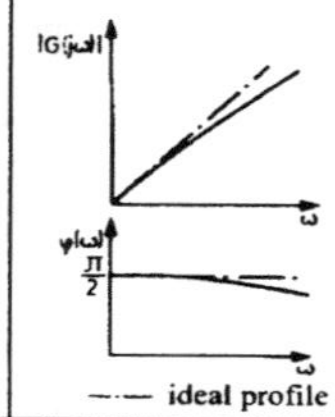	Differentiator introduces unstable phase shift by $\pi/2$; amplitude proflie not ideal	Cannot be avoided in simple RC systems having $T_1 = CR$.

10.3. Sensors determining robot's state parameters

To control the end-effector of the robot we must know its state parameters. These parameters include the positions and velocities of end-effectors as well as the forces and moments acting in various points on the robot's arm.

The application of position and velocity sensors robot designers to use drive systems containing one or two feedback loops (position feedback, or combined position and velocity feedback). These drive systems provide the accuracy needed to perform programmed tasks. Force and torque sensors are installed to protect the manipulator structure against overloading, or to introduce adaptative control for some operations.

10.3.1. Position sensors

There are many and varied designs of position sensors, such as **resistive, inductive, capacitive and optoelectronic** sensors. **Potentiometric sensors** are an example of resistive sensors, that transform linear or angular displacement into variable resistance proportional to the displacement. The resistance causes changes in the output voltage ($U_o = k\varphi$ and $U_o = kx$) in the measuring system (Figures 10.3a, b, c).

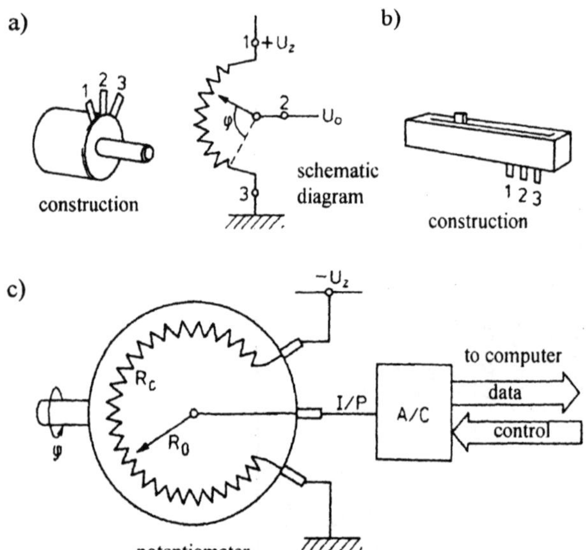

Fig. 10.3. Potentiometric sensors: a) angular; b) linear; c) connections between the sensor and the computer; U_z - supply voltage, U_w - output voltage, x - actual linear displacement, φ - actual angular displacement, A/C - analog-to-digital converter, R_o,R_c - resistance values.

Very important features of potentiometric sensorrs are their resolution and durability. Theoretically measuring systems with potentiometric sensors do not require amplifiers, but in practice amplifiers are used to normalize the characteristics of the position measuring system. Figure 10.3c shows how an angular position sensor is connected to the computer [10.3].

Position may also be measured by **inductive sensors,** utilizing the principle of the changes in self-inductance or mutual inductance caused by displacement. There are many designs of this type of sensors; the most common use the transformer principle of operation. **Resolvers** and **inductosyns** are widely applied.

The **resolver** is a transformer-type angular position transducer consisting of a stator with two excitation windings shifted in phase by $\pi/2$, and a rotor with one output winding, the ends of which touch the slide rings fitted on the resolver shaft. Its construction is similar to that of a very small two-phase AC motor (rotor diameter 10÷12 mm, rotor length 20÷30 mm) fitted with very precise bearings. Resolvers usually transform the rotation angle into a phase of the voltage induced in the output winding. Figure 10.4 shows their operating principle. Figure 10.4a illustrates the operation of a resolver with sine excitation voltages and sine output voltage, Figure 10.4b shows the operation of a resolver with square excitation voltages and square output voltage. In both examples the quantity representing the angular displacement is the phase of induced voltage. Coefficient k_r is the resolver's ratio.

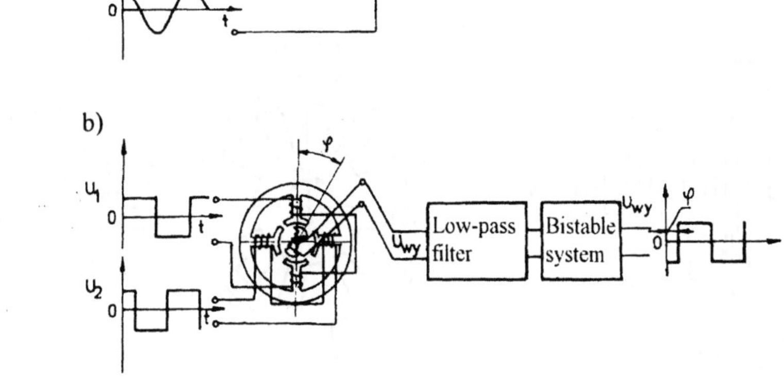

Fig. 10.4. Resolver with phase output, principle of operation: a) excitation voltage and sine profile output voltage, b) excitation voltage and square profile output voltage.

As digital measurements are simpler, and it is easier to compare phase shifts of square signals than of sine signals, in practice the square configuration is more often used. More details concerning the application of resolvers as position sensors can be found in [10.1].

Inductosyns are transformer-type linear or angular position transducers. The construction of an inductosyn is illustrated in Figure 10.5. The windings shown in the drawing are printed circuits and feature high precision of shape. Inductosyns are more precise transducers than resolvers.

Digital methods, applied to increase the acuracy of linear and angular displacement measurements, can be classified as either *incremental measurements* or *absolute measurements*. In both these methods the full range of variability of the measured quantity is divided into a finite number of identical subranges, called quantization intervals. The objective is to determine within which quantization interval the cosidered value falls.

The main component of the incremental measuring system is a **linear (bar-type)** or **angular (disc-type) pulse emitter** (Fig. 10.6), the length or circumference of which is divided into a number of segments of the same width. When the disc rotates, or the linear mask moves, two transducers (photoelectric or based on the Hall-effect) shifted by the half of the quantization interval width are activated.

Using two transducers we can determine either the direction of disc rotation, or the direction of linear mask displacement. For optical pulse emitters, the width of quantization intevals ranges from 0.1 to 0.01 mm.

The main component of an absolute measuring system is a **linear (bar-type)** or **angular (disc-type) code-bearing device**. A specific optical, magnetic or electric pattern is assigned to each quantization interval a, representing this interval as a digital code. In industrial practice optical bars and discs are generally preferred.

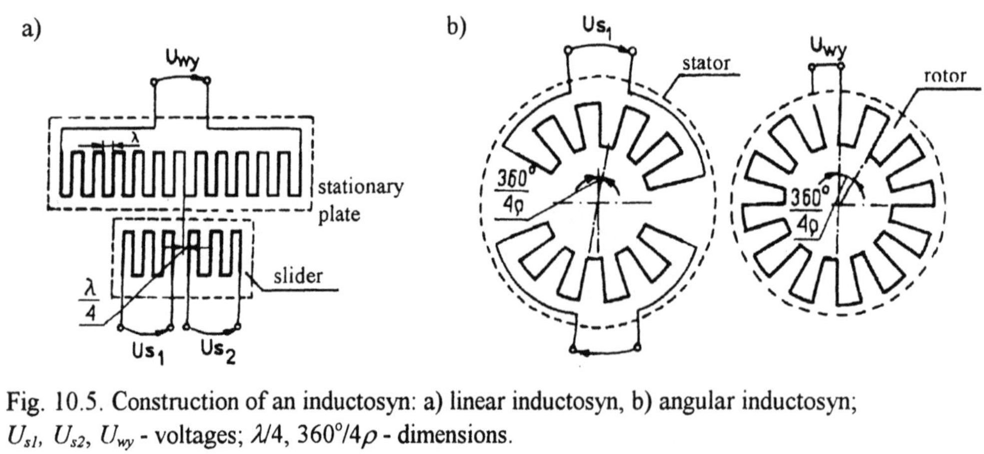

Fig. 10.5. Construction of an inductosyn: a) linear inductosyn, b) angular inductosyn; U_{s1}, U_{s2}, U_{wy} - voltages; $\lambda/4$, $360°/4p$ - dimensions.

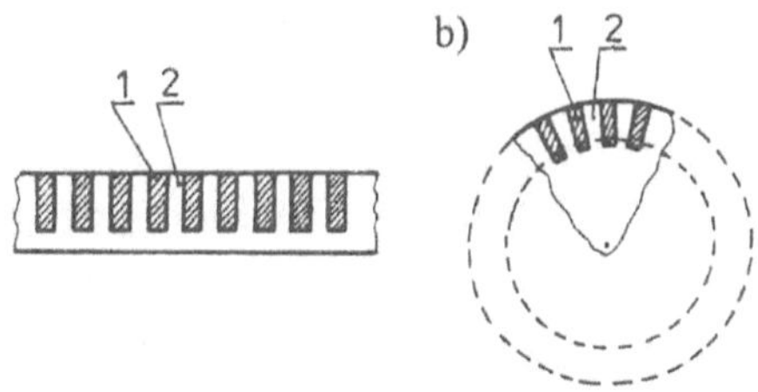

Fig. 10.6. Position sensors - pulse counters: a) linear pulse emitter, b) angular (disc) pulse emitter; *1* - is a transparent element in the case of optical sensors, a permanent magnet in the case of magnetic sensors; *2* - an opaque element in case of optical sensors; a paramagnetic material in the case of magnetic sensors.

Figure 10.7 shows binary code and the gray code bars. The resolution of code bar, or disc is usually about 0.012% of the full measurement range. When higher resolutions are required, measurement systems using two or three code bars or discs coupled by decimal mechanical gears are used.

In comparing incremental and absolute measurement principles, it is important to remember that in incremental measurements, the transition from one quantization interval to the next always generates the same signal, irrespective of the actual position of the bar or disc. More information can be obtained from absolute measurements, as each quantization period is assigned to it a unique digital code.

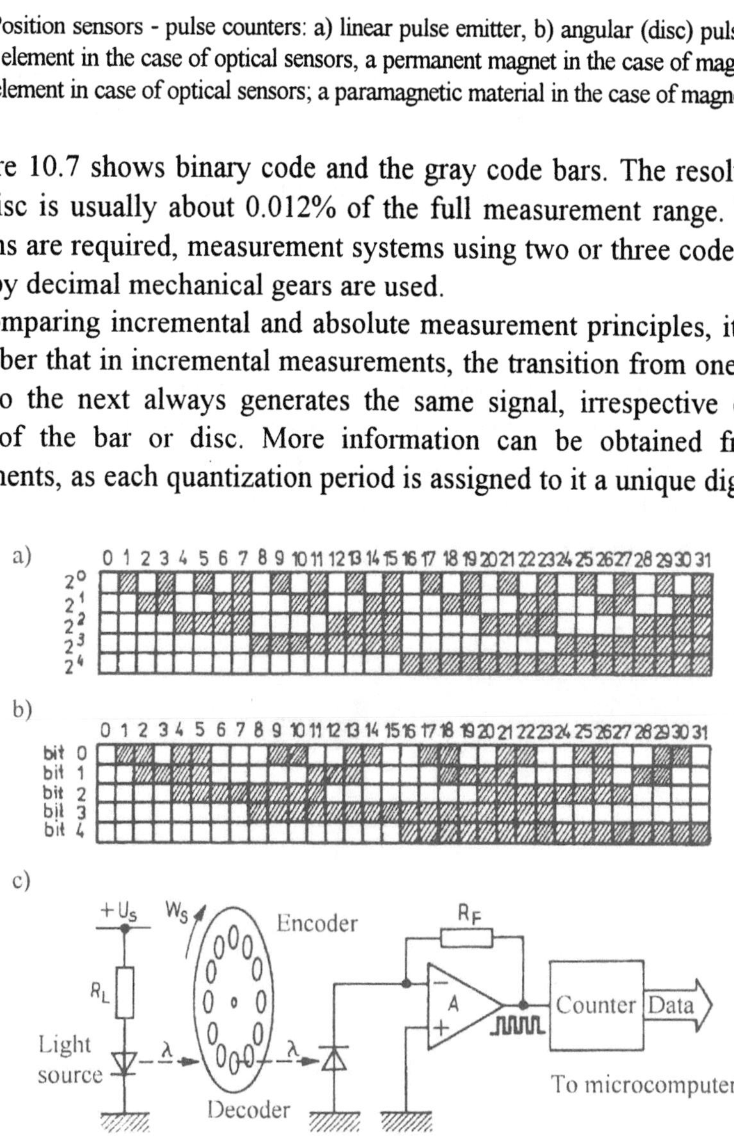

Fig. 10.7. Code bars for absolute measurements: a) binary code, b) gray code; c) velocity sensor based on the optical encoder (light interrupter); U_S - supply voltage; R_L, R_F - resistances.

10.3.2. Velocity sensors

Velocity sensors are used in robots to increase the accuracy in maintaining the programmed velocity of the end effector. **Tachometric generators** using **optical incremental encoders** are the velocity sensors most commonly used. They measure velocity proportional to shaft velocity directly. **Position sensors with an electronic differentiator** (e.g. $\omega = \mathrm{d}\alpha/\mathrm{d}t$) form another group of sensors. Figure 10.7 c shows an example [10.3].

Other concepts of velocity sensors combining a position sensors and a velocity sensor in one element are also known. If the measuring element has the form of bar pulse emitter or disc pulse emitter, the velocity is determined on the basis of field shift frequency. The Japanese painting robot "Mitsubishi-Ivata" uses a velocity sensor of this design.

10.3.3. Force sensors

Force sensors are used in robots in devices for protection from mechanical overload, in velocity control systems, and in mass-measuring systems, etc. A very common type of sensors is the **strain gauge**. Strain gauges are used to measure force by measuring the deformation of an elastic part (to which the strain gauge is bonded) under a specific load.

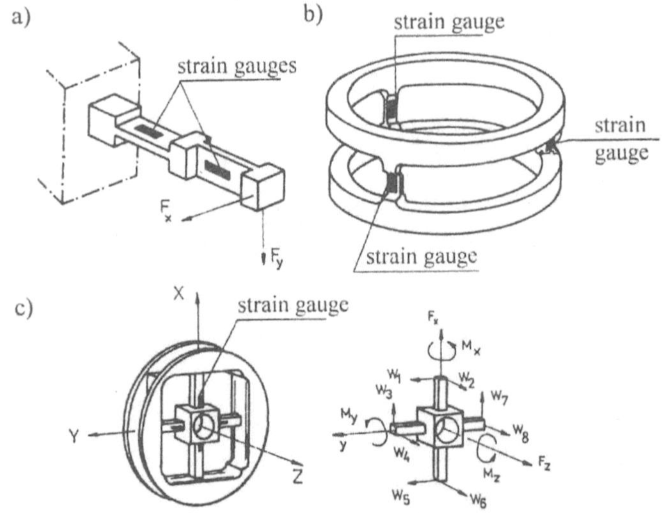

Fig. 10.8. Construction of force and torque transducers using resistor-type strain gauges: a) strain gauges in orthogonal, planar arrangement, b) strain gauges on a ring, in spatial arrangement; c) strain gauges in spatial, Maltese cross arrangement.

Many different designs of strain-gauge force sensors are used in a bridge arrangement. Some of that arrangements can be used to measure one, two, or more force components in order to protect a mechanical system from overloading. Force sensors are also used to control the velocity of objects being transported by the robot depending on their mass.

The sensor illustrated in Figure 10.8a is used to measure force and torque components in a planar system of Cartesian coordinates. The sensors shown in Figures 10.8b and 10.8c are designed to measure force and torque components in space [10.1, 10.3].

We may take the Maltese cross sensor as an example to describe how force and torque components are determined. Sixteen strain gauges, four on each cross arm, have been used to build this sensor. This produces eight output signals, W_1 through W_8. On the basis of these signals three force components (F_x, F_y, F_z) and three torque components (M_x, M_y, M_z) (acting in the place where the sensors are installed) are detemined. The range of the Maltese cross sensor described above is from 0.5 to 300 N (Figure 10.8c).

10.4. Sensors determining the state of the robot's surrounding

The modern robot is to some extent an adaptive and autonomous system operating in a changing environment. To have it perform its required functions, closed loop control systems must be applied, in which control signals are generated using signals representing the changing conditions in the robot's surroundings. The ability to recognize the state of robot's surroundings depends on the kind of sensors the robot is equipped with. These can be single sensors, or groups of sensors providing data for the data processing systems. They replace in some way human receptors. Important sensors in modern robot designs are **proximity detectors, touch sensors, slip sensors, grasping force sensors,** and **machine vision systems**. Robotic sensors are usually installed on or near the gripper in different points of the workstation, as required by the robotized manufacturing process.

10.4.1. Proximity detectors

Proximity detectors are devices used to measure the distance between the sensors and the workpiece to be grasped, or to detect obstacles along the robot's trajectory in order to avoid collisions. As the proximity detector is small in size, it is easily filted in the robot's gripper. The sensing range of the detector is usually from 10 to 200 mm. When smaller distances must be detected, special detectors covering a range from 1 to 20 mm can be employed.

There are many types of proximity detectors; these most commonly used in robots are **inductive sensors, ultrasound sensors and optical sensors. Pneumatic and microwave proximity detectors** also exist, but present disadvantages which prevent a wide range of applications.

Inductive sensors register the changes of the magnetic parameters when the sensor is approaching the object. The sensing range falls within 5 and 40 mm. Although inductive sensors are durable and reliable, they can be used only to detect metallic objects, which is a significant drawback. One typical application of inductive sensors is the measurement of the distance between the end-effector of a welding robot and two plates welded at a specific angle (Figure 10.9).

The **ultrasound proximity detector** consists of two basic parts: an ultrasound signal transmitter and a signal receiver. A reflected ultrasound signal can be used to detect objects, and to measure distance, as well as to determine the direction and velocity of moving objects by using the Doppler effect. The minimum dimensions of an object that can be detected by means of ultrasound sensors are 3×3×40 mm, while the maximum sensing range is up generally to 3 m; there are also models operating up to 10 m. The relatively low resolution of ultrasound sensors results from the significant length of ultrasound waves.

Ultrasound proximity detectors are used in robotics mainly as range-finders, serving to detect objects before grasping them, and to detect and circumvent obstacles on the manipulator's trajectory [10.5].

Optical sensors constitute a large group of proximity detectors. **Laser detectors** and **photometric detectors** furnish the best examples here.

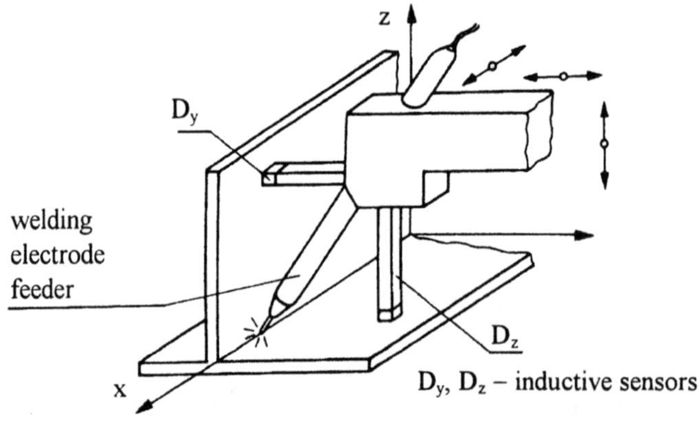

Fig. 10.9. Application of inductive sensors to trace the contact line of welded plates

The **laser detector** uses an optical system to operate by directing a beam of coherent radiation in a particular phase on the object grasped by the robot. The signal reflected from the object is detected by the receiver lens, and directed to the photodetector system.

Laser detectors measure distance by one of two methods:

- by using pulse-modulated radiation (see Figure 10.10a); the distance is determined by the number of pulses reaching the receiver within a given time, starting from when the light beam is directed onto the objects;
- by using continuous radiation; see Figure 10.10b, the distance is determined by comparing the phase of the laser-generated radiation with the phase of radiation received by the receiver.

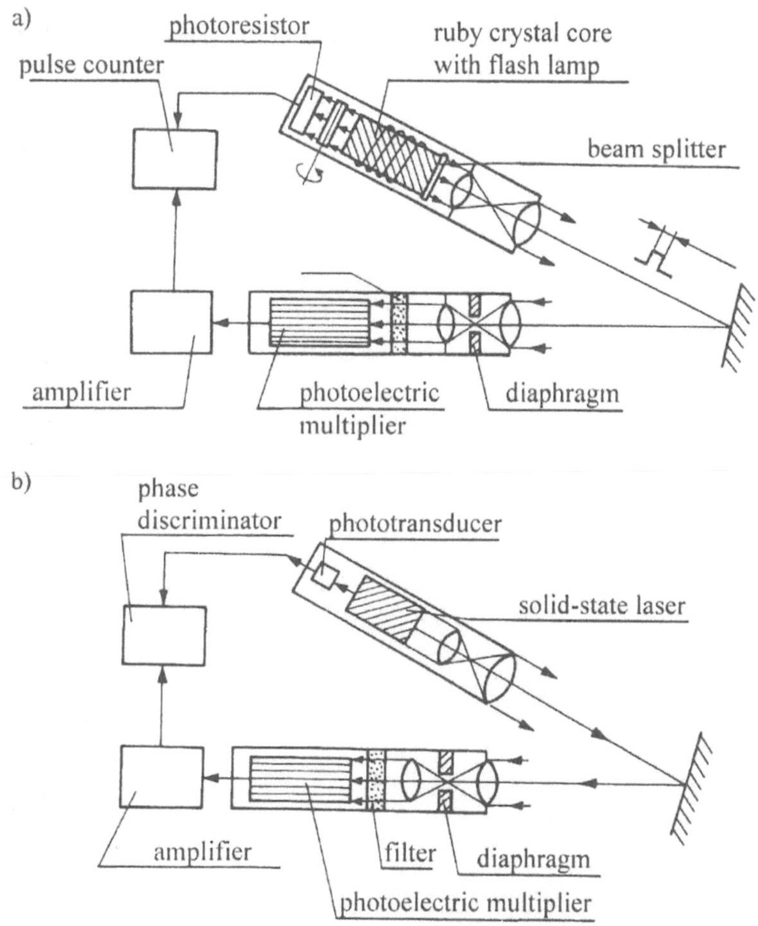

Fig. 10.10. Laser range-finder, using: a) pulse modulated, b) continuous radiation.

Distance determined by laser detectors can achieve accuracy of 0.5 mm. In older detectors He-Ne lasers were used, but solid-state GeAlAs lasers emitting infrared radiation are now employed.

Photometric sensors consists a large group of optical proximity detectors used in robots. The simplest example of the photometric sensor is the **reflective transoptor** (Fig. 10.11). This transoptor is an optically coupled pair of elements: a radiation source (usually a light-emitting diode) that serves as the input circuit of the transoptor, with a photodetector (usually a photodiode, or phototransistor), as the output circuit. The radiation source and the photodetector are usually mounted in a common housing and galvanically isolated. The optical coupling is provided by the reflection of the source radiation from the surface of the object being detected. The common part of two cones, the source radiation cone and the photodetector sensing field cone (shown in Fig. 10.11) determines the photometric detector sensing range of up to 200 mm, measured between the detector head and the reflective surface.

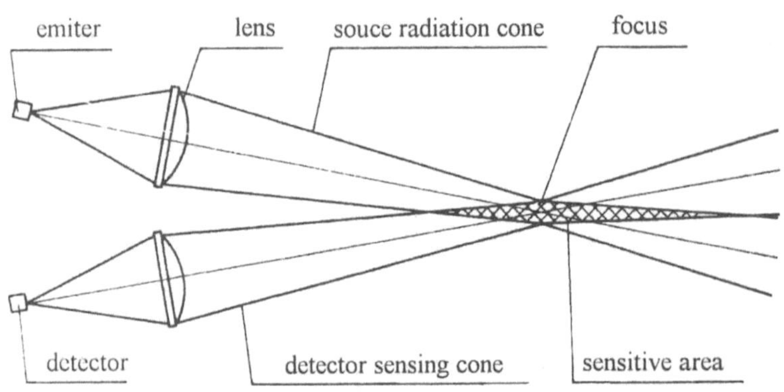

Fig. 10.11 Reflective transoptor.

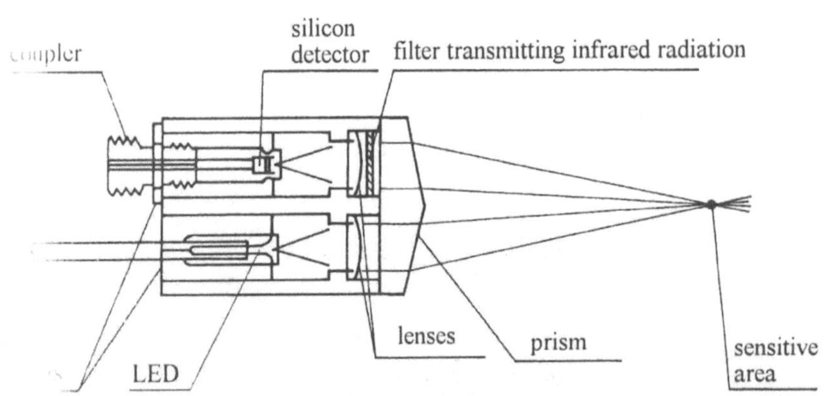

…entiometric proximity detector, using lenses and prism; LED: light-emitting diode.

The resolution of detectors within the sensing range is about 1 mm. When higher resolution is required, it can be obtained by reducing the sensing range. Figure 10.12 shows a photometric proximity detector designed for mounting on the robot's gripper.

A new element in photometric proximity detectors is the application of **fiber optics**. The concept of a sensor using these fibres is shown in Figure 10.13. This sensor, in addition to all the advantages of the photometric sensor, allows highly sensitive electronic parts to be moved from the gripper to a safer site, a smaller detector head, and a detector more resistant to ambient conditions (temperature, humidity) and to external electromagnetic fields.

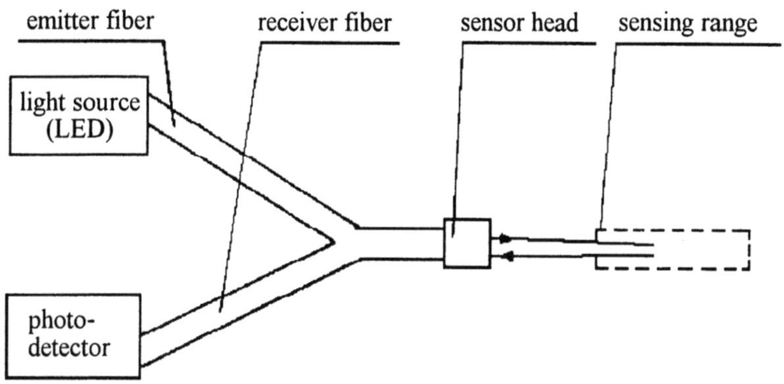

Fig. 10.13. Proximity detector using fibre optics technology.

10.4.2. Touch sensors

Touch sensors are located on the inner side of the gripper's fingers and detect the contact between the gripper and the object. A number of touch sensors may constitute of a touch matrix, called an "artificial skin". This matrix can be used to recognize the shape of the object from the contour of the part of the object touching the gripper surface. It can also detect the position and orientation of the manipulated object in the gripper. Figure 10.14a shows a robot gripper fitted with touch pads; Figure 10.14 illustrates the principle of operation of the sensor, and Figure 10.14c shows an example of a touch sensor fitted with an ultrasound sensor and used to follow the contour of the object.

Also used in robots are **microswitch touch sensors** and **continuous operation touch sensors**. Simple touch sensors having switch characteristics use microswitches. An example of this type of sensor is presented in Figure 10.14d. The sensor has a small flexible plate that presses the microswitch when the sensor touches an object. Because of the small dimensions and relatively wide assortment of microswitches, they are commonly used in touch sensors.

Continuous operation touch sensors are employ inductive transducers (needle sensors), or elastomers, which change their properties with deformation, and utilize the **elastomeric resistance**. This effect can be described as the change in resistance between two electrodes applied to the elastomeric material in respect to the external pressure exerted on the electrodes. Higher pressures cause lower resistances and higher currents flowing through the electrodes. An example of a touch sensor exploiting elastomeric resistance is shown in Figure 10.15.

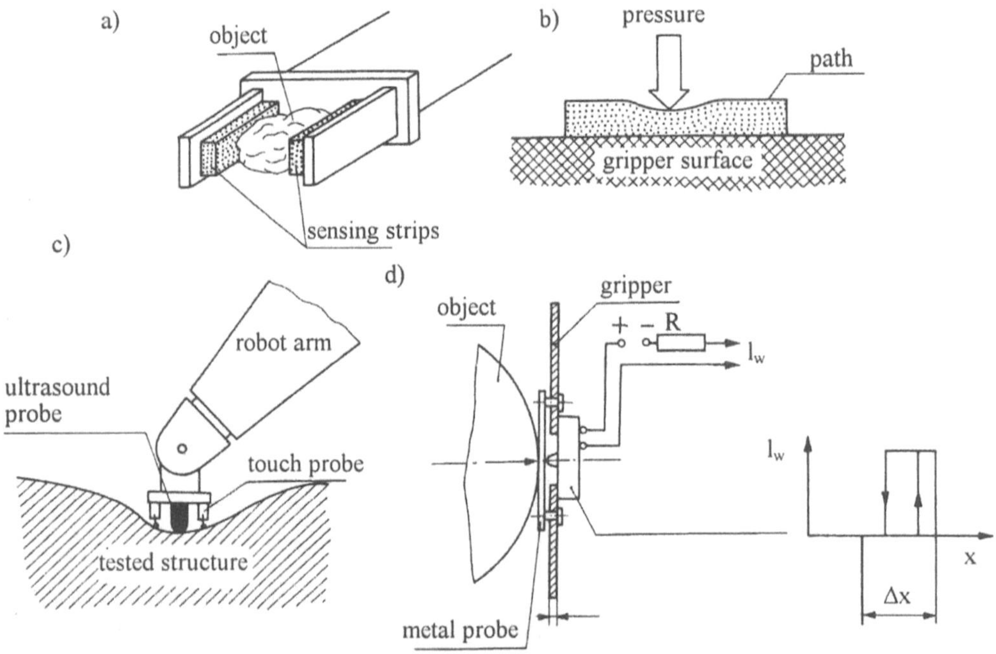

Fig. 10.14. Touch sensor: a) gripper with touch pads; b) principle of operation; c) touch sensor used to measure contour changes; d) sensor with microswitch.

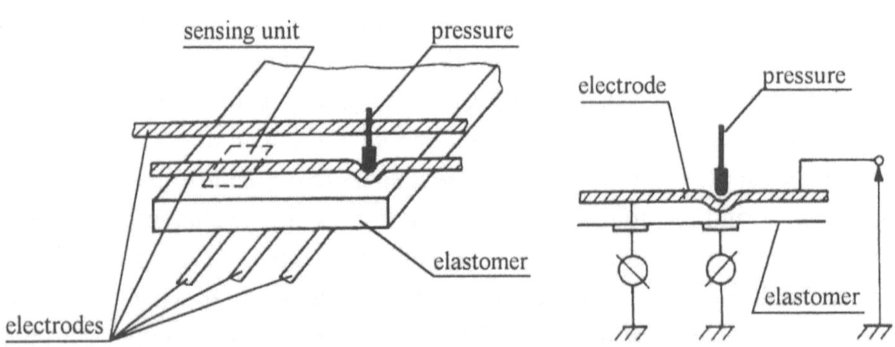

Fig. 10.15. Touch sensor utilizing elastomeric-resistance effect

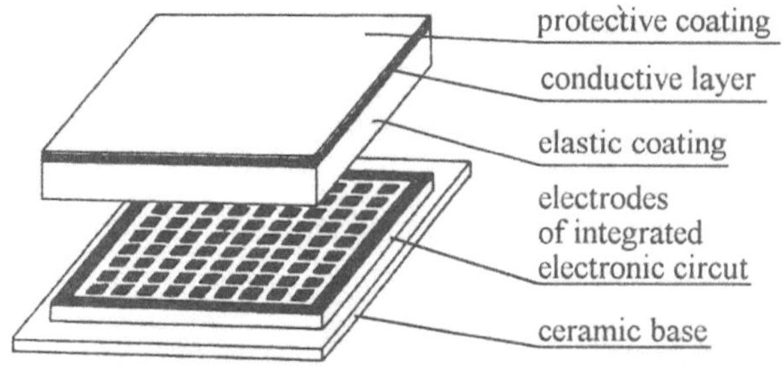

Fig. 10.16. Touch matrix constituting an "artificial skin" designed for mounting on the inner sides of the gripper's fingers

Touch sensors using various elastomers have come very common in robotic applications. A sensing unit measures only 4×4 mm, allowing the construction of matrices to constitute an "artificial skin" (Figure 10.16). The sensing range of these touch sensors extends from 2 to 50 N/m^2.

10.4.3. Grasping force sensors

Grasping force sensors are used to measure the force the gripper exerts on the grasped object. Measurements of this kind are usually made by means of **strain gauge sensors**, similar to the sensors used to measure forces and torques acting in the robot's arm (section 10.2.3).

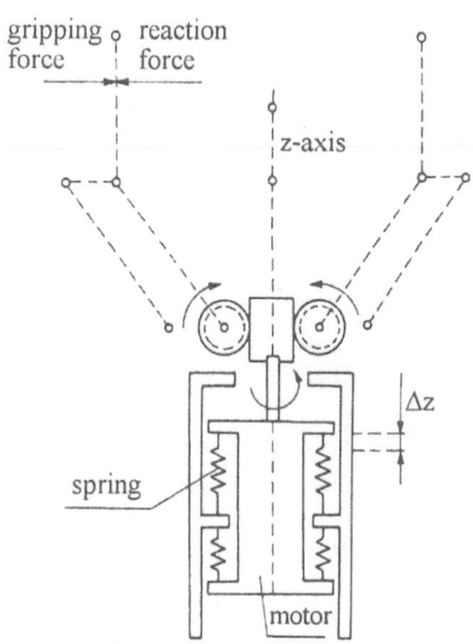

Fig. 10.17. Gripper fitted with a grasping force measuring system using springs and potentiometers (Δz is measured by means of a potentiometer).

Another measurement method and sensor design can be used to determine the grasping force. In this solution the fingers of the gripper are installed in the gripper frame on springs (Figure 10.17). The springs are linked with potentiometers measuring spring deformations, and the grasping force is calculated by processing the output signals from the potentiometers.

10.4.4. Slip sensors

The problem of proper adjustment of the grasping force arises mainly when grasping unknown objects, or objects belonging to a determined set of workpieces, but selected in a random manner. In the most general case the grasping force can be adjusted by subsequent trials of grasping and lifting the object, having the gripper of the manipulator apply increasingly greater grasping forces, until the object can be securely lifted. Determining the proper grasping force is the key issue of this method. In existing grippers the reliability of the grasp has been determined so far by controlling the slip, i.e. the displacement of the grasped object relative to the gripper's fingers.

Various slip sensors are shown in Figure 10.18. The magnitude and direction of slip can be also determined by analyzing the changes in the output signals of touch sensor matrices attached to the inner sides of the gripper's fingers.

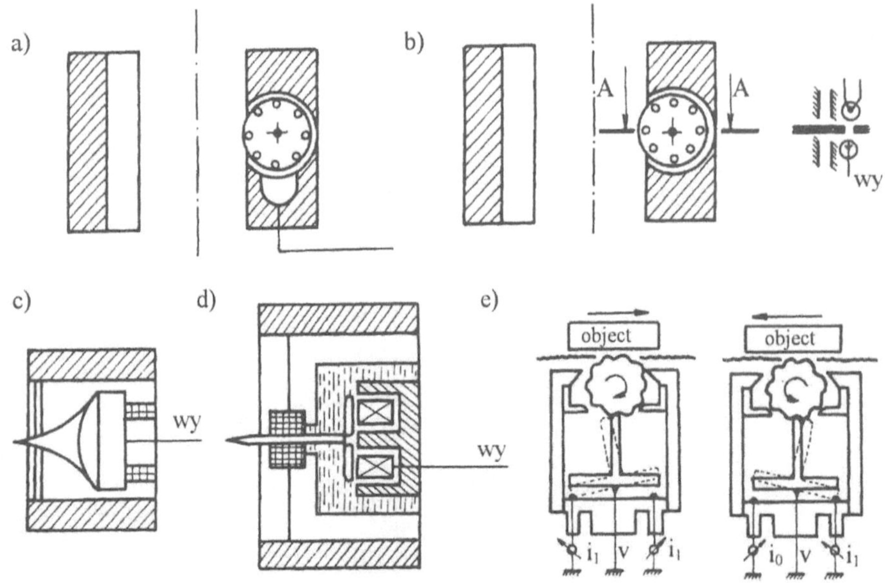

Fig. 10.18. Construction of slip sensors; a) rotating coil (magnetic measurement); b) rotating roller (photoelectric measurement); c) needle (electric measurement); d) blade (inductive measurement); e) rotating ball with elevated contact points (electric measurement).

10.4.5. Sensors and machine vision systems

A vision sensor collects information about the robot's surroundings by optical means. These sensors together with a data processing system (usually incorporating a computer) forms a **machine vision system**.

The most important tasks performed by vision systems are:

- object recognition,
- storage of the description other object and its surroundings,
- determining object orientation relative to the environment,
- controlling the state of the objects, i.e. measuring object parameters and comparing them with the data stored in the computer memory, allowing for the pre-set tolerances.

The vision system`s basic function of recognition and identification of objects is the initial condition for the other operations performed by the robot, such as object inspection, assembly, or sorting. In other words, the function of the vision sensor is to transform the representation of the workspace (or more precisely - the primary image, usually supplied by optical transducers) into signals used to adapt the operating equipment for further processing (the optical signals are usually transformed into electrical signals).

The primary image at each point of the workspace can be described using a parameter of light intensity (brightness), which in case of analogue images is a non-negative, continuous (or continuous in sections) function. The function may be equal to zero in sections where there is no lighting.

The image representation obtained at the input of the vision sensor can be classified as:

- two-dimensional: for each point (x, y) of the object surface the brightness function $s(x, y)$ is defined,
- three-dimensional: described, for example, by a pair $(s(x, y), r(x, y))$, where $r(x, y)$ is the distance between the sensor and the corresponding point of the robot workspace.

The vision sensor has the following drawbacks:

- nonlinearity of the excitation current with respect to the brightness,
- inertia (the image cannot follow rapid movements of objects),
- occasional over-exposure,
- noise that causes image disturbances (thermal effects).

Machine vision systems use various criteria to identify objects:

- coordinates of the gravity center or mass center of the object;
- moment of inertia of one projection of the object;
- area of one projection of the object;

- number of holes in the planar image of the object;
- angles between the edges of the object projection, length and number of edges;
- coordinates of the object gravity centre and coordinates of the points of intersection between the object contour and four concentric circles, the center of which is also the center of gravity of the object;
- colour of the object.

Solid-state image transducers (CCD or CJD) and **laser scanners** are the vision sensors most commonly used in robotics. In practice, the widest range of applications employ by **vision systems consisting of a camera and an image processing system with a computer incorporated,** such as the Univision system produced by Unimation, or the English Nelson II system, or the Autovision system produced by Automatics Incorporated.

Figure 10.19a illustrates the operation of the machine vision system equipped with a camera, showing an artificial vision system used to recognize parts during assembly [10.14].

Figure 10.19b shows the process of teaching the robot to recognize two-dimensional objects by showing it the different parts from different sides. Two tv-cameras collect the information: one studies the object to be built, and the other observes the table on which all parts which will be used as subassemblies are laid out. Other parts, not necessary in the assembly process, can be also put on the table. Special programs activate the manipulator, which starts assembly according to a determined order. The figure shows the successive stages of the teaching and recognition process. The first step is the location of the object, the second, the detection and location of its center of gravity, the third, its description in polar coordinates. Further steps include teaching, recognition, storage, comparison, and classification. In the teaching stage, the image registered by the camera is transferred to the computer. An appropriate program detects the contours of the sample and locates its center of gravity in the center of the coordinate system. Next, the coordinates of characteristic points are registered and stored in the memory, and in this manner a model of the object is generated. In the comparison stage this description is compared with all the models stored in the memory. During analysis, the reference angle is adjusted to fit the orientation of the objects.

In recent machine vision systems, solid-state one-dimensional (in the form of lines), or two-dimensional (in the form of matrices) image processors are used. In reply to the image projected on the screen processor. These generate a series of vectors of digital information representing the brightness of discrete points of the image being processed.

New technologies can be used to produce matrices (e.g. CCD-type), in which the individual photosensitive elements are spaced by a hundredth of millimetre, and one line in the module may contain up to 2^{11} (2048) elements. Cameras using these matrices do not require additional image discretization. In comparison with the earlier VIDICON camera, which was an electron-tube type image processor, these new cameras feature a digital output signal, higher accuracy, and a higher scanning rate, as well as greater mechanical durability.

Machine vision systems used in robotics contain a varied number of vision sensors. The image of the object being recognized is two-, or three-dimensional depending on the number of sensors and the placement of cameras [10.7].

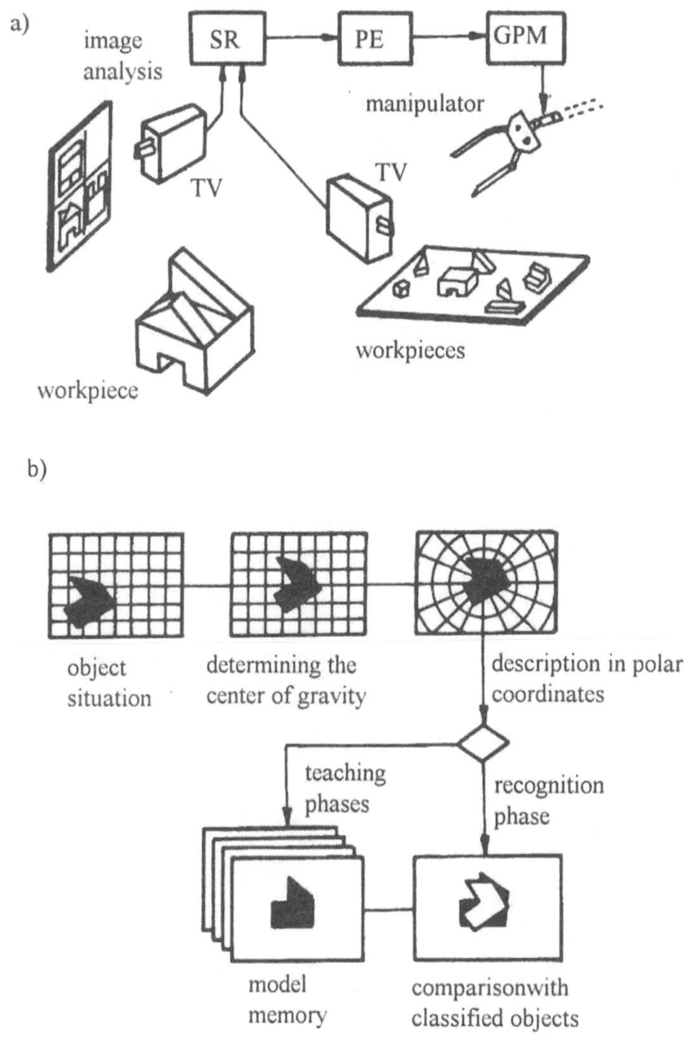

Fig. 10.19. Manipulator equipped with senses and sight (as described in [10.4]): a) principle of operation (SR - recognition system, PE - processing programs, GPM - generation of manipulator program), b) teaching process.

Planar images can be represented in binary form, or in grey levels (up to 64 levels). To obtain a three-dimensional image of the object, in order to measure its position and orientation in space, and properly identify it, vision systems using two cameras are applied.

Cameras used in robotic vision systems usually located in the gripper, or on the manipulator arms. When it is necessary to have a more complete image of the workspace, additional cameras are fitted in the robot's surroundings.

Laser scanners are also used as vision sensors in robotic vision systems. The most common are: scanners measuring distance by the triangulation method, and scanners measuring distance by measuring the time between the pulse emitted by the scanner and the pulse reflected back to the scanner from the object. An example of the former is shown in Figure 10.20. Laser scanners are active vision sensors as they have a built-in source of radiation, and the scene being observed does not require lighting (which is instead necessary when using camera-based vision systems).

Older designs of laser-based vision systems used He-Ne lasers of large dimensions. More modern systems use solid-state lasers, which provide a higher power output, but are much smaller and cheaper. An additional, very important advantage is that the length of the infrared waves emitted covers the range of maximum sensitivity of the receiving system (silicon photodetector). Laser scanners are very expensive measuring systems, and this fact limits their range of application considerably. Laser scanners which will give a slightly more affordable performance in terms of accuracy and resolution, but still be suitable for automatization and robotization applications are now under development.

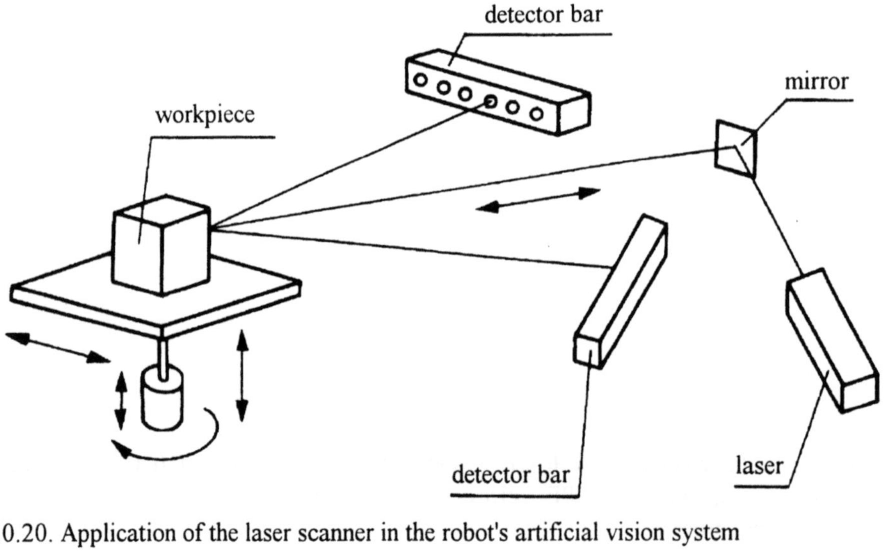

Fig. 10.20. Application of the laser scanner in the robot's artificial vision system

A new component of vision systems are **optical fibers**. Their first application in robotics was tested at the ITT Research Institute in Chicago. A **fiber optic sensor** was made, the head of which consisted of 16 optical fibers arranged in four groups of four fibers, and mounted above the grasping mechanism of the robot. The sensor was used to identify gears handled by the robot's arm, and to determine their position and orientation before grasping. Another example of the application of fiber optics in vision systems is a sensor in the form of a fiber strip attached to the finger of the gripper (Fig. 10.21). When the grasping mechanism is extended, the strip (perpendicular to the direction of motion) scans the object located between the fingers. A binary two-dimensional image of one of the object surfaces is obtained by having a source of light appropriately located on the opposite finger of gripper.

Ordered and non-ordered **fiber optics bundles** are also used in studies on artificial vision systems o. The former are used to transfer the image from inaccessible places to a remote camera lens. The latter are used to provide lighting for the objects examined by traditional, camera-based artificial vision systems.

In addition to the machine vision systems described above, new systems capable of identifying objects by analyzing their color are being studied. The criterion of color is especially important when the objects are of the same shape, have a similar surface texture, and differ only in color. Moreover, recognizing the object by its color significantly shortens data processing time.

A Japanese system for identifying color consists of a camera, a set of colour filters installed on a rotary disc, and a computer. This vision system has been used to analyze the image of a tower built of blocks of different colors, touching and overlapping each other.

Machine vision systems using colour tv-cameras are also being studied. The flowchart of these systems is shown in Figure 10.22. The signal from the camera is amplified, and split into R-G-B signals corresponding to the basic colors: red, green, and blue.

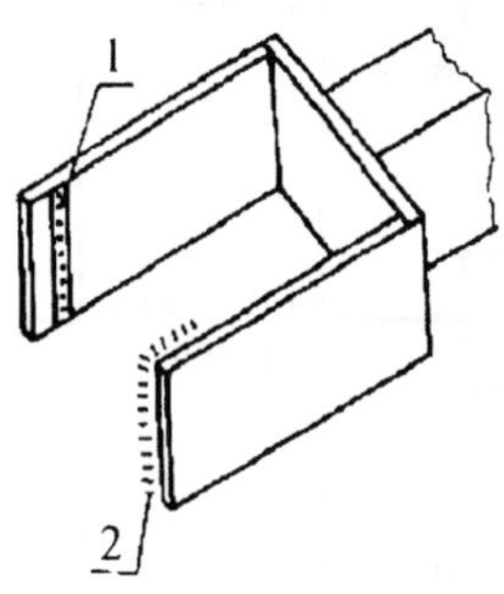

Fig. 10.21. Fiber optics vision sensor in the form of fibre strip attached to the fingers of the robot's gripper.

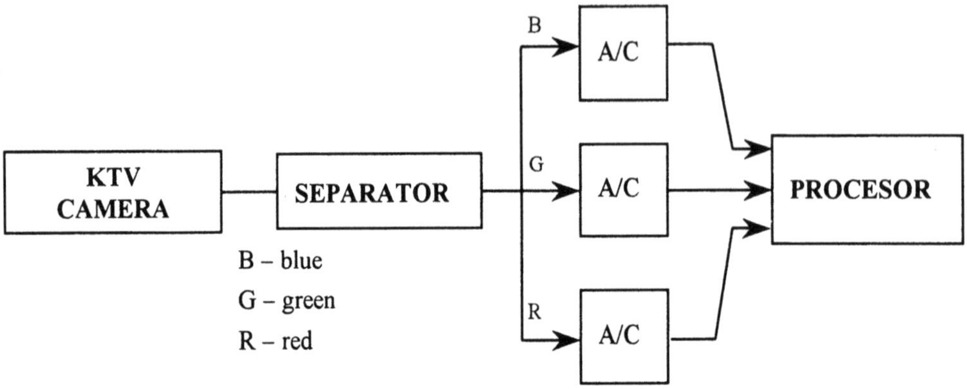

Fig. 10.22. Flowchart of a machine vision system based on colour analysis by means of colour tv-camera

The analysis of the scene starts with an "order" to find a particular color coming from the processor. Next, an image containing only elements of the "ordered" color is separated from the general image, the parameters of the objects featuring the selected color are measured, and the appropriate commands sent to the robot control system.

A vision system utilizing a fiber optics color sensor has been developed at the Warsaw Technical University (Fig. 10.23) [10.5, 10.8].

The issue of machine vision is further discussed in section 12 "Robot task planning".

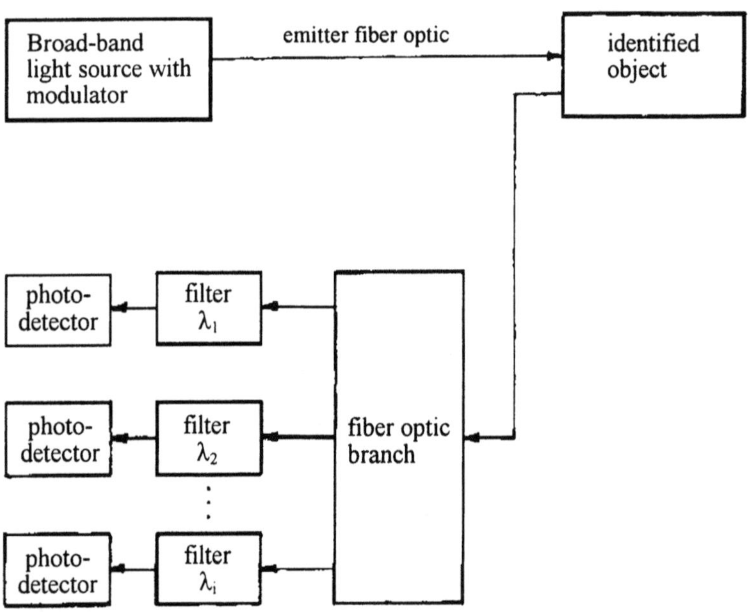

Fig. 10.23. Flowchart of a fiber optics vision processor used to identify object by their colours

A broad-band source of light emits a white modulated light, which passes through the emitter fiber and illuminates the object being identified. The object reflects the maximum quantity of radiation, the wavelength of which corresponds to the object color. The reflected light is registered by the receiver fiber and transmitted through the branching fiber and an appropriate interference filter to the photodetector surface. The electrical signal generated by the photodetector is used to control the grasping mechanism of the robot.

Of all the different sensors used in robotics, the largest amount of information is supplied by vision sensors. Vision systems are accompanied by a very complex (in terms of hardware as well as software) data processing system, unlike any other sensor system used in robotics. The direct reason for this is the fact, that the vision sensor supplies the robot sensing system with a large quantity of analysed data of varied structure. A very high number of images representing the same object but viewed from many different points and accompanied by various configurations of the robot's surroundings are generated.

So much data requires long processing times and a large processing system memory, but only properly processed information can be used to control a robot's movements. At the current level of development the whole processing system cannot operate in real time. Even in laboratory tests numerous limitations and constraints concerning the number of objects, and their position and orientation, as well as the diversity of their shapes, must be imposed on the system. Thus it seems more practical, especially in industry, to use the limited vision systems that analyse only a restricted scope of the robot workspace and a limited number of objects, according to an accepted protocol. Today we must say that the problems connected with the application of machine vision are more complex than expected [10.7].

10.5. References

[10.1] *Doeblin E. O.*: Measurement systems. New York, McGraw-Hill Book Comp. 1976.

[10.2] *Morecki A., Nazarczuk K.*: Description of dynamic measurements of mechanical quantities. Part I. Theoretical Bases. Publ. of Warsaw Univ. of Technology, Warsaw 1981, (in Polish).

[10.3] *Ruocco S. R.*: Robot sensors and transducers. Holstead Press, John Wiley & Sons, 1987.

[10.4] *Morecki A.*: Binic Manipulators. Warsaw, Polish Sc. Publ. 1976.

[10.5] *Marszalec E., Marszalec J. Morecki A.*: Theoretical and Experimental Investigations of Optical Fibre Refflective Sensors for Robotics, RoManSy 6. Proceed. of the Sixth CISM IFToMM Symposium on Theory and Practice of Robots Manipulators. Cambridge Massachusetts, Ed by A.Morecki, G.Bianchi and K.Kędzior, the MIT Press, 1987.

[10.6] Metrology. Theoretical Bases. Warsaw, WKŁ 1988 (in Polish).

[10.7] *Pugh A.*: Robot Sensors. Vol.1 Vision. Vol.2 Tactile and non-vision. Campton, Bertford (U.K.), IFS Publ. 1986.

[10.8] *Marszalec E., Marszalec J.*: Theoretical and experimental study of fibre optic sensors for robotics. Doctoral Diss., Publ. of Warsaw Univ. of Technology, Warsaw 1986.

[10.9] Conference papers, Advanced Robotics: Beyond 2000, the 29th ISR, 27th April–1st May 1998, N. E. C. Birmingham (UK).

Part Three

CONTROL OF MANIPULATORS AND ROBOTS TASK PLANNING

11. CONTROL OF MANIPULATORS

We describe only a few of the methods used in the field of manipulator control here [11.10]. The function of a control system is, given reference trajectory $Q(t)$, to determine these signals controlling the actuators of the manipulator that allow the manipulator to follow this trajectory. They are usually computed using the information supplied by feedback sensors. A typical control system for manipulator links is illustrated in Figure 11.1, where

Z is the generator, supplying reference configuration signals,
S, the control system (controller),
M, the manipulator,
Q, the actual manipulator configuration,
C, the set of sensors measuring position and generalized velocities, and
E, the difference between the reference configuration and the actual configuration (servo error).

It is difficult to predict the effects of disturbances acting on the manipulator during its motion because of the discrepancies between the reference configuration and the actual configuration. It is especially difficult to evaluate the influence of different types of friction in actuator bearings, as there are no accurate models of this phenomenon. If we treat manipulator control as a task of motion trajectory planning, three cases can be identified:

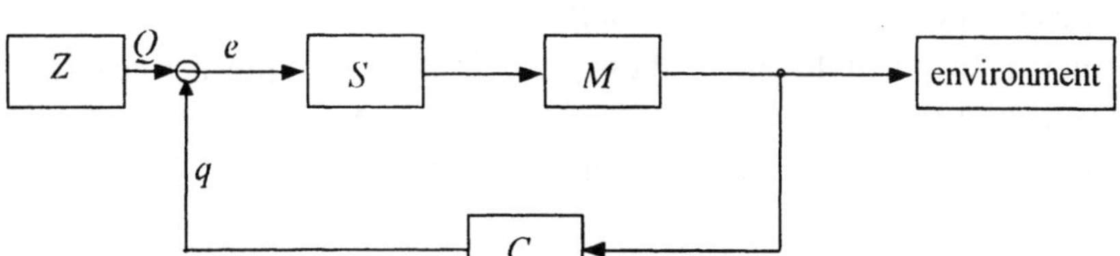

Fig. 11.1. Typical manipulator control system

1. Separate motion control for each manipulator link (local control):
- servomechanisms the: PUMA robot for example),
- torque compensation,
- time-optimal control,
- control with a variable structure,
- independent linear control with feedback
2. Programmable motion control in the Cartesian space:
- velocity control,
- acceleration control,
- control with force feedback
3. Adaptive control:
- adaptive control with reference model,
- self-adjusting adaptive control,
- adaptive control with simple compensation of disturbances,
- adaptive control of programmable motion.

A selection of these three types of control are described here.

11.1. Control equations

Dynamic equations for manipulators are usually nonlinear functions of generalized coordinates and their derivatives. In the general case, they can be written in the following form (see chapter 5):

$$\mathbf{u} = \mathbf{M}(\mathbf{q})\ddot{\mathbf{q}} + \mathbf{C}(\mathbf{q}, \dot{\mathbf{q}}) + \frac{\partial \mathbf{G}(\mathbf{q})}{\partial \mathbf{q}} \tag{11.1}$$

where

$\mathbf{q}$ is generalized coordinate, $\mathbf{q} \in \mathbf{R}^n$;

$\mathbf{M}(\mathbf{q})$, the $(n \times n)$ matrix of inertia moments about the axes of the manipulator`s kinematic pairs;

$\mathbf{C}(\mathbf{q}, \dot{\mathbf{q}})$, the n-dimensional vector of centrifugal forces and Coriolis forces acting on manipulator links, and

$\mathbf{G}(\mathbf{q})$, the potential energy of the manipulator.

Friction may also be included in these equations. Assuming a linear friction model with a constant-term $(n \times n)$ matrix $\mathbf{T}$ of friction coefficients, and adding term $\mathbf{T}\dot{\mathbf{q}}$ to the right side of equation (11.1), we obtain modified dynamic equation

$$\mathbf{u} = \mathbf{M}(\mathbf{q})\ddot{\mathbf{q}} + \mathbf{C}(\mathbf{q}, \dot{\mathbf{q}}) + \mathbf{T}\dot{\mathbf{q}} + \frac{\partial \mathbf{G}(\mathbf{q})}{\partial \mathbf{q}} \tag{11.2}$$

This equation describes a nonlinear dynamic system of the second order. Let us assume that the terms $\mathbf{M(q)}$, $\mathbf{C(q,\dot{q})}$, $\mathbf{T\dot{q}}$ and $\partial\mathbf{G(q)}/\partial\mathbf{q}$ accurately represent the real manipulator (i.e. that accurate link dimensions, inertia matrices, and friction coefficients are known).

Important properties of the $\mathbf{M(q)}$ matrices, namely their non-singularity, are utilized in designing the control systems,. In this case the general theory of systems [11.2] tells us that for the system described by (11.2) there is a feedback

$$\mathbf{u} = \mathbf{M(q)v} + \mathbf{C(q,q)} + \mathbf{T\dot{q}} + \frac{\partial \mathbf{G(q)}}{\partial \mathbf{q}} \tag{11.3}$$

where $\mathbf{v}$ is a new control which globally linearizes the equation of robot dynamics and transforms it (by equating (11.2) and (11.3)) to a second order linear system (Fig. 11.2)

$$\mathbf{\ddot{q}} = \mathbf{v} \tag{11.4}$$

describing the manipulator with the new control $\mathbf{v}$.

For this control chart based on dynamic equations the task of finding a controller can be formulated. The controller must, for example, minimize the servo error $e = \mathbf{Q} - \mathbf{q}$. This goal can be achieved by using the PD controller described by the equation

$$\mathbf{v} = \mathbf{\ddot{Q}} + \mathbf{K}_p(\mathbf{Q} - \mathbf{q}) + \mathbf{T}_d(\mathbf{\dot{Q}} - \mathbf{\dot{q}}) \tag{11.5}$$

where

$\mathbf{K_p}$ is the diagonal $(n \times n)$ matrix of proportional gains; and

$\mathbf{T_d}$, the diagonal $(n \times n)$ matrix of differentiation constants

Figure 11.3 shows a complete manipulator control system using PD controller.

Looking at relationships (11.4) and (11.5), we see that the servo error equation is a uniform, second-order differential equation

$$\ddot{e} + \mathbf{K}_p e + \mathbf{T}_d \dot{e} = 0 \tag{11.6}$$

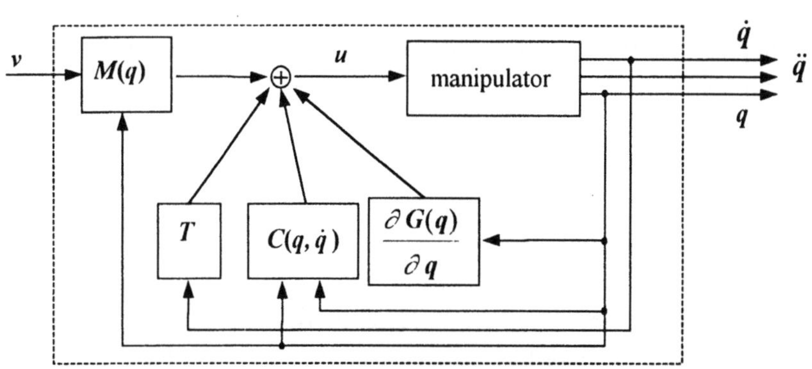

Fig. 11.2. Globally linearized dynamic equation of the manipulator (according to [11.1, 11.2])

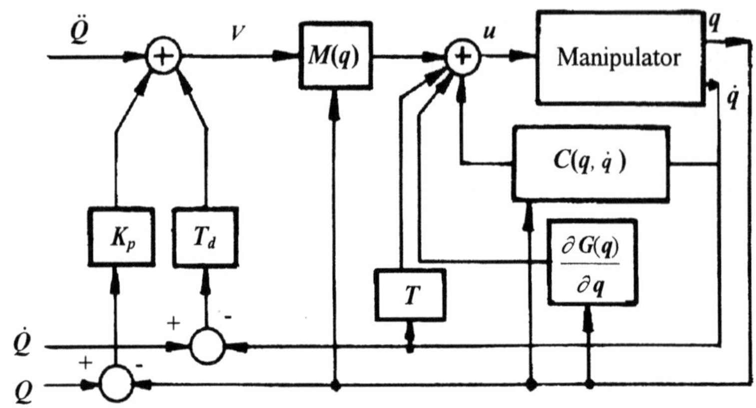

Fig. 11.3. PD-type manipulator control structure

the characteristic equation of which, for the i-th component of the error vector, is described as

$$s^2 + T_d^i s + K_p^i = 0 \tag{11.7}$$

where

K_p^i, T_d^i are, respectively, the proportional gain and differentiation constants of the i-th component of the error vector; and $i = 1, \dots, n$.

To fulfil the necessary and sufficient condition for the stability of the linear control system shown in Fig. 11.3, all roots

$$s_1^i = \frac{T_d^i}{2} + \frac{\sqrt{(T_d^i)^2 - 4K_p^i}}{2}$$

and

$$s_2^i = -\frac{T_d^i}{2} + \frac{\sqrt{(T_d^i)^2 - 4K_p^i}}{2}$$

must have negative real parts (roots located on the left side of the imaginary axis). From inequalities

$$\begin{cases} re(s_1^i) < 0 \\ re(s_2^i) < 0 \end{cases} \tag{11.8}$$

one can find appropriate constants.

As a result of the unavoidable disturbances caused by the inaccuracy of the dynamic equations applied (e.g. an unsuitable friction model) the use of a PD controller introduces a steady-state servo error. We can compensate for this error by using a PID controller in the arrangement shown in Fig. 11.4.

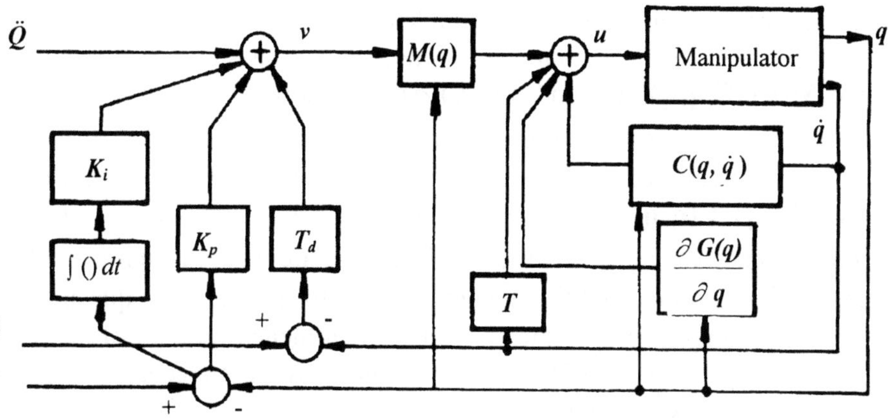

Fig. 11.4. PID-type manipulator control structure

The control equation in this case can be written as

$$\mathbf{v} = \ddot{\mathbf{Q}} + \mathbf{T}_d(\dot{\mathbf{Q}} - \dot{\mathbf{q}}) + \mathbf{K}_p(\mathbf{Q} - \mathbf{q}) + \mathbf{K}_i\int_0^t(\mathbf{Q} - \mathbf{q})dt \tag{11.9}$$

where $\mathbf{K}_i$ is the diagonal $(n \times n)$ matrix of the controller`s integration constants.

Thus the servo error equation can be written as

$$\dddot{e} + \mathbf{T}_d\ddot{e} + \mathbf{K}_p\dot{e} + \mathbf{K}_i e = 0 \tag{11.10}$$

Diagonal matrices $\mathbf{T}_d$, $\mathbf{K}_p$ and $\mathbf{K}_i$ ensure the stability of the control system shown in Fig. 11.4. Recapitulating, the control method described requires real-time execution of feedback procedures linearizing dynamic equations. The computational complexity of these procedures strongly influences control frequency, and consequently control quality. In most of the manipulator control systems currently in use, in equations

$$\mathbf{u} = \mathbf{M}(\mathbf{q})\mathbf{v} + \mathbf{C}(\mathbf{q}, \dot{\mathbf{q}}) + \mathbf{T}\dot{\mathbf{q}} + \frac{\partial \mathbf{G}(\mathbf{q})}{\partial \mathbf{q}} \tag{11.11}$$

it is assumed that

$$\mathbf{M}(\mathbf{q}) = \mathbf{I},$$

where $\mathbf{I}$ is a $(n \times n)$ identity matrix; and

$$\mathbf{C}(\mathbf{q}, \dot{\mathbf{q}}) = \mathbf{T}\dot{\mathbf{q}} = \frac{\partial \mathbf{G}(\mathbf{q})}{\partial \mathbf{q}} = 0$$

The result is a very simple and rapid control system, which treats each kinematic pair as an independently controlled object. But when we apply these simplified control systems, it is difficult to evaluate their accuracy, since in the control equation no interaction between individual joints is taken into account. If we do not consider these interactions, additional errors are introduced, which must be eliminated by using PID controller.

The linearized equations of manipulator dynamics, on which the control systems described above are built, are determined by assuming that the accurate values of $\mathbf{M}(\mathbf{q})$, $\mathbf{C}(\mathbf{q},\dot{\mathbf{q}})$, $\mathbf{T}\dot{\mathbf{q}}$ and $\partial\mathbf{G}(\mathbf{q})/\partial\mathbf{q}$ are known.

In practice, these values are given with certain errors, the system of equations (11.4) becomes a coupled system, and the issue of the precise identification of the values of $\mathbf{M}(\mathbf{q})$, $\mathbf{C}(\mathbf{q},\dot{\mathbf{q}})$, $\mathbf{T}\dot{\mathbf{q}}$ and $\partial\mathbf{G}(\mathbf{q})/\partial\mathbf{q}$ or their adaptation in the control algorithm becomes a significant problem. But these are issues which extend beyond the scope of this book. If the reader wishes to pursue the question of the identifications manipulator parameters and their influence on control quality, he is referred to [11.3], [11.4] and [11.7]. One of the possible concepts of the adaptive control system is shown in Figure 11.5.

The flowchart of the control system shown in the figure is also based on the equation of manipulator dynamics. The system contains an adaptive module which modifies manipulator parameters until the servo error e is completely eliminated, on the basis of continuous recording of the manipulator state (link positions, generalized velocities) and the servo error. Such system should "know" its dynamic properties. For further discussion of the issues of adaptive control, see [11.5], [11.6], [11.7] and [11.10].

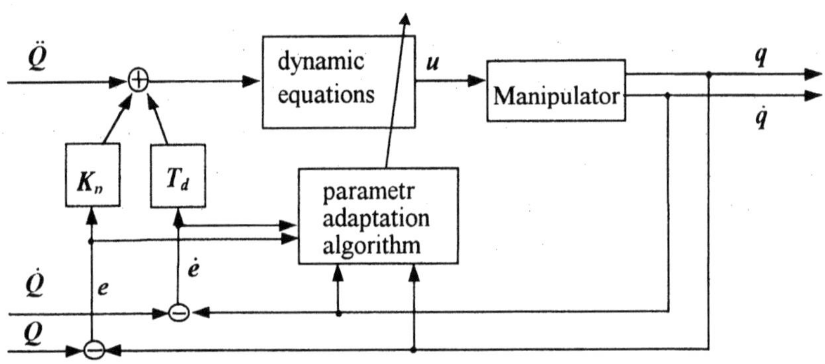

Fig. 11.5. A concept of adaptive control system

11.2. Model-reference adaptive control

To describe control using a reference model [11.8] we write the dynamic equations (11.1) in the following form:

$$M(q)\ddot{q}+C^1(q,\dot{q})\dot{q}+G^1(q)q=u \tag{11.12}$$

where (by definition) $C^1(q,\dot{q})\dot{q}=C(q,\dot{q})$ and $\dfrac{\partial G(q)}{\partial q}=G^1(q)q$.

Introducing the *2n*-dimensional state vector $x = (q \quad \dot{q})^T$, dynamic equation (11.12) can be expressed as a state equation:

$$\dot{x} = A_p(x) + B_p(x)u \tag{11.13}$$

where $A_p(x) = \begin{bmatrix} 0 & \mathrm{I} \\ -M^{-1}(q)G^1(q) & -M^{-1}(q)C^1(q,\dot{q}) \end{bmatrix}$, $B_p(x) = \begin{bmatrix} 0 \\ M^{-1}(q) \end{bmatrix}$

Matrices $A_p(x)$ and $B_p(x)$ have the dimensions of $(2n)\times(2n)$ and $(2n)\times n$ respectively. The main concept of adaptive control with a reference model is to have such synthesis of control as will force the dynamic system (11.13) to behave in a certain desired way, as determined by the reference model. On the basis of the structure of equations (11.13), the reference model is selected so as to ensure its stability and linearity, as well as its invariability with respect to time. The form of the model is

$$\dot{y} = A_M y + B_M r \tag{11.14}$$

where

y is the *2n*-dimensional state vector of the reference model;

r, the *n*-dimensional control vector of the reference model;

$$A_M = \begin{bmatrix} 0 & \mathrm{I} \\ -\Lambda_1 & -\Lambda_2 \end{bmatrix}, \qquad B_M = \begin{bmatrix} 0 \\ \Lambda_1 \end{bmatrix}$$

Λ_1, Λ_2 are $(n\times n)$ diagonal matrices with positive terms.

Control vector *r* equals *Q*.

The difference between state vector *x* and reference vector *y* yields the state error, which is used in the adaptive algorithm. The objective of the algorithm is to maintain this error as close to zero as possible. The adaptive algorithm has been built to fulfil the asymptotic stability requirement for the system described above, using the Lyapunov stability criterion. This approach assumes the control method

$$u = -K_x x + K_u r \tag{11.15}$$

where K_x, and K_u are, respectively, $(n)\times(2n)$ and $(n\times n)$ matrices of gain coefficients, which are time variables. By defining $x = (x_1 \quad x_2)^T$, $K_x = \begin{bmatrix} K_{x_1}{}^T & K_{x_2}{}^T \end{bmatrix}$, a closed loop control system (11.13) can be obtained:

$$\dot{x} = A_s(x)x + B_s(x)r \tag{11.16}$$

where

$$A_s(x) = \begin{bmatrix} 0 & \mathrm{I} \\ -M^{-1}(q)\left(G^1(q)+K_{x_1}\right) & -M^{-1}(q)\left(C^1(q,\dot{q})+K_{x_2}\right) \end{bmatrix}; \text{ and}$$

$$B_s(x) = \begin{bmatrix} 0 \\ M^{-1}(q)K_u \end{bmatrix}$$

Suitable selection of the K_x and K_u matrices permits "adjustment" of the system (11.16) to the reference model (11.14). To do this, the following *2n*-dimensional state error vector is defined:

$$E = y - x$$

Hence

$$\dot{E} = A_M E + (A_M - A_s)x + (B_M - B_s)r \tag{11.17}$$

Matrices K_x and K_u must fulfil the condition

$$\lim_{t\to\infty} E(t) = 0 \tag{11.18}$$

This problem is solved by using the Lyapunov theorem of stability. An additional function V is defined:

$$V = E^T PE + tr\left((A_M - A_s)^T F_A^{-1}(A_M - A_s)\right) + tr\left((B_M - B_s)^T F_B^{-1}(B_M - B_s)\right) \tag{11.19}$$

where P, F_A, and F_B are $(2n)\times(2n)$ positively defined matrices; and $tr(\ )$ denotes the matrix ( ) trace. Differentiating both sides of (11.19) and using (11.17) we obtain

$$\dot{V} = E^T(A_M P + PA_M)E + tr\left((A_M - A_s)^T\left(PEx^T - F_A^{-1}\dot{A}_s\right)\right) + tr\left((B_M - B_s)^T\left(PEr^T - F_B^{-1}\dot{B}_s\right)\right) \tag{11.20}$$

The Lyapunov theorem of stability tell us that condition (11.18) is fulfilled when $\dot{V} < 0$. This yields the conditions:

$$\begin{aligned} A_M{}^T P + PA_M &= -W \\ \dot{A}_s &= F_A PEx^T \\ \dot{B}_s &= F_B PEr^T \end{aligned} \tag{11.21}$$

where $\boldsymbol{W}$ is a positively defined matrix.

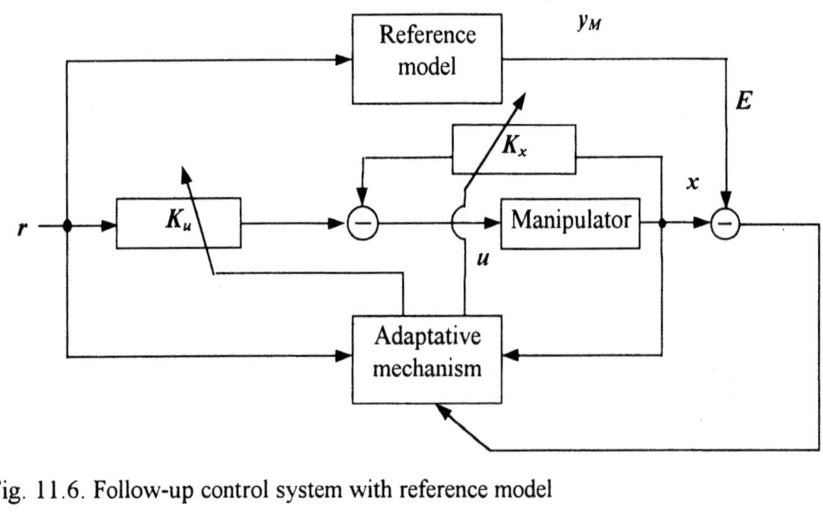

Fig. 11.6. Follow-up control system with reference model

Assuming that $A_s \cong A_M$, $B_s \cong B_M$ and $\|\dot{A}_p\| << \|A_p\|, \|\dot{B}_p\| << \|B_p\|$ we obtain

$$\dot{A}_s \cong B_p \dot{K}_x$$
$$\dot{B}_s \cong B_p \dot{K}_u$$

Taking these assumptions into consideration, we obtain an adaptive algorithm (Fig. 11.6.) for the K_x and K_u matrices:

$$\begin{aligned} \dot{K}_u &= K_u B_M^+ F_B E r^T \\ \dot{K}_x &= K_u B_M^+ F_A P E x^T \end{aligned} \qquad (11.22)$$

where B_M^+ is a $(n)\times(2n)$ B_M pseudoinverse matrix; and $B_M^+ = \left(B_M^T \quad B_M\right)^{-1} B_M^T$.

Matrix differential equations (11.22) build an adaptive mechanism and ensure the asymptotic stability of the approach.

11.3. Sliding control of manipulators

Sliding control`s (Utkin [11.9]) feature discontinuities in time described by

$$u_i(x\,,\,t) = \begin{cases} u_i^+(x,t) & if \quad s_i(x) > 0 \\ u_i^-(x,t) & if \quad s_i(x) < 0 \end{cases} \qquad (11.23)$$

where

$s_i(x) = 0$ is i-th component of n switchable hypersurfaces in the state space

$$s(x) = 0 \qquad (11.24)$$

where s: $\boldsymbol{R}^{2n} \rightarrow \boldsymbol{R}^n$; and $i = 1{:}n$.

The sliding mode is applied on switching hypersurfaces $s_i(x)=0$ when the robot state trajectory moves towards these hypersurfaces. In this case the state vector "slides" over the surface described by equation $s_i(x) = 0$. The condition for using the sliding mode on the i-th hypersurface is the following inequality:

$$\lim_{s_i \to 0} s_i \dot{s}_i < 0 \qquad (11.25)$$

true in the neighbourhood of hypersurface $s_i(x) = 0$. In the sliding mode the system described by the state equation (11.13) meets the conditions of

$$\begin{aligned} s_i(x) &= 0 \\ \dot{s}_i(x) &= 0 \end{aligned} \qquad (11.26)$$

Switchable hypesurfaces usually take the form of hyperplanes:

$$s(x) = R\begin{pmatrix} e \\ \dot{e} \end{pmatrix} = 0 \qquad (11.27)$$

where

R is a $(n\times(2n))$ matrix, usually having the form of $R=[C_s \quad I]$, and

$C_s = \mathrm{diag}(c_1, c_2, \ldots, c_n)$; $c_i > 0$; and $e = q - Q$.

In the case of time-dependent switchable surfaces $s(x, t)$, condition (11.25) can be written as

$$\lim_{s_i \to 0} s_i \dot{s}_i < -\Psi(|s_i|) \tag{11.28}$$

where Ψ is the increasing function of its argument.

A dynamic system in the sliding mode satisfies equation (11.26). Relationships (11.13) and (11.24) yield

$$\dot{s}(x) = \frac{\partial s}{\partial x}\dot{s} = G_s A_p(x) + G_s B_p(x) u_{eq} = 0 \tag{11.29}$$

where $G_s = \frac{\partial s}{\partial x}$, and u_{eq} is known as the equivalent control (Utkin [11.9]).

Matrix $G_s B_p(x)$ is non-singular, and thus from (11.29) we can derive the control function u_{eq}, implying the ideal sliding mode:

$$\dot{x} = \left(I - B_p(x)\left(G_s B_p(x)\right)^{-1}\right) A_p(x) \tag{11.30}$$

Control function u_{eq} is an equivalent of the average control function value u, maintaining the state vector, x, on the switching surface $s(x)$. The actual control function contains a low-frequency component Δu (vibration)

$$u = u_{eq} + \Delta u \tag{11.31}$$

$$\Delta u_i(x, t) = \begin{cases} \Delta u_i^+(x,t) & if \quad s_i(x) > 0 \\ \Delta u_i^-(x,t) & if \quad s_i(x) < 0 \end{cases}$$

In the case of precise knowledge of the $\left(G_s B_p(x)\right)^{-1}$ factor, the discontinuous control function becomes $\Delta u = -\left(G_s B_p(x)\right)^{-1} \mathrm{diag}(w_1, \ldots, w_n)\ \mathrm{sgn}(s(x))$, where $w_i > 0$; $i = 1{:}n$. Hence, taking into account (11.26), we obtain

$$\dot{s} = \frac{\partial s}{\partial x}\dot{x} = G_s A_p(x) + G_s B_p(x) u_{eq} + G_s B_p(x)\Delta u \tag{11.32}$$

and finally

$$\dot{s}_i = -w_i\ \mathrm{sgn}(s_i(x)) \tag{11.33}$$

From equations (11.33) we know that $s_i \dot{s}_i < 0$; thus the sliding mode condition (11.25) is fulfilled. In most practical applications matrix $\left(\widehat{G_s B_p(x)}\right)^{-1}$ is usually estimated by matrix $\left(\widehat{G_s B_p(x)}\right)^{-1}$. In the case when for matrix

$$\left(G_s B_p(x)\right)\left(G_s B_p(x)\right)^{-1} = I + \delta I$$

where

$\delta I = \left|\delta I_{ij}\right|_{n\times n}$, the following condition is met [11.1]:

$$1+\delta I_{ii} > \sum_{\substack{j=1\\ j\neq i}}^{n}\left|\delta I_{ij}\right|\frac{w_j}{w_i} \tag{11.34}$$

then for the estimated matrix $\left(\widehat{G_s B_p(x)}\right)^{-1}$ the sliding mode condition (11.25) is fulfilled as well. The main drawback of the sliding mode strategy is the discontinuity of the control function on the switching hyperplane (11.27). As a consequence, the control function may introduce theoretical "vibration" with infinite frequency. Such a control system is difficult, or even impossible, to implement. To avoid this problem, the discontinuous function sgn( ) is replaced by a continuous function $conti(s_i) = \frac{s_i}{|s_i| + \delta_i}$, where $\delta_i > 0$ is a certain coefficient. When $\delta_i = 0$, function $conti(s_i)$ is reduced to $sgn(s_i)$. On the other hand, when δ_i is large, function $conti(s_i)$ approaches zero, i.e. the discontinuous control component vanishes. Suitable selection of δ_i may significantly improve the control process. Coefficient δ_i is assumed as [11.11]):

$$\delta_i = \delta_{i0} + \eta_i \tag{11.35}$$

where

$\eta_i = k_i\left|\int_0^t s_i dt\right|$, $\delta_{i0}, k_i > 0$ are given numbers.

It is assumed that component η_i appears in formula (11.35) if inequality $|s_i| < \varepsilon_i$ is true, when ε_i is a given small number. When $|s_i| \geq \varepsilon_i$ control function Δu is written as

$\Delta u = -\left(\widehat{G_s B_p(x)}\right)^{-1} diag(w_1,\dots,w_n)\mathrm{sgn}(s(x))$, ($\eta_i$ components are zeros).

This mechanism is convenient when the robot follows a given trajectory. In the neighbourhood of trajectory Q (i.e. $|s_i| < \varepsilon_i$), the control function u is continuous and ensures vibration-free motion. In event of a sudden change in trajectory or strong external disturbances ($|s_i| \geq \varepsilon_i$), the control is switched to discontinuous control.

11.4 Robust control of manipulators

Robot control methods have recently appeared [11.12], that do not require estimation of the kinematic parameters of robots. This new approach significantly simplifies and accelerates the construction of trajectory following controllers. It does not require detailed knowledge of the robot's dynamics. As a result, a model more general than the one described in section 11.1 can be considered:

$$M(q)\ddot{q}+C^{1}(q,\dot{q})\dot{q}+\frac{\partial G(q)}{\partial q}+F_{t}(\dot{q})+F_{z}(t)=u \tag{11.36}$$

where: $F_t(\dot{q})$ denotes static or dynamic friction forces; and $F_z(t)$ are the external disturbances acting on the robot (e.g. temperature, etc.). It is assumed that $F(\dot{q})$ are Coulomb friction forces, and that $F_z(t)$ are limited in time. By introducing function $x_s=\dot{Q}-\beta e$, where, as usual, $e=q-Q$; and β is a constant positive number, dynamic model (11.36) may be expressed as a trajectory tracking error

$$\mathrm{M}\begin{pmatrix}\dot{e}\\ \ddot{e}\end{pmatrix}+\mathrm{C}\begin{pmatrix}e\\ \dot{e}\end{pmatrix}=u-\mathrm{L}(q,\ \dot{q},\ x_s,\ \dot{x}_s,\ t) \tag{11.37}$$

where

$$\mathrm{M}=[\beta M(q)\quad M(q)],$$

$$\mathrm{C}=\left[\beta C^{1}(q,\dot{q})\quad C^{1}(q,\dot{q})\right],\ \text{and}$$

$$\mathrm{L}=M(q)\dot{x}_s+C^{1}(q,\dot{q})x_s+\frac{\partial G(q)}{\partial q}+F_t(\dot{q})+F_z(t)$$

For this kind model, an adaptive PD controller is used to follow trajectory Q (Figure 11.7.):

$$u=-(k+\zeta)(\dot{e}+\beta e) \tag{11.38}$$

where k is a certain positive number; ζ is an adaptive function derived from the formula:

$$\zeta=\frac{d(3+\phi_a)\hat{a}}{\|\dot{e}+\beta e\|+\varepsilon},$$

$d\geq 1$ is an arbitrary real number,

$$\phi_a=\|\dot{x}_s\|+\|x_s\|\|\dot{q}\|+\|\dot{q}\|;\ \dot{\hat{a}}=-\sigma\hat{a}+gd\frac{\|\dot{e}+\beta e\|^{2}(3+\phi_a)}{\|\dot{e}+\beta e\|+\varepsilon};$$

$\sigma>0$ is a certain parameter (e.g. function of time) introduced to ensure that function $\hat{a}(t)$ is limited;

$g>0$ is a selectable real number influencing the precision of trajectory tracking;

ε is a coefficient given by $\varepsilon=\dfrac{\nu}{d(3+\phi_a)}$; and

ν is small positive number (or function of time).

With these assumptions it can be shown that control system with controller (11.38) is stable according to the Lyapunov criterion. When ν and σ have the form:

$$\nu = \frac{1}{(1+t)^{\alpha}},$$

$$\sigma = e^{-\nu_1 t}, \quad \alpha > 1, \quad \nu_1 > 0,$$

a control system with controller (11.38) is asymptotically stable. (The control function described by (11.38) is a continuous function in the time domain, contrary to the function described in the previous section.)

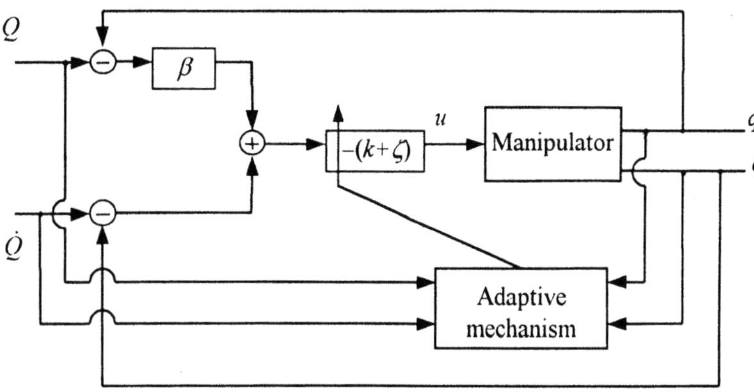

Fig. 11.7. Robust control system

11.5. References

[11.1] *Craig J. J.*: Introduction to Robotics. Mechanics and Control. Addison - Wesley Publ. Comp. 1989.

[11.2] *Wonham W. M.*: Linear multivariable control - a geometric approach. New York, Heidelberg, Berlin, Springer - Verlag 1979.

[11.3] *Lee C. S. G., Lee B. H.*: Resolved motion adaptive control for mechanical manipulators. Trans ASME, Jnl of Dyn. Syst., v. 106, 1984, pp. 134-142.

[11.4] *Koivo A. J., Guo T. M.*: Adaptive linear controller for robotic manipulators. IEEE Trans. Aut. Contr., vol. 28, 1983, pp. 162-171.

[11.5] *Craig J. J., Hsu P., Sastry S. S.*: Adaptive control of mechanical control. Proc. IEEE Cont. Rob. Aut., 1986, pp. 190-195.

[11.6] *Vukobratovic M., Stokic D., Kircanski N.*: Towards to nonadaptive and adaptive control of manipulation robots. IEEE Trans. Aut. Contr., vol. 29, 1984, pp. 841-844.

[11.7] *Abdallach C., Dawson D., Dorato P., Jamishidi M.*: Survey of robust control for rigid robots. IEEE Control System Mag., vol. 11, No. 2, 1991, 24-30.

[11.8] *Nicosia S., Tomei P.*: Model Reference adaptive control algorithms for industrial robots. Automatica, vol. 20, 1984, pp. 635-644.

[11.9] *Utkin V. I.*: Sliding modes and their applications. Moscow: Mir, 1978.

[11.10] *Slotine J. J., Li W.*: Applied nonlinear control. Englewood Cliffs, NY: Prentice Hall, 1991.

[11.11] *Harashima F., Hashimoto H., Maruyama K.*: Practical robust control of robot arm using variable structure system. In Proc. IEEE Rob. and Aut., 1986, pp. 532-539.

[11.12] *Song Y. D.*: Adaptive motion tracking control of robot manipulators-non-regressor based approach. In Proc. IEEE Rob. and Aut., 1994, pp. 3008-3013.

12. ROBOT TASK PLANNING

12.1. Introduction

Robot programming is particularly difficult in the case of complicated tasks requiring complex operations in three-dimensional workspaces where those operations are also coordinated by the information from sensors. And even for relatively simple tasks performed by currently produced industrial robots, the cost involved in their programming can be comparable to the price of the robot itself. Consequently, it is only natural that methods for simplifying robot programming are a priority issue. Treating a robot's operations only in terms of results on the objects of manipulation seems to be one way of solving this problem. For instance, it is easier for a user to define an operation: "place the pivot in the hole" than specify the sequence of manipulator motions required for achieve that result.

To perform the task, specified in this simplified manner, in the workspace, the robot must have information about the objects to be found there, the relations (usually geometrical) between them, the objects it is made of itself, and the admissible actions implying changes in those relations. In other words, a robot must have information about: geometric models (representations) of the workspace, as well as of itself; kinematic and dynamic models (equations) of the manipulator, together with the conditions of their application set, for example, by design limitations, the performances of the actuators used and, in general, by the ambiguity of the task performed.

Apart from those models, a robot must be fitted with sensors of touch, and vision sensors (usually recording two-dimensional images of the workspace) which make it possible to recognize objects of manipulation during operation planning and updating of the model of the workspace. In the event of changes in the workspace (the introduction of new objects, of obstacles), the information obtained must be transformed into the current form of the model of the workspace. This process usually involves the reconstruction of a three-dimensional workspace, done on the grounds of the two-dimensional images obtained from vision sensors.

Knowledge of the elements listed above enables the robot to:

- obtain (by means of vision sensors), and analyze images of the workspace, as well as create its geometric model;
- assign and understand the task assigned by a user in natural language;
- perform that task by generating elementary operations, that is, a finite sequence of manipulator trajectories (expressed as values of the relative rotations and

translations of the joints corresponding to a certain trajectory of a gripper motion, with time as a parameter) that realize that task.

Robots capable of performing the above functions are called the third generation robots, and are discussed in chapter 4. According to the defining characteristics defining third generation robots, they meet the basic aim which of the simple programming of the task assigned by a user. The analysis of all functional elements of the third generation robot is a very broad issue. Only its main function, that of **planning the elementary operations** required to perform a given task, is considered here. The user-robot communication module, the objective of which is syntactic analysis and semantic interpretation of a natural language after converting speech into a language of symbols, is omitted, as are the vision module that analyzes the information obtained from TV cameras and the teaching module used to generalize the working plans that have been generated. The reader who wishes to explore these subjects is further referred to [12.24].

A simplified operation flowchart of a robot analyzing the task assigned by a user, is shown in Figure 12.1.

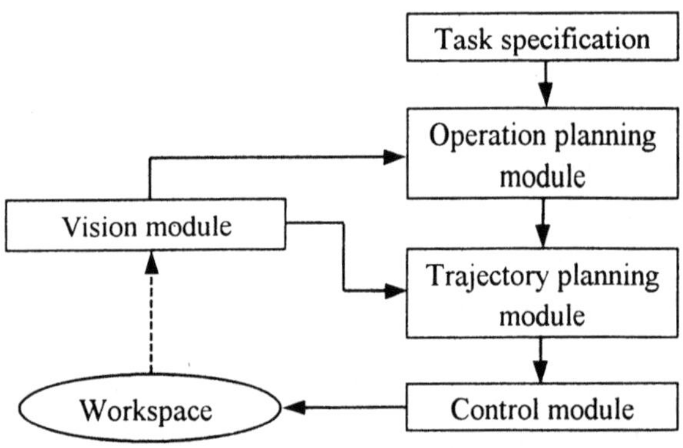

Fig. 12.1. Simplified operation flowchart of a robot performing a task assigned by a user

The user`s. specification of the task is transmitted to the operation planning module which, with the input of the vision module that recognizes the objects of manipulation and their location in the workspace, generates a finite sequence of elementary operations. In the successive step, the appropriate trajectories of the manipulator and gripper are determined in the operation planning module, taking into account information from vision sensors (e.g. obstacles encountered). The trajectories are then passed on to a control system which generates the control signals activating the manipulator actuators that move the gripper and manipulator along the trajectory. Every change in the manipulator position affects the workspace, together with the vision module. The robot has completed its task when it has carried out all the elementary operations assigned, which will include operations of the type: grasp an object and move it (or move and release) to a certain place. Vision analysis gives the robot only partial information about the possibility of realizing these operations; it simply identifies the object to grasp, and its position and orientation in the workspace.

In order to grasp an object, the position and orientation of the gripper (relative to the base system) which is to the grasping must be generated.

12.2. Grasp planning

Selection of the gripper configuration with which to grasp the recognized objects is usually not obvious. It depends on:
- the design parameters of gripper,
- the shape of the object to be grasped, and
- the presence of other objects in the vicinity of that object.

Figure 12.2 shows a parallel jaw gripper for which the grasp orientation is determined by three unit vectors (v_z, v_p, v_o), where: v_z is a position vector, v_p , an approach vector, and v_o , an orientation vector. They determine a coordinate system attached to the gripper.

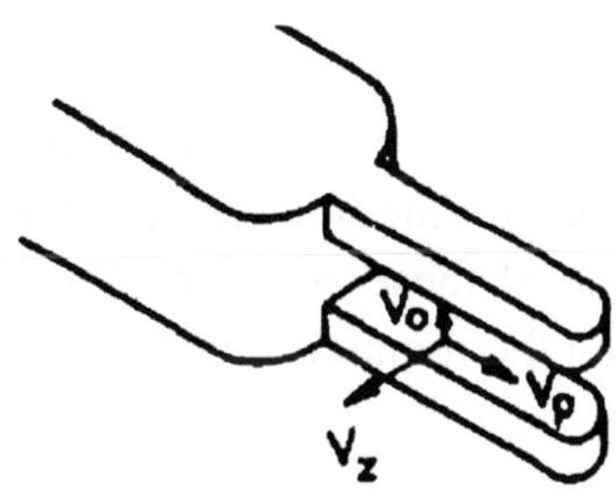

Fig. 12.2. Model of a gripper and notation of parameters

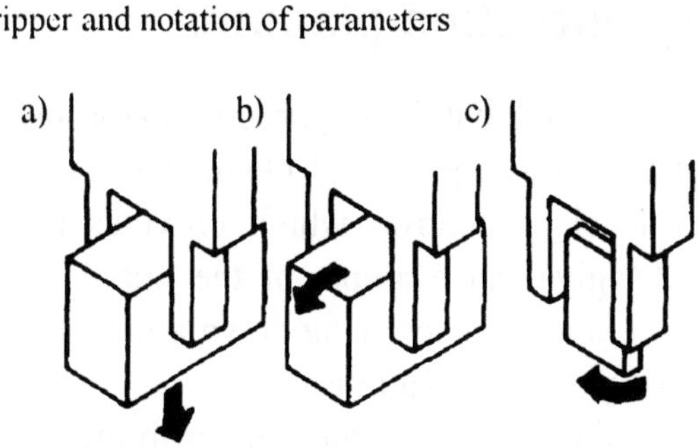

Fig. 12.3. Possible types of displacements of an object in the gripper

The orientation and position (i.e. the operational point) of the gripper are generated by:

1. selecting the faces of the object that are approximately parallel to each other. The normal vector of the planes containing these faces defines the orientation vector v_p.
2. producinga finite sequence of approach vectors for each pair of these faces. They are determined by measuring slip resistance of the object in the gripper and its twist resistance when closing the gripper, as well as by obstacles in the vicinity (their influence on the grasp will be discussed in later chapters).

Additional limitations are imposed on grasp selection result by the fact that:

- the gripper jaws must come in contact with the object,
- the object must be stationary with respect to the gripper when grasped.

Figure 12.3 shows the possible movements between the object and gripper.

All these criteria are applied to determine the approach vector v_p . In the case of objects with simple shapes located in simple workspaces, the vector v_p is selected basing on the basis of axes of the symmetry of the objects to be grasped.

Given vectors v_o and v_p , a vector v_z can be found according to the relationship

$v_z = v_o \times v_p$, where $\times$ denotes a cross product.

In sum up, the result of grasp planning is the gripper configuration expressed by a matrix

$$T_0 = \begin{bmatrix} v_p & v_o & v_z & r_c \\ 0 & 0 & 0 & 1 \end{bmatrix} \tag{12.1}$$

where r_c is a vector from the base system to the gravity center of the object of the grasp.

Data, obtained from the vision module, about positions of objects and possible grasp configurations for some of these are then used in the process of elementary operation planning and to plan the required trajectory.

12.3. Elementary operation planning

In the introduction we stated that the required task (for instance, an assembly) is decomposed by the robot into a sequence of elementary operations, these are then delivered to the lower level of control, which provides for performance of the primary task. Figure 12.4 gives an example of the initial situation of the task of putting a washer on a bolt, and then screwing down a nut. In other words, the robot has to achieve the final situation, shown in Fig.12.5.

The robot can change the position of the gripper in the workspace, turning it as well as closing and opening its jaws. A set of facts in which the given are: the initial situation, the final situation, and a set of possible robot actions constitutes a **problem situation**. Its solution depends on the generation of a sequence of elementary operations that will result in the final situation, as in the example presented in Fig. 12.5. The actual (or planned) execution of an elementary operation changes the real situation in the workspace (or its model). Consequently, a set of different situations can be generated by carrying out the above operation Clearly, not all of these operations can we employ for every situation. For instance, it is not possible to grasp an object when another object is already held by the gripper. Generalizing this example, we shall try to describe more formally a problem situation to be presented to the robot given the task to solve it [12.35].

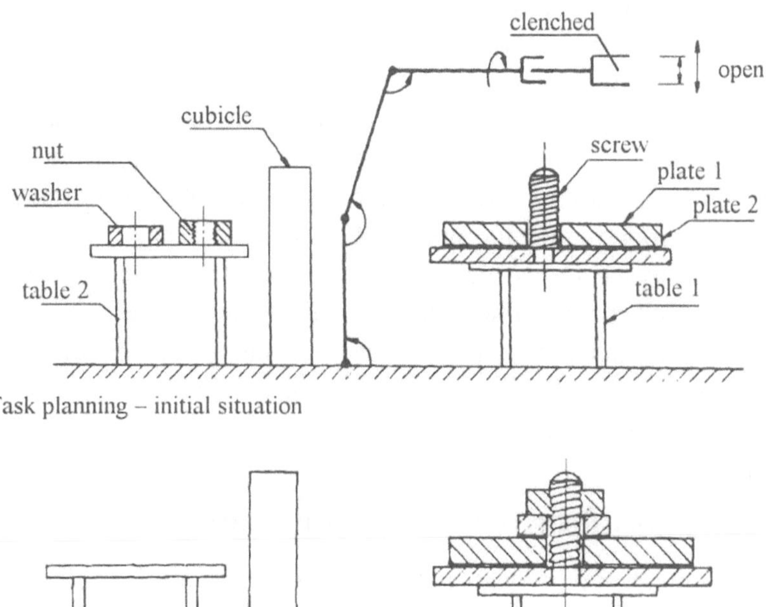

Fig. 12.4. Task planning – initial situation

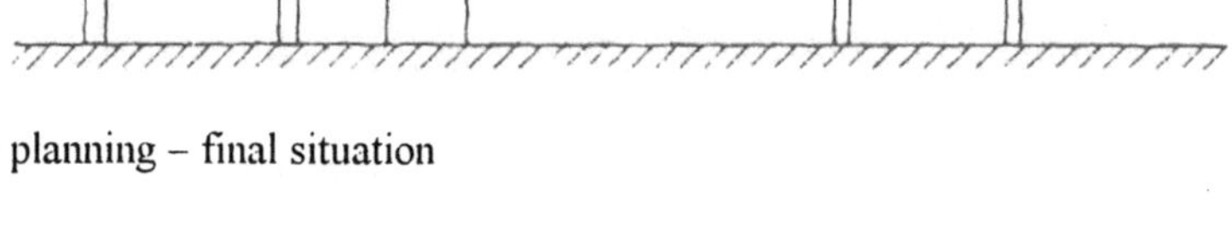

Fig. 12.5. Task planning – final situation

The above description contains the basic notion of the state of a problem situation, meant as a formally written set of relationships (representations) the truthfulness of which is determined by the given problem situation. Formally, what is called a problem situation consists of four quantities in the form (S_s, S, F, S_t), where: S_s ($S_s \subset S$) is a set of initial states; S, a set of possible states of the problem situation; S_t ($S_t \subset S$) , a set of final states; and F, a set of operators transforming particular states into other states. Any

Any operator $f \in F$ is a function in the form $S_f \to S$, where S_f ($S_f \subset S$) is an admissible domain of the operator f. When $s \in S_f$, we say that the operator f is applicable in state s.

In the example presented set S_s contains one element, represented by the situation in Figure 12.4. Set S_t also contains one element, a description of the final situation, as shown in Fig. 12.5, while set F consists of the three elementary operations required together with their applicability conditions, S_f operations for any $f \in F$. The solution of the problem situation is a finite sequence of such operators f_1, ..., f_l, that $f_i \in F$, where $i = 1{:}l$, $f_l(\ldots(f_2(f_1(s)))\ldots) \in S_t$ and $s \in S_s$.

The above assumption, regarding the operators, holds if the following relationships exist: $s \in S_{f_1}$, $f_1(s) \in S_{f_2}$, ..., $f_l(\ldots(f_2(f_1(s)))\ldots) \in S_{f_{l+1}}$.

Sometimes certain constraints are imposed on the selection of operators. For instance, a sequence of operators that are optimal in some sense is required.. Heuristic search methods are usually employed to solve such tasks. The method of solving of a problem situation (S_s, S, F, S_t) is called the **heuristic searching method** if, in each operation step, it finds all the admissible operator applications for a given actual state, and the order of examining the states, and of applying those operators depends on the properties of the previously examined states. The analysis of the example, as well as the formal definition of the problem situation and its solution indicates, first of all, the need to formalize notions (proper representation of states and operators) introduced into the robot. Representation using calculus of the first order predicate, which will be presented in the next chapter, seems to be the most satisfactory of formulating problems of this type.

12.3.1. Calculus of the first order predicate

Predicate calculus is a mathematical theory particularly well suited for the formalization of descriptions of problem situation states, and of operators with their applicability conditions. The fundamental element of predicate calculus is an alphabet which consists of the following set of symbols:

- punctuation marks (),
- logical connectives ~ (no), $\vee$ (or), $\wedge$ (and), $\Rightarrow$ (if, then)
- quantifiers:
  $\forall$ universal quantifier,
  $\exists$ existential quantifier,
- symbols of variables x_k, $k = 1, 2, \ldots$,
- n-argument function symbols f_k^n,
  where $k \geq 1$, $n \geq 0$, and f_k^o is the symbol of a constant,

- n-argument predicate symbols p_k^n,
  where $k = 1, 2, ..., n \geq 1$.

In further examples, we shall, for convenience, write u, v, w, x, y, z instead of x_k, a, b, c, d instead of f_k^o; f, g, h, φ instead of f_k^n where $n > 0$; and P, Q, R, S instead of p_k^n.

Using the elements of the alphabet, various finite sequences of them can be built.

Some of these sequences , for example $(P(x) \vee Q(y)) \rightarrow R(z)$, are of greater interest to us than others, such as $\rightarrow a \vee f(x, y)$. Those which will be of interest to us are:

1. Terms which can be determined recurrently in the following way:
- any symbol of a variable, or of a constant is a term,
- if $t_1, t_2, ..., t_n$, where $n \geq 1$, are terms, then $f_k^n(t_1, ..., t_n)$ is a term too,
- an expression is a **term** if it is composed according to the above rules. For instance, the expression $f(a, g(x))$ is a term.
2. Elementary formulas (atoms), defined as follows:
   If p_k^n is a predicate symbol and $t_1, ..., t_n$ – terms, then $p_k^n(t_1, ..., t_n)$ is an **atom**.
3. Formulas that are built recurrently:
- an atom is a formula,
- if A and B are formulas and x is a variable, then each of the expressions $(\sim A)$, $(A \vee B)$, $(A \wedge B)$, $(A \Rightarrow B)$, $(\forall x A)$ and $(\exists x A)$ is also a formula.

An expression is a **formula** when it is composed according to the above rules. Examples of formulas are: $P(x)$, $Q(y) \vee R(a, z, h(u))$.

In order to bestow essence on a formula, it is interpreted as a theorem which concerns the subject domain considered.

The first interpretation of a formula consists of a non-empty set D called a subject domain together with the given values of all the constant, function and predicate symbols, occurring in the formula, according to the following rules:

- to any symbol of a constant, a certain element of set D is attributed,
- to any n-argument function symbol, a certain function from D^n to D is attributed, where $D^n = \{(x_1, ..., x_n); x_1, x_n \in D\}$,
- to any n-argument predicate symbol, a certain function from D^n to $\{T, F\}$ is attributed, where T and F denote True and False, respectively,
- if a formula is not an atom, then its value is calculated recursively, starting from the values of component formulas, noting that if A and B are formulas with known values, then the values of formulas $\sim(A)$, $A \vee B$, $A \wedge B$, $A \Rightarrow B$ are presented in the truth table 12.2.
- the value of formula $(\forall x)A$ equals T if A equals T for each $x \in D$, otherwise formula $(\forall x)A$ equals F,

- the value of formula $(\exists x)A$ equals T if A equals T for at least one $x \in D$, otherwise the value of formula $(\exists x)A$ is F,

Table 12.2. Values of formulas A, $A \vee B$, $A \wedge B$, $A \Rightarrow B$

A	B	$\sim A$	$A \vee B$	$A \wedge B$	$A \Rightarrow B$
T	T	F	T	T	T
F	T	T	T	F	T
T	F	F	T	F	F
F	F	T	F	F	T

As an example we shall consider the formula:

A: $((\forall x)(P(x) \Rightarrow Q(f(x), a))$

Formula A contains a constant a, a one-argument function symbol f, a one-argument predicate symbol P, and a two-argument predicate symbol Q. We then interpret formula A for the following data:

domain $D = \{1, 2\}$

values of a:

a
1

values of f:

$f(1)$	$f(2)$
2	1

values of P and Q:

$P(1)$	$P(2)$	$Q(1, 1)$	$Q(1, 2)$	$Q(2, 1)$	$Q(2, 2)$
F	T	T	T	F	T

We now evaluate the value of formula A.

If $x = 1$, then:

$P(x) \Rightarrow Q(f(x), a) = P(1) \Rightarrow Q(f(1), 1) = T.$

If $x = 2$, then:

$P(x) \Rightarrow Q(f(x), a) = P(2) \Rightarrow Q(f(2), 1) = T.$

Since the expression $P(x) \Rightarrow Q(f(x), a)$ is true for all $x \in D$, then formula A is true in this interpretation.

Clearly, some formulas can be true or false depending on the interpretation chosen.

A formula A is *consistent* if and only if the first interpretation is such that A assumes the value T in I. In this case we say that I is a model for A.

A formula A is a *tautology* if and only if for each interpretation I it assumes the value T.

A formula A is *inconsistent* if and only if for each interpretation I it assumes the value F.

The following examples are of consistent and inconsistent formulas, and tautologies:

$P(a) \Rightarrow \sim((\exists x)\, P(x))$ (consistent)

$((\forall x)\, P(x) \Rightarrow (\exists x)\, P(x))$ (tautology)

$(\forall x)\, P(x) \wedge ((\exists x) \sim P(x))$ (inconsistent)

In the process of task planning a robot must often answer the question of whether in the current state of a problem situation a certain operator can be used. Hypothesizing an affirmative answer to that question, the robot, applying what is called **mechanical** (formal) **theorem proving** (later discussed in this chapter), verifies this self-imposed hypothesis. The essential element of the formal method of proving theorems is a notion of **inference rules** we shall attempt.

Formula B is a consequence of formulas $A_1, ..., A_n$ (written as $\frac{A_1,...,A_n}{B}$) when for each interpretation for which formula $A_1 \wedge A_1 \wedge ... \wedge A_n$ is true, B is also true. Formulas $A_1, ..., A_n$ are called **premises** or **axioms,** and formula B is their **conclusion**.

The following theorems are true:

Th. 12.1.

$\frac{A_1, A_2, ..., A_n}{B}$ if and only if formula $(A_1 \wedge A_1 \wedge ... \wedge A_n) \Rightarrow B$ is a tautology.

and

Th. 12.2.

$\frac{A_1, A_2, ..., A_n}{B}$ in and only if formula $A_1 \wedge A_1 \wedge ... \wedge A_n \wedge \sim B$ is inconsistent.

Theorems 12.1 and 12.2 are quite important: from them we derive the proof that a certain formula is a logical consequence of a finite set of formulas is equivalent to the proof that a certain formula, related to these, is inconsistent.

This routine is used in mechanical theorem proving. To verify whetheror not a given formula is inconsistent, it is more convenient to express it in a certain specific form.

Formula $A_1 \vee A_1 \vee ... \vee A_n$ is called a **disjunction** of formulas $A_1, A_1, ..., A_n$. An atom or its negation is a letter. A disjunction of letters is called a **clause**.

Formula F is in a normal conjunctive form when F can be expressed as

$$F\colon (Q_1 x_1)\, (Q_1 x_1) ... (Q_1 x_1)\, M \qquad (12.2)$$

where

$Q_i x_i$ assumes the form of $\forall x_i$ or $\exists x_i$, and

M is a conjunction of clauses containing no quantifiers.

The following formula, in which a is a constant, is in normal conjunctive form:

G: $(\forall x)(\exists y)(\exists z)(P(x)\vee Q(y))\wedge(R(x, z)\vee P(a))$.

Every formula can be converted into the above form, using the rules for formula transformations of the first order predicate calculus.

Formula (12.2) is in standard Skolem form if it does not contain existential quantifiers.

The process of elimination of existential quantifiers is shown below. We assume that formula F is in the form of (12.2), and that Q_r, when $1\leq r\leq n$, is a existential quantifier. When there is no universal quantifier ($\forall$) to the left of Q_r, a new symbol of a constant, different from the rest of the symbols of constants included in M, is introduced, c, for instance, and all the x_r in M are exchanged for c, and the $(Q_r x_r)$ is canceled from F. When $Q_{s_1}, Q_{s_2}, \ldots, Q_{s_m}$ denotes the set of all the universal quantifiers occurring to the left of Q_r, that is $1\leq s_1 < s_2 \ldots < s_m < r$, a new m-argument function symbol, different from the rest of function symbols of formula F is introduced, f for instance, all x_r are exchanged for $f(x_{s_1}, x_{s_2}, \ldots, x_{s_m})$, and $Qr\,xr$ is canceled.

This process is repeated for any consecutive existential quantifier, until they are all eliminated.

The last formula obtained in this way is a **standard Skolem form**, also called, for simplification, a standard form F. Constants and functions used to exchange existential quantifier variables are called **Skolem constants,** or **functions**. As an example, we take the formula

$$(\exists x)(\forall y)(\forall z)(\exists u)(\forall v)(\exists w)(P(x, y, z, u, v, w)\vee Q(b))\wedge R(c)$$

The corresponding standard form is

$$(\forall y)(\forall z)(\forall v)(P(a, y, z, f(y, z), v, g(y, z, v))\vee Q(b))\wedge R(c)$$

where

a is a Skolem constant; while f and g are Skolem functions.

Since all the variables of the standard form formula are bound by universal quantifiers, and the order in which they occur is not important, they are omitted for the sake of simplification, yielding a quantifierless standard form formula.

In mechanical theorem proving as well, the set of clauses is treated as a conjunction of all the clauses in this set. This convention allows us to express the quantifierless standard form as a set of clauses. For example, quantifierless standard form of the last example be expressed by as a set of clauses of the form

$$\{(P(a, y, z, f(y, z), v, g(y, z, v))\vee Q(b))\wedge R(c)\}$$

A clause, that does not contain letters is empty, and an empty clause is assumed to be an inconsistent formula. We denote it by $\Box$. Theorem 12.3 is true [12.36]:

Th. 12.3

Let S denote a set of clauses that represent the quantifierless standard form of formula F. Then formula F is contradictory only when S is a contradictory set. According to the convention, set S is contradictory when the conjunction of clauses that form the set is contradictory.

12.3.2. Resolution

In the process of mechanical theorem proving it is often necessary to "adjust" certain formulas (represented by sets of clauses) to each other in order to prove the theorem.

Determining the terms to be substituted for variables in order to form identical expressions (formulas) is called **unification**.

An arbitrary substitution can be expressed as a set of ordered pairs in the form

$$s = \{t_1/v_1, \ldots, t_n/v_n\} \tag{12.3}$$

where

t_i is a term to be substituted for the variable v_i; and

n is the number of terms substituted.

The result of substituting s in formula A we denote as As. As an example, we take the formula $P(z, f(w), b)$ for which the following substitution has been applied:

$$s = \{x/z,\ g(y)/w\}$$

The new formula takes the form $Ps = P(x, f(g(y)), b)$.

We say that a set of formulas $\{A_i;\ i = 1{:}n\}$ is unifiable if there is a substitution s for which $A_1s = A_2s = \ldots = A_ns$. In this case s is an unifier because it reduces set $\{A_i;\ i = 1{:}n\}$ to one element. For example, the substitution $s = \{f(a)/x,\ b/y\}$ unifies set $\{R(f(y), x),$ $R(f(b), f(a))\}$ yielding, as a result, a single element of the form $R(f(b), f(a))$.

The notion of a minimal unifier (the most general unifier), which will be useful later, can be explained in a simple example:

Let us assume a given set of formulas: $\{P(x, y), P(a, y)\}$. The substitution $s = \{a/x,$ $b/y\}$ unifies this set, but in a sense it is not the simplest one, as the second substitution $s = \{a/x\}$ unifies this set as well. This second substitution is called a minimal unifier.

Let C_1 and C_2 denote two clauses (called premise clauses) that do not contain common variables. Let B_1 and B_2 be letters contained in C_1 and C_2, respectively. Let us assume that B_1 and $\sim B_2$ have the minimal unifier s.

A resolution [12.36] is an inference rule of the form

$$\frac{C_1, C_2}{(C_1 s - \{B_1 s\}) \cup (C_2 s - \{B_2 s\})} \tag{12.4}$$

The deduction of the resolution inference rule is called a **binary resolvent**.

Let us now consider the examples $C_1 = P(x) \vee Q(x)$ and $C_2 = \sim P(a) \vee R(x)$. Since the variable x appears in C_1 as well as in C_2, it is exchanged in formula C_2. We let, for example, $C_2 = \sim P(a) \vee R(y)$, and have $B_1 = P(x)$ and $B_2 = \sim P(a)$.
B_1 and $\sim B_2$ have a minimal unifier $s = \{a/x\}$. Hence:

$$C_1s - \{B_1s\} = \{Q(a)\},\ C_2s - \{B_2s\} = \{R(y)\} \text{ and}$$

$$(C_1s - \{B_1s\}) \cup (C_2s - \{B_2s\}) = \{Q(a), R(y)\} = Q(a) \vee R(y).$$

The last equality results from the previously accepted convention. The resolution is proved by the truthfulness of the given formula:

- reducing it to the standard form,
- contradicting opposing formulas,
- entering an empty resolvent □ by applying a resolution.

As an example we shall execute a resolution proof of the theorem: "Show that the trapezoid's opposite inner angles determined by its diagonal are equal" (Figure 12.6).

In order to prove this theorem, we shall axiomatize it: Let $T(x, y, u, v)$ denote a predicate, and the points x, y, u, v determine a trapezoid with the upper left vertex x, the upper right vertex y, the lower right vertex u and the lower left vertex v. We then let $P(x, y, u, v)$ denote a predicate, the interval xy is parallel to the interval uv , and $E(x, y, z, u, v, w)$ denote a predicate, $\angle xyz$ equals $\angle uvw$, to obtain the following axioms:

A_1: $(\forall x)\,(\forall y)\,(\forall u)\,(\forall v)\,(T(x, y, u, v) \Rightarrow P(x, y, u, v))$, the trapezoid definition.

A_2: $(\forall x)\,(\forall y)\,(\forall u)\,(\forall v)\,(P(x, y, u, v) \Rightarrow E(x, y, v, u, v, y))$, the angles determined by the intersection of two parallel lines by a non-parallel one are equal.

A_3: $T(x, y, u, v)$ – points a, b, c, d form a trapezoid.

On the basis of these axioms we wish to prove that

$$A_1 \wedge A_2 \wedge A_3 \Rightarrow E(a, b, d, c, d, b)$$

is a true formula. According to the resolutive proving method we contradict the last formula and try to prove that formula

$$A_1 \wedge A_2 \wedge A_3 \wedge \sim E(a, b, d, c, d, b)$$

(see Th. 12.1, 12.2) is inconsistent.

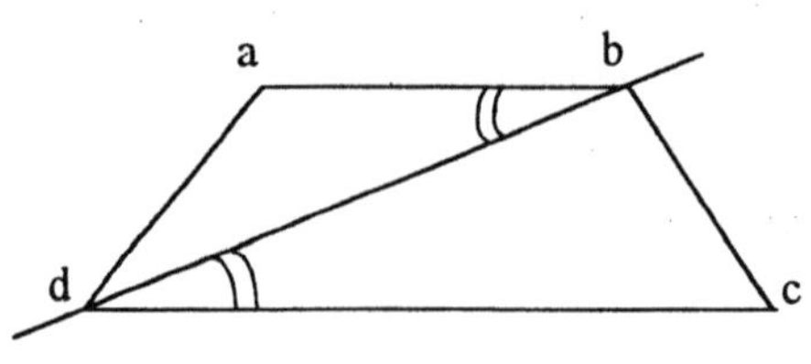

Fig. 12.6. Trapezoid from the above example

In order to do this, we transform axioms A_1, A_2, A_3 and the contradiction of the theorem to the quantifierless standard form

$S = \{\sim T(x, y, u, v) \vee P(x, y, u, v),\quad \sim P(x, y, u, v) \vee E(x, y, v, u, v, y),$
$T(a, b, c, d),\quad \sim E(a, b, d, c, d, b)\}$

Now, using the resolution method, we prove that set S is inconsistent.

(1)	$\sim T(x, y, u, v) \vee P(x, y, u, v)$	the clause from S
(2)	$\sim P(x, y, u, v) \vee E(x, y, v, u, v, y)$	the clause from S
(3)	$T(a, b, c, d)$	the clause from S
(4)	$\sim E(a, b, d, c, d, b)$	the clause from S
(5)	$\sim P(a, b, c, d)$	the resolvent from (2) and (4)
(6)	$\sim T(a, b, c, d)$	the resolvent from (1) and (5)
(7)	$\Box$	the resolvent from (3) and (6)

Since the last clause derived from S is an empty clause $\Box$, we conclude that set S is inconsistent, and, consequently, the original theorem is true. The successive steps of the proof are shown in the form of a tree in Figure 12.7.

This tree is called a deduction tree. The tree of the proof from set S of clauses is a downward growing tree in which a clause from S is attributed to each initial node, and a resolvent of the clauses taking part in the resolution to each successive node. One of the algorithmic methods of determining an empty clause is called **linear resolution**. Given set S of clauses, as well as the clause $C_0 \in S$, an inference, which is presented graphically in Fig. 12.8, and is called a linear proof from set S of clauses with a top clause C_0, where

1. for $i = 0{:}n\text{-}1$, C_{i+1} is the resolvent of C_i (called a central resolvent) and B_i (called a side resolvent), noting that
2. for $i = 0{:}n\text{-}1$, $B_i = C_j$ for some $j < i$ or $B_i \in S$.

It is easy to show that the tree in Figure 12.7. represents a linear resolution proof with a top clause $\sim E(a, b, d, c, d, b)$.

As can be seen, linear proof theorem proving is the equivalent of searching for node $\Box$ in the tree. We now present, in mechanical proving language, one of the many algorithms [12.35], [12.36] of this search: a heuristic algorithm A^* [12.24], which considerably reduces the number of resolvents created while searching for clause $\Box$. For this purpose, let us assume that we have a function with values in the real positive number set given in the form

$$f(C_i) = g(C_i) + h(C_i), \tag{12.5}$$

where:

C_i is a resolvent obtained from the top clause after a i-th process of resolving;

$g(C_i) = i$, $h(C_i)$ is a certain heuristic function which is a measure of the distance of resolvent C_i from the terminal resolvent $\Box$.

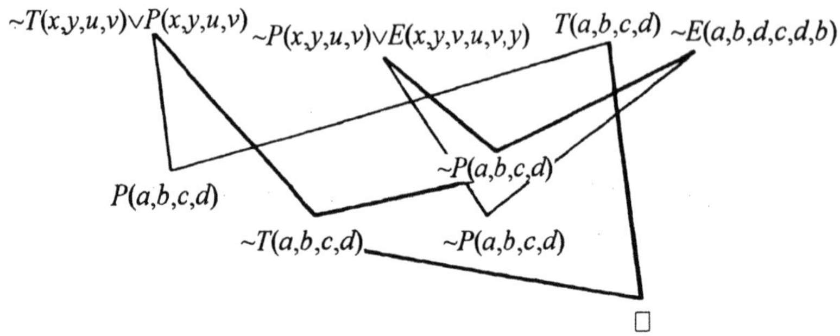

Fig. 12.7. Deduction tree

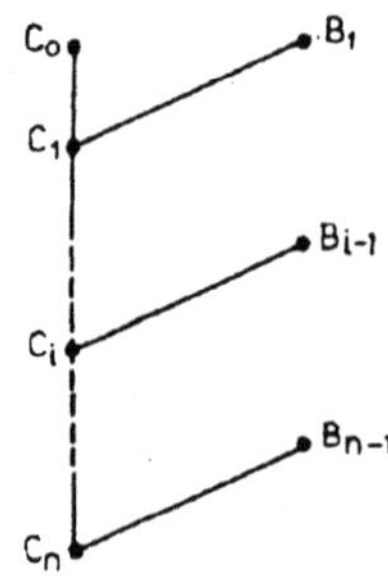

Fig. 12.8. Operation diagram of linear resolution

The heuristic function can be stated in the form of an expression

$$h(C_i) = w_1 h_1(C_i) + w_2 h_2(C_i), \tag{12.6}$$

where:

w_1 and $w_2 \geq 0$ are constant coefficients,

$h_1(C_i)$ is the "length" of clause C_i measured by the number of literals it contains, and $h_2(C_i)$ is the degree of complexity of clause C_i (for example, the greatest number of compositions of functions occurring in C_i).

Algorithm A* uses two lists OPEN and CLOSED, which are empty in the beginning of its course, and an input set of clauses from set S.

The following notation is introduced:

C_0 is the top clause from set S, and

:=, an operation of attribution in the usual pascal notation.

The course of the algorithm is:

1. OPEN:= OPEN $\cup$ $\{C_0\}$.
2. If OPEN is an empty list, finish the algorithm – the proof does not exist. Otherwise compute: $f(C) = \min_{C' \in \text{OPEN}} \{f(C')\}$
3. If $C = \Box$, CLOSED:= CLOSED $\cup$ $\{C\}$ and finish the algorithm. The goal has been reached.

4. Otherwise: CLOSED:= CLOSED $\cup$ $\{C\}$, OPEN:= OPEN – $\{C\}$.
5. Find all clauses in set S which can be resolved with clause C. If there are no such clauses, go to step 2. Otherwise perform the resolution of these clauses with clause C. Let R_1, R_2, ..., R_m denote the resolvents obtained (in the graph notation the last operation is equivalent to generating from node C all its admissible successors, that is nodes R_1, R_2, ..., R_m). Compute $f(R_i)$ for $i = 1{:}m$. If $R_j \notin$OPEN for a certain $j \leq m$, then OPEN:= OPEN $\cup$ $\{R_j\}$ while if for a certain $k \leq m$, $R_k \in$CLOSED, then compare the newly computed value of $f(R_j)$ to the value already attributed to R_j and denoted by f_s. If $f_s > f\{R_j\}$, then OPEN:= OPEN $\cup$ $\{R_j\}$ and CLOSED:= CLOSED – $\{R_j\}$. Go on to step 2.

In the case when set S is inconsistent, algorithm A^* will find the proof the least possible number of steps (depending upon the heuristic function used).

In elementary operation planning it is often necessary to prove a theorem in which some of the variables appearing in it are treated differently (these special variables correspond to certain operator parameters). In order to perform such proofs, the process of unification must be modified as follows:

- terms which can be substituted for variables are: variables, constants, parameters and functional terms without variables;
- terms which can be substituted for parameters are: constants, other parameters, functional terms with neither variable parameters nor Skolem functions;
- the substitution of whatever for a parameter in a certain formula must be performed in the whole given set of formulas as well.

In the next subchapter the resolution method of theorem proving will be used to plan the elementary operations of a robot.

12.3.3. Generating an elementary operation plan

We take example of the planning system STRIPS to describe the method of elementary operation planning addressing the performance of a task (final situation) assigned by the user.

By operation planning system we mean an algorithm which is used to have its input information that concerns:

- a description of an initial problem situation,
- a set of operator descriptions (models of operation),
- a description of a final problem situation,

and whose output is a plan (or information that a plan cannot be obtained). By M we denote a set of all the possible states of the problem situation. By plan is understood a directed graph that satisfies the following conditions:

1. each arc is associated with a certain operator $f_i \in F$ – a set of operators (in a general case an operator f_i can be parametrized by a set of parameters $f_i(p_1^i, \ldots, p_k^i)$, where p_j^i is the j-th parameter of an i-th operator, $j:=1:k$);
2. a formula G_k, describing a certain subset M_k in the space of states M, is associated with each node n_k;
3. arcs associated with only one operator come from a single node;
4. for each node n_k, the corresponding set M_k is contained in the admissible domain of operator f_j associated with the arc (arcs) coming from node n_k;
5. the graph has a node n_i describing the state of an initial situation. A plan that satisfies an additional condition: on the plan there is a path linking node n_i with node n_l describing the state of the final problem situation, is called a general plan.

As we have said before, the state of the problem situation is a set of facts concerning the objects to be found there, as well as, the relationships between them. The state is described by pointing the set of formulas written in the standard normal form. Below, there is a description of the state of a simple situation, as shown in Figure 12.9, *in(A, r)* denotes a formula in which object A is in position r relative to system OX_1X_2. We established earlier that an operator models an elementary operation. Operators are represented in the STRIPS system by a list consisting of three sub-lists. The first describes conditions that must be satisfied in a given state for the operator to be used. This is so called the precondition list (formula). The other two are descriptive lists.

The first of these, called a delete list, contains formulas which cease to be true after employing the operator. The second, an add list (formula), contains facts which become true after employing the operator. The main goal, the final state of the problem situation, is represented, like other states, in the form of a set of conditions describing it. The work of the planning system proceeds according to the algorithm shown in the diagram in Fig. 12.10, in which the following notation has been assumed: G is the current (top) aim; LC, a list of top goals; W, a current node in a graph; M, a current description of a state (a model) (Fig. 12.10).

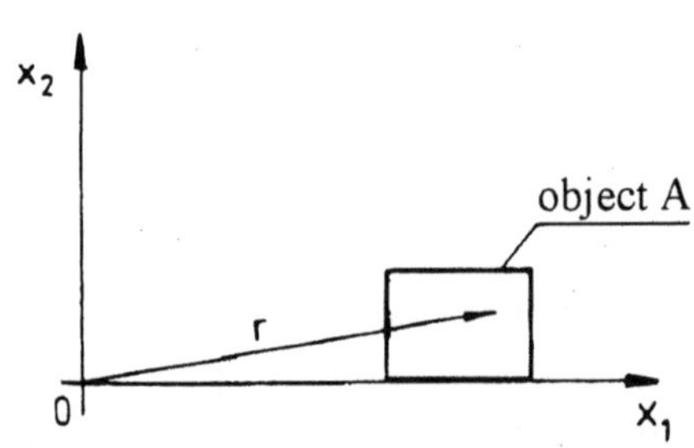

Fig. 12.9. Example of scene description that results of an operator activity on a given state.

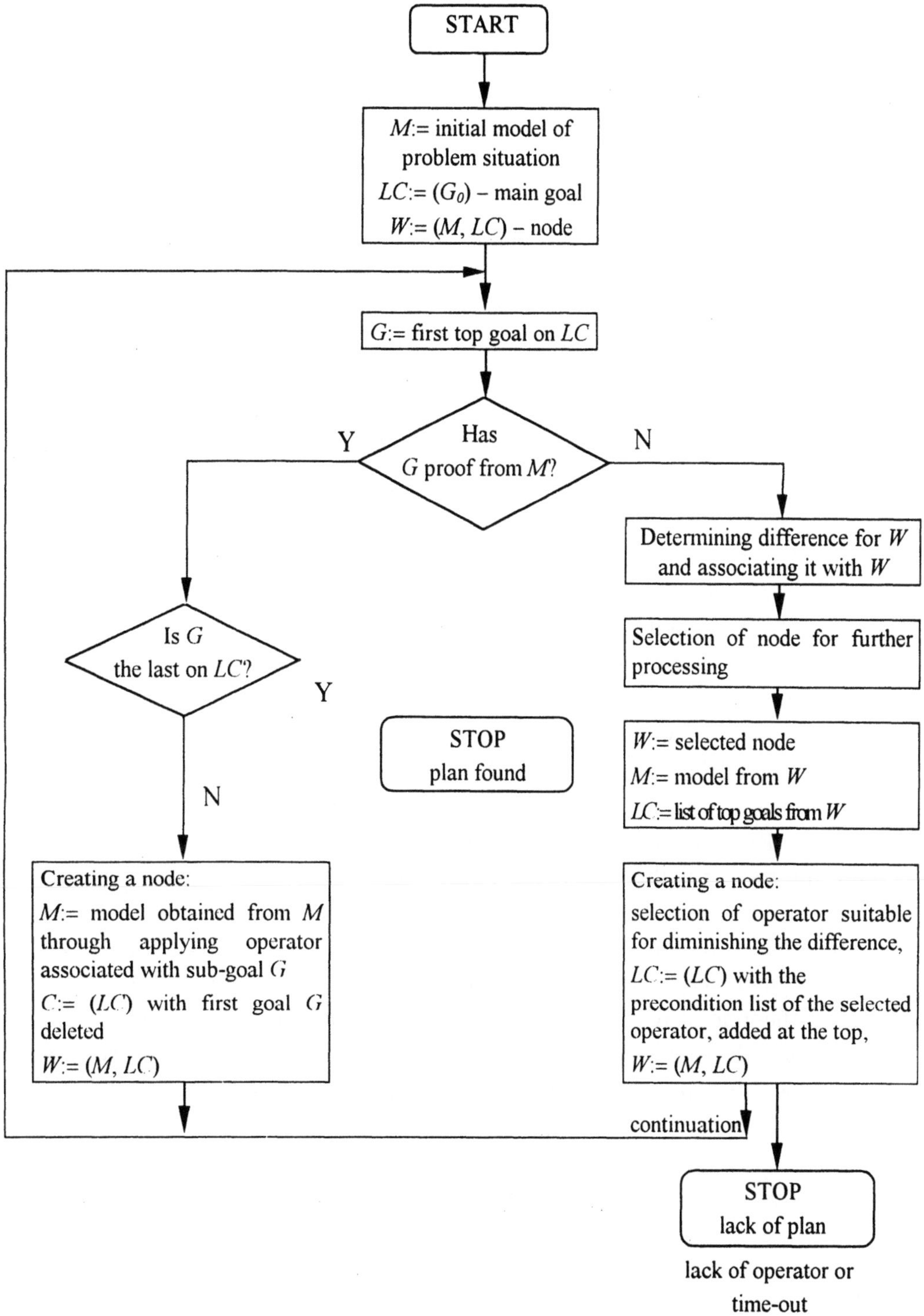

Fig. 12.10. Flow chart of system STRIPS

During operation, the system forms the search tree whose nodes have the form $(M,(G_i, G_{i-1}, ..., G_0))$, where M is always a model of a problem environment obtained from the initial model employing all the operators already used on a given branch of the tree, while a list of top goals $(G_i, ..., G_0)$ consists of indirect aims with the help of which the system will try to achieve its main goal (in the expansion of the given branch). The first top goal from the list is always considered first; in the case when its achievement prove impossible, a new sub-goal is generated (and added at the beginning of the list).

The course of the algorithm, in brief, is: we begin with an attempt to prove that the main goal formula G_0 is satisfied in the initial model M_0. At the same time an initial tip (root) of the search tree is created in the form: $(M_0,(G_0))$. The theorem proving subsystem acting on the principle of resolution is called to carry out the proof. If it is successful, the system ends its work, and the plan generated contains no operator. In the non-trivial case (failure of proof), a difference in the form of an unfinished proof of the theorem G_0 from M_0 is attributed to node W (how to find that difference will be discussed later). Then an operator, which system considers useful for decreasing the difference, i.e. continuing the proof (how that operator is chosen, will be described later), is selected.

Obviously, after the first algorithm loop there is only one node, one difference, and this is the only one that can undergo decreasing. In subsequent cycles there will be several such nodes on the tree, and the selection of the operator must be preceded by the selection of the node for which the operator is to be chosen. The selection of the node is based on a certain heuristic function that evaluates what chances the particular nodes stand of finding the plan. When the node and then a suitable operator for reducing the difference related to the node have been chosen, the system sets itself sub-objective of that operator's application by creating a new node whose first goal is the operator's precondition (when there are several operators, all of these are included in the tree, forming independent branches). The algorithm resumes its theorem proving, and again attempts to prove that the first sub-goal from the list derives from the model. If it fails to do so, the whole above procedure is repeated: the difference is determined, the node associated with the difference, a new node selected for further processing, an operator (operators) selected and a new sub-goal generated.

If, instead, it is successful (in the case discussed that would mean that the operator chosen could be used immediately), other operations are performed: a simulation is carried out with the operator applied for the model in the node considered. This creates a new model which, with the preceding list less the first goal (that concerned the preconditions of the operator just used) forms a new node.

Therefore, the list`s description of the state M can be called the current model. It is updated with the introduction of each successive operator. The algorithm returns to the theorem proving phase, and considers the next goal on the list. The subsequent action depends on the result of the proof attempt. The job is done when the last sub-goal (i.e. the main goal) is achieved, or when no suitable operator can be selected for any difference on the tree of unexpanded nodes or, for practical reasons, when the time allotted or computation is exceeded. Clearly, we can speak of a success only in the first case.

The difference is determined when the theorem proving procedure does not find the proof for set $M \cup \{\sim G\}$ (does not generate an empty clause) within the specified time. That is, the algorithm fails to generate the desired empty clause. Considering the proving process as a decision tree, it is this incomplete decision tree, i.e. the tree's set of clauses itself, that the STRIPS system regards as this node's difference.

The selection of operators proceeds as follows: we have a current model M of a problem situation, the first (tip) goal on the list and the current difference D between them in the form of a set of clauses from M, G and their resolvents. From the accessible operators we must choose the one that can be expected to be helpful in decreasing of the difference. The conditions of operator selection can be comprised in a short algorithm in which during the first step the clauses from the difference D are compared to the clauses on the operator`s add lists. If the clause containing the contradiction of the position predicate appears in the difference, then during the first step of the algorithm all the operators associated with motion (as their add lists contains clauses describing new positions) will be selected. The second step of the operator selection algorithm employs the proving procedure to establishing whether the clauses from the add lists of the operators selected in the first round will actually enable the generation of new resolvents in difference D. If this is the case, then the specified operators with the parameters already adjusted (not all parameters must be changed) are considered as suitable for diminishing difference D.

Note that a single operator can cause different results in diminishing the difference depending upon the substitution made in parameters.

Let us illustrate the algorithm with an example that is not too complicated

Given the initial problem situation shown in Figure 12.11 and the desired final situation shown in Figure 12.12, the robot's task is to move the object to position c.

The model of the initial situation can be written in the form

$$M_0 = \begin{cases} \sim ingripper \\ in\left(gripper,\, a\right) \\ in\left(object,\, b\right) \end{cases}$$

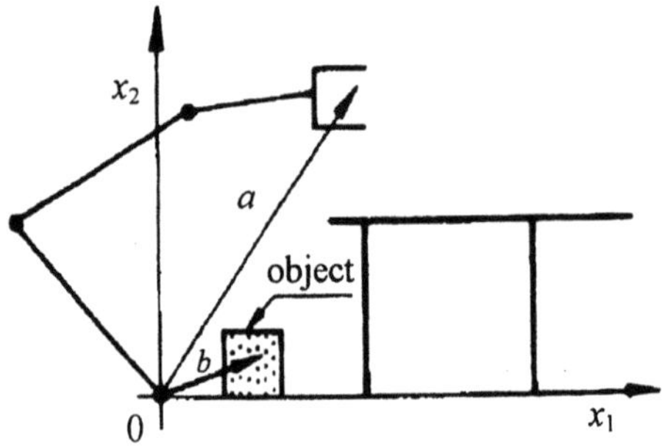

Fig. 12.11. Initial situation

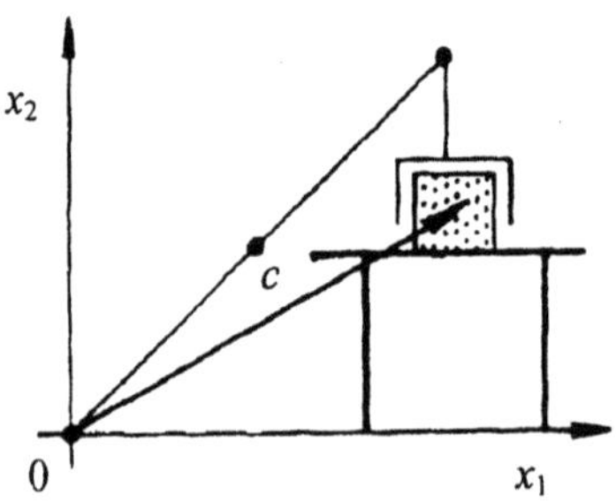

Fig. 12.12. Desired final situation

where:

ingripper is a predicate of the value T if the gripper is holding the object;

in (gripper, a) is a predicate denoting that the gripper is in position a ;

in (object, b) is a predicate stating that the object is in position b;

while the goal formula corresponding to the final situation is:

$G_0 = in\ (object,\ c) \wedge \sim ingripper$

The perators (models of operation) available for the robot are:

1. move (t, u) – displacement of the empty gripper from position t to position u,
   - a precondition list – p.l.: ~ *ingripper* ∧ *in (gripper, t)*,
   - a delete list – d.l.: ~ *ingripper* ∧ *in (gripper, t)*,
   - an add list– a.l.: ~ *ingripper* ∧ *in (gripper, u)*,
2. carry (x, v) – displacement of the gripper holding the object from position x to position v,
   - p.l.: *ingripper* ∧ *in (gripper, x)* ∧ *in (object, x)*,
   - d.l.: *ingripper* ∧ *in (gripper, x)* ∧ *in (object, x)*,
   - a.l.: *ingripper* ∧ *in (gripper, v)* ∧ *in (object, v)*,
3. pickup (z) – grasping the object in position z,
   - p.l.: ~*ingripper* ∧ *in (gripper, z)* ∧ *in (object, z)*,
   - d.l.: ~*ingripper* ∧ *in (gripper, z)* ∧ *in (object, z)*,

- a.l.: *ingripper* ∧ *in* (*gripper*, *z*) ∧ *in* (*object*, *z*),

4. putdown (*y*) – releasing the object in position *z* of the gripper,
   - p.l.: *ingripper* ∧ *in* (*gripper*, *y*) ∧ *in* (*object*, *y*),
   - d.l.: *ingripper* ∧ *in* (*gripper*, *y*) ∧ *in* (*object*, *y*),
   - a.l.: ~*ingripper* ∧ *in* (*gripper*, *y*) ∧ *in* (*object*, *y*).

System STRIPS begins by creating an initial node (M_0, (G_0)) and attempting to prove G_0 with M_0 (i.e. to prove that the set $M_0 \cup \{G_0\}$ is inconsistent), using the rule of resolution.

After forming the deduction tree

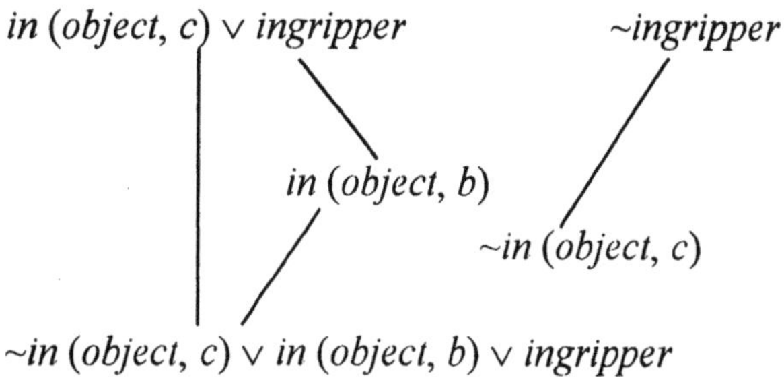

the system abandons work on the proof, saving the incomplete tree as a difference for node (M_0, (G_0)). Its review of the operators has yielded only the one selected in the first step of the operator procedure *carry* (*x*, *v*). The selection was made using the predicate *in* (*object*, *v*) from the operator add list. As can be seen, different substitutions of parameter *v* give different results. Let us assume that the substitution *v* = *c* has been made.

Thus, the operator takes the form

op1: *carry* (*x*, *c*).

The preconditions of this operator are

G_1 = *ingripper* ∧ *in* (*gripper*, *x*) ∧ *in* (*object*, *x*).

Having created set $M_0 \cup \{G_1\}$ we form a new deduction (resolution) tree.

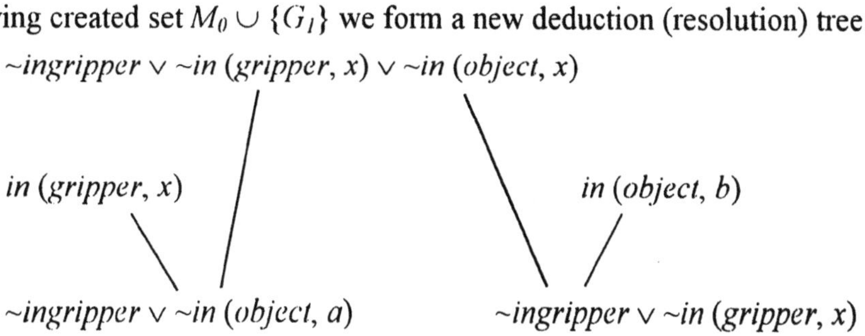

Since an empty clause has not been generated, the above tree is treated as a difference for the second node of the form of (M_0, (G_1, G_0)). The search graph now created takes the form

$(M_0, (G_0))$

$(M_0, (G_1, G_0))$

For further analysis we may once more choose the first node (there are also other cases of operator *carry*) or take the second one. Let us assume that the second node is chosen (we would be able to affect the selection of the node after creating a node) selection heuristic function which has not been introduced). Review of the operators yields two operators: *move* and *carry* which turn out to be useful in the next step. For simplicity, let us assume that operator *move* has been selected as first. Taking into account a substitution $u = b$ in the predicate *in* (*gripper*, *u*) we choose the operator

op2: *move* (*t*, *b*).

We create a new sub-goal

$G_2 = in\ (gripper,\ t) \wedge \sim ingripper$

and a new node $(M_0, (G_2, G_1, G_0))$ in the graph tree:

$(M_0, (G_0))$
$(M_0, (G_1, G_0))$
$(M_0, (G_2, G_1, G_0))$

The proof G_2 from M_0 associated with this node proves successful.
After two steps of resolution we have:

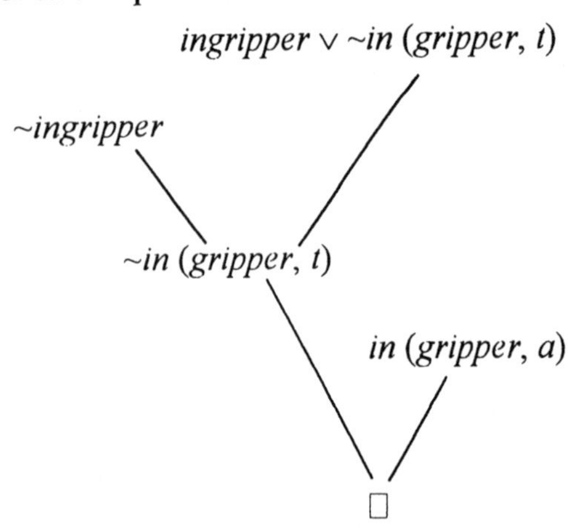

Hence, *op2* can be used for M_0. Having carried it out according to the delete and add lists we have

$$M_1 = \begin{cases} \sim ingripper \\ in(gripper, b) \\ in(object, b) \end{cases}$$

as well as a newly created node $(M_0, (G_1, G_0))$. Proof G_1 from M_1 fails to give a result

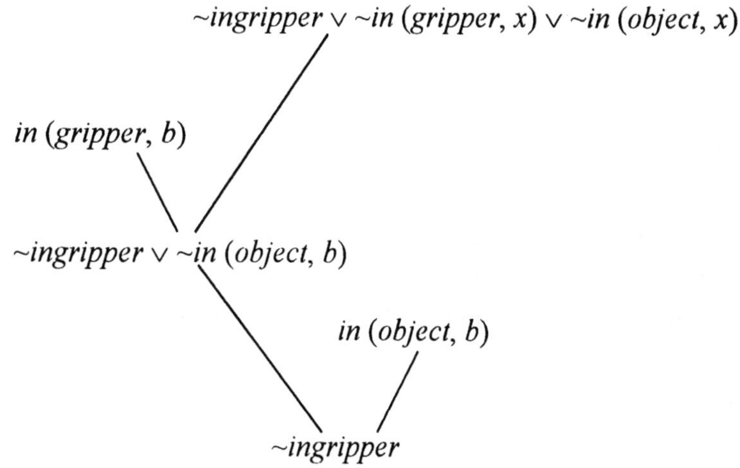

The review of the operators for the above difference yields operator *pickup* with a substitution $z = b$

op3: *pickup* (b).

The operator's preconditions yield a new sub-goal

$G_3 = \sim ingripper \wedge in\ (gripper,\ c) \wedge in\ (object,\ c)$

for which the proof from M_1 is a success. We then create a model of the problem situation after applying *op3*. It takes the form

$$M_2 = \begin{cases} ingripper \\ in(gripper, b) \\ in(object, b) \end{cases}$$

Proof G_1 from M_2 is not difficult now

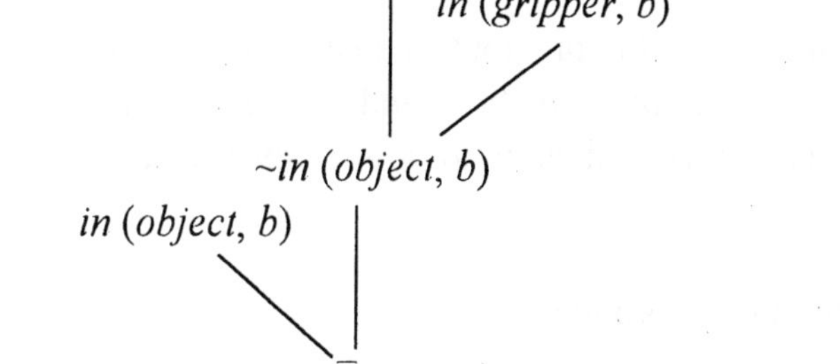

and *op1* can be applied in M_2 yielding

$$M_3 = \begin{cases} ingripper \\ in(gripper,c) \\ in(object,c) \end{cases}$$

Proof G_0 from M_3 is not successful. Operator *putdown* (y) is selected with a substitution $y = c$ and a corresponding sub-goal

$$G_4 = wgripper \wedge w\,(gripper,\, c) \wedge w\,(object,\, c)$$

Proof G_4 from M_3 is succ7essful. As a result, after using *op4* we have

$$M_4 = \begin{cases} \sim ingripper \\ in(gripper,c) \\ in(object,c) \end{cases}$$

The proof of the goal formula G_0 in model M_4 is immediate. The complete search graph takes the form

$(M_0, (G_0))$
$(M_0, (G_1, G_0))$
$(M_0, (G_2, G_1, G_0))$
$(M_1, (G_1, G_0))$
$(M_1, (G_3, G_1, G_0))$
$(M_2, (G_1, G_0))$
$(M_3, (G_0))$
$(M_4, (G_4, G_0))$
$(M_4, (G_0))$

and the solving sequence of operators is:

op2 move (a, b)
op3 pickup (b)
op1 carry (b, c)
op4 putdown (c)

In real problem situations model M_0 contains a number of formulas and in order to generate a plan it is necessary to use the appropriate heuristic functions, to eliminate the many redundant cases.

Combining grasp planning (discussed in the previous chapter) with elementary operation planning yields the sequence of gripper positions and orientations, needed to carry out the task. This information is delivered to lower level control for trajectory planning.

Exercises

1. Reduce to the normal form the following formula:

$$(\forall x)(P(x) \Rightarrow (\exists z)(Q(x, z) \vee R(x)))$$

2. For $D = \{1, 2\}$ and formula

$$P(a) \Rightarrow \sim(\exists x)P(x)$$

indicate one interpretation for which this formula has a model, and a second one for which it is false.

3. Reduce the following formulas to the standard form:

a) $((\forall x)(P(x) \Rightarrow (\exists y)(\forall z)(Q(y, z))$

b) $((\forall x)(P(x) \Rightarrow (\exists y)P(y)$

4. Let F_1 and F_2 be formulas in the form:

$$F_1\text{: } ((\forall x)(P(x) \Rightarrow Q(x))$$

$$F_2\text{: } \sim Q(a)$$

Using the resolution method prove that $P(a)$ is a conclusion from formulas F_1 and F_2.

5. Find all the resolvents of the formulas below:

a) $C = \sim P(x) \vee Q(x, x)$, $\quad D = \sim Q(a, f(a))$

b) $C = \sim P(v, z, v) \vee P(w, z, w)$, $\quad D = P(w, h(x, x), w)$

12.4. Trajectory planning

Planning a trajectory is a matter of finding a time-parametrized curve in the joint space. That curve connects the initial configuration with the desired final configuration of the manipulator. Sometimes a trajectory can be found with the knowledge of the gripper path in the robot base system.

To perform the process of trajectory planning, we must have (Fig. 12.13):

- a geometrical description of the manipulator, gripper and the objects in the workspace, obstacles in particular;
- kinematic and dynamic equations for the manipulator, which determine its working capabilities;
- the performance index used according to which an optimal trajectory is determined.

Trajectory planning (Figure 12.13) is one of the most difficult tasks in robotics and, consequently, is usually divided into two main phases: spatial planning of the trajectory, described by joint coordinates, with a corresponding gripper path in the workspace, connecting the gripper's initial position with its final one; and its time-parameterizing (the phase of planning that takes time into account).

Although the particular stages (indicated in blocks) of the planning process are shown separately, in practical applications sometimes they can coincide.

The block for converting the description of the workspace is usually introduced to determine the algorithms for the spatial planning of the trajectory (see below). Its existence is a consequence of the fact that the workspace and manipulator kinematic equations are described in different coordinate systems. As the workspace where the robot performs the manufacturing operations usually contains other objects, or obstacles, a major issue in the spatial trajectory planning block is the problem (which we leave to latter discussion) avoiding the collision of manipulator links with these objects.

Knowledge of the geometrical description of the workspace, and of the kinematic and dynamic equations of the manipulator inspires two general approaches to the spatial planning of its trajectory. In the first, the geometric description of the workspace and the manipulator is usually transformed into its joint space coordinate (also called the joint space) by the kinematic equation.

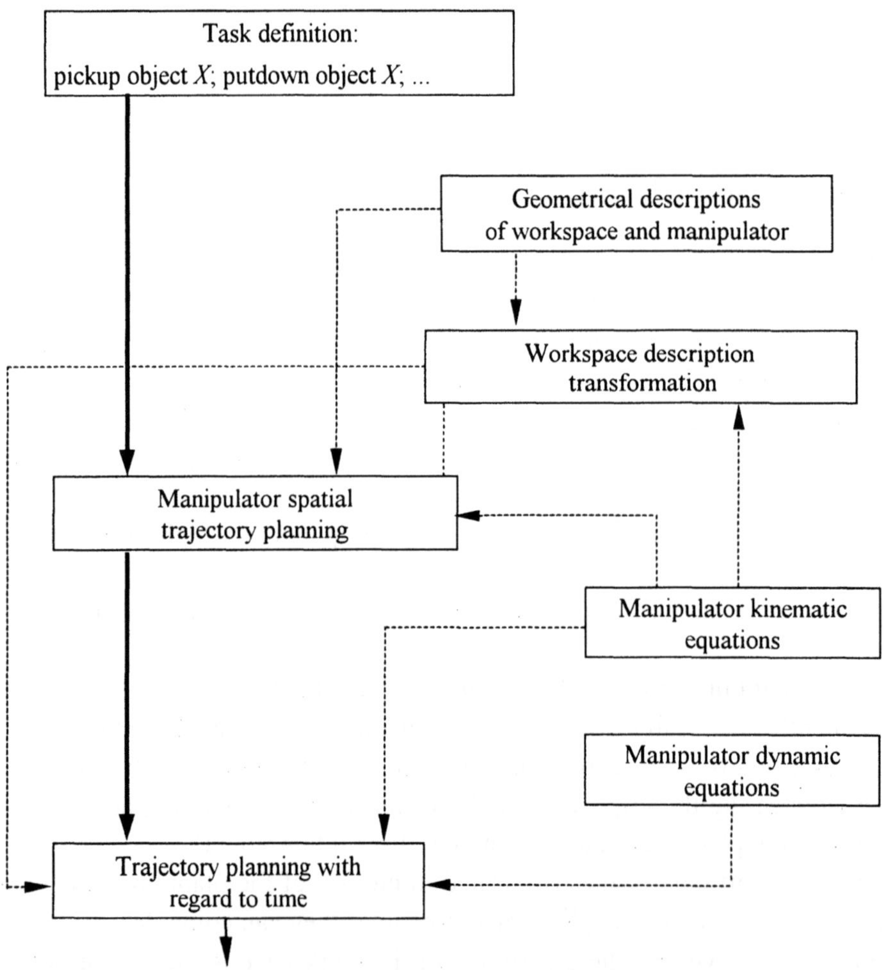

Fig. 12.13. Flow chart of trajectory planning

Transforming the obstacles (objects, given in the base system) in the workspace, we obtain a set of points given in joint space coordinates called **spatial configuration obstacles**. The manipulator configuration, regarded as a system of linked, rigid bodies, is described in the joint space as one element (point). The sum of the configuration obstacles determines the free space in the joint space, that is, a **set of collision-free manipulator configurations**.

In this way the problem of manipulator trajectory planning in the workspace is reduced to the problem of determining the collision-free motion of a point within the free configuration space. The solutions used differ in the ways in which this free space is represented, and in their search within it.

The second approach to manipulator trajectory planning (in which the separation between the spatial trajectory planning phase and the time-parameterization of the trajectory is less clear) is realized in the workspace. Since the free configuration space is not known explicitly in this case, constraints expressed in both the joint space coordinates (regarding mechanical limitations) and the workspace (regarding obstacles) are imposed on the planned motion of the gripper and manipulator links.

The result of spatial trajectory planning is a curve (most often a quasi linear one) representing a set of collision free manipulator configurations (corresponding to the successive phases of the gripper motion to achieve a final position) and connecting the manipulator initial configuration with the final one. The next step is time-parameterization of the curve determined (trajectory planning with regard to time). This problem is usually solved by assuming a certain criterion (usually time minimization) according to which the trajectory is to realized. The division of trajectory planning methods is illustrated in Fig. 12.14.

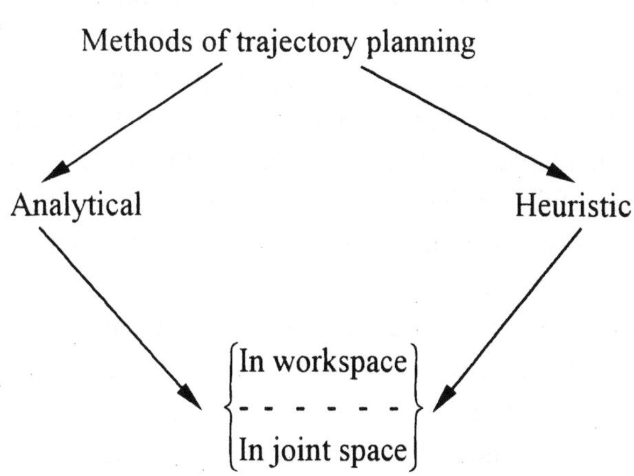

Fig. 12.14. Division of trajectory planning methods

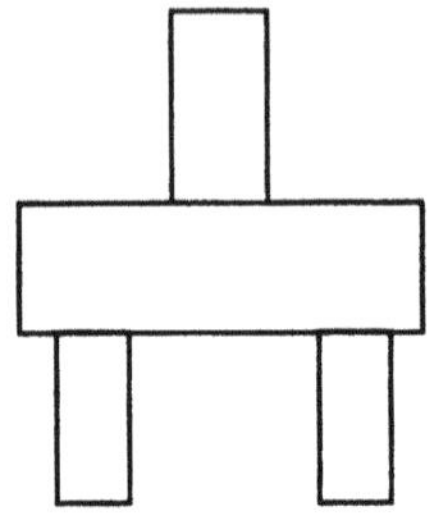

Fig. 12.15. Two-dimensional gripper representation made of elementary figures

An important issue of trajectory planning, affecting the complexity of the algorithms used, is the geometric description of the workspace, and of the robot itself. Proper selection of this description can significantly simplify the planning method.

The descriptions most commonly used in practice are

- A representation based on the elementary solid geometry [12.1, 12.2] in which, given original set of elementary solids (described analytically) given, such as segments of straight lines, cubicoids, spheres, cones, cylinders, constituting the lowest level of the workspace representation, and given some operations (sums of solids and complements of one solid in another), new solids (a higher level of the workspace representation) can be created. Figure 12.15. shows a two-dimensional representation of a gripper as of a sum of rectangles.
- A representation of the workspace through oct-trees [12.3, 12.4] which show the successive decompositions of space (originally given in the shape of a cube) into certain sub-spaces. Initially the whole space is represented by a single node (a root) of the tree (a particular case of a graph). When this space is fully occupied by a workspace object (an obstacle, the manipulator link), or entirely free, the tree root is not decomposed any further, but labeled "full" or "empty", respectively. A partially occupied, or empty space, is labeled "mixed" and divided into eight identical octans which become successors of the root in the tree. This process is repeated until all the nodes in the tree are homogeneous, i.e. they are all tagged "full" or "empty" (or a given threshold is reached). Figures 12.16 and 12.17 show two-dimensional workspace and its representation as a quadruple tree (in this particular case).
- A representation of the workspace as packed with spheres [12.5, 12.6]; the objects in the space are approximated by spheres of different radiuses. The result of this sphere approximation of the workspace is a graph whose nodes are the centers of the spheres and whose branches constitute a link between two adjacent spheres (Fig. 12.18).

- A representation of the workspace based on polyhedrons [12.7, 12.8, 12.9]; in the most general case, objects are approximated by convex polyhedrons (or a group of them) specified by their boundary planes, the intersections of which determine the polyhedron edges, whole the intersections of the edges determine the nodes (Fig. 12.19).

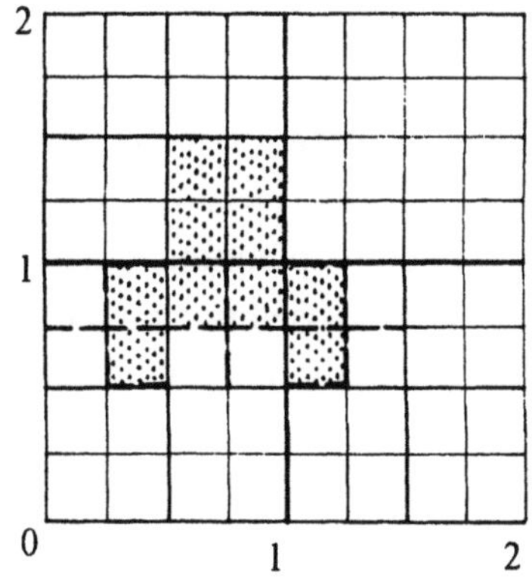

Fig. 12.16. Example of two-dimensional workspace

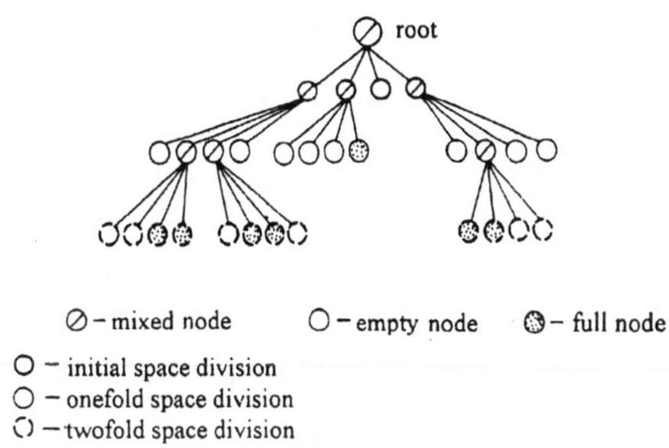

Fig. 12.17. Representation of the workspace from fig. 12.16 as a model tree

Fig. 12.18. Sphere approximation of workspace

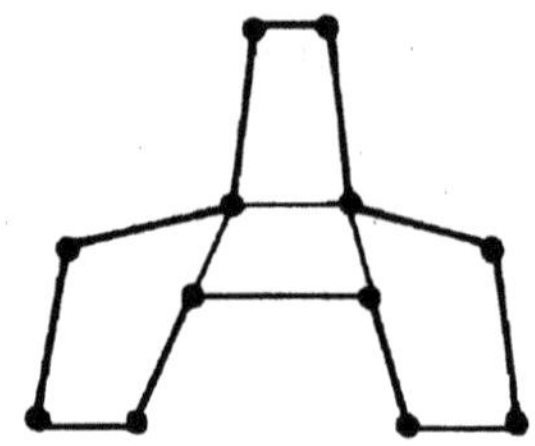

Fig. 12.19. Approximation of the workspace as convex polygons

Usually the objects in the workspace are expressed using the coordinate system attached to the stationary robot base (base coordinate system). The successive elements in the manipulator trajectory planning process provide the manipulator's kinematic and dynamic equations discussed in chapter 5.

Knowledge of the kinematic equations together with the geometric description of the workspace, and of the manipulator itself makes it possible to detect potential collisions (between the manipulator and obstacles), an important element in the block of spatial trajectory planning.

12.4.1. Collision detection in the workspace

The algorithms of collision detection used in practice depend heavily on the geometric description of the workspace employed. When the obstacles, manipulator links, and gripper are represented by elementary solid geometry, by spheres or polyhedrons, the issue of collision detection can be reduced to the solution of a certain task of nonlinear programming. For that purpose, the following elements are assumed to be given:

a set constituting the representation of an object in the workspace

$$\mathbf{A} = \{x = (x_1, x_2, x_3) \in \mathbf{R}^3 : A(x) \leq 0\} \tag{12.7}$$

and a set expressing another object in this workspace (Fig. 12.20)

$$\mathbf{B} = \{y = (y_1, y_2, y_3) \in \mathbf{R}^3 : B(y) \leq 0\} \tag{12.8}$$

where $A(x)$, $B(y)$ are scalar functions.

Checking whether these two objects collide is the equivalent of minimizing a function of the form

$$(x^* - y^*)^2 = \min_{x,\, y}\{(x - y)^2\} \tag{12.9}$$

where

$$(x - y)^2 = \sum_{i=1}^{n} (x_i - y_i)^2$$

with the constraints $x \in \mathbf{A}$, and $y \in \mathbf{B}$.

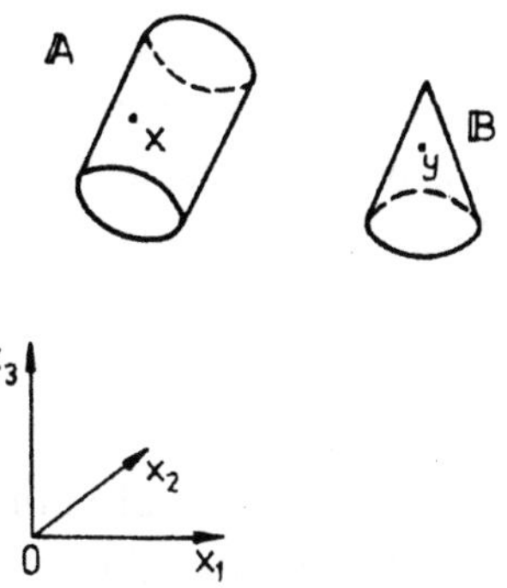

Fig. 12.20. Examples of objects of elementary solid geometry

When $(x^* - y^*)^2 = 0$, we assume a collision between **A** and **B;** when $(x^* - y^*)^2 > 0$, those objects do not collide.

In particular, when the workspace is represented by convex polyhedrons (Figure 12.21), we have

$$(x^* - y^*)^2 = \min_{x,\, y}\{(x - y)^2\} \tag{12.11}$$

$$\{<a, x> + b \le 0\}$$
$$\{<c, y> + d \le 0\}$$

where

$< , >$ is the dot product of vectors;

$$<x, y> = \sum_{i=1}^{3} x_i y_i$$

$x = (x_1, x_2, x_3)$

$y = (y_1, y_2, y_3)$

$<a, x> + b \le 0$ and $<c, y> + d \le 0$

– are object boundary plane equations that can be solved by efficient methods (for instance, using Lemke's algorithm) [12.10].

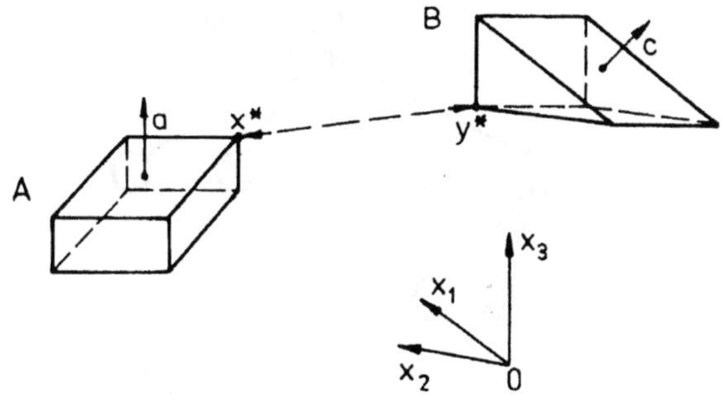

Fig. 12.21. Examples of convex polyhedrons

Collision detection in workspaces represented by oct-trees is very simple. Determining whether a given point in a workspace is an object (representing an element of the manipulator link), or an obstacle that belongs to a "full" space, or a free workspace depends on the recursive division of the tree and the subsequent determinations of the octants to which this point belongs (i.e. determining this space's value intervals to which the coordinates of the point belong). When the octant is full – a collision occurs (Figures 12.22 and 12.23).

A relatively general and simple algorithm of collision detection for all types of representations of the workspace [12.11] uses the condition of collision between a cubicoid (object **A**) and a sphere (object **B**). This condition can be expressed through the conjunction of inequalities (Fig. 12.24):

$$\begin{aligned} |x_1'| &\le \frac{a}{2} + r \\ |x_2'| &\le \frac{b}{2} + r \end{aligned} \qquad (12.12)$$

We assume that a collision between *A* and *B* occurs if both of the above inequalities are satisfied (Fig. 12.25). To use this method of checking for the collision of two cubicoids *A* and *B* (Fig. 12.26), a sphere is circumscribed on one of them.

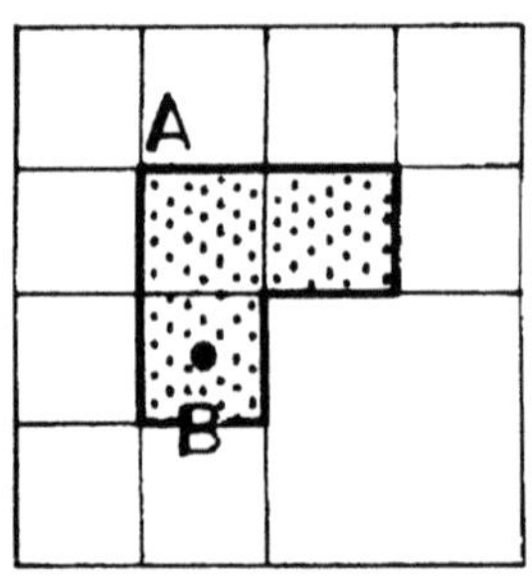

Fig. 12.22. Collision of objects for a quadruple tree representation

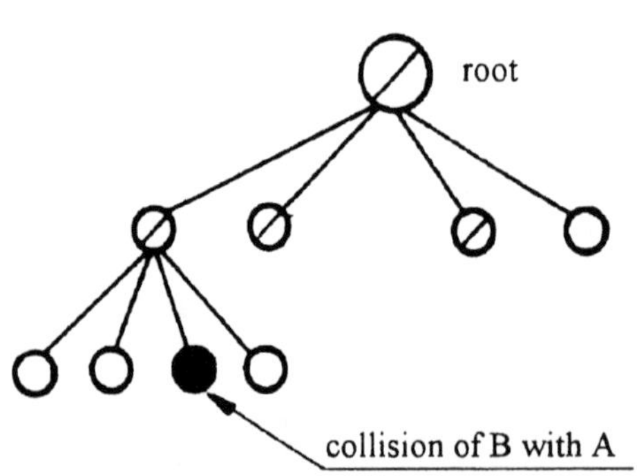

Fig. 12.23. Example of quadruple tree corresponding to fig. 12.22

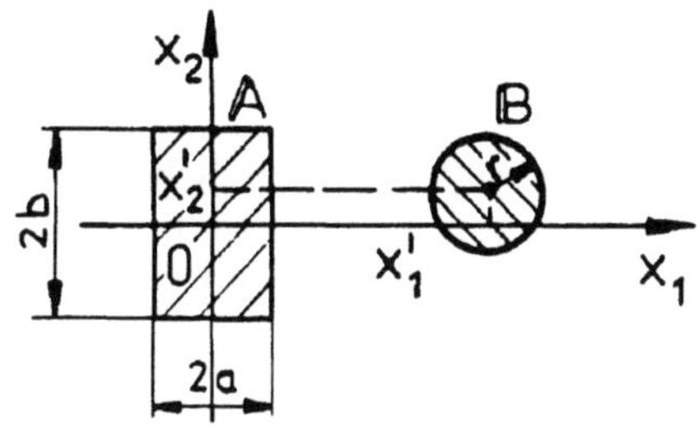

Fig. 12.24. Undetected collision between **A** and **B**

As this is a rough approximation of object *B*, the algorithm uses second collision detection procedure, proceeding as follows: the two dimensions of cubicoid *B* are divided by 2, dividing *B* into four smaller cubicoids. To each of these we circumscribe spheres, after calculating the radiuses, and then expression (12.12) is used to check each sphere for the collision condition. When a collision is detected between the smaller spheres and object *A*, the division procedure is repeated and the collision condition (12.12) checked again for the subsequently divided parts of the smaller cubicoids (Figure 12.27).

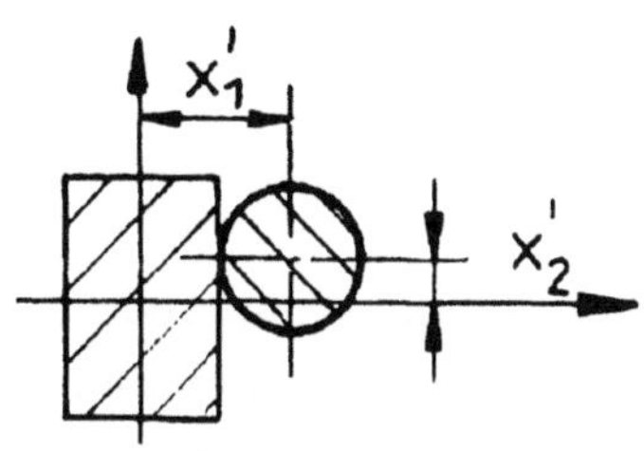

Fig. 12.25. Detected collision

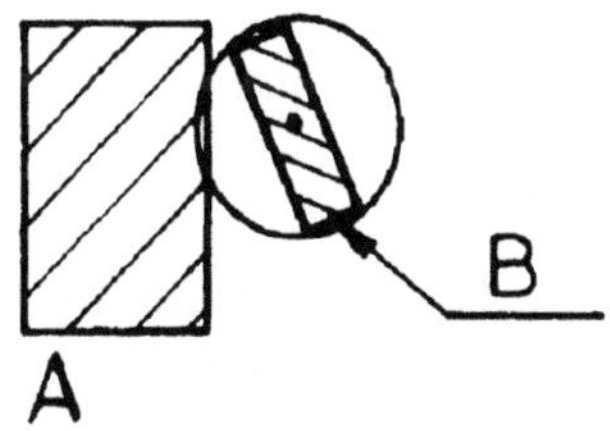

Fig. 12.26. Collision detection by means of a sphere circumscribed on cubicoid

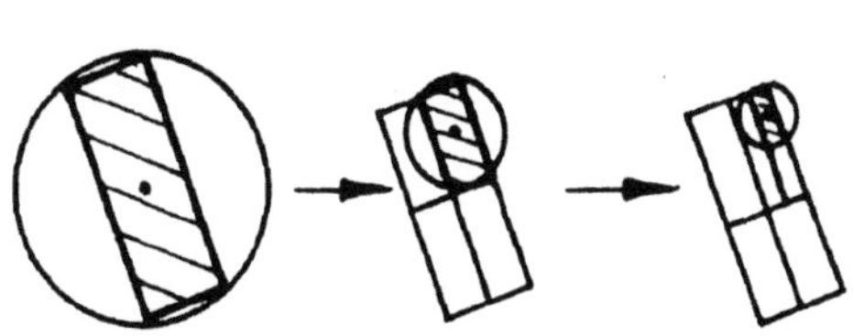

Fig. 12.27. Division of object **B**

The criterion for stopping the procedure is the value of the minimum radius of circumscribed sphere. When the minimum is achieved for the condition of the collision of the smallest spheres with object *A*, it is assumed that objects ***A*** and ***B*** collide with each other. If the approximation of the robot parts and obstacles by single cubicoids and spheres is insufficient, their number is increased for a more accurate approximation of these objects.

12.4.2. Methods of spatial trajectory planning

When we have the geometrical description of the workspace, manipulator links and gripper, the collision detection algorithms, and the robot`s kinematics equations, the process of the collision-free trajectory planning can be performed. The trajectory connects the initial configuration $q_s = (q_s^1, q_s^2, ..., q_s^n)$, where n is the number of kinematic pairs (revolute and/or prismatic), and $n \leq 6$, with the given final configuration $q_f = (q_f^1, q_f^2, ..., q_f^n)$. The manipulator motion must proceed in such a way that the mechanical limitations are not exceeded, that is

$$q_{min} \leq q \leq q_{max}$$

where

$$q_{min} = (q_{min}^1, q_{min}^2, ..., q_{min}^n)$$

$$q_{max} = (q_{max}^1, q_{max}^2, ..., q_{max}^n)$$

q_{min}, q_{max} are the lower and upper limits imposed on the manipulator joint space coordinates; and q is a current manipulator configuration. The inequality of vectors is meant in Pareto's sense, in terms of coordinates. In addition, manipulator links and gripper ones must not collide with objects in the workspace.

The methods of solving these problems (the most representative ones, in our opinion) are presented in the next sub-chapters.

12.4.2.1. Dynamic programming

Application of the method of dynamic programming requires the introduction of the criterion according to which an optimal trajectory is determined, this can be the trajectory length (expressed in joint space coordinates) given in the form

$$I = \int_{q_s}^{q_f} \|dq\| \tag{12.13}$$

where

$$\|dq\| = \sqrt{\sum_{i=1}^{n} \left(dq^i\right)^2},$$

$\|dq\|$ is the norm of the differential of the desired trajectory's length,

$dq = (dq^1, dq^2, ..., dq^n)$

In order to use the method of dynamic programming, the joint space coordinates are discretized with an increment of Δ^i for the i-th joint space coordinate. Consequently, we take into consideration only the discrete values of the joint space coordinates when determining the manipulator trajectory

$$q^i(k_i) = q^i_{\min} + k_i \Delta^i \tag{12.14}$$

where

$$k_i = 0{:}N_i;\ n_i = \left[\frac{q^i_{\max} - q^i_{\min}}{\Delta^i}\right],$$

$[x]$ is an integer of x; and $i = 1{:}n$,

generating a descretised space in the form

$$Q_d = \{q \in \boldsymbol{R}^n{:}\ q = (q^1(k_1), q^2(k_2), ..., q^n(k_n));\ 0 \le k_i \le N_i;\ i = 1{:}n\} \tag{12.15}$$

We assume that configurations q_s and q_f belong to set Q_d.

In view of possibile obstacles in the workspace, we must check, for each configuration of the above mentioned set, whether the manipulator link positions corresponding with this configuration collide with any of those obstacles, using, for example, the methods mentioned in the previous chapter, together with the kinematic equations. In the case of a collision, this configuration is not taken into account in set Q_d. Consequently to take advantage of the dynamic programming method, we must transform the obstacles expressed in rectangular coordinates into joint space coordinates through the manipulator kinematic equation. An example of this, regarding for a planar manipulator with two revolute joints is shown in Figures 12.28 and 12.29.

For the purposes of the dynamic programming method, for each configuration $q \in Q_d$ we introduce a set of its possible changes in the form [12.12]

$$\{q' = q + \Delta : \Delta \in Z\} \tag{12.16}$$

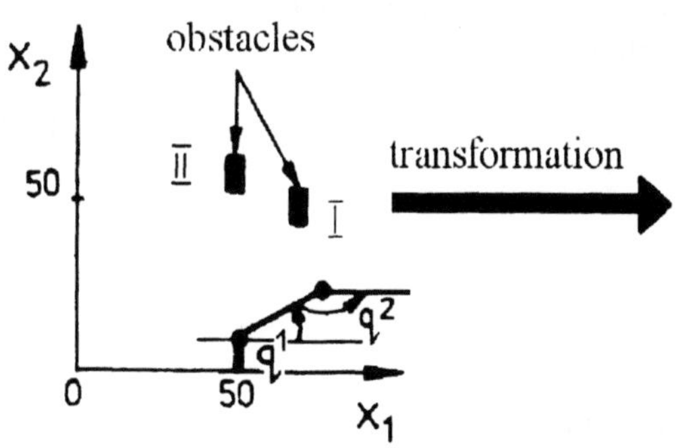

Fig. 12.28. Example of a workspace with obstacles

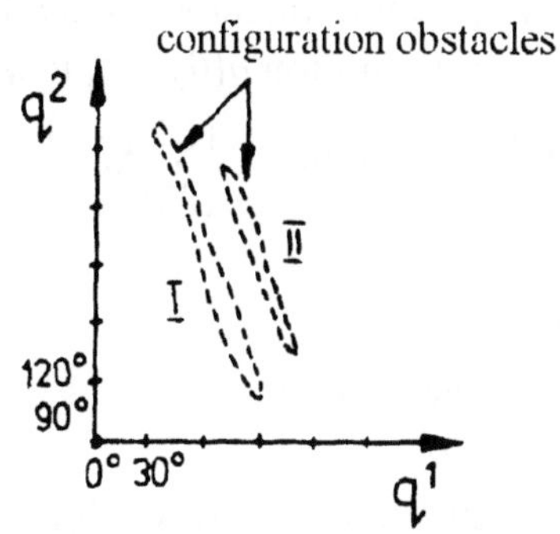

Fig. 12.29. Configuration of obstacles corresponding to Fig. 12.28

where

$$Z = \{(l_1\Delta^1, l_2\Delta^2, ..., l_n\Delta^n): \sum_{i=1}^{n} = (l_i)^2 \geq 0; l_i = 0 \text{ or } l_i = 1 \text{ or } l_i = -1; i = 1:n\}$$

In the equation $q' = q + \Delta$ the compound Δ can be interpreted as a control vector. As a result of its action configuration (state) q changes into configuration (new state) q'. In the case of $n = 2$, sets Z and $\{q' = q + \Delta ; \Delta \in Z\}$ are expressed in the form of the relationship

$$Z = \{(-\Delta^1, -\Delta^2), (-\Delta^1, 0), (-\Delta^1, \Delta^2), (0, -\Delta^2), (0, \Delta^2), (\Delta^1, -\Delta^2), (\Delta^1, 0), (\Delta^1, \Delta^2)\} \quad (12.17)$$

With the discretization of the joint space, the integral performance index (12.13) is reduced to a sum

$$I = \sum_{k=0}^{N-1} \|q(k+1) - q(k)\| \quad (12.18)$$

where:

$q(0) = q_s$; $q(N) = q_f$;

N is the number of elementary division intervals of the trajectory, determined in the optimization process.

Applying the equation

$$q(k) = q(k-1) + \Delta(k-1), \quad (12.19)$$

where $\Delta(k-1) \in Z$,

relationship (12.18) can be written in the form

$$I = \sum_{k=1}^{N} \|\Delta(k-1)\| \quad (12.20)$$

Thus, our task is to find the sequence of controls $\Delta(0)$, $\Delta(1)$, ..., $\Delta(N-1)$ that allows the manipulator to pass from the initial to the final configuration without collision, together with the minimizing criterion (12.20) (the trajectory length).

For all admissible configurations $q(k) \in Q_d$ (in the k-th step of the process of finding an optimum trajectory, $k \leq N$) we denote by $\| q(k)\|$ the length of the shortest path from configuration $q(0) = q_s$ to $q(k)$, that is

$$I(q(k)) = \min_{\Delta(0), \Delta(1), ..., \Delta(k-1)} \left\{ \sum_{i=0}^{k-1} \|\Delta(i)\| \right\} \quad (12.21)$$

where

$\Delta(i) \in Z$

with a boundary condition

$$I(q(0)) = 0 \quad (12.22)$$

Bellman's equation [12.13] for $I(q(k))$ takes the form

$$I(q(k)) = \min_{\Delta(k-1)} \{I(q(k-1)) + \|\Delta(k-1)\|\} \quad (12.23)$$

Our task is to solve it for all $k = 1{:}N$. On the basis of the known value of $I(q(0))$, we begin minimization of the above mentioned criteria with $k = 1$. In this case for each admissible value of control $\Delta(0) \in Z$, configuration $q(1)$ is determined on the basis of equation (12.19). The finite values of index $I(q(1))$, corresponding to those configurations, together with controls $\Delta(0)$ are then saved. The operations described are repeated in successive steps $k = 2, 3, \ldots, N$. The consequence of this iteration is the increase in the number of configurations $q \in Q_d$ for which optimal controls have been determined from the initial configuration. At the beginning of the optimization the number N is unknown. It is established by performing the optimization steps until the connection between the initial and the final configuration is found, and the performance index corresponding to all the configurations determined is finite. With the satisfaction of these conditions, the optimisation of index (12.20) ends. During computation to determine the optimal values of the control index in the k-th step, a configuration $q(k)$ may be achieved, that does not belong to set Q_d because it could cause a collision, or exceed mechanical limitations. In this a case, a value $\|\Delta(k-1)\| = \infty$ is attributed to the corresponding control, and this forbidden direction is ruled out during minimization. The result of the optimization performed is a sequence of controls (Fig. 12.30)

$$\Delta_{opt}(0),\ \Delta_{opt}(1),\ \ldots,\ \Delta_{opt}(N-1),$$

which pass the manipulator from configuration $q(0) = q_s$ to configuration $q(N) = q_f$ along the shortest trajectory without a collision.

The method presented above was employed to find an optimal trajectory for the manipulator shown in Figure 12.28. For the numerical computation the following data have been assumed:

$$q_s = (30°, 150°), \quad q_f = (90°, 210°), \quad q^1_{min} = q^2_{min} = 0°;$$
$$\Delta^1 = \Delta^2 = 2.25°, \quad N_1 = 80, \quad N_2 = 120.$$

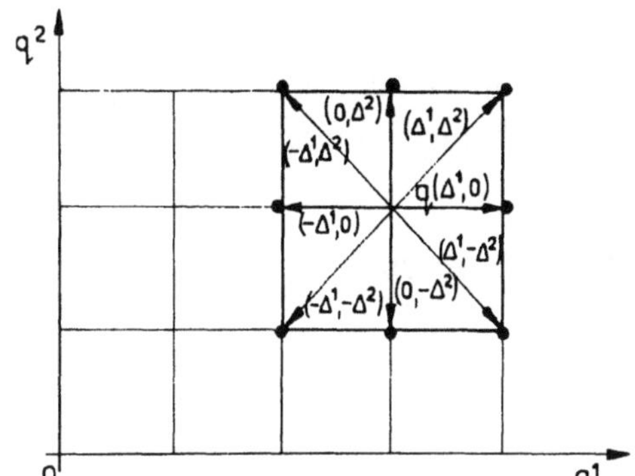

Fig. 12.30. Admissible infinitesimal motions of the manipulator from Fig. 12.28

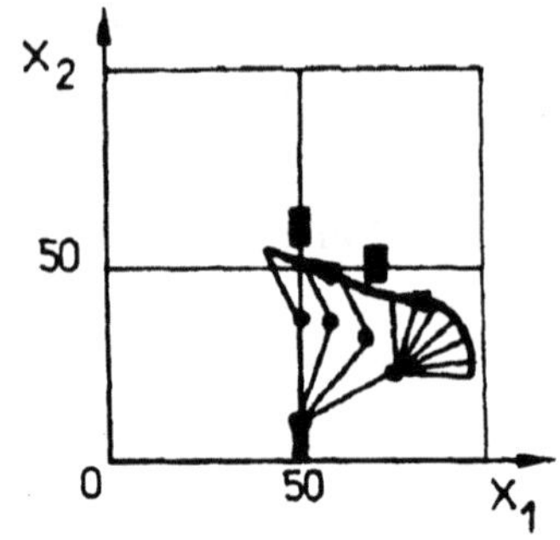

Fig. 12.31. Manipulator trajectory in rectangular coordinates

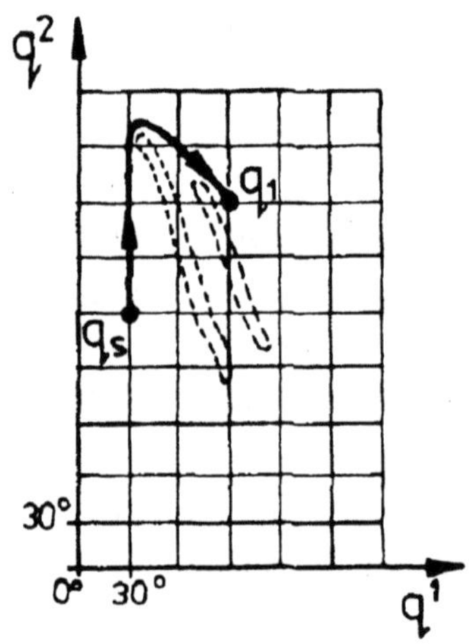

Fig. 12.32. Trajectory in joint space, corresponding to the trajectory in Fig. 12.31

Set Z is given by expression (12.17).
The results of the numerical simulations are presented in Figure 12.31 and 12.32. The dynamic programming method required $N = 69$ optimization steps to find an optimal trajectory.

This method, regardless of its generality, is, in practice, used for trajectory planning for manipulators with three joints at the most, due to its computational complexity.

12.4.2.2. Artificial potential method

The approach using the artificial potential field [12.14, 12.15, 12.16] seems to be an efficient tool for planning collision-free manipulator trajectories. In the case of non-complex workspaces (e.g. with a small number of obstacles) and when the robot is to perform rather simple manufacturing operations, this method can even be used for on-line manipulator control.

The method entails, for the purpose of avoiding collisions, the introduction of a local potential field in the neighbourhood of the obstacle. Thus, the manipulator points lying inside those fields are subjected to repulsive forces. The obstacles are described by the following relationships:

$$Q_j = \{x: B_j(x) \le 0\}, \tag{12.24}$$

where:

$j = 1{:}p$ is a number of obstacles,

$x \in \mathbf{R}^3$ is a point belonging to j-th obstacle,

$B_j(x)$ is a scalar function describing the boundary of j-th obstacle.

The equations of the local potential fields of these obstacles (Fig. 12.33) can be expressed in the form

$$V_j(x(q)) = \begin{cases} C_j\left(\dfrac{1}{d_j(x(q))} - \dfrac{1}{d_j^0}\right)^2 & \text{for } d_j(x(q)) \le d_j^0 \\ 0 & \text{for } d_j(x(q)) > d_j^0 \end{cases} \tag{12.25}$$

where:

$d_j(x(q))$ is a distance of point $x(q)$ from j-th obstacle,

$d_j(x(q)) = \min\limits_{x \in Oj} \{\|x(q) - x\|\}$; and

d_j^0 is a given distance, reducing the influence of the potential field;

$x(q)$ is a point belonging to the edge of a link of the manipulator whose position in the workspace, with configuration q, is determined by means of the kinematic equations;

C_j is positive coefficient (Fig. 12.33).

From relation (12.25) it can be inferred that the tips of manipulator are subjected to repulsive forces only if their distance from the obstacle is not greater than d_j^0. The closer the manipulator to the obstacle, the more rapidly the value of potential function $V_j(x(q))$ increases. To prevent breaching the mechanical limitations during the motion of manipulator links, a potential function (Fig. 12.34) is introduced in the form

$$V_m(q) = \sum_{i=1}^{n} C_i'\left(V_i^d(q) + V_i^g(q)\right) \tag{12.26}$$

where

$$V_i^d(q) = \begin{cases} \left(\dfrac{1}{S_i(q)} - \dfrac{1}{S_i^0}\right)^2 & \text{for } S_i(q) \le S_i^0 \\ 0 & \text{for } S_i(q) > S_i^0 \end{cases}, \qquad S_i(q) = q^i - q^i_{\min}$$

$$V_i^g(q)=\begin{cases}\left(\dfrac{1}{W_i(q)}-\dfrac{1}{W_i^0}\right)^2 & \text{for } W_i(q)\le W_i^0\\ 0 & \text{for } W_i(q)>W_i^0\end{cases},\qquad W_i(q)=q^i_{min}-q^i$$

S_j^0, W_j^0 are given distances limiting the influence of the potential field; and

C' is a positive coefficient (the enhancement of the potential).

To reach the desired final configuration q_f, the manipulator is also subjected to constant attractive forces

$$V_a(q) = C''(q_f - q)^2, \tag{12.27}$$

where

C'' is a positive constant (the enhancement of the attractive force); and

q is a current manipulator configuration.

Assuming that the M points are placed on manipulator links with sufficient density, the total potential function for the given configuration q takes the form

$$V(q)=\sum_{i=1}^{M}\sum_{j=1}^{p}V_j(x_i(q))+V_m(q)+V_a(q) \tag{12.28}$$

where

$x_i(q)$ is the i-th point on the manipulator (Fig. 12.33).

The manipulator motion`s is the result of the combined forces, and is executed so as to minimize function (12.28), that is

$$V(q_{(k+1)}) = \min_q \{V(q)\}, \tag{12.29}$$

where $q_{(0)} = q_s$; k - k-th phase of minimisation.

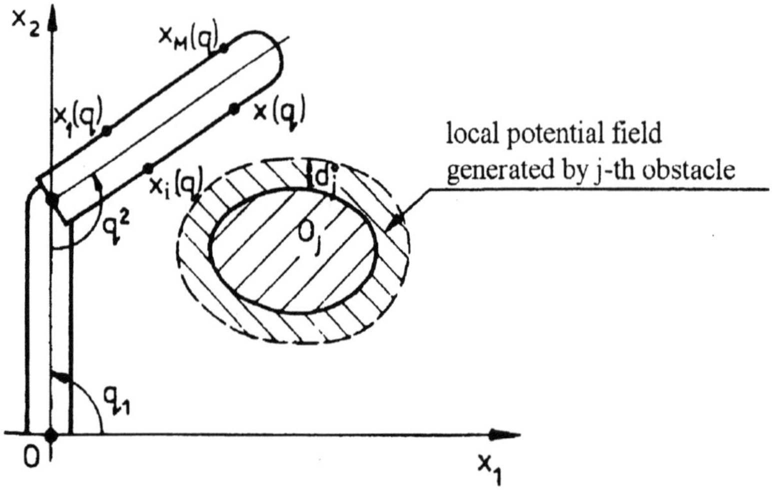

Fig. 12.33. Example of local potential fields

The result of the minimization of the above function is the configuration

$$q_{(k+1)} = q_{(k)} - \alpha \frac{\partial V\left(q_{(k)}\right)}{\partial q} \tag{12.30}$$

where α is a step length.

Finding the gradient of function $V(q_{(k)})$ for each obtained configuration entails M_p tasks of mathematical programming in the form:

$$d_j(x_i(q)) = \min_{x'} \{\| x_i(q) - x' \|\} \tag{12.31}$$

$$B_j(x') \leq 0$$

where $i = 1{:}M$; and $j = 1{:}p$. If this method is to be used for collision-free trajectory planning on-line, the manipulator links and obstacles must be approximated as simple elementary solids (intervals, spheres, cubicoids) which make it possible to solve tasks (12.31) with analytical methods. The potential function method does not require transforming obstacles into joint space. One fundamental drawback to this method is its locality, due to the existence of local minima of the potential function. Consequently, the manipulator may jam, halt before reaching the final configuration. Dynamic programming, instead, is global in this sense, it does not present this drawback.

12.4.2.3. Harmonic function method

The method using harmonic functions, presented in [12.40] and [12.41], does not have the disadvantage. In this case the artificial potential in the locality workspace is induced by harmonic functions, which are solutions of Laplace equations. Moreover, harmonic functions have a very convenient property for the planning of collision-free trajectories: they satisfy a maximum principle which says that a non-constant harmonic function reaches its smallest and its greatest value on the boundary of its domain. The basic concepts of trajectory planning with the use of harmonic functions are presented here in the two-dimensional joint space of what is called a point robot (certain generalizations concerning manipulators are possible; e.g. those presented in [12.41]). This approach assumes a homogeneous flow of liquid in the workspace, in the positive direction of axis $0q^1$, for instance (Fig. 12.35) given by the equations:

$$\begin{aligned} \Phi\left(q^1, q^2\right) &= -Uq^1 \\ u\left(q^1, q^2\right) &= \frac{\partial \Phi}{\partial q^1} = -U \\ v\left(q^1, q^2\right) &= \frac{\partial \Phi}{\partial q^2} = 0 \end{aligned} \tag{12.32}$$

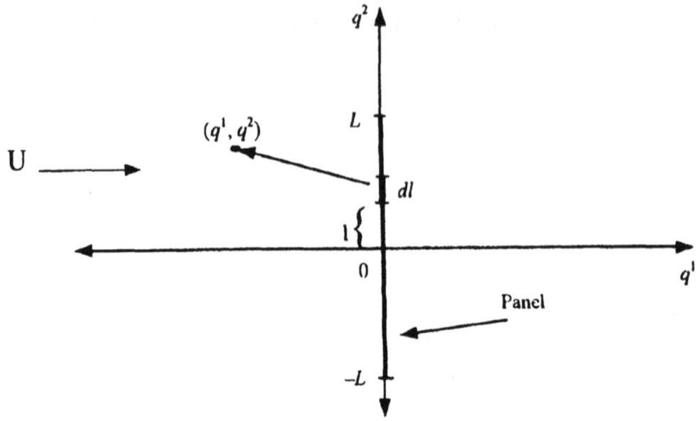

Fig. 12.35. Graphic illustration of the harmonic function method

where U describes the force (power) of the homogeneous flow; while $u(\cdot,\cdot)$ and $v(\cdot,\cdot)$ denote the components of the liquid flow velocity. Obstacles are represented by intervals called panels. The potential from a single obstacle (panel) at point (q^1, q^2) of space induced by sources in a small element dl of the panel located at point $(0, l)$ is expressed as

$$d\Phi_p = \frac{\lambda dl}{2\pi} \log\left((q^1)^2 + (q^2 - l)^2\right)^{\frac{1}{2}} \tag{12.33}$$

where $\lambda < 0$ is called the force (power) of the source (the repulsive force of the panel differential acting on a particle located at (q^1, q^2)). The induced potential function of the whole panel is

$$\Phi_p(q^1, q^2) = \frac{\lambda}{2\pi} \int_{-L}^{L} \log\left[(q^1)^2 + (q^2 - l)^2\right]^{\frac{1}{2}} dl \tag{12.34}$$

The magnitude of coefficient λ affects the flow profile while avoiding the panel. If an obstacle is formed by many panels (Fig. 12.36), the potential at an arbitrary point (q^1, q^2) caused by panel j equals:

$$\Phi_{p_j}(q^1, q^2) = \frac{\lambda_j}{2\pi} \int_{-L_j}^{L_j} \log\sqrt{(q^1 - q_j^1)^2 + (q^2 - q_j^2)^2}\, dl_j \tag{12.43}$$

Collision avoiding depends on displacing the point robot, from its current position (q^1, q^2), to the desired final position (configuration). For this purpose we need a potential that attracts the robot to a final configuration in which a global minimum is reached (the only one, considering the properties of harmonic functions). This potential can be expressed by the relationship

$$\Phi_f(q^1, q^2) = \frac{\lambda_f}{2\pi} \log\sqrt{(q^1 - q_f^1)^2 + (q^2 - q_f^2)^2} \tag{12.36}$$

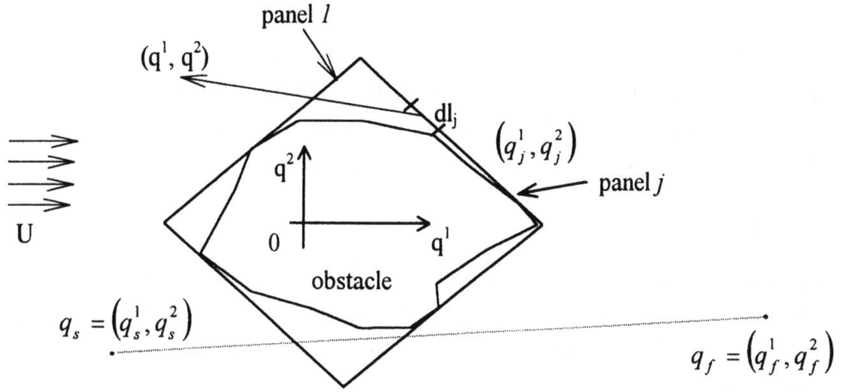

Fig. 12.36. Illustration of trajectory planning

where $\lambda_f > 0$ is the coefficient of the force attracting the robot. The potential of the homogeneous flow, which tends to displace the robot from its initial position (configuration) $q_s = \left(q_s^1, q_s^2\right)$ to its final one, takes the form:

$$\Phi_{hom.} = -U\left(q^1 \cos\alpha + q^2 \sin\alpha\right) \tag{12.37}$$

where $\alpha = ar\,\mathrm{ctg}\dfrac{q_f^2 - q_s^2}{q_f^1 - q_s^1}$

The global potential generated by m panels (obstacles), the final configuration, and the homogeneous flow equals:

$$\begin{aligned}\Phi_{gl.}\left(q^1, q^2\right) &= \Phi_{hom.} + \Phi_f + \sum_{j=1}^{m} \Phi_{p_j} = -U\left(q^1 \cos\alpha + q^2 \sin\alpha\right) + \\ &+ \frac{\lambda_f}{2\pi} \log\sqrt{\left(q^1 - q_f^1\right)^2 + \left(q^2 - q_f^2\right)^2} + \\ &+ \sum_{j=1}^{m} \frac{\lambda_j}{2\pi} \int_{-L_j}^{L_j} \log\left(\left(q^1 - q_j^1\right)^2 + \left(q^2 - q_j^2\right)^2\right)^{\frac{1}{2}} dl_j\end{aligned} \tag{12.38}$$

Having selected the proper coefficients, U, λ_f and λ_j, we obtain the conditions for realizing the desired trajectory. The robot`s motion is defined by the equations:

$$\begin{aligned}\dot{q}^1 &= \frac{\partial \Phi_{gl.}(q)}{\partial q^1} \\ \dot{q}^2 &= \frac{\partial \Phi_{gl.}(q)}{\partial q^2}\end{aligned} \tag{12.38a}$$

Although the method of harmonic functions appears very attractive, considering the possibility of its implementation on-line, it too presents certain shortcoming. First, in more complex workspaces it induces what are called stagnation points, that is points at which the robot halts as a result of a combination of different potentials. However, a stagnation point is not at one of a local minimum. Moreover, if a non-point robot (manipulator) is concerned, the harmonic function method may lead to structural local minima in which reaching the final point for a given potential is impossible. In this event the robot control strategy must be changed.

12.4.2.4. Penalty function method

The approach using a penalty function [12.17, 12.18] combines the spatial trajectory planning phase with its time-parametrizing. The manipulator trajectory is determined by a set of functions which minimize a certain quality index generally expressed in the form

$$I(q)=\int_0^T F(q,\dot{q},u)dt \tag{12.39}$$

where:

T is the trajectory duration, which may or may not be;

$F(q,\dot{q},u)$ is a certain cost function, usually assuming non-negative values;

u is a control associated with joint space coordinates by the robot dynamic equations which can generally be written in the form $u = f(q,\dot{q},\ddot{q})$; and

$f(q,\dot{q},\ddot{q})$ is a non-linear vector function of variables q, $\dot{q}$ and $\ddot{q}$.

When $F(q,\dot{q},u)=1$ and the time is not set, we have to determine a time-optimal trajectory, that is a trajectory such that the time it takes the manipulator to pass from initial configuration q_s to final one q_f, maintaining motion constraints, is minimal. The minimization of the energy consumption is achieved when $F(q,\dot{q},u)= \langle u, u \rangle$ and time T is given. The constraints imposed on configurations and manipulator velocities in the initial and final instances usually take the form

$$\begin{cases} q(0)=q_s \\ q(T)=q_f \\ \dot{q}(0)=0 \\ \dot{q}(T)=0 \end{cases} \tag{12.40}$$

Maintaining the notation from the previous chapter, the conditions of collision avoidance in the workspace and respect the mechanical limitations at any instance of manipulator motion are expressed by the equations below:

$$\begin{cases} d_j(x_k(q(t)))-d_j^0 \geq 0 \\ q^i(t)-q^i_{\min} \geq 0 \\ q^i_{\max}-q^i(t) \geq 0 \end{cases} \tag{12.41}$$

where: $i = 1{:}n$, $j = 1{:}p$, $k = 1{:}M$

Equations (12.39)÷(12.41) describe a problem of dynamic optimization that is complex and difficult to solve (considering the inequality constraints). The **penalty function method** is an efficient method to apply here, in terms of computation. It maintains the inequality-type constraints while slightly increasing the value of the performance index (12.39) by forcing the solution away from the boundary set described by relationships (12.40) and (12.41).

Assuming that a solution to this dynamic optimization task exists, the general form of the performance criterion can be reduced to the expression

$$J(q)=\int_0^T \widetilde{F}(q,\dot{q},\ddot{q})dt \tag{12.42}$$

where:

$$\widetilde{F}(q,\dot{q},\ddot{q})=F(q,\dot{q},f(q,\dot{q},\ddot{q}))+\beta\sum_{i=1}^{n}\left[k_1\left(q^i-q^i_{\min}\right)+k_2\left(q^i_{\max}-q^i\right)\right]+$$
$$+\beta\sum_{m=1}^{M}\sum_{j=1}^{p}k_3\left(d_j\left(x_m(q)\right)-d_j^0\right);$$

β is a positive coefficient establishing a compromise with the performance index (12.39);

$k_l(x)$ is a penalty function with the general form of (Fig. 12.37)

$$k_l(x)=\begin{cases} d_l\left(\dfrac{x}{b_l}-\dfrac{b_l}{x}-2\right) & \text{for } 0<x<b_l \\ 0 & \text{for } x\geq b_l \end{cases}$$

d_l is a coefficient describing the magnitude of the penalty,

b_l is a given threshold value of variable x with the exceeding of which the penalty function is activated; and l = *1:3* (Fig. 12.37).

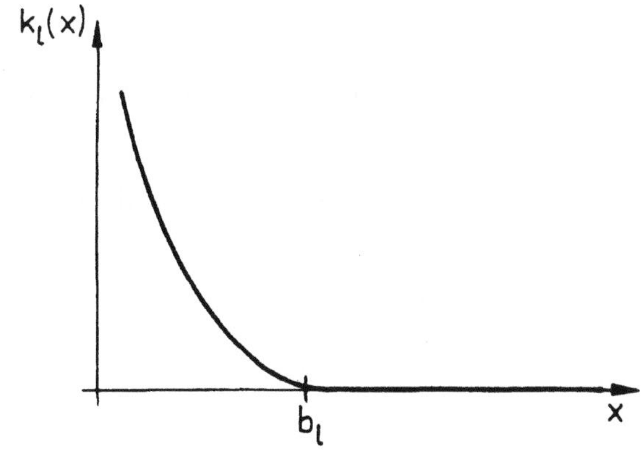

Fig. 12.37. Chart of penalty function

The method of the penalty function reduces the original optimization task to one of finding a trajectory that minimizes the performance index (12.42). In a general case as stated above, an analytical solution can not be found. Consequently, approximate solutions must be sought. With the use of the generalized Euler's method [12.19], the optimal trajectory can be found, for example in a set of functions

$$q(t)=\sum_{i=1}^{N}[C_i]\varphi_i(t) \tag{12.43}$$

where:

$\varphi_i(t)$ is the n-dimensional time function which is given, a so called **base function** (spline, polynomial);

$[C_i]=\operatorname{diag}\left(c_i^1,c_i^2,\ldots,c_i^n\right)$ is a $(n\times n)$-dimensional diagonal matrix the elements of which are subjected to the optimization process, and

N is a number which describes the accuracy of the approximation of the optimum trajectory by base functions $\varphi_i(t)$.

Coefficients c_j^i in matrices $[C_i]$ are not independent. They must satisfy the system of $4n$ linear equations, a result of constraints (12.40), that is:

$$\begin{aligned}
&\sum_{i=1}^{N}\operatorname{diag}\left(c_i^1,\ldots,c_i^n\right)\varphi_i(0)=q_s\\
&\sum_{i=1}^{N}\operatorname{diag}\left(c_i^1,\ldots,c_i^n\right)\varphi_i(T)=q_f\\
&\sum_{i=1}^{N}\operatorname{diag}\left(c_i^1,\ldots,c_i^n\right)\dot\varphi_i(0)=0\\
&\sum_{i=1}^{N}\operatorname{diag}\left(c_i^1,\ldots,c_i^n\right)\dot\varphi_i(T)=0
\end{aligned} \tag{12.44}$$

from which it is usually possible to determine $4n$ coefficients depending on the remaining $(N-4)n$ ones. We denote the independent coefficients by $\bar{c}_i^j$, where $j = 1{:}n$; and $i = 1{:}(N-4)$. As a result, the elements of k-th diagonal matrix from equation (12.43) can be expressed as linear functions of $\bar{c}_i^j$, further denoted as

$$a_k^j=a_k^j\left(\bar{c}_1^1,\bar{c}_2^1,\ldots,\bar{c}_i^j,\ldots,\bar{c}_{N-4}^n\right)$$

Taking these functions into account, the desired trajectory assumes the following form:

$$q(t)=\sum_{i=1}^{N}\operatorname{diag}\left(a_i^1,a_i^2,\ldots,a_i^n\right)\varphi_i(t) \tag{12.45}$$

By substituting the right side of equation (12.45) in the performance index (12.42) the problem of its minimization is reduced to the task of the static optimization of the function with $(N-4)n$ variable coefficients $\bar{c}_i^j$ written below:

$$J\left(\bar{c}_1^1, \bar{c}_2^1, \ldots, \bar{c}_i^j, \ldots, \bar{c}_{N-4}^n\right)=$$

$$=\int_0^T \tilde{F}\left(\sum_{i=1}^N \operatorname{diag}\left(a_i^1, a_i^2, \ldots, a_i^n\right)\varphi_i(t), \sum_{i=1}^N \operatorname{diag}\left(a_i^1, a_i^2, \ldots, a_i^n\right)\dot{\varphi}_i(t), \sum_{i=1}^N \operatorname{diag}\left(a_i^1, a_i^2, \ldots, a_i^n\right)\ddot{\varphi}_i(t)\right)dt \tag{12.46}$$

Using standard algorithms of optimization of multi-variable functions [12.20], we can determine the values of the coefficients we are looking for.

In order to initiate the process of optimization of function (12.46), we must know the initial values of coefficients $\bar{c}_i^j$. These can be obtained on the grounds of a certain admissible trajectory that meets constraints (12.40), (12.41).

The initial values of the coefficients are selected in the process of approximation of the given admissible trajectory by base functions $\varphi_i(t)$. The method of the penalty function makes it possible to efficiently determine the optimal trajectory in complex workspaces, which may contain many obstacles. However, in order to employ this method we must to know the initial admissible trajectory, and finding this may turn out to be difficult and more time consuming than the optimization of function (12.46).

12.4.2.5. Heuristic planning methods

The algorithms of spatial trajectory planning presented in the preceding chapters were based on the use of analytical methods only. **Heuristic planning methods,** instead, up-date the direction of the searching for a trajectory on the grounds of information coming from environment. The methods applied [12.21, 12.22, 12.23] differ in their representations of joint space and their manner of searching. They are used, in principle, for manipulators with no more than three kinematic pairs. The application of heuristic methods usually requires that workspaces be transformed into joint spaces.

Assuming, for further simplification, that $q_{\min}^i = q_{\min}^j$ and $q_{\max}^i = q_{\max}^j$, where: $i \neq j$; $i, j = 1:n$, we discretize the latter in $(2N)^n$ elementary cubes, where:

$2N = \dfrac{q_{\max}^i - q_{\min}^i}{\delta}$; and δ is the assumed discretization accuracy.

A proper discretization of the workspace is determined by the relationship

$$\varepsilon = \max_{q_{\min} \le q \le q_{\max}} \left\{ \left\| J(q)\Delta q \right\| \right\}$$

where

$J(q) = \left[\dfrac{\partial f(q)}{\partial q}\right]$ is the Jacobian matrix of kinematic equations $f(q)$; and

$\Delta q = (\delta, \delta, \ldots, \delta)$ –is a vector of equal n-coordinates.

We represent the free joint space by octal trees (the advantages and disadvantages of which were discussed in preceding chapters). To do so, we must, for each elementry cube, check that the configurations contained in it do not cause a collision:

$$\Delta q = \{q_c + (\delta q_1, \delta q_2, ..., \delta q_n):-\delta/2 \leq \delta q_i \leq \delta/2; i = 1:n\}$$

where q_c is the centre of the cube (Fig. 12.38).

This can be done in the following manner (Fig. 12.38):

1. The manipulator is set in configuration q_c.
2. The dimensions of all the manipulator`s and the gripper's links are increased by $\varepsilon/2$ reflecting approximately the volume of the real, not enlarged, sweep of the manipulator in the workspace, while changing all the configurations from elementary cube Δq.
3. Using, for example, the collision detection method presented at the end of chapter 12.4.1, we check whether this enlarged manipulator collides with obstacles. If it does, then the label "full" is attributed to the cube being considered; otherwise, the label "empty" is assigned.
4. Each successive elementary cube is tested in this way. Then the procedure is ended.

As a result of this reiteration of the procedure we obtain the highest level of the division of the joint space representation, which we shall call the N-th level. On the basis of this we determine the joint space representation of level $N-1$. To do so we consider a cube formed by eight mutually adjacent N-th level cubes. If all the cubes of the N-th level are labelled "full" ("empty"), then the same label is attributed to the cube at level $N-1$. Otherwise, the label "mixed" is attributed to the cube at level $N-1$. This process is repeated until we obtain the joint space representation at level zero which, as we know, constitutes the tree root (Figure 12.39).

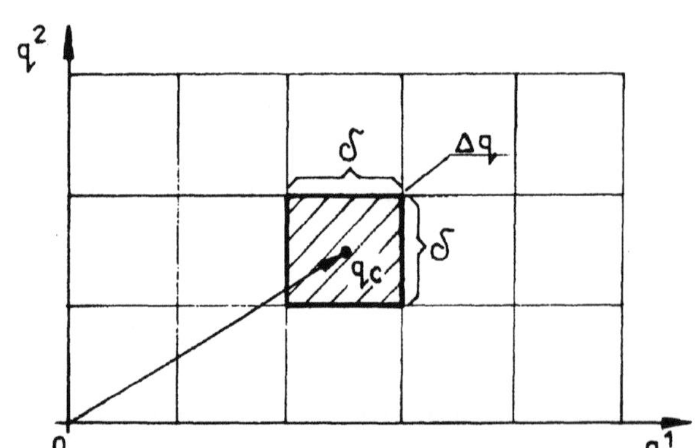

Fig. 12.38. Forming an elementary square

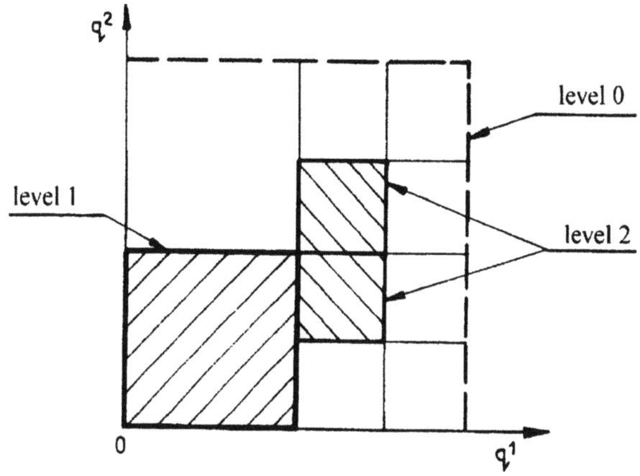

Fig. 12.39. Example of forming a joint space representation on the basis of the oct-tree for $n=N=2$

In order to determine a trajectory in the oct-tree from the initial configuration q_s to the final configuration q_f (under the assumption, not reducing the generality of the considerations, that they coincide with centers of certain cubes in the oct-trees) in the shortest possible computation time, three *search techniques* are combined: *hypothesis and verification*, *hill climbing* and *algorithm* A^*. The first determines a hypothetical trajectory linking the current configuration with the final one (e.g. a straight line) and then verifies that this curve does not collide with obstacles (does not pass through cubes labelled "full"). The second technique uses the *hill climbing* algorithm [12.24]. The search for the trajectory is carried out in a graph obtained by linking the centers of all the adjacent cubes of the oct-tree labelled "empty". The algorithm employs a cost function whose value at any node (configuration) of the free joint space graph is proportional to the Euclidean distance of this node from the given final configuration and equal to infinity for any collision configuration.

When the algorithm is run, the manipulator is displaced to the configuration with the smallest cost function from among the adjacent configurations. This technique is very fast because it takes advantage of local information concerning the direction of the manipulator displacement. However, if for a certain configuration the corresponding value of the cost function is smaller than all the values at the neighbouring nodes, the *hill climbing* algorithm jams.

The third technique employs algorithm A^* but with suitable functions g and h. It also uses the representation of an oct-tree in the graph form. As we know from the previous considerations, algorithm A^* performs a global search in this graph. Consequently, it takes into account various trajectories during the search, and is slower than the *hill climbing* algorithm. However, it offers the strong advantage of always finding the trajectory connecting configuration q_s with q_f (if only one is possible).

When we attribute to the graph arcs weights equal to the Euclidean distances between the configurations determining these lines, and assume that the simple Euclidean distance between the current manipulator configuration and the final one is a heuristic function, the determined trajectory will be also be optimal (of minimum length).

The techniques described above can be combined for the quickest determination of a possible collision-free trajectory (not necessarily optimal). Figure 12.40 shows how this combined heuristic method works on the example of a joint space represented by a quardruple tree.

Hill climbing and A^* algorithms are used to search the graph. The initial and the final configuration are located in squares 12 and 5, respectively. The *hill climbing* algorithm is activated; first it finds its way through squares 12 and 9 to square 10, where it gets "tuck". Consequently, algorithm A^* is applied to find a path through squares 10, 8 and 2.

In the second step, the *hill climbing* algorithm is employed again and reaches configuration q_f through squares 2, 3, 4 and 5. For the example presented we may conclude that if the hypothesis and verification method were used, a solution could be found immediately by drawing a straight line linking the center of square 2 with configuration q_f. The combined method described above determinies a quasi-linear, collision-free spatial trajectory. Current research concentrates on using this planning method in the workspace; the objective is to eliminate the time-consuming process of finding transformations of these spaces.

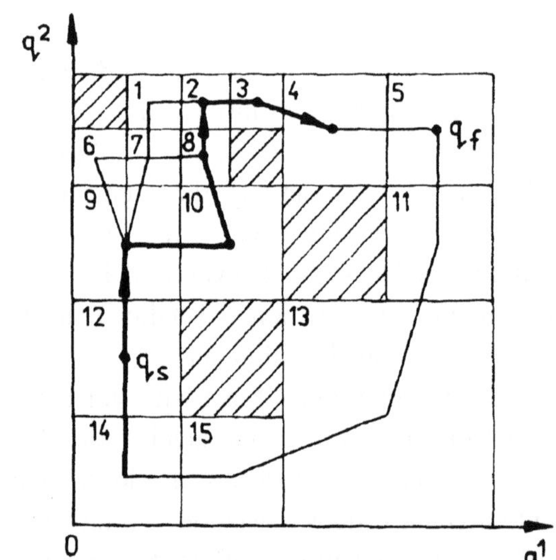

Fig. 12.40. Example of operation of combined heuristic method

In sum, the spatial trajectory planning results in an ordered set of collision-free configurations (curves in the joint space or in discretized approximation), containing the initial and final configurations.

12.4.3. Time-parametrization of spatial trajectories

Manipulator motion in the workspace is enabled by control signals, certain functions of time, sent to its actuators. The values of these signals can be determined, if we know the time profiles of the joint space coordinates. Consequently, the spatial trajectories (discrete, continuous) obtained in preceding chapter must be subjected to time-parameterization in order to achieve the desired manipulator motion in the workspace.

In the next chapters we describe parameterization methods using two representations of a spatial trajectory.

12.4.3.1. Spline functions

We assume a discretized spatial trajectory, found by one of the methods presented in the preceding chapter, and given in the form

$$\{q_j : q_j = (q_j^1, \ldots, q_j^n)\},\ j = 0:m;\ (q_o^1, \ldots, q_o^n) = q_s;\ (q_m^1, \ldots, q_m^n) = q_f \tag{12.47}$$

The time functions $Q^i(t)$ of i-th joint space coordinate are determined with the conditions

$$Q^i(t_j) = q_j^i \tag{12.48}$$

where $i = 1{:}n; j = 0{:}m$; and t_0, t_1, ..., t_m is a non-decreasing sequence of instances of time $[0, T]$ within which the motion takes place; $t_o = 0$; $t_m = T$.

Additional conditions that ensure subsequent correct manipulator control may be imposed on these functions. Apart from that, the problem arises to find the total time T required to move the manipulator from configuration q_s to configuration q_f.

One method for solving these problems uses the information about the capabilities of the actuators employed in the manipulator, and solves the interpolation task, that is, the task of determining a configuration for any instant of the motion (and not only at the instances t_j). To determine value t_j we can use the relationship

$$t_j = t_{j-1} + \max_{1:i:n} \left\{ \frac{|q_j^i - q_{j-1}^i|}{\dot{q}_{\max}^i} \right\} \tag{12.49}$$

where $\dot{q}_{\max}^i$ is a maximum joint space velocity of i-th actuator.

Quantity $\dot{q}_{\max}^i$ describes the capability of the i-th actuator which is to ensure sufficiently small tracking errors of coefficient $Q^i(t)$. The interpolation task is solved using spline functions [12.25].

A **spline function** of the l order is a polynomial of the l order defined in any of the intervals $[t_{j-1}, t_j]$, where $j = 1{:}m$, satisfying the additional condition of continuity of the derivatives up to the $(l-1)$-th order at points $t_1, t_2, \ldots, t_{m-1}$. In practice, spline functions of the third order ($l = 3$) are used.

The function below, which interpolates the values of the i-th joint space coordinate in interval $[t_{j-1}, t_j]$, satisfies continuity conditions up to the second order in points $t_1, t_2, \ldots, t_{m-1}$

$$
\begin{aligned}
Q^i(t) = {} & M^i_{j-1}\frac{(t_j - t)^3}{6\Delta_j} + M^i_j\frac{(t - t_{j-1})^3}{6\Delta_j} + \\
& + \left[\frac{q^i_j - q^i_{j-1}}{\Delta_j} - \frac{\Delta_j}{6}\left(M^i_j - M^i_{j-1}\right)\right](t - t_{j-1}) + q^i_{j-1} - M^i_{j-1}\frac{(\Delta_j)^2}{6}
\end{aligned}
\tag{12.50}
$$

where $t \in [t_{j-1}, t_j]$, $\Delta_j = t_j - t_{j-1}$; M^i_j is what is called the moment of function $Q^i(t)$; and $j = 1{:}m$.

Moments $M^i_0, M^i_1, \ldots, M^i_m$ in equation (12.50) are determined on the grounds of the continuity conditions of the first derivative of function $Q^i(t)$ at points $t_1, t_2, \ldots, t_{m-1}$, yielding a system of equations in the form

$$
\frac{\Delta_j}{6}M^i_{j-1} + \frac{\Delta_j - \Delta_{j+1}}{3}M^i_j + \frac{\Delta_{j+1}}{6}M^i_{j+1} = \frac{q^i_{j+1} - q^i_j}{\Delta_{j+1}} - \frac{q^i_j - q^i_{j-1}}{\Delta_j}
\tag{12.51}
$$

where $j = 1{:}(m-1)$.

These are $m-1$ equations for $m+1$ moments. The conditions for the joint space velocities at the ends of the trajectory $\dot{Q}^i(0) = \dot{Q}^i(T) = 0$ yield two additional equations for determining moments $M^i_0, M^i_1, \ldots, M^i_m$:

$$
\begin{cases}
M^i_0\dfrac{\Delta_1}{3} + M^i_1\dfrac{\Delta_1}{6} = \dfrac{q^i_1 - q^i_0}{\Delta_1} \\[2ex]
M^i_{m-1}\dfrac{\Delta_m}{6} + M^i_m\dfrac{\Delta_m}{3} = \dfrac{q^i_{m-1} - q^i_0}{\Delta_m}
\end{cases}
\tag{12.52}
$$

Consequently, moments M^i_j can be determined from the following system of $m+1$ linear algebraic equations with a band matrix of the form:

$$
\begin{bmatrix}
2 & \lambda_0 & & & 0 \\
\mu_1 & 2 & \lambda_1 & & \\
 & & & & \\
 & & \mu_{m-1} & 2 & \lambda_{m-1} \\
0 & & & \mu_m & 2
\end{bmatrix}
\begin{bmatrix}
M^1_0 \\ M^1_1 \\ \\ \\ M^1_m
\end{bmatrix}
=
\begin{bmatrix}
b^1_0 \\ b^1_1 \\ \\ \\ b^1_m
\end{bmatrix}
\tag{12.53}
$$

where

$$\lambda_0 = \mu_m = 1; \quad b_0^i = \frac{6(q_1^i - q_0^i)}{(\Delta_1)^2}; \; b_m^i = \frac{6(q_{m-1}^i - q_m^i)}{(\Delta_m)^2}; \quad \mu_j = \frac{\Delta_j}{\Delta_j + \Delta_{j+1}};$$

$$\lambda_j = 1 - \mu_j; \quad b_j^i = 6\left[\frac{q_{j+1}^i - q_j^i}{\Delta_{j+1}} - \frac{q_j^i - q_{j-1}^i}{\Delta_j}\right] \Big/ (\Delta_{j+1} - \Delta_j); \text{ and } j = 1{:}(m-1)$$

No general method, such as Gauss's elimination method [12.25], is required to solve this system of equations The particular form of the band matrix makes it possible to determine moments M_j^i recursively according to the following algorithm. In the first step the values of auxiliary variables p_j^i and r_j^i are calculated:

$$p_j^i = -\lambda_j / (\mu_j p_{j-1}^i + 2)$$
$$r_j^i = \frac{b_j^i - \mu_j r_{j-1}^i}{\mu_i p_{j-1}^i + 2} \tag{12.54}$$

where $r_{j-1}^i = p_{j-1}^i = 0$, $j = 0{:}m$; and $i = 1{:}n$.

Having determined p_j^i and r_j^i, the successive moments M_m^i, M_{m-1}^i, ..., M_0^i are then calculated:

$$\begin{cases} M_m^i = r_m^i \\ M_j^i = p_j^i M_{j+1}^i + r_j^i \end{cases} \tag{12.55}$$

where $j = (m-1){:}0$; and $i = 1{:}n$.

Using spline functions for the parameterization of a spatial trajectory is a simple, efficient, and, consequently, rapid way to solve equations (12.53). Another advantage of employing spline functions for interpolation of the joint space coordinates is the manipulator controls are generated in the form of continuous time functions (the first and second derivatives are continuous functions) which can be used to implement manipulator controls in real time with feedback. A certain shortcoming of this method is the possibility of exceeding the limitations imposed on the admissible manipulator controls, as the dynamic equations are not taken into account. As a result, the use of spline functions requires sufficiently dense discretization of interval $[0, T]$. A method avoids this disadvantage at the cost of greater computational complexity, is presented in the next section.

12.4.3.2. Phase space method

We further assume that the manipulator spatial trajectory is given in parameterized form

$$q(s){:}\ 0 \le s \le s_f \tag{12.56}$$

where s is a parameter describing the length of the trajectory, measured from the initial configuration; $q(0) = q_s$ and $q(s_f) = q_f$.

If the trajectory is given by a finite sequence of configurations, it is easy to obtain its representation in the form of (12.56), using one of the interpolation methods (for example, spline functions).

The phase space method [12.26, 12.27] uses the robot`s dynamic equation (see the previous section) given by the set of equations

$$\langle M_i, \ddot{q} \rangle + \left\langle \dot{q}, \left(\frac{\partial M_i}{\partial q} - \frac{1}{2} \frac{\partial M}{\partial q^i} \right) \dot{q} \right\rangle + \frac{\partial G}{\partial q^i} = u^i \tag{12.57}$$

where

M_i is the i-th row of inertia matrix M;

G, the manipulator potential energy; and

u^i, the i-th control signal (e.g. the i-th generalized moment).

For simplification, we shall assume that the control limitations resulting from the capabilities of manipulator actuators are constant, and their absolute values equal to one another:

$$u^i_{\min} \le u^i \le u^i_{\max} \tag{12.58}$$

where $u^i_{\min}$ and $u^i_{\max}$ are lower and upper limitations for i-th control, respectively; and

$$\left| u^i_{\min} \right| = \left| u^i_{\max} \right|$$

Substituting derivatives

$$\dot{q} = \frac{dq}{ds} \dot{s} \quad \text{and} \quad \ddot{q} = \frac{dq}{ds} \ddot{s} + \frac{d^2 q}{ds^2} (\dot{s})$$

(12.57) yields dynamic equations expressed as a function of parameter s and its derivatives with regard to time:

$$A_i(s)\ddot{s} + B_i(s)\dot{s}^2 + C_i(s) = u^i \tag{12.59}$$

where:

$$A_i(s) = \left\langle M_i, \frac{dq}{ds} \right\rangle; \; B_i(s) = \left\langle \frac{dq}{ds}, \left(\frac{\partial M_i}{\partial q} - \frac{1}{2} \frac{\partial M}{\partial q^i} \right) \frac{dq}{ds} \right\rangle + \left\langle M_i, \frac{d^2 q}{ds^2} \right\rangle; \; C_i(s) = \frac{dG}{dq^i}.$$

Inequalities (12.58), considering relations (12.59), constitute pairs of parallel planes in phase space $(\ddot{s}, \dot{s}, s)$ of the form

$$\begin{cases} A_i(s)\ddot{s} + B_i(s)\dot{s}^2 \le u^i_{\max} - C_i(s) \\ A_i(s)\ddot{s} + B_i(s)\dot{s}^2 \ge u^i_{\min} - C_i(s) \end{cases} \tag{12.60}$$

By solving these inequalities for a fixed value of s we obtain a set of admissible manipulator motions contained in phase plane $(\ddot{s}, \dot{s}^2)$ (Figure 12.41).

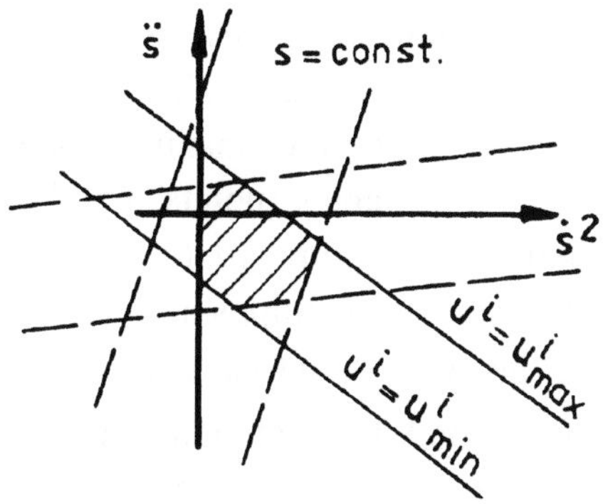

Fig. 12.41. Area of admissible manipulator motions

For any velocity $\dot{s}$ there are usually two corresponding acceleration limits $\ddot{s}$. The maximum velocity may have a single value (as shown in Figure 12.41) or an infinite number of them if $A_i(s) = 0$. The set of the values of admissible maximum velocities forms what is called a **velocity limit curve** in space $(s, \dot{s}^2)$ (Fig. 12.42).

The manipulator's motion along the trajectory given by relationship (12.56) is possible only below the area limited by this curve. If it causes the intrusion into the area above the limit curve, the trajectory realised by the manipulator will be deflected from the one given by equation (12.56), because it is impossible for the actuators to achieve the required joint space velocities.

The dashed curves in Figure 12.42 denote manipulator motions with the maximum possible acceleration, which is determined in phase space $(\ddot{s}, \dot{s}^2)$. The dotted curves denote motion with the least acceleration. Curves with such properties are called **extremals**. Curves, the first of which begins at point $(0, 0)$, while the second one ends at point $(s_f, 0)$ are called **boundary extremals**. With the curve of velocity limits known, we shall demonstrate a very simple algorithm for the parameterization of trajectory (12.56) which will not, in general, time-optimal (for time-optimal parameterization, see [12.26]) but the manipulator controls will not exceed the admissible values.

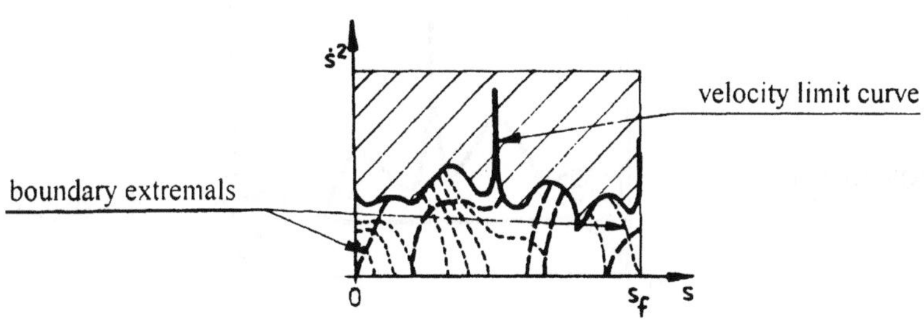

Fig. 12.42. Example of velocity limit curve

In the first step of the algorithm the lowest value of the velocity limit curve in interval $[0, s_f]$ is determined (point *a* in Figure 12.43).

Then for this value we find the maximal and minimal accelerations corresponding to the boundary extremals (points *b* and *e* in Figure 12.43). These values are usually lower than those at points $(0, 0)$ and $(s_f, 0)$ because as the velocity increases, the areas of admissible accelerations usually shrink (see Figure 12.41). In the next step we determine the curves in phase space $(s, \dot{s}^2)$(the first of which begins at point $(0, 0)$, while the second ends at point $(s_f, 0)$) corresponding to these accelerations as well as to the points of the intersection of these curves with the straight line of the lowest velocity limit (points *d* and *c* in Figure 12.43), called **switching points**. The manipulator moves in phase space $(s, \dot{s}^2)$ along the curve determined by points $(0, 0)$, $(s_f, 0)$ and the switching points 0, d, c and s_f (Fig. 12.43).

The total duration of the motion is described by the relationship

$$T = \int_0^{s_f} \frac{ds}{\dot{s}} \tag{12.61}$$

Given T and knowing the lowest value of the velocity limit, we find the switching instances and eventually searched function $s = s(t)$ which is substituted in relationship (12.56). The trajectory searched is expressed as $Q(t) = q(s(t))$.

If the boundary extremals intersect in the area below the lowest velocity limit (Fig. 12.44), we assume that the manipulator moves according to these extremals. In this case the algorithm determines the parameterization for which the obtained trajectory is time-optimal.

The phase space method is generally used to form algorithms for a time-optimal trajectory. Readers interested in some algorithms of time-optimal trajectory parameterization are referred to [12.26, 12.27, 12.28].

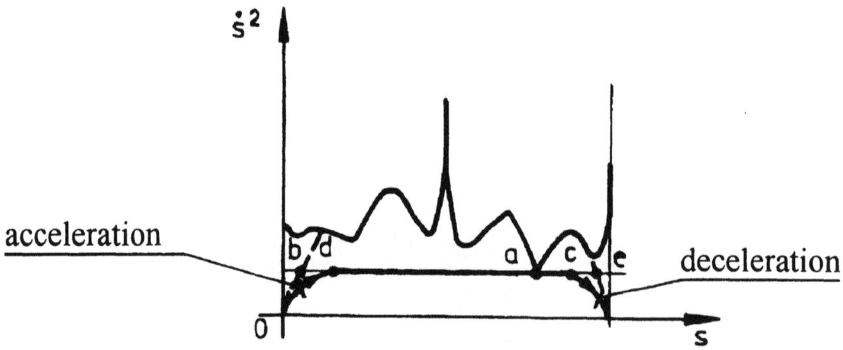

Fig. 12.43. Determining the lowest values of the limit curve

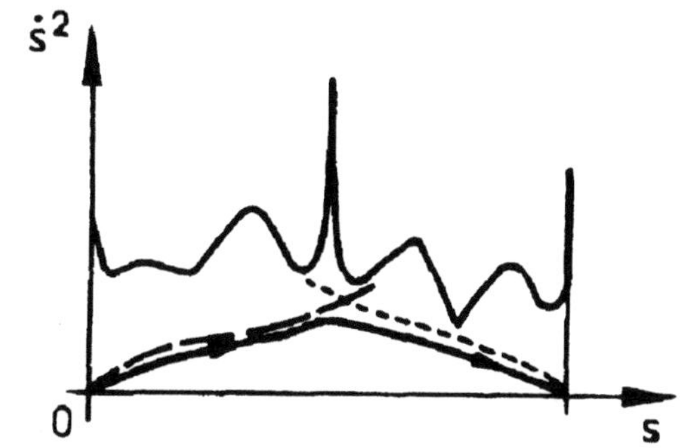

Fig. 12.44. Manipulator motion along the extremals

The fundamental advantage of phase space is that it reduces the task of finding n functions of time for joint space coordinates to the one-dimensional problem of finding of function $s(t)$. It can also be used when the limitations u^i_{min} and u^i_{max} are functions of q and $\dot{q}$. The fact that the method takes into account the dynamic characteristics of actuators implies the time parameterization of trajectory (12.56) for which the corresponding controls will not exceed the admissible values. The cost of these advantages is greater complexity of the method.

The stage of time parameterization ends trajectory planning. We have the result obtained time function $Q(t)$, where $t \in [0, T]$, now called the **reference trajectory,** which at each instance of time determines the admissible configuration, which should be realized by the manipulator as it moves from the initial configuration q_s to final configuration q_f.

Exercises

Do inequalities (12.12) present an accurate condition of collision detection, or it is possible that the objects will not collide, with the conditions satisfied.

Let us assume that there are two rectangles, described by coordinates of their successive vertices, given on the plane. Find the computationally efficient algorithm of detection of these polygons, which differe from those presented in chapter 12.4.1.

Program in PASCAL a method of trajectory planning using dynamic programming. The heading of the procedure should have the form

```
procedure UNKOLMPD (n: Integer; qs, qf, qmin, qmax: tab;
                    rowkin: tabk; obst: tabp; var traj:tabow);
```

where:

n is the number of kinematic pairs;

qs, qf, qmin, and qmax are initial and final configurations and lower and upper limitations of vector q, respectively;

rowkin is a table containing the link kinematic parameters;

obst is a table describing obstacles (point coordinates, plane equations describing the boundaries of obstacles); and

traj is a table of a mobile spatial trajectory.

Modify the algorithm of trajectory planning using the potential function method for the case in which the initial and final positions of the gripper are given without determining q_s and q_f.

Using the phase space method, write a simple algorithm for the time-optimal (approximately) planning of the trajectory given by equation (12.56).

12.5. References

[12.1] Baer A., Eastman C., Henrion M.: Geometric modelling: a survey. Comp. Aided Design, 11, 1979, 75-145.

[12.2] Rechiqua A.A.G.: Representations for right solids theory, methods and systems. ACM Comp. Surveys, vol.12, 4, 1980, 437-464.

[12.3] Jackings C.L., Tanimoto S.L.: Oct-trees and their use in representing three-dimensional objects. Comp. Graph. Image Process., 1980, 249-270.

[12.4] Meagher D.: Geometric modelling using octtree encoding. Comp. Graph. Image Process., 1982, 129-147.

[12.5] Bajcsy R., Solina F.: Sphere packing algorithm for sparce 3-D points. Technic. Report. Univ. Pensylvania, Philadelphia, 1984, 1-16.

[12.6] Mohr R., Bajcsy R.: Packing volumes by spheres. IEEE Trans. on Patt. Anal. and Mach. Intell., 1983, 111-116.

[12.7] Fangeras 0.0. and all.: Polyhedral approximation of 3-D objects without holes. Comp. Vision, Graph. Image Process., 1984, 169-183.

[12.8] Henderson T.: Efficient 3-D objects representations for industrial vision systems. IEEE Trans on Patt. Amal. and Mech. Intell., 1983, 609-618.

[12.9] Shirai Y., Suwa N.: Recognition of polyhedra with a range finder. Proc. 2nd IJCAI, 1971, 80-87.

[12.10] Zoutendijk C.: Mathematical programming Methods. North Holland, Amsterdam, 1976.

[12.11] Gilles I.W. Universal computer program for seeking a collision free path for an industrial robot. Proc. 15-th ISIR, 1985, 935-942.

[12.12] Gerke W.: Collision-free and shortest paths for industrial robots found by dynamic programming. Roboter-Systeme, 1985, 42-52.

[12.13] Bellmann R.: Dynamische Programmierung und selbstandpassende Regel-processe. R.Oldenbourg Verlag, Munchen, Wien, 1967, (there is a Russian translation as well as the original English version).

[12.14] Khatib 0.: Real-time obstacle avoidance for manipulators and mobile robots. Int. Journ. Rob. Research. 1986, 90-98.

[12.15] Krogh B.N.: A generalized potential field approach to obstacle avoidance control. Proc. Cont. Rob. Res. 1984, MS84-484.

[12.16] Ochocimski D.E., Golubiev Y.F.: Mechanika i upravlenye dvizenyem avtomaticeskogo sagayuscego apparata. Nauka, Moskva, 1984.

[12.17] Bobrow I.E.: Optimal robot path planning using the minimum-time criterion. IEEE J. Rob. Aut. , vol.4, 1988, 443-450.

[12.18] Gilbert E.G., Johnson D.W.: Distance functions and their application to robot path planning in the presence of obstacles. IEEE J. Rob.Aut.,vol.1,1985,21-30.

[12.19] Elsgolc L.E.: Diffieriencalnye uravnienia i variacionnye iscisleniye. Nauka, Moskva, 1965.

[12.20] Findeisen W., Szymanowski J., Wierzbicki A.: Theory and numerical methods for optimization. Polish Sc. pub., Warsaw, 1977 (*in Polish*).

[12.21] Lozano-Perez T.: A simple motion – planning algorithm for general robot manipulators. IEEE J.Rob. Aut., vol.3, 1987, 224-238.

[12.22] Herman N.: Fast three dimensional, collision free motion planning. Proc. IEEE Conf. Rob. Aut., 1986, 1056-1063.

[12.23] Gouzenes L.: Strategies for solving collision-free trajectories problems for mobile and manipulator robots. Int. Journ. Rob. Research, vol.3, 1984, 51-65.

[12.24] Nilsson N.J.: Principles of artificial intelligence. Tioga, Palo Alto, 1980.

[12.25] Stoer J.: Introduction to numerical methods. Polish Sc. Pub., Warsaw, 1979 (*in Polish*).

[12.26] Pfeifer F., Rainer J.: A concept for manipulator trajectory planning. IEEE J. Rob. Aut., vol.3, 1987, 115-123.

[12.27] Pfeifer F.: Manipulator trajectory planning and control. Proc. IFAC Theory of Robots, 1986, 325-330.

[12.28] Shin K.G., McKay N.D. Minimum time control of robotic manipulators with geometric path constraints. IEEE Trans. Aut. Contr., vol. 30, 531-541.

[12.29] Wong E.K., Fu K.S.: A hierarchical orthogonal space approach to three dimensional path planning. IEEE J. Rob. Aut. , vol.2, 1986, 42-53.

[12.30] Gibson J.J.: The senses considered as percepted Systems. Honghton Mitten co, Boston, 1966.

[12.31] Helmholtz H.L.:Handbook of Physiological Optics. JPC Southall,Dover, 1962.

[12.32] Roberts L.G.: Machine perception of three-dimensional solids, Optical and Electro-optical Information Processing. MIT Press, 1965.

[12.33] Coiffet P.: Interaction avec L'Environnement. Hermes Publish, 1981.

[12.34] Fu K.: Syntactic methods in pattern recognition. Academic Press, 1979.

[12.35] Popov E.V., Firdman G.R.: Algoritmiceskye osnovy intieliektualnych robotov i iskustviennogo intielekta. Nauka, Moskva, 1976.

[12.36] Chang C.L. Lee P.C.: Symbolic logic and mechanical theorem proving. Academic Press, 1973.

[12.37] Fu K.S., Gonzales R-C., Lee C.S.G.: Robotics, Control, Sensing, Vision, and Intelligence. McGraw-Hill 1987, Russ. trans. Izd. Mir, Moskwa 1989.

[12.38] C.I. Connolly J.B. Burns, R. Weiss.: Path planning using Laplace's Equation. In Proceedings of the 1990 IEEE International Conference on Robotics and Automation, IEEE, May 1990, 2102-2106.

[12.39] Kim I.O., Khosla P.:Real-time obstacle avoidance using harmonic potential functions, IEEE Trans. Rob. And Aut., vol 8, no. 3, 1992, 338-349.

Part Four

ROBOTICS - SELECTED ISSUES

13. ROBOTS AND MANIPULATORS FOR UNDERWATER OPERATION

13.1. Introduction

Torpedos are the oldest underwater vehicles. The first, made Britain in 1863 can be considered as the first underwater robot. In 1953 Riebekoft designed a robot for archaeological works in the Mediterranean Sea, using underwater television and remote control. In 1975 the RCV-225 appeared on the market; it is still used today for civil and military operations.

There has been a growing interest in the resources of the Atlantic Ocean in the last fifty years, especially since crude oil deposits were found at low depths offshore. In the seventies this interest intensified because it is possible to extract minerals, mainly manganese concretions, from depths of up to 6000 meters. In the first period of development (the fifties and sixties) the main technical means of deep sea penetration were manned underwater vehicles. Some of them were equipped with manipulators, at first very simple, to take ground samples or to pick up sunken objects. The first underwater manipulator was installed in 1961 in the manned bathyscaph "Trieste" [13.6].

In the seventies unmanned underwater sea vehicles began to be developed. These vehicles have all the characteristics of robots: mobility, capability of manipulation, the technical ability to observe the environment, and sometimes the ability to make decisions autonomously in typical, repeatable situations.

13.2. Description of selected design features

Because of the conditions in which they operate, most of manipulators used in underwater robots are similar to the anthropomorphic manipulators used, for example, in nuclear technology. Such manipulators can perform various underwater tasks requiring a gripper and changeable tools. They are located at the front of the vehicle, accompanied by spotlights, TV cameras and/or an underwater detection system (sonar) for observation of the sea bed and the surrounding water (Fig. 13.1.) There are also manipulators for quite different purposes, for example:

- arms to hold the robot near the wreck or hydroengineering structure, while another manipulator performs the actual task (Fig. 13.2),
- arms with terminating in vacuum cups to fix the robot when working near flat walls,
- manipulators-grippers with limited mobility, for lifting large, heavy objects (e.g. sunken torpedos),
- manipulators for fixing a detachable gripper to a selected, sunken object (Fig. 13.3),
- manipulators operating TV cameras, photo cameras, underwater detection systems or spotlights.

Table 13.1 lists examples of several typical manipulators used in underwater technology. The number of degrees of freedom does not reflect the mobility of the gripper. These manipulators run on hydraulic power (hydraulic .parameters are given in the last two columns).

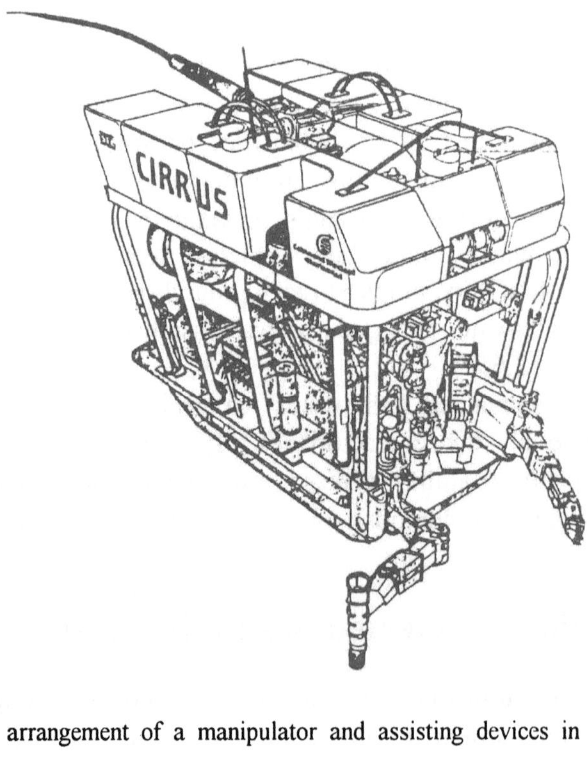

Fig. 13.1. Typical arrangement of a manipulator and assisting devices in an underwater robot [according to 13.5]

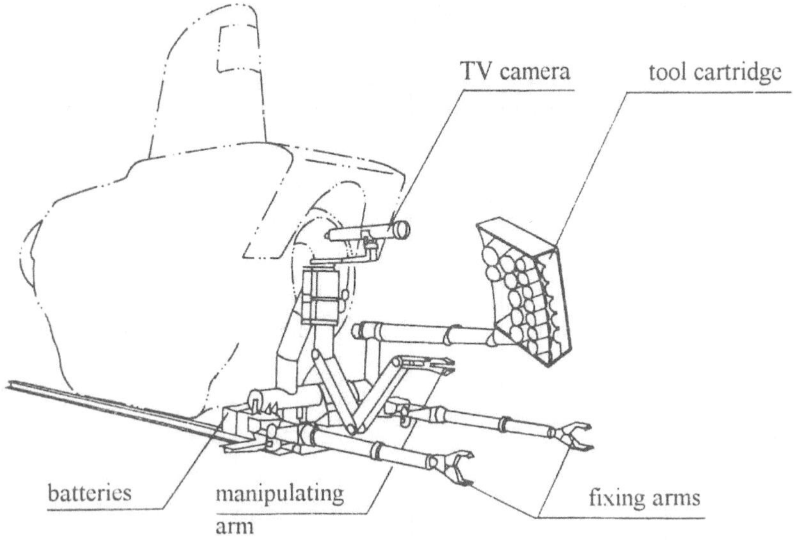

Fig. 13.2. Bow section of the Alvin underwater vehicle. Note the two arms holding the robot near the place of manipulation [according to 13.4].

Manipulators designed to execute underwater tasks, such as object lifting, rope cutting, surface cleaning, hydraulic tubing and cable joining, drilling, usually have four to five degrees of freedom, which (combined with the proper mobility of the manipulator itself) allows the manipulator to achieve any position and orientation of the tool in the workspace.

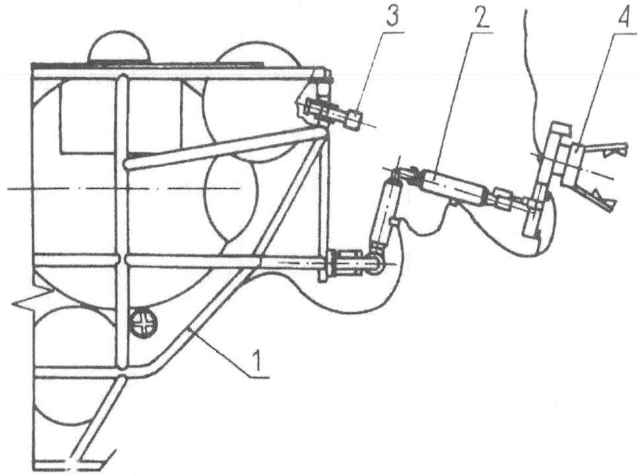

Fig. 13.3. Manipulator with gripper that can be detached after grasping the selected object [according to 13.2]: 1 - bow section of the vehicle, 2 - main manipulator, 3 - TV camera, 4 - detachable gripper.

Table 13.1. Specifications of manipulators used in selected robots of human-operated underwater vehicles; p - push-button control, j - joystick control, m/s - master/slave.

Manipulator name Manufacturer	Mass [kg]	Operating depth limit [m]	Weight in water [N]	DOF	Operating Range [m]	Mass [kg]	Control system	Hydraulic power pressure [MPa]	Hydraulic power delivery [dm^3/min]
Model 150 General Mills, USA	-	-	-	5	0.98	215	p	-	-
UWM 1.6/50 SAL Jugner Marine, Sweden	130	700	-	7	1.58	500	m/s	21.0	30.0
UWM 2.5/80 SAL Jugner Marine, Sweden	250	-	-	7	2.50	800	m/s	21.0	30.0
Beaver Autonetics Division, USA	-	600	-	8	1.80	215	p	-	-
Titan 7F Schilling, Great Britain	-	600	80	6	2.05	1200	m/s	17.0	13.5
Titan 5F Schilling, Great Britain	-	1000	60	4	1.80	1200	p	17.0	13.5
Norson Norson Power Ltd.	17	1000	110	6	1.12	160	j	5.8	2.25
Cirrus CWMarine Great Britain	-	-	-	5	1.30	500	m/s	-	-
DSRV USA	-	-	-	5	2.60	230	p	-	-
UT 004 Slingby Eng, Great Britain	-		-	7	2.00	-	m/s	-	-
NPG 600 IO Szczecin Technical Univ. Poland	150	600	1080	6	2.50	500	p	32.0	-
AG5 Marine Academy, Poland	90	500	350	5	1.80	300	p	16.0	-

More expensive manipulators with six to seven degrees of freedom are less common, although they are capable of performing tasks when the robot itself is not moving and (in the case of 7-DOF manipulators) of avoiding potential obstacles positioned in the workspace. This classification does not consider gripper or tool mobility, sometimes treated as an arm with an additional degree of freedom.

The fewer degrees of freedom of some very commonly used manipulators require the application of a suitable drive system, sometimes very complex, to provide the manipulator with dynamic positioning capabilities in the water. The drive system of this kind of robot usually consists of several fixed, or rotary nozzle propellers situated in different points of the hull.

In respect of the structure of the kinematic chain, underwater robots use manipulators with a suitable number of degrees of freedom and ranges of angular displacement in each joint that allow for robot arrangement of auxiliary equipment. The workspace should be placed low on the front of the robot as the majority of manipulations are performed near the bottom. The design of the workspace should also contemplate camera observation and lighting by spotlights. The kinematic chain of most robots contains only revolute joints, the first with a vertical axis.

The lifting capacity of manipulators used in underwater vehicles and robots ranges from 250 to 1000 N, and is sufficient for operating with typical underwater tools. In more recent constructions although, naturally, limited there has been a slight tendency to increase this force. Practical experience shows that under water the manipulator can not manipulate objects weighing less than 10% of its nominal lifting capacity [13.1]. Consequently, two or more manipulators of various (sometimes very high) lifting capacity are often installed, especially on large underwater ships. For example, project [13.2] requires a manipulator with a lifting capacity of 10 kN.

The average length of the whole kinematic chain of current underwater manipulators ranges between 1.0 and 2.0 meters. The longest manipulator ever built, installed on the DSRV (see Table 13.1), is 2.6 meters long and can lift a load of 230 N of load.

The arm length and lifting capacity of underwater robot manipulators are limited by the requirement of robot stability, which is easy to achieve when the vehicle is resting on the bottom, but is extremely difficult when it is afloat and being positioned dynamically. The characteristics of the manipulator are functions of the power output and configuration of the drive system. The length and dimensional proportions of the arms depends on the required workspace, which depends in turn on the shape of the bow section of the vehicle, where the manipulator, spotlights and TV camera used to observe the workspace are usually housed. Appropriate arm dimensions, the shape of the workspace and the location of spotlights and camera are often a decisive factor in underwater operations. An example of successful arrangement of manipulators and auxiliary equipment in the front of the vehicle is the manned vehicle Alvin (Fig. 13.4).

Most of the manipulators installed for sea exploration are prepared to cutt off and abandon cables in the event of collision or interference from an obstacle, wreck, etc, so that the vehicle will not become entrapped.

In underwater manipulators two types of actuators are used in arm drive systems: (more commonly) hydraulic and (occasionally) electric. The latter are an inheritance from designs of terrestrial manipulators.

Hydraulic actuation under water has several advantages over an electric drive. The most important is its resistance to the corrosive action of sea water. Manipulators with hydraulic actuators also feature good dynamic behaviour (as the ratio of driving torques to inertia forces is higher), overload resistance, and simple control.

In these manipulators the valve may be located in the arms, or, in a stationary case, at the beginning of the first arm. For example, in the Sligsby manipulator (Fig. 13.5) two hydraulic lines (one supply and one return line) are installed inside the arms, and two valves are located in the joints and controlled electrically by wires laid inside the arms.

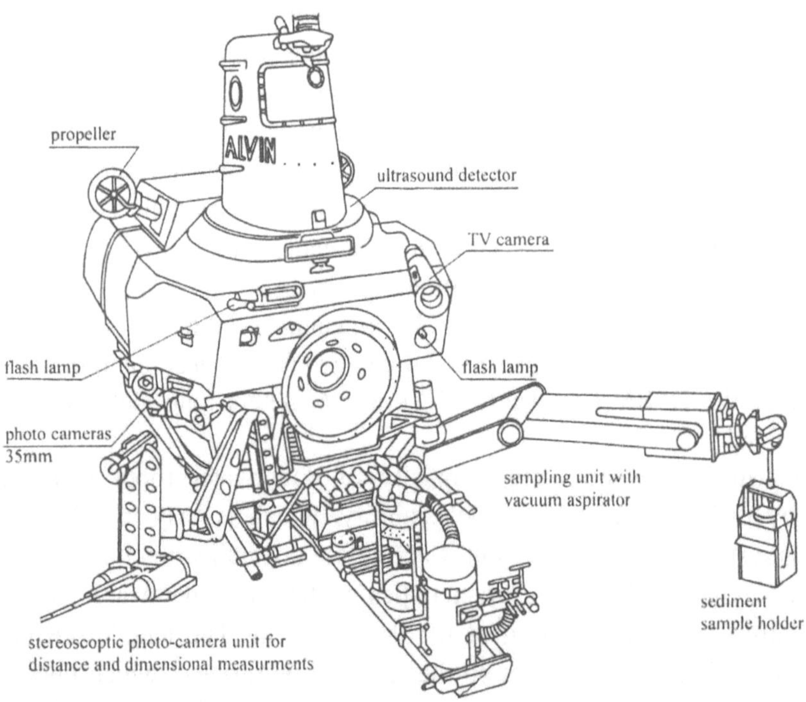

Fig. 13.4. Arrangement of manipulators and auxiliary equipment in the front part of the manned vehicle [according to 13.4]

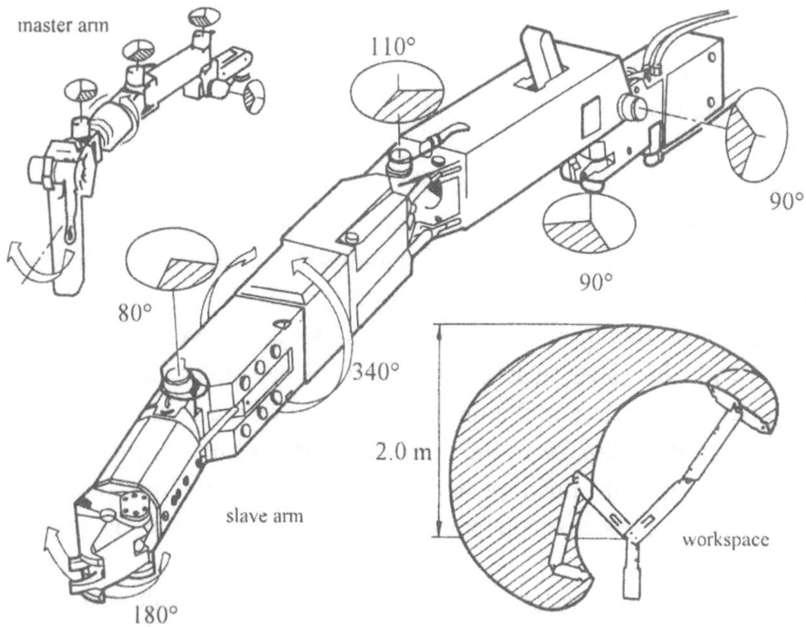

Fig. 13.5. Sligsby Inc. manipulator [according to 13.4]

In the Perry manipulator [13.1] the valves are located in an external case, and the seven actuators are serviced by seven supply lines. The hydraulic fluid returns from the valves through another seven return lines. Inside the arms the fluid passes through rigid pipes and in the joints through flexible hoses. The number of supply lines may be reduced by half if sea water is used as the hydraulic medium. This solution was first presented in a manipulator built by the Navy Electronics Laboratory in 1964 [13.6]; it is currently being employed by Osel company.

Electric actuators (the most common model is a high-rpm motor combined with reduction gear having a 1:1000 reduction ratio) are seldom used, as they are sensitive to corrosion in sea water, and pressure compensation is a difficult problem. An electric drive was applied for the main manipulator, for example, in the manned underwater vehicle Alvin [13.6]. Since it develops less power than hydraulic drive, it is usually employed to control the spotlights, cameras, underwater detection systems, etc.

13.3. Underwater robots: manipulator control

Most of manipulators of underwater robots and manned vehicles are controlled by open control systems with operator feedback. The operator is located in the underwater vehicle, or on the mother ship, where observes the action on a TV screen, taking decisions about further steps in the process. The following modes of manipulator control may be specified:

- push-button control;
- master-slave control;
- master-slave control with force feedback.

The first control mode is time-consuming and often requires additional feedback (by measuring the positions of manipulator arms), which increases costs and the complexity of the manipulator structure.

Much simpler in operation, but also much more expensive are master-slave type manipulators, with the slave structure installed on the underwater robot, or in the front of underwater vehicle and the master structure, which is kinematically similar, installed on the mother ship, or in the crew compartment of the underwater vehicle. The operator observes the position and orientation of the slave mechanism, and maneuvers the master mechanism appropriately, which motion is copied by the slave system.

Often the master structure can cannot be housed in the small space of the crew compartment of the underwater vehicle, and is replaced by a joystick, combining the features of push-button and master-slave control systems. Miniaturized master-slave position transmitters are also used in underwater technology, although these are usually less precise. An example of master structure miniaturization is the solution applied in the Titan SF manipulator made by Schilling: the transmitter is enclosed in a small suitcase (Fig. 13.6) [13.3].

Joystick control, together with the advantage of miniaturization, presents apparent disadvantages, as it is more difficult to use and does not provide the most precision.

Fig. 13.6. Control case of the Titan 5F manipulator [according to 13.4]

An important improvement of control systems has been the introduction of force and torque feedback, providing information about the forces and torques acting on the slave system. They are effects of reaction forces generated during operation on the object of manipulation. If the robot is floating in water, these reactions significantly disturb the positioning process, but the measurement of these forces and torques facilitates positioning and task execution. In a General Electric manipulator the information about the loading of the slave structure is obtained by measuring the pressures in actuator supply lines. This information is then converted into torques developed by the electric actuators of the master structure, which can be sensed by the operator. Another important virtue of such system is manipulator elasticity, i.e. its compliance to increased loading, impacts, operator mistakes, etc. This feature protects the manipulator against damage.

13.4. Effectiveness of the underwater operations carried out by robots

Underwater operations can be broken down into simple tasks to be performed by the manipulator. The robot's ability to perform these tasks is influenced by many factors, including the dimensions of the manipulator arms and the number of degrees of freedom of the manipulator, its lifting capacity, manoeuvrability, stability, visual and acoustic object observation systems, the control system of the arms, etc.

Table 13.2 shows a conventional classification of several typical underwater operations broken down into manipulation tasks. Table 13.3. shows the relationship between the ability of the manipulation system to perform tasks of different complexity and the control system applied. From these tables it can be reasoned that the method of closing the control system significantly improves the robot's performance. One must nevertheless remember that the application of manipulators with an increased number of degrees of freedom and using force feedback entails a considerable increase in costs.

Consequently, selecting the appropriate manipulator for the underwater robot requires a thorough analysis of the frequency and complexity of the tasks peformed. If the robot is to perform very complex tasks relatively seldom, a simpler and less expensive manipulator should be applied. Then more complex tasks can be performed exploiting the robot's manoeuvrability and extending the amount of time allotted in execution of the task.

The time of execution for some underwater operations may be very long. It is assumed that an expensive and complex master-slave manipulator with force feedback is about ten times slower than a frogman performing the same task, but still ten times faster than a simple push-button controlled manipulator. The conditions of observation of the object on which the work is carried out plays an important role. Poor visibility resulting from unsufficient lighting or turbid water may prevent the execution of a specific task.

Table 13.2. Break-down of some typical operations performed by underwater robots

Underwater operations	Manipulator tasks										
	1	2	3	4	5	6	7	8	9	10	11
Removing debris from the bottom	■	■	■	■							
Installation and dismantling of hydroengineering systems		■	■		■	■			■		■
Lifting operations		■	■				■		■		
Connecting and disconnecting cables and ropes	■	■	■								■
Mounting anodes on underwater structures							■	■	■	■	
Opening and closing valves	■							■	■		
Burying cables and pipelines under the bottom	■	■	■	■	■						
Operations are following: 1 - Grasping; 2 - Lifting/Lowering; 3 - Pushing/Pulling; 4 - Sucking/Flushing; 5 - Cementing; 6 - Welding; 7 - Driving bolts; 8 - Drilling; 9 - Unscrewing bolts; 10 - Cutting and sawing; 11 - Connecting cables.											

Table 13.3. Influence of the control system on the the ability of the manipulator system to perform tasks of different, increasing complexity

Task type	Control system applied		
	1	2	3
Observation	■	■	■
Lifting/Lowering	■	■	■
Module replacement	■	■	■
Rope cutting/Debris removal		■	■
Structure cleaning			■
Using tools/Opening and closing valves			■
Connecting ropes and hydraulic lines			■
Joining pipeline flanges			■
Drilling and tapping			■
Complex assembly and disassembly			■
Control systems: 1. Simple, 4-DOF manipulator with push-button control; 2. 7-DOF master-slave manipulator; 3. 7-DOF master-slave manipulator with force feedback			

Table 13.4 gives the results of the studies on the influence of visibility conducted by Hackman in 1982. The time requires for task execution using various control systems and can also be compared.

Other test results concerning task time distribution (Wernli 1985) illustrate the specificity of underwater manipulation. Alvin, the underwater vehicle tested was integrated with a tool cartridge, TV camera and spotlights (Fig. 13.2).

Table 13.5 reveals that manipulation takes only one-third of task time, the rest of the task time is used by the operator to plan the next step and correct the position of the camera or spotlights. Thus it is frequently need that a simple manipulator with few degrees of freedom, but operated by a well-trained operator would be the best solution. This reasoning is confirmed by a study on the ocean exploration market, where together with extensively commercialized complex master-slave manipulators with 7 or 8 DOF and force feedback, simple and effective robots and manned underwater vehicles with simple manipulators can also be found.

Consequently in designing of underwater manipulation systems the following guidelines should be:

- precise identification of possible tasks (as suggested by the authors of [13.2, 13.4, 13.6]); the manipulator should be designed to perform these tasks;
- a compromise between expensive, master-slave manipulators with force feedback and simpler, cheaper manipulators with fewer degrees of freedom and greater manoeuvrability of the manipulator base (carrier);
- control systems should be developed mainly to cope with the situation under water and to be with apply master control systems selecting pre-programmed operation to be executed without any human interference.

Several recent designs of underwater vehicles with various manipulation systems are described below.

Figure 13.7 presents an interesting comparison of operating and maintenance costs for different models of underwater vehicles. Figure 13.8 shows an autonomous inspection vehicle for inspecting underwater objects.

Table 13.4. The influence of visibility on the ability of the system to perform underwater tasks (flange joint disassembly test)

Visibility	Task execution time [s]	
	master-slave control	master-slave control with force feedback
Good	122	88
Sufficient	236	130
Poor	573	251
Very poor	∞	1180

Table 13.5. Underwater robot worktime distribution [s]

Task	Operation	
	without tools	with tools and force feedback
Decision making	50	37
Movements of arms	33	30
TV camera adjustment	17	11
Work with tools	-	22
Total	100	100

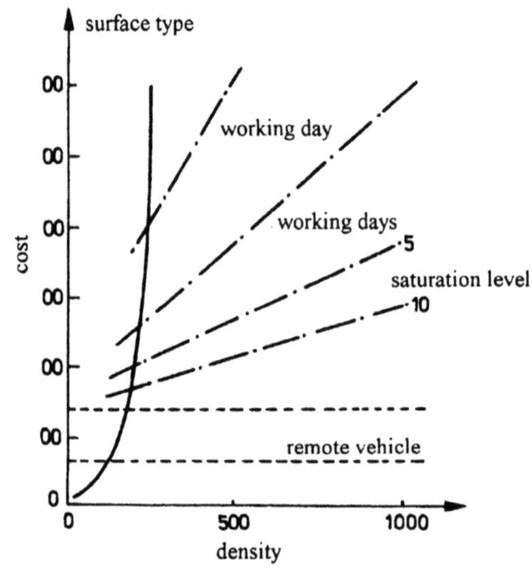

Fig. 13.7. Comparison of costs of operation for different models of underwater vehicles.

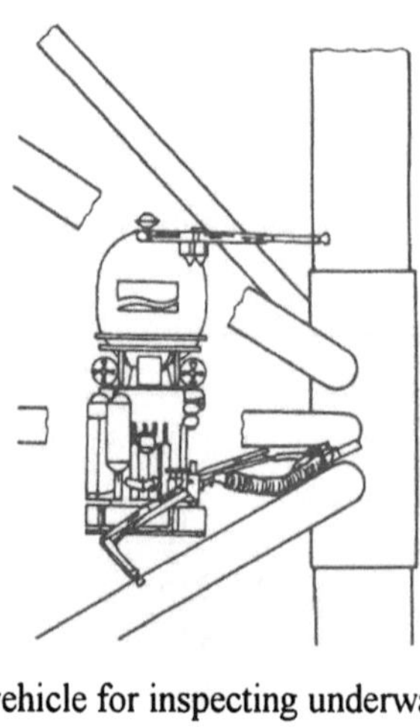

Fig. 13.8. Autonomous inspection vehicle for inspecting underwater objects [according to 13.7]

Figure 13.9 shows the Titan underwater system, a multistage system for controlling underwater vehicles equipped with manipulators. Figure 13.10 shows a towed vehicle with standard equipment (Marine Physical Laboratory, Scripps Institution of Oceanography, San Diego, California).

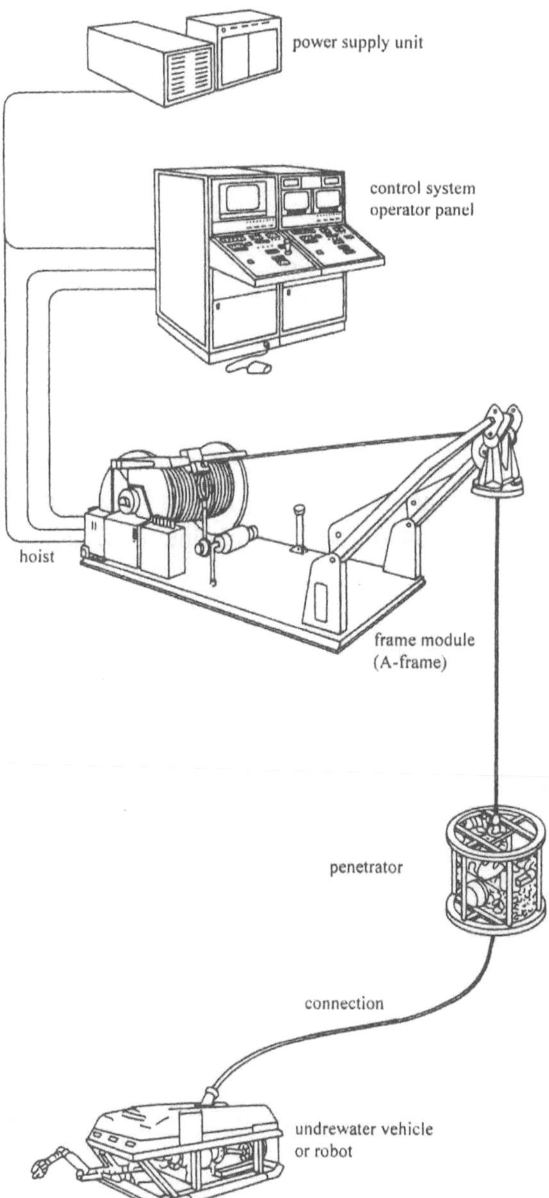

Fig. 13.9. The Titan underwater system [according to 13.7]

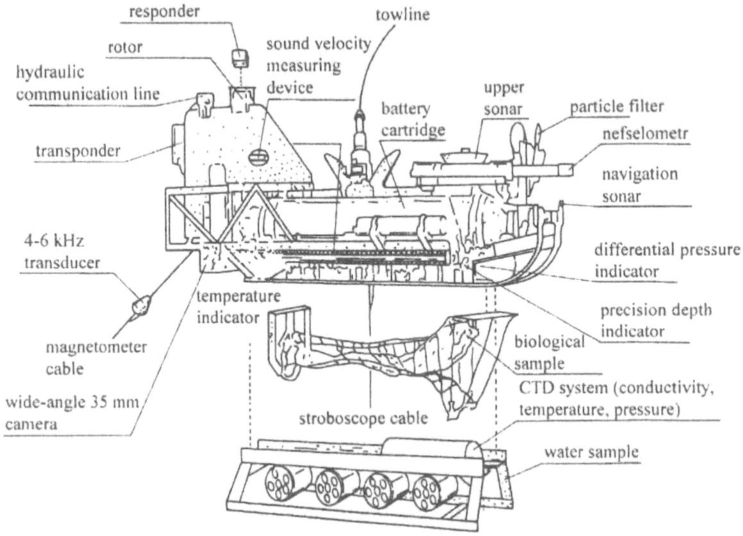

Fig. 13.10. Towed vehicle with standard equipment [according to 13.7]

The last two examples are a cable-guided underwater vehicle, the *sea dog* (British Telecom Inter. Southampton, UK)(Fig. 13.11), and the SCAN inspection robot (Underwater Maintenance Co. Ltd., Southampton, UK)(Fig. 13.12).

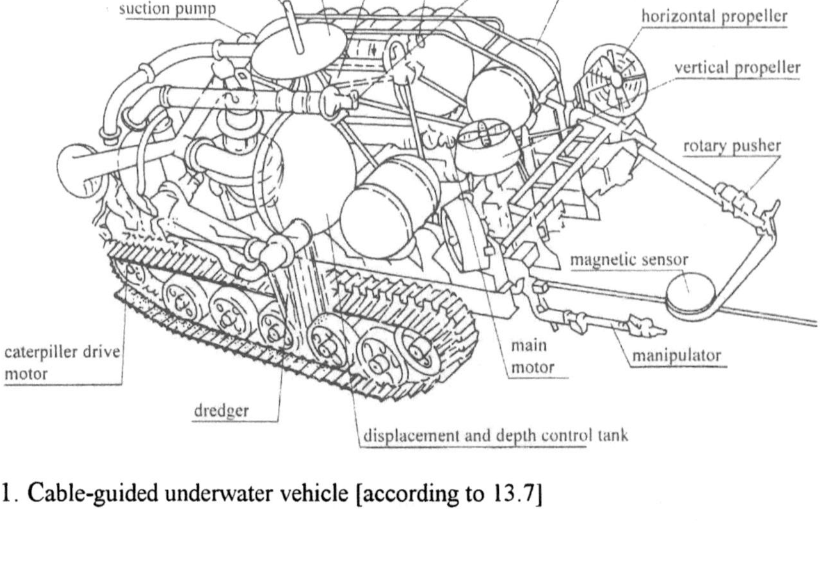

Fig. 13.11. Cable-guided underwater vehicle [according to 13.7]

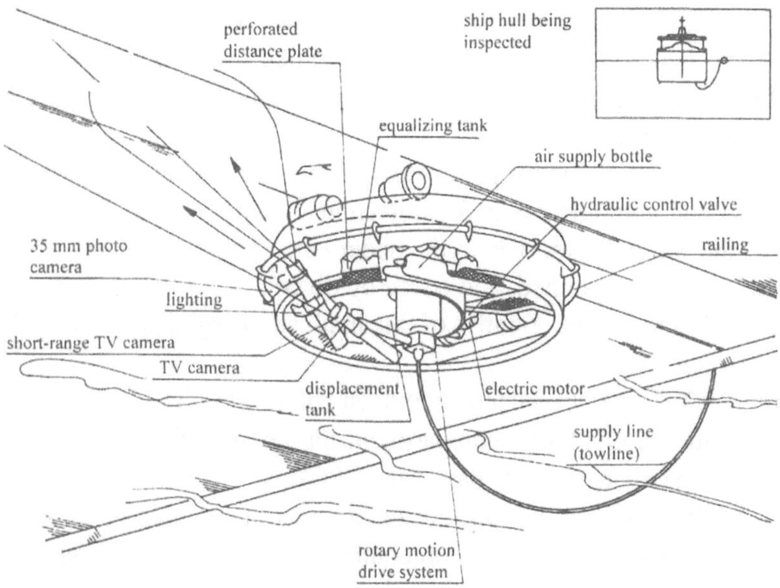

Fig. 13.12. SCAN ship hull inspection robot [according to 13.7]

We anticipate further development of manipulators and robots for underwater research in the near future. Interesting studies on robots for cleaning ship hulls are currently being carried out in Poland.

13.5. Underwater Robotics in Poland

13.5.1. Early developments

Direct underwater investigations using divers method began in Poland in 1939 under the supervision of R. Wojtusiak. The main goal of these investigations was to study the fauna and flora of Puck bay and other areas of the southern Baltic sea. These investigations were continued by Professor Wojtusiak's team from the Jagiellonian University in 1947-50 and 1960. In 1953 the divers from the Institute of Fishing in Gdynia under chairmanship of P. Ciszewski started underwater investigations in classical diving suits. To avoid problems with the cables connecting the diver on the bottom with the ship new French equipment was introduced in 1956. Direct observations of fish were made employing a vehicle connected with the ship by cable. In the seventies more advanced underwater vehicles and some underwater laboratories (bases) were designed: DELFIN I,

DELFIN II, MEDUZA I, MEDUZA II, and together with the buoy caisson bell GENUOR for geological research. Poland joined the underwater international program for the acceleration of progress in the botinautical field in the North-East Atlantic Ocean and Baltic Sea.

Table 13.6 shows the main parameters of the underwater bases, the MEDUZA lab was used for geological and hydrotechnical experiments. In the years 1970 through 77 the single man, wet type underwater vehicles BŁOTNIAK 1A and BŁOTNIAK 1B were designed, and tested by divers.

Table 13.7 shows the family of manned submergibles built and tested in different situations.

Figure 13.13 shows the general arrangement of the LTS-7 „GRZES" two-man submergible with the GRP pilot compartment for a depth of 200 meters. In 1977 the prototype was tested in Gdańsk bay. The NPG-600 is of the manned vehicle for depths up to 600 meters. It was designed to collect geological and biological samples. Two configurations were contemplated: ball + cylinder and ball + ball. It was equipped with two manipulators with hydraulic actuators. The gripper was of the parallel type (0-150 mm) with force about 800 N. It was connected to the front wall of the vehicle. The control system is of the copying type.

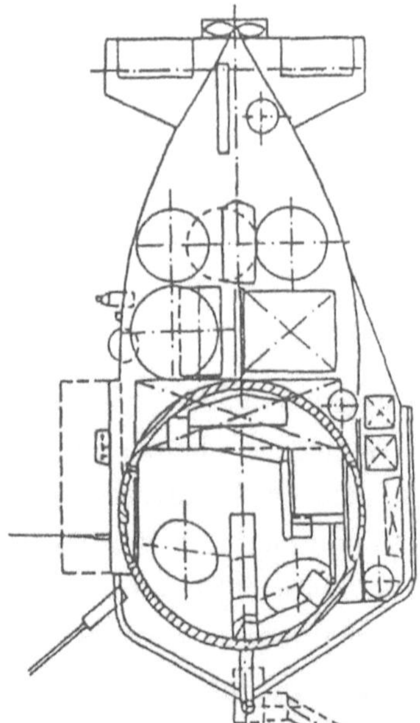

Fig. 13.13. General arrangement of LTS-7 „GRZEŚ", two-man submersible with GRP pilot compartment for 200 m

Table 13.6 Underwater bases

Base name	Mooring place	Acceptable max diving depth	Application working depth	Crew	Time of working		Power supply			Communication	TV	Main dimensions			Weight	Decompression	Carriage	Employment place
					Normal	Hard operating conditions	Manner	Electric energy	Respiration mixture			Length	Width	Height				
		[m]			days							[m]			kg			
Meduza I	Gdynia	50	24	2	3	–	from land	–	37 % O_2 63 % N_2	diving telephone	–	3,6	2,2	1,8	3000	–	–	Kłodno Lake
Meduza II	Sopot	60	24	2	6	–	from land from ship	–	air	diving telephone	–	4,49	3,34	2,8	8000	in underwater base	tow	Gulf of Gdańsk Baltic Sea

Table 13.7. Manned underwater vehicles

Vehicle name	Maximum diving depth	Speed		Autonomy			Service	Lookout men	Useful load	Length	Weight
		working	top speed	standard working	emergency normal	emergency max					
	[m]	[kn]		person-hour	hour	person-hour			[kg]	[m]	[kg]
Delfin I	35,0	tow		4,5	–	4,5	1	–	–	2,60	500
Delfin II	60,0	tow		4,0	–	4,0	1	1	–	3,45	1500
Błotniak	50,0	3,5	5,0	10÷12	–	–	1	1	–	2,70	3500
GTS-7 Grześ	200	–	2,5	8,0	32	–	1	1	150	4,00	3000
NPG-600	600	2,3	–	10,0	72	–	2	2÷3 divers	600÷800	8,80	27 600

13.6. Present developments

Table 13.8 shows the family of underwater submergibles designed in the years 1985-1990. A small, modular, remotely operated vehicle CORAL for 400 meters was developed in the Department of Underwater Technology. After laboratory tests the CORAL ROV system was used for underwater observation of the PETROBALTIC drilling rig. The design, technological and operational experience gained from the CORAL ROV permitted modifications and development of new types of underwater vehicles [13.3, 13.12, 13.14, 13.15].

The CORAL ROV System consisting of unmanned underwater vehicle (Fig. 13.14) and surface equipment, allows:

- real-time documentation of observations,
- simple manipulation,
- searching and identification to a depth of 400 meters.

At present, a new, lighter, and smaller version of CORAL AT that is more powerful and faster and provides better data transmission be prepared. The remotely controlled underwater vehicle MONA-400 was designed at the Department of Informatics and Automatics of Technical University of Szczecin. The main purpose of this vehicle is the inspection of the sea bottom, and the identification and location of the objects found. MONA-400 has a modular configuration and comes in different models. At present a new version allowing motion programming is in preparation.

The underwater vehicle TUM-100 was designed for the transportation of monitoring devices [13.1]. An overall view of this vehicle is shown in Figure 13.15 [13.14, 13.15, 13.16, 13.17, 13.18, 13.19].

Fig. 13.14. CORAL ROV system

Table 13.8. Unmanned underwater vehicles

Underwater vehicle	Produced by	Basic task	Working depth	Main dimension L×B×H	Weigh	Payload	Speed	Propulsion system	Power suply	Supply conduit	Equipment	Serwis
			[m]	[m]	[kg]	[N]	[m/s]			[m]		person
AITS-1	Technical University of Szczecin 1985	experimental vehicle	10	1,48×0,83×0,75	144	150	0,2	direct current motors	Accumulator 12V, 170 A·h	25	Video camera, reflectors	1
MUNA	Technical University of Szczecin 1988	inspection, target search, defects diagnosis, executing of underwater's small work	400	2,50×1,00×1,00	500	500	3,09	6 induction asynchronous squirrel-cage motors	3×380 V 50 Hz	–	TV camera, 2×150W reflectors, azimuth stabilization head-up, depth stabilization, depth gauge, gyro, log	3
KORAL 100÷400	Technical University of Gdansk 1990	inspection, observation	100 ÷ 400	0,60×0,60×0,70	90	150	1,5	5 electric-motors	–	100 or 400	Color or monochrome video camera, 3×100 W reflectors, depth gauge, echo sounder, magnetic compass, photo camera, sonar,	2
NUR	Technical University of Szczecin 1990	inspection, diver work helping, carriage	300	4,5×2,5×2,7 with grab 5,2×2,5×2,7	7130 with welding device 7630	12350	1,5	7 hydraulic motors	3×380 V 50 Hz	450	2 TV cameras, 2×100 W reflectors, portable camera, 650 W portable reflector, log, depth gauge, echo sounder, sonar, gyro, transponder	4
TUM-100	Technical University of Szczecin 1990	inspection, observation	100 or 600	1,00×0,80×0,63	70	–	1	4 direct current motors	220 V 50 Hz	100 or 600	TV color camera, photo camera, 2×150 W reflectors, photo-flash lamp, echo sounder, compass, magnetometer.	2

The underwater vehicle TUM-100 was designed for the transportation of monitoring devices [13.1]. An overall view of this vehicle is shown in Figure 13.15 [13.14, 13.15, 13.16, 13.17, 13.18, 13.19].

The underwater vehicle NUR was designed in 1990. The main purpose ot this unit is inspection, transport and support of deliveries. The general view of this vehicle equipped with a manipulator is shown in Figure 13.16.

Figure 13.17 shows the tug HOLONUR adapted to complex underwater searches using a miniature sonar device. The dimensions are 1300x25 mm, its max. velocity is 0.9-1.1 m/s, and its weight in water, -0.1 daN, allowable distance 5000 meters.

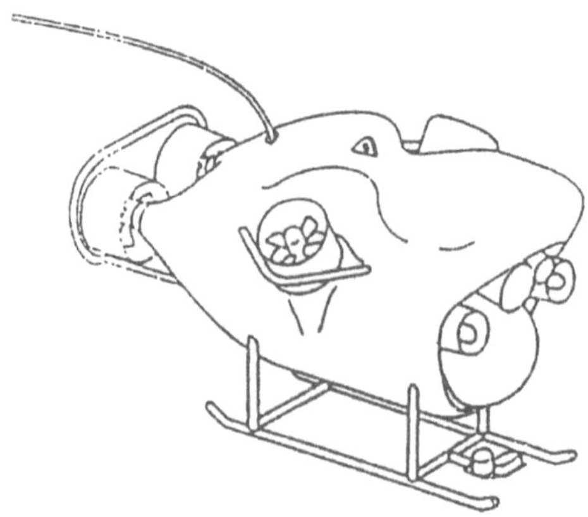

Fig. 13.15. Vehicle TUMMY-100

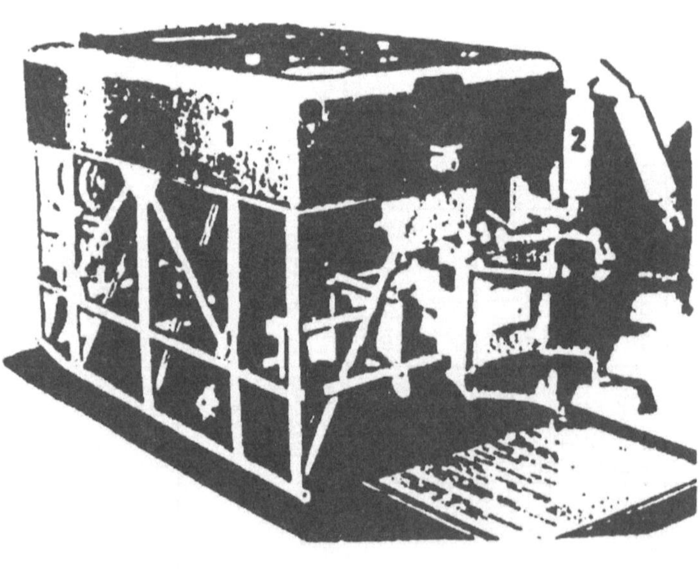

Fig. 13.16. The underwater vehicle NUR

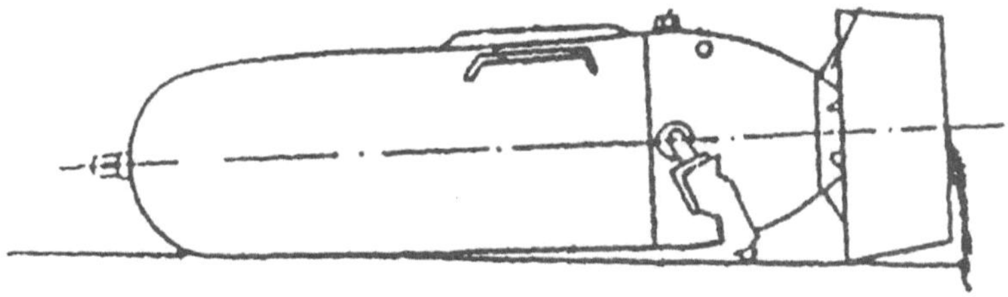

Fig. 13.17. The HOLONUR tug

13.7. Ship hull cleaning underwater robot [13.20]

13.7.1. Main goal of this design

One of the main reasons for ship dry docking is to eliminate marine growth that has accumulated on the sides and flat bottom of the ship. This growth causes great resistance to the passage of the hull through the water, reducing the speed of the vessel, and requiring extra power from the main engines to maintain a reasonable schedule, which in turn results in greater fuel consumption. Toxic paint, the most common anti-fouling agent used to destroy a marine growth, is an expensive product and must be applied in dry dock. Considering such factors as ship maintenance service and the toxicity of the anti-fouling paint, the average time on which fouling appear can be estimated at six months. Dry docking should then be scheduled. Moreover, dry cleaning is more difficult to perform than wet, since the fouling quickly firms out of the water.

The cheapest and most convenient alternative to regular dry docking is underwater (UW) cleaning. When carried out while the ship is loading or discharging, underwater hull cleaning requires no extra lay-up time and does not interfere with cargo operations or other shipping operations. After cleaning, 95 percent of the ship's original speed is regained.

13.7.2. General idea

There are underwater cleaning systems controlled either by divers or an electronic equipment. Figure 13.18 shows the structural scheme of the mobile cleaning tool of the robot. The numbers indicate: the bristle brush head (1), driving gear of arms (2), drive wheel (3), driving gear (4), electrohydraulic rotary joint (5), frame (6), and ballast tanks (7), (8).

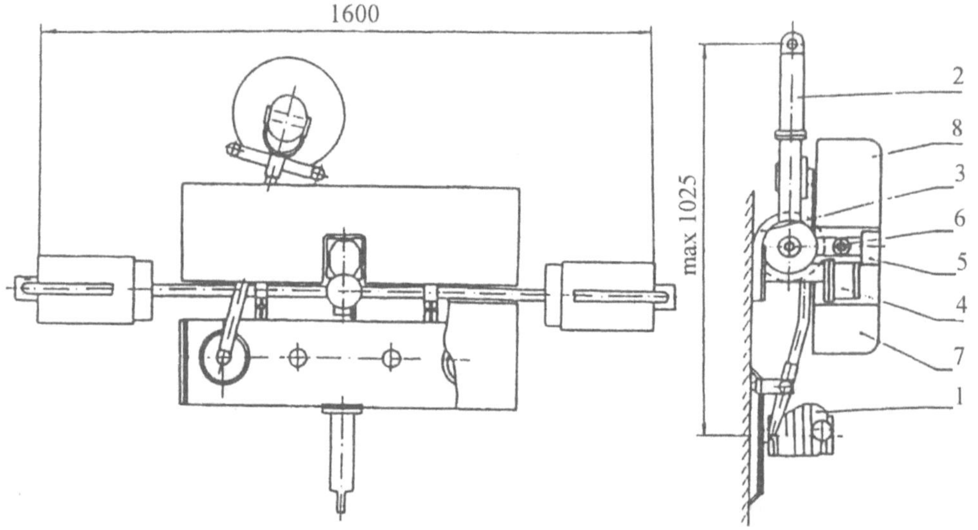

Figure 13.18. Structural scheme of the mobile cleaning tool of the robot

The tool contains two wheels driven by axles propelled by a hydraulic engine. The wheels are controlled by two electromagnetic clutches and gripped to a cleaned hull on the principle of magnetic power attraction.

Two rotating brushes driven by hydraulic engines are mounted on two swinging arms which are propelled by a hydraulic servomotor. The brushes scrape of the marine growth The bristles lie flat and smooth on the vessel's surface producing a shaving effect that does not injure the underlying coat of paint. The system may also be supplied with the appropriate television or camera and used for underwater inspection.

The working movement of the arms of the cleaning heads is synchronized with the rate of the robot-tool's movement. The tool can be directed by applying the clutch and brake of the left or right wheel respectively.

An electrically controlled electromagnetic distributor controls the mechanical movement of the hydraulic engine, and switches on and off the cleaning heads.

Some technical data: hydraulic actuators 50 l/min, 16 MPa; electromagnetic force for connecting with the body -2.3 kW, cleaning efficiency 20-30 m/min at a velocity of 15-20 m/min; width of the cleaning track, 1.6 m.

13.7.3. Concept of robotization

It is assumed that the cleaning robot will be able to operate irrespective of surface weather-conditions, place of mooring (sheltered area, or open sea), or underwater visibility and drifts (up to 4 knots). It should be a multi-task system for both small and large vessels, and it should operate as a mobile supply station for the manual cleaning of small surfaces (a rudder for instance) as well.

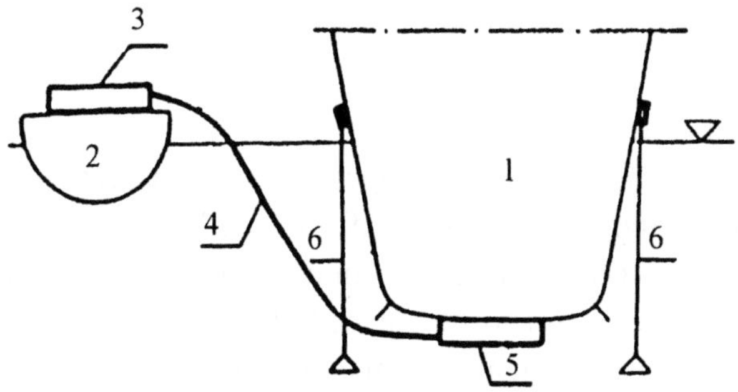

Fig. 13.19. General concept

The general concept of the robot system is shown in Figure 13.19. The robot tool equipped with cleaning heads and a drive is moving on the underwater surface of the ship`s hull (1). The cleaning tool (5) of the system is supplied with all the necessary sensors and actuators, as well as with a local microprocessor controller. An electrohydraulic turning joint and a set of electrical and hydraulic leads (4) serve as a central system (data and power-supplying) base connecting the working tool with a host computer (3) placed on an assistant vessel (2). A hydroacoustic subsystem is applied to localize the current position of the tool in relation to the vessel's hull. The markers (6) of the localization subsystem are hanging down on both sides of the ship.

The robot tool is put into the water using the small crane on the assistant boat. Note that the design of the two ballast tanks of the tool makes them unsinkable. When the cleaning brushes start to work at the moment the tool itself starts to move towards the hull of the vessel providing the tool with an appropriate orientation. The movement results from the action of small propellers which are commonly used to press the tool onto the surface being cleaned. When the tool is close to the vessel, the electromagnets are switched on so as to press the tool against the side of the vessel. Now the tool is able to move over the hull's surface using its hydraulic drive. All the tool's operations are controlled by an operator using the host computer console. The set of working movements of the cleaning tool over the surface concerned (a side or a flat bottom of the hull), are selected by the system operator; they compose the working cycle. One longitudinal working movement produces a clean strip of 1.8 meter. When the tool reaches the end of the assignated area it is turned around and comes back. The essential curvature of the hull is cleaned manually by a diver (Fig. 13.20).

The tool of the robot system can also be used for inspection, when equipped with a TV camera. The following issues must be considered in designing the system:

- current tool-position identification
- local control of the cleaning tool
- central computer-aided robot control.

Taking into account possible changes of the vessel's position during the cleaning, caused by wind, waves, and drift, all earth-based position-identification systems (the gyro compass, for instance) must be eliminated. The base should be related to the hull of the vessel to be cleaned. The solution proposed here uses an underwater distance - measurement system composed of several response devices.

The following functional operations must be performed by the local robot-tool controller:

- mechanical operations (forward movement, turning, start, stop of cleaning operation, switching the electromagnets on and off and the like);
- auxiliary operations (sensing operations, such as detection of curvatures and obstacles, indication of depth and level, the measurement of the cleaned surface, periodic responder calling, time delay measurement, as well as on/off auxiliary manipulations e.g. TV camera and light control);

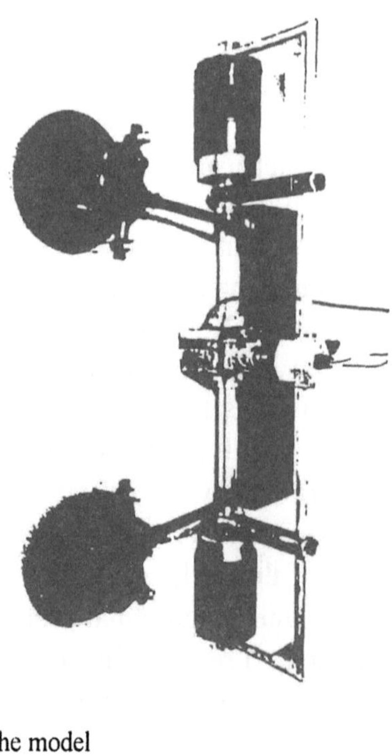

Fig. 13.20. General view of the model

- communication operations (service of the host computer messages, and on-going supply of information about the current robot-tool state, including data for position calculation, to the host).

The host computer is responsible for the following tasks:

- receiving and displaying the data of the current status of the cleaning process;
- interpretation, registration and transmission of operator commands (teach - in);
- preparing and editing reports on the cleaning operation.

13.8. References

[13.1] *Rowiński L.* et al.: Design and Construction of Abbysal Vehicles. Ship-Building Institute of Technical University of Gdañsk (manuscript), Gdańsk 1985 (*in Polish*).

[13.2] Proposal and Design of Manipulators for Lock-Out Abbysal Vehicles. Team work under the guidance of *H. Szymański* for the Ship-Building Institute of Technical University of Szczecin. Warszawa 1989 (*in Polish*).

[13.3] Under Water Manipulators. SAL Junger Marine. Hemvarnsgatan, Sweden, 1981.

[13.4] Undersea Equipment. Norson Power Ltd., Glasgow 1983.

[13.5] *Liddle D.*: Cirrus-New Underwater Cable, Burial, Location and Repair Vehicle. Int. Underwater System Design, 1/2 1986.

[13.6] Ocean Technology. Team work edited by *J.F. Brahtz*. Sea Publishers, Gdañsk 1974 (*in Polish*).

[13.7] Intern. Encyclopedia of Robotics: Applications and Automation. Vol. 1, Ed. by *R.C. Dorf*, Wiley 1988 V-Undersea Robots.

[13.8] *Ostachowicz W., Rożnowski G.* et al.: Robot for Underwater Operation. Study ordered by IPPT PAN/CPBP 02.13, manuscript, 1986 (*in Polish*).

[13.9] *Ostachowicz W., Rożnowski G.* et al.: Development of a Method and Technical Assumptions of Industrial Robot Application to Underwater Works. Overview of Equipment for Underwater Ship Hull Cleaning. Study ordered by MERA-PIAP/CPBR 7.1, manuscript, 1987.

[13.10] *Ostachowicz W., Rożnowski G.* et al.: Robot for Underwater Works at Ship Hull - Concept of Robot Positioning and Control. Study ordered by MERA-PIAP/CPBR 7.1, manuscript, 1988.

[13.11] *Kitowski Z., Morecki A., Ostachowicz W.*: Underwater Robotics in Poland. Proceed. of the 24th ISIR, 1993, Tokyo, Japan.

[13.12] *Rowiński L.A., Narewski M.*: Underwater Technology Development in Ship Research of Technical University of Gdañsk. Proc. Black Sea'90 Conference, Varna, Sept. 17-21, 1990.

[13.13] Exploitation Experiments with Remotely Controlled Underwater Vehicle KORAL. Report (*in Polish*).

[13.14] *Orlowska A., Skorski W., Matejski M.*: Remotely Controlled Instrumentation Carrier for Depths down to 400 m. Proc. of 5th Conference on Design and Construction of Ocean Technology Objects. Szczecin 1989 (*in Polish*).

[13.15] *Bednarski L.*: Abbysal Vehicle for Divers. Proc. of 5th Conference on Design and Construction of Ocean Technology Objects. Szczecin 1989 (*in Polish*).

[13.16] *Rowiński L.*: Plastic Vessel for Underwater Observation. Ship Building, 1978 (*in Polish*).

[13.17] *Bednarski L.*: NPG-P0 - Design and Proposal. Technical University of Szczecin, 1989 (*in Polish*).

[13.18] *Bednarski L.*: Remotely Controlled Underwater Carrier for Technical Equipment - a Concept. 13th Scientific Session of Ship Builders, Gdañsk 1988.

[13.19] *Graczyk T.*: Unmanned, Remotely Controlled Abbysal Vehicles - Designs and Applications. Ship Research Institute of Technical University of Szczecin, 1991.

[13.20] *Ostachowicz W., Rożnowski G.* et al.: Robot for Underwater Tasks by Ship Body. Under CPBR 7.1, Gdańsk 1988.

14. INDUSTRIAL ROBOTICS

Industrial robotics deals with the application of robots and manipulators for welding, casting, painting, surface protection, machining, assembly, and many other industrial processes which are physically demanding, hazardous, or monotonous.

In recent years the application of industrial robots has expanded beyond the engineering industries, and penetrated the mining, ship building, agriculture, transport, and construction, telecommunications, chemistry, and forestry sectors. Today robots are increasingly employed in underwater and in space, as well as in the service sector.

14.1. Basic terms. Systematics of robots

We assume the following definition of robot [14.1, 14.2]: A robot is a machine that can be programmed to perform tasks involving manipulative or locomotive actions under automatic control (ISO/TR 8373 3.3).

We can divide robots into four types.

1. Sequence robot

Definition: A robot having a sequence control function (Fig.14.1).

Note: Sequence control (ISO 2806 3.1.07) is a system of control in which a state of machine movements occurs in a desired order; the completion of one movement is not specified by numerical data.

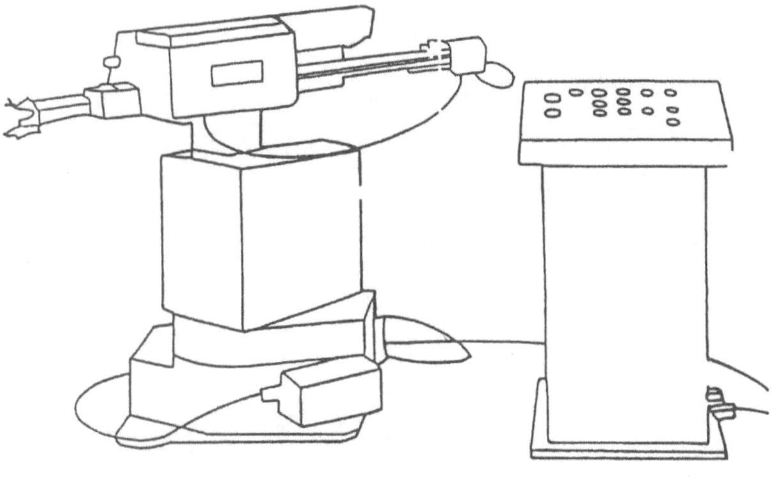

Figure 14.1. Sequence control robot [14.9]

The operations of the robot are binary, i.e., on/off, start/stop, and the trajectory of an operation between its two binary end conditions is not under program control. A typical type of sequenced robots is „Non-servo, PTP".

2. Trajectory operated robot (Playback and CNC robot)

Definition: A robot which performs a controlled procedure whereby two or more controlled motions operate in accordance with instructions that specify the next required position (normally achieved through interpolation) and the velocity required to reach it. The velocity is varied for each axis motion so that the desired path is generated (Fig.14.2 and Fig.14.3).

These include categories of robot often identified as Playback, CNC or Continuous Path.

Notes:

1. Playback robot (ISO, TP.8373 3.4). The robot can repeat a task program entered in the remote teach programming.
2. CNC (ISO 2806 3.1.3). A numeric control system, in which a dedicated, stored program computer is used to perform some, or all of the basic numeric control functions.

3. Adaptive robot

Definition: A robot having sensory control, adaptive control, or learning control functions (Fig.14.4).

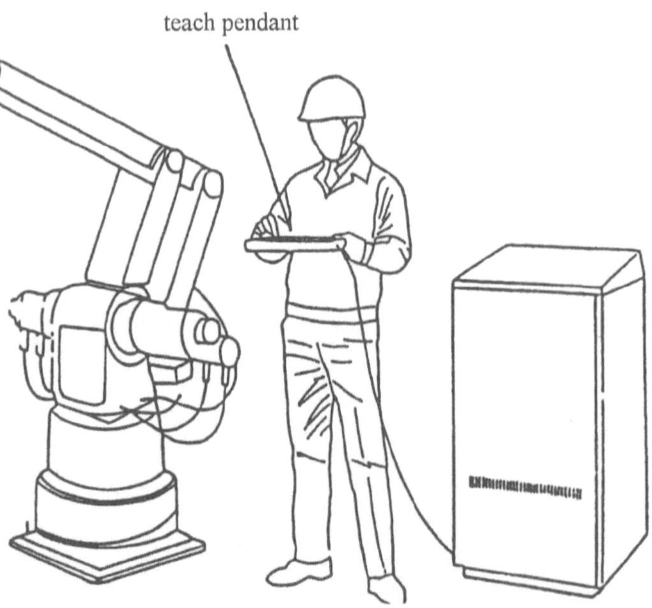

Figure 14.2. Playback robot [14.9]

Figure 14.3. CNC robot [14.9]

Examples of these functions are the correction and change of behavior applying sensory information, or past experience; the planning and scheduling of tasks or jobs, and training and learning by iterated behavior. A typical Adaptive Robot is a vision equipped robot that can correct positioning when selecting parts, or an arcwelding robot equipped with a continuous beamtracking function.

Notes:

1. Sensory control (ISO/TR 8373 6.3.3): A control scheme where the robot motion or force is adjusted in accordance with the output of external sensors.
2. Adaptive control (ISO/TR 8373 6.3.4): A control scheme that adjusts control system parameters to cope with conditions detected during the process.
3. Learning control (ISO/TR 8373 6.3.5): A control scheme that automatically used experience obtained during a previous cycle to adjust control parameters and/or algorithms.

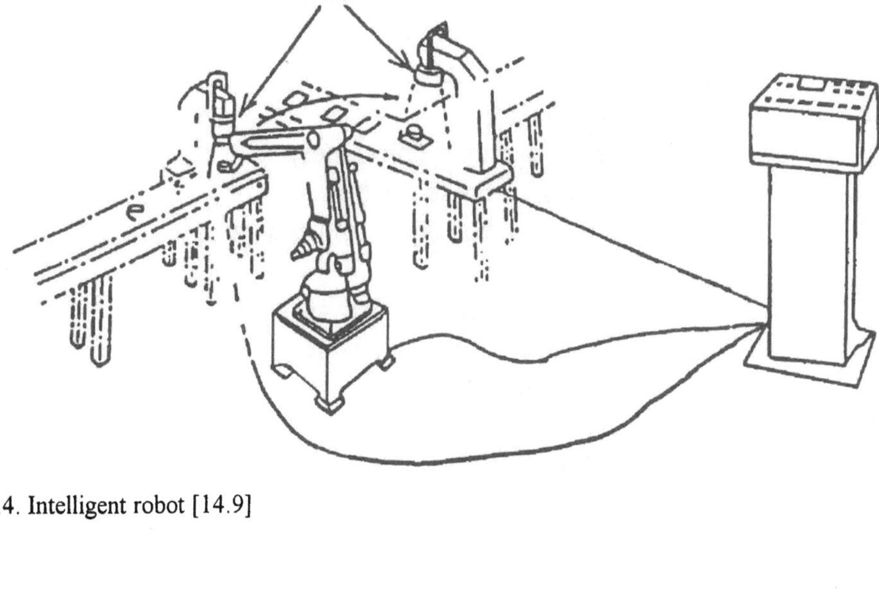

Fig.14.4. Intelligent robot [14.9]

4. Teleoperated Robot

Definition: A robot that can be remotely operated by a human operator; its function extends the human`s sensory-motor functions to remote locations (Fig.14.5).

These include categories of robot often identified as teleoperating robots, or, computer-aided teleoperators, with simple or sophisticated controls, and commands communicated along physical (wires, conduits, or pipes), or non-physical (wireless) channels. Devices that use only mechanical linkages are excluded.

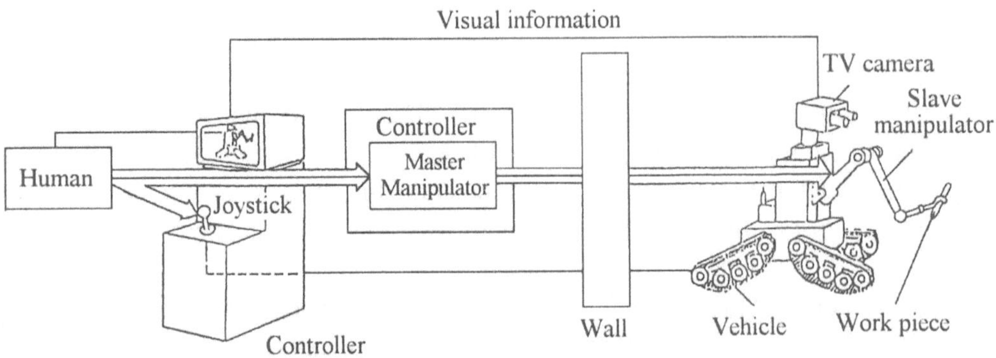

Fig. 14.5. Teleoperating robot [14.9]

14.2. Classification of industrial fields and applications

We have distinguished the following manufacturing industries [14.1]

ISIC Class

211	Iron and steel
212	Non-ferrous metals
22	Metal products
23	Machinery
23	General purpose machinery
232	Special purpose machinery
233, 234	Electrical equipment and apparatus
235, 236	
237	Precision / optical instruments
24	Transportation equipment
241	Shipbuilding
242	Railway
2431	Motor vehicles
2440	Aircraft

2491	Motor cycles
2491	Bicycles
2499	Industrial vehicles
2499	Other transportation equipment
31	Food / beverages
32	Textiles / leather products
33	Food / wood products
34	Paper / paper products / printing
35	Chemical / chemical products
36	Petroleum products / coal products
37	Rubber products
372	Plastic products
38	Ceramic / stone products
	Other

and non-manufacturing industries

ISIC Class

0	Agriculture / forestry / fishing
1	Mining
4	Electricity / gas / water supply
4	Construction / building
893	Research / development
93	Other

The classification by application in manufacturing industries is shown below

IFR Class

000	Unspecified
110	Casting
111	Iron and steel casting
112	Die casting
119	Other
120	Forging
130	Plastic molding
140	Heat treatment
150	Stamping
160	Welding
161	Arc welding
162	Spot welding
163	Gas welding
164	Laser welding

169	Other
170	Painting
180	Machining
181	Machine loading/unloading
182	Mechanical cutting
183	Grinding/deburring/polishing
189	Other
190	Cutting
191	Laser cutting
192	Water jet cutting
199	Other
200	Assembly
201	Mechanical assembling/attachment
202	Inserting/mounting
203	Bonding
204	Soldering
205	Sealing/gluing
209	Other
210	Palletizing / packaging
220	Measurement/inspection/testing
230	Material handling
240	Education/research
900	Other

14.3. Brief history

Five eras can be distinguished in the development [14.3, 14.4, 14.8, 14.24]

1. Pre - electronic
2. First steps
3. Euphoria
4. Disillusionment
5. Strong recovery

The pre-electronic era stretches back many centuries. Of it we conserve mechanical curiosities such as a Chinese four legged machine (dated from the third century), the sixteen century story of the Golem, that of Frankenstein in the nineteen century story, next R.U.R., a 1921 play about a robot, Robbie - the robot from Isaak Asimov story, and the Star Wars - a movie.

The next era, first steps of electronics, began in the early 1950`s, after the development of a remote manipulator system by the Atomic Energy Commission (G.M.).

Industrial robotics as we know it today began with George Du Vall and Joseph F. Engelberger who brought the first UNIMATE to market in 1957. Removing parts from the casting machine was the first industrial application. Over ten years passed before UNIMATION turned a net profit on into investment. Cincinati-Milacron, Autoplace, Praab and several manufactures then entered the market.

The first industrial robot arrived to Europe in 1968. Robotics research was began in the 1960`s through Darpa`s funding of the artificial intelligence labs at MIT, Stanford University and Stanford Research Institute. In the late 1960`s and early 1970`s NASA began funding work in robotics at the Jet Propulsion Lab (J.P.L), Ames. Industrial robots not did become really profitable until the late 1970`s, when automotive manufactures began to use robots for spray painting and spot welding.

Next the Euphoria was started. General Motors joined with Fanuc – GMT, and Automofix was founded, G.E. Teamed with robotics companies from Japan, Italy and Germany, and Westinghouse, IBM, and Bendix entered the market. Over 500 companies in Europe, Japan and USA rushed to join in what promised to be a new multibillion industrial revolution built on robot labor. Analysts were predicting 35% per year growth through the 1980`s.

Until the end of 1981, the prediction of the 35% growth rate seemed correct. But in 1986 the era of disillusionment set in. In 1986 and 1987 sales fell by 35%. In 1988 the market was equally dismal. The analysts asked themselves what had gone wrong? The diagnosis was that the technical difficulties had been masked by the fact that the really easy tasks, like spray painting, spot welding, and press loading were done first. When this market became saturated, robotics ran into a technology barrier.

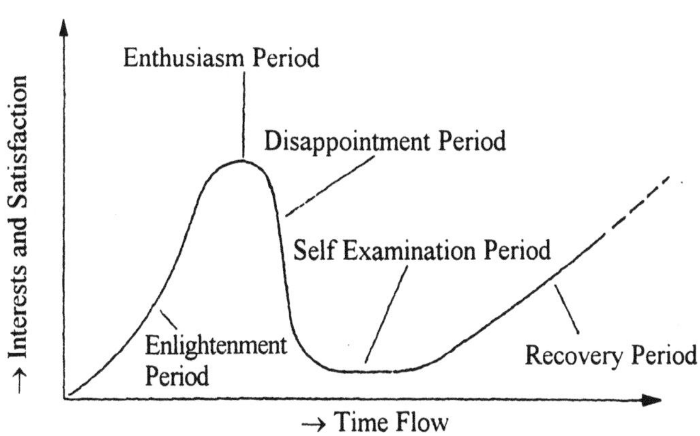

Fig.14.6. General process of robot introduction and diffusion into industry [14.5]

To optimistic predictions of exponential growth, based on robot assembly (by 1988, 50% of the labor in small component assembly would be replaced by automation). Next by 1990 the development of sensory techniques would enable robots to approximate human capability in assembly. Figure 14.6 presents the eras as seen in 1989/1990. Progress had resumed. But between 1996 and 1997 another recession took place.

14.4. Strong recovery in the engineering industries world-wide across the globe

Engineering industries are in a phase of recovery. They have come out of a deep recession, which had its trough in 1992 (in 1991 in the United States, arriving in Europe in 1993). Engineering production had surged by over 6% on a worldwide basis (by 10,5% in the United States, while in Europe it stopped at 4% by 1994, see Figure 14.7). In Europe were harder hit by the recession than other regions: production fell 4,2% per year between 1990 and 1992 and 6,4% in 1993. The United States, on the other hand, has seen booming growth in the engineering industries since 1992.

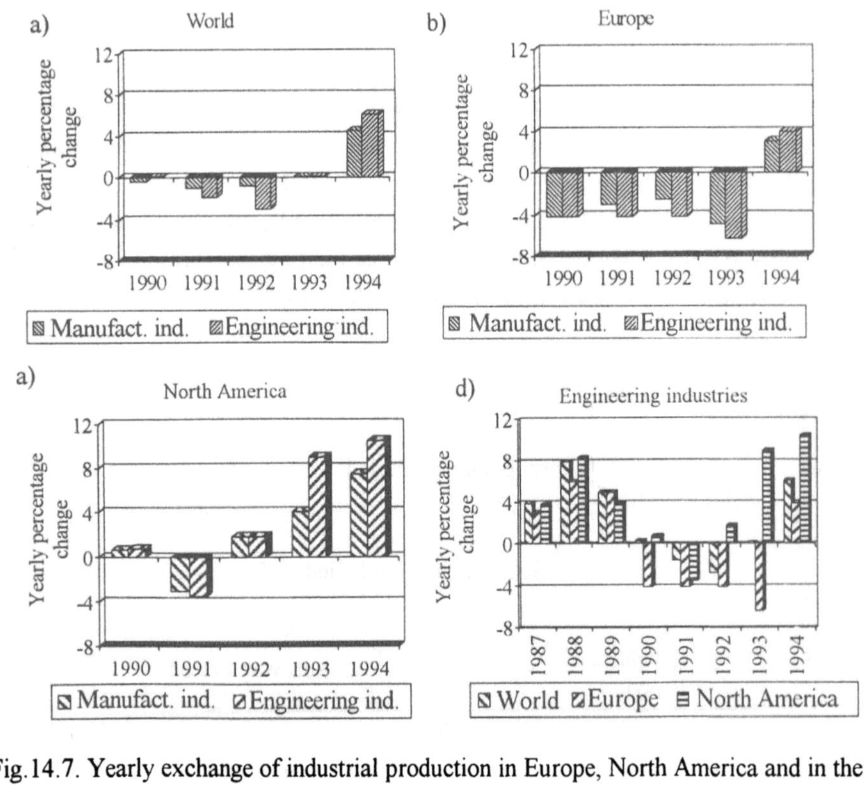

Fig.14.7. Yearly exchange of industrial production in Europe, North America and in the World

In periods of recession the engineering industries normally show less growth than manufacturing as a whole. This, however, is more than compensated for during recovery, when the engineering industries expand significantly faster than the rest of manufacturing. This phenomenon is clearly registered in Fig. 14.7.

Yearly percentage change in the production in manufacturing industries (ISIC Rev.2:3) and engineering industries (ISIC Rev.2:38) in the world (Fig. 14.7a), Europe (Fig. 14.7b) and North America (Fig. 14.7c). Yearly percentage change in engineering production, 1987-1994 (Fig. 14.7d).

14.5. Performance of the engineering industries in individual countries

When analyzing the indices for the manufacturing industry, it is clear that the recession which hit industry so hard in 1992 and 1993 was over in 1994 in almost all the ECE countries. Of 28 reporting countries all (except Belarus, Moldova, the Russion Federation, Turkey and the Ukraine) had a higher rate in 1994/93 than in 1993/92. Denmark, Finland, Ireland, Sweden and Australia saw production in the manufacturing industry increase by between 10% and 13%. In the United States growth reached 6%, while it registered to between 4% and 5% in Germany, France and the United Kingdom. After falling by 6,5% in 1992/91 and by 4,5% in 1993/92 modest growth of just under 1% was obtained in Japan, indicating that the country was not yet on a path of solid recovery.

The greatest drop was recorded in Moldova and the Ukraine, where production fell by almost 30%. After a fall of 13% in 1993 in the Russian Federation, production fell by 22,5% in 1994.

After several years of falling production the countries in central Europe finally experienced an increase. In Poland, which had already increased its production in 1992 by 4,6%, growth reached an impressive 8,4% in 1993.

The engineering industries in all the 27 reporting countries, except for Belarus (where production fell by14%), and Moldova, the Russion Federation, Turkey the Ukraine had a growth rate in 1994/93 higher than that in 1993/92.

The highest growth, 21%, was recorded in Finland followed by Sweden with 18%, Ireland with 16%, Slovenia and Australia with 15%, Canada with 14%, Denmark with 13%, Spain with 12% and the United States with just under 10%. For all these countries the growth rates were significantly higher for the engineering industry, than for manufacturing on the whole.

Japan only recorded growth of just under 2%. The 1994 production was over 10% lower than the 1990 level. The engineering industries in France, Germany, and the United Kingdom saw growth of between 4% and 6% [14.10]. As we know the economic recession causes the permanent shut-down of some machinery. During recovery old equipment is often replaced by new, more efficient machines. But even those industrial robots installed before recession are treated as useful tools. That is the reason that number of units decline more slowly than other machines. In any case the annual growth and decline of robots influenced more production than global investments. Another important economic factor in production of industrial robots (together with the general economic conjuncture) is the labor market, from the point of view of labor costs and levels of employment. The cost of labor is compared with the application cost of robot. Analysis shows that the price of the robot be decreased in recent years. The average price of industrial robots was 82000 USD in 1995, and 108000 USD in 1991. In 1991 we saw the highest differences in prices: 79000 USD in the U.K, and 131000 USD in Italy. In 1995 - 65000 in U.K. and 88000 in U.S.A. There has been a gradual increase in labor costs (Fig.14.8).

The comparison of these two factors gives us what is called the Golden Cross Point (Fig. 14.8). Crossing this point, we accelerate the introduction of robots.

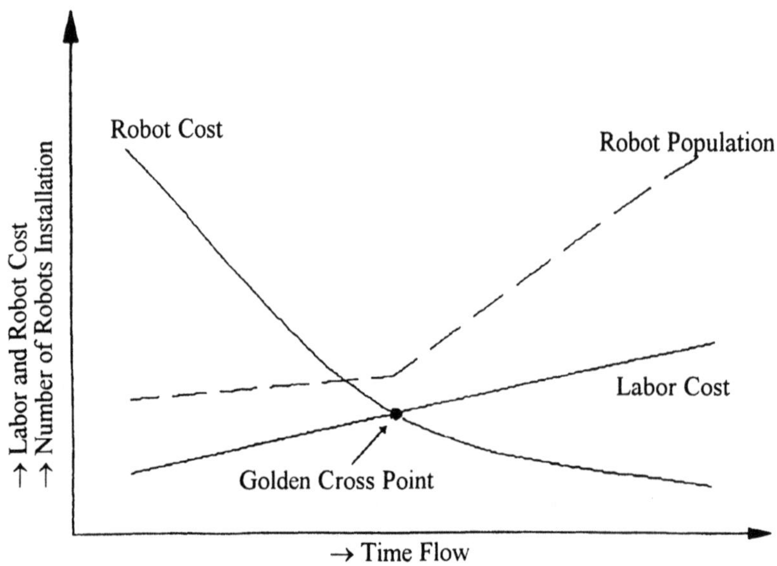

Fig.14.8. Golden cross point of robot diffusion [14.5]

14.6. Statistics of robots

The latest statistics (1983-1996) and forecasts to 1999 [14.11] shows that robots have now become a symbol of industrial automation in the most advanced form.

Table 14.1. Estimated operational stock of industrial robots [14.1]

Country	Number of units						% change
	1990	1991	1992	1993	1994	1995	1995/94
Japan	274 210	324 895	349 458	368 054	377 025	387 290	2,7
United States	38 980	42 250	45 610	50 454	57 088	66 286	16,1
Germany	27 320	32 760	37 550	41 415	45 340	51 375	13,3
Italy	12 200	14 090	16 410	18 735	20 593	22 963	11,5
France	8 350	9 510	10 420	11 295	12 292	13 276	8,0
United Kingdom	5 940	6 550	7 030	7 476	8 123	8 314	2,4
Big Six	**367 000**	**430 055**	**466 478**	**497 429**	**520 461**	**549 504**	**5,6**
Austria	1 150	1 430	1 670	1 872	2 125	2 325	11,0
Benelux	1 680	1 920	2 480	2 900	3 400	4 000	17,6
Denmark	470	520	540	568	622	672	8,0
Finland	810	940	1 030	1 129	1 223	1 398	14,3
Norway	490	500	500	504	486	477	–1,9
Spain	2 160	2 670	3 430	3 874	4 266	4 913	15,2
Sweden	3 340	3 420	3 650	3 677	4 177	4 459	6,8
Switzerland	1 510	1 670	2 010	2 450	2 927	3 390	15,8
West-Europe – 8	**11 610**	**13 070**	**15 310**	**16 974**	**19 226**	**21 668**	**12,7**
Czechoslovakia	7 060	7 060					
Hungary	199	229	237	247	113	128	–13,3
Poland	520	620	630	642	629	618	–1,7
Slovakia				113	139	164	18,0
Czech. Rep.				200	300	400	33,3
Slovenia				133	160	180	12,5
East-Europe – 6	**7 779**	**7 909**	**867**	**1 335**	**1 341**	**1 490**	**11,1**
Australia	1 430	1 550	1 630	1 739	1 776	1 840	3,6
Rep. of Korea	2 990	4 070	6 060	9 155	13 705	19 991	45,9
Singapore	1 620	1 900	2 220	3 970	7 960	11 924	49,8
Taiwan	1 290	1 690	2 220	2 727	3 277	3 849	17,5
Asia – 4	**7 330**	**9 210**	**12 130**	**17 591**	**26 718**	**37 604**	**40,7**
USSR	64 204	65 000	55 000	45 000	40 000	35 000	–12,5
Other Countries	1 420	1 810	2 210	2 850	3 350	3 900	16,4
Grand total	**459 300**	**527 100**	**552 000**	**581 179**	**611 096**	**649 166**	**6,2**

Sources: ECE, IFR and national robot associations.
Note: The operational stock of industrial robots is calculated as the last 12 years of accumulated sales except for those countries which have reported data for the operational stock.

Together with computerized, numerically controlled machine tools, automated guided vehicles, and host computers of various kinds, robots form the centerpiece of the computer-integrated manufacturing system [14.11]. Table 14.1 shows the estimated operational stock of industrial robots.

The world total is some 650000 units, 60% of which in Japan, 10% in the U.S.A, and 15% in Germany, Italy, France and the U.K. The highest percentage growth (more than 40%) registered in 1995 in four Asian countries, with lower growth in western countries (12,7%), and the lowest in the biggest producing countries (5,6%). In the former Soviet Union, in recession lost 12,5% of robots.

The prognosis for the next years (1996 - 1999) is shown in Table 14.2.
Table 14.4 shows the 1995 operational stock of robots, listed by industrial sectors. The picture described above has not changed significantly. The importance of the transport equipment industry for the robot market is still quite evident in the motor vehicle producing countries, except for Japan where is the second largest user. In nine countries, the transport equipment industry accounted for more than 25% of the total robot stock with a record 57% in Spain. Table 14.3 shows the estimated operational stock of industrial robots in the industrialized countries.

Table 14.2. Estimated operational stock of industrial robots at year end 1990-1995 and forecasts 1996÷1999 [14.11]

Country	Number of units			
	1996	1997	1998	1999
Japan	406 500	438 000	469 000	498 250
Germany	72 800	79 500	87 500	99 200
Italy	57 300	63 300	68 700	76 100
France	25 400	28 100	30 700	33 300
United Kingdom	8 600	9 000	9 700	10 700
Big Six	**631 500**	**678 500**	**740 100**	**815 700**
East-Europe – 8	**24 600**	**27 900**	**31 500**	**35 500**
West Europe – 6	**1 700**	**2 000**	**2 400**	**2 900**
Asia – 4	**51 300**	**68 300**	**89 800**	**115 300**
Former USSR	30 000	30 000	30 000	30 000
Other countries	4 700	5 800	7 300	9 200
Grand total	**743 800**	**812 600**	**901 100**	**1 008 600**

Sources ECE, IFR and national robot associations.
Note: The operational stock of industrial robots is calculated as the last 12 years of accumulated sales except for Japan which reports data on operational stock.

Table 14.3

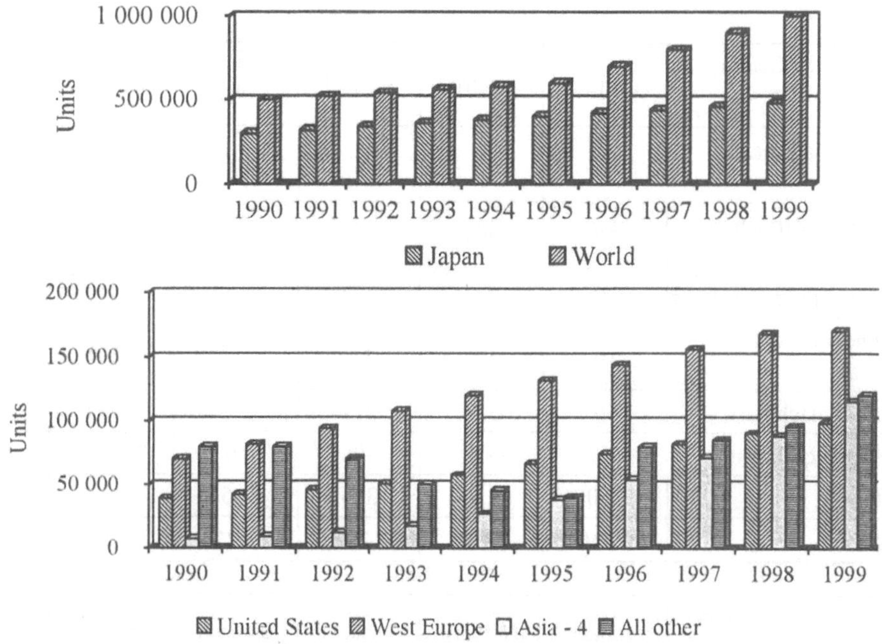

In Japan and Austria, the electrical machinery industry is the largest user of robots, while in Finland, Hungary, and Norway the leading user of robots to fabricated metal products industry.

We calculate that 65% to 80% of all robots are concentrated in the engineering industries (machine tools, electrical machinery, transport, and precision equipment). Figure14.9 shows the estimated operation stock of industrial robots at year-end.

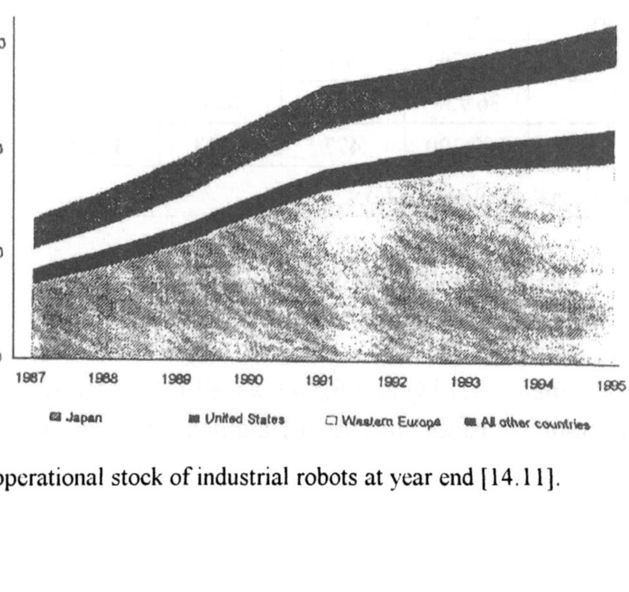

Fig.14.9. Estimated operational stock of industrial robots at year end [14.11].

Table 14.4. Operational stock of robots at the end of 1995 by industrial branches and countries. Number of units.

Branch \ Country	Australia	Austria	Finland	France	Germany	Hungary
Food, beverages and tobaco	178		62	463	1 189	
Chemicals, petroleum, plastics	27	298	223	2 130	4 964	6
Fabricated metal products	642	320	392	179	4 727	37
Machinery except electrical	16	405	83	1 776	3 468	6
Electrical machinery	60	533	217	753	3 561	8
Transport equipment	657	320	106	6 921	21 993	16
All other	260	256	315	1 055	11 473	55
Total	**1 840**	**2 130**	**1 398**	**13 276**	**51 375**	**128**

Branch \ Country	Japan	Norway	Spain	Sweden	Taiwan	United Kingdom
Food, beverages and tobaco	4 120	12	25	13	63	113
Chemicals, petroleum, plastics	50 181	27	261	109	128	1 509
Fabricated metal products	19 071	236	304	994	201	641
Machinery except electrical	36 206	38	687	644	299	409
Electrical machinery	138 213	42	180	268	641	507
Transport equipment	102 761	16	2 821	1 710	1 887	3 961
All other	36 738	105	634	721	630	1 144
Total	**387 290**	**477**	**4 913**	**4 459**	**3 849**	**8 314**

Sources: IFR and national robot associations.

The number of robots per 10000 employees in 1994 is shown in Figure 14.10. Table 14.6 shows the density of supply for all types of robots and for advanced robots.

The relationship of the highest and lowest coefficient of density of all types of robots (Hungary: Japan = 1:148) and for advanced robots (Slovakia: Japan = 1:524). The worldwide supply of robots in 1995 by application areas and countries is shown in Table 14.6.

Table 14.5. Estimates of the total 1995 stock of operational robots (all types of robots), the stock of advanced robots (trajectory-operated and adaptive robots only), densities of all type of robots and densities of advanced robots [14.11]

Country	**All types of robots**	**Advanced robots**	**Robot densities**	
			All types of robots	**Advanced robots**
Australia	1 804	1 564	16,7	14,2
Austria	2 359	1 274	31,0	16,8
Benelux	4 000	3 200	23,1	18,4
Denmark	672	672	14,5	14,5
Finland	1 398	1 383	37,9	37,5
France	13 276	12 652	32,4	30,9
Germany	51 375	46 238	57,7	52,0
Hungary	128	97	1,7	1,3
Italy	22 963	19 059	49,7	41,3
Japan	387 290	322 613	251,5	209,5
Norway	477	384	18,4	14,8
Poland	618	428	2,9	2,0
Rep. of Korea	19 991	15 993	64,3	51,4
Singapore	11 924	2 969	335,9	83,6
Spain	4 913	3 930	20,4	16,3
Sweden	4 459	4 013	59,5	53,6
Switzerland	3 390	3 051	43,3	39,0
Taiwan	3 849	3 079		
Former USSR	35 000	6 580	26,8	5,0
United Kingdom	8 314	6 651	17,4	13,9
United States	66 286	59 657	35,7	32,1
Slovakia	164	20	3,7	0,4
Czech Rep.	400	240	3,5	2,1
Slovenia	180	101	7,7	4,3
Other Countries	3 900	1 950		
Total	**649 166**	**517 799**		

Sources: UN/EC, OECD, IFR and national robot associations.

a/ Estimates for certain countries.

b/ Not comparable with other countries.

Note: Densities are defined as the number of robot per 10000 people employed in the manufacturing industry (ISIC 3, rev.2). Data for the robot stock refer to the year 1995, whether these data are reported by national robot associations or estimated by the ECE secretariat. The year for employment data is the most recent year available which in the best case is 1994 and in the most cases 1993 or 1992. The sources for employment data are the ECE/ITD data bank and OECD.

Table 14.6. Worldwide supply of robots in 1995 by application areas and countries. Numbers of units [14.11]

	Australia	Austria	Denmark (a)	Finland	France	Germany
Welding	101	79	26	34	421	2228
Arc welding	40	62		33	195	790
Spot welding	60	17			224	1438
Dispensing	4	22		3	32	315
Painting	1	0		3		222
Sealing/gluing	3	22				93
Machining	23	87		49	167	1145
Loading/unloading	19	62		37	129	863
Mech. cutting/grinding/ deburring/polishing	2	11		12	17	282
Special processing	4				12	
Assembly	19	24		66	116	866
Palletizing/packing	44	7		11	71	436
Material handling	18		39	6	243	907
Others	1	57		51	333	1439
Total	**214**	**276**	**64**	**212**	**1384**	**7335**

	Hungary	Italy	Japan (b)	Norway	Poland	Singapore (c)
Welding	11	609	667	9	6	119
Arc welding	11		229	9		119
Spot welding			424			
Dispensing	4	174	482	3	1	120
Painting	4		109	3		120
Sealing/gluing			591			
Machining		636	2068	5		330
Loading/unloading			1137			385
Mech. cutting/grinding/ deburring/polishing			669	2		245
Special processing		59	76	1		4
Assembly		213	4836			6717
Palletizing/packing			793	8		43
Material handling		1400	3426			289
Others		130	858	15	8	398
Total	**15**	**3120**	**10265**	**41**	**14**	**8020**

	Spain	Sweden	Taiwan	United Kingdom	United States	Korea
Welding	400	172	282	365	5099	2668
Arc welding	156	87	211	164	1632	1032
Spot welding	244	84	71	201	3467	1424
Dispensing	57	14	34	92	1020	248
Painting	29	12	34	51	306	248
Sealing/gluing	28	2		41	714	
Machining	39	132	66	50	2967	
Loading/unloading	32	94	42	43		
Mech. cutting/grinding/ deburring/polishing	7	31	24	7	306	
Special processing	10	2		3		
Assembly	12	11	22		918	1620
Palletizing/packing	23	65				
Material handling	71	19	154	93		273
Others	218	46	72	189	204	1537
Total	**830**	**461**	**630**	**792**	**10198**	**6336**

Sources: IFR and national robot associations.
a/ Includes all kinds of handling operations, machining and special processing.
b/ Net increase in stock between 1995 and 1994.
c/ Data for both 1994 and 1995.
Note: Data for Japan are not yearly supply data but net increase in the operational robot stock which explains why some of the data are negative. The actual gross supply of assembly robots, for instance, amounted units in 1995 to 15974. Moreover, as be noted that not all subgroups under welding, dispensing and machining are included, is the aggregations of the subgroups do not always add up to the main groups.

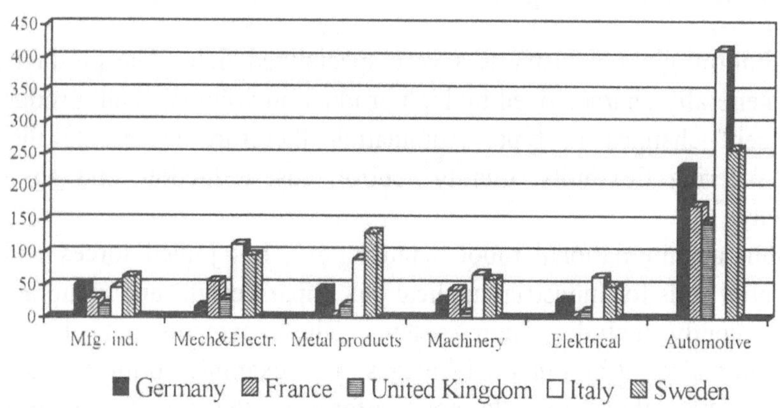

Fig.14.10. Number of robots per 10000 employees in 1994.

Welding is still the dominating application, except in Japan, Italy and Singapore. The demand for industrial robots generally follows the same cyclical pattern as that of other investment goods. However, compared with machine tools, which are extremely sensitive to changes in economic activity, there are some differences. This is clearly seen in Figure 14.11, which presents time series data for the consumption of machine tools and robot investment in the U.S.A.

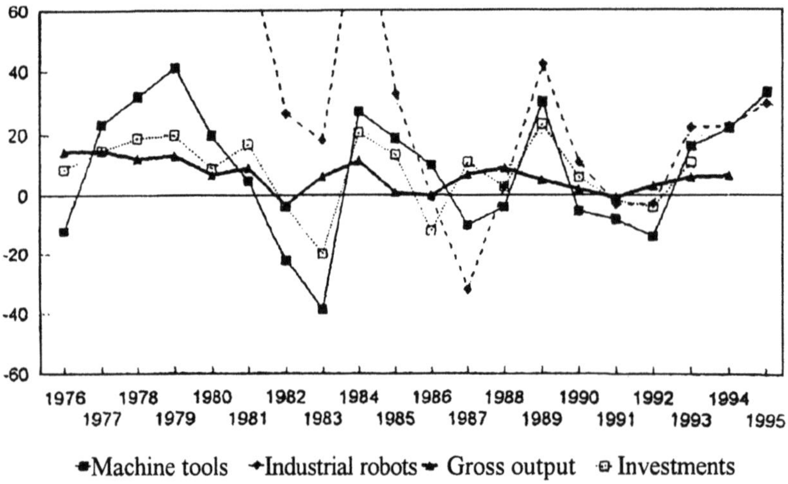

Fig.14.11. Machine-tool and robot investments in the U.S.A. [14.10, 14.11]

14.7. Examples of the latest applications

14.7.1. Conditioning in the hollow glass industry

The hollow glass industry is a very specialized field. The product packaging lines are generally characterized by high production volumes, high production rates, and frequent changes in type. Automation therefore represents the optimum solution, offering flexibility, quality control, cost reduction, and an increase in productivity.

Staubli, an international robot manufacturer, has joined forces with Sipar to provide robot cells for industrial applications. Sipar designs and manufactures cells, where frequently include vision systems. These systems not only handle the products, but also offer other advantages. For example, prior to packaging the cardboard boxes, those that do not conform to the specific requirements are rejected. It is also possible to check the product`s position in the box.

Other of the installations developed by the two companies one operates at Schott-Zwiesel in Germany. This 600-employee company specializes in the manufacture of stem glassware. Sipar has developed and integrated a robotised cell consisting of one Staubli RX90 robot with a vision system.

The inverted robot picks up the stemware from the feeding conveyor belt and deposits them in boxes. The stemware can be packed in a horizontal or vertical position. After washing the stemware is relatively aligned in the opposite conveyor belt. The vision system identifies each part (Fig.14.12). The system began operations in May 1996 on three-shift system, and is capable of packing 40 glasses a minute.

Fig.14.12. Staubli robot handling fine glassware [14.12]

14.7.2. Polishing operation in the car industry

Figure 14.13. shows the 4-axes robot used for car polishing. This process is performed using a special system of control.

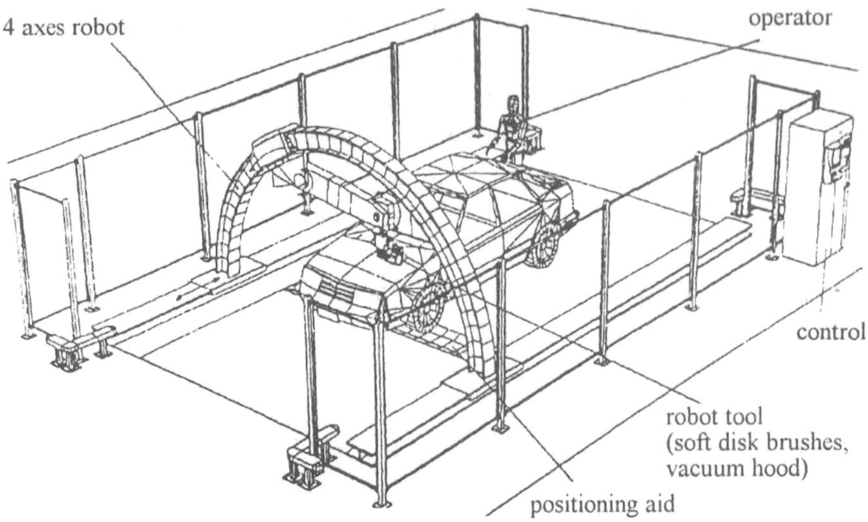

Fig.14.13. Robot for car polishing [14.7]

14.7.3. Drain cleaning and maintenance

In Germany alone, roughly 40 million drains of streets, courts, airstrips etc. have to be cleaned regularly. Up to now this task is performed manually. The lifted by a hook and the basket is taken out and cleaned manually. Figure 14.14 show the robot for drain cleaning and maintenance.

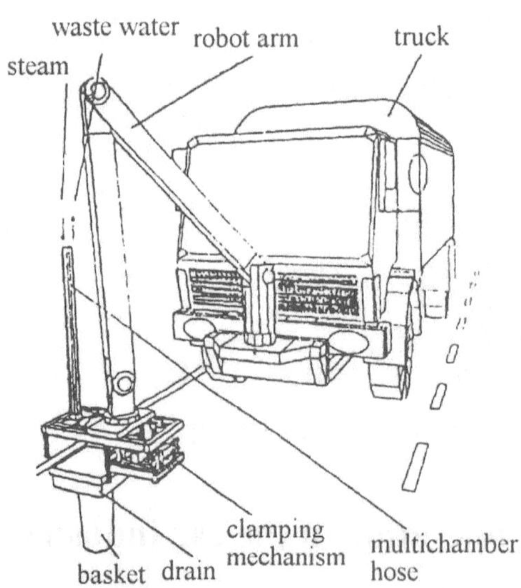

Fig. 14.14. Robot for drain cleaning and maintenance

14.7.4. The redundant heavy manipulator

The redundant heavy manipulator was developed at the Institute of Mechatronics, University of Duisburg, Germany, with an industrial partner. The redundancy manipulation in narrow environments with enough flexibility to perform task. It can load up to 500 kg in a range of eight meters. The integrated sensors allows supervision and assistance of movements. It will be used in tunnelling to fix cc elements on the wall and roof [14.7].
The redundant manipulator MEYCO Robojet on a mobile platform with a sprayed concrete aggregate. Applications are found in tunnel construction. The robot con allows uniform spraying onto unknown surfaces. The accuracy of the thickness of the concrete is carried out through a laser scanner. At the beginning, the shape of the tunnel is measured and at the end of the process its quality is checked. The advantage is that the amount of concrete bouncing back is reduced and the performance of the process is improved. (Source: MEYCO Equipment, MBT Switzerland AG) [14.7]

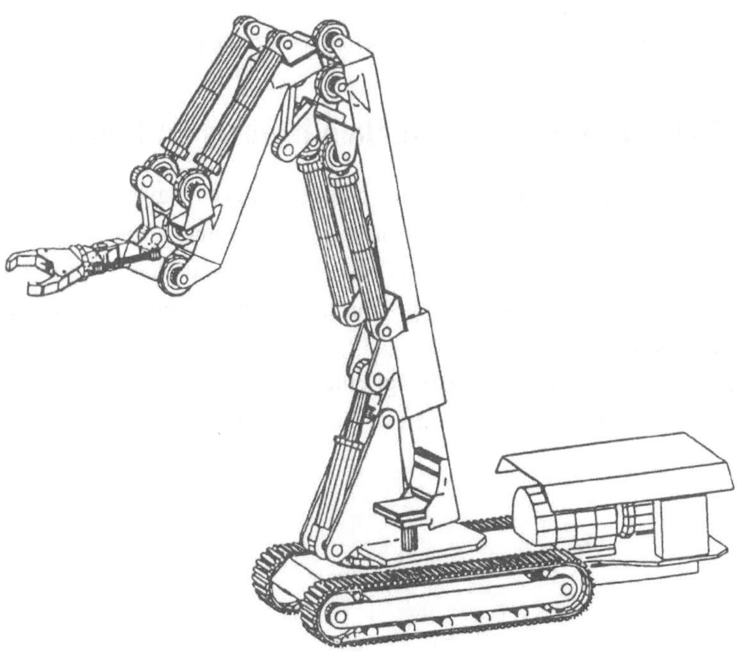

Fig. 14.15. The heavy redundant manipulator

14.7.5. Robot for sheep-shearing and picking of apples

Figure 14.16 shows the system with manipulator used for sheep-shearing. The details of hydraulic and electric equipment are omitted.

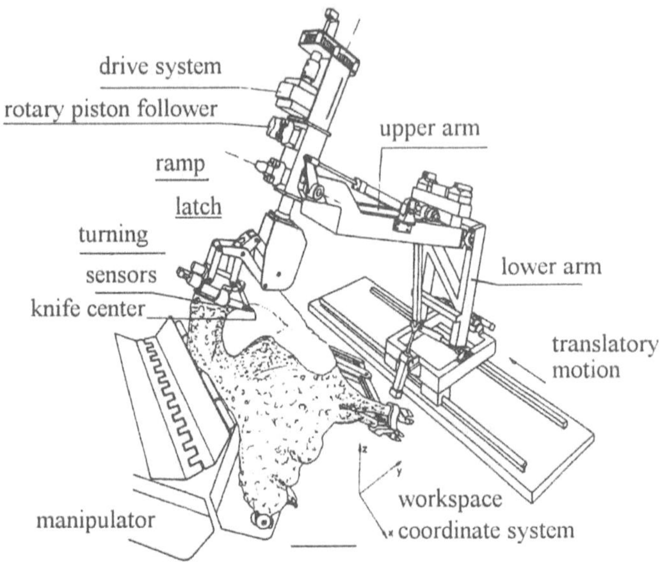

Fig. 14.16. The machine ORACLE for sheep-shearing (according to ASAE and SME) [14.7]

14.7.6. Robot for cutting the chicken breasts and picking of apples

Cut into "spear" shaped portions, the chicken breasts have been cut to meet the specifications provided by the customer, minimize over shipment, and to provide consistent product shape and quality. Each breast has been scanned and shape optimized with order make-up programmed prior to processing. The portioner can be programmed to create a variety of cut configurations.

It is expected, that in the next years agriculture, forestry and fruit-farming will be the serious users of robots (Fig. 14.17). More examples can be find in [14.23]

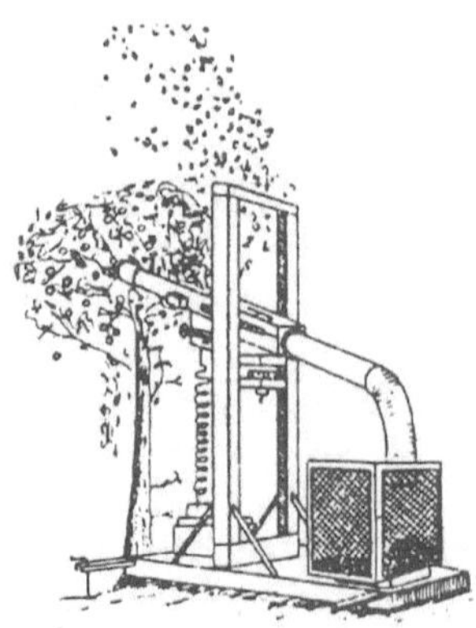

Fig. 14.17. The French robot for picking of apples [ASAE]

14.8. Development of industrial robotics in Poland

Research and application of industrial robots dated from the sixties and seventies in Poland [14.25, 14.17, 14.18, 14.19, and 14.21]. Some selected examples of applications in the years from 1975 to1996 are described briefly.

14.8.1. The RIMP robot and its applications

Figure14.18 shows an example of the application of six RIMP robots to service a light press (1MN) line. The robot used to service the first press, P1, picks the small plate from the container, and fixes it into the robot gripper. The workpiece is then transported to press, P2. When the last operation is finished at press P6, the gripper of the last robot places the workpiece in the container. Both the press line and the robot line are controlled by microcomputers. With the application of the robot line the number of workers can be reduced by ten workers in two relays with a 20% increase in productivity [14.17, 14.18, 14.19, 14.20].

The latest installations in Poland are described in [14.22].The most modern press shop is situated in Tychy, at the FIAT Auto Poland (FAP) car factory, where robots are used for the automation of small, medium and heavy press lines. Each type of line is automated using different kind of robots. One of these solutions, automating the medium-size press line, provides for the use of universal, electrically driven robots, i.e. standard IRb-60 robots, developed by PIAP (the Industrial Institute of Automation and Measurements) (Fig.14.27).

The complex consists of the following equipment:

- 1 host PLC controller - type Simatic S5 155U, with ready to-use, checked connection to a remote PC or networking system for long term production data processing;
- 7 Irb-60 (+ 1 Irb-60 as a reserve) robots made by PIAP;
- press line operator console with indicators and push-buttons, visible along the whole line;
- sensor system;
- auxiliary equipment:
  - input rotary magazine - feeder of blanks, equipped with pneumatically driven manipulator;
  - custom designed pneumatically driven gripper with vacuum cap-lifters;
  - rail track carriages, pneumatically driven, for removing robot manipulators from the press line;
  - safety barriers system,

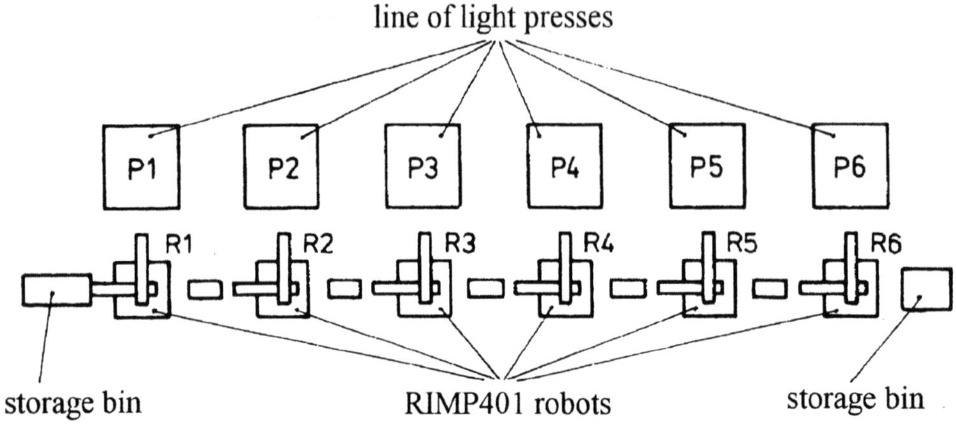

Fig. 14.18. The robotized line of light presses in the automotive industry.

- output magazine with two containers.

The press line host PLC tasks are as follows:

- coordination of robot - press auxiliary interactions, and, most important, press activation and enabling of robot loading/unloading;
- full time diagnostics regarding the presses, robots and auxiliaries, including an emergency switch-off of robots, and presses either by segment or the whole line;
- automatic supervisory testing; the communication between the robots and the controller automatically is checked before the start of the operation;
- data processing:
  - process visualization on the PLC monitor;
  - monitoring of the press line with instantaneous output;
  - recording of shifts` outputs;
  - calculation of the current utilization coefficient for the whole line and technical availability coefficients for three groups of equipment: automation (robots, PLC), presses, and auxiliaries.

The robots` user programs are composed of three main parts:

- automatic testing producers executed automatically under the host PLC, checking at every start-up of the press line:
  - robots - controller I/O exchange,
  - grippers movement,
  - master positions of the robots` manipulators,
  - number of robots working programs and drawpieces,
  - vacuum cap-lifters and piece presence sensors,
  - positions of the presses` tools,

- working programs for loading and unloading the presses,
- service, customized automatic procedures, enabling service staff to locate any failure in the equipment.

The main requirement were:

- instantaneous output: a minimum of eight pcs/min,
- flexibility - the ability to produce from one to three different kinds of drawpieces
- at the same time on the line,
- working staff per shift - max. 2 persons,
- positioning accuracy (of pieces in the presses) with a tolerance of + 1 mm.

All these requirements were met.

The most demanding requirement was that of instantaneous output, but actually an output of up to even 9 pieces/minute was obtained. Many time-consuming experiments, tests and programming were needed to achieve this break through came with the special positioning of the six robots\` manipulators, permitting a substantial reducing of the main rotation angle. Grippers were equipped with small, pneumatically driven rotating motors to compensate for this reduction.

The press line run automatically served by one operator on each shift. Before automation, seven people per shift were needed. To fulfil the flexibility requirement, the line may be divided into 2 to 3 sublines, with separate blank input/output devices for every subline. The host PLC controller is programmed accordingly.

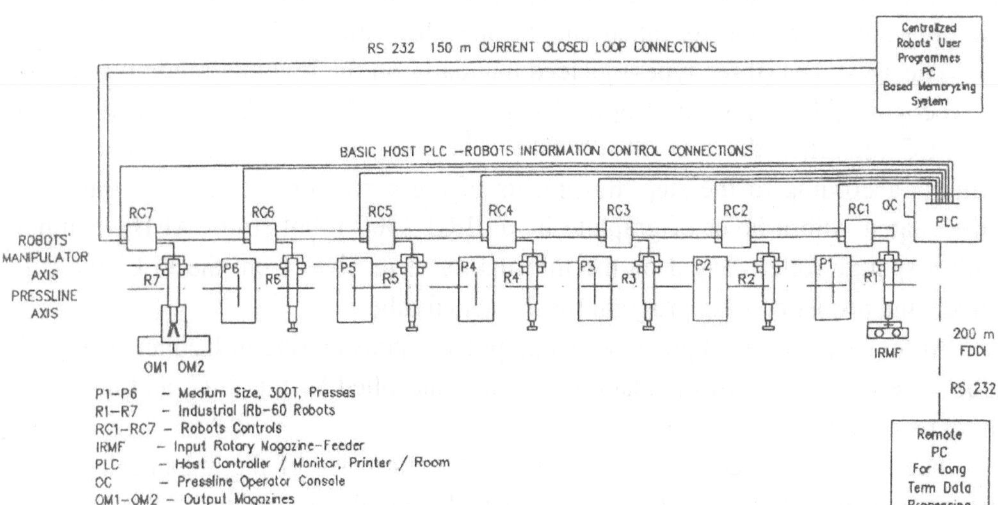

Fig. 14.32. The lay-out of the robotized-automated medium-size press line at FIAT-AUTO Poland, Tychy [14.22]

During operation magnetic tape IRb-60 cassettes were found to be very ineffective for memorizing user programmes. The write-in time from tape recorders to the seven robots on the line took too long.

Consequently, it was crucial to modernize the memories of the automated press line in order to increase the actual utilisation of the line:

- the memories containing the inner robots` user programs were changed and modernized by installing a hybrid chip, with a very long back-up time memory, in robots` control cabinets,
- an IBM PC - based centralized system for reading, collecting, editing, modifying, checking, storing on the FD, reading and writing user programs from/to robots was introduced. The robots are connected with a computer by a 150 meter loop RS 232 current interface, creating a simplified version for programming the robots off-line.

This significantly increased the utilisation coefficient of the automated press line.

A robotized line for graphite coating of TV - tubes

In January, February and March 1993, at Thomson-Polkolor Factory near Warsaw, three robotized lines for the graphite and lacquer coating of TV-tubes came successively on stream. These three lines were designed, built, and put into operation in nine months. This very short period of time caused many technical and organizational problems. One of which was the lack of time for testing the various technological solutions and choosing the best ones for this unique process. It was therefore decided to use equipment from well-known, reliable European companies. But, it was still necessary to design and build customized equipment, select the proper sensors for the whole line, and create user programs for robots and the lines PLC (Fig. 14.28).

The TV-tubes (three types), placed manually on the line in a random sequence, are recognized by a system of optical sensors. Using the signals from these sensors, the PLC identifies the type of TV-tube. The information is sent to all the work stations according to the step-movements of the conveyor. At the first station the tube coated with a layer of graphite by an IRB 1500 robot (from ABB), equipped with a sponge roller. Based on the information it receives from the PLC, the robot selects the proper coating program for the actual tube.

At the next station, a pneumatic manipulator sprays a ring of lacquer on the TV-tube. The movements of the manipulator are controlled by the PLC, and depend on the type of tube.

The last station automatically measures the resistance of the graphite layer. The measured value of the resistance is compared with the set value of this device to determine whether the TV-tube is satisfactory, or faulty. This information is sent to the PLC and is shown as a green or red lamp in front of the operator unloading the tubes.

A controller of the line (a C200H produced by OMRON), except cooperation with all working stands, collects production information (the amounts of satisfactory and faulty tubes) for every type of tube and supervises the correct functioning of equipment. The PLC warns the line service by visual and acoustic alarms. The production is interrupted depends on the type of fault. The line also has a conveyor, two large tunnel ovens for heating the TV-tubes, a lacquer pump, the equipment to supply the graphite, and various peripheral equipment (platform, mechanical barriers, light barrier, etc.).

A main user requirement was to achieve an efficiency of 180 tubes/hour on every line, while maintaining the quality of the graphite layer. Because the robot was the slowest element of the line, the question of its user's programs was very important (see Fig.14.28). After a preliminary start-up, an efficiency of about 158-165 tubes/hour (depending on the type of tube) was achieved.

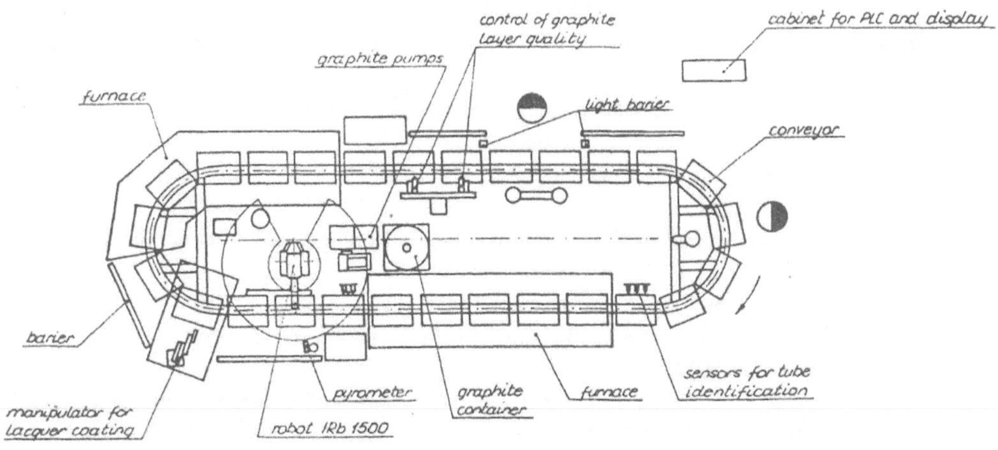

Fig.14.33. Robotized line for the graphite coating of TV-tubes [14.22]

14.9. References

[14.1] Classification of Industrial Robots. IFR 1990

[14.2] IFR Robot Statistics. Documentation Package Revison, Jau.1997.

[14.3] Conogigliaro L.: Trends in the robot industry (Revisited: Where are we now?). 11th ISIR, Chicago 1983

[14.4] Morecki A., Ekiel J., Fidelus K.: Cybernetic Systems of Limbs Movements in Man, Animals and Robots. Ellis Morwood Chichester, Warsaw, PWN 1984.

[14.5] Hasegawa Y.: Down of Asian Robot Age and Key Issues. Proc. of 26th ISIR, Oct 1995, Singapore, pp.101-108.

[14.6] Morecki A.: One Men's View. The Industrial Robots, Vol.6, No 2, 1980.

[14.7] Proceed. of ISIR, 1974, 1976, 1977, 1978, 1979, 1980, 1981, 1982, 1983, 1984, 1985, 1986, 1987, 1988, 1989, 1990, 1992, 1993, 1994, 1995, 1996.

[14.8] Morecki A.: Present state and perspectives of industrial robotics. 3rd Polish Conf. on Robotics, Vol.1, ITC Wroclaw 1990, pp.5-15 (in Polish).

[14.9] Morecki A.: The problems of present robotics. Bulletin PIAP, No 2 166/93, pp.3-31 (in Polish).

[14.10] World Industrial Robots. 1996, Statistics 1983-1996. Forecosts to 1999. IFR, United Nations, New York - Geneve 1996.

[14.11] World Engineering Industries and Automation Performance and Prospects 1994-1996. United Nations, New York - Geneve 1996.

[14.12] Robotics Newsletter, No 24, Dec. 1996/Jan.1997.

[14.13] Schraft R.D., and Merlinger A.: The Relation Men/Robot - A brief review and outlook on social and technical aspects. Proc. 27th ISIR, Milan 1996, pp.2-7.

[14.14] Schraft R.D., Hogele M.: Methods and tools for an efficient design of service robot application. Proc. of the 26th ISIR, Singapore 1995, pp.71-78.

[14.15] Morecki A., Kaczanowski S.: Central Programme of Applied Investigations - Industrial Robots. Prc. of the 22nd ISIR, Detroit 1991.

[14.16] Scientific projects. Grants in the field of robotics. Bulletin PIAP, Warsaw 1994

[14.17] Morecki A., Buc J.: Development of Robotics in Poland in 1977-80. Proc. of 11th ISIR, Tokyo 1981.

[14.18] Morecki A. and others. Is Compactness and Simplicity Needed in Robotics. Proc. 14th ISIR, Gothenburg 1984.

[14.19] Morecki A., Buc J.: From experience of Polish robotics. Proc. 2nd Yugoslav. Symp. on Applied Robotics 1981.

[14.20] Morecki A.: New Branch of Education in Poland - Robotics and Automation. Proc. 19th ISIR 1988.

[14.21] Robotics Research and Applications (Selected Topics presented at the 1st, 2nd and 3rd Polish National Conf. on Robotics). Edited by A. Morecki, W. Muszyński and K. Tchoń, Sc. Tech. Publ., Warsaw 1992.

[14.22] Chosen Aspects of Building Robotized Lines (two practical examples). Preprints of 2nd IFAC/IFIP/IFORS Workshop, Intelligent Manufacturing Systems INS`94, Vienna 1994.

[14.23] Engelberger J.F.: Robotics in Service. The MIT Press, Cambridge, Massachusetts 1989.

[14.24] Albus J.S.: Robotics where has it been?. Where it is going. Robotics and Autonomous Systems, 1990, No 6, pp.199-219.

[14.25] Karlsson J.M.: A Decade of Robotics. Geneve 1991.

[14.26] Khatib O.: Mobile Robotic Manipulation. Proc. of 26th ISIR, Singapore 1995, pp.7-12.

[14.27] Morecki A.: The contribution of IFToMM in the next decade to the theoretical and applied robotics. Proc. of 27th ISIR, Milan 1996, pp.11-15.

[14.28] Buć J., Morecki A., Study and Applications of the Industrial Robots in Poland. Proceedings of the 9th ISIR, March 13-15, 1979, Washington D.C. USA

[14.29] Proceedings of the '93 ICAR, Nov. 1-2, 1993, Tokyo, Japan

15. REHABILITATION MANIPULATORS

15.1. Introduction

Statistical data identifies about 10÷20% of the population in the developed countries is disabled, half of them requiring the assistance of special devices to substitute or support diminished manipulation and locomotion functions. Together with developments in recent years, new trends address the design and application of different types of manipulators, which enable the patient to help in contact the outside world although confined at home and permit him to carry out some working activity. A brief description of control systems of single and multi-functional prosthetic devices is given here.

We also describe two large rehabilitation systems created, one in Germany and one in England to assist tetraplegia patients. In view of the interesting properties of living organism control systems, motion control systems are illustrated as well.

15.2. Control of motion

Knowledge of the human control system is important not only to physiologists and physicians but to investigators and developers of artificial limbs as well.

The control of motion requires the cooperation of physiologic mechanisms that enable us to perform a feasible and concrete motion task in accordance with a motion program. The feasibility of a task depends on the capabilities of the organism in terms of force, speed and power, while actual results are also determined by the degree with which the person has mastered his motion habits, in other words, by his level of training and development [15.1]. Motion's coordination by nerve impulses sent to the muscles, and must be based on information about the values of external forces.

Moreover, studies on the cooperation of muscles show that there is no clear relationship between the degree of muscle excitation U/U_{max} (where U, U_{max} denote current and maximum muscle electroactivity) and the torque developed by the muscle in the joint.

These facts, together with the 250 degrees of freedom in the joints, and the number of actons in the near 600 muscles of the human body, make the process of motion control dependent on the stream of information gathered by receptors which pinpoint the biomechanic parameters of human motion, as well as those of the surrounding environment.

15.2.1. General system of motion control

N. A. Bernstein (1957) proposed a general model of the control of voluntary movements (Fig. 15.1) consisting of the following systems [15.1]:

1. a motor effector whose operation is controlled according to assumed parameters;
2. a program yielding the values of desired parameters;
3. a receptor which detects the actual values of parameters and hands them over to a comparing mechanism;
4. a comparing mechanism which determines the difference between the actual and the desired values;
5. a decoder, that is a device which decodes the comparing mechanism data as a consequence of the correction signals given to a controller by the feedback loop;
6. a controller which controls the effector according to the parameters obtained.

The programming device can be either of the stabilizing, or the tracking type, which means that it can yield a program with fixed, or variable values. The human body has set programmed parameters for most vegetative functions, such as body temperature, sugar content in the blood, and blood pressure. Motion parameters and those stabilizing the particular intervals of the body are programmed in the follow-up system.

According to Bernstein's assumptions this control depends fundamentally on the follow-up-manner comparing of the desired value (*Sollwert SW,* from the German), which is worked out by the man through learning, as a motion program, with the current values of motion parameters (*Istwert IW* , from the German). Consequently, muscles are excited by the nervous system (controller) not according to a fixed routine but with constant adjusting (ΔW) to the situations encountered so that the final result enables achievement of the motion task.

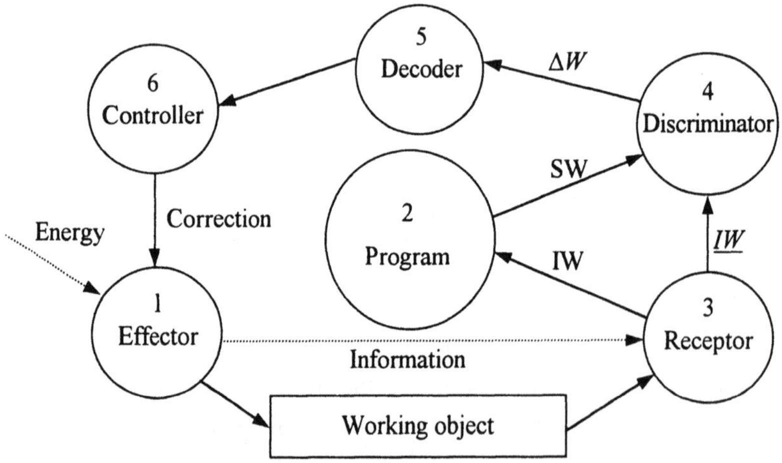

Fig. 15.1. System model of the control system of human movement proposed by Bernstein [15.1, 15.18]

The comparison of values *SW* and *IW* in the discriminator makes it possible to:
1. determine the correction impulses;
2. accept the executed motion phase;
3. determine a new value of *SW* if the motion is ineffective, or impossible.

The time Δt it takes for information to pass from the receptor to the effector equals approximately 0.07÷0.12 s, according to Bernstein. The control process is performed with a frequency of 8÷14 Hz. This frequency has quite a general character and value Δt can be regarded as a biological time unit: the α rhythm in EEG and muscle tremor have the same frequency. It is also close to the lower threshold of audibility and the upper threshold of recognition of single visual images.

Value Δt constitutes the lower threshold of human reaction time. Man probably evaluates the velocity of his body on the grounds of so called "fresh traces", that is recorded values of ΔW within the time interval Δt, so that

$$v_{\acute{s}r} = \frac{\Delta W}{\Delta t} \tag{15.1}$$

Taking into account value Δt and the possibilities of movement control connected with it, the following three types of movements can be distinguished (Morecki, Ekiel, Fidelus – 1967):
1. *involuntary movements* (*reflexes*) – realized on the lowest levels of the central nervous system (the spinal cord) according to the pattern of the reflex loop, that has no feedback.
2. *ballistic movements* – of a duration of 0.1÷0.2 s., which means they are too short to control during the motion; and are controlled ante factum. Examples are: throwing or hitting some object, or a single rebound after a jump.
3. *continuous movements* – controlled in facto, i.e. during their execution; corrections are input without delay to have *SW* in accordance with *IW*.

Modelling and controlling involuntary and ballistic movements, which are low precision movements, poses no great difficulty since they are performed more or less according to the "hit or miss" rule. It is continuous movements which are of the greatest importance, and they require complicated control systems.

15.2.2. Hierarchic systems of muscle control

The nervous system has many concentrations of motion, associative and feeling cells. These are the nerve centres spread along the spinal cord from the lowest levels to the highest ones in the cerebral cortex. Below the cerebral cortex large concentrations of neurons form what is called the basal nuclei. There are also a number of nuclei associated with the motion system in the central part of the brain, on the spinal cord extension: from the top of the spinal cord up to the thalamus nerve cells interconnected by means of dendrites form a mesh, called a reticular formation. Another very important element of the nervous system is the cerebellum, located in the rear part of the brain. However, as the functions of different centers of the brain, as well as the way they are interconnected, are not fully known, all the models discussed are purely hypothetical.

A simplified diagram of the interconnections of the motion system is shown in Figure 15.2. It is the cerebral cortex that global information about external surroundings, obtained mainly from telereceptors (hearing, vision, smell), is analyzed and synthesized. On the basis of this analysis and considering the desired function, a criterion of optimum motion is determined. Subcortical centers establish, or select muscle synergies adequate to the criterion. These are movements of the muscle cooperation and their components aimed at performing a certain motion task, for example maintaining the position of the body, walking, running, jumping, speaking and writing. Consequently, the synergies form the basic "motion vocabulary" of a given person, with the excited muscles being the letters.

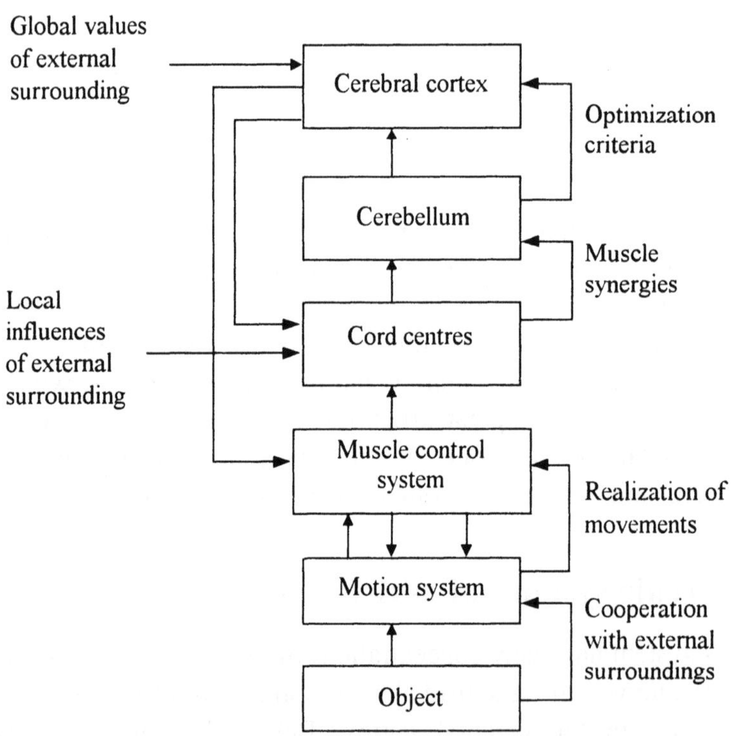

Fig. 15.2. Hierarchic model of control of motion system proposed by Braines and Świerczyński [15.1]

Subcortical centers control only the basic groups of synergies, and influence the cooperation of lower levels of the nervous system. The spinal centers obtain information about the local values of the external surroundings as well as the inner parameters of motion. They affect the realization of concrete movements and spinal cord reflexes through the motion system which influences the working objective and the external surroundings. A more detailed structural system of the interconnections of the basic centers of the motor control system is shown in Figure 15.3.

Let us look now at that part of the control system which is responsible for generating signals controlling the work of the muscle actuators. The most important centers and interconnections of the motor system are presented in Figure 15.3, which shows the motor system as a five level hierarchy, consisting of:

I. A set of feedback loops involving muscle actuators with appropriate sensing elements: this level we shall call peripherical-spinal level.
II. A segment of the spinal cord participating in the coordination of cooperating muscles.
III. A Set of supraspinal centers involving the reticular formation, vestibular system and cerebellum. It is accepted usually that this assembly forms a coordinating and corrective system.
IV. The subcortical nuclei forming the basal ganglia, including the thalamus,.and participates in the programming, coordination and creating of stereotypes.
V. Cerebral cortex – the highest coordination and decision center.

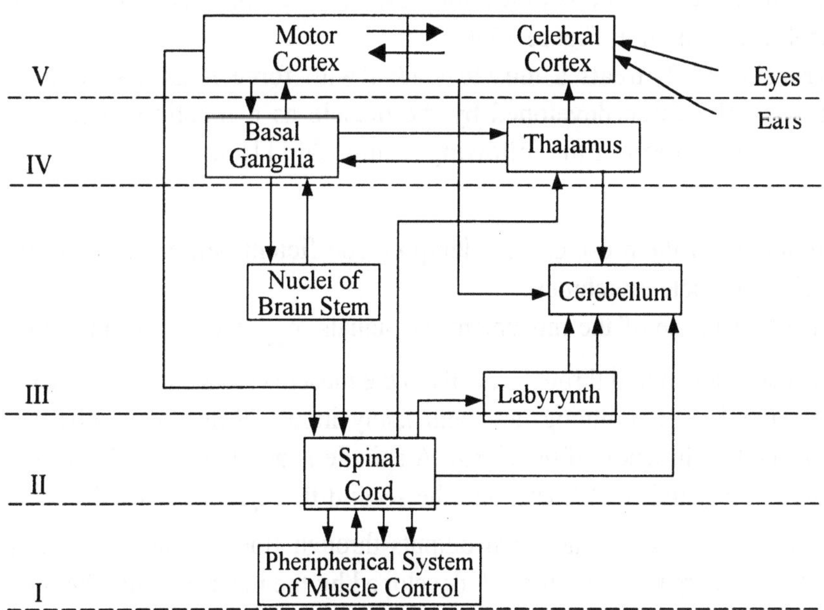

Fig. 15.3. Hierarchic control system and more important subsystems proposed by Gawroński [15.2]

From the hierarchic control system two general **control principles** can be inferred:

1. **centralization** – realized by the higher levels, especially as concerns the meaningful part of the motion,
2. **autonomy of control** (especially on the lower levels of the nervous system) – associated with the conditions of maintaining a certain stability of motion and the position of the body (homeostasis).

The centralization of the control process causes a two stage muscle excitation. Some 60÷70 ms after the movement decision has been taken, the central command is sent to the muscle actons to stabilize certain parts of the body, or move the center of gravity of the body. After 100÷130 ms, the excitation of the muscles responsible for performing a given task starts. The simplest criterion of movement control is called the **principle of reciprocal enervation**, in other words mutually opposition. However, as various studies show, its application for the control of arbitrary movements is rather limited. Nevertheless, it can be used to explain the activity of feedback loops on the spinal cord level.

At present, the cooperation between antagonistic muscles, i.e. causing movements in opposite directions (for instance flexion and extension in the elbow joint), is relatively widely recognized. The principle of reciprocating movements is based on Sherrington's rule that all signals causing the contraction of a given muscle dull the antagonistic muscle at the same time, and vice versa. The model of this system of interconnections constitutes the lowest level in hierarchic system of muscle control; it is shown in Figure 15.4.

R. Gawroński [15.2] treats a muscle loaded with force P as a system of the second order, and the force developed by the muscle as a function of the control signal x presents in the form of the following relationship [15.2]:

$$F(x) = P + \lambda \dot{l} + m\ddot{l} \tag{15.2}$$

where l is the length of the muscle, λ, a damping coefficient, and m, a mass reduced to the point of the muscle attachment.

During the first phase of the movement the signals $x_{\alpha exr}$ and $x_{\alpha fl}$ from the central nervous system are transmitted through α cells to the muscle causing it to contract. An α cell can be regarded as a summing unit with many inputs. Renshow's cells make it possible to change the direction of the signal. A spindle is an element with two inputs,: the first through γ ($x_{\gamma ex}$ and $x_{\gamma fl}$) fibers, and the second through changing the length of muscle fibers (l_{1ex}, l_{2ex}, l_{1fl}, l_{2fl}), and two outputs through fibers a_1 and a_2. The input signal for the Golgy's organs is the force F developed by the muscle, while the output is fibers connected by synapses with the α cells of antagonistic muscles.

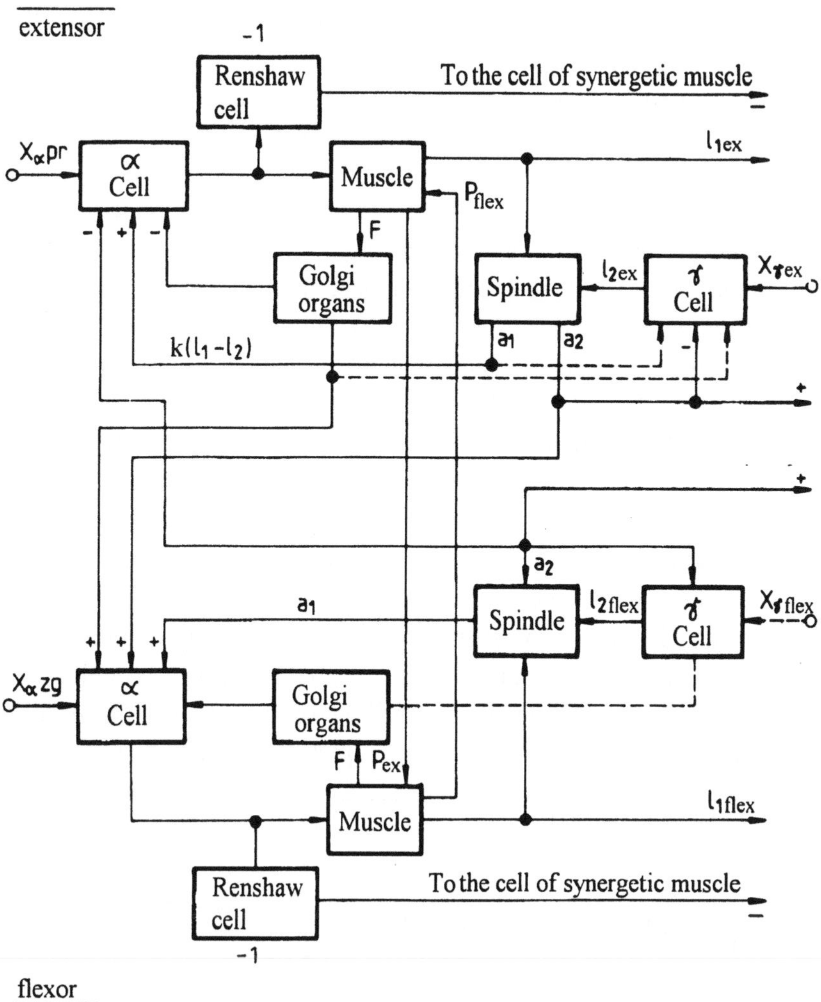

Fig. 15.4. Block diagram of antagonistic muscle control proposed by Gawroński [15.22, 15.8]

The second phase of the movements is controlled by system γ of contracting of the spindle ends. The difference between lengths l_1 and l_2 causes the excitation of fibers a_1. The signals from a_1 are delivered to the α cells as well as to the muscle, which contracts until the difference between l_1 and l_2 is zero. When the muscle produces a considerable force, the Golgy's organs are activated and send signals enervating the activity of the α cells (negative non-linear feedback). The remaining interconnections lead to synergistic and antagonistic muscles. As for the asymmetric system of antagonistic muscle control, the additional muscle load (p_{fl}, p_{ex}) due to the simultaneous action of the antagonistic muscle has also been taken into account, while the biomechanical parameters in the joint have not been considered in this model.

Outputs l_{lex} and l_{lfl} assume that the change in muscle length causes the motion in the joint, this being the reason for the lack of force output. In reality the force is the cause of the movement, while the change in muscle length is the effect.

This situation, in which the biomechanical parameters of muscles and a joint are taken into account first, while the parameters of the nervous system are treated as one object with many inputs and outputs, is shown in Figure 15.5. The input for the nervous system is the information from the telereceptors (muscle receptors which provide information about the change in length l subject to the angular position α), sinew receptors (information about the muscle force F_m), joint receptors (information about the change of angles α),and contact receptors (information about forces applied to the body links). Nervous motion stimuli I on the output of the nervous system excite the pair of antagonistic actons into action. The acton, as a part of a muscle, or a muscle with certain functions exerts force (F_m) on the links forming the joint. From this output we obtain information in the form of electromiogram U and the prioproception data $l(\alpha)$ and F_m. A muscle acton is also characterised by its physiological crosssection p and radii r. Force F_m and the external torques M_z determine the position in the joint and the velocities and accelerations $\dot{\alpha}$ and $\ddot{\alpha}$ which, considering masses and moments of inertia of the links, determine the kinematic and dynamic motion parameters of the particular links and of the whole body. As we can see, the modelling of the system of muscle control, maintaining certain feedback circuits, is very complex, difficult, and ambiguous [15.1].

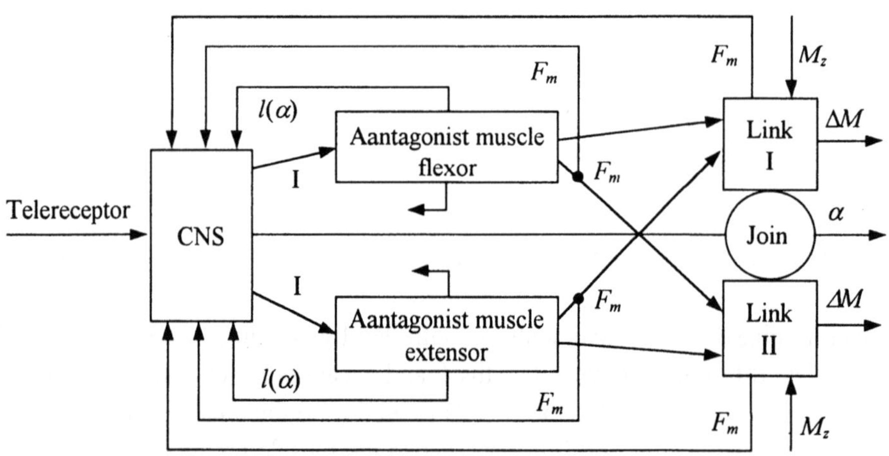

Fig. 15.5. Control bloc diagram proposed by Fidelus and Morecki

15.2.3. Perfecting the motion control process

There are a number of possible functional models of the nervous system. All of these can model a transfer function of the "black box" of certain biological phenomena. Such models usually do not take many functions into account, but by modelling a phenomenon we can obtain a "criterion of unsuitability" for a given model. If a model reflects some function well, this function may, but does not necessarily have, such a characteristic in reality. Otherwise, we know that the functional model is inadequate.

In biology, apart from the question of how and why a given process proceeds, we should ask what for? In other words, modelling functions of living organisms requires stating the goal of the modelling.

L. W. Czchaidze (1965) has presented a flow chart of the information circulating in the organism when mastering (the goal) new movements (Fig. 15.6). It shows two control circuits. The first, whose structure makes it possible to use an external feedback, controls the meaningful part of the motion task; the second, whose structure ensures that the information circulates exclusively inside the organism, controls the realization of the executive part of the motion task. The meaningful part of the task determines "what for" and "what" to do, so it decides the tactics, while the executive part determines "how to do this" which means the technique of a given movement.

When mastering unknown movements the external circuit controls the lower levels through the higher ones. Because this is not its usual function, the movement is stiff, uneconomical, and of limited accuracy. The visual control of such a movement is slow, and absorbs the actor`s attention. However, after learning the parameters of the given spatial motion, the organism itself works out a certain motion program according to which it can activate the inner control system. This encompasses the lower levels of the nervous system which now act faster and more accurately in the encountered situation. Using information about the state of the motion system (priopriorecеption), a program allowing more accurate and automatic movement performing is formulated.

A pathology provides us with the proof of this reasoning. A man or animal deprived of the functions of the higher levels of the brain (e.g. the cerebral cortex) is unable to learn any new movement. They can, however, perform movements that were known before. The situation is similar when the lower levels of the central nervous system have been damaged, for example in tabes dorsalis, a person can not perform movements automatically. It can move, supported by visual feedback, but his movements are very unsteady, like those of an apprentice instructed by his master.

Czchaidze's model separates the functions of the lower levels of the control system and regards the spinal cord as a purely motion system. His model also ignores reflexes on the spinal cord level.

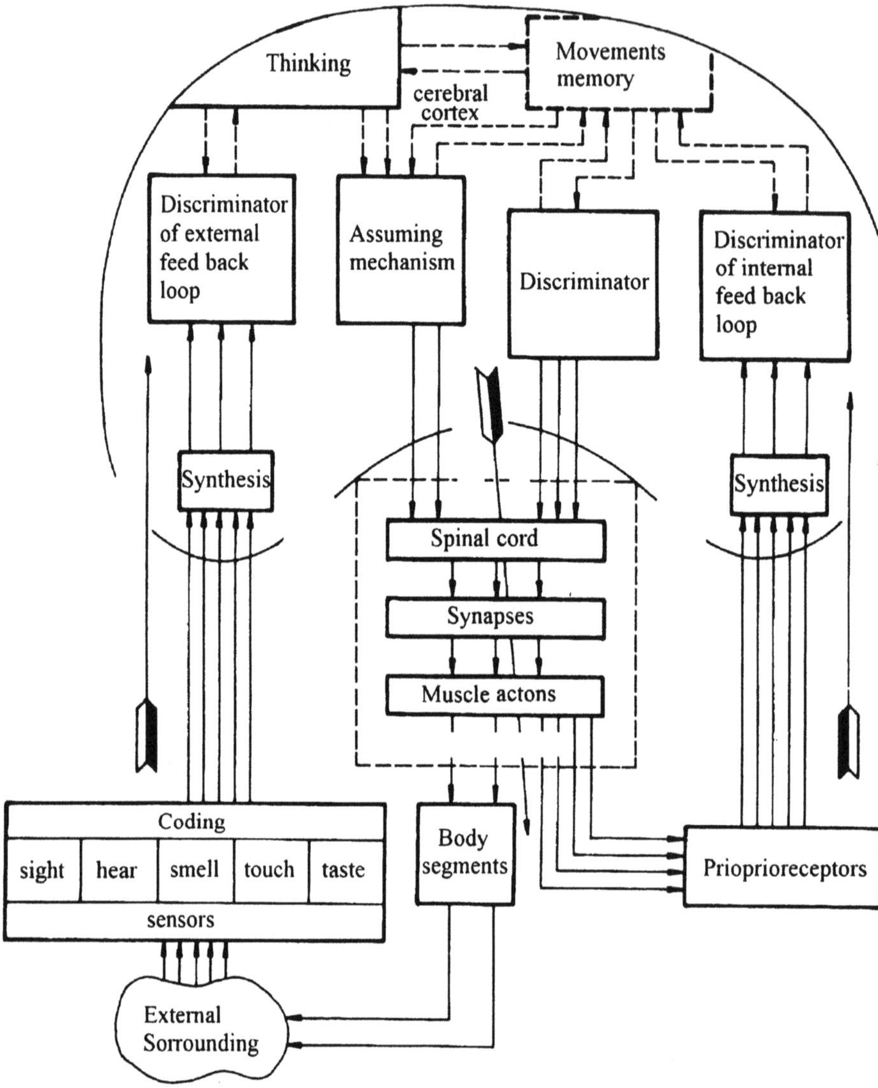

Fig. 15.6. Flow chart of presumable information circulation during voluntary movement control proposed by Czchaidze [15.1]

Fidelus, Ekiel and Morecki have proposed (1970) an expanded model of the control of the man's movements and reflexes (Fig. 15.7). The motion task can be programmed both by thinking, which imposes the criterion of the motion goal and by the use of motion memory, on which the movement optimization criterion depends. The circulation of information: thinking, the motion program, the movement memory and a programming device, makes it possible to imagine movements without having to perform them.

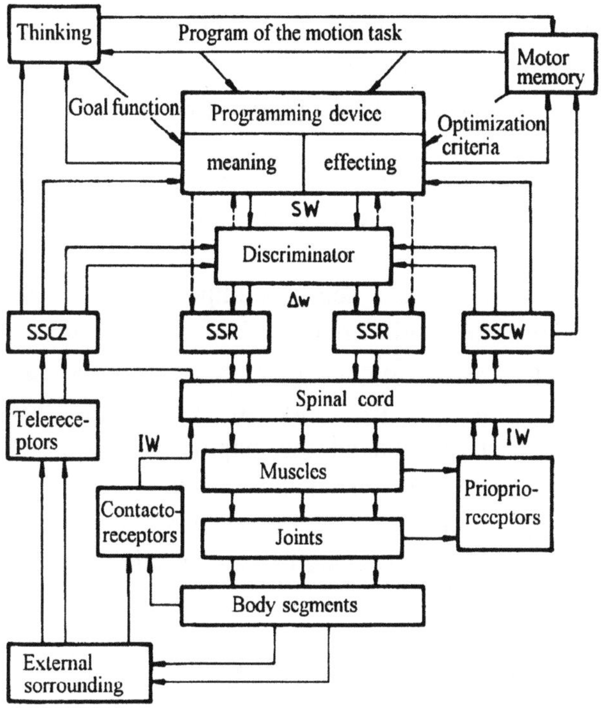

Fig. 15.7. Flow chart of human movement control proposed by Fidelus, Ekiel and Morecki [15.1]

Each motion must go through the comparing mechanism (discriminator) which compares our motion representation (*SW*) with the actual movement (*IW*). This model also has two feedback circuits providing the meaningful and executive part of the motion program. This enables the realisation of reflexes through the contacto/receptors and prioprioreceptors on the spinal cord level.

In the initial phase of learning a movement the organism can not determine the difference values ΔW. However, by using its motion memory it can and must give the desired values (*SW*) directly to the motion centres (*SSR*) which excite the muscles according to already mastered muscle synergies. In Figure 15.7 this path is indicated by the dashed lines.

In this way the control organ takes advantage of both external and internal feedback. In our opinion, without the latter the movement would be impossible.

The learning process also assumes the direct transfer of information to both the person's consciousness (thinking) and to the motion memory.

In this model we can distinguish three phases of forming movement habits, that is perfecting the processes of movement control. The first phase assumes reducing the excessive number of degrees of freedom in order to render the movements in the joints more economical and adequate.

In the second phase, the degrees of freedom and external forces are more widely used. An optimum variant of motion is sought by taking into consideration the individual qualities of the person and motion parameters (l, v, p), leading to the automation of movements.

In the third phase the most efficient external forces for variable conditions of external surrounding are purposefully evoked. Consequently, the movement habit is fully automated [15.1].

From this short outline we can see how important a role the central nervous system plays as a controller and "organizer" of the complicated human motion organ. In the control process different criteria must be satisfied at the same time, such as those concerning the trajectory to realize, with its velocities and accelerations, or in the case of the hand, the shape of an object, the stability of grip, energy consumption, and the information flow (Nightingale, 1979). This substantiates the concept in which the central nervous system (*CNS*) is regarded as a hierarchic system capable of repeating learned and optimized tasks.

15.3. Control of prostheses of arm

In this chapter we present some aspects of the mechanics and control of the prpstheses of people with partial or total amputation of the arm.

The flow chart of the exchange of information between a patient, prosthesis and the control system is shown in Figure 15.8.

The prosthesis of the upper limb is a bionic anthropomorphic (or close to anthropomorphic) mechanism, partially substituting the missing functions and located directly on the stump (Morecki, Ekiel, Fidelus, 1989). The difference in the way the natural and artificial limbs are controlled stems from the difference between the configurations of their control systems (Fig. 15.3 and Fig. 15.8) [15.3].

As it has been said, the operation of the joint-muscle system is based on an hierarchic type of pheriphery control system. The amputation of part of the limb leaves the amputee's system incomplete, with α and γ loops of no use and no signals in the feedback loops. This situation is illustrated in Figure 15.9 (amputation removes the part of the diagram to the right of the dotted line (A)[15.4]).

In the case of damage at the neck level caused by an illness or pathology, the control loop for all levels is disabled (damage to the afferent and efferent fibers, section B in Fig. 15.9) (Scede, 1980). In the amputee of one limb, simple functions are usually taken over by the voluntary limb. Activities (tasks) requiring two limbs can sometimes be performed using the healthy limb together with the stump, in which case new strategies have to be developed through learning.

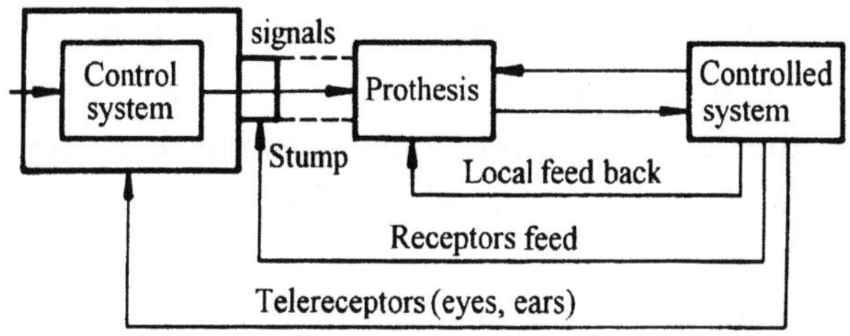

Fig. 15.8. Flow chart of prosthesis control system

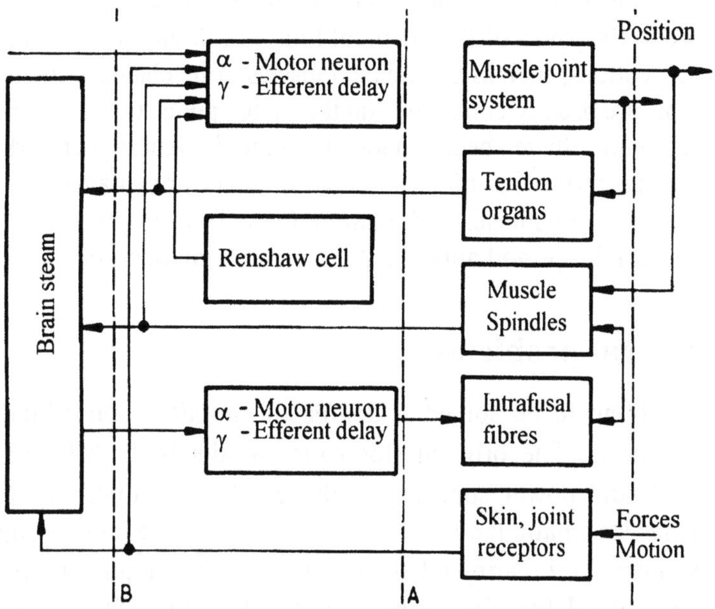

Fig. 15.9. Damage to control system of voluntary movements [Scede 1980]

A prosthesis can substitute the lost part of the limb, but it can never replace the damaged control loops.

15.3.1. Stucture and control systems of prostheses

Before a patient is provided with a prosthesis, orthosthesis, implanted stimulator, or medical manipulator, several factors must be determined:

- the patient's capability of movement and the possibilities of its control,
- the patient's real needs, and his current abilities,
- the patient's psychic and emotional statue; his readiness to cooperate with the rehabilitation device,
- the patient's ability to conform to new working conditions (change of the job), family situation, and finances.

Since the main goal of using a prosthesis, or manipulator is the partial substitution of lost functions with the cooperation of a living organism, it becomes necessary to decide the degree of anthropomorphism of these devices, both in terms of shape and function. As far as a control and feed system are concerned, a certain degree of anthropomorphism is recommended. An anthropomorphic shape is advisable in the geometrical-structural as well as psychic-aesthetic sense, while functional anthropomorphism concerns kinematic and dynamic properties.

The issue of the type of actuator (electrical, hydraulic, or pneumatic) and its location relative to the joints of the artificial limb is still an open matter.

With few exceptions, such as pneumatic muscles or mechanical-chemical actuators, technical actuators are usually employed and the same applies to feed sources.

Various attempts have been made to model neural networks (adapting, learning), and some properties of these networks may advantageously exploited in the control systems of prostheses and manipulators. Two properties in particular are interesting: feedback loops and antagonistic and synergic operation.

15.3.2. Technical problems

The first problem to be solved is degree of simplification of the engineering model of a prosthesis. The original (lost) part of the limb differed (in material, structure, control, and power supply) but the prosthesis should substitute the lost functions. Various kinematic models of the prosthesis of the human upper limb, that is a hand with fingers, a forearm and an arm, are shown in Figure 15.10.

The forearm model has five degrees of freedom ($w = 5$), including the wrist joint (Fig. 15.10a). The prosthesis of the whole limb has eight degrees of freedom. Experience in prosthesis design indicates that a reasonable model should have eight degrees of freedom at the most but not less than four (Fig. 15.10b). Figure 15.10c shows a simple prosthesis of a hand with three fingers.

The second problem refers to the selection of the method and system of control. This is a difficult problem, and it is being solved in various ways, usually unlike the control principles of living organisms in which the nervous system sends signals to all relevant muscles at the same time. The drive delay of 30÷40 ms in the control signal distribution results from the need to stabilize some functions. Such a system does not seem feasible today. Consequently, the two solutions used in control systems are:

- *sequential control* allowing the activation of many degrees of freedom with a minimum of control signals (1 or 2) and employing suitable logic units (Fig. 15.11a),
- *parallel control* of many degrees of freedom using specialized processors and sets of programs (Fig. 15.11b).

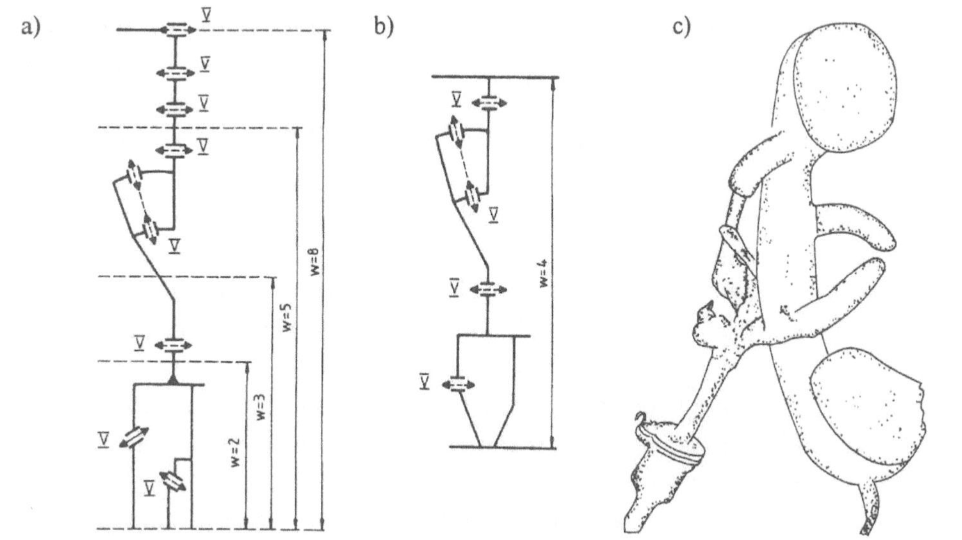

Fig. 15.10. Kinematic chains of the prosthesis of the man's upper limb: a) system with $w = 8$, $w = 5$, $w = 3$ and $w = 2$, b) simplifies system with $w = 4$, c) prosthesis with three fingers

The use of external control systems began in the fifties (pioneers: Marquardt in Heidelberg, and Aldersoni Kiessling in the USA, both in 1955).

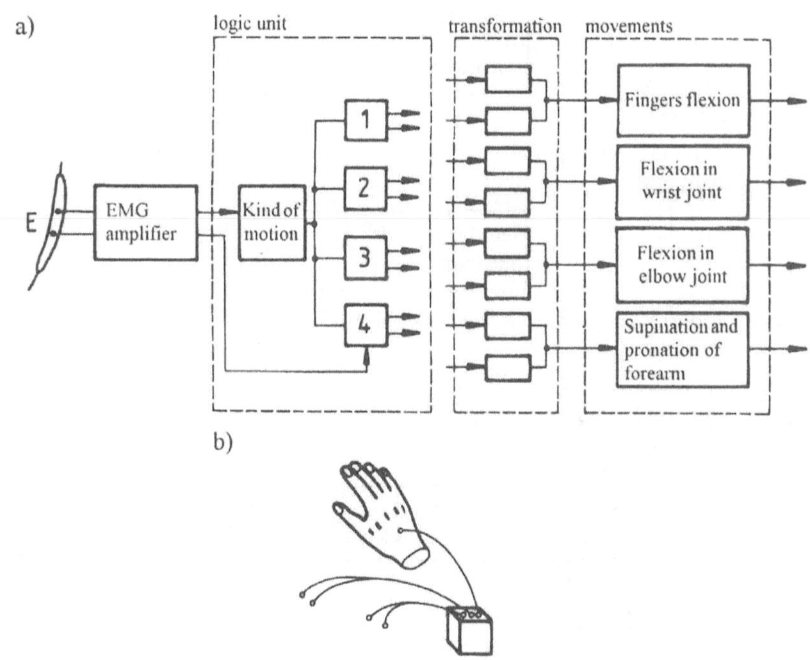

Fig. 15.11. Bioelectrical control systems of prosthesis: a) sequential system, b) parallel system

Bottles with liquid CO_2 for controlling pneumatic actuators and electrical batteries are used as a source of energy. Each of the systems mentioned poses different problems of control.

A quite common control system is the bioelectrical (Bottomley, 1970). The use of an electomiogram (EMG) as a control signal date back one 40 years. The first experimental prosthesis, with bioelectrical control, was developed by the team of Battye, Nighting and Whillis (1955) at Guy's Hospital in London [15.7]. The prosthesis with bioelectrical control developed by Kobryński and others in 1961 [15.1] was widely used for people with limb amputation below the elbow joint.

Figure 15.12 is the flow chart of a system developed in 1957, in which the electrical activity of the antagonistic muscles – flexor and extensor of the forearm is used to control the grasping motion (opening and closing of the artificial hand). Necessary transmissions were applied in this system.

Different varieties of multi-purpose prostheses with bioelectrical control of the 0-1 type have been developed over the last forty years. Independently, systems with continuous control such as the bioelectrical hand constructed by Bottomey in 1970, presented in Figure 15.13 have appeared on the scene. In this system a backlash generator, two feedback loops (for the velocity and force) and a force sensor were applied.

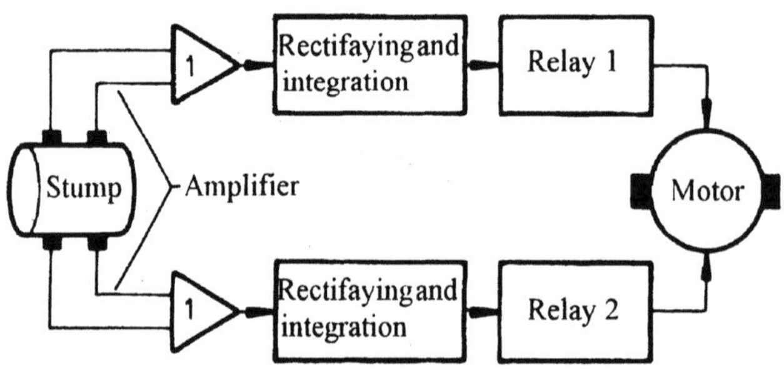

Fig. 15.12. Prosthesis control system developed by Kobryński [15.1]

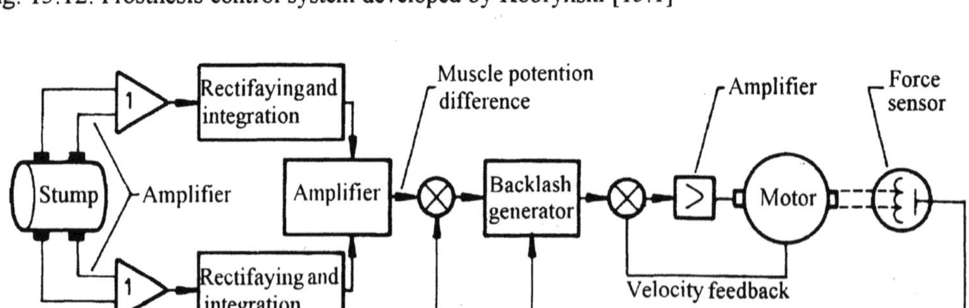

Fig. 15.13. Prosthesis control system developed by Bottomley [15.1]

P r o b l e m t h r e e, also connected with the issue of control, concerns the selection of proper sensors and feedback loops. Only a few systems are fitted with sensors (Fig. 15.13), and most of them operate as open systems, using their vision to control the behavior of the system.

P r o b l e m f o u r concerns to the feed system. Both electrical batteries and liquid gas (CO_2) are exhausted after a relatively short time, requiring recharging or refilling. This causes certain difficulties for the patient.

P r o b l e m f i v e concerns the every day cooperation between patient and prosthesis. It is known from experience that only about 40 per cent of amputees use their prostheses every day. Various technical problems, service difficulties, or simply using of the healthy limb mean that a considerable number of patients do not use their prostheses all the time.

To sum up the issue of patient-prosthesis cooperation we examine those aspects of control that seem to be crucial factors in the usefulness of bioelectrical prostheses. Current trends tend to increase the number of functions provided by a prosthesis and their reliability. An analysis of the existing methods of control of multi-purpose prostheses of the upper limbs, made in 1984, divides them into:

- multi-channel systems,
- program systems,
- recognizing systems.

A recognizing system, which acts on the principle of combination of the bioelectrical signals that determine the realization of a certain movement, seems to be the most promising of these systems. Progress in microprocessor techniques has brought marked development in microprocessor systems for prosthesis control in recent years. One of the first of these systems was produced in the seventies (Lyman, Freedy, 1971) at the Biotechnology Laboratory of UCLA. The aim of this project was to develop a microprocessor control system for a prosthesis with three degrees of freedom (a recognizing system), together with information optimization for the purpose of sensor feedback [15.8].

Figure 15.14 shows the hardware elements of this system. As a result of the first experiments the ability to control eight functions has been achieved (flexing and extending the forearm, supination and pronation of the wrist joint, opening and closing of the hand, the combination of flexing the forearm and pronation of the wrist joint, and the combination of extending the forearm and supination of the wrist joint and rest).

The control process assumes a constant read-out of the vector of signals obtained from electrodes attached to the patient's body and their transformation into motion signals. Figure 15.14a shows the flow chart for the organization of the control system.

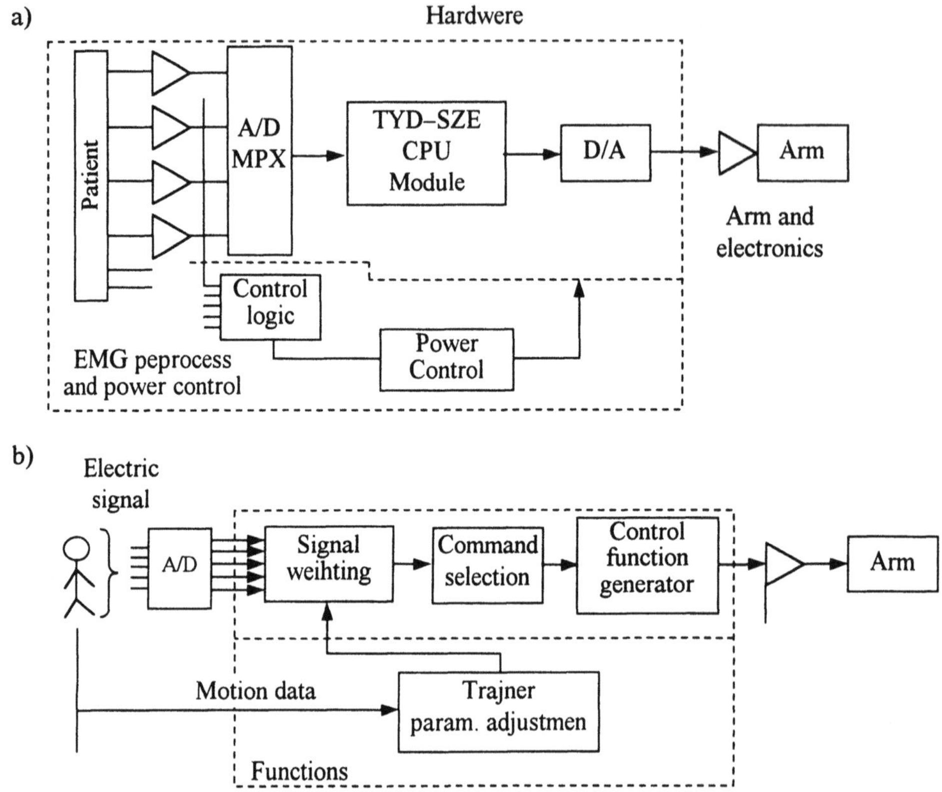

Fig. 15.14. Computer control of prosthetic manipulator arm: a) flow chart of functional organisation of control system , b) processor system [15.8]

The system consists of three units: a bioelectrical signal preprocessor with power control, computer equipment, and the mechanical and electronic part of a prosthesis.

The transmission of a program from a computer depends on various simplifications aimed at obtaining adequate speed at low power. The main part of the system (Fig. 15.14b) is a processor and memory (ROM and RAM).

15.4. Rehabilitation manipulators and robots

In recent years, bioengineering has been moving in a new direction, independently of prosthetic-orthotics, that of **rehabilitation manipulators and robots**. We can distinguish three types of manipulators:

- directly attached to the patient,
- free standing,
- attached to an electrical wheelchair.

Various control systems are employed, using, for example, the head, tongue, chin, voice, etc.

A *rehabilitation manipulator* is a bionic mechanism, partly anthropomorphic, designed for the partial realization of the motion of the upper limb (Fig. 15.15a).

A *manipulative-pedipulative system* is a device of the bionic mechanism type, partly anthropomorphic, designed for the partial realisation of the motion of the upper and lower limbs (Fig. 15.15b, Morecki, 1984).

Various systems of rehabilitation manipulators for aiding disabled patients have been developed in recent years (T.B. Sheridan, D.R. Whitney, J. Lyman, A. Freedy, B. Roth, Roessler, M.R. Hillman and others, 1970-90).

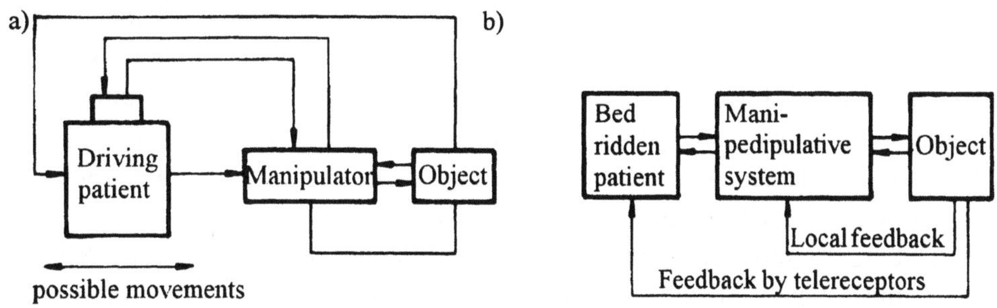

Fig. 15.15. Flow charts of rehabilitation manipulators a) patient-manipulator flow chart, b) patient-manipulative-pedipulative system flow chart

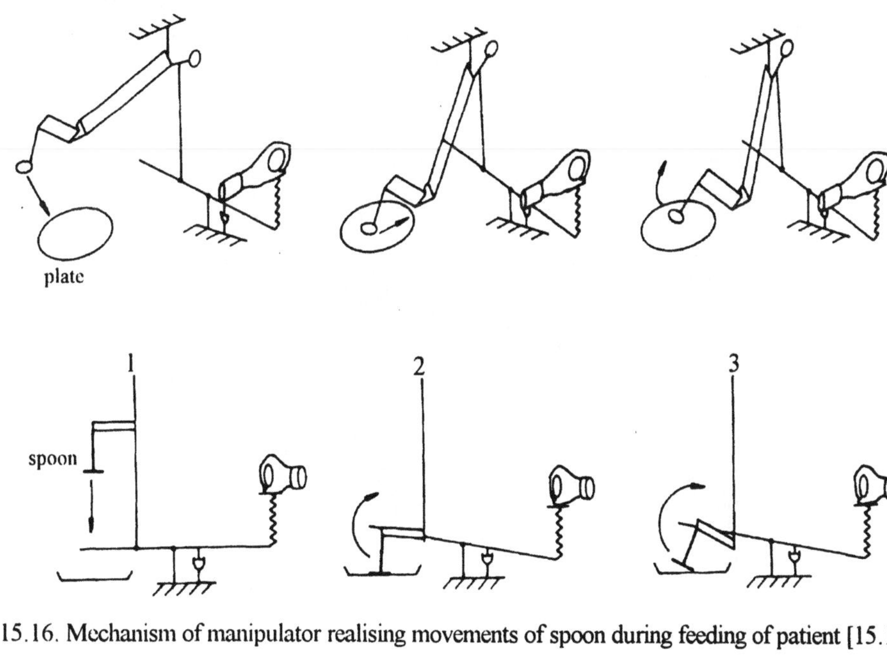

Fig. 15.16. Mechanism of manipulator realising movements of spoon during feeding of patient [15.13]

Figure 15.16 shows a manipulator adapted to feed a patient while Figure 15.17 shows this device during the feeding activity. A spoon is attached at the end of a parallelogram mechanism. The subsequent Figures (Fig. 15.16) show the successive positions of the mechanism. The arrows indicate the successive motion phases (1, 2 and 3). After the spoon has touched the plate, it performs a rotational movement dipping one side into the food. During its rotation it can be moved horizontally across the plate for a span of 100 mm.

The supporting mechanism moves horizontally. After taking the spoon out of the plate, it reverts to the vertical position, with the parallelogram mechanism holding.the spoon in position while moving it towards the mouth. The mechanism enables rotation about a vertical axis allowing the spoon to move along different trajectories on the plate, and at the sides, to collect the remnants of the food. The mechanism is appropriately protected to ensure adequate hygiene. It has been tested by four patients suffering from motion pasy (ataxy) at Frenchay Hospital, Bristol, England. Laboratory examinations have also been carried out to check the mechanism's response to different frequency input [15.13]. In some models the patient controls the manipulator`s movements with his leg [15.15].

Figure 15.18 shows a manipulator with computer control. The arrows indicate possible movements of the manipulator, while the dashed lines represent the trajectories realized in taking the food from the plate and transporting it to the patient's mouth (P_3) [15.6].

Another variety of rehabilitation manipulators is located on an electrical wheelchair. The patient controls its movements by means of various parts of his body, by moving his head, or leg, for example, or by voice.

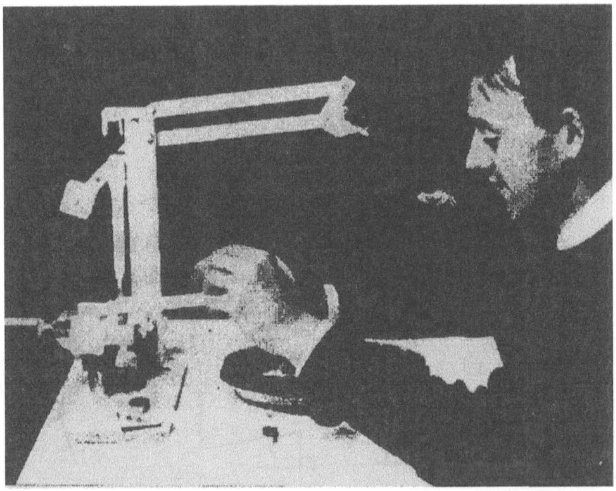

Fig. 15.17. Control of mechanism position [15.13]

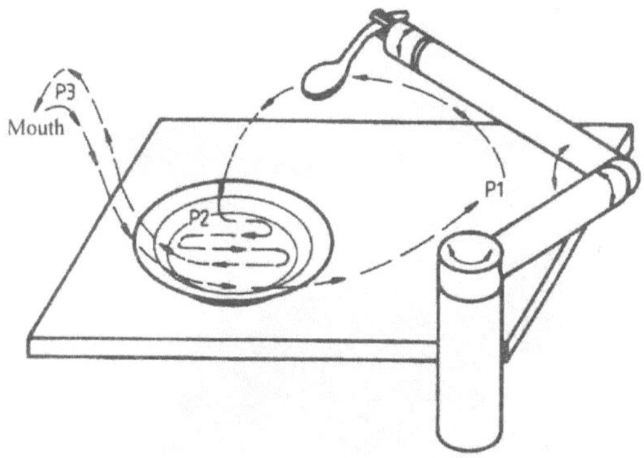

Fig. 15.18. Manipulator with computer control

Figure 15.19 shows a manipulator with many degrees of freedom. The arrows indicate possible movements. The manipulator presented in Figure 15.20 [15.10] is an example of a voice-controlled system.

Figure 15.21 [15.14] shows another variety of a manipulator. This is a manipulator attached directly to a patient, located on his head and controlled by slight movements of his head or finger. This is particularly helpful in cases of bilateral amputation. It has unquestionable advantages, such as low mass and a simple control system, which render it competitive with stationary prostheses and manipulators.

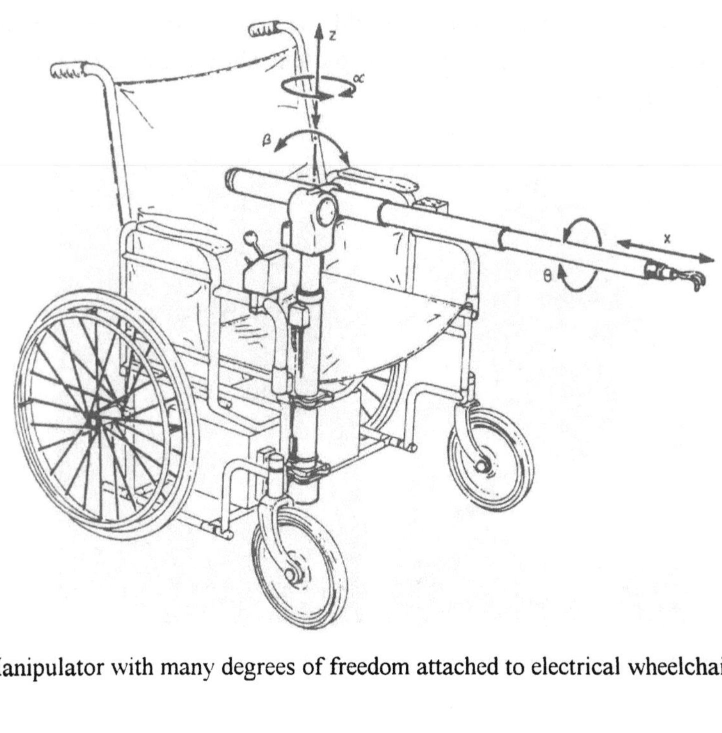

Fig. 15.19. Manipulator with many degrees of freedom attached to electrical wheelchair [15.6]

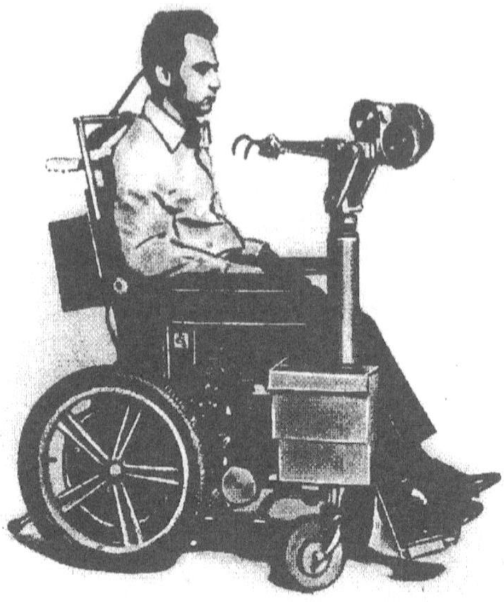

Fig. 15.20. Manipulator with voice control installed on electrical wheelchair [15.10]

On the grounds of research regarding spinal cord injury conducted at the Department of Mechanics of Stanford University, as well as at the Veteran Hospital in Palo Alto, the use of the SMART ROBOTIC ARMS manipulator (Leifer, Roth, 1978) [15.11] has been proposed. Three questions have been asked: what the signals controlling the robot's movements should be like, what the communication-interaction between the robot's output and patient should be like, and what should be the price of such a device.

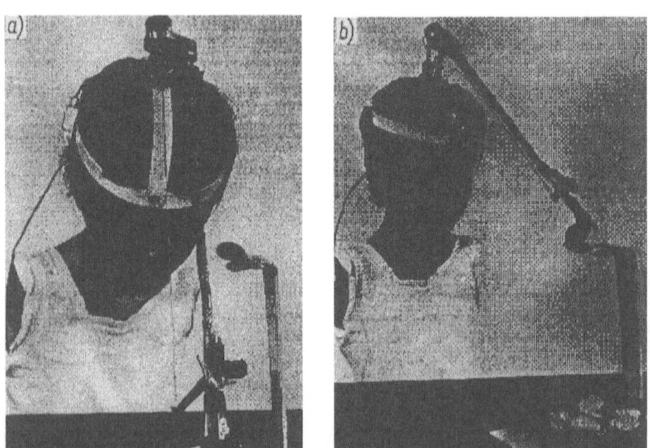

Fig. 15.21. Manipulator attached to patient's head [15.14]

Figure 15.22 shows such a system. It consists of an executive microcomputer, programs, and manipulator with seven joints, each controlled by a processor. The movement of the head, chin, eyes, or voice can be used as the output signal. A graphic terminal allows the patient to communicate with the manipulator.

The capabilities of the control in an open, closed, or hierarchic system have also been considered. The estimated cost of such a system would be about 8000 US dollars.

Figure 15.23 presents a full system with computer control. An interesting trend in the development of rehabilitation manipulators is the rehabilitation system fitted with a manipulator for tetraplegics (Fig. 15.24).

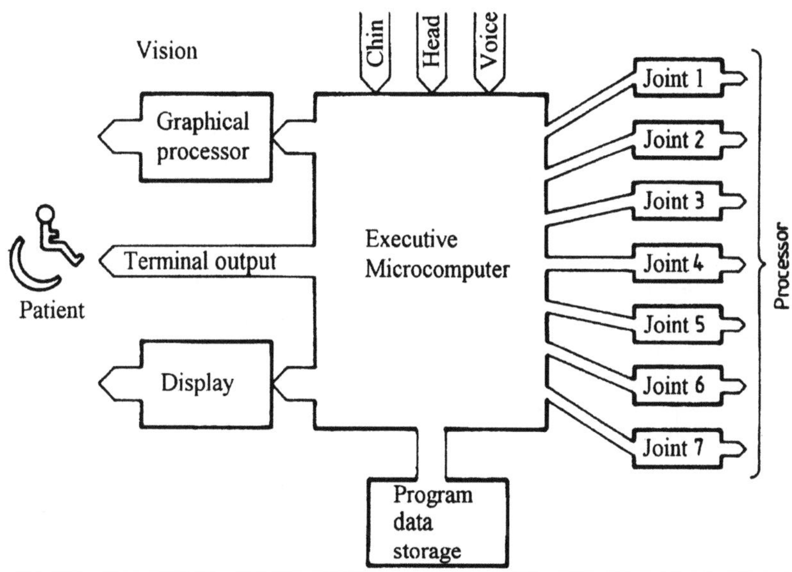

Fig. 15.22. Manipulator control system proposed at Stanford University [15.11]

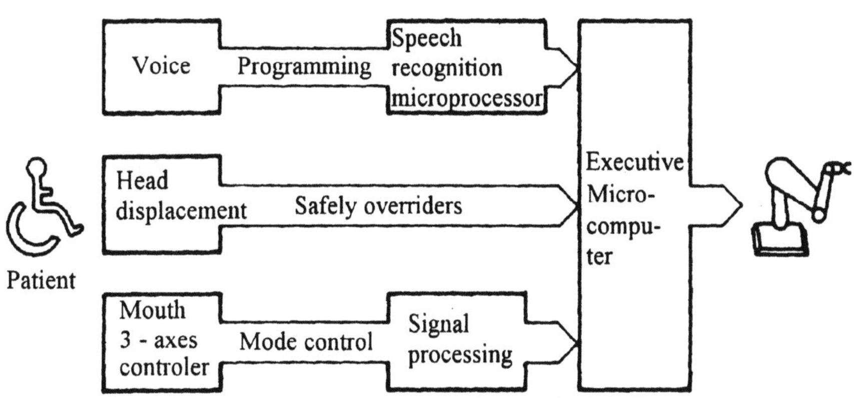

Fig. 15.23. Developed control system of Stanford manipulator

Figures 15.24a and 15.24b show a system of this kind, developed by Roesler and Paeslack [15.12] while Figures 15.24c and d present a SPARTACUS system [15.17], consisting of a properly prepared work station, manipulator and special equipment. The manipulator is a system with five degrees of freedom, which enables it to manoeuvre in the service space. The control system is of the end point control type, which means that a certain movement of a lever corresponds to the movement of the manipulator's end point in certain direction. This solution seems preferable to the individual control of each joint, and simplifies the control process. The Spartacus system assumes control by means of the head (Fig. 15.24d).

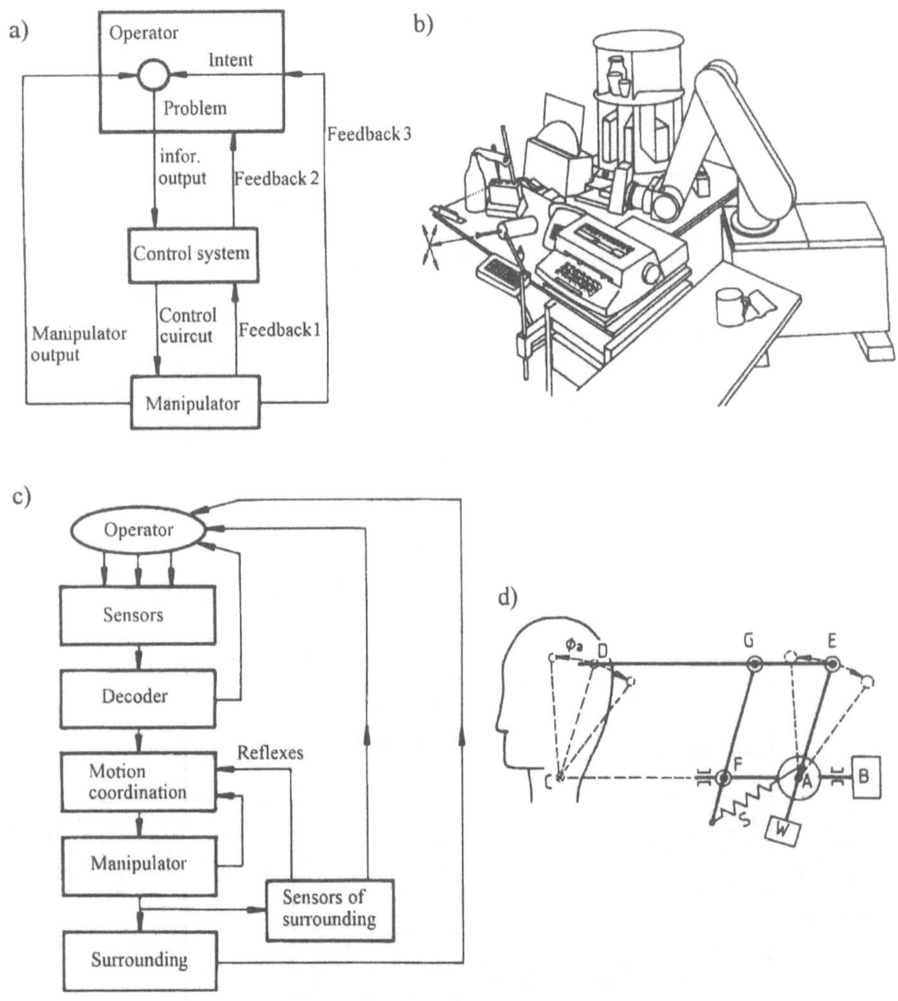

Fig. 15.24. Medical manipulator for tetraplegics [15.12]: a) control system, b) general view, c) flow chart of SPARTACUS type control, d) control by means of the head

Another example of a work station fitted with a manipulator developed by a team of engineers and physicians from England [15.13], is presented in Figure 15.25. A prototype system has been developed using an inexpensive Atlas manipulator with five degrees of freedom and grippers with stepper motors (the cost is about 2000 UK pounds). The station is 9.7×0.9 m. The various tasks for the manipulator to perform have been distributed in a circle around it. The Arcon BBC computer has been used as its control. The menu is displayed on the screen; the user selects a symbol through a two-switch input and resolving the motion direction.

The system has been tested in hospital conditions on a sample group of eight patients with high spinal cord injuries. On the basis of these tests, a new version of the system, shown in Figure 15.25, has been designed. The dimensions of the work station have been reduced for domestic conditions and adapted to a patient using an electrical wheelchair. The system's hardware is located in a desk (Fig. 15.25a). The Atlas robot has been replaced by a SCARA type robot (Fig. 15.25b). The gripper is positioned in the horizontal plane. The arm and elbow joints are driven by rotary motors. Figure 15.26a shows the vertical and horizontal actuators employed. The whole manipulator can be lifted by means of a vertical actuator. In this device DC motors as well as increment type visual encoders in the feedback loop have been used. The control system is similar to that applied in the prototype system [15.5].

Finally, we shall briefly discuss the concept of the so called "alive manipulator" developed by Polish scientists [15.15, 15.16].

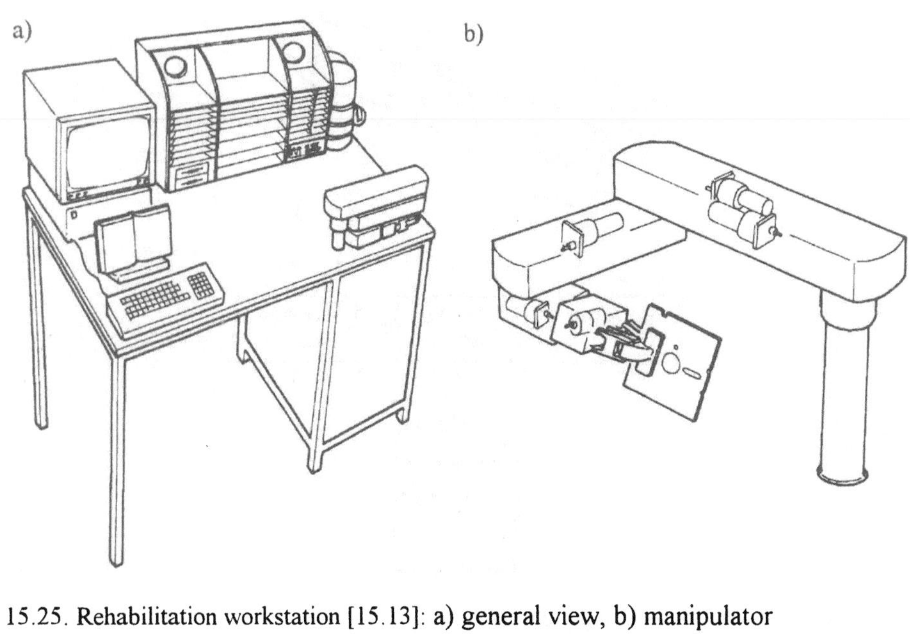

Fig. 15.25. Rehabilitation workstation [15.13]: a) general view, b) manipulator

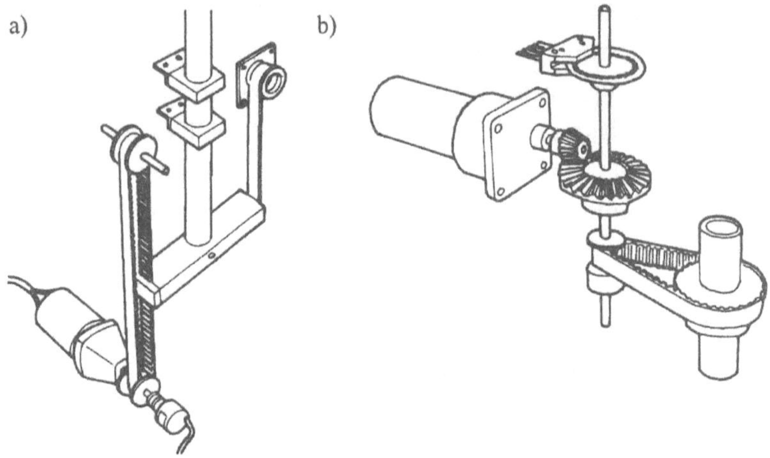

Fig. 15.26. Actuators of SCARA type manipulator used in the work station from Fig. 15.25 [15.13]: a) horizontal, b) vertical

The diagram of a hybrid system for assisting prehension motion is shown in Figure 15.27. An injury of a spinal cord at the C5-C6 level precludes the ability to perform the prehension motion.

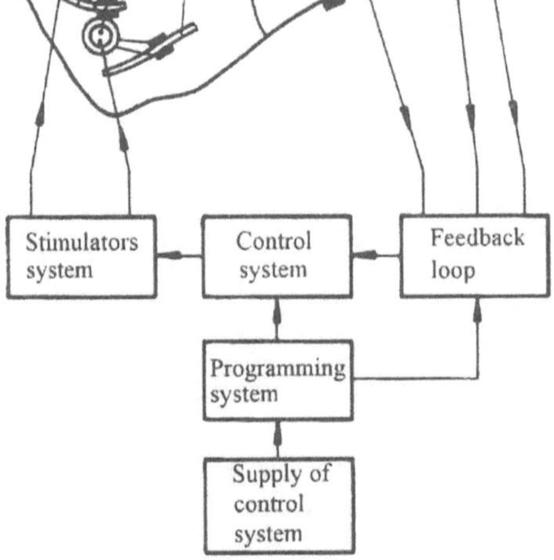

Fig. 15.27. Hybrid system for assisting prehension motion [15.1]

The essence of the project lies in the simultaneous cooperation of the external orthosthesis system (3) with a system of implanted stimulators (5) and electrodes attached to the nerves (4). The system of converters consists of converters of position (2) and force (1), and a sensors (1'). The position converter, which is a precise potentiometer, allows control of the opening and closing of the hand. The force converter, during the contact with an object, makes it possible to control the value of the force necessary to pick up the object.

A full system, presented in Figure 15.27, consists of feeding, programming and control systems, as well as feedback loops. Stimulators are attached to a radial and pheripheral nerve of the disabled limb which provides selective excitation of certain muscle groups. The system makes it possible to adjust such stimulation parameters as an amplitude, its duration, and the pause of the stimulation pulse. The course of the stimulation program is as follows: 5 s stimulation, 30 s pause – 3÷5 times within 5 min and then a break of 55 min. Such a program was run for 3÷4 hours every day for 4÷6 months. The duration of the stimulation pulse was 0.7 ms, with frequency 45 Hz. In many cases a (8÷10)-fold increase in force (from 2 N to 20 N) was observed, and this was maintained for many years.

Figure 15.28 shows the successive operations of performance of the prehension motion. They include three phases, as well as operations *A* and *B*. In the first phase, the device is set in an initial position. Flexors and extensors are not stimulated. In the second phase, the hand is open (the fingers are straightened). The stimulation of the flexors is switched off and that of the extensors switched on.

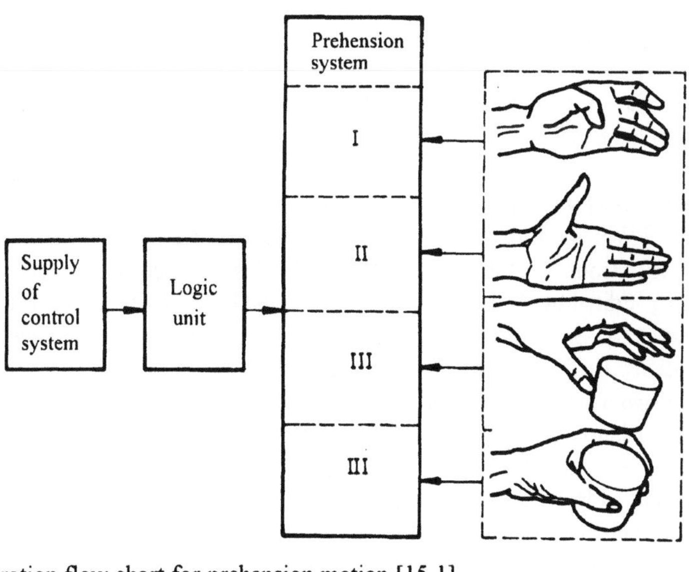

Fig. 15.28. Operation flow chart for prehension motion [15.1]

The third phase includes two successive operations:

A. extensor stimulation is switched on and flexor stimulation switched off, so that the fingers approach the object with controlled velocity,

B. the object is lifted and the stimulation adjusted to maintain a firm grasp (subject to the object's weight)

In Figure 15.29a the test bench is shown and in Figure 15.29b operation *A* and preparations for operation *B*. The feedback provides a gradual, fluent increase in force, and the closing of grasp.

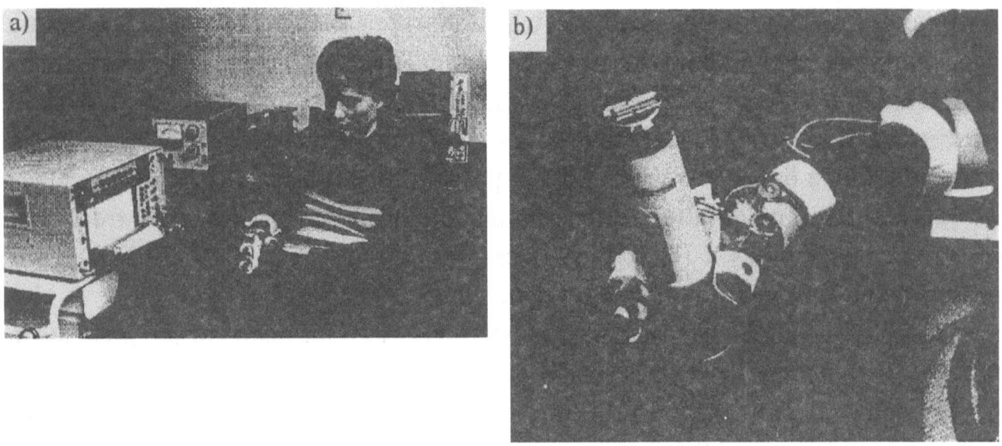

Fig. 15.29. "Live" manipulator [15.1]: a) patient at test bnch, b) grasping and lifting of object

15. 5. References

[15.1] *Morecki A.*: Bionic manipulators. Warszawa, PWN 1976, (in Polish).

[15.2] *Gawronski R.*: Problems of bionics in large systems. Warszawa Wyd. MON 1975 (in Polish).

[15.3] *Morecki A., Ekiel J., Fidelus K.*: Cybernetic systems of limb movements in man, animals and robots. PWN and Ellis Horwood Limited Publ. Chichester 1984.

[15.4] *Morecki A.*: Control Aspects of Artificial Hand, Control Aspects of Biomedical Engineering. Edited by. N. Nałęcz, Published for the IFAC by Pergamon Press, 1987.

[15.5] *Broadhurst M.J., Stammers C. W.*: Mechanical feeding aids for patients with ataxia: design considerations. J. of Biomedical Engineering. Vol.12, No 3, May 1990.

[15.6] *Ramey R.L.*and others: Manipulator control by Minicomputer. Colloques IRIA 1978. Int. Conf. on Telemanipulation for the physically handicapped. Rocquencourt France, 4-6 Sept. 1977.

[15.7] *Bottomley A.R.*: The Control of the upper limb. In D.C. Simpson (Ed.) Modern trends in Biomechanics - I Butterccorths, London 1970.

[15.8] *Lyman I.* and others: Control of an artificial limb with several degrees of freedom of motion. Bulletin of Prosthetics Research. Dept of Medicine and Surgery V.A.Washington, 1977.

[15.9] *Taylor H.J.*: The Development of two Wheelchair Manipulator Systems. Colloques (IRIA). Int. Conf. on Telemanipulation for the physically handicapped. Rocquencourt 4-6 Sept.1978.

[15.10] *Bejczy A.K.*: Voice Command System For Motion Control. Colloques IRIA. Int. Conf. on Telemanipulation for the physically handicapped. Rocquencourt 4-6 Sept.1978.

[15.11] *Leifer L.* and others: Robotic aids for the severely disabled - needs assesment. Colloques IRIA. Int. Conf. on Telemanipulation for the physically handicapped. Rocquencourt 4-6 Sept.1978.

[15.12] *Poeslack V., Roesler H.*: Medical Manipulators - general considerations on the design criteria of manipulative technical aids for quadriplegics. First CISM-IFTOMM Symposium on Theory and Practice on Robots and Manipulators. Vol.1. Udine Springer Verlag. 1974.

[15.13] *Hjlinann M.R.* and others: Development of a robot arm and workstation for the disabled. J.of Biomedical Engineering. Vol.12, No 3, May 1990 Butterworths.

[15.14] *Ober J.K., Leonhard J., Ogórkiewicz A.*: On the Head Manipulator for Bilateral Shoulder Disarticulated Children. Proceed. on Theory and Practice of Robots and Manipulators Ed. by A.Morecki, K.Kedzior, PWN – Elsevier Scientifics Publishing Company, 1977.

[15.15] *Morecki A., Weiss M., Kiwerski J. and Paśniczek K.*: A New Method for Forcing Lost Grasping Functions of Extremities by Use of on Orthotic Manipulator Combined with Implanted Stimilators of Nerves. Int. Conf on Medical Devices and Sport Equipment. Century 2-mF.CA August 18-20, 1980 ASME. Ed. by T.E.Shoup and J.G.Thacker.

[15.16] *Morecki A.* and others: Substituting of Upper Human Extremities Functions –where we are going? Lecture Notes in Medical Informatics Europe 1981. Third Congress on the European Fed. of Medical Informatics. Proceed. Tolouse, France, March 9-13, 1981.

[15.17] *Gujttet I.* and others: The SPARTACUS telethesis: Manipulator Control and Experimentation. Colloques IRIA. Int. Conf. on Telemanipulation for the physically handicapped. Rocquencourt 4-6, Sept. 1978.

[15.18] *Morecki A.*: Control Aspects of Artificial Hand. IFAC Control Aspect of Biomedical Engineering, 1987.

16. ELASTIC MANIPULATORS

16.1. Basic definitions and classifications

Elastic manipulators are manipulators designed to operate in workspaces containing obstacles, as well as limited workspaces. During positioning of the end-effector at a certain point of the manipulator's workspace, the arm of such manipulator must be able to take various, sometimes very complex shapes, and the spatial orientation of the end-effector must be independent of the position of that point.

The arm mechanism of such a robot must have a sufficient number of degrees of freedom, and cannot contain rigid elements the lengths of which are comparable to the distances between obstacles [16.1].

Elastic manipulator arm structures may be divided into two main classes. The first includes mechanisms composed of deformable links, the other includes mechanisms containing rigid links; the linear dimensions of which are small compared with arm length. The required number of degrees of freedom is obtained by linking movements of selected groups of links.

Two further structural modifications can be specified within the second class of mechanisms. In the first modification the manipulator is an open kinematic chain containing identical kinematic pairs with a geometrical or force closure by wire systems; the motion in particular groups of pairs is linked by the drive system. The other possibility is a chain of closed subchains. The motion of link groups is linked by the mechanism structure, not by the drive system.

First scientific studies on snake-like locomotion dates back to the thirties. First models of elastic manipulators appeared in the sixties and was called **snake-like manipulators** [16.2]. The later version was named **tensor arm** [16.3]. Next generations of snake-like mechanisms were constructed in the seventies [16.4]. In the seventies and eighties other modifications of elastic manipulators appeared, namely spine-type manipulators and elephant trunk manipulators.

Elastic manipulators of today may be divided into three large groups:

- snake-like manipulators,
- elephant trunk manipulators,
- spine-type manipulators.

On the next pages of this chapter various designs, their properties and some structural, kinematic and dynamic aspects of such manipulators are described in detail.

16.2. Snake-like manipulators

The manipulator shown in Figure 16.1a is a model worked out in the Laboratory of Artificial Intelligence at Stanford University by L. Leifer and V. Schneiman [16.2]. This mechanism was named „Orm", Norwegian for „Snake", and resembling to English word „Arm".

The„Orm" manipulator was developed in [16.3] into the „Tensor arm" manipulator (see Fig. 16.1b). From the structural point of view this mechanism is a multi-DOF manipulator composed of short links; the motion of this manipulator is similar to continuous motion, and its shape is like that of a snake. "Tensor arm" was considered that is the optimum solution for underwater operations.

The „Orm" model shown in Figure 16.1a has six moving links, and a pneumatic drive system. The kinematic pairs are of the Hook's joint type, allowing independent rotations about two axes. Binary type actuators are used in the drive systems, meaning that they may be fully opened or fully closed (0-1), without intermediate positions. Thus the actuator may be controlled by one bit only and there is no need to apply a D/A converter in the control system. In each link 8 actuators are installed around the link circumference, equally spaced and equally distant from the point of intersection of the axes. This allows 8 configurations when only one actuator is moved. Thus a 6-link manipulator has 8^6 configurations. This manipulator may also use stepping motors instead of actuators. To provide general usefulness of such manipulator, the number of 8^6 configurations should ensure manipulation capabilities in the given workspace.

It has been estimated that a 24-link system may, on the average, reach points spaced by 0.00004 inch in a 50.000 cubic inch workspace.

The dynamic control of such a manipulator is not easy. The transmission of motion is achieved by means of wires. The main disadvantage of this manipulator is the number of actuators, valves and wires required. The manipulator shown in Figure 16.1 was patented by V.C. Anderson and R.C. Horn (US patent no. 3497083, 1970). By using direct drive this type of manipulator may find many applications coping with situations (like moving through an opening) in which typical manipulators are limited.

Interesting studies on kinematic analysis and zoological experiments concerning snake-like locomotion were performed by the Japanese researchers Hirose and Umetani [16.4]. They produced a model of snake-like vehicle named „**robot**".

The snake`s body acts as a „foot" when moving over the ground, as a „hand" when wrapping itself around an object, and as a „body" when moving from one branch onto another on the tree. The authors consider the snake as a multi-functional robot, regardless of its simple shape.

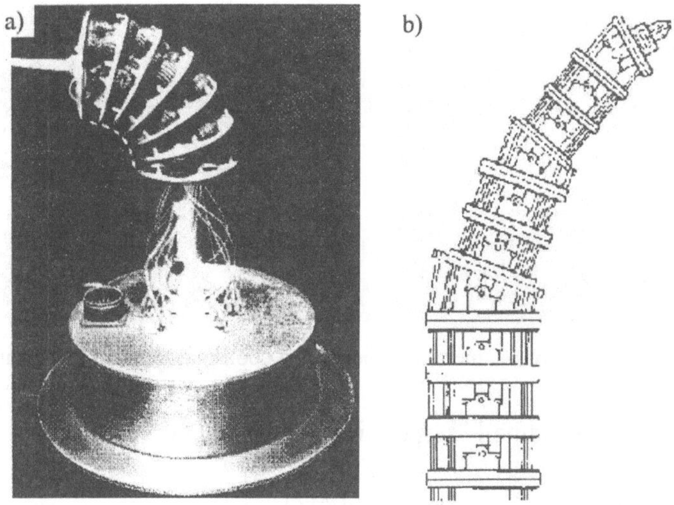

Fig. 16.1. Snake-like manipulator [according to 16.2 and 16.3]: a) Model built at Stanford University; b) „Tensor arm" manipulator.

The studies gave kinematic relationships concerning the snake-like locomotion, which were verified by experiments on a snake, Elaphe Quadrivirgate. Next a special type of locomotion which the authors called "sine-type locomotion" was defined. It is optimal for fast movements and the movements adapt to the terrain. Finally, a mechanical model composed of links driven by electric actuators and performing artificial snake-like locomotion was built.

The authors identified two types of curves representing the sliding motion of a snake: clotoid (Cornu spiral) and "snake curve".

Figure 16.2 shows the curves (clotoid segments) of snake-like locomotion in a coordinate system. The equations of these curves are:

$$\begin{bmatrix} x(s) \\ y(s) \end{bmatrix} = l\sqrt{\frac{\pi}{2\alpha}} \begin{bmatrix} \cos\alpha & -\sin\alpha \\ \sin\alpha & \cos\alpha \end{bmatrix} \begin{bmatrix} 1 & 0 \\ 0 & -1 \end{bmatrix} \begin{bmatrix} C\left(\frac{1}{l}\sqrt{\frac{2\alpha}{\pi}}s\right) \\ S\left(\frac{1}{l}\sqrt{\frac{2\alpha}{\pi}}s\right) \end{bmatrix} \tag{16.1}$$

where:

$C(*)$ and $S(*)$ - Fresnel integrals,

l - distance along the spine axis from O to P,

α - angle,

s - auxiliary variable.

This equation was obtained on assumption that muscles contract and expand monotonously and the bending angle changes proportionally along the spine axis.

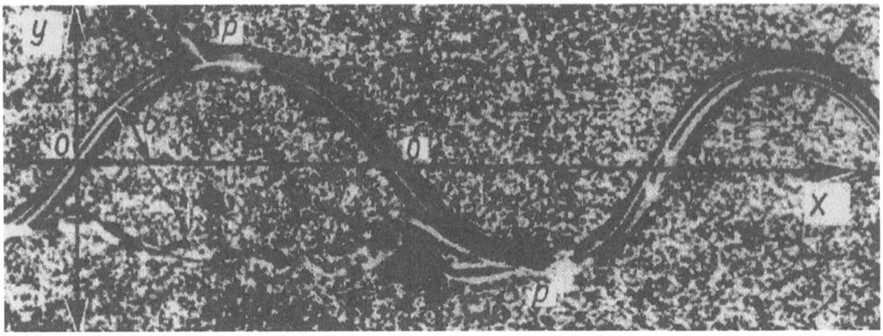

Fig. 16.2. Motion registered by a camera and shown in a coordinate system [according to 16.4].

Another shape was named "serpenoid", ie. "snake curve":

$$\left.\begin{aligned} x(s) &= sJ_0(\alpha) + \frac{4l}{\pi}\sum_{m=1}^{\infty} J_{2m}(\alpha)\sin\left(m\pi\frac{s}{l}\right) \\ y(s) &= \frac{4l}{\pi}\sum_{m=1}^{\infty}(-1)^{m-1}\frac{J_{2m-1}(\alpha)}{2m-1}\sin\left(\frac{2m-1}{2}\pi\frac{s}{l}\right) \end{aligned}\right\} \tag{16.2}$$

where $J_n(*)$ are Bessel's functions of first kind.

The tangent force may be determined from the following formula:

$$f_{tOP} = -\frac{a\alpha}{l^2}\int_0^l \frac{dF_m(s)}{ds} s\,ds \tag{16.3}$$

where:

a is the average distance between the joint and the muscle, and

F_m the muscle force along s-axis determined on the basis of physiological assumptions.

This equation assumes that the muscles contract in such a way that the curvature of the spine line changes according to the sine function.

In order to verify their properties, these curves were compared with curves observed for real snakes. The results indicate that the curves obtained from equations (16.1) and (16.2) are similar to observed curves (Fig. 16.3). The snake curve seems superior from the physiological point of view.

Kinematic analysis of snake-like locomotion was performed on a model of a kinematic chain containing rotational joints (Fig. 16.4). Antagonistic muscles causing the body to bend were replaced by actuators connected to both sides of each pair of links. Figure 16.4 shows a situation when f_{mi} is the contracting force on the antagonistic muscle side, $f_i = a/\delta s\, f_{mi}$ acts on joints J_{i-1}, J_i and J_{i+1}. The resultant force acting on joint J_i results from muscle contraction forces f_{mi-1}, f_{mi} and f_{mi+1}.

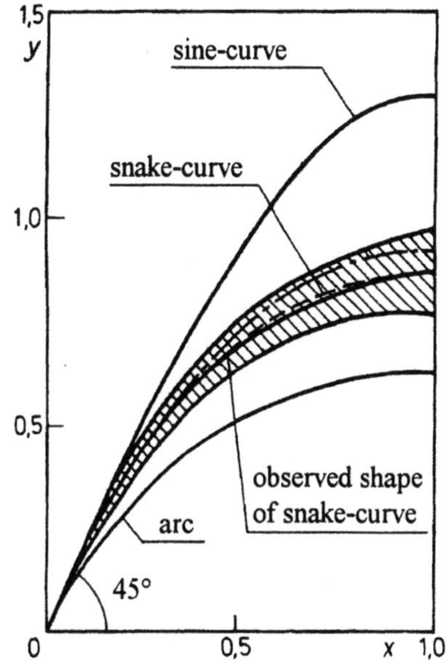

Fig. 16.3. Comparison of theoretical curves with real curves of a sliding snake. The hatched area denotes deviations from the mean value [according to 16.4].

The normal component of force may be determined in the same way. A series of zoological experiments were conducted to verify the proposed equations. The muscle force was measured by means of electromyography (Fig. 16.5) [16.4]. On the basis of these experiments the values of α and α_0 (the lower limit of curvature angle) describing this type of locomotion were determined. Several models were built in order to illustrate the snake-like vehicle and its locomotion.

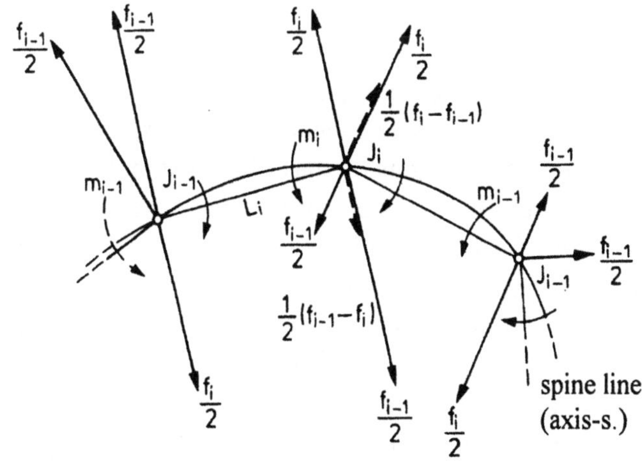

Fig. 16.4. Forces of contracting muscle acting on joint J_i [according to 16.4].

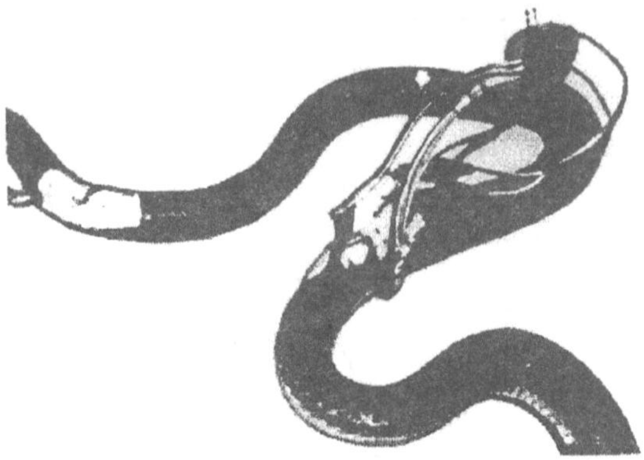

Fig. 16.5. Electromyograph electrodes and force sensors installed on a snake (Elaphe Quadrivirgate) [according to 16.4]

Figure 16.6 shows a mechanical model of a snake-like vehicle. The shape was established by determining the angle α and wave amplitude l. The sliding speed was controlled by varying the time (Δt - adjusted by changing amplitude A and frequency f of the function generator). If the unit length is 10 cm, the locomotion speed of locomotion may be written as

$$V = \frac{10}{\Delta t} \tag{16.4}$$

the amplitude, l, may be expressed as

$$l = \frac{2.5}{f \cdot \Delta t} \tag{16.5}$$

and the angle, α, as

$$\alpha = 1.79 \frac{A}{f \cdot \Delta t} \tag{16.6}$$

In the experiment the following set of parameters was used:
$\Delta t = 0.69$ s, $f = 0.12$ Hz, $A = 1.5$ V, which yields $V = 17$ m/s, $l = 35$ cm, $\alpha = 40^{\circ}$.

In the seventies other modifications of snake-like manipulators appeared. Some of these are shown in Figures 16.7a ÷ d; one of the models is shown in Figure 16.8.

The kinematic chain of this manipulator contains coaxial and coplanar joints which are fifth-class kinematic pairs. Those joints may be replaced by differential mechanisms, or by coaxial joints alone. It can easily be proved that the motion capabilities of this modified mechanisms remain unchanged. The shape of the described mechanism depends strictly on the position of each segment. The manipulator is loaded not only by external load, but by its own weight as well. Thus its structure must be as light and rigid as possible.

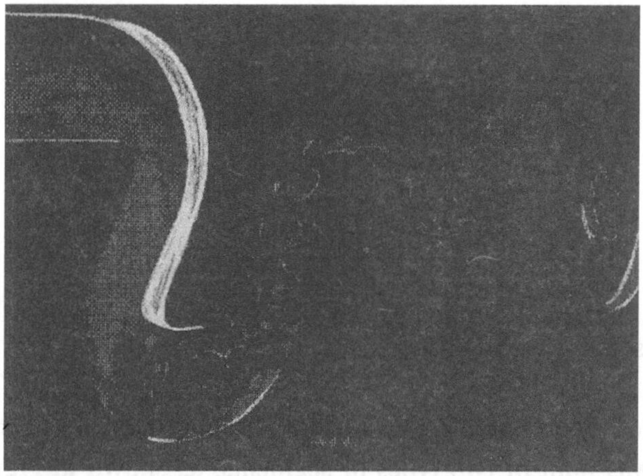

Fig. 16.6. Mechanical model of a snake [according to 16.4] and its locomotion capabilities

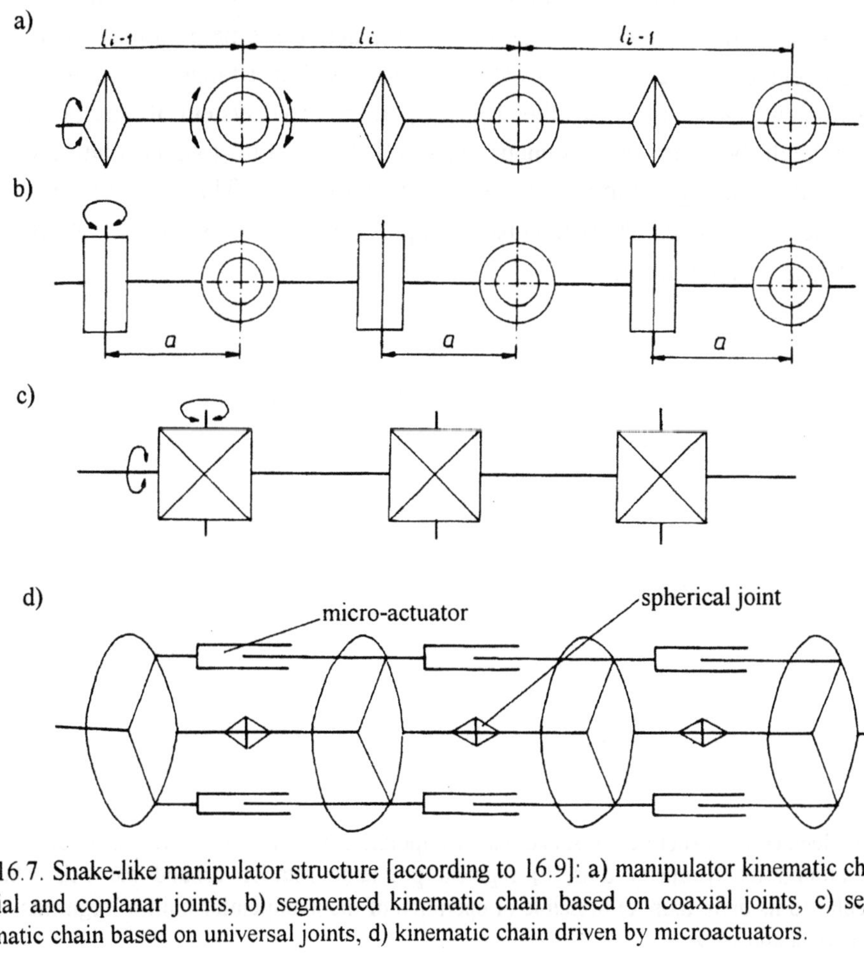

Fig. 16.7. Snake-like manipulator structure [according to 16.9]: a) manipulator kinematic chain with coaxial and coplanar joints, b) segmented kinematic chain based on coaxial joints, c) segmented kinematic chain based on universal joints, d) kinematic chain driven by microactuators.

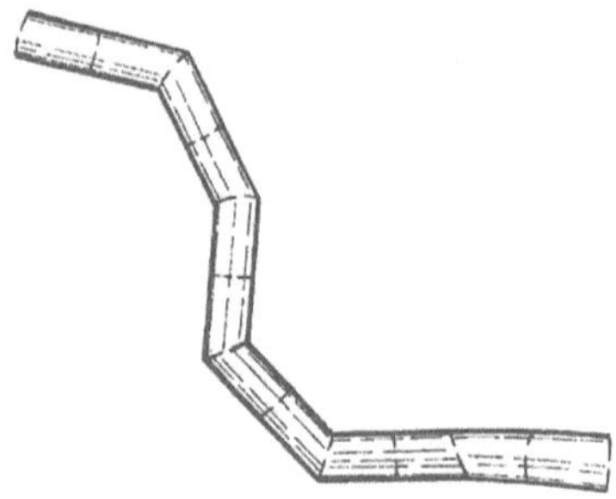

Fig. 16.8. Model of segmented manipulator

The application of a torsional skew active wire mechanism proved to be effective. The mechanism was built of two types of segments, as shown in Figures 16.9a and b. The first type of segment provides rotation about its axis of all segments situated above the segment. The second type provides rotation of the symmetry axis about the rotation axis. In this way all the kinematic chain above the active segment can sweep from the axis of this segment. The angle of rotation changes within a range of -2α to 2α. This mechanism significantly increases the efficiency of the drive system. It may be designed so that the angle between the symmetry axis and the rotation axis is relatively small. This means that the light components are sufficiently rigid. This type of design also presents other advantages. A mechanism composed of those components maintains a smooth surface when deformed.

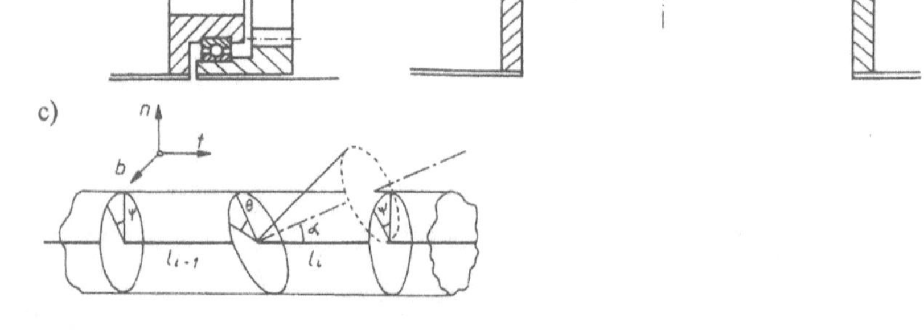

Fig. 16.9. Segmented structure of snake-like manipulator: a) segment providing rotation of all segments around its symmetry axis, b) segment providing rotation around the axis which is perpendicular to its drive axis, c) principle of operation of the manipulator built of segments shown in Figures 16.9a and 16.9b

The structural components of the mechanism are hollow, giving the structure of high stiffness and permitting the other systems to be installed inside, protected against potential damage. In order to achieve high motion capabilities the mechanism must be built of sufficient number of elements (Fig. 16.9c).

16.3. Manipulators with shape memory actuators

When constructing miniactuators the main structural component is made of an alloy with shape memory. In these actuators electrical energy is converted to thermal energy by means of heating coils, and this energy is used to heat elements which are able to change their shape [16.5].

A typical and one of the most effective alloys that changes its shape with its temperature is NITINAL-55.

NITINAL-55 is an alloy containing 53÷57% of nickel and titanium. Its property of „shape memory" means that an element made of NITINAL in specific conditions is able to regain its original shape after heating up to a particular temperature. The component remembers its original shape after special heat treatment, including the following phases:

- shaping the component,
- placing the component in the mold and soaking it at temperature of 482°C,
- cooling in the mould.

Now, if the component removed from the mold and distorted, it regains its original shape, when heated to the temperature of transformation. The temperature of transformation strongly depends on the composition of the alloy, ranging from -50°C to 160°C and even droping to -238°C if the alloy also contains cobalt. A component made of NITINAL is capable of overcoming some resistance when returning to its original shape. The process of deformation and its reverse may be repeated many times. The shape memory of NITINAL components is the result of what are called backward stresses, depending on the alloy temperature and the initial deformation of the sample.

The ability to exploit this property to perform mechanical work was tested by NASA_GOODYEAR company. These tests consisted in tensioning a wire by means of a weight suspended on it, and heating the wire with additional weight. The mechanical work W was determined using the following formula

$$W = \frac{F^2 l}{2EA} \tag{16.7}$$

where l is the length; F the force; A, the cross-sectional area, and E the modulus of elasticity.

The specimen was able to perform maximum work when the initial deformation was 7%. NITINAL also presents high corrosion resistance owing to its high nickel content.

16.3.1. Construction of light zoomorphic NITINAL actuators

The principle of operation of these actuators is simple. The actuator returns to its original (remembered) shape after deformation using thermal energy and converting it to mechanical energy. The simplest way to heat the actuator to the required temperature is by applying a voltage to the actuator, causing a suitable current to flow through its active elements.

The thermal balance can be written in the form

$$Q = cm\Delta T$$
$$Q = UI\Delta T \tag{16.8}$$

where Q is the heat energy; c, the specific heat; m, the mass, ΔT, the temperature difference; U, the voltage, and I, the current.

The simplest method of cooling the actuator is air cooling.

A component made of NITINAL is capable of carrying 500 N of load. Consequently small and light actuators can develop considerable forces.

The disadvantages of NITINAL actuator are:

- its low efficiency compared with that of the Carnot engine,
- its slow action,
- the high precision required during production of the alloy.

These drawbacks exclude NITINAL actuators from applications where they are the main actuators, although several prototypes using only these actuators have been built (and in Poland). Professor Hirose has suggested using NITINAL actuators as auxiliary ones (Figures 16.10 and 16.11).

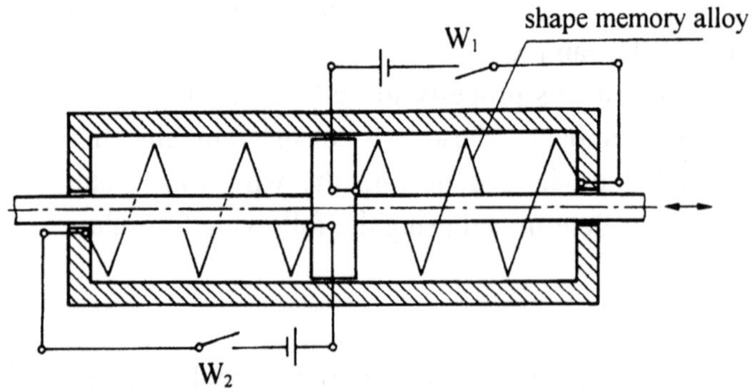

Fig. 16.10. NITINAL actuator

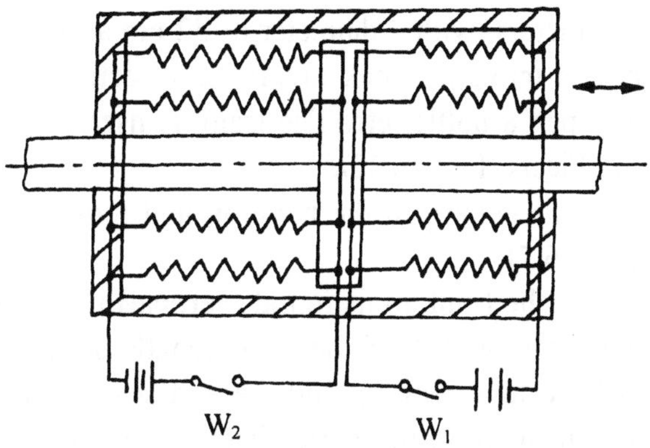

Fig. 16.11. NITINAL actuator with parallel system of springs supplied as serial system.

The actuator shown in Figure 16.11 consists of a piston moving in a cylinder. The piston is maintained in equilibrium by two NITINAL springs. When both switches *W1* and *W2* are in the open position, both springs are in the state of permanent elastic deformation. The actuator operates in the following manner:

If we close switch *W1* and leave switch *W2* open, the first spring will conduct electric current, which causes it to reach (after a certain time) the temperature of transformation. The thermal energy accumulated in the spring will be partially converted to mechanical work, i.e. the spring will try to return to the initial shape given it in the process of heat treatment, moving the piston rod to the left and increasing the elastic deformation of the second spring. The first spring can also to perform some work to overcome external forces. When switch *W1* opens, the current stops flowing through the first spring, its temperature decreases, the modulus of elasticity decreases as well, and the second spring attempts to decrease its elastic deformation. As a result, the actuator returns to its equilibrium position. Displacement in the opposite direction is caused in a similar way. The springs may be connected in parallel in order to increase the force exerted by the actuator. In this case an interesting power supply circuit is used (see Figure 16.11).

Several variations of this type of actuator have been built. One of them, for example, can perform a rotational motion. Another can act as a leaf spring in which the distance between its ends changes.

16.4. Trunk-of-elephant manipulator

The anatomical studies [16.6] for this type of manipulator were limited to an analysis of the system of locomotion (muscular system). The physiological description concerned only the motion of the trunk. On the average, the Indian elephant's trunk is 2.4 meters, with a diameter of 32 cm at the head and of 8 cm at the tip. The elephant can move its trunk in various directions, it can pick a needle from the ground, and open a bottle, etc. The trunk is used to transport food to the mouth and to pick up objects (Fig. 16.12a).

The anatomic analysis employed the notation and terminology used in human anatomy (Fig. 16.12b) [16.7].

Figure 16.12b shows the trunk in its initial (vertical) position. It can be seen as three segments: the base ($1/4 \div 1/3$ of the total length of the trunk), the middle segment ($1/2 \div 1/3$ of the total length), and the end segment ($1/4 \div 1/3$ of the total length). The total length of the trunk varies between 1.5 m (Indian elephants) to 2.4 m (African elephants).

The trunk has no skeleton, thus consequently its motion is continuous. The muscular system can be divided into an internal subsystem, containing the internal muscles, fat and connective tissue, and an external subsystem, containing the external muscles. The internal subsystem acts as a skeleton, while the external one acts as a drive system. The internal subsystem also serves as a cylindrical core. It is surrounded with layers of various thickness. Inside the core two nasal channels run parallel to the frontal surface (of cartilage), the diameter of which varies along the trunk length.

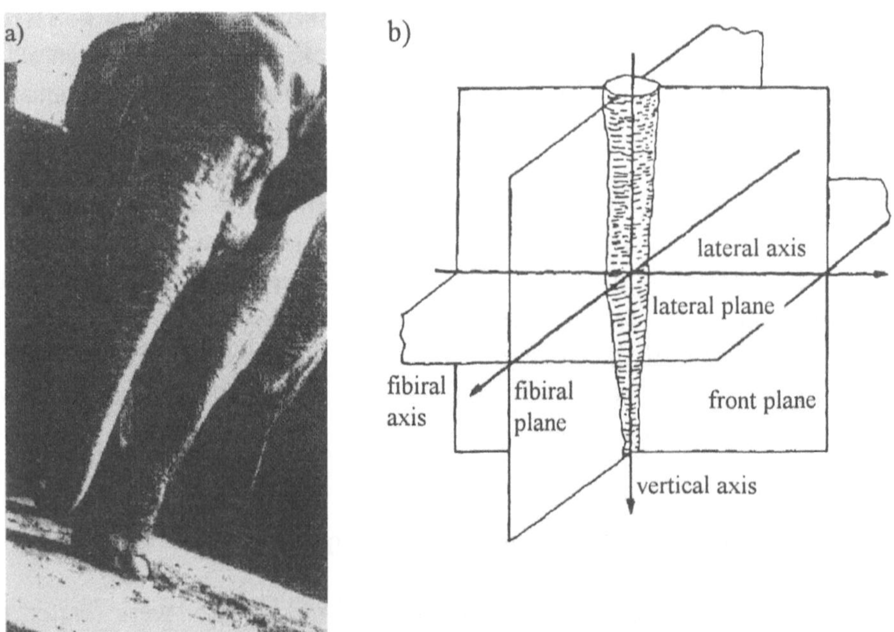

Fig. 16.12. Elephant trunk and its model: a) vertical position of the trunk; b) model and notation

Figure 16.13 shows a cross section of a specimen of elephant trunk prepared in the Laboratory of Anatomy of the Agricultural Academy in Warsaw [16.6].

By analyzing several trunk segments the cross-sectional model shown in Figure 16.14 was conceived. Areas 1 to 3 are internal muscles, fat and connective tissue; 6 represents the neural fiber; 7, the nasal channels; 4 and 5, the external muscles running the whole length of the trunk. They are composed of 30÷40 cm long interconnected sections. At the tip of the trunk, the muscles are twisted, allowing the elephant to perform twisting movements. Two issues were studied in detail during modeling: the kinematic structure of the model, which would emulate trunk motion, and the loading of the model and its drive system, in order to provide adequate control. In the continuous model the core is a chain (series connection) of elements having discrete masses of m_i and stiffness coefficients of kx_i, ky_i, and kz_i. The model consists of three segments, like the actual trunk. The first segment contains three masses and three cores; the second, six masses and six cores; and the third, twelve masses and twelve cores. The connections between masses and cores are rigid (Fig. 16.15a).
The motion capabilities of the model depend on the stiffness ratios for the particular segments. For the first segment the following relationships are maintained:

$kx_i \leq ky_i$ and $kx_i \leq kz_i$

Fig. 16.13. Cross section of elephant trunk at the base level (near the head) [according to 16.6]

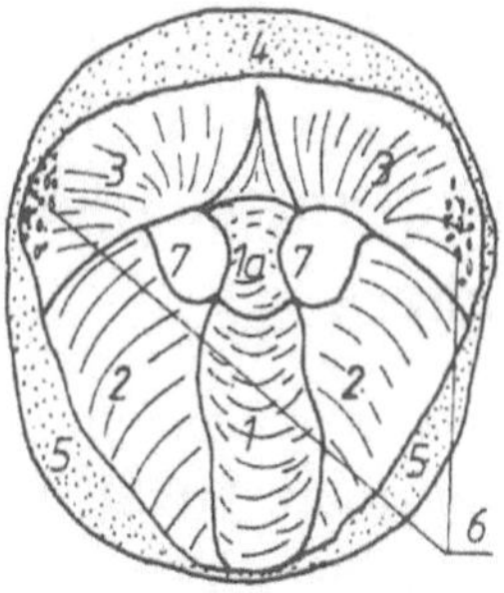

Fig. 16.14. Model of the cross section: 1-5 - various layers, 6 - neural fibers, 7 - nasal channels

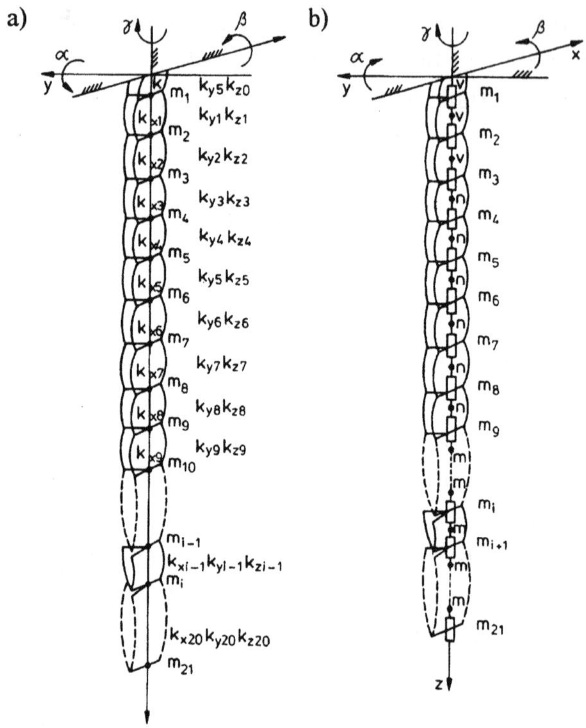

Fig. 16.15. Physical models of the trunk: a) continuous model, b) discrete model

for the second segment

$kz_i \geq kx_i$ and $kz_i \geq ky_i$

and for the third, all stiffness coefficients kx_i, ky_i and kz_i have the same values.

The discrete model is shown in Figure 16.15b. It also consists of three segments. The segments contain correspondingly three, six and twelve joints and rigid components. The first segment contains fifth-class kinematic pairs; the second, fourth-class pairs; and the third, third-class pairs. The total number of degrees of freedom is 51. The drive system is similar to the muscular system, providing for trunk motion in various configurations.

16.4.1. Continuous model of the elephant trunk proposed by American researchers

In recent years American researchers have proposed an experimental-theoretical continuous model of the elephant trunk [16.8]. In 1985 Croner and Wainwright tested a specimen of African elephant trunk and built the section shown in Figures 16.16a and 16.16b.

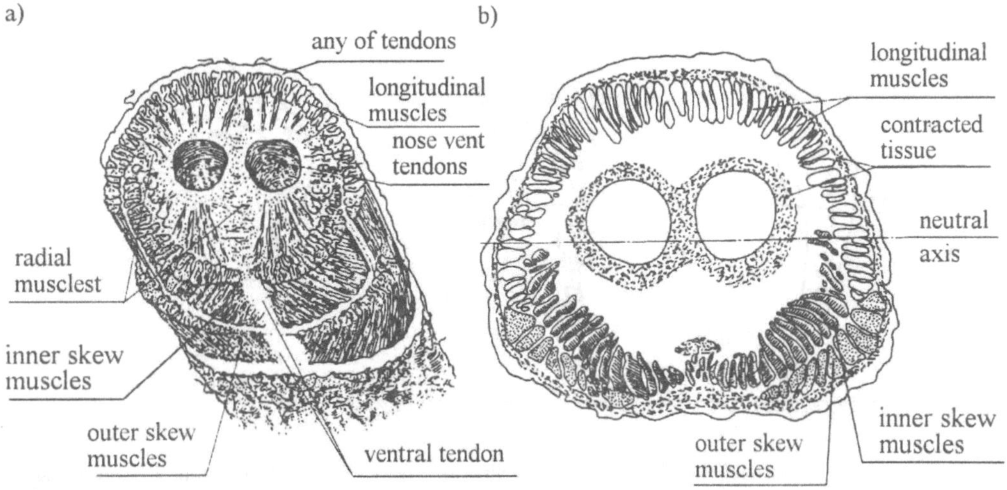

Fig. 16.16. Trunk models worked out by American researchers [according to 16.8]; a) elephant trunk cross-section, b) muscle arrangement

This model reconstructs the trunk cross-section (Fig. 16.16b) on the basis of earlier studies of Bors and Pauli (1908). Wilson and others found [16.8] that this cross-section corresponds to cross-section A-A (Fig. 16.17). Successive measurements supplied additional knowledge regarding angles, the average density of the trunk (γ = 1180 kg/m^3), and the weight of the trunk segment between cross-section A-A and the tip of the trunk.

Two components of strain were measured for an elephant lifting a weight located at the tip of the trunk: uniform trunk contraction ε_n and maximum bending strain ε_x. The measurements were taken on a mature, female African elephant in a Washington zoo. Before lifting the weight the initial length of the trunk was about 1.5 meters; its diameter at the head was 32 cm, and at the end, 8 cm. Ten markers 1 cm in diameter were painted along the central line of the trunk. Using a 16 mm camera, two situations were photographed at a distance of 12 meters: the trunk in repose at its maximum length, and the position when its tip was grasping a with a 26.2 kg weight.

Figure 16.17 shows three situations illustrating the process of lifting the weight (Fig. 16.17a, b and c). For these situations the uniform trunk contraction ε_n was determined. For a given position of the trunk the marker moves from the initial position l_i by Δl in order to reach another position corresponding to the shortening of the trunk (Fig. 16.17c). The results are illustrated in Figure 16.18 (straight lines are obtained by using linear regression). The trunk curvature radius was also measured, giving ρ_0 = 1.18 m and ρ_l = 1.67 m.

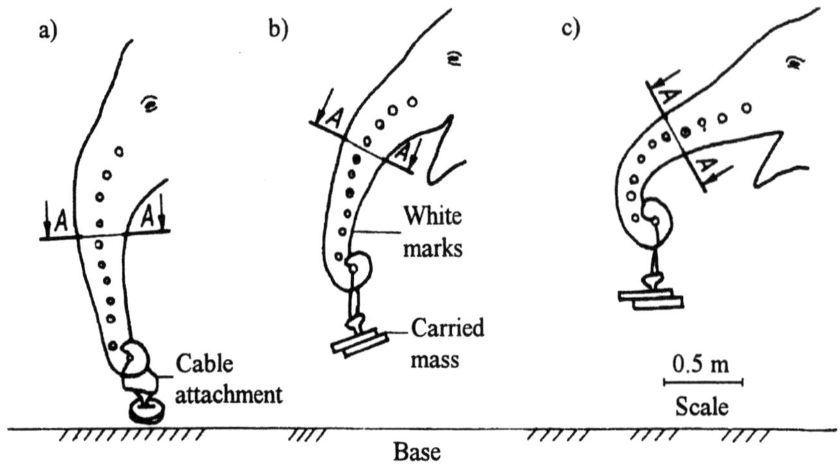

Fig. 16.17. Lifting a weight by the trunk [according to 16.8]; a) initial position, b) intermediate position, c) final (upper) position

Figure 16.19 shows changes in the total longitudinal stress ($\sigma_u + \sigma_x$) as a function of total longitudinal strain ($\varepsilon_u + \varepsilon_x$).

As said, Wilson assumed that the cross-section shown in Figure 16.16a corresponds to the cross section A-A shown in Figure 16.17. On the drawing of Bors and Pauli he then measured the angles θ formed between oblique muscles with the longitudinal axis of the trunk. The angles corresponding to the internal and external muscles are 29.1 and 34.1 degrees.

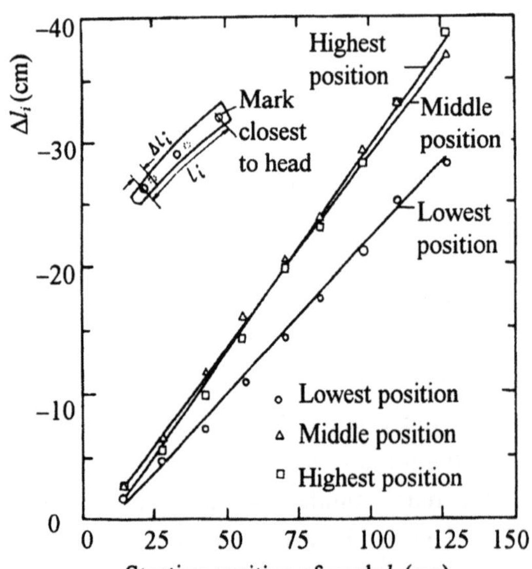

Fig. 16.18. Marker displacement vs its initial position (according to Fig. 16.17)

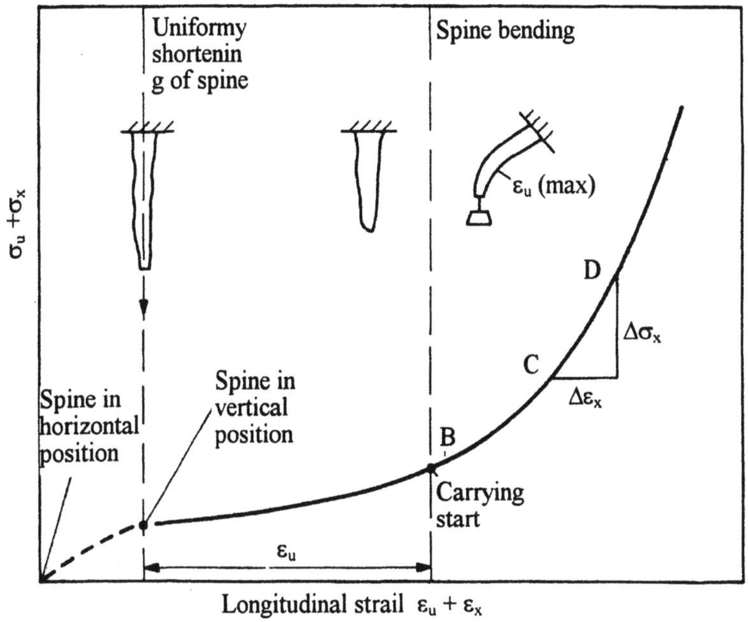

Fig. 16.19. Changes in the total longitudinal stress as a function of total longitudinal strain [according to 16.8]

Table 16.1 contains data collected from these measurements and calculations. A continuous trunk model was then proposed (Fig. 16.20). Assuming that the longitudinal strain ε_x resulting from the bending of the trunk to the curvature radius of ρ has a linear distribution from the upper to the lower side, we obtain (Fig. 16.20a)

$$\varepsilon_x = -\frac{y}{\rho} \tag{16.9}$$

Figure 16.20b shows the distribution of the strain as the sum of ε_x and ε_u.

When the assumptions conform to the classical theory of elasticity, the normal strain ε_0 of each oblique muscle at the angle of θ equals:

$$\varepsilon_0 = \varepsilon_x \cos^2\theta \tag{16.10}$$

For minor changes in trunk position we obtain:

$$\Delta\sigma_x = -ka\, \Delta\varepsilon_x, \quad \Delta\sigma_\theta = -E_\alpha\, \Delta\varepsilon_\theta \tag{16.11}$$

where:

$\Delta\sigma_x$ is normal stress increment for the combination of tendons and longitudinal muscles,

$\Delta\sigma_\theta$, the normal stress increment for oblique muscles,

$\Delta\varepsilon_x$, $\Delta\varepsilon_0$ are the corresponding strain increments, and

k_a is the actual trunk stiffness.

Table 16.1. Main variables, values and symbols

Quantity	Symbol and value
Trunk cross-section A-A, Fig. 16.20a	$D = 24$ cm
Displacement m_t	$d_0 = 5.0$ cm
Displacement m_t	$d_l = 8.0$ cm
Displacement m_p	$e_0 = 9.29$ cm
Displacement m_p	$e_l = 13.27$ cm
Marker positions, Fig. 16.18	l_i , Δl_i
Loading	$m_p = 26.2$ kg
Trunk mass between trunk end and A-A cross-section	$m_t = 16.37$ kg
Number of longitudinal muscles in A-A cross-section	$M = 78$
Number of torsional muscles in A-A cross-section	$N = 72$
Trunk density	$\gamma = 1180$ kg/m^3
Twist angle of internal muscles	$\theta = 29.1°$
Twist angle of external muscles	$\theta = 34.1°$
Curvature radius of A-A section	$\rho_0 = 118$ cm
Curvature radius at A-A cross-section	$\rho_l = 118$ cm
Area of all longitudinal muscles (54.6 cm^2) and tendons (67.8 cm^2)	$A_X = 112.4$ cm^2
Area of all internal (46.2 cm^2) and external oblique muscles (67.8 cm^2) Fig. 16.16	$A_X = 71.9$ cm^2
Effective modulus of the trunk, equation (16.14)	$E\alpha = 0.938\times10^6$ N/m^2
Moment of inertia of the cross-section of all muscles and tendons relative to neutral axis	$I = 6045$ cm^4
Neutral axis position, Fig. 1620	$z = 12.0$ cm
Change in maximum stresses in longitudinal muscles	$\Delta\sigma_x = 2.27\times10^6$ N/m^2
Change in maximum stresses in internal oblique muscles	$\Delta\sigma_\theta = 1.57\times10^6$ N/m^2
Change in maximum stresses in external oblique muscles	$\Delta\sigma_\theta = 1.67\times10^6$ N/m^2
Change in maximum strains in longitudinal muscles	$\varepsilon_x = 0.0242$
Change in maximum strains in internal oblique muscles	$\varepsilon_\theta = 0.0168$
Change in maximum strains in external oblique muscles	$\varepsilon_\theta = 0.0178$
Uniform stress (lower), Fig. 16.28	$\varepsilon_u = -0.244$
Uniform stress (intermediate) , Fig. 16.28	$\varepsilon_u = -0.315$
Uniform stress (upper) , Fig. 16.28	$\varepsilon_u = -0.330$

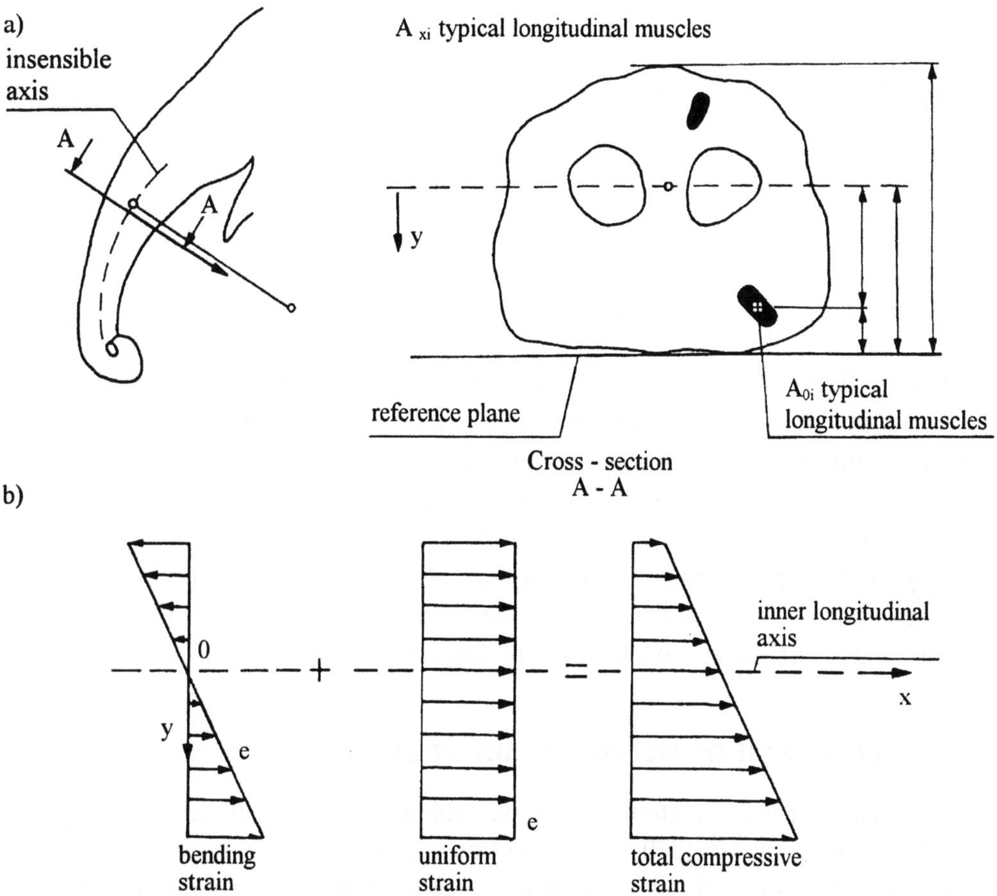

Fig. 16.20. Computational model [according to 16.8]: a) cross-section A-A; b) strain distribution.

The position of neutral axis $y = 0$, and the plane along which the strain of the muscles is zero has been determined from the condition of equilibrium of forces, yielding

$$\int_{A_x} y dA_x + \int_{A_0} y \cos^4 \theta dA_0 = 0 \tag{16.12}$$

The distribution of strains can be calculated from the following formulas

$$\Delta\sigma_x = -k_a \frac{y}{\rho} \quad \text{and} \quad \Delta\sigma_\theta = k_a \frac{y}{\rho} \cos^2 \theta \tag{16.13}$$

When the arrangement of muscles and tendons in the cross-section of the trunk is known, and the neutral axis (Fig. 16.20) can be found, equation (16.12) may be solved using numerical methods.

When the position of the neutral axis and the curvature radius ρ are known, the distribution of bending strain can be determined by equation (16.9). On this basis increments of bending strain and stiffness can be found.

The stiffness is defined as

$$k_a = \frac{\bar{\rho}}{I}\left[m_t g(d_1 - d_0) + m_p g(e_1 - e_0)\right) \tag{16.14}$$

where

$$\bar{\rho} = \frac{\rho_0 - \rho_1}{\rho_0 \rho_1} \tag{16.15}$$

and

$$I = \int_{A_x} y^2 dA_x + \int_{A_0} y^2 \cos^4 \theta dA_0 = 0 \tag{16.16}$$

By adding these results we can calculate the stiffness k_a and, then the strain increment and strain distribution, when, of course, the geometry, mass, muscle and tendon loading for the given cross-section are given.

On the basis of Figure 16.16 the area and moments of 150 muscles (78 longitudinal muscles, 51 internal oblique muscles and 21 external oblique muscles have been determined. Using the relationships (16.12), (16.14) and (16.15) the position of the neutral axis $z = 12$ cm (Fig. 16.20), the moment of inertia of the cross-section $I = 6045\ \mathrm{cm}^4$, and stiffness $k_a = 0.938\times10^6\mathrm{N/m}^2$ have been found.

16.4.2. Technical model of the manipulator

As a result of these anatomical and physiological studies on the real object, the elephant trunk, and the modeling of its motion capabilities using its muscular system as the drive system, a flexible manipulator arm of the elephant trunk type with hydraulic drive has been proposed.

The manipulator described below is a stationary machine. It consists of three main assemblies: the arm, the body, and the hydraulic power unit, together with the control system.

The manipulator arm is built of three segments: upper, middle and lower. The segments of the manipulator are stacked one on the other (see Fig. 16.21a). Each segment consists of a different number of elastic elements (1). These elements are partitioned and linked by circular separators (4, 9, 10, 15, 16). Subsequent segments consist of four, two and one elements (1) counting from the bottom. Thus the whole arm consists of 7 elements. The control system allows each segment to move independently. The nominal range of 20° spring deflection was assumed (during design of the drive system). The maximum spring deflection permitted by the construction is 90°. At this deflection the spring coils on the compressed side touch one another. Elastic elements in each segment are prestressed. The elements (1) of prestressed segments have following dimensions: the upper one, 56 mm, the middle one, 50 mm; and the lower one, 44 mm.

As the preliminary tension of each segment is different, different spring reaction forces are obtained, compensating to some extent the increase in the loading of the lower segments for unfavorable arm configurations.

The upper segment (Fig. 16.21b) consists of one elastic element (1), a coil spring located between separators (3) and (4). The spring is prestressed by wires (6). The prestressing force is adjusted by the limiter (2). The wires causing the elastic elements to move are situated inside the springs at 90° intervals around the circumference. Each two opposite wires are connected to a lever (7) so that the movement of the lever in one direction causes the wires to move in opposite directions.

One arm of the lever is connected to the piston rod (8) of a double-acting hydraulic actuator. The lever swivels on a bolt mounted into the support (2). The actuator housing is attached to the body. Two assemblies containing the actuator and the lever provide for the motion of one segment by means of four wires. Wires of the upper segment start from the separator (3). Under the separator (4) the wires (6) are armored (5). The armor is attached to the separator (4) at the bottom and to the support (2).

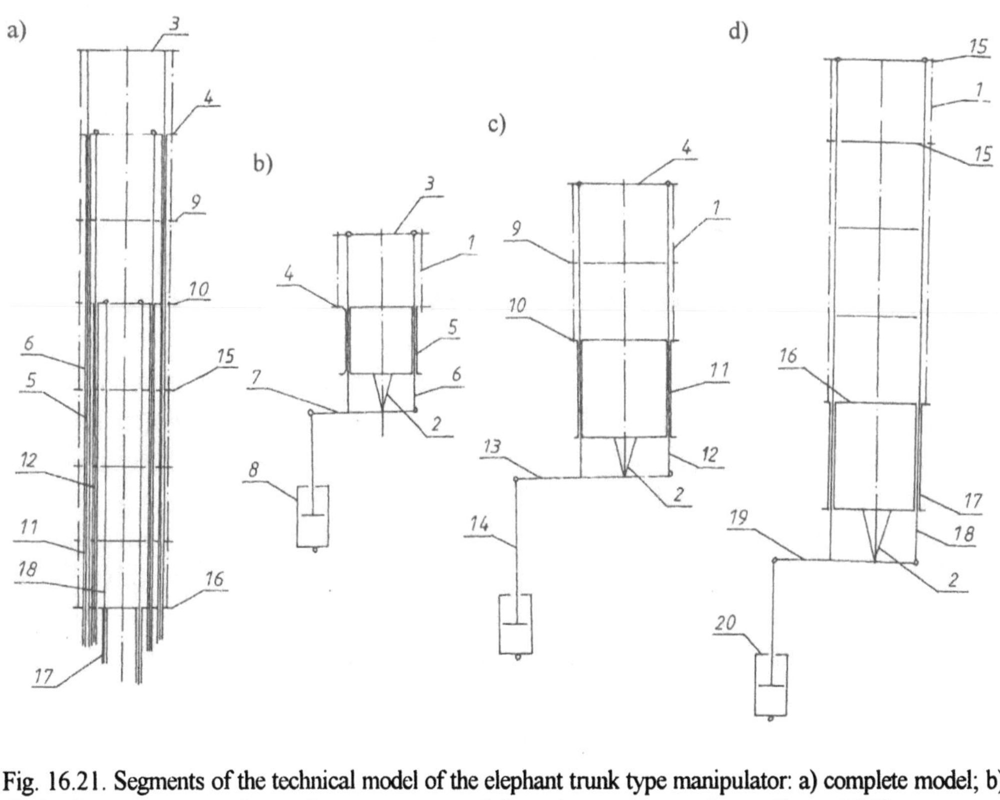

Fig. 16.21. Segments of the technical model of the elephant trunk type manipulator: a) complete model; b) single-element segment; c) two-element segments; d) four-element segment [according to 16.1]

The middle segment (Fig. 16.21c) consists of two elastic elements (1). It is situated between the separators (4) and (10). The springs are separated by a separator (9). The motion is generated in the same way as in the upper segment. The segment is driven by two actuators by means of two levers (13) attached to the support (2). The levers are connected to four wires (12). Between the separators (4) and (10) the wires are exposed while between the separator (10) and the support the wires are hidden in the armor (11).

The lower segment (Fig. 16.21d) consists of four elastic elements (1). It is situated between the separators (15) and (16). The springs are joined by three separators (15). The motion is generated in the same way as in the upper and middle segments. The segment is driven by two actuators (20) by means of levers (19) attached to the support (2). The levers are connected to four wires (12). Between the separators (15) and (16) the wires are exposed and run through the holes in the separators (15). Between the separator (16) and the support (2) the wires are hidden in the armor.

16.4.3. Manipulator drive system

The drive system is shown together with the body column in Figure 16.22a. The arm is driven by six actuators (two actuators driving each segment). Pairs of actuators provide independent bending in one plane. By combining these motions it is possible to obtain intermediate configurations. The actuators are connected to the bottom plate by means of joints. Actuator piston rods are attached to the lever assembly. The levers transmit the motion from the actuators to the wires. The lever assembly pivots on a bolt attached to the support. Bolts in the support provide for wire tension adjustment and for setting the initial position of the arm. The adjustment is within the range of about 30 mm. The lever support is fixed to the frame which is the main structural component of the manipulator body. In the upper part of the support a plate is mounted to which the manipulator arm is bolted. Armored wires run through the holes in the plate to the levers, to which they are attached. The body frame is bolted to the lower plate. This plate is a base plate for the actuators, frame and panel with solenoid valves.

The actuators are powered externally. The panel, with hydraulic connections, is placed on the side of the body. The panel also contains solenoid valves. The actuators are connected to the panel with rigid tubing. The hydraulic fluid is discharged from the system through the solenoid valves and collecting manifold. Figure 16.22b shows a general view of the manipulator.

Figures 16.23a and b show the model of manipulator in two positions. Figure 16.23a illustrates the motion of the upper segment, and fig. 16.23b, the motion of the whole manipulator.

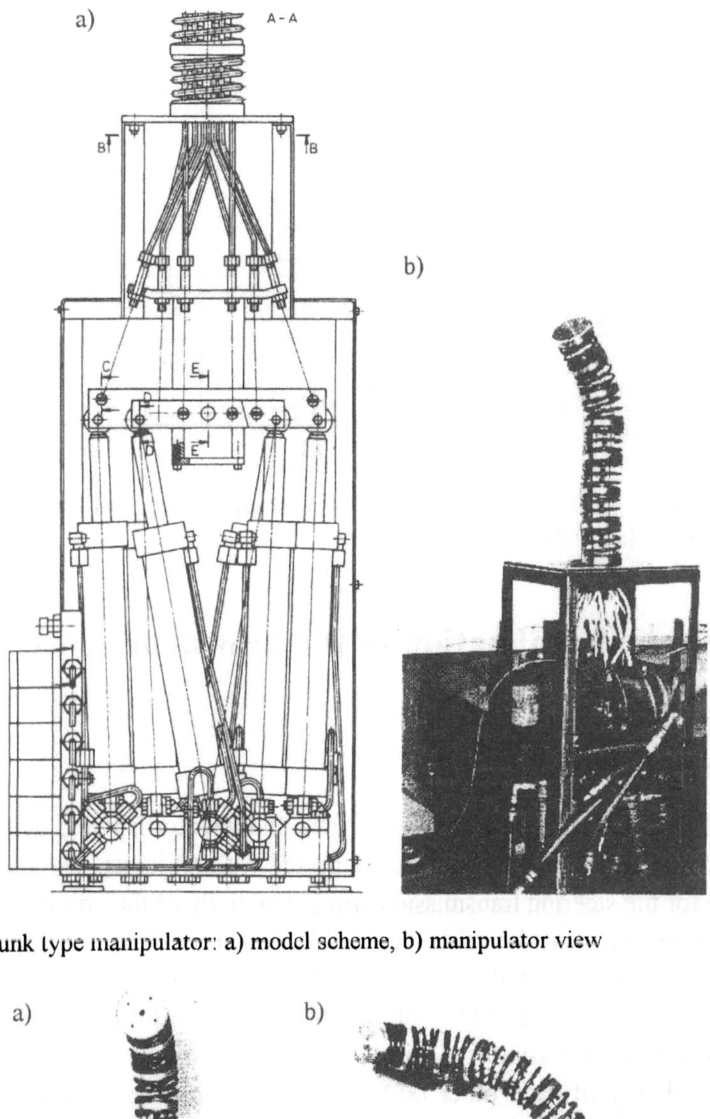

Fig. 16.22. Trunk type manipulator: a) model scheme, b) manipulator view

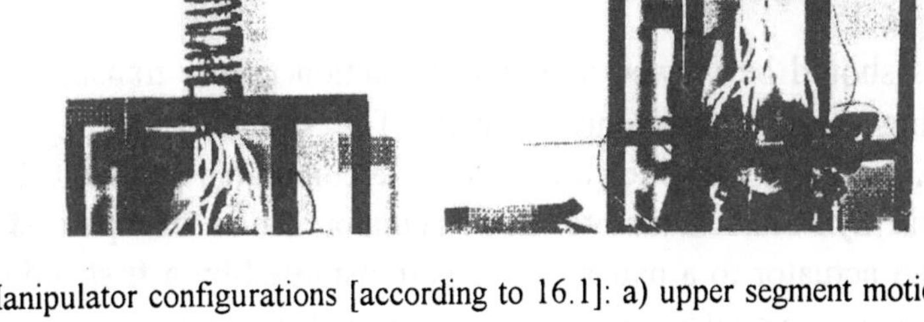

Fig. 16.23. Manipulator configurations [according to 16.1]: a) upper segment motion; b) motion of all segments.

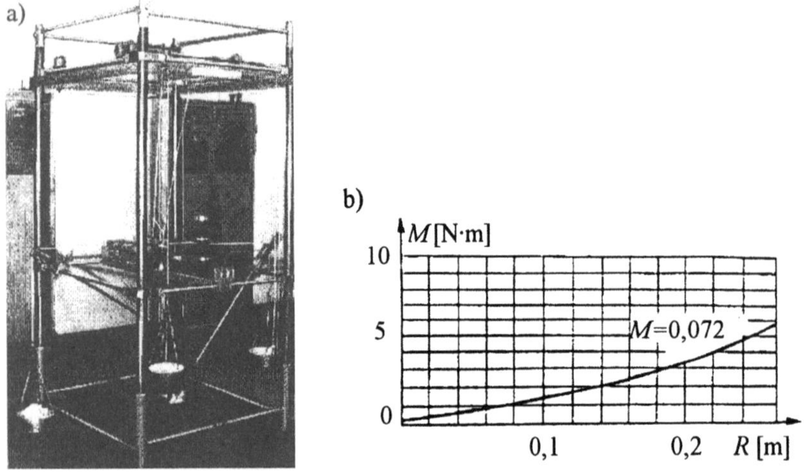

Fig. 16.24. Testing the springs included in the manipulator segments: a) test bench; b) torque M as a function of radius R [according to 16.9]

16.5. Technical realization of the new arm

16.5.1. New solution

A new arm of the elephant trunk type elastic manipulator is made of seven flexible elements (coil springs). The elements are connected without joints. The rigid separators act as a guide for the steering transmission wires. The body of the arm is hollow and the transmission wires are located outside the flexible elements [16.6].

An original position of the arm is similar to the anatomical position of the elephant trunk. The wrist member is simply directed toward the ground [16.16].

Anatomical analysis identified three parts (base, middle and wrist) which are driven independently of each other by two pairs of transmission wires (Fig 16.25). The transmission wires in any pair are controlled independently. The shortening of one of the wires in a pair causes an equal elongation of the other (linear dependence) [16.15].

The wires should be viewed as unilateral action power transmission. In other words, a spring element may only be bent. The straightening of an element is realized by exploiting the elastic strain energy accumulated inside it. The wires are actuated by six hydraulic actuators. Each actuator drives one pair of wires. The motion from an actuator to a pair of wires is transmitted by a first order lever. The hydraulic actuators are controlled by six proportional valves USEB6.

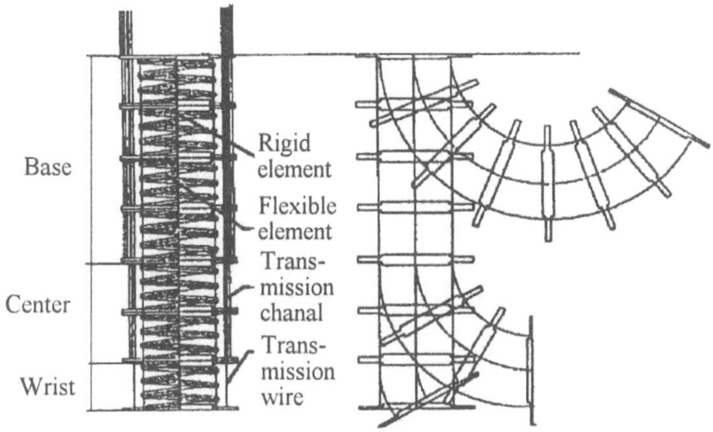

Fig. 16.25. Design diagram and motion possibilities of the elephant trunk type manipulator of new solution.

The control system of a valve causes proportional control of actuator velocity. It a multistage control system. The first stage is an open-loop control utilizing a computer program and two D/A converter cards AX5414 (fig 16.26).The second stage is an analogue closed-loop.

In this system the input signal is the voltage from the first stage and the output is the voltage to controlling proportional valves. Feedback is realised by sensors of separator position in the valves.

16.5.2. Description of the problem

In the elephant trunk type elastic manipulator each element (coil spring) tipically has six degrees of freedom: it may be compressed twisted and bent in three directions of Cartesian space.

The design of this manipulator is based on the assumption, regarding control of the flexibility of the segment, that control is possible with insufficient constraints. If we use open-loop control, the load attached to the elastic manipulator must be known. In closed-loop control a method of measuring the orientation is needed [16.16].

The main goal of the construction of the experimental manipulator is the verification the possibilities of feedback for that structures of the arm.

The second goal is the verification of the theoretical work regarding the dynamics and kinematics of the arm. In reality, the manipulator members are loaded by external forces produced by its deflection and torsion. In the elastic manipulator two kinds of external forces interfere with the control of the structure.

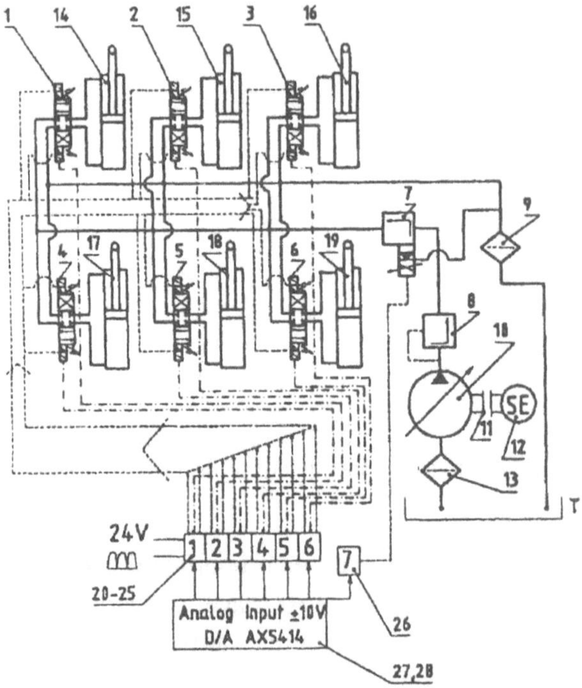

Fig. 16.26. Hydraulic scheme of the drive of the experimental stand. 1÷6 - proportional valves, 7 - proportional control valve, 8 - safety valve, 9,13 - filters, 10 - pump, 11 - coupling, 12 - electric motor, 14÷19 - hydraulic actuators, 20÷26 analogue control cards, 27,28 - D/A converter cards.

1. The transverse force that causes the flexural bulking of an element
2. The torsional force that causes the torsional deflection of an element.

Because of this a stiffness ratio of the element must be taken into consideration in the kinematics and dynamics equations. The stiffness of each element is defined by a matrix of stiffness ratios.

The effect of an action of generalised external forces on the orientation vector is described by the formula

$$[Q]=\begin{bmatrix} [K_1] & 0 & 0 & 0 \\ 0 & [K_2] & 0 & \vdots \\ 0 & 0 & \ddots & 0 \\ 0 & \cdots & 0 & [K_n] \end{bmatrix} \cdot [l_i] \tag{16.16}$$

Each element of the stiffness matrix is a non-linear function dependent on its current orientation. This makes description of the natural stiffness of elements difficult. A platform on four spring supports has been accepted as a theoretical model of the manipulator element. The springs have non-linear characteristics. The transverse flexibility of an element is also represented by a four-spring element. One spring element in the platform center represents the torsional flexibility. Theoretically the flexibility of the manipulator segment should be equivalent to the inverse of the matrix of the element's stiffness ratio.

$$[c]=\begin{bmatrix} c_1 \\ c_2 \\ c_3 \\ \vdots \\ c_p \end{bmatrix} \equiv [K]^{-1} \tag{16.17}$$

The matrix of stiffness may be calculated using a computer, but the theoretical functions must be verified experimentally.

16.5.3. Description of the measuring system

The experimental stand consists of three subsystems:

1. Mechanical arm
2. Hydraulic motion system
3. Control system

It is designed so that the following values can be controlled:

1. Forces in the transmission wires
2. Forces in the hydraulic actuator
3. Position of the hydraulic actuator
4. Changes of length in transmission wires
5. Pressures in the hydraulic system

We can also control the orientation of the manipulator independently, using the stereogrametrical method. Tension forces and pressures are measured in static condition. The remaining values can be measured dynamically with a sampling frequency of 200-1000Hz. This is sufficient because the natural frequency of manipulator vibration is at the most 20 Hz.

The following dependencies are of special interest:

1. Change in the length of wires as a function of forces in the actuators
2. Change in the deflection angle as a function of control forces
3. Orientation of the arm as a function of control forces

During investigation some basic characteristics were measured.

All results were obtained using a 16 channel A/D converter card PCL 818H and processed using the STATISTICA computer program.

16.5.4. Motion analysis of the manipulator arm

16.5.4.1. Geometrical analysis

The full bending angle of the manipulator link is limited to 30 degrees in each direction. We can assume that the arm is an axially-symmetrical element with equal mobility in each direction. The base segment can bend 120 degrees; the middle segment, 90 degrees; and the end, 30 degrees. Because the bending of each link is independent, the number of possible configurations is infinite.

For geometrical analysis, the global reference coordinate frame was located at the central point of the first rigid element of the arm (Fig. 16.27).

The coordinate frames are affixed to each link. Each one is located at the bottom invert. The movement of the arm is fully defined by describing the positions, velocities, and accelerations of eight points. Seven of these are the origins of $i+1$-th coordinate frames, the eighth is the center of the last (ninth) rigid element. The geometry of a single link is described by the Denavit-Hartenberg notation.

Contrary to manipulators with rigid links, the value of the joint angle for this manipulator is constant. Link length, and the offset and torsion angle are variable.

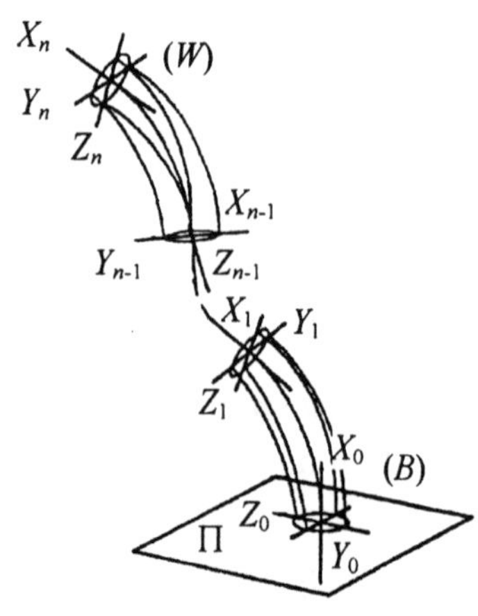

Fig. 16.27. The location of the coordinate frames affixed to the manipulator arm.

16.5.4.2. The manipulator model

Each link of the manipulator is loaded by external forces. During operation the control forces in the wires, reaction forces, gravity force, and generalized external force and torque all act on each element. The reloaded manipulator is additionally loaded by forces exerted by the material transported. The latter forces are of the stochastic type, and should be included in the equations of motion. For liquid and bulk materials such equations are well known.

Each link of the manipulator has six degrees of freedom. It is driven by four actuators (control wires) which are coupled pairwise. Each element is underconstrained, with a deficiency of constraints equal to four. Because the elements are elastic, the lack of constraints is partially compensated by the mechanical energy accumulated in the initially tensioned springs.

The fact that the elements are underconstrained may cause the torsion of the element in the case of shear force or torque. Low values of shear forces and torsional torque are transmitted due to the enhanced stiffness caused by the wires and their armoring.

Each manipulator element can be modeled as a platform supported by four elastic supports (Fig. 16.28). Because the supports are fixed to the base, the stiffness of the element is modeled by the two elastic elements in the plane normal to the axis of the manipulator. Torsion of the link is represented by a single elastic element located in the center of the rigid insert. The stiffness coefficients form an equivalent matrix to the real deformability matrix of an element of the manipulator. The damping elements represent damping caused by Coulomb friction resulting from control wires rubbing over passages.

The elasticity and damping coefficient functions were determined by FEM and verified experimentally.

16.5.4.3. Computer simulation

The computer simulation program dealt with the static and dynamic analysis of a single element and of a group of elements. The main goal of the procedure was to determine the stiffness equivalent matrix, and prepare the model for the analysis of direct and inverse dynamics problems of the elastic manipulator, including the forces caused by the material transported.

The FEM results show that the relationship between the bending angle and the control forces is nonlinear. It results from the nonlinear distribution of the winding angle of the turns of the spring when it is bent (Fig. 16.29).

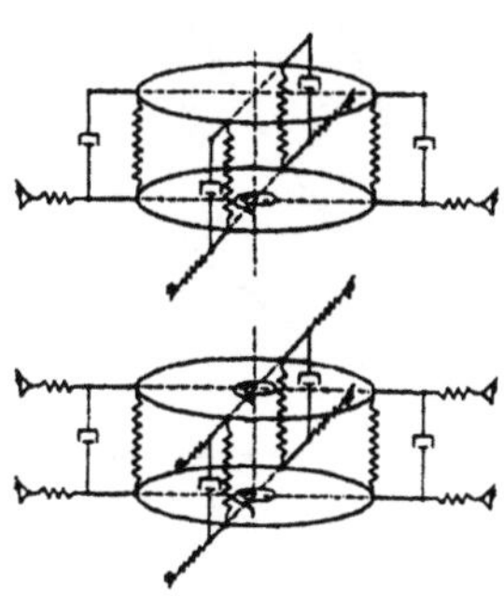

Fig. 16.28. Physical model of the elastic manipulator.

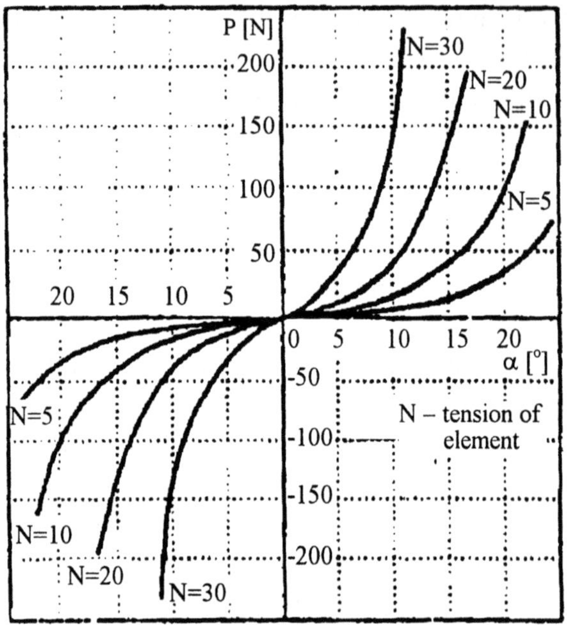

Fig. 16.29. Relationship between the bending angle and the control forces.

The vibration damping is represented by the relationship between reaction forces in the holes and the link bending angle. This relationship is quasilinear. It is nonlinear for small angles of bending, because of the backlash in the holes (Fig. 16.30).

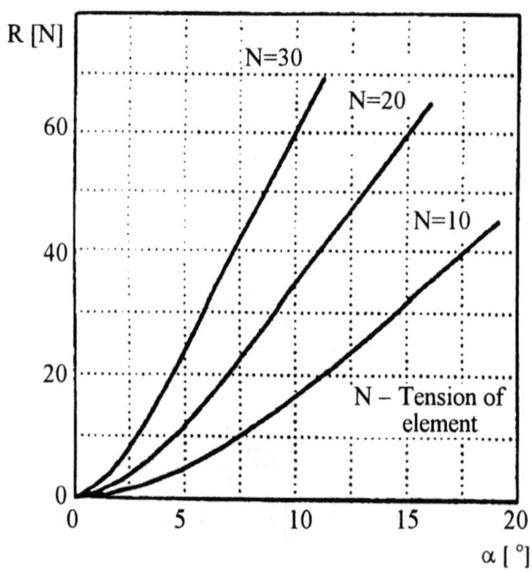

Fig. 16.30. Reaction forces in the holes of a rigid link as a function of the bending angle.

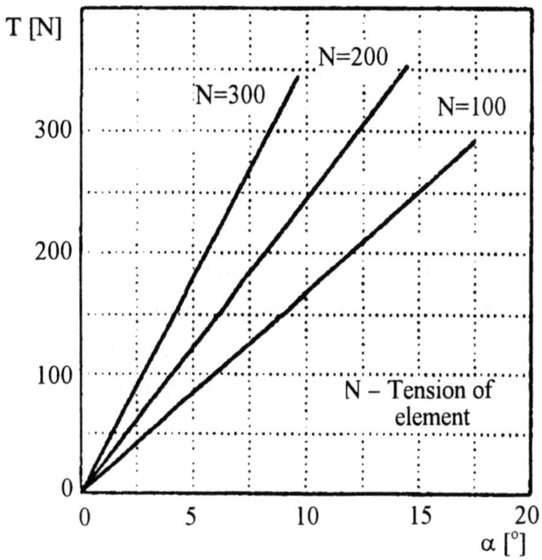

Fig. 16.31. The coefficient of elasticity of the element

The coefficient of elasticity of the element during bending is a linear function. The force during the simulations was directed transversally. The tangent of the angle of inclination of the curves is different for different values of the initial tension.

The last of the functions investigated was the response time to the input force step function. This differs for different bending angles. The characteristics of this function probably result from the friction caused by the motion of the control wires in the holes (Fig. 16.32) [16.17].

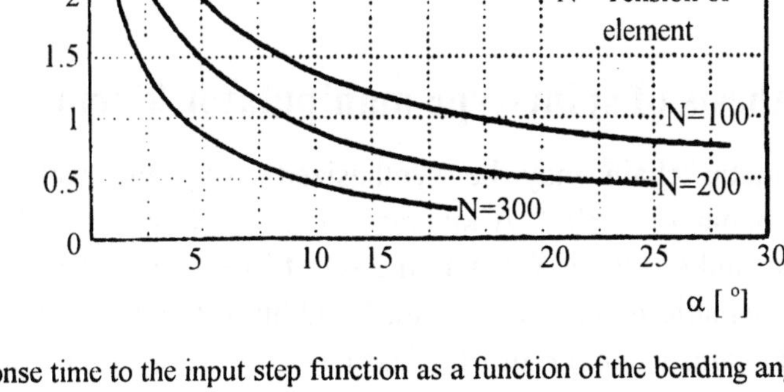

Fig. 16.32. Response time to the input step function as a function of the bending angle.

16.5.4.4. Possible applications

The loading and unloading of bulk and liquid materials can be performed by a manipulator with automatic or remote control. Elastic manipulators can operate in a workspace with obstacles and with access through small holes. Manipulators with the structure of a long joint, built of short links connected to each other by joints or elastic links connected rigidly are useful for such a task. The application of a manipulator of a given type to the transportation of bulk materials depends on its design. This kind of manipulator must have a hollow arm and the possibility of installing the material handling channel there. In our opinion the manipulators with rigid links currently in use do not fulfil these requirements.

16.6. Spine-type manipulator

16.6.1. Design and construction of the manipulator

A multi-link segment of a spine-type manipulator is shown in Figure 16.33. It represents a closed kinematic chain consisting of fifth-class revolute kinematic pairs and second-class pairs providing three elementary rotations and one translation. Figure 16.33a presents a general view of the segment and Figure 16.33b shows a detailed scheme of the kinematic chain in which the number of moving links $n = 6$ (numbered respectively 1, 2, 3, 4, 5 and 6), the number of fifth-class kinematic pairs $p_5 = 6$, and the number of second-class kinematic pairs $p_2 = 2$ [16.1]. Figure 16.33c is the side view of the manipulator.

By applying the general mobility formula for spatial mechanisms

$$W = 6n - 5p_5 \, i \, 2p_2 \qquad (16.17)$$

one can obtain $W = 36 - 30 - 4 = 2$

This means that the segment shown in Figure 16.25a can be driven by two actuators alone.

16.6.2. Synthesis of spine-type manipulator kinematic chain

This kinematic chain may be synthesized using the well-known formula describing mechanism mobility. A kinematic chain is first selected from the set of manipulator arm links. We let S denote the set of links and $P = UP_i$ ($i = 1, 2, ..., 5$) the set of all kinematic pairs in the chain. In addition there are sets (K_j n_k) of links forming closed chains (loops) r_i of i-th class pairs ($i = 1, 2, ..., 5$).

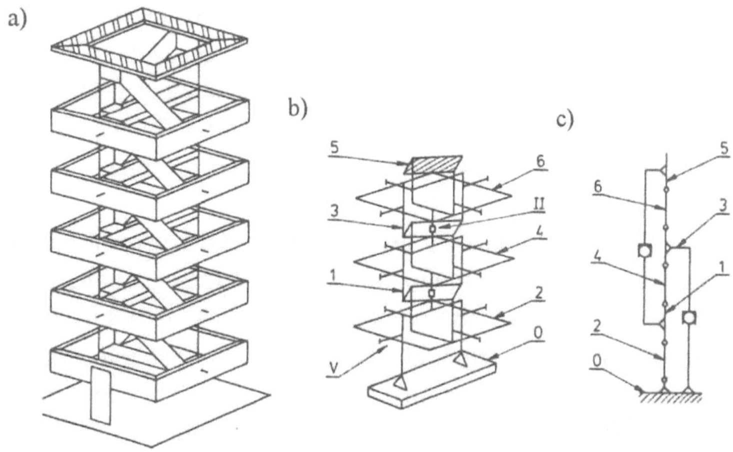

Fig. 16.33. Kinematic chain of spine-type manipulator [according to 16.1]: a) general view of the segment; b) kinematic chain; c) side view.

Let us assume that $Kj \subset S$ ($j = 1, 2, ...$) and that S is ordered with respect to j.

Now let us consider a special case in which all K have the same kinematic structure. As the chain S is connected, for each j, the product $K_j \,\Pi\, K_{j+1} \neq 0$. Assuming that some links may belong to several closed chains (loops), we determine cardinality α_c:

$$\alpha_c = \text{card}\,(K_j \,\Pi\, K_{j+1} \,\Pi ... \Pi\, K_{j+c})$$

Let R_{ij} be a set of i-th class pairs within the closed chain j. As $\alpha_c \geq 1$, at least two pairs belong to two consecutive chains. Let β_{ip} denote cardinality

$$\beta_{ip} = \text{card}\,(R_{i,j} \,\Pi\, R_{i,j+1} \,\Pi ... \Pi\, R_{i,j+p}) \quad (i = 1, 2, ..., 5)$$

Assuming α_c and β_{ip} the cardinality of P may be written as

$$P_i = \text{card}\,(p_i) = m(r_i - \sum_{p=1}^{n} \beta_{ip})$$

for $m = 1, 2, ...$ and $i = 1, 2, ..., 5$, where $n = \sum_{c=1}^{m} \alpha_c$

On the condition that within one group there are no passive constraints and that by joining two groups no additional constraints are imposed on chain S, the formula determining the group mobility is:

$$w = \sigma\left[m\left(n_k - \sum_{c=1}^{m} \alpha_c\right) - 1\right] - \sum_{i=1}^{5} i\left[m\left(r_i - \sum_{p=1}^{n} \beta_{ip}\right)\right] \tag{16.18}$$

Let S be referred to as the segment of mobility w and K be referred to as C-class cage, if these groups fulfil equation (16.18) for $k = 1$ and $k = 2$ and $m = k\alpha$, where α is the smallest $\alpha_c \neq 0$ and C i the greatest α_c.

This equation may have many solutions for w, nk, ri, α_c and β_{ip}. The simplest way of finding these solutions is by testing various combination of these parameters.

The solution does not give the full kinematic chain of the mechanism. A suitable version may be implemented its kinematic and dynamic properties. The mobility of a segment equals 2 regardless the number of cages. The mechanism shows no passive constraints.

The arm of the manipulator is usually a mechanism composed of several segments connected in a series. Kinematic analysis of the arm consists in testing its motion capabilities by analyzing its workspace as a set of points determining the possible positions of the orthonormal coordinate system attached to the end-effector and by analyzing the possible orientation of this system. This analysis also permits the determination of the structure, displacements and velocities of the drive system components that provide for the execution of the desired motion.

A kinematic description of k-th pair of links (i) and (j) may be obtained by attaching orthonormal coordinate systems Π^i and Π^j to these links. The relative positions of the links are described by 7 coordinates:

$$q_k = [x_k\, y_k\, z_k\, a_1\, a_2\, a_3\, a_4]_k^{\mathrm{T}} \tag{16.19}$$

where a_i (i = 1, 2, 3, 4) denote Euler parameters for the rotation of Π^j relative to Π^i. A_k is the matrix of this rotation.

Vector $c_k = [x_k,\ y_k,\ z_k]^{\mathrm{T}}$ describes a translation of Π^j relative to Π^i. The parameters of rotation $a_k = [a_1\, a_2\, a_3\, a_4]^{\mathrm{T}}$ satisfy the equation

$$f_k(a_k) = a_1^2 + a_2^2 + a_3^2 + a_4^2 - 1$$

resulting from the definition of these parameters; moreover coordinates q_k satisfy another equation $g_k(q_k) = 0$ defined by the class and type of kinematic pair (i, j).

The closure condition for the cage loop consisting of m kinematic pairs may be written in the coordinate system Π^1 of the link number (1), then referred to as the base link, in the form of equations:

$$c_1 + A_2 c_2 + A_2 A_3 c_3 + \ldots + A_2 A_3 + A_m c_m = 0 \tag{16.20}$$

Moreover, rotation of zero magnitude of the base link adds the following equation:

$$\operatorname{diag}\,[A_1 A_2 \ldots A_m - E] = 0 \tag{16.21}$$

Now let us write the coordinates of the cage belonging to segment s as vector q_s:

$$q_s = [q_1\, q_2 \ldots\, q_{ns}]_s^{\mathrm{T}} \tag{16.22}$$

The system of equations (16.18), (16.19), $f_k(a_k) = 0$ for k = 1, 2, ... , l_0 and $g_k(q_k)$ = 0 for k = 1, 2, ... , m where l_0 is the number of rotational pairs in the cage may be written as $g_s(q_s) = 0$. Independent coordinates q_s of the cage (segment s) are those q_s for which the Jacobian matrix $\left[\dfrac{\partial g_i}{\partial q_j}\right]$ is nonsingular. By equating the Jacobian matrix to zero for each segment independent variables q of manipulator arm can be determined. The assumption that there is no constraints in the cage can be also verified in this way.

Among the links belonging to a cage one link (p) can be selected which belong to two cages within the segment. This is the intermediate link of the cages. For the purpose of numbering intermediate links within one segment let us assume that the link bearing the highest number is the intermediate link between two consecutive segments. The origin of the orthonormal coordinate system Π^1 attached to (p) has the same coordinates c_p in the base systems of related cages.

Similarly, the matrix A_p of rotation transforming Π^p into Π^1 is equal to $A^m{}_p$. The base system of segment (s+1) $\Pi^1{}_{s+1}$ may be rotated and translated with respect to the intermediate system $\Pi^p{}_s$. Thus by denoting this translation in $\Pi^p{}_s$ system as $\boldsymbol{r}$, and the rotation as $\boldsymbol{P}$ ($\boldsymbol{r}$ and $\boldsymbol{P}$ are constant), we obtain translation $\boldsymbol{d}_s$ of the system $\Pi^1{}_{s+1}$ expressed in the Cartesian coordinates of system $\Pi^1{}_s$:

$$\boldsymbol{d}_s = \boldsymbol{b}_s + \boldsymbol{C}_s \boldsymbol{r}_s \tag{16.24}$$

where $\boldsymbol{C}_s = \boldsymbol{P}_s \boldsymbol{B}_s$.

The rotation transforming $\Pi^1{}_{s+1}$ into $\Pi^1{}_{s+1}$ is described by matrix $\boldsymbol{C}_s$. By using a notation similar to the description of translations and rotation of segment base systems relative to the manipulator base system, translation c_e of coordinate system Π_e attached to the end-effector and its rotation A_e around the base coordinate system $\Pi^1{}_0$ of the manipulator can be written as

$$c_e = d_1 + C_1 d_2 + C_1 C_2 d_3 + \ldots + C_1 C_2 \ldots C_{n-1} d_n \tag{16.25}$$

$$A_e = C_1 C_2 \ldots C_n$$

where n denotes the number of segments in the manipulator.

The method of describing the position of the manipulator having a "cage" structure in its internal coordinates $\boldsymbol{q}$ is the basis for determining the manipulator workspace. By finding the functional relationship $f(q,u) = 0$ where u = $[u_1\ u_2 \ldots .\ u_r]^T$ coordinates of the drive system components are determined. By analyzing the Jacobian matrix with respect to variables u it is possible to verify the independence of these variables and in this way verify wheter the drive system provides full motion capabilities [16.1].

16.6.3. Model of the manipulator and control system

Figure 16.34a shows an example of computer animation [16.10] of the manipulator segment, Figure 16.34b shows the animation of three segments, and Figure 16.34c shows a model of the segment with two hydraulic actuators and part of the drive system.

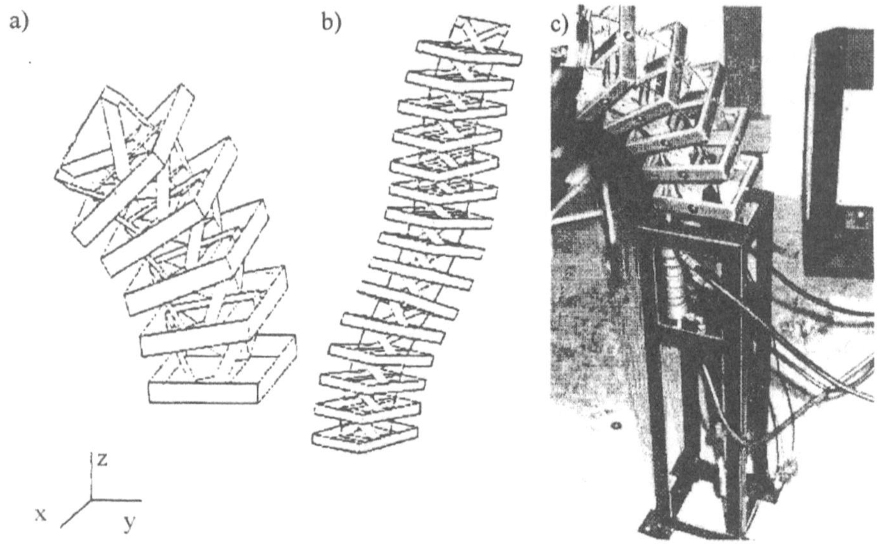

Fig. 16.34. Spine-type manipulator: a) animation of the segment; b) animation of three segments; c) view of the manipulator with two hydraulic actuators.

The manipulator is driven by two hydraulic actuators (Fig. 16.34c). Each actuator is controlled by a Moog-type electrohydraulic servovalve. The position of manipulator end-effector is measured in the feedback loop by transformer-based position sensors of the PELTRON type. Control signals are sent by an IBM PC. Position signals from the manipulator are transmitted to this computer through A/D converters. An electronic device between the manipulator and the drive system contains amplifier and corrector, as well as the system receiving and transforming signals from the position sensors (Fig. 16.35a).

The position v_1 of the transducer probe is the input signal for the PELTRON transducer. Signal x_1 is the output signal of this transducer adjusted by the transforming system. The PELTRON transducer with the signal transforming system may be treated as an inertialess element. The difference between the amplified control signal v_2 controlling the position of the manipulator and the signal x_1 from the position sensor is treated as a position error of the manipulator. The servovalve is controlled by voltage x_3 supplied from its amplifier. Both amplifiers in the control system may be treated as inertialess elements [16.11].

The Moog-type servovalve is a two-position hydraulic amplifier. The first amplifier stage is an amplifier of the nozzle-diaphragm type, in which the movement of the diaphragm (1) is controlled by current flowing through the coil (2) situated in the field generated by a permanent magnet (3) (fig. 16.35). The second stage is a spool-type flow control valve.

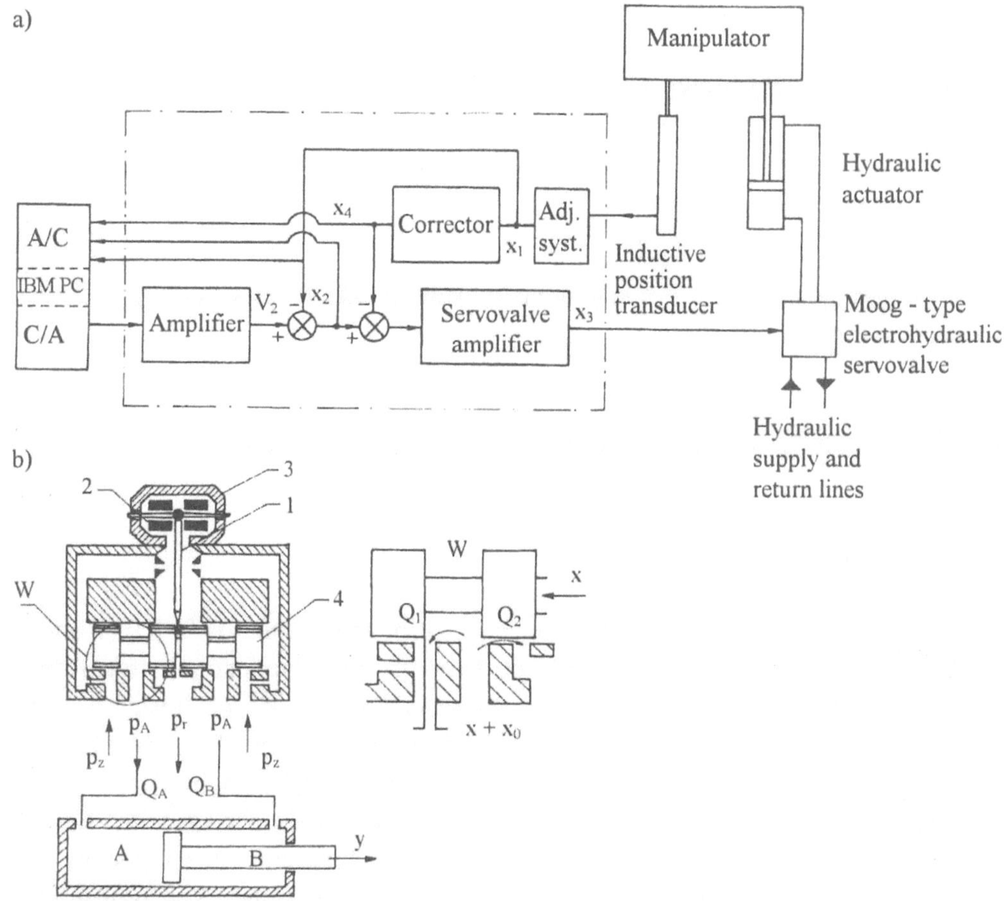

Fig. 16.35. Manipulator control system: a) diagram of one control path [according to 16.11]; b) Moog-type servovalve.

Internal feedback in the servovalve is obtained by linking the movements of the spool (4) with diaphragm movements. The output signals of the servovalve are the rates Q_A and Q_B of the oil flow through the ports connected with chambers A and B of the actuator. The static equilibrium of the signals results from the equilibrium of forces acting on the diaphragm and the spool. For certain conditions it can be assumed that the pressure in one actuator chamber equals the supplied pressure p_z and the pressure in the second chamber equals return pressure p_r. The pressures are regarded as constant. The relationship between the displacement of the control valve spool $x(t)$ and the current $i(t)$ may be described by the equation experimentally verified

$$\dot{x} = \frac{k}{T} i \tag{16.26}$$

The time constant T and gain coefficient k of the amplifier are determined separately for each valve.

Relationships describing the flow rate Q_i of oil flowing through the ports of the control valve, the coefficient of hydraulic losses ξ and pressures p_A and p_B in chambers A and B of the actuator can also be found. Without detailed considerations we can compute the actuator force

$$F = S_A p_A - S_B p_B - F_T \mathrm{sgn}(\dot{y}) \tag{16.27}$$

where S_A and S_B are actuator piston areas on sides A and B respectively, while F_t is the friction force.

16.7. Existing designs of elastic manipulators

16.7.1. SPINE manipulator

Volvo uses a spine-type manipulator on its production line. Its SPINE-Robotics has flexible arm with seven degrees of freedom, and the end-effector reaches a velocity of 1.5 m/s. The degrees of freedom in this case do not result from the number of rigid links and types of joints, but from the number of drives. The robot has 50m^3 of serviceable workspace. The maximum lifting capacity of the arm is 5 kg, but optimum operation is ensured when the load does not exceed 3 kg. The manipulator is driven by a combined electric and hydraulic drive system with feedback control providing high positioning accuracy (±2 mm). The transmission mechanism contains only four wire lines, and is built from biconvex elements touching one another in one contact point (Fig. 16.36a).

Figure 16.36b shows the construction of the SPINE manipulator, Figure 16.36c shows the workspace with its dimensions and Figure 16.36d shows the prototype of the manipulator performing painting operation.

The mechanism shown in Figure 16.37a can be mounted on the end-effector of a painting robot. It is driven by a rotary actuator (θ) and a linear actuator (x) (Fig. 16.37b). Figure 16.37c illustrates the kinematic chain of the mechanism, together with the drive system. The drive system provides for the positioning of the mechanism, which takes the shape of a curved segment lying on the plane containing the rotation axis A and the curved segment $\rho(\lambda_i)$ itself (fig. 16.37b).

The mechanism shown in Figure 16.38a performs a function similar to that of the system illustrated in Figure 16.38. This system has the curvilinear shape of an arc of radius ρ, which assumes various positions in plane P.

The system is driven by four hydraulic actuators and pretensioned wire lines. If ρ and θ are given, only two variables are used to control the mechanism.

The manipulator contains a certain combination of simple modules composed of a particular number of segments. The rotational joints are in the form of pivots.

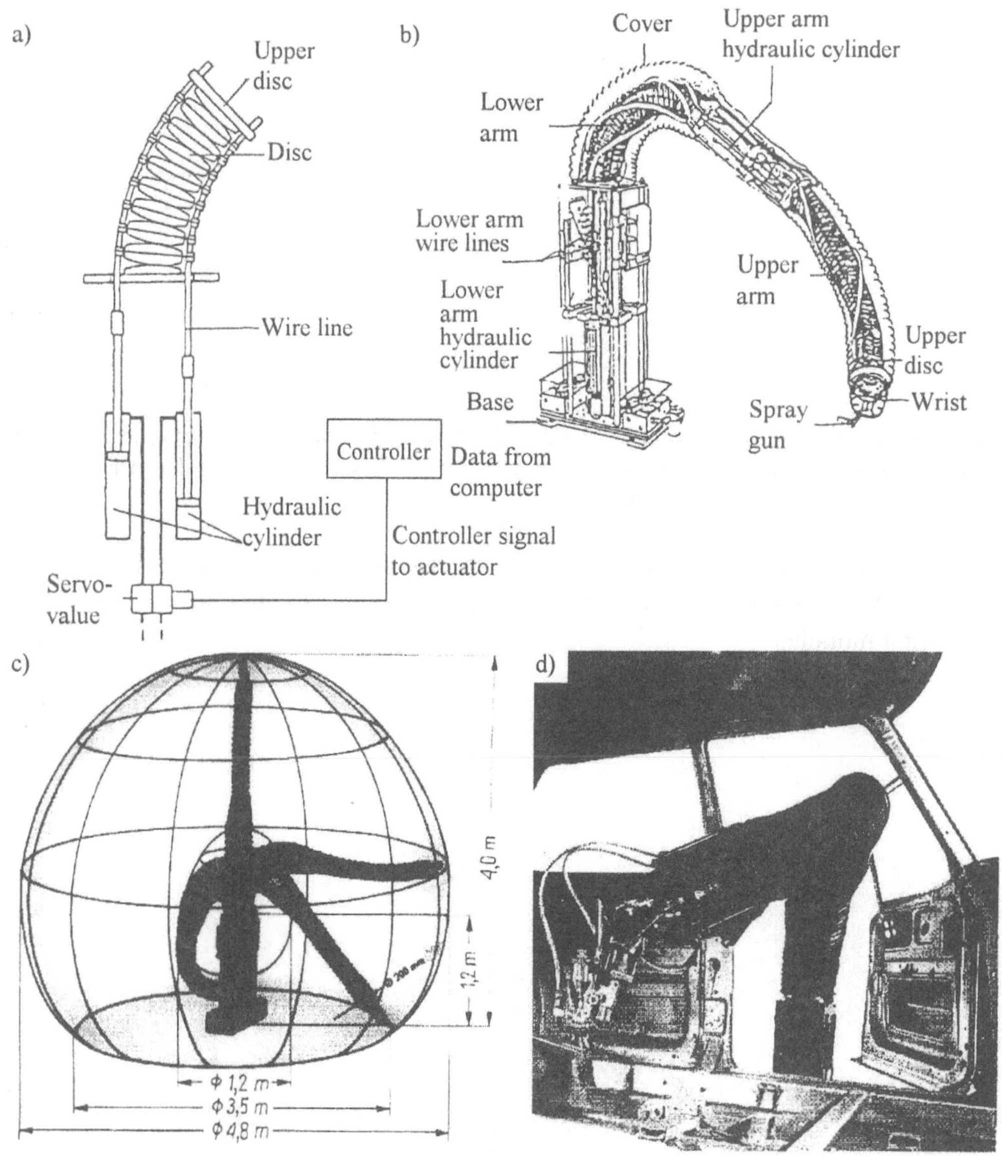

Fig. 16.36. SPINE Robotics robot [according to 16.12 and 16.13]: a) SPINE robot drive mechanism (according to Komatsu Ltd); b) robot construction; c) robot workspace; d) prototype of the manipulator.

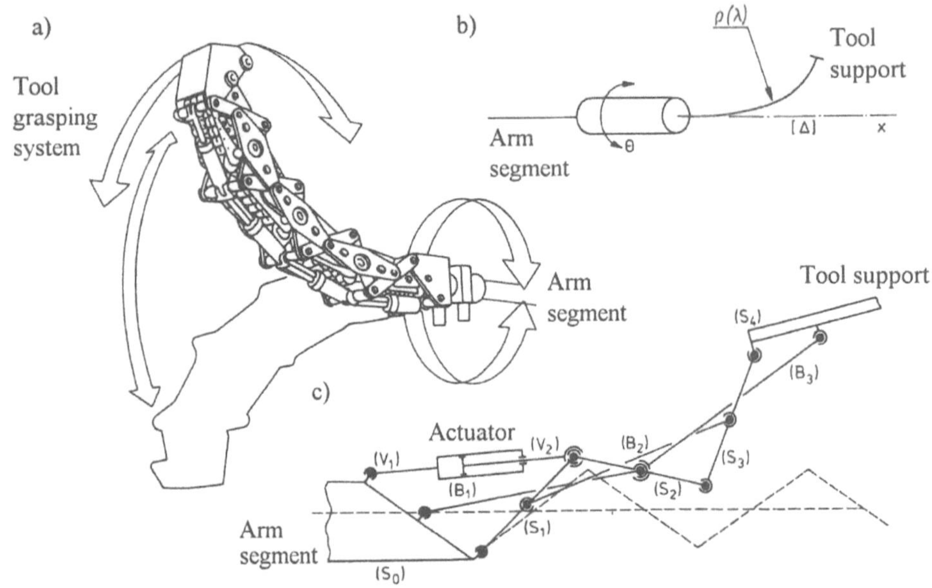

Fig. 16.37. Manipulator arm segment: a) Nitro-Nobel MEC mechanism [according to 16.14]; b) motion capabilities; c) kinematic chain of a two-module control mechanism for one actuator: S_i, B_i - ball joints; V_1, V_2 - rotational joints.

The simplified kinematic diagram is shown in Figure 16.38b (without wire lines). In this drawing S_7 denotes the input shaft, and the S_1 , the intermediate shaft of the first module.

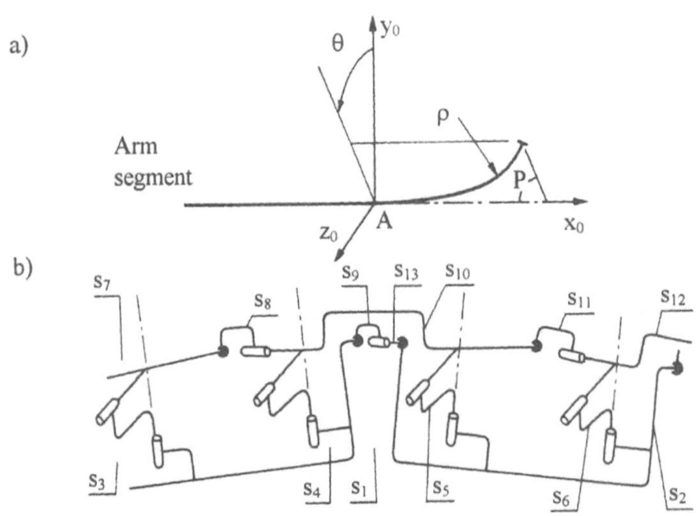

Fig. 16.38. ACMA-Renault painting robot [according to 16.14]: a) parameters describing robot mechanism; b) simplified kinematic diagram of the robot.

Figure 16.39a shows the mechanical structure of the flexible arm and Figure 16.39b shows the general view of the robot with the segment of the mechanism installed.

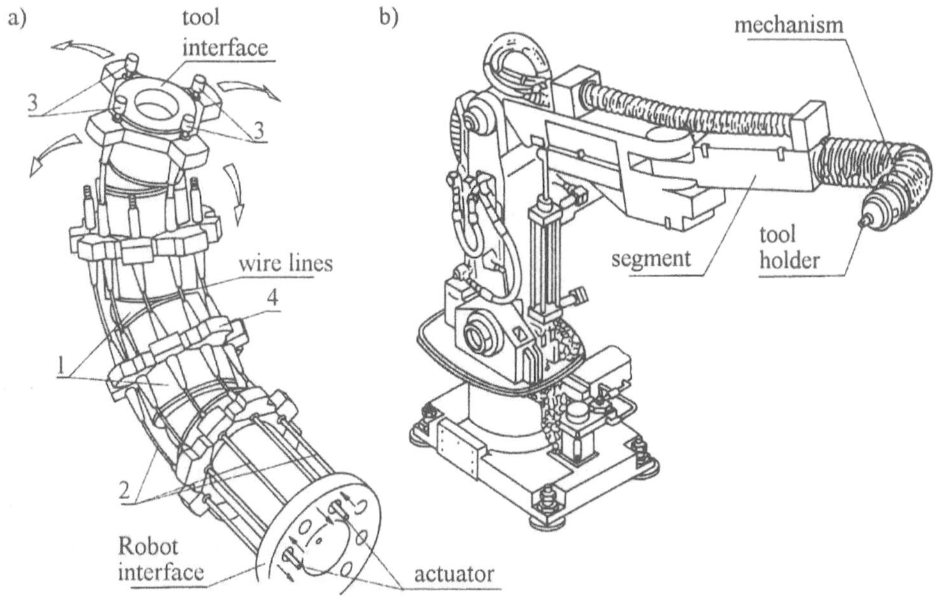

Fig. 16.39. Robot with a flexible painting mechanism [according to 16.14]: a) external structure; b) overall view of the robot.

16.7.2. Manipulators composed of pressurized pipe segments

In the years 1986-1988 studies were conducted on the mechanics of orthotropic pressurized pipe segments emulating specific movements of octopus tentacles.

Morgan (1986, 1987) presented a description of experimental flexible limbs of first generation; Wilson (1988) published a detailed description of the experiment. In the study [16.15] the issues of mechanics and positioning of highly flexible manipulators were disscussed.

Figure 16.40 shows the manipulator limb made of ten pressurized pipe segments, bending under the influence of internal pressure. Five of the elements illustrated in this drawing are reinforced on one side, and five on the opposite side (thick black line). When the pressure is suitably adjusted, the end-effector of the manipulator acts as a gripper lifting the weight (Fig. 16.40b) and transporting it to a given point on the plane.The single pipe segment is made of a reinforced elastic polymeric compound with nonlinear properties (e.g. polyurethane) which bends proportionally to the internal pressure and loading. One of the possible designs of these elements is illustrated in Figure 16.41.

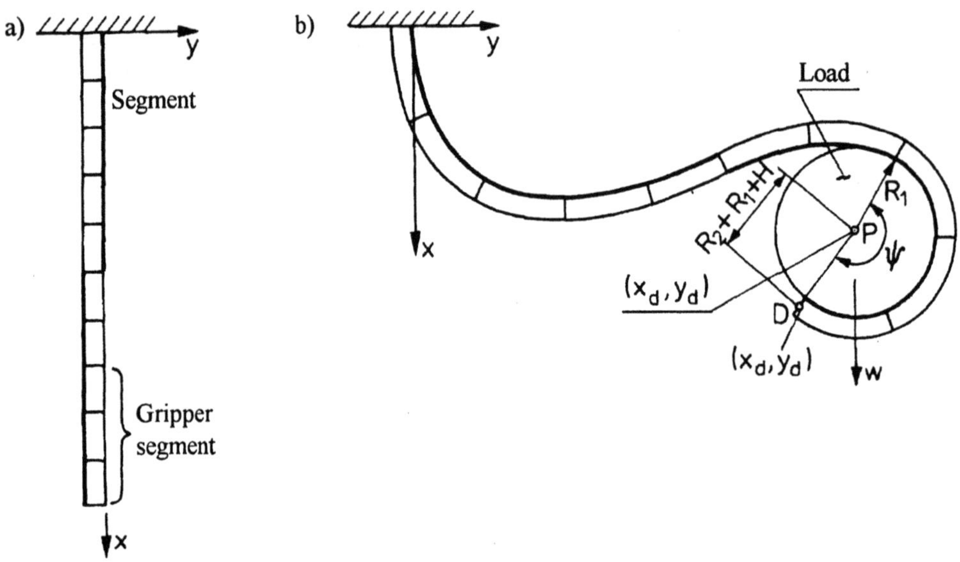

Fig. 16.40. Robot arm composed of 10 pressurized pipe segments: a) at rest; b) lifting the weight.

The study [16.15] gives a detailed analysis of the mechanical properties of this component. It is assumed that stress distribution takes the form of

$$\sigma = A\varepsilon + B^{1/n} \tag{16.28}$$

When the segment has smooth wall, A, B and n are material constants.

If the values of A, B, n and E, and segment geometry are konwn, it is possible to determine $y = y(x)$, the neutral axis, and its curvature radius ρ for the value M_C of torque applied to the end of the segment. The method of analysis of a multi-segment limb is given in [16.14]. The recurrent formula that determines the internal pressure in static equilibrium conditions is

$$M_{ti} = (p_i - p_{i+1})\pi R_2^2(R_1 + H_1) \tag{16.29}$$

where

M_t is the moment applied to the end of the segment;

R_1, the outside radius of the segment;

H_1, 1/2 of the segment wall thickness;

p_i, p_{i+1}, the internal pressure.

The following is an example of an adaptive program for the calculation of this system, in the case when all elements are identical, have identical characteristics and each of them is initially pressurized. Elements 1 through 5 are situated on the right side, elements 6 through 10 are situated on the left side. The given data are: $n = 0.2$; A = 13.8 N/cm^2; B = 34 N/cm^2; E = 2.07 N/cm^2; R_1 = 1.33 cm, R_2 = 1.21 cm, H = 0.0381 cm, L = 5.08 cm, b = 1.27 cm, Mc = 42.9 N·cm.

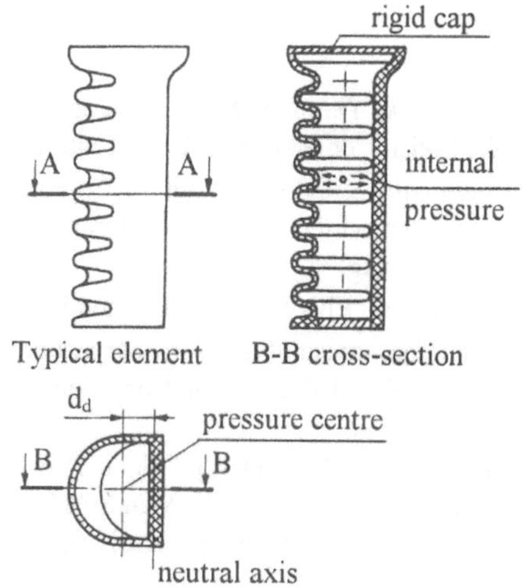

Fig. 16.41. Bellows-type segment. US Patent No 3, 981, 528 [according to 16.15].

Figure 16.42 shows the main dimensions of the segment. The resulting values are $H_1 = 0.0190$ cm, $\rho = 6.09$ cm, $y = 2.85$ cm, $x = 1.29$ cm. In this case the internal pressure required to produce the desired torque is $p = 6.93$ cm. We have assumed that the weight of 8.9 kN and 3,92 cm in diameter was to be lifted from the position defined as ($x_1 = 36.6$ cm, $y_1 = 5.29$ cm) to ($x_f = 20.3$ cm, $y_f = 21.6$ cm) (Fig. 16.34b).

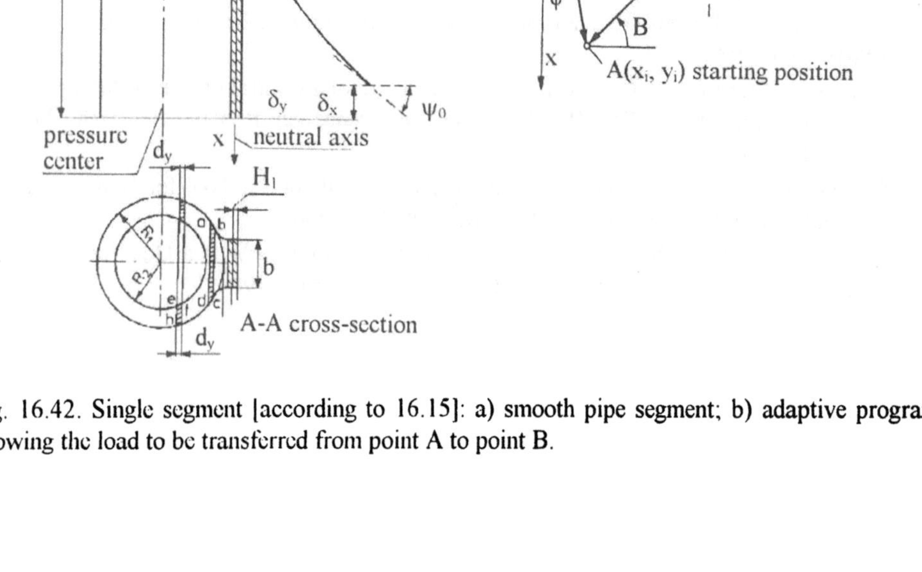

Fig. 16.42. Single segment [according to 16.15]: a) smooth pipe segment; b) adaptive programme allowing the load to be transferred from point A to point B.

16.8. References

[16.1] *Buśko Z., Frączek J., Kamiński D., Morecki A.*: Designing and Modelling of Elastic Manipulators. Proc. of XII Polish Conference on TMM, Bielsko-Biała, 1989.

[16.2] *Roth B., Rastegor J., Scheinman V.*: On the Design of Computer Controlled Manipulators. First CISM-IFtoMM Symposium on the Theory and Practice of Robots and Manipulators. Vol. 1, Udine 1974, Springer-Verlag.

[16.3) *Thring M.W.*: Robots and telechirs, Ellis Horwood Limited Publishers. Chichester, 1983.

[16.4] *Umetani Y., Hirose S.*: Biomechanical Study of Serpartine Locomotion. First CISM-IFtoMM Symposium on Theory and Practice of Robots and Manipulators. Vol.1, Udine 1974, Springer-Verlag.

[16.5] *Hirose S., Ikuta K., Umetani Y.*: A New Design Method of Servoactuators based on the shape memory effect. Proceedings of the Ro.Man.Sy'84; The Fifth CISM-IFtoMIM Symposium. Edit. by A. Morecki, G.Bianchi, K. Kędzior, Kogan Page, London. Hermes Publishing, 1985.

[16.6] *Malczyk G., Morecki A.* Elastic Manipulator of the Elephant Trunk Type. Biocybernetics and Biomedical Engineering, 1987, Vol.7, Nol-4, pp 155-168.

[16.7] *Morecki A., Ekiel J., Fidelus K.*: Bionics of motion. Warsaw, PWN 1971.

[16.8] *Wilson J. F., Muhajan U., Wainwright S.A., Croner L.J.*: A Continuum Model of Elephant Trunks, Manuscript (presented of ICAR, Ohio, Columbus, USA, 1989).

[16.9] *Cieślak R.*: Elastic Manipulator - Design and Static Analysis. M. Sc. Thesis. Warsaw University of Technology, Warsaw, 1990.

[16.10] *Frączek J.*: Computer method of dynamic analysis of kinematic chains of manipulators. Doctoral dissertation, Warsaw Technical University, Warsaw 1990.

[16.11] *Morecki A., Busko z., Czerwinski M., Frączek J.*: Model of robot with elastic manipulation system. MERA-PIAP, CPBR 7.1, 1989.

[16.12] Spine-Robotics, Sweden 1983.

[16.13] Spine News, Sweden 1984.

[16.14] *Lhote F., Kaufmann J.M., Andree P. and Taillord J.P.*: Robots, Components and Systems. Robot Technology. Vol. 4, Kogan Page, London, 1984.

[16.15] *Wilson J.F., Manajan V.*: The Mechanics and Positioning of Highly Flexible Manipulator Limbs. Jnl of Mechanisms, Transmissions and Automation in Design. Transactions of the ASME. June 1989.

[16.16] *Cieślak R., Morecki A.*: Technical Aspects of Design and Control of Elastic Manipulator of Elephant Trunk Type. RISME 5th World Conference on Robotic Research. Proceed. Cambridge USA, 1994.

[16.17] *Cieślak R., Morecki A.*: Elephant Trunk Type Manipulator - a Tool for Bulk and Liquid Material Transportation. Proceed. 27 ISIR, 1995, Singapore.

[16.18] *Morecki A.*: Robotics System - Elastic Manipulator Combined With a Quadruped Walking Machine. 7th CISM-IFToMM Symp. Ro'Man'Sy-88, Hermes 1988.

17. WALKING MACHINES

17.1. Basic definitions and types of walking machines

The walking machine can be defined as "a technical device designed to perform functions similar to the locomotion of animals and insects. The locomotion of the walking machine is of the discrete type and may be performed using one, two, three, four, six, eight and more legs - pedipulators – to walk, run and jump over a hard surface"[17.3].

Research on walking machines seen rapid development in recent years, due in large part to the introduction of new generations of computers. Without computer control fast coordination of leg and body motion and efficient avoiding of obstacles would be impossible.

The availability of small computers and progress in computer controlled machines (especially industrial robots) have made it technicaly possible to design walking machines. The usefulness of such machines is clear as a means of transport over the roadless terrains and for operations on sea and river bottoms (in the construction of drilling rigs and bridges, etc.) There is also a demand for mobile industrial robots, featuring a high degree of autonomy.

Walking machines may be useful when operating in environments harmful to the man, such as inspecting installations in nuclear power plants, and fire fighting. Prototypes of walking machines confirm the significance of research work in this direction. However, investigation of walking machines is still at the laboratory stage.

The majority of walking machines are six-legged machines based on insect leg structure and insect locomotion, due to the fact that principles of insect locomotion are well known [17.3] and easy to imitate. However, insects move slowly, and their ability to avoid obstacles and their maneuverability cannot be compared to that of four-legged mammals (especially horses).

Complex control rules must be developed to apply the principles of locomotion of four-legged animals. In particular, we must solve the problem of positional stability when moving over ground with a non-uniform profile. Developments in information technology and the increase in the computational speed of electronic systems allow us to think that four-legged walking machines emulating mammals will be more effective than six-legged machines. The four-legged machine can be fitted with a simpler drive system (with fewer driven elements and less power consumption) and the number of control signals required for the drive system is lower.

Studies on four-legged machines emulating four-legged mammals were, however, been abandoned after the first experience was gained with computer control in the sixties when electronic technology was still yet in its initial phase of development [17.5]. At present research on various models of six-legged and four-legged machines is conducted; the legs of these machines are modeled on spider legs; and the machines move very slowly.

The design of walking machines requires great familiarity with the principle of locomotion in insects. These principles can be limited to six (Wilson) [17.3]. The locomotion of insects is easily emulated as there is a one-to-one correspondence between the speed of locomotion and the sequence of leg movements. Four-legged animals are quite another case. In spite of years of study and the development of various hypotheses, the locomotion of such animals cannot be described as a set of simple rules. However, these studies have produced significant energy savings and the capability of reaching high locomotion speeds on various terrains. Gait, defined as the sequence of leg movements, is the subject of many research programs. Some of these studies concern the number of gait patterns, depending on the number of walking machine legs [17.6, 17.7].

Independently of the studies on six-legged and four-legged machines, various models of anthropomorphic two-legged robots (bipeds) [17.2, 17.8, 17.10, 17.11], and one-legged jumping machines, together with two-legged running machines, and recently four-legged jumping machines [17.4, 17.10] have been built. Various models of these machines are described later, in this chapter.

As has been said, currently known and studied walking machines have one (jumping machines), two (bipeds), four (quadrupeds), six (hexapeds), eight, and more legs. Various models of crawling (snake-like) machines have been built as well.

Figure 17.1 shows an one-legged jumping machine. Various sensors, valves and control units are attached to the leg and body. The leg terminates in a foot. The rotation of the foot is controlled by a servovalve generating torque on the "hip" joint of the machine. The leg is rotated during the "flight" phase. Effective control algorithms [17.4], [17.10] must be applied to ensure stable flight and landing.

Figure 17.2 shows a model of the WL-12 two-legged walking robot designed and built at the Waseda University in Tokio. Each of the lower legs has six degrees of freedom, while the body has three. The balancing mass (8) m_p is 30 kg; the mass of the robot $m = 107$ kg; the height of the robot $h = 1.8$ m. The robot is driven by electrohydraulic servo-drive system with rotary hydraulic motors in the rotational joints, and by a linear motor in the prismatic joint. The robot is equipped with a suitable potentiometric system, built into the feedback loop, for the control of rotation angles and angular velocities. The execution of one step takes 1.3 s, and the length of the step is 0.3 cm [17.11].

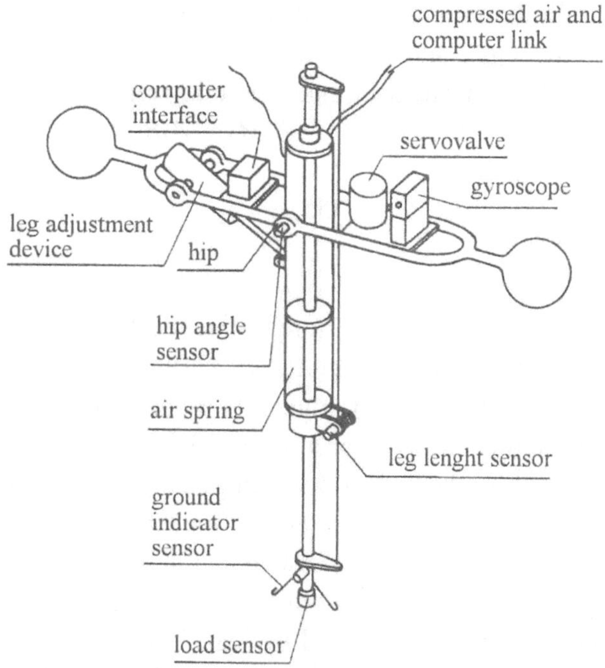

Fig. 17.1. One-legged jumping machine (kangaroo) in the fibular plane (University Carnegie Mellon) [according to 17.10].

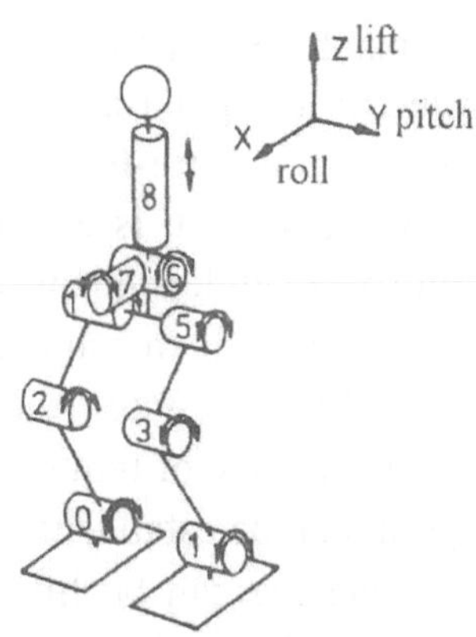

Fig. 17.2. Kinematic chain of WL-12 two-legged walking robot: 0÷7 - links; 8 - balancing mass [according to 17.11].

17.2. Four-legged walking machines

P. L. Tschebyschev (1821-1884) designed a walking mechanism and made its model. The kinematic diagram of Tschebyschev's "walking foot machine" is shown in Figure 17.3. During the forward movement platform $CC'C'_1C_1$ rests first on legs 1 and 4, and then on the legs 2 and 3. The legs end in large, flat feet.

The dimensions of the mechanism links are following:

$$A_1B_1 = B_1C = B_1M_1 = A_2B_2 = B_2C = B_2M_2 = A_3B_3 = B_3C_1 = B_3M_3 = A_4B_4 = \\ = B_4C_1 = B_4M_4 = 1,$$

$$A_1C' = A_2C' = A_3C'_1 = A_4C'_1 = 0.355,$$

$$CC' = C_1C'_1 = 0.785,$$

$$A_2A_4 = A_1A_3 = C'C'_1 = 0.634.$$

These link dimensions lead to four straight-line mechanisms connected together so that their cranks form a parallelogram $A_1A_2A_3A_4$ (link A_1C' is fixed to link A_2C', and link $A2C'_1$ is fixed to link $A_4C'_1$). Figure 17.3 shows trajectories of points M_3 and M_4 in relative motion with respect to platform $CC'C'_1C_1$. This trajectory resembles a curve traced by the end of a walking man's foot. When points M_1 and M_4 are on the straight segments of their relative trajectories, the mechanism stands on legs *1* and *4*, while legs *2* and *3* are elevated and move forward. When points M_1 and M_4 end its rectilinear motion, the feet of legs *2* and *3* come into contact with the ground. The mechanism now is supported by legs *2* and *3*, while legs *1* and *4* are lifted and start to move forward, together with the platform [17.10].

But the first four-legged walking machine was a Chinese design of the third century (Fig. 17.4). In 231 a famous Chinese commander Zhu Ge-Liang used wooden four-legged walking machines for the transportation of food (flying horse) [17.12]. In 1964 General Electric designed a machine (AMS) called "the walking truck". Each leg of this machine is a three-link pedipulator 2.3 m long performing planar motion. The machine weighs 1.5 t, is 3.5 long and 3 m high. The operator sitting in a special cabin controls 12 independent drives simultaneously and is able to control the length and pace of gait, which can reach speeds of 10 km/h. The cabin is inconveniently situated several meters above the ground and from this height the operator must take into account the scale of motion and the profile of the terrain.

An optimum gait program can be established on the basis of horse or cow gait analysis. Assuming that only one leg can be lifted at a time (in order to provide stability), six possible gaits can be obtained:

1-2-3-4	1-3-2-4	1-4-3-2
1-2-4-3	1-3-4-2	1-4-2-3

where: 1 - left front leg, 2 - right front leg, 3 - right rear leg, 4 - left rear leg.

The slow gait of the horse or cow is of the 1-3-2-4 type, which provides maximum stability. Another four-legged walking machine was proposed by R.B. McGhee of Ohio State University [17.13]. This is a model of the walking horse controlled by a computer, instead of an operator.

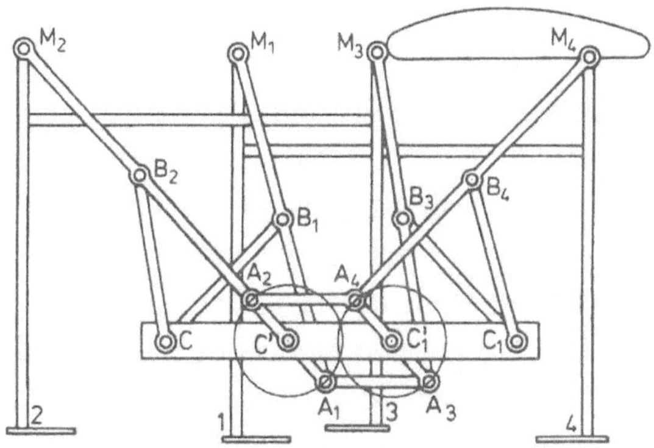

Fig. 17.3. Kinematic chain of Tschebyschev's "walking foot machine" [according to 17.10].

Fig. 17.4. Chinese four-legged walking machine (year 231) [according to 17.12].

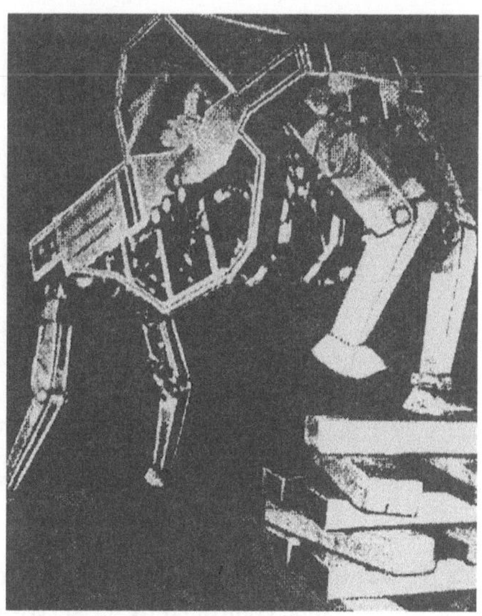

Fig. 17.5. Walking truck [according to 17.5].

Figure 17.6a is a general view of „Pony", a walking machine designed by American and Yugoslavian researchers; Figure 17.6b shows one of the models of four-legged walking machines designed in Japan [17.14]. Each leg of the machine (Fig. 17.6a) has a hip joint (2 DOF) and a knee joint (1 DOF). The four legs are driven by 12 servo-motors. Three gyroscopes, providing information about angular velocities about three axes, were used as the base sensors. The robot can turn without angular errors and move along a straight line.

Figures 17.7a and 17.7b present two models of four-legged walking machines, one with horse-type legs, the other with spider-type legs [17.16]. Figure 17.7c shows the relationship between power consumption in the support and in the transfer phase for each.

Figure 17.7d shows an interesting solution of a four-legged walking machine designed by Hirose. The leg design allows the machine to climb stairs, while the body maintains a horizontal position [17.7].

In 1983 studies on MK-4 walking machine in the Section of Robotics and Biomechanics of the Institute of Aviation Technology and Applied Mechanics at the Warsaw University of Technology were started. The legs of the machine were modeled on toe-walking mammals [17.10].

For the initial analysis it was assumed that the machine moves using the quadruped crawl technique. This gait features a fixed sequence of leg movements, and the machine is always supported by at least three legs. This gait may be observed in horses at slow pace (0.7÷1.7 m/s) [17.3, 17.5].

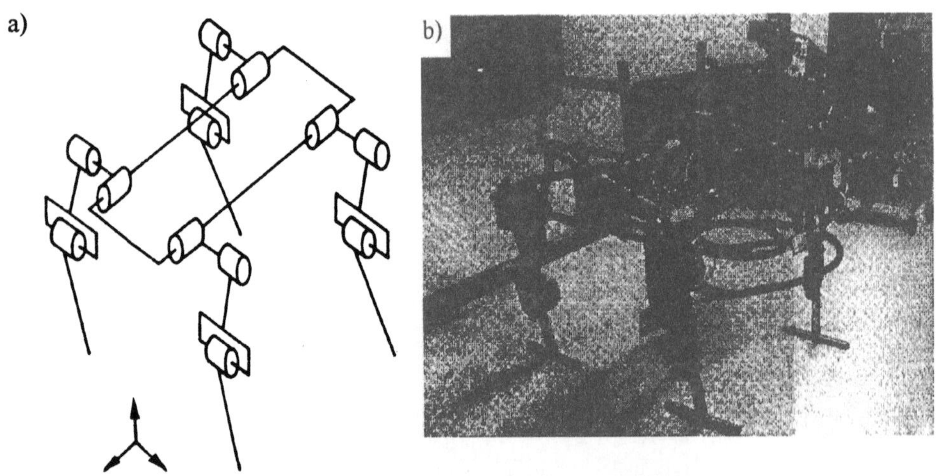

Fig. 17.6. Four-legged robots (quadrupeds): a) the„Pony" robot designed by American and Yugoslavian researchers [17.32]; b) a machine controlled by 12 servo-motors, designed by Japanese researchers [17.14].

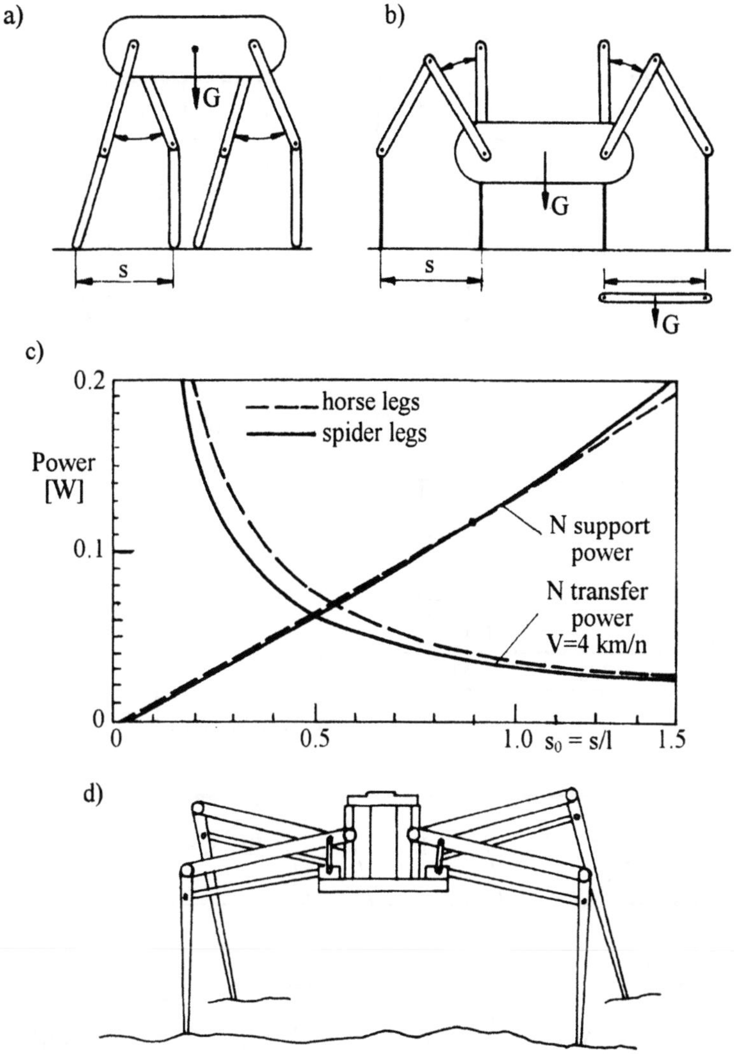

Fig. 17.7. Models of four-legged robots designed by Japanese researchers [17.14]: a) model with horse legs; b) model with spider legs; c) relationship between power in the support phase ε_{pod} and in the transfer phase ε_{przen}; d) four-legged Hirose machine [according to 17.16]

Figure 17.8 shows the subassemblies and model of a four-legged walking machine [17.7]. The geometry of the legs of this walking machine is modeled on the legs of toe-walking mammals, such as the rabbit, dog, or horse. The legs are assumed to be planar linkages having two degrees of freedom and driven by double--acting hydraulic cylinders. Figure 17.8a shows the kinematic chain and dimensions of the legs. Figure 17.8a$_1$ shows the ultrasonic position sensors in the leg. Some issues of the control of this machine are described later in the chapter.

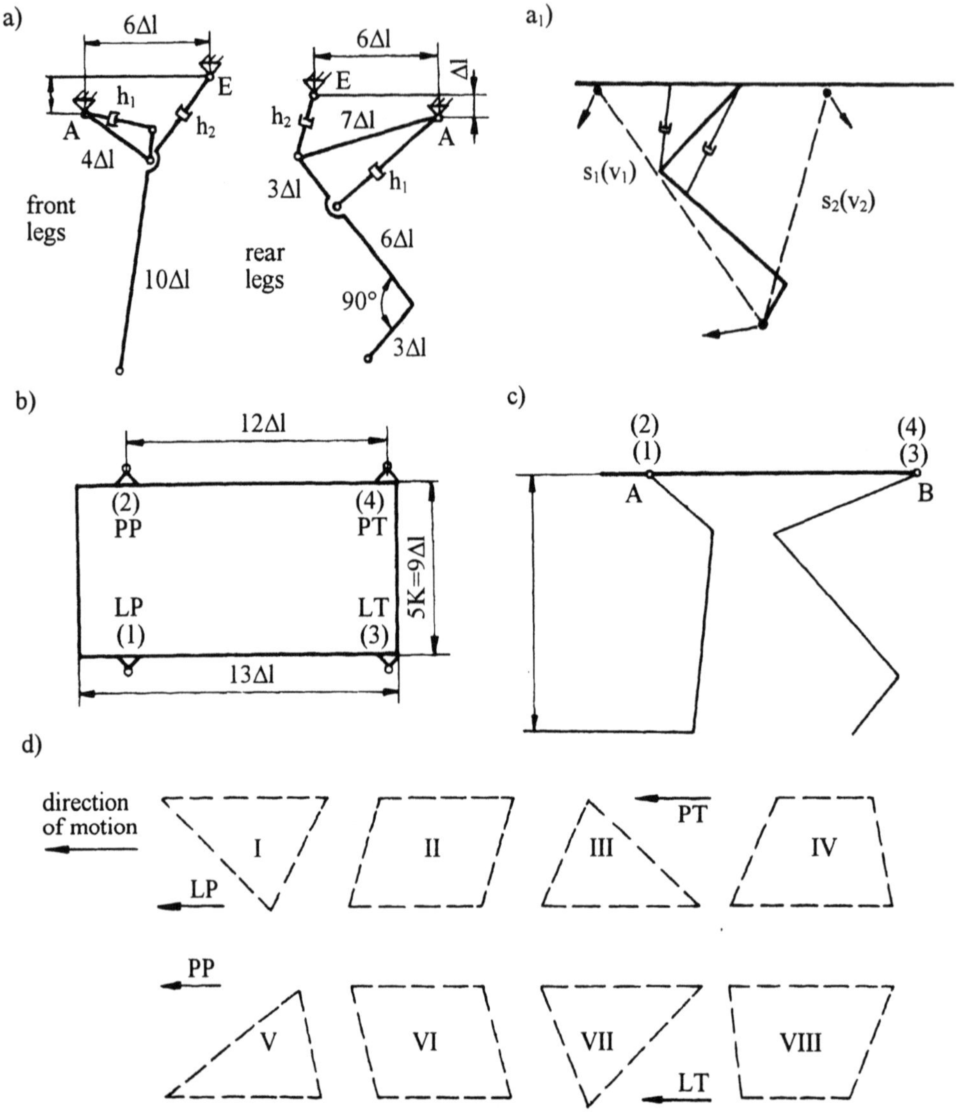

Fig. 17.8. Models of four-legged walking machine designed by Polish researchers [according to 17.10]: a) kinematic chain and leg dimensions of the MK-4 machine (Δl =H/12), $a_{1)}$ - machine leg with position sensors; b) top view of the machine; c) side view of the machine: S_k - machine body width, L - machine body length, H - machine height, Δ - leg attachment points, (1), (2), (3), (4) - leg numbers.

To better illustrate the dimensional proportions, the dimensions of particular components are given as multiples of Δl, where Δl = H/12; H is the height of the platform (machine body) above the ground. The platform dimensions are proportional to leg dimensions: the platform length is assumed to be 12Δl and the width, 9Δl. Figure 17.8b shows the top view of the machine body and Figure 17.8c, the side view of the machine. The numbers at the points of the legs denote the assumed numbering of the legs, namely:

- left front leg LP - number 1;
- right front leg PP - number 2;
- left rear leg LT - number 3;
- right rear leg PT - number 4.

The MK-4 machine is also assumed to a quadruped crawl technique, maximizing the static stability margin. The sequence of leg movements is:

- left front leg LP - number 1;
- right rear leg PT - number 4;
- right front leg PP - number 3;
- left rear leg LT - number 3 (Fig. 17.9).

The phases of leg movements are separated by short periods when the machine is supported by all four legs. Figure 17.8d shows the sequence of support polygons.

Figure 17.10 shows the points of support during locomotion. For the assumed gait the three-legged support phases have the same duration. Four-legged support phases all have an equal duration, but are considerably shorter than in three-legged support phases.

The process of leg transfer combined with the movement of the center of gravity relative to the leg ends is as follows:

Let T denote the duration of one sequence of machine leg movements and K, the step length. Let us divide the period T into 12 sub-periods $T/12$, and consider the positions of leg ends and machine body at the end of each sub-period. The relative positions of legs and body in the three-legged support phase denoted by I in Figure 17.8 are shown in Figure 17.10a. Points denoted by numbers 1, 2, 3 determine the positions of the leg ends after each sub-period (1 $T/12$, $2T/12$, $3T/12$). Figures 17.10b and 17.10c show the leg support points in phases III, V and VII (Fig. 17.8).

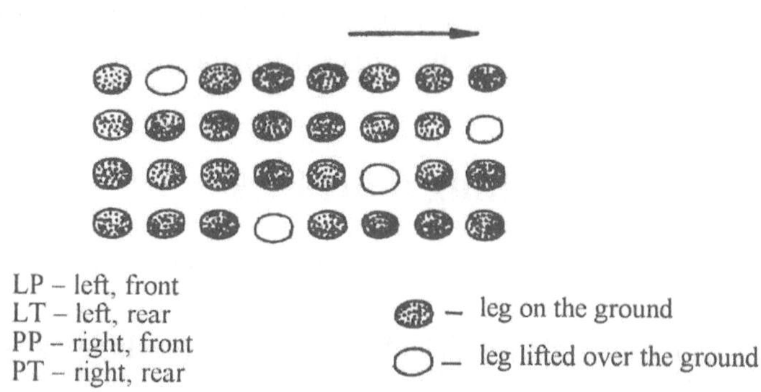

Fig. 17.9. Leg transfer sequence

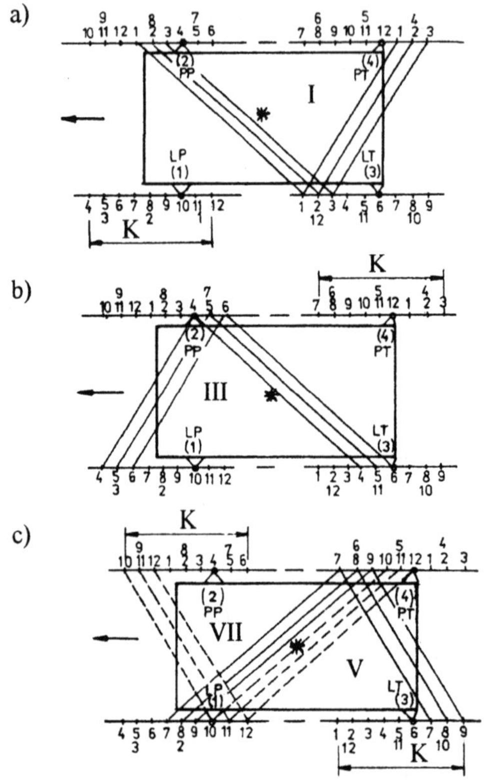

Fig. 17.10. Leg support points during locomotion: a) three-support phase I; b) three-support phase III; c) three-support phases V and VII; (1), (2), (3), (4) - leg numbers, 1,2,3, ..., 12 - numbers of sub-stages, K - step length, Δ - leg attachment points, * - vertical projection of the gravity center.

The legs of the machine are planar mechanisms operating in the vertical planes perpendicular to the body (platform) surface. The positions of leg ends are determined relative to the vertical projections of their points of attachment, denoted in the figure by symbol of Δ.

As is shown in Figure 17.10, the vertical projection of the gravity center (denoted by *) always lies within the relevant support triangle.

17.3. Six-legged walking machines

The first studies of six-legged walking machines were undertaken, in several research centers around the world, in the seventies [17.6, 17.9, 17.15].

Figure 17.11 shows the model of a six-legged machine called Hexapod. This machine was 1.3 m long, and 1.4 m wide. It weighed (including its electronic systems) 100 kg.

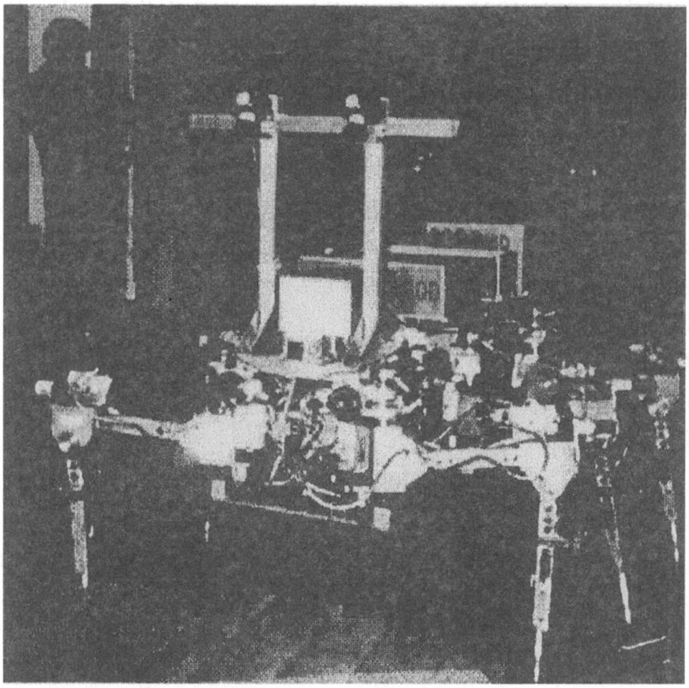

Fig. 17.11. The Hexapod six-legged machine [17.10]

Six-legged machines present the complex problem of how to determine the maximum number of gait schemes. Studies conducted by A.B. Bessonow and others [17.18] have led to the relationships given below:

$$W = \sum_{k=1}^{N} \frac{1}{k}\left[k^N - C_k^1 (k-1)^N + C_k^2 (k-2)^N + \ldots + (1)^{k-1} C_k^{k-1} 1^N\right] \qquad (17.1)$$

where N is the number of legs, and C, the number of combinations

$$C_k^i = \frac{k!}{(k-i)!\,i!} \qquad (17.2)$$

Relationship (17.1) includes all possible combinations of leg functions, such as in jumping (all six legs are simultaneously lifted into the air), and also those combinations when one, or two legs rest on the ground. The general formula for the number of gaits in which less than three legs are in contact with the ground is the following

$$W^i = \frac{N(3N-1)}{2} + 1 \qquad (17.3)$$

Relationship (17.3) is correct for $N \geq 5$. For lower values of N the value of W^i must be found directly. Let W^{st} denote the number of all statically stable gaits

$$W^{st} = W - W^i \qquad (17.4)$$

If $N = 6$, then $W_6^{st} = 1030$. If $N = 8$, then $W_8^{st} = 94\,493$. With such a high number of possible gaits, it is difficult to choose the correct one.

The assumption of leg transfer symmetry reduces considerably the number of possible gaits: as reduction ratio is about 50, the number of gaits is 24. Assuming the principle of wave gait, the number of statically stable gaits is reduced to 7. These results can not be applied without limitations to the machines with greater number of legs.

For specific limitations the number of statically stable gaits for eight-legged machine equals 4. These are the patterns:

a) 16258374| 1
b) 17358264| 1
c) 14725836| 1
d) 16385274| 1

For example, in case a) after lifting leg 1 (after 1/8 of the cycle), the next leg lifted is leg 2, etc.

Table 17.1 gives the number of possible gaits, depending on the number of legs and the value of W^{st}.

Table. 17.1. The number of possible gaits, subject to the number of legs

N	W	W^i	W^{st}
1	1	1	0
2	2	2	0
3	6	6	0
4	26	20	6
5	150	36	114
6	1 082	52	1 030
7	9 366	71	9 295
8	94 586	93	94 493
9	1 091 670	118	1 091 552
10	14 174 322	146	14 174 376

In the last 20 years many different models of walking machines have been built. Figure 17.12 shows a six-legged machine built by M. Ignatiev in 1973, at the Leningrad Aviation Institute [17.7]. The machine was equipped with adaptive control. The control system was a three-level system (see Fig. 17.30). The machine had a round body and leg resembling spider legs.

In 1975 the research team directed by D.E. Ochocimski [17.6] of the Moscow Institute of Aviation Mechanics built a walking machine named T-2 (Fig. 17.13). The control system of the machine (see Fig. 17.31) implemented five functions.

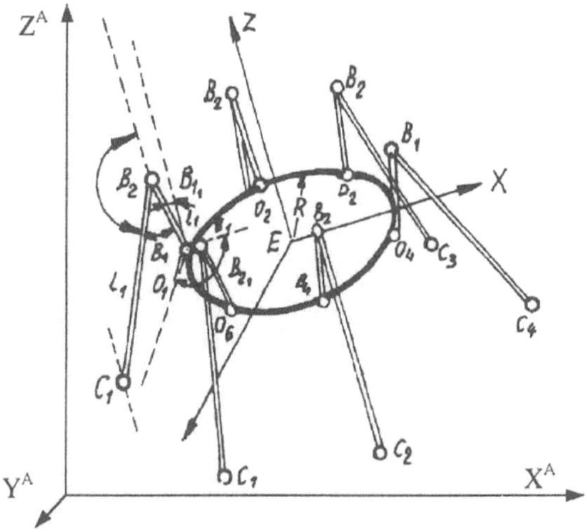

Fig. 17.12. Six-legged machine built by Ignatiev [17.7].

A six-legged machine constructed in one of Japanese robotics laboratories [17.21] is illustrated in Fig. 17.14, which also shows the arrangement of the legs. Each leg of the machine is driven by three motors and controlled by a microprocessor system. The machine is also equipped with force sensors and a compass. The operator can control all six legs simultaneously. The machine is designed for underwater inspection.

One of the more interesting designs is also the six-legged machine built by Odetics (USA) [17.10].

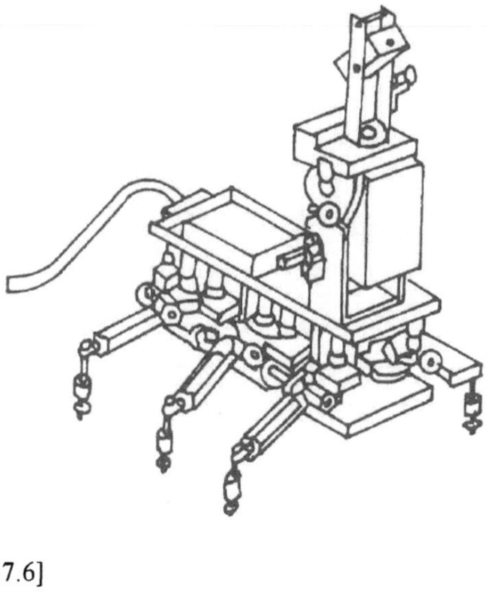

Fig. 17.13. T-2 machine [17.6]

Fig. 17.14. Six-legged walking machine built at the Robotics Laboratory, Machinery Division, of the Port and Harbour Research Institute of the Japanese Ministry of Transport [17.21].

An interesting design of six-legged machine has been proposed by Bessonov (Fig. 17.15). In this design all the joints are prismatic. As the authors maintain, in all the machines built to date the kinetic energy is lost twice during each gait cycle,

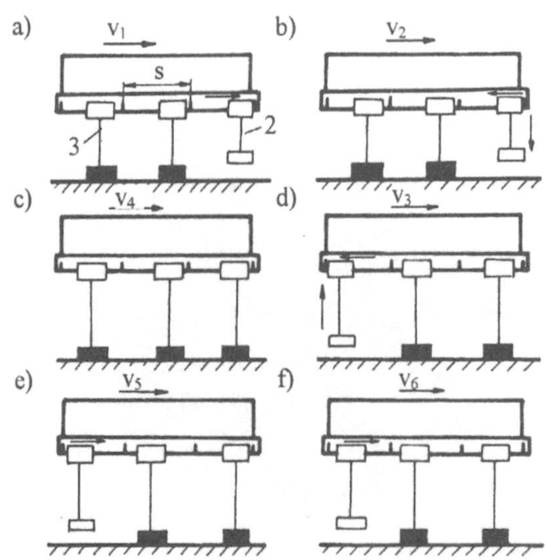

Fig. 17.15. Six-legged machine designed by Bessonov [17.18].

a solution is proposed in which each leg of the machine is stopped at the end of the transfer phase (for example leg no. 2 in Fig. 17.15a).
The sense of the velocity vector then changes, and the leg is lowered (Fig. 17.15b). When the leg touches the ground (Fig. 17.15c), the next leg starts its cycle. The operation of leg 3 is similar (Fig. 17.15d). The leg loses the contact with the ground and is stopped (Fig. 17.15e). Then, after the change of direction of motion the leg accelerates reaching transfer velocity (Fig. 17.15f) [17.18].

The most advanced six-legged machine is the ASV machine designed and built at Ohio State University under the supervision of R. McGhee and K.J. Waldron [17.17]. In the late eighties this machine passed successfully laboratory and field tests. The machine is shown in Figure 17.16.

From the 122 880 possible gait schemes the sequence 1-3-5-2-4-6 was selected, for its locomotion symmetry. The types of gaits, which can be implemented on the machine, were then determined.

Figure 17.17a shows an example of walking down a step with significant body pitch angle. Figure 17.17b shows the machine walking over a narrow wall.

The leg design was proposed after a thorough analysis of various solutions, e.g. four-bar linkage, seven-bar linkage, etc. Finally, the leg was designed as a pantograph mechanism (Fig. 17.18a and 17.18b). The legs are driven by hydraulic actuators. Figure 17.18b shows the view of the leg from its right and left side. The pantograph mechanism ensures that the horizontal actuator causes horizontal movement of the foot. The first drawing from the left shows the situation of the vertical axis relative to which the leg and the actuator are transferred.

a)

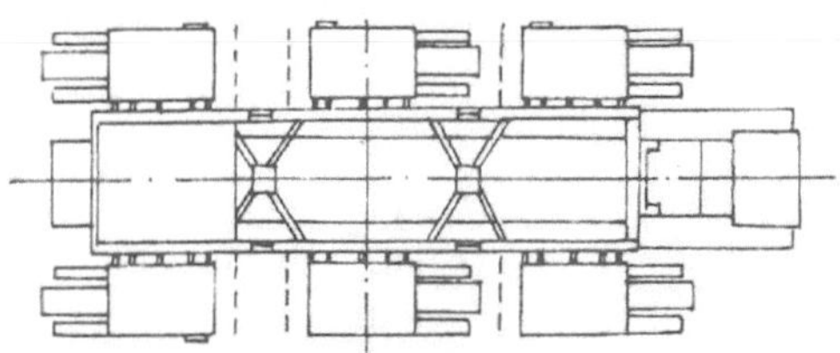

b)

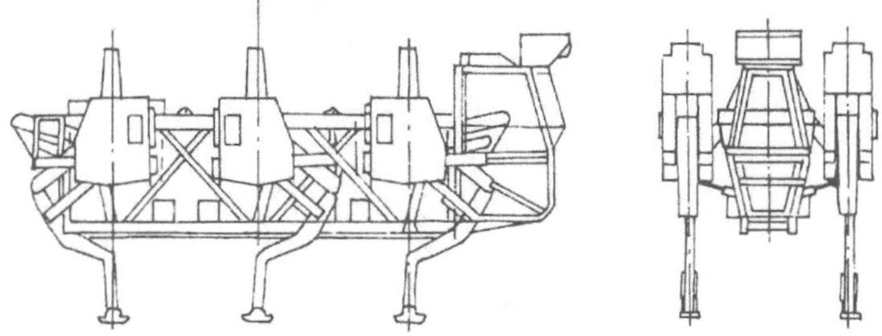

Fig. 17.16. Six-legged walking machine (ASV) [according to 17.17]: a) walking down the step with variable machine body pitch angle; b) walking over a narrow wall.

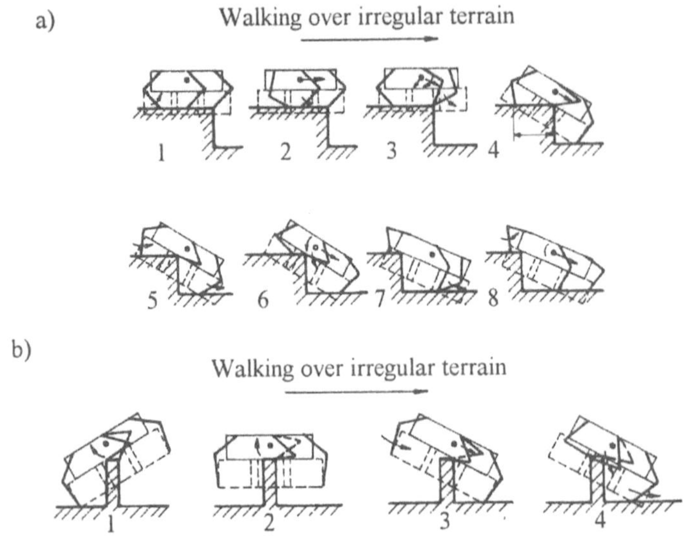

Fig. 17.17. Mobility of the six-legged machine (ASV) [17.17]: a) walking down a step with significant body pitch angle, b) walking over a narrow wall.

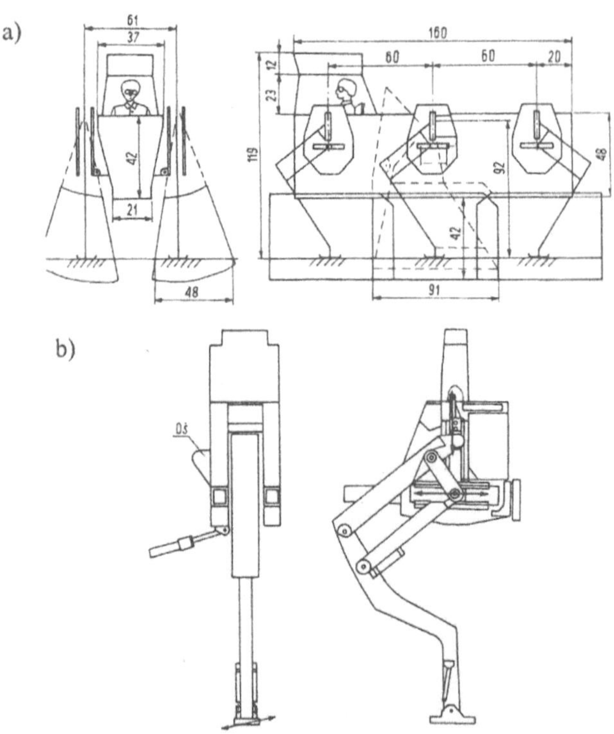

Fig. 17.18. Machine legs [17.17]: a) the pantograph legs of the ASV machine including dimensions in meters; b) a single leg

The pantograph mechanism by actuating three joints of the leg causes three motions: parallel to the longitudinal axis of the machine, perpendicular to the leg plane and rotational motion about the horizontal axis. The machine took its first steps in December 1985. The main dimensions and technical data are given in Table 17.2.

Table. 17.2. Final dimensions and performance of ASV machine

Length	5.2 m
Width	2.4 m
Height	3.0 m
Load	220 kg
Velocity	2.25 m/s
Maximum velocity	3.60 m/s
Leg height	1.8 m
Leg lateral displacement	±20°
Vertical step	2.1 m
Horizontal displacement	2.7 m

The machine is equipped with a system of sensors that enable it to operate on varying terrains. The system measures 82 variable analogue signals which are fed to the control computer in feedback loops. Six additional analogue channels allow the operator to control the machine by means of a joystick. The computer control system is based on the Intel 86/30 processor.

Fig. 17.19. General view of the ASV machine [17.17].

Each computer contains 8086 processors, 8087 coprocessors, 128 kB of memory, 16 A/D ports and 8 D/A ports. The system also contains many auxiliary devices. The output of the computer system contains 18 analogue channels transmitting control signals to the actuators. An additional channel controls the four-cylinder motorcycle engine developing 50 kW of power. The flywheel accumulates 0.25 kWh of energy. Figure 17.19 shows the machine in the laboratory.

17.4. Eight- and more-legged walking machines and special vehicles

Figure 17.20a shows an eight-legged machine. No further closer data is available concerning this machines, other than its name „The Moonwalker". It was built by Space-General Corp. and designed for walking on the moon`s surface. The machine body is supported by eight widely spaced legs grouped at the corners of the body.

Another interesting vehicle was designed by M.W. Thring of Queen Mary College in London [17.19]. The body is equipped with two caterpillars, each fitted with 20 legs. This vehicle can move over a soft terrain and walk up stairs, carrying loads of 70-80 kg (Fig. 17.20b).

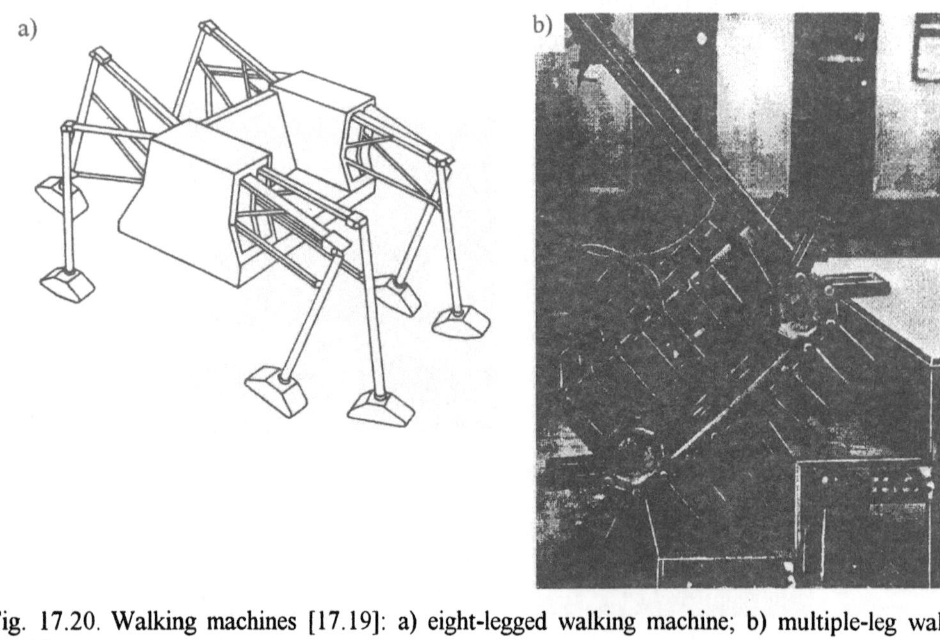

Fig. 17.20. Walking machines [17.19]: a) eight-legged walking machine; b) multiple-leg walking machine.

A separate group of machines are those designed for snake-like locomotion. A model of snake-like vehicle has been built (Fig. 17.21) on the basis of biomechanical studies carried out by Y. Umetani and S. Hirose of Tokio University [17.16]. It is 2 m long and consists of 20 links driven by servomechanisms. New versions of this kind of vehicle have been built in recent years.

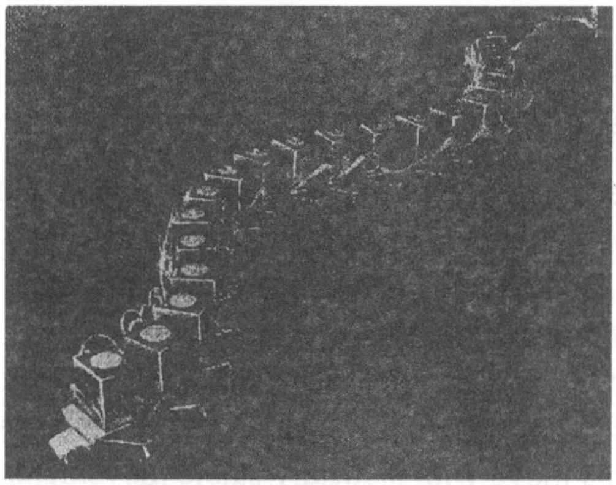

Fig. 17.21. Machine designed for snake-like locomotion [according to 17.16].

a)

b)

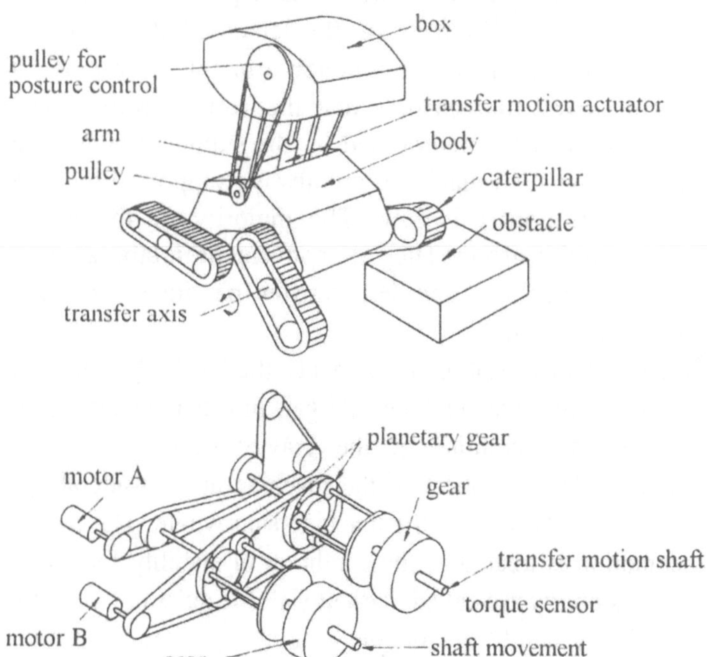

Fig. 17.22. HELIOS-II, designed and built by Japanese scientists: a) the quadratrack vehicle; b) the drive system

Figure 17.22 shows an interesting vehicle built recently by Japanese researchers [17.20]. This is a mobile robot with four caterpillars fitted with legs. It is capable of moving in variable terrain and over obstacles.

17.5. Motion control of walking machines

17.5.1. Achieving position stability

Currently designed machines ensure **static,** or **dynamic locomotion stability.** Control of the motion of statically stable machines prevents the overturning of a motionless machine. Dynamically stable machines maintain then equilibrium by balancing. Statically stable machines have at least four legs (and more commonly six), while the dynamic stability may be achieved by machines with one to four legs.

Stability also depends on the geometry of the limbs, and on the way the machine moves. Insects move with statically stable movements. People and animals such as cats, dogs, or horses maintain dynamic stability when walking(excluding very low-speed movements). Machines with static posture stability are called **static crawlers**. Irrespective of differences in the details of the design and of control strategies these machines all have one common principle of motion. Statically stable machines move in a way that ensures that the vertical projection of the gravity center lies within the polygon determined by the support points (support polygon) [17.16]. This means that such machines are always supported by at least three legs, when the velocity of motion is low. The majority of the machines currently constructed belongs to this group. They are convenient in laboratory investigation of terrain sensing, gait selection, and leg transfer parameters (such as leg stroke, positions of support points, etc.).

Determining the sequence of leg movements, the leg stroke, etc. is a complex task. For machines of the static crawler type, the gait model is generated assuming quasi-static motion, when accelerations of the gravity centers are minor, and when accelerations of individual members of the machine are considerably lower than the acceleration of the gravity center of the machine as a whole. The condition of stability during such locomotion is treated as the condition of stability for a motionless object. Gaits ensuring such stability are called **statically stable gaits** [17.4, 17.6].

The static stability condition for the walking machine is formulated using the notion of „frozen" configuration of the machine: the machine is treated as a rigid body and the configuration includes the body and legs locked relative to the body in the instant of motion considered.

The walking machine is statically stable at instant if its "frozen" configuration is in static equilibrium. This statement assumes that the only forces acting on the machine are the force of gravity and reaction forces from the ground. [17.6].

This condition is usually expressed in the form most practical for gait description.

The machine walking over a level terrain is statically stable in any instant time if the vertical projection of the center of gravity of its "frozen" configuration on the plane of motion lies within the support polygon [17.5, 17.6].

The study [17.16] gives a proof for a theorem stating that the static stability of the machine with forces acting on it will be maintained if the projection of the center of gravity is not closer to the edge of the support polygon than the given distance ξ.

In the modeling of a statically stable gait a given *margin of static stability* must be maintained. This margin is usually defined as the minimum distance of the vertical projection of the center of gravity from the edge of the support polygon. This distance is measured along the direction of motion and during motion, The evaluation of the desired margin takes into account the influence of inertia forces, external forces, and the effects of all simplifications and computational error on the direction of the resultant force.

The definition of the dynamic stability of a walking machine given by M. Vukobratovic is [17.8]: " .. let height H denote the distance between the gravity center of the machine and the ground level and ... given is a set of coordinates permitting precise determination of machine position. *Dynamic stability is a fact, when there is a closed domain R containing a line segment representing the machine height and trajectory of coordinate changes when no disturbances are present.* Domain *R* has the following feature: when a disturbance *D* acts on the machine, the trajectories of coordinate changes will return to domain *R* in a finite time. *D* is a class of allowed disturbances.

17.5.2. Execution-level control

Tasks implemented in the control system can be divided into tasks concerning the planning of walking machine motion (real-time generation of the gait model) and tasks related to execution control.

The system of control on the execution level usually applies principles for the control of the drive system of the machine (e.g. electric motors, hydraulic actuators), aimed at the sufficiently accurate execution of the given gait scheme. On the execution level, control is usually implemented in a closed system (Fig. 17.23), and the control principles depend on the gait modeling principle applied.

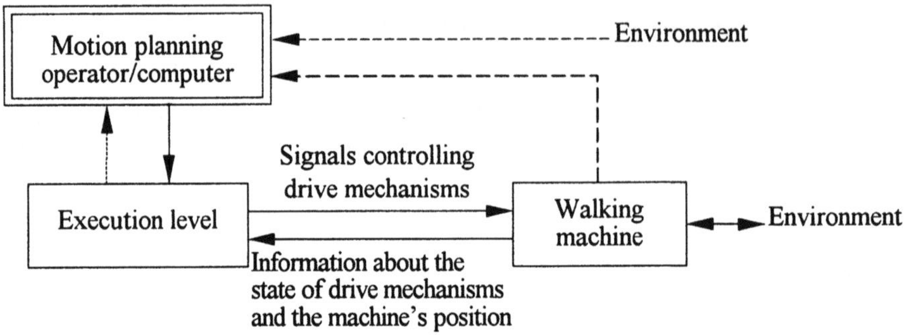

Fig. 17.23. General flow chart of control of walking machines: feedback in the case of closed system control

The first historical principle was **finite state control** (Tomovic 1961, Hemai 1976, McGhee 1977). Control states were defined for the actuators driving the legs for executing in a specific sequence. For example, the following states were defined for electric motors situated in the knee and hip joints, free lifting of the leg, lowering of the leg, pushing the body forward, and pushing the body backward. The feedback in this case usually consisted in checking the states of limit switches or encoders. Transition to the subsequent state was effected when the required position in the previous state had been reached.

Another control principle was **model reference control**. On the execution level this control principle caused the legs to follow given trajectories. These trajectories were usually described as functions of angles measured in the joints (knee and hip). On the grounds of this approach a control system was designed minimizing the following index [17.5]:

$$M = K_p\,(\theta - \theta_{poż}) + K_V\,\dot{\theta} \tag{17.6}$$

where:

θ is the actual (measured) value of angle,

$\theta_{poż}$ is the desired value of angle (taken from the reference model), and

K_p, K_V are constants.

Dynamic phenomena were not taken into account. In quick changes of the reference trajectory coordinates, the execution level of the control system was found to be too slow. For this, and other reasons, the modeling of the whole machine dynamics was proposed, including real-time determination of the driving torque on the basis of the gait scheme and dynamic model. This control option was called **algorithmic control.** Walking machines of the static crawler type, currently designed, are equipped with control systems with a reference model and minimization of control quality coefficients on the execution level. Machines with dynamic stabilization use algorithmic control.

17.5.3. Higher levels of control systems with a reference model

The connections between actions executed by the control system are shown in Figure 17.24. This is a hierarchical system. Actions on the highest level (path and target point selection and motion initialization) are performed by the operator. Actions on lower levels are performed by a suitably programmed control system. The operation of the execution systems also requires various interfaces (e.g. D/A converters) between the control system (computer) and execution mechanisms.

The first model of a machine with dynamic stability was an automatically controlled inverted pendulum [17.4, 17.8]. Studies on the motion of this pendulum permitted the implementation of an one-legged jumping machine [17.4]. A four-legged running machine is currently tested. The control system coordinates the motion of legs, treating them as a single, virtual leg, and the control principles are similar to those of an one-legged machine. The inverted pendulum model is usually the basis for the derivation of control principles for dynamically stable machines.

Two-legged machines are a significant group of dynamically stable walking machines. The control of the motion of these machines is designed using the results of experiments and modeling on human gait. Various techniques are employed, such as the linearization of equations of motion, and stabilization of the posture using the measurements of reaction forces. Extensive information on this subject can be found in [17.2] and [17.8].

The strategy of motion control is an important factor in the designing of a walking machine. Control strategy determines the type of locomotion (gait scheme), machine stability, and power consumption of the system.

Artificial neural networks, which can be taught the process of motion control, are a modern alternative to traditional control systems.

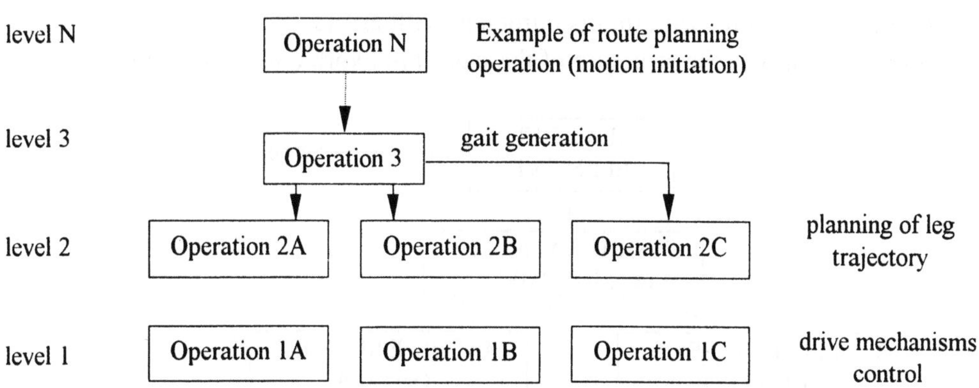

Fig. 17.24. Operations performed by a control system [17.7]

As defined by I. Alexander (1991) the *artificial neural network* is a system that can store knowledge gained experimentally. The operation of such a network resembles the operation of the brain, because the knowledge is collected by training, not by programming, and the network may generate relationships which cannot be described analytically.

The six-legged walking machine LAURON [17.22] has a control system based on neural networks. This machine, built at the University of Karlsruhe (Germany) moves with speeds of 2 to 4 kph. The machine measures 70 cm in height, and weights about 9 kg.

The control system has a hierarchical structure. The lowest level contains two controllers for the phases of the lifting and lowering of the legs. Each controller is built on a neural network. One controls all the legs being lifted (transferred); the other one controls all the legs in contact with the ground (supporting the machine).

When the leg being transferred finds an obstacle, the controller changes the trajectory of the leg end to avoid the obstacle.

The next level of the control system is responsible for the coordination of leg movements. It synchronizes the movements of the individual legs in order to execute the desired gait scheme and leg movement sequence.

The highest level contains a *reactive element,* also based on a neural network. The task of this element is to select the gait scheme and direction of machine walk using information read from the sensors.

Each neural network of the control system has been taught to perform its specific task. Teaching consisted in supplying the reference standards to the inputs and generating the corresponding output states. The network, by proper adjustment of its internal parameters, learned the correct generation of outputs for the given inputs. The main advantage of a properly taught neural network is that it can correctly generate outputs for inputs not used in the teaching process. The network can also operate correctly with inputs burdened with errors.

Figure 17.25 shows the structure of the neural network control system.

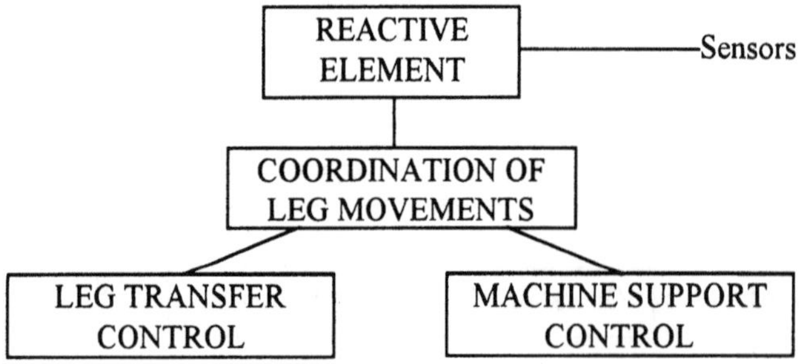

Fig. 17.25. Structure of the control system based on neural networks

17.5.4 Mathematical model of a static crawler

17.5.4.1. Assumptions

The four legged walking machine MN-4 (Fig.17.26) is an example of static crawler. Each leg is driven by two linear hydraulic actuators (h_1, h_2) and can move only in one plane. The mathematical model [17.7] of the machine was developed to facilitate description of gait. Special attention was paid to the description of actions of the control system.

The following assumptions were made:

- the control of motion is realized without feedback (from the environment; information about the state of the environment is deterministic),
- the model describes a machine of the static crawler type (the described machine moves with statically stable gaits),
- model reference control is used (with digital controller),
- the machine's model was divided into three parts: the mechanical structure (legs, body, and actuators), control, the interface between the mechanical structure and the control,

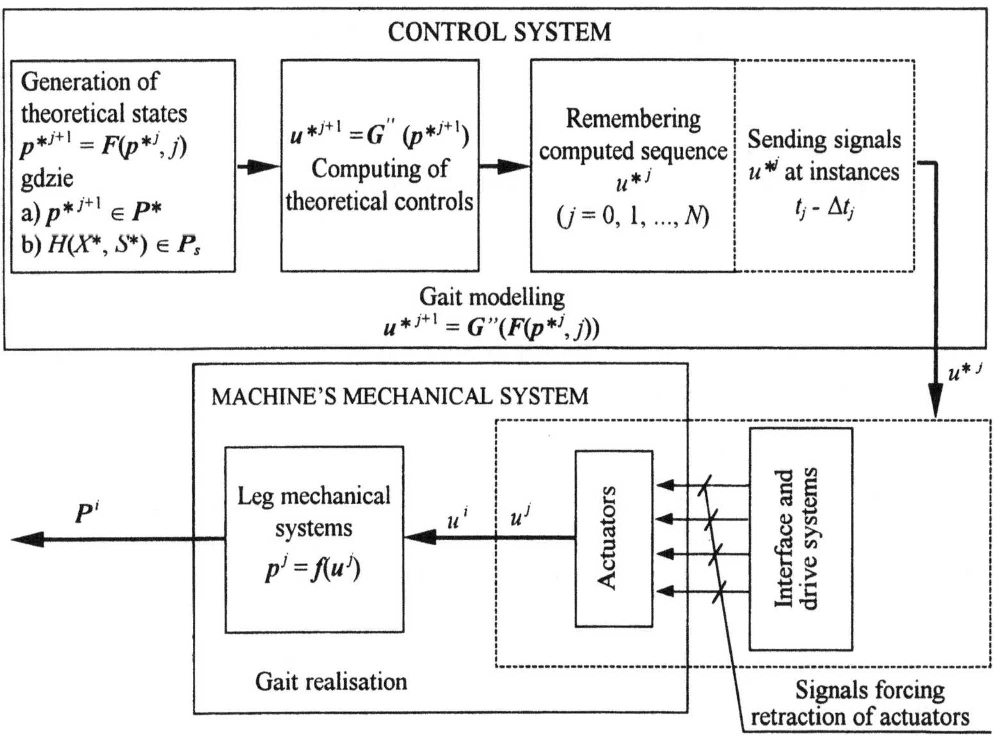

Fig. 17.26. Mathematical model of the walking machine

- the mechanical structure executes the gait, whereas controls organize the functioning of the mechanical structure, so that the realization of the gait is possible,
- the control system sends the control signals in discrete instants of time.

17.5.4.2. Mechanical structure

The mechanical structure is the controlled object. Its goal is the execution of gait. The adequately defined state of the mechanical part is called the *real state.* The *real state* of the i-th leg at instant f is defined by a set of coordinates univocally describing the position of the leg end. This state is denoted by:

$$\boldsymbol{p}_i = \boldsymbol{p}_i(t),\ i \in \{1, 2, 3, 4\} \tag{17.7}$$

where $\boldsymbol{p}_i \in \boldsymbol{P}_i$

$\boldsymbol{P}_i$ is a bounded set of values of p_i (resulting from construction constraints). The real state of the leg is represented by the coordinates of its tip end, expressed in a Cartesian coordinate frame fixed to the body, so:

$$\boldsymbol{p}_i = \boldsymbol{p}_i(t) = (x_i(t), y_i(t), z_i(t)) \tag{17.8}$$

The *real state* of the walking machine is described by the states of its legs:

$$\boldsymbol{p} = \boldsymbol{p}(t) = (p_1(t), p_2(t), p_3(t), p_i(t)) \tag{17.9}$$

where $\boldsymbol{p} \in \boldsymbol{P}$, $\quad \boldsymbol{P} = P_1 \times P_2 \times P_3 \times P_4$

The position of all actuators is called the *real control* of the mechanical system

$$\boldsymbol{u} = \boldsymbol{u}(t) = (u_{11}, u_{12}, u_{21}, u_{22}, u_{31}, u_{32}, u_{41}, u_{42}) \tag{17.10}$$

where:

$\boldsymbol{u} \in U, \quad U = U_{11} \times U_{12} \times U_{21} \times U_{22} \times U_{31} \times U_{32} \times U_{41} \times U_{42}$

U_{ki} – is the set of position of k-th actuator fixed to the i-th leg

$(k \in \{1, 2\}, \quad i \in \{1, 2, 3, 4\})$

The real state of a walking machine is a function of $\boldsymbol{u}$:

$$\mathrm{p} = f(\boldsymbol{u}) \tag{17.11}$$

The function $f(\boldsymbol{u})$ describes the relationship between the actuator`s position and the positions of leg ends, as realized by the mechanical system.

17.5.4.3. Control system

The state of the machine calculated for the instant t_j of simulation time is called the *discrete theoretical* state and denoted by $\boldsymbol{p}^{*j} = \boldsymbol{p}^*(t_j)$. The theoretical control signals $\boldsymbol{u}^{*j}$ are related to the theoretical states in the following manner:

$$\boldsymbol{u}^{*j} = G''(\boldsymbol{p}^{*j}) \tag{17.12}$$

The form of function G'' depends on the geometry of the legs. The control system causes a statically stable gait. If we let:

$$X^* = (\boldsymbol{p}^{*0}, \boldsymbol{p}^{*1}, \boldsymbol{p}^{*2}, \dots, \boldsymbol{p}^{*N}) \tag{17.13}$$

be any trajectory of discrete theoretical states of the walking machine,

$$S^* = (S^*(t_j)) \tag{17.14}$$

is a sequence of disturbances changing the conditions of the modeled motion at instants t_j of simulation time ($j \in (0, 1, 2, ..., N\}$).

The disturbances are interpreted as factors, such as variations in the slope of the terrain, causing changes in the static stability margin. A function $H(X^*, S^*)$ of constraints must to be introduced. Let D_s^* be a non-empty subset of a set of theoretical states, for which the conditions asserting the static stability of gait introduced by the constraints functions are fulfilled. This can be written as:

$$H(X^*, S^*) \in D_s^*, \tag{17.15}$$

for $D_s^* \subset \boldsymbol{P}^*, D_s^* \neq 0$

The functioning of the control system (*generation of gait model*) can be described as:

$$\boldsymbol{p}^{*j+1} = F(\boldsymbol{p}^{*j}, j), j \in \{0, 1, ...\} \tag{17.16}$$

Function F is such that the states calculated according to (17.16) fulfill the following conditions:

a) $\boldsymbol{p}^{*j} \in P^*$, the condition constraining the values of state variables, resulting from the constraints imposed on motions of leg ends depending on the machine's construction,

b) $X^* = (\mathbf{p}^{*0}, \mathbf{p}^{*1}, \mathbf{p}^{*2}, ..., \mathbf{p}^{*N})$, the condition of static stability of gait:
$H(X^*, S^*) \in D_s^*$,
where:
$D_s^* \subset \boldsymbol{P}^*, D_s^* \neq 0$

The gait modeling task consists in calculating a set of functions F such that the states $\boldsymbol{p}^{*j}$, generated according to relationship (17.16), fulfill the constraint conditions a) and b).

17.5.4.4. Model of the machine

The control system sends the digital control signals $\boldsymbol{u}^{*j}$ at predefined intervals of time. The signals containing information about $\boldsymbol{u}^{*j}$ are transmitted at instants $t_j - \Delta t$ of real time. Δt is the anticipated interval of time lag between initiating the transmission of signals $\boldsymbol{u}^{*j}$ and realization of state $\boldsymbol{p}^j = \boldsymbol{p}(t_j)$ of the mechanical system. Taking into account the accepted models of the machine's systems (control system, interface and mechanical system), the real state $\boldsymbol{p}^j$ of the machine at instant t_j can be described as a superposed function of the theoretical state $\boldsymbol{p}^{*j}$ modeled in the control system:

$$\boldsymbol{p}^j = f(G'(G''(F(\boldsymbol{p}^{*j})))) \tag{17.17}$$

The form of function F depends on the gait rules (when conditions a) and b) are fulfilled); function G'' is defined by the geometry of the legs. The functions F and G'' can be represented in an analytical form. Functions G' and f depend on the method of generating control signals of the executive level, the design of the machine, the current conditions of motion (environmental influence), and the operating particulars of the mechanical system. Analytical definition of functions G' and f is impossible, but if necessary, certain approximations can be obtained. The real states p^j differ from theoretical states p^{*j}. The task put before the control system is: modify function G' (when the position of the actuators are measured) or the superposition $f(G')$ (when coordinates of leg ends are measured), so that states p^j will he sufficiently close to states p^{*j}.

17.6. Control system of the MK-4 machine: realization

17.6.1. Structure of the control system

The system has five levels of control to provide the detail required by motion task specification.

At the lowest (*joint*) level the motions of all links of the limb are controlled. At the next (*leg*) level the positions of leg ends will suffice. At the still higher (*gait*) level the gait formula (a coordinated method of placing the limbs) must be supplied. At the fourth (*body*) level the trajectory of the walking machine's body must be specified. At the highest (*task*) level only the task to be executed by the walking machine must be indicated. A single function of each of the control levels enumerated can be decomposed into a sequence of functions of a lower level. Information regarding the current state of the machine is fed back to the joint level of the control system (Fig. 17.27).

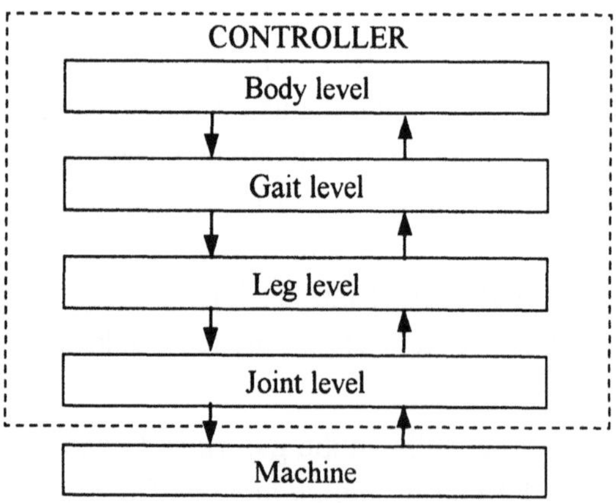

Fig. 17.27. Hierarchical structure of walking machine controller

We began building the control system at the gait level, specifying the task of the control system as: move to a point in the environment (the trajectory of the body was fixed; in initial study it was a straight line). The control system's task was to find the possible gait and evaluate its parameters. After the values of leg-end coordinates and actuator lengths were computed.

17.6.2. Gait level of the control system

17.6.2.1. Features of the modeled motion

The machine's gait is described by a set of *gait parameters* (e.g. step length) which depend on the type of gait modeled, on the state of the environment, and on the conditions of motion formulated by the operator. To date the principles of gait parameter selection developed, take into account:

- variations in the slope of the terrain, but the angle of slope must be in one plane (the machine can be higher in front than in back and vice versa),
- variations in the speed of the machine's motion (in particular, stopping and starting),
- motion along straight lines.

In situations when a gait with a fixed sequence of leg transfers cannot be realized, a free gait is modeled. During the process of free gait planning the control program searches the tree of permissible machine-motions several times, detrmining a sequence of motions leading to leg positions such that the predefined type of gait can be continued.

17.6.2.2. Evaluation of gait parameters

The type of gait is identified by a predetermined sequence of leg transfers, and its parameters are necessary for the construction of the gait model. The values of parameters of statically stable motion depend on the following factors:

M, the geometrical parameters of the machine (body length, height, distance of center of gravity from the ground, minimum and maximum swing of the legs backward and forward, measured at the point of connection with the body),

d, the parameters resulting from the relationships between the dimensions type of gait (the distance between footsteps of the legs measured along the diagonal of the body),

X_C, the coordinates of the machine's center of gravity (in the considered reference frame).

The predefined conditions of motion:

V – velocity,

M_s, acceptable static stability margin,

S, environmental influences (the changing slope of the terrain).

The values of all gait parameters can be expressed as functions of arguments represented the elements of the above mentioned list of factors. Moreover, all the other parameters depend on the single parameter of the *length of step K*. This is measured between two consecutive footprints left by the same leg. During a constant type of gait the steps of all legs are identical in length. Step length K, for a statically stable gait is such that it fulfills the relationship:

$$K \in \bigcap_{i=1}^{n} f_i \qquad (17.18)$$

where

f_l: $K \leq f_l(M, d, X_c, M_s, S)$, $l \in \{1, 2, ..., n\}$

f_m: $K \geq f_m(M, d, X_c, M_s, S)$, $m \in \{1, 2, ..., n\}$, $m \neq l$

The form of functions f_i is determined by stating the static stability condition for the motion and taking into consideration the coordinates of leg ends. This condition is formulated for all of the gait phases identified. The number of inequalities depends on the type of gait and the considered constraints of leg motion space. The algorithm determining relationships (17.18) is relatively simple, but time consuming.

After calculating, according to (17.18), the values of parameter K, the remaining gait parameters are evaluated:

$$GP = F_c(M, d, X_c, M_s, S, K) \qquad (17.19)$$

where GP is the set of remaining gait parameters (excluding K)

Different GP sets can be constructed, because there different quantities can be used in the description of gait. The form of function F_c depends on the set selected.

We define function F_c for the following set of GP:

T is the gait period (the duration of one sequence of leg transfers),

X_{pmax}, X_{tmin} are the maximum leg swings at the points of the legs connections to the body (X_{pmax} is the foreleg swing, X_{tmin} , the hind leg swing),

X_p, X_t are leg coordinates at instants when the static stability margin is equal to M_s (X_p , for the foreleg; X_t , for the hind leg),

r_0 T is the time when the leg remains on the ground during one sequence of leg transfers.

These parameters, together with step length K, suffice for the creation of a univocal description of the gait.

17.6.3. Leg level of the control system

Having determined the gait parameter values, the machine's theoretical states are generated. The rule of generating the theoretical states is formulated in the following manner:

$$\text{Let } N = \frac{T}{\Delta t} \tag{17.20}$$

where Δt is the simulation time interval between the consecutive instants of theoretical state generation, and T is the gait period. The following relationship is created:

$$\boldsymbol{p}^{*j+1} = \boldsymbol{p}^{*j} + w^j, j = 0, 1, 2, ..., N \tag{17.21}$$

where:

$$\boldsymbol{p}^{*0} = \boldsymbol{p}^{*N},$$
$$\boldsymbol{p}^{*j} = \boldsymbol{p}^{*j}(K, GP),$$
$$w^j = (K, GP)$$

The sequence of generated states determines the walking machine's statically stable motion.

17.6.4. Joint level of the control system

17.6.3.1. The control principle: indirect method

The position of the machine's leg is evaluated on the basis of the distance between the leg-end and two fixed points of the body. The distances are measured by ultrasonic range finders (Fig. 17.28). The calculation of the values of leg-end coordinates on the basis of the distances measured is fairly easy (deduced from the Pythagorean theorem).

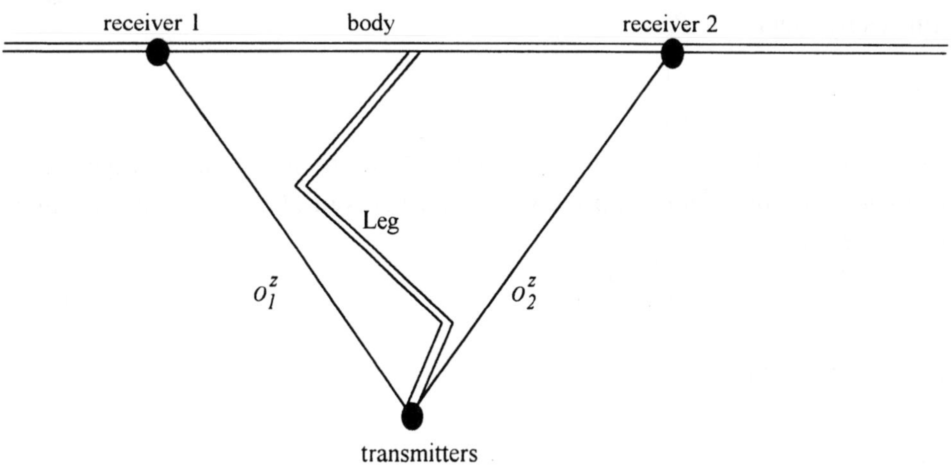

o_1^z, o_2^z – measured distances

Fig. 17.28. Ultrasonic range finder system

Values o_1, o_2, the distances between the transmitters and the receivers of ultrasonic waves, are the output quantities of the system generating the model of the machine's gait. These values are the values set for the executive level of the control system. During the control process the positions (lengths) of actuators are set incrementally. In each step of the control process, after comparing the measured values o_1^z, o_2^z with the set values o_1, o_2 the actuator positions (lengths) which have to be achieved in the next step of control, are evaluated. The proposed control method does not require evaluation of the leg's kinematic model. Nevertheless, this model was used to determine the indirect control rule.

17.6.4.2. Relationships

Let us introduce the following denotations:

h_1, h_2 are the lengths of the prismatic actuators 1, 2 driving the leg,

x, z, coordinates of a leg-end, expressed in a Cartesian coordinate frame of reference attached to the body of the machine,

o_1, o_2, distances of ultrasonic receivers from the transmitters.

The lengths of actuators are related to the coordinates of the leg-ends (direct kinematic problem):

$$h_1 = \phi_1(x, z), \qquad h_2 = \phi_2(x, z) \tag{17.22}$$

Using the vector notation we obtain

$$H = \Phi(x_z), \qquad \left(H = \begin{bmatrix} h_1 \\ h_2 \end{bmatrix}, \quad \Phi = \begin{bmatrix} \phi_1 \\ \phi_2 \end{bmatrix}, \quad x_z = [x \quad z]\right) \tag{17.23}$$

The distances of ultrasonic receivers from transmitters are related to the leg-end coordinates by functions Ψ:

$$0 = \Psi(x_z), \qquad O = \begin{bmatrix} o_1 \\ o_2 \end{bmatrix}, \qquad \Psi = \begin{bmatrix} \psi_1 \\ \psi_2 \end{bmatrix} \tag{17.24}$$

The evaluation of the total differentials of functions Φ and Ψ with respect to x_z allows us to state the relationship between increments of H and O, and increments of coordinates x_z:

$$dH = \frac{\partial \Phi}{\partial x_z} dx_z, \qquad dO = \frac{\partial \Psi}{\partial x_z} dx_z \tag{17.25}$$

The above relation implies that:

$$\frac{dH}{dt} = J_1(x_z)\frac{dx_z}{dt}, \qquad \frac{dO}{dt} = J_2(x_z)\frac{dx_z}{dt} \tag{17.26}$$

where J_1, J_2 are the Jacobian matrices. During the control process, the displacement of actuators is determined by comparing the set values o_1, o_2 and the measured values o_1^0, o_2^0. To allow the determination of the control values and actuator displacements, in relation to the measured values O, we must to find the functions $\frac{dH}{dt}(\frac{dO}{dt})$. Taking into account (17.26) we obtain:

$$\frac{dH}{dt} = J_1(x_z)J_2(x_z)^{-1}\frac{dO}{dt} \tag{17.27}$$

The matrix $J_2(x_1)$ is non-singular.

17.6.4.3. Control law

In the control process the following function is minimized:

$$B = \gamma_1(o_1 - o_1^0) + \gamma_2(o_2 - o_2^0) \tag{17.28}$$

where γ_1, γ_2 are constant weighting coefficients. We assume that the desired velocity is equal, in each control step, to the difference between the desired displacement (o_1, o_2) and the measured displacement (o_1^0, o_2^0) divided by an adequate time increment:

$$\left(\frac{o_1 - o_1^0}{\Delta t}, \frac{o_2 - o_2^0}{\Delta t}\right) = \frac{\Delta O}{\Delta t} \tag{17.29}$$

The control signals, determining the desired changes of H depending on $\frac{\Delta O}{\Delta t}$, are transmitted every Δt units of time. The time Δt is called the time of realization of a single control step. During each step the actuators are displaced by ΔH, so they must move at a velocity of $\frac{\Delta H}{\Delta t}$.

During the control we take into account the relationship $\frac{\Delta H}{\Delta t}(\frac{\Delta O}{\Delta t})$. If formula (17.27) were used directly the computation required each time would be excessive (the Jacobian matrix $J_1(x_z)$ contains inverse trigonometric functions). But detailed analysis of relationship (17.27) allows us to identify the linear factors of $J_1(x_z)J_1(x_z)^{-1}\frac{dO}{dt}$ that determine the sign of expression. Moreover, we have found that the absolute value of the terms $\frac{dO}{dt}$ is dominant in the right hand side of expression (17.27). Finally, in the control we consider:

$$\frac{\Delta H}{\Delta t} = sign\left(linear\left(J_1(x_z)J_2(x_z)^{-1}\frac{\Delta O}{\Delta t}\right)\right)\left|\frac{\Delta O}{\Delta t}\right| \tag{17.30}$$

where *sign(linear...)* is the sign of linear terms of the argument, and $|\frac{dO}{dt}|$ is the absolute value of the argument. The values o_1^0, o_2^0 are measured during the i control step. These values are the basis for the evaluation of (17.30) for the $i+1$ step. The next values of o_1, o_2 are transmitted when the absolute values of differences ΔO become smaller than certain constant values, or, if this does not happen, when k control steps have been performed.

The classical control method (utilizing the kinematic model of the leg directly) makes use of the following relationship instead of formula (17.29):

$$\frac{\Delta H}{\Delta t} = \frac{\Phi(x_z) - \Phi(x_z(o_1^0, o_2^0))}{\Delta t} \tag{17.31}$$

where $\Phi(x_z)$ is the desired value of actuator displacement, and $\Phi(x_z(o_1^0, o_2^0))$ the real value, computed by taking into account the measured values o_1^0, o_2^0 and introducing them into function Ψ (17.23), (17.24). The values of $\frac{\Delta H}{\Delta t}$ computed with the help of formula (17.31) are the values set for the *i-1* control step, where the values o_1^0, o_2^0 are the results of distance measurements obtained in the *i-1* control step. The values of $\Phi(x_z)$ transmitted from the hierarchically higher levels of the control structure remain unchanged until the absolute value of the difference $\Phi(x_z)$ – $\Phi(x_z(o_1^0, o_2^0))$ drops, below a certain constant. If this condition is not satisfied, during k control step, the newly set values $\Phi(x_z)$ are transmitted. Before experimenting with the real machine, simulation trials were conducted to compare the proposed method with the classical method. The results obtained proved that in the case of leg-end motion from point to point there is no significant difference in the trajectories obtained. When a predetermined trajectory is followed, the indirect control method results in greater deflection from the specified trajectory points. These deflections can be considerably reduced by increasing the number k of steps k performed for a single set of values $\Phi(x_z)$. As the values of Ψ^1 (17.24) and $\Phi(x_z)$ (17.23), which contain inverse trigonometric functions, do not have to be computed repeatedly, we could expect that the indirect method be much more efficient than the classical method. The slowness of the classical method, resulting from its computational load (which resulting turn in a small number of control steps for a single value of $\Phi(x_z)$) may actually introduce greater errors than indirect control.

17.6.5. Conclusions

The method of gait description and its parameter determination can be used for any statically stable motion. The method of theoretical state generation mentioned above yields a relationship between the machine's geometric parameters (body length, machine height, the range of motion of each leg) and the gait parameters. These relationships are instrumental in the design of a machine performing an assumed type of motion, or when the types of motion executed by a given machine have to be determined.

The initial investigations of the real control system allowed us to assume that the indirect control method would be effective. This method did not prove less accurate then the classical control method in following a given trajectory, while computations required in a single control step, it are much less time consuming. The identification of the properties of the hydraulic control path is very important, as it affects control accuracy. Introducting the autocorrection of the characteristics of this path is a significant advance. The feedback path (the ultrasonic distance measurement system) should take into account the possibility of statistical errors (single disturbance, systematic error).

17.7. Computer controlled machines

17.7.1. Devices with constant gait program

The first machine to walk under the computer control was the Phoney Pony, also called the Californian Horse. Its designers, R. B. McGhee and A. A. Frank, presented this walking machine in 1966. The device, of the height of a small pony and weighting 50 kg, was connected to a computer. The computer sent signals controlling the electro-motors actuating the legs, which performed sequences of identical movements. So Phoney Pony was a computerized version of a walking machine with a constant gait program.

The work of R.B. McGhee and his team on quadruped gait specification was terminated in 1973 with the presentation of a supervisory remote control of the Californian Horse (Fig. 17.29). The device could now perform four different types of motion: crawling, walking, trotting and galloping. The software of the control computer consisted of two elements (Fig. 17.29):

- a generator of kinematic commands,
- a generator of forces and torques.

The appropriate type of motion was chosen, depending upon the velocity selected by an operator, by the kinematic command generator. The generator of forces and torques converted the coordinates of leg-ends given by the kinematic command generator into signals controlling the machine's drive system. Sensor made it possible to "trace" the positions of the ends of the machine's limbs. Despite this increase in its motion capabilities, Phoney Pony remained a device with a constant gait program; the change of gait could be triggered only by the generator.

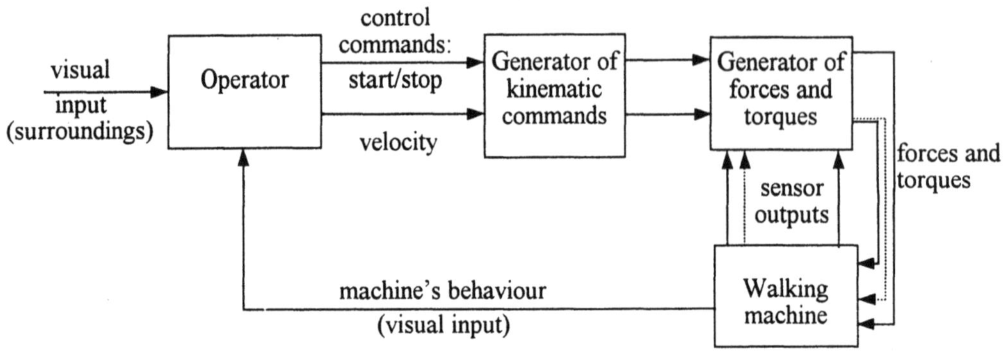

Fig. 17.29. Flow chart of Californian Horse's control system [17.13]

Another machine with a constant gait program is the quadruped device presented in 1975 by K. Ikeda, K. Tagushi, and S. Matsumoto from the Mechanical Engineering Company. The device, which had pneumatic drive, could move only with a four-legged crawling type of gait. The legs were so simple in their design that the motion could be divided into proper vertical and horizontal displacements. The control program simply generated sequences of these displacements. The design of this device and the concept of its control system was the result of a trend towards maximum simplification of the control algorithm, and played a considerable role in the research on methods for minimizing operating costs such as, for example, further presented attempts to minimize the expenditure of energy for motion. The assumed concept of a constant program of gait entailed no contribution to the common search for ways of increasing the motion capabilities of walking machines.

A six-legged machine T-1, designed by Russian scientists in 1975, moved only with constant velocity. Built by the team of D. E. Ochocimski, this machine was an experimental piece in a research project to expand motion capabilities, and was used to evaluate the efficiency of programmed control of leg movements [17.6]. In this case, the control device was an analogue machine.

17.7.2. Machines with adaptive gait programs

A. Simple experimental devices

The computer, used as the control system, made it possible to expand of gait programs. Since the seventies, machines that can, to certain degree, adjust their movements to the terrain, have been designed. New notions of *terrain adaptability* and *adaptive control* have appeared. Some scientists have investigated various control concepts using simple experimental devices; the models of walking machines are designed. This is the course of research taken by Ochocimski and his team. In 1971 they presented the concept of a multi-level control system. Particular levels were to realize the following functions [17.6]:

- selection of body trajectory (the highest level),
- selection of the type of motion (backward, forward, turning),
- selection of step points,
- leg movement coordination,
- leg movement control (the lowest level).

The performance of the lowest control level has been verified in a model, called ricksha, built in 1973, – a machine with two legs and two wheels. A control system was used here to determine the way of generating the gait type, as well as finding an effective method for converting the coordinates of the leg-ends in quantities that could be interpreted by the leg actuating system. The second version of this device, with two legs and an underslung body, was used to observe the position of the body and leg behavior while moving on uneven terrain. Those observations have been taken into account in designing operations at higher levels of the control system, as well as in determining methods of telereception in walking machines.

The results of these experiments were exploited in the construction of two six-legged walking machines: the T-1 and T-2 machine presented above (Fig.17.13).

B. Walking machines – laboratory prototypes, models

One of the earliest model designed was the six-legged walking machine ESA UNO. M. Petternella and S. Salinari worked on this device at University of Rome, from 1969 to 1979. The dedicated control system of ESA UNO had a hierarchic structure, with the following functions attributed to the different levels:

- decision making (at the highest in hierarchy) – corresponding to a driver who chooses the body motion direction and velocity,
- an algorithm (at the next level), responsible for the selection of gait type within the realm of the machine's motion capabilities and dependent on the power of the leg drive motors,
- gait coordination (at the third level), controlling limb movements performing the gait during transition phases, and when changing direction of motion,
- dynamics (at the fourth level), converting the leg-ends motion trajectory into the signals controlling the servo-mechanisms of the executing part,
- a servo-mechanism (at the fifth level), the machine's mechanical structure and its drive system (motors).

The design of the system's structure is shown in Figure 17.30. Work on the software for its two lowest levels was completed in 1973. The work on ESA UNO was abandoned year later [17.7], leaving no data regarding methods for realizing the functions of the remaining levels of the control system. Nevertheless, this project is one of the first to take into consideration the task of adapting to the conditions of the terrain to be done by an appropriate cooperation between the levels of coordination and dynamics.

The notion of an adaptive multi-legged vehicle appeared for the first time in 1973 in the description of a six-legged walking machine designed by the team of M. Ignatiev at the Leningrad Aviation Institute. Ignatiev designed a three level control system for an adaptive machine, to provide for the following tasks:

- the third (*highest*) level determined motion velocity and direction, as well as the type of gait; it cooperated with an operator;
- the second level was responsible for the realization of the gait, that is, it coordinated leg movements and worked out the data needed to achieve a given type of gait,
- the first level generated signals for the execution mechanisms.

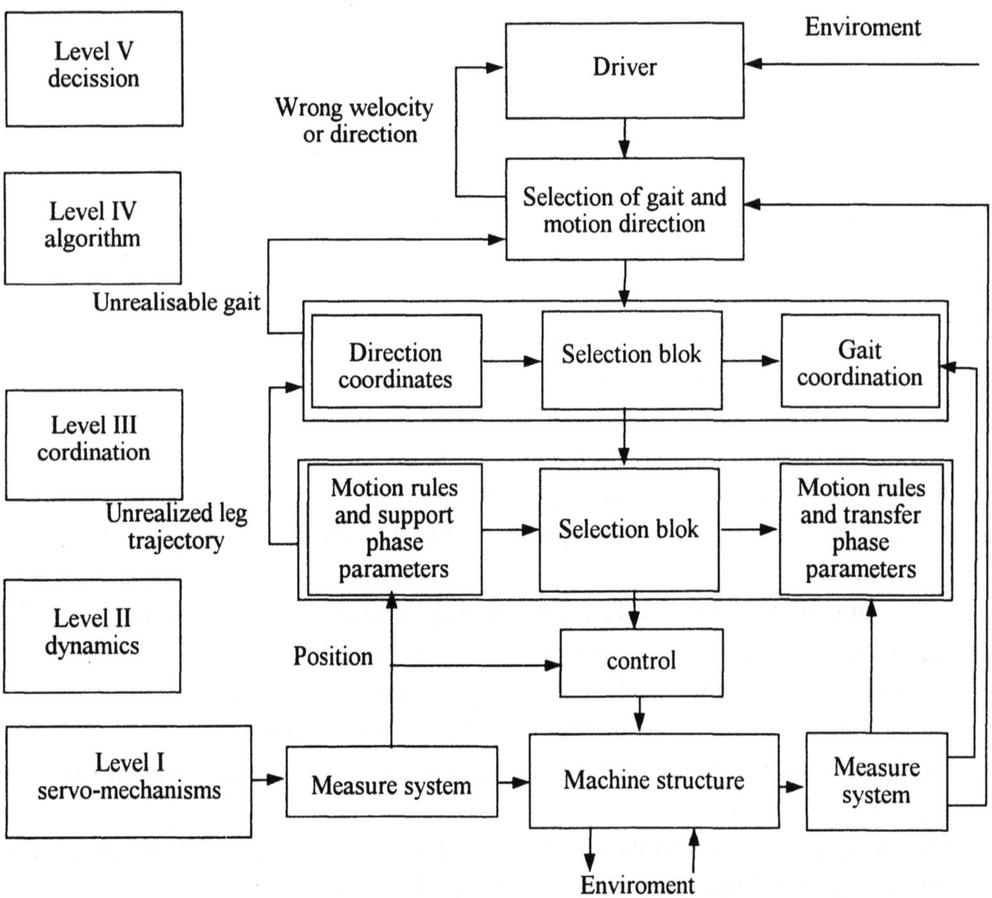

Fig. 17.30. Multi-level flow chart of the control of ESA UNO machine [17.7]

The adaptation to terrain conditions was achieved by practice relied on maintaining the machine's center of gravity on a certain trajectory. This meant that in overcoming terrain obstacles the legs had to bend in a way that kept the body parallel to the horizontal plane. This was done by the second level of the control system. The machine's body was round, and its legs resembled those of a spider. This exceptional design made it easier to realize the assumed adaptation concepts. Ignatiev presented the machine's description in the form of a mathematical model, closely connected with the machine's design (e.g. the body shape, the number of legs). The adaptation in the computational sense relied on modification of the model's parameters (Fig. 17.31).

The next version of the control system, presented in 1972 in the T-2 machine (Fig. 17.32) was more versatile. This machine was built by the team of Ochocimski, working at the Moscow Institute of Aviation Mechanics.

The operation of the control system was determined on the grounds of an analysis of the machine's kinematic model. This kinematic description can be applied to any multi-legged machine. Moreover, as it was later seen, it allows further expansion of the control system. The presented version of the control system provided for the five functions, (listed for the design of the experimental devices).

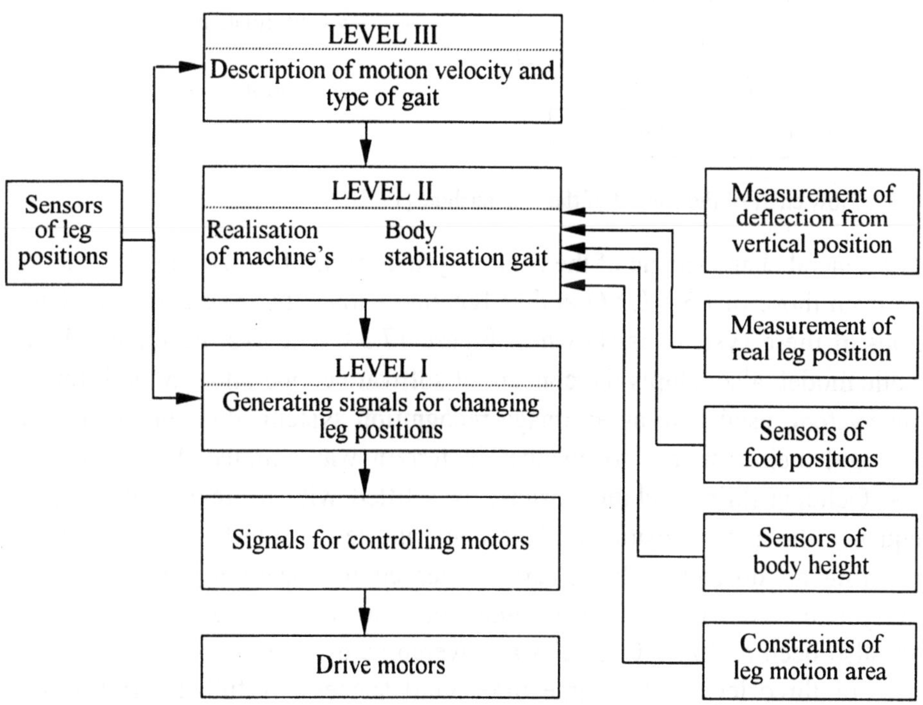

Fig. 17.31. Three-level flow chart of the machine control designed by M. Ignatiev [17.7]

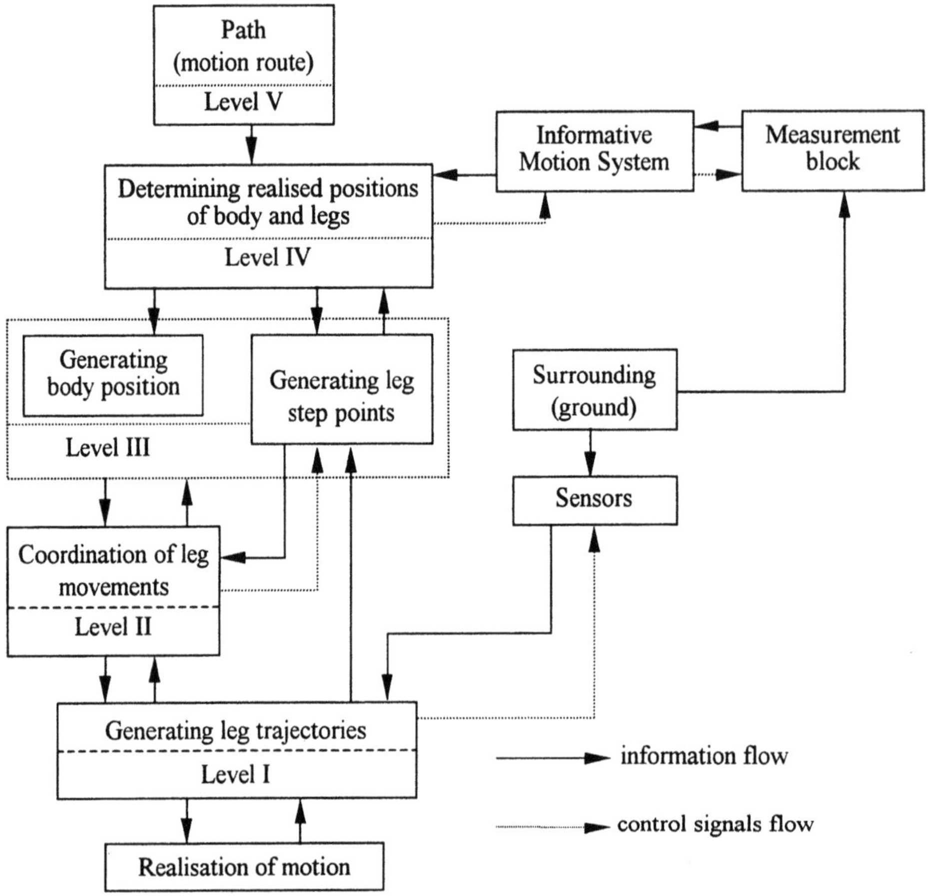

Fig. 17.32. Flow chart of the control of the T-2 machine [17.7]

The general characteristic of the control system of the T-2 machine does not differ greatly from those of ESA UNO and of Ignatiev's machine. The correspondence of functions of these systems is shown in Figure 17.33. The T-2, beside employing a kinematic model, also adopts the concept of software cooperation in the Informative Motion System, used here as an image recognition system. Previous devices were "blind", and their response to changes in terrain was stimulated by a system of sensors. Ochocimski's machine, instead, was fitted with a range finder, and as a consequence the actual positions of the body and legs could be determined before encountering an obstacle. This clearly increased the operating efficiency of the control system. Work on the T-2 machine continues today, and the concept of free gait has also been realized. Obstacles are overcome by repeated attempts of moving one, two or three legs at the same time, until success. A full description of the research of Ochocimski's team can be found in [17.6].

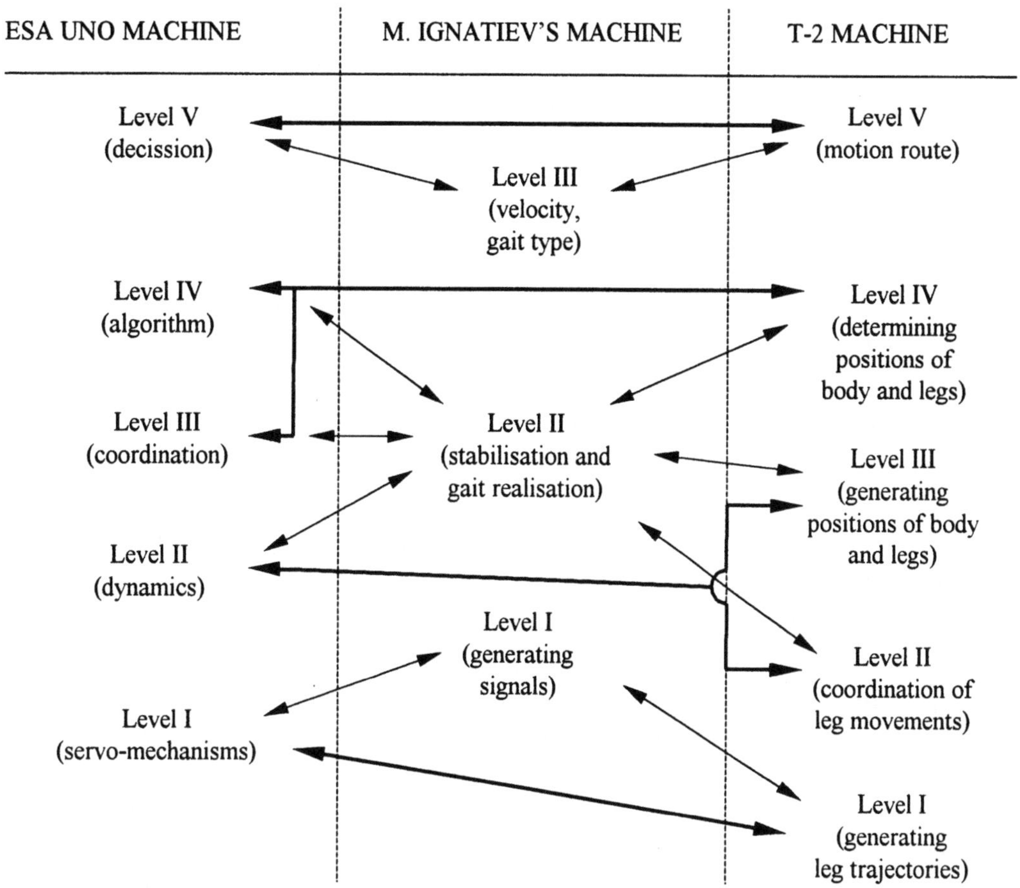

Fig. 17.33. Correspondence of functions of particular control system levels [17.6]

A review of other walking machines reveals control systems that similar in terms of structure. This is only natural fact since the process of generating the gait of a walking device is a consequence of a series of certain operations, such as generating signals for execution elements, motion stabilization, and other. However, the methods of realizing those operations may be different. We shall not present only control system flow charts, but attention to important, new solutions.

The six-legged MEWALK walking machine is an example of attempt to radicaly simplification of the control system by selecting an appropriate design of the machine and its capabilities of movement.

In 1984 by Kaneko and his collaborators, from the Mechanical Engineering Laboratory of Tsukuba, presented two versions of their machine, Mark I and Mark II [17.7] which used a gait with a fixed routine. Overcoming obstacles relied on "sensing", by sensor systems, a leg is "blocked" on the ground and was lifted until it is not blocked any more, while maintaining the body in a horizontal position. The designers felt that such motion capabilities are in many cases sufficient, and quickness of a simple control system (which determines gait speed) is of great advantage. However, many potential applications, such as exploring planets, or supporting the movement of people with disabled limbs, entails high motion efficiency. This is demonstrated by common trends in the research conducted in this field.

The six-legged OSU Hexapod machine combines the expanded motion capabilities with optimization of energy consumption and an increase in control system efficiency. This machine with digital control was activated for the first time in 1977 [17.17]. The control system of OSU Hexapod has since been considerably modified. Other important information imbedded into the control software, apart from the model of leg geometry were the transmittances of the elements of the drive system (electric motors, signal converters and so on). These data permit a control of the velocity of leg movements and modification of their trajectories such that maximum energy saving is possible [17.17].

Analysis of computational processes was led to the replacement of the original control computer with two operating in parallel as much computation as possible [17.17]. Combining the issue of energy consumption analysis together with that of multi-processor control resulted in a more energetic efficient machine at greater velocities [17.17]. Further research has concerned the expansion of motion capabilities. The machine has been fitted with a vision system, and a newly developed algorithm for the motion entailed in surmounting obstacles. Developed in 1983, the so called follow-the-leader rule (FTL) had the legs located on the same side of the body on exactly step in the same places as the appropriate front leg [17.17]. This principle reduced surmounting of minor obstacles to planning the motion of the front legs only. The body remains in a horizontal position during the maneuver. Considering the scope of the research carried out on the OSU Hexapod, we can conclude that despite the non-universal concept of this machine description, it is of great practical significance, as the important issues in machine operation have been thoroughly discussed here. The results of this research are currently being applied to design a machine for practical applications.

The problem of energy efficiency has also been considered in the PV II and TITAN III machines, built at Tokyo Technology Institute. S. Hirose and his team analyzed of energy expenditures while determining its mechanical design. Consequently, they chose a quadruped machine, its legs resembling spider's legs (the concept of the pantograph mechanism – PANTOMEC) and electric motors as drive units. The PV II and TITAN III are fitted with range-finders, one short-range, for immediate leg position planning, and other long-range for general planning of the motion trajectory. The body kept in horizontal position while by passing obstacles. Recently simulation study has addressed on the issue of moving over a surface with so-called forbidden zones [17.5, 17.6]. PV II and TITAN III machines are supposed to present – according to the designer's assumption – the concept of energy efficient adaptive machines [17.5].

A specialized programming language was been introduced for the first time in the control of the six-legged H1 and H2 machines. J. Kessis and his colaborators from Paris University have developed the concept of LP4.5 language (Walking Robot Plan Language), dedicated to walking machine gait programming [17.7]. Beside the typical control system functions, an additional function, set at the highest level, called here the Level of Intelligence and Plan Generating, has been developed for the H1 and H2 machines.

The task of this level is the interpretation of the machine's behavior, programmed in LP4.5 language. For example, a typical command in the program may be on meaning "move with a certain gait for *n* time units". At highest level of the control system the means of realizing this command are developed considering data on terrain configuration obtained from a visual input. Study of the control of H1 and H2 machines is still in an experimental phase. Considerable difficulties arise in developing the rules of adaptation to the terrain while realizing the strategy written in LP4.5. But the control concept presented here is a break through increasing not only the motion capabilities, but also the autonomy of walking machines. The authors of the project plane to fit walking machines fitted with speech recognition devices, which will make command remote possible.

17.7.3. Current research trends

Today there is a tendency to use methods of artificial intelligence in control systems. They are employed in recognition of terrain configuration (visual systems), as well as in the selection of motion strategy (expert systems). This evolution of control system is connected with the attempts to develop a universal walking machine programming language, of which the LP4.5 language mentioned above is an example. Its authors have announced their intention of creating a more versatile one, WRL (Walking Robot Language), and OWL language [17.1] was developed by I.E. Satherland from Carnegie-Mellon University in 1983. This expansion of control systems decreases the dependence of walking machines on the human operator.

Certain research groups see expansion of motion capabilities trough the analysis of motion dynamic phenomena. Kinematic models commonly used in control systems are completed with dynamic description.

Of these two trends in expanding control systems, the former one is more popular. The analysis of dynamic phenomena is very complicated and, in the case of multi-legged machines, a completely new domain. Consequently the idea of increasing machine autonomy by the use of artificial intelligence methods together with specialized programming languages, seems to be the easier solution, and with greater hopes of success.

Other currently considered issues regard the design of the walking machines. The problem of optimum multi-processor control and of the optimum number of legs (with regard to control) is dealt with here. For example, different concepts of multi-processor networks can be found in the control of the OSU Hexapod, and of the six-legged ODEX I machine, as well as in the project for the ASV-84 control.

Multi-processor networks, currently in use, include a main processor and many subordinate processors forming systems of varying architecture [17.5]. The configuration of multi-processor networks depends on the functions attributed to particular processors and is chosen for a machine design.

The problem of optimization of the number of legs have found no univocal solution. S. Hirose and Y. Umetani came to the conclusion that four-legged machines should be built for the following reasons:

- The increase in the number of legs improves:
  a) motion stability (ease of maintaining stability),
  b) adaptability to an uneven terrain, and
  c) energy savings while displacing a single leg.
- The decrease in the number of legs causes:
  a) simplification of the drive system (the number of drive elements decreases as well as energy consumption), and
  b) simplification (smaller number) of control signals.

Hirose and Umetani hold that the above mentioned facts allow us to set the optimum number of legs between four and eight. However, they emphasize, as technological limitations in the realm of drive units are crucial, the best number of legs is only four.

Ochocimski and Platonow claim, for their part, that theoretical premises and the results of practical studies indicate that the reduction of the number of legs to four, instead of six, considerably limits motion velocity and ability to adapt to uneven terrains. They also believe that the alternative to six legs is two, which reduces drive elements and mass, but poses new problems such as those of: dynamically stable control and flexibility of body motion.

The advantage of six-legged machines presented here indicates the tendency to simplify control rules (stability, adaptation) at the expense of the complexity of execution systems.

17.8. References

[17.1] *Donnes M.D.* Control of Walking. Local Control and Real Time Systems. Post Doct. Dissertation. Carnegie Mellon University, 1984.

[17.2] *Bielecki W.W.*: Two-legged locomotion: dynamical modeling and control methods.(in Russian) Moskva, "Nauka" 1984.

[17.3] *Morecki A., Ekiel J., Fidelus K.*: Cybernetics systems of Limb Movements in Man, Animals and Robots. Polish Scientific Publ.,Warsaw 1984.

[17.4] *Reibert M.H.*: Legged robots. Communication of the ACM, 1986, Vol.20, No 6, p.499-514.

[17.5] *Todd D.J.*: Walking Machines. An introduction to legged robots. Kogan Page, 1985.

[17.6] *Ochocimski D.E., Golubiev Ju.T.*: Miechanika i upravienije dvizenja avtomaticeskovo sagajuscevo apparata. Moskva, "Nauka" 1984.

[17.7] *Zielińska T.*: Gait modelling of the 4-legs walking machine. Doct. Diss., Warsaw Univ. of Technology 1986.

[17.8] *Vukobratovic M.*: Legged Locomotion and Anthropomorphic Mechanisms. Mihailo Pupin Institute. Belgrad 1975.

[17.9] *McGhee R.B., Orin D.E.*: A mathematical of Joint positions and torques in Legged Locomotion Systems. Proced. Theory and Practice of Robots and Manipulators. Ed. by A. Morecki, K. Kędzior, PWN – Elsevier Publishing Company, 1977.

[17.10] *Morecki A.*(editor): Basic Problems of Contemporary Technics. Vol. XXV, Robotics. Polish Scientific Publ.,Warsaw 1987 (in Polish).

[17.11] *Takanishi A., Egusa Y.E., Tochizawa N., Takega T. and Kato I.*: Realization of Dynamics Biped Walking Stabilized with Trunk Motion. RoManSy'7.Proc. of the Seventh CISM-IFToMM Symposium on Theory and Practice of Robots and Manipulators. Ed. by A.Moreckl, G.Bianchl and K. Jaworek, Hermes, Paris1990 .

[17.12] *Morecki A., Oderfeld J.*: Theory of Machines and Mechanisms. Warsaw PWN 1987 (*in Polish*).

[17.13] *McGhee R.B. and Orin B.E.*: An Interactive computer control system for a quadrupedix robot. First CISM-IFToMM Symposium on Theory and Practice on Robots and Manipulators. Vol. 1, Springer Verlag, 1974.

[17.14] *Emura T. and, Arakova A.*: A Study on Walking Robots Controlled with Attitude Sensors. Advanced Robotics, 1989, Ed. K.J. Waldron, Springer-Verlag 1989.

[17.15] *Bessonov A.P., Goncharov S.S., Umnov N.V.*: Evaluation of the Top Limit Velocity for Walking Robots; RoManSy'7 Proc. of the Seventh CISM-IFToMM Symposium on Theory and Practice of Robots and Manipulators. Ed. by A. Morecki, G. Bianchi and K. Jaworek, Hermes, Paris 1990.

[17.16] *Hirose S., Umetani Y.*: Some Considerations on a Feasible Walking Mechanism as a terrain Vehicle. Proc. of the 3rd Symposium on Theory and Practice of Robots and Manipulators. Ed. by A.Morecki, G.Bianchi, K. Kędzior, PWN – Elsevier Scientific Publishing Company, 1980.

[17.17] *Song Shin-Min and Waldron K.J.*: Machines that Walk: the Adaptive Suspension Vehicle, The MIT Press, Cambridge, 1989.

[17.18] *Bessonov A.P., Umnov NV.*: Choice of Geometric Parameters of Walking Machines. Proc.. on Theory and Practice of Robots and Manipulators. Edited by A. Morecki, K. Kędzior, PWN - Elsevier Scientific Publ. Company, 1977.

[17.19] *Thring M. W.*: Robots and talechirs. Ellis Horwood Lmtd, Publ.-Chichester, 1983.

[17.20] *Hirose S., Aoki S. and Michyahe J.*: Design and Control of Quadro-Truck Crawler Vehicle HELIOS-II. Preprints Eight CISM-IFToMM Symposium on Theory and Practice of Robots and Manipulators, 2-6 July 1990, Cracow

[17.21] *Akizowo J., Iwasaki M., Nemoto T., Asukura 0.*: Development on Walking Robot for Underwater Inspection. J. Waldron (Ed.). Advanced Robotics. Springer-Verlag.

[17.22] *Jlg.W., Berns K.*: A learning architecture based on reinforcement learning for adaptive control of the walking machine LAURON. Robotics and Autonomous Systems, no.15, pp. 321-334, 1995

[17.23] *Ritzman R. E.*: The Neural Organization of Cocroatch Escape and its Role in Contex Dependent Orientation. Biological Neural Networks in Inverteberate Neuroethology and Robotics. Edit. R. D. Beer, R. E. Ritzman, T. McKenna, Academic Press Inc. 1993.

[17.24] *Dean J.*: A Model of Leg Coordination in the Stick Insect, Carausius morosus. Biological Cybernetics, No 64, s. 393–411, Springer Verlag 1991.

[17.25] *Dean J.*: A Model of Leg Coordination in the Stick Insect, Carausius morosus. Biological Cybernetics, No 66, s. 335–355, Springer Verlag 1992.

[17.36] *Zielińska T.*: Walking Machine as a Mechatronic Device. Int. J. of Intell. Mechatronics, Design and Production. Vol. 1, No 3, s. 156–170, 1995.

[17.27] *Bogutsky A. V.*: Control of Walking Machine on Extra–Complex Terrain. Proc. of European Control Conf.`93, s. 1261–64.

[17.28] *Probir K. P., Jarayan K.*: Generation of Free Gait – a Graph Search Approach. IEEE Trans. On Robotics and Automation, Vol. 7, No 3, June 1991, s. 299–305.

[17.29] *Morecki i in.*: Robotics System: Elephant Trunk Type Elastic Manipulator Combined with Quadruped Walking Machine. Proc. of the 2nd Int.Conf. on Robotics and Factories in Future. Springer Verlag 1988.

[17.30] *Muller-Wilm U. et all*: Kinematic Model of a Stick Insect as an Example of a Six–Legged Walking System. Adaptive Behavior, Vol. 1, No 2, s. 155–170, The Massachusets Institute of Technology 1992.

[17.31] *Halme A. i in.*: Terrain Adaptive Motion and Free Gait of a Six Legged Walking Machine. Proc. Euromech`307 „Walking Machines", Duisburg 1993.

[17.32] *Vukobratovic M.*: Legged Locomotion Robots and Anthropomorphic Mechanism. Michailo Pupin Institute, Beograd 1976.

[17.33] *Formalski A.M.*: Peremiescenije antropomorfnych miechanismow. Moskwa, „Nauka" 1982.

[17.34] *Kato I.*: Bulletin of Sciences and Engineering Research Laboratory. No 62, Waseda Univ, 1973 (Special Issue of Wabot).

[17.35] *Kato I.*: Bulletin of Sciences and Engineering Research Laboratory. No 112, Waseda Univ, 1985 (Special Issue of Wabot-2).

[17.36] *Bay J. S., Hemami H.*: Modelling of a Neural Pattern Generator with Coupled Nonlinear Oscillators. IEEE Trans. On Biomedical Engineering, Vol. BME-34, No 4, s. 297–306, 1987.

[17.37] *Brogan W. L.*: Modern Control Theory. 3rd edition, Prentice Hall Inc. 1991.

[19.38] *Buschman A., Frik M.*: Nonlinear Pattern Generators for Legged Locomotion. Euromech 307 „Walking Machines", Duisburg 1993.

[17.39] *Cohen A.H., Rossignol S., Griller S.*: Neural Control of Rhythmic Movements in Vertebrates, J. Eiley and Sons, New York 1988.

[17.40] *Formalski A.*: Impulsive Control for Antropomorphic Biped. RoManSy`10, Edited by A. Morecki, G. Bianchi, K. Jaworek, Springer Verlag 1995, s. 387–393.

18. MICROROBOTICS

18.1. Basic terms and systematics

A micromechanism is a device handling micro objects and itself micro sized, i.e. much smaller than the size of ordinary machines [18.1].

MM consists of a man-operated control C, a drive source S, an actuator for positioning (A_p), and an actuator to perform the operation (A_w) (Fig.18.1). Here are different classes of micromechanism: fixed MM, floating MM, and self-running MM.

The first investigations in the field of micromechanisms (MM) began in the seventies; the first models arrived in the eighties [18.1, 18.3]; and the first International Symposium on Micromechanisms took place in July 1993 [18.2]. The present state of development of MM is shown in Figure18.2. There are here three generations of MM:

- First generation: MM having dimensions from several mm to one mm. These are devices such as the wrist watch, which contain a motor and gears of outer dimensions of 0,5 mm. Devices of this generations are based on the conventional technology of precision mechanisms.
- Second generation: devices having dimensions from 1mm to several μm. Their development is based on the technology of semiconductor materials.

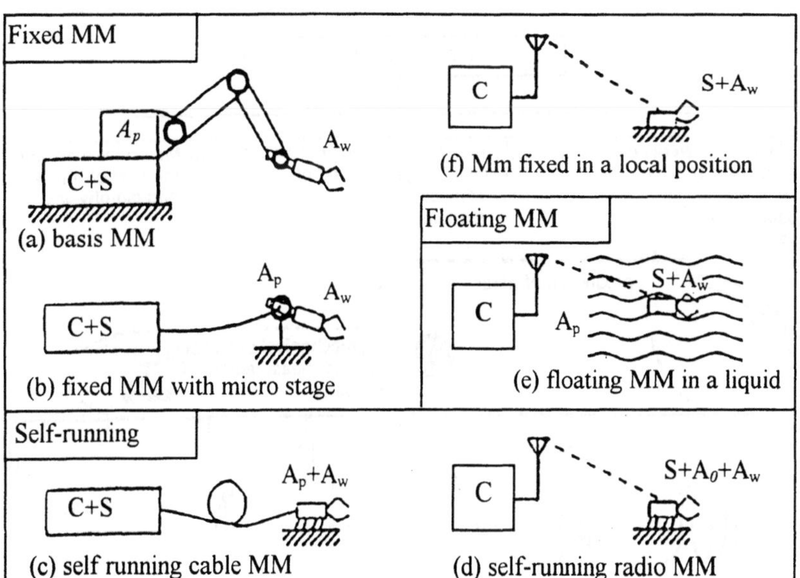

Fig. 18.1. Classification of MM by type (after [18.2])

Third generation: devices ranging from several μm to bacteria dimensions. The further development of this generation depends strongly on investigation of living organisms. It is expected that future development of MM will be based on molecular technology. Japanese researchers maintain that all machines of less than few mm in size can be made even smaller.

18.2. Present state of research

The dimensions shown in Figure18.2 range from 10^{-3} to 10^{-9} m. The principles of their functioning and research methods change with their size. The middle level in Figure18.2 shows the present situation of MM as a function of dimensions. The wrist watch now belongs to the smallest machines of every day use. The smallest bearing of the electrostatic motor on a silicon wafer is one tenth of a milimeter. The mechanical transformation and measurement of parameters of the size of one tenth of micrometer (μm.) is also possible.

Level (I) in Figure18.2 deals with micromechanisms based on conventional technology. The step motor and microturbine with an outer diameter of 2 mm also belong to this level. Other examples are the spherical shells a few μm in diameter produced from different materials for use in the chemical industry, the ultrathin belts of diameters of 1 - 4 μm., and amorphous film 1 - 2 μm thin used in semiconductor industry. It is assumed that first generation MM will continue to evolve in this way.

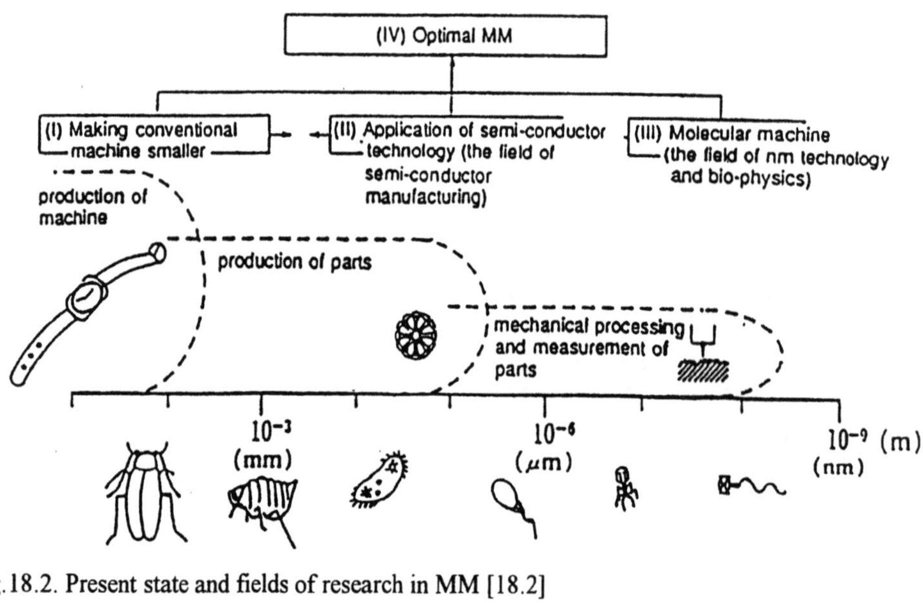

Fig.18.2. Present state and fields of research in MM [18.2]

The current level (II) in Figure18.2 is based on semiconductor technology. Lithography and vacuum technology are used here. The technical examples realized on this way are the accelerometric oscillator of the dimensions of a few μm. As the wafer, fluid element, and biochamber for cell fusion, electrostatic motor, air turbine and different sensors for the measurement of physical and chemical parameters (polycristalline materials), level III deals with the class of mobile machines. In this field work is concentrated on bioactuators and artificial bacteria, and the results promise be useful for the design of new mechanisms.

There is also a fourth level (IV) concerned with what are called optimal mechanisms. Here investigation is concentrated on the components of MM and the possibility of producing them using micromaterials and microtribology.

18.3. The properties of MM

Research [18.1, 18.2] on the moving models has indicated that:

- dimensional analysis by length is important in comparing the achievements of the MM with different principles of action;
- simple MM can be designed to fulfill the assumed requirements. For example, the eel model moving in silicone oil at the speed of 24 μm/s allows to investigate the speed of living organisms or machines in the water and air (Fig.18.3). The three-dimensional space model (Fig.18.4) can be used to investigate the horizontal surface.
- there is a relationship between the dimensions (l), velocity $v(l^1)$ and Reynolds number $Re(l^2)$ of moving devices (Fig.18.5);
- method of the error analysis must be refined for the design of MM because of the difficulties in producing very small parts.

The research also identified the salient characteristics of MM:

- the influence of the dimensions of the developed forces. Different forces have different influence (L^n) according to their action. For example, the three-dimensional arch model (Fig.18.4) stopped working when $L < 3$ mm; the electrostatic resistance force of the air acting on this MM body was bigger than the driving force (L^2) produced by the elastic deformation of the materials with which it was built (Fig.18.4).
- increased strength of materials. It is known that unit crystals or amorphous materials are 100 to 1000 times stronger than polycrystalline materials of the some chemical consistency because of the block of the crystalline barriers. These properties can be exploited in the design of MM.
- amplification of surface features. Because the surface effect is greater than that of the volume, chemical reactions or heat phenomena become more important. These

phenomena are useful in building mechano-chemical and heat driven actuators, which are not used at present because of their low velocities. It is expected such a microtribology phenomena like absorption forces in contact or slippery.

- decrease in manufacturing accuracy. It is difficult to achieve accuracy E in the production of the different parts (equal to e_x/L - 10^{-3} at present), but on the other hand an accuracy of E - 10^{-4} meets the requirements of typical machines.
- the relationship between dimension and speed of travel must be taken into account in the design of MM. The Reynold`s number $Re(L^2)$ is equal to 1 by L = 1 mm, its value decreases from 1 in the square of the length L by L - 1, which means that the principles of MM motion differ from those of machines of conventional dimensions (see Table 18.1).

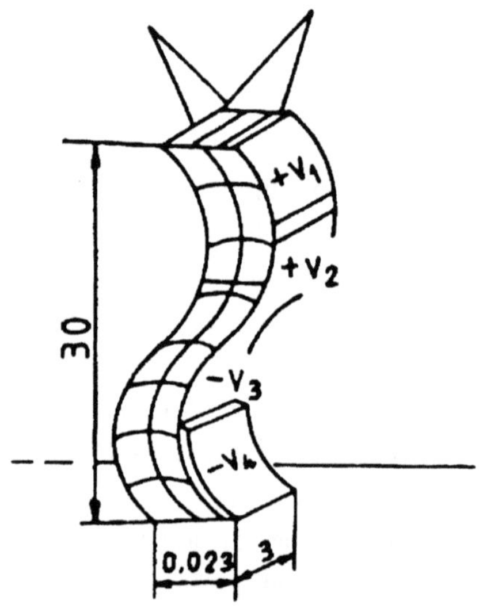

Fig. 18.3. The model of artificial eel for measurements in the liquid [after 18.2]; v_i - the electrodes.

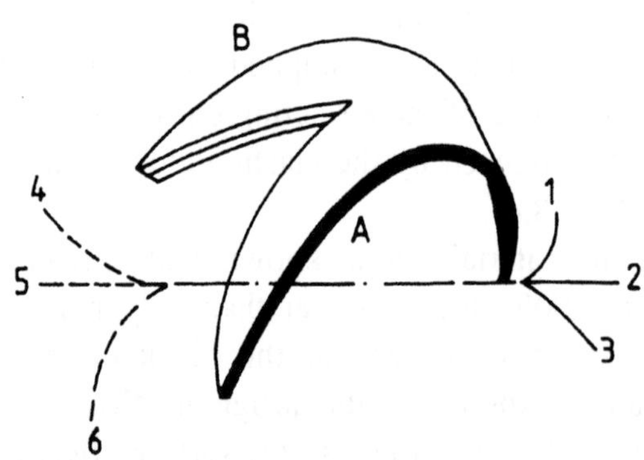

Fig. 18.4. The 3-dimensional space model for the investigations on horizontal surface [after 18.7].

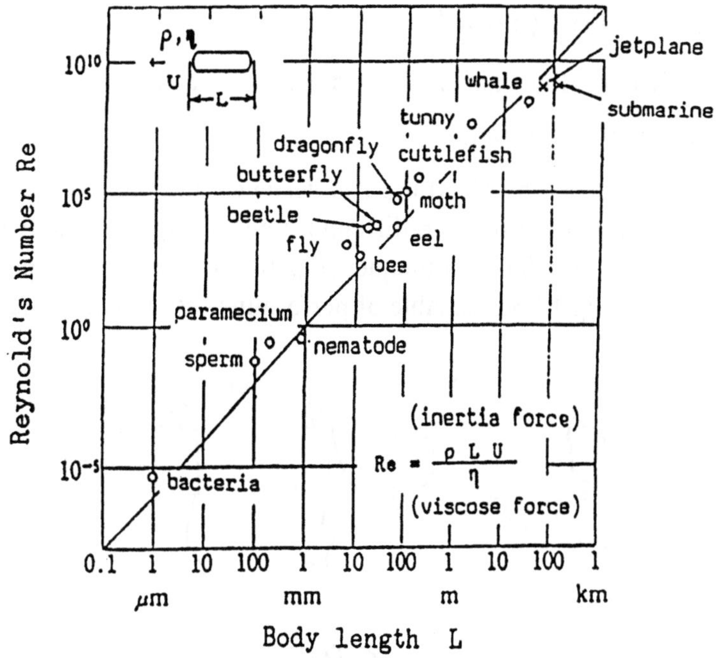

Fig.18.5. Relationship between dimensions (L) and Reynold`s number Re (L^2)

Table 18.1 Traveling speed on moving things

Traveling: circumstances	Traveling velocity		
	U mm/s.	Dimension	Dispersion
In water	U = L		
On surface On air 1	U = 10 L	∞ L	1/10 ~ 10 X
On air 2	$U = 10^3 L^{0.5}$	$\infty L^{0.5}$	1/10 ~ 100 X
X – Dispersion			

18.4. Mobile microrobots

There are:

- walking robots,
- mobile vehicles,
- running robots,
- flying robots,
- -swimming robots

We shall briefly discuss their properties and design.

18.4.1. Microwalking robots

These include two-, four- and six legged walking machines.

18.4.1.1. Biped walking robots

The laboratory headed by H. Miura has designed and built different models of biped robots [18.19]. On the fig.18.6 different types of biped are shown. These are robots such as fongo (Fig.18.6a), bilboquet (Fig.18.6b), the inverted pendulum type (Fig.18.6c) and stilt (Fig.18.6d) capable of performing fast and accurate movements.

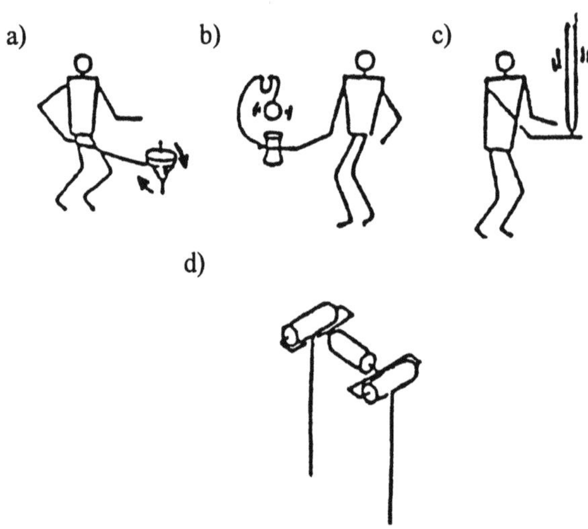

Fig. 18.6. Examples of the intelligent biped robot built in Miura's laboratory [after 18.19]

18.4.1.2. Four-legged micro-walking robot driven by electromagnetic force

The drive of this micro-walking robot is based on mutual interaction between a constant magnet field, the magnetic field produced by an electromagnet, and the force of the spring (see Fig. 18.7). It consists of two independent parts connected by a coupling. These parts can move independently. The constant magnet is connected to the left part of the body and, the coil, which creates the electromagnet to the right part.

The distance between the two parts of the body is fixed by the coil spring surrounding the magnet and electromagnet. There are two possible kinds of movement. The first takes place when $F_s > F_m$, meaning that the spring force dominated. When $F_m > F_s$, the magnet is the dominating force. In the letter case, when the electrical current is off, the foreleg and handle are stationary, and the interval between them is minimum, because the repulsion from the spring F_s is smaller than magnetic force F_m (Fig.18.8a).

When the electrical current is switched on, the sum of forces $F_{em} + F_s - F_m$ becomes greater than force F_m, and the foreleg and hindleg move in the direction of force F_s. Since the friction force of handle N_b is greater, than that of the foreleg N_a ($F_s + F_m - F_m + N_b$), the foreleg moves from B_1 to B_2 (Fig.18.8b). The other part is fixed.

When the electrical current is cut off, the foreleg and handle contract toward the center of the machine (Fig.18.8c). Since friction force N_b is now smaller than N_a ($N_b - N_a$), the hindeleg moves from A_1 to A_2.

These three phases of motion described above create so called step of the robot. By repeating these steps the robot is able to move in one direction.

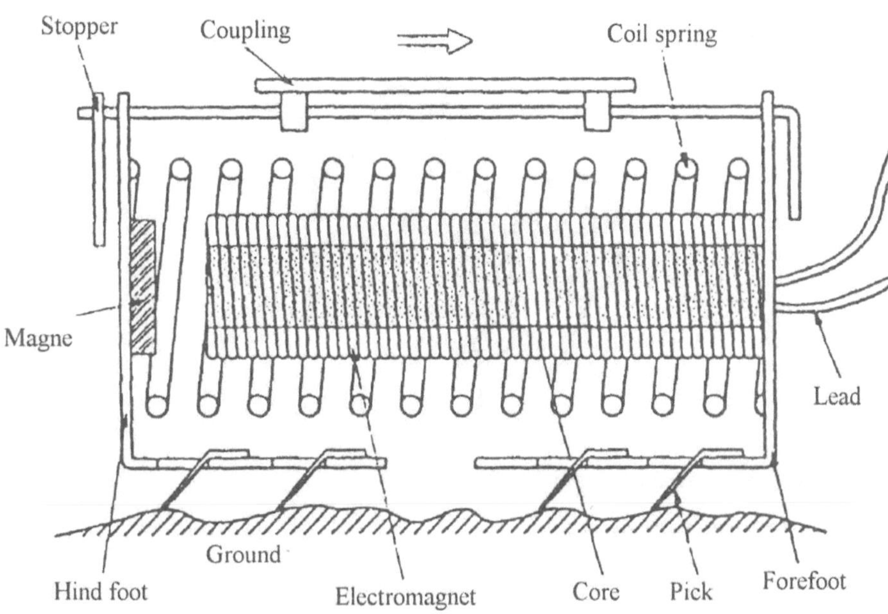

Fig. 18.7. Overall view of a four legged robot driven by electromagnetic force [18.8, 18.20, 18.23]

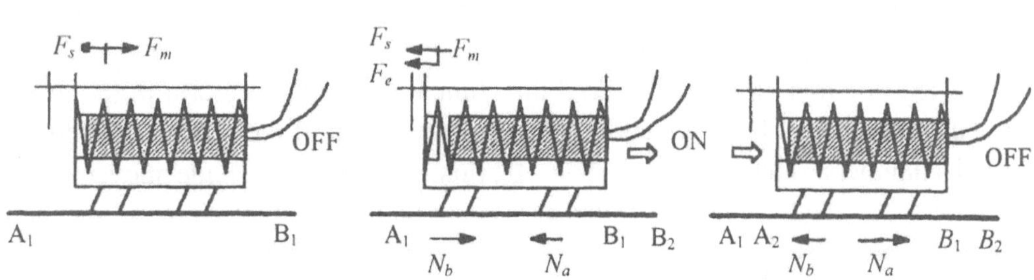

Fig. 18.8. The principle of walking

During the design process one of the most critical problems was the selection of proper servoactuator. The final electromagnet was 10 mm long, and 10 mm in diameter. Another problem was the design of the legs. In addition there were the torsion motions of the two parts of the robot during the movement. These were eliminated by special connections. The energy supply also created a problem. A special experimental rig was built and used for experiments. The influence of the surface on the robot motion and the influence of the wires were also tested.

The experiments showed that the relationship between the current and its frequency was about 8 Kz (the regular - 4 Hz). The possibility of moving the robot on a slope was not investigated.

Figure 18.9 shows the four-legged robot. This robot is dynamically stable in motion. In the swing phase of the gait two of the legs are in contact with the ground.

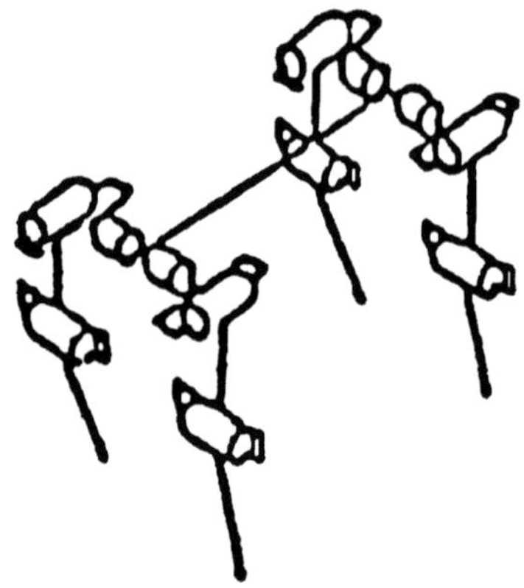

Fig. 18.9. Four legged micro-robot [after 18.19]

18.4.1.3. Six legged machines

The limbs of a six legged walking robot are similar to those of a mammal, and the gait is realized by controlling the „flexion" movements of the elastic legs. By elastic here we mean (Fig.18.10) a tube of rubber fibers the deformation of which can be obtained by controlling the flow of air or liquid. Two types of six legged walking robot are shown in Figure18.11.

The leg deforms easily in the axial direction and not at all deform in the radial one. Deformation can be obtained by independent control of the pressure inside the internal chambers. The leg has 3-DOF (flexion, extension and side movements). The diameter of the leg changes from 4 to 12 mm, its length, from 20 to 50 mm, and its mass, from 0,3 to 8 g, the flexion force is 0,15÷1 N. Figure 18.11a shows a six-legged robot walking on a finger. The six legs of the second type of robot (Fig.18.11b) are of the overhang type. Each leg has a diameter of 2 mm, and is

12 mm long. The dimensions of the first six legged robot described are 12*10*12 mm, while the latter's 12*20*7 mm.

The air passes through 18 tubes with a internal diameter of 0,2 mm each. In general, 12 tubes are enough to control movement (back and forth and rotating); the speed of the gait is about 20 cm/min and the load, 300 mg for the four-legged, and 100 mg for the six-legged machine. This class of robots includes devices with a light drive, electrostatic drive, ultrasonic motor, piezoelectric element, electromagnetic force, and air turbine drives.

Fig.18.10. The flexible actuator (FMA with diameter of 1 mm) [after 18.3]

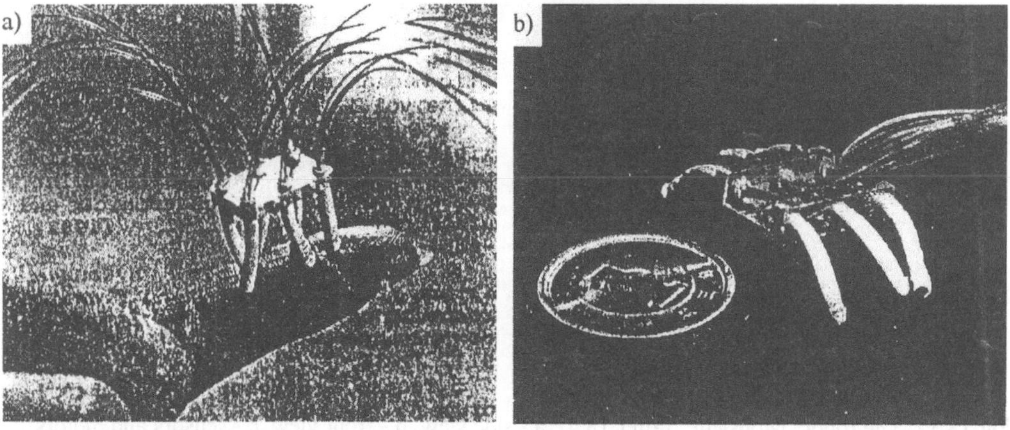

Fig.18.11. Micro-walking robots with six legs

18.4.1.4. Light driven mobile robot

This machine is known as „Monsieur", and will be the first micromachine used for industrial purposes. It consists in a central unit (CPU-IC), condenser, two drive units, and a pair of photosensors, that enable an independent drive in the direction of the light. It consists of 98 parts, the majority from the Seico Epson Corporation watch.

In 1991 this robot was received and in 1994 was included at the Guinness book as the smallest robot in the world in 1994 [after 18.4]. As said, „Monsieur" can move independently in the direction of the light. It is loaded for successive trips by antennas and tail electrodes. The machine moves forward at a speed of 1,8÷14,7 mm/s; and can negotiate slopes of up to 5 degrees, its dimensions are 12,4*11*10,8 mm, operating time is 5 min (if already loaded, 3 min), loading time 30 s , mass 4,3 g; its 98 parts are the equivalent of two watches; and it is made in sterling silver (Fig.18.12) [18.4].

A two pool stepping motor, of the kind usually employed in ladies` watches, was used as the ultrasmall drive. The magnet of the rotor has a diameter of 1 mm, a thickness of 0,4 mm, and a mass of 2 mg, wire coils of 13 μm, a resistance of 2,3 Ω. The stator is made of permalloy.

In Figure18.12, a front view, showing the location of sensors and motors, and an overall view of „Monsieur" are presented.

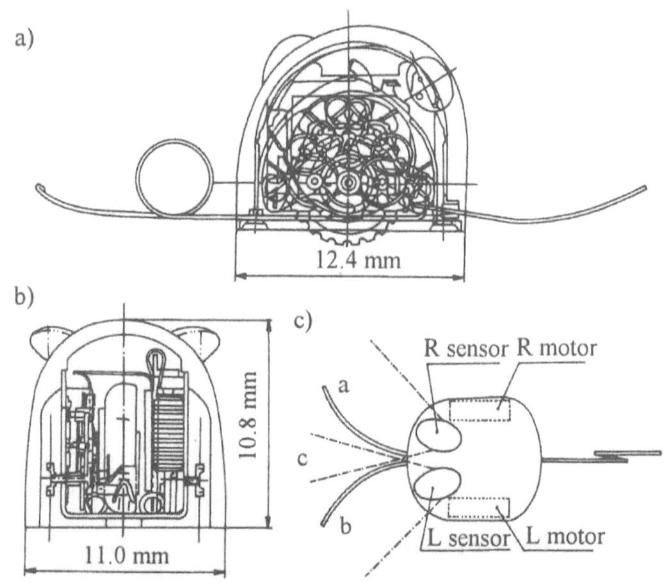

Fig.18.12. The robot „Monsieur" [after 18.4]: a - side view, b - front view, c - sensors and motors

18.4.1.5. Hermes walking robot

The Hermes robot is the most recent of a long line of walking robots offered by IS Robotics. It was developed for a museum display, showcasing unique insectoid micro-rovers. Building on IS's experience in prototyping rovers for planetary missions, Hermes was designed to be a small, rugged and powerful robotic exploring tool [18.26]

Hermes is a six-legged walking robot (Fig.18.13) incorporating high powered processing, rich sensing, and high mobility in a compact body. These features make it an ideal support in research into high degree-of-freedom systems and cooperative behavior. The compact body makes it suitable for use as an extension vehicle deployed from a parent craft, or for tasks that require mobility in tight spaces. This, and its ability to carry a video camera, make it a useful tool for remote inspection tasks.

Walking with six-two-degree-of-freedom legs, Hermes easily scrambles over rough terrain. Force sensing in each leg, bump sensors, and IR proximity sensors provide information that the robot can use to negotiate obstacles. A pitch and roll inclinometer allows the robot to adjust to rough terrain.

The robot is supplied with a code which utilizes all these sensors in its exploring demonstration. A serial cable can be attached to the direct robot by supervisory commands from a host computer. A detachable LCD display can be used to monitor the state of the robot during programming. An optional color video system can be mounted that transmits a robot's eye view of the world.

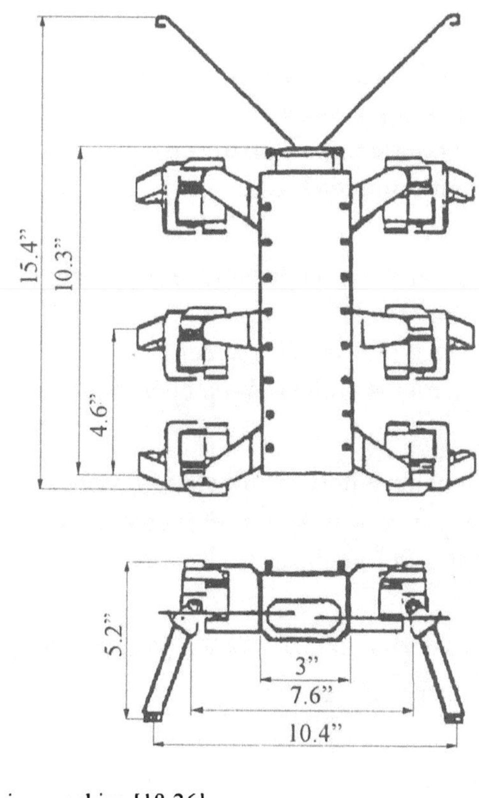

Fig.18.13. Hermes walking machine [18.26]

The machine consist of:

Mechanical part

- 12 high-torque ball bearing servo motors (0,72 Nm)
- Durable aluminium construction
- Can clear up to 2`` vertical steps
- 6 easy-to-replace leg modules;

Sensors

- Collision detection on each leg
- 2 forward-mounted whiskers for obstacle detection
- 3-bit support force detection on each leg
- 4 element in board IR proximity sensors
- Pitch and roll inclinometers;

Processors

- 1 Motorola 68332 microprocessor 16,78 Mhz
- with 1 MB battery-backed SRAM and 64 KB EPROM
- and operating system, diagnostics and demo program;

Programming Environment

- Behavior Language/subsumption compiler supplied
- This LISP-like AL Language allows user to add and modify behaviors
- Currently supported on Macintosh platforms
- Graphical based monitoring interface;

Input/Output Devices

- Programmable tone and duration piezo-electric buzzer
- 16 large red LED_5 (8 per side)
- Detachable Super Twist (STN) LCD screen
- Serial line communications with hook-up to Macintosh;

Configurations

- Rechargeable NiCd battery-powered operation with 30 min autonomy
- 72`` tether provides power and direct serial communication through
- a slip ring for continuous robot use

User options

- High resolution color CCD video camera with video transmission
- system for use in either configuration

Dimension, weight (total)

- Length: 0,39 m, width: 0,26 m, height: 0,14 m, weight: 2,5 kg

Host computer

- Power Macintosh 7500/100, 16 MB RAM, 1000 MB HD,
- CD 4 x drive, video card on board, 14`` monitor, keyboard

Locomotion patterns

- six legs, each has two degrees of freedom; the machine moves at a constant velocity of 0,028 m/s,
- the machine gait is similar to insect wave gait, always 4 or 5 legs in the support phase, one stepping pattern.

Figure18.14 shows the gait diagram, the sequence of the legs and changes in the support polygon.

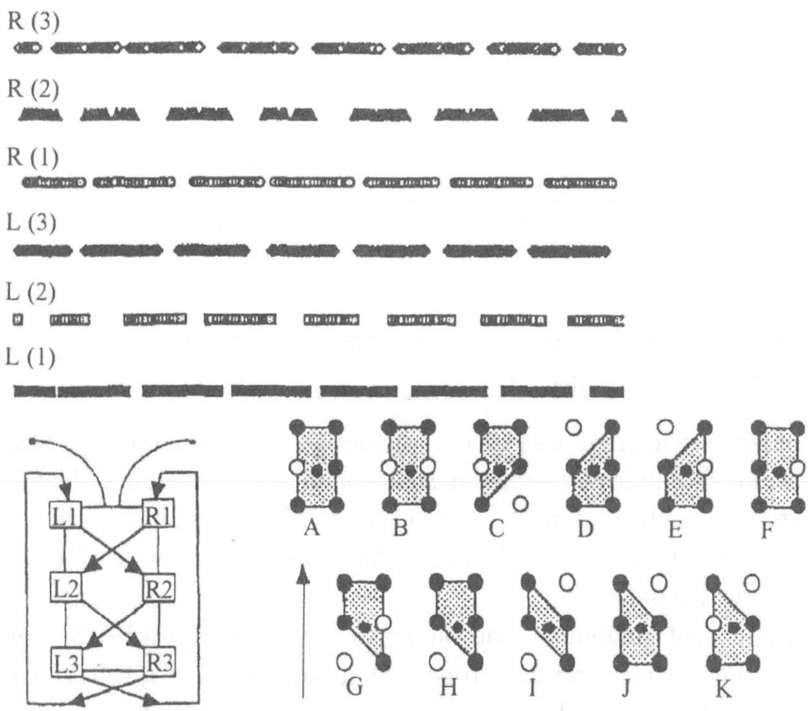

Fig.18.14. The diagrams of walking

18.4.1.6. Robot with a caterpillar drive

This robot can move on vertical walls. It consists of the linear actuators (SMA type), caterpillar drive and ratchet-and-pawl wheel system. The actuators are 0,05 mm in diameter and 17 mm long; the traveling speed is 1 mm/s, and the dimensions are 10*10*7 mm.

Figure 18.15 presents the mobile robot of the caterpillar type. The first experiments have shown that this vehicle is rather slow, and has difficulties moving on the vertical, but designers are now trying to overcome these difficulties.

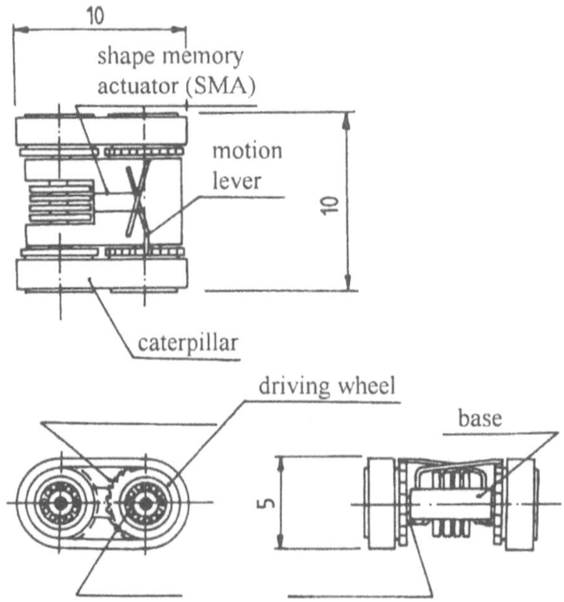

Fig.18.15. Mobile robot with caterpillar drive (dimensions and components) [18.5]

18.4.1.7. Mobile robot with ultrasonic motor

The piezoelectric effect was used in the design of this micromotor. High energy from ultrasonic vibrations caused by mechanical torsion of the piezoelectric element is converted through friction to driving force. This motor has the following properties: small dimensions, high torque, high efficiency, low level of noise, and high positioning accuracy.

The basic configuration of the motor is shown in Figure18.16; Figure18.17 shows the four wheel running robot "Yamakoshi Jivo" (1991), which has two independently controlled ultrasonic motors of a diameter of 10 mm, one on the left side, and one on the right.

This vehicle can run freely, changing direction and speed because of proper control of the motors. The ultrasonic motor improves the parameters of this vehicle, compared with the solution based on the stepping motor.

The technical data of this robot are: voltage - 6 V; torque - 1,0 gcm, angular velocity - 6000 rev./min, mass - 2,0 g. A new version of "Yamakoshi Jivo" was presented in 1992 [18.6].

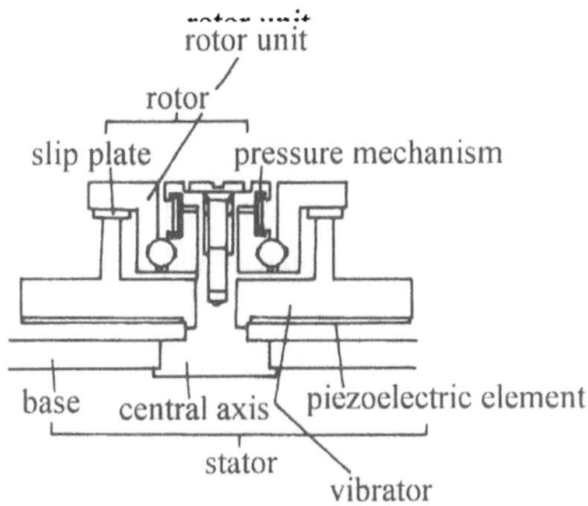

Fig.18.16. Basic configuration of the ultrasonic motor [18.6]

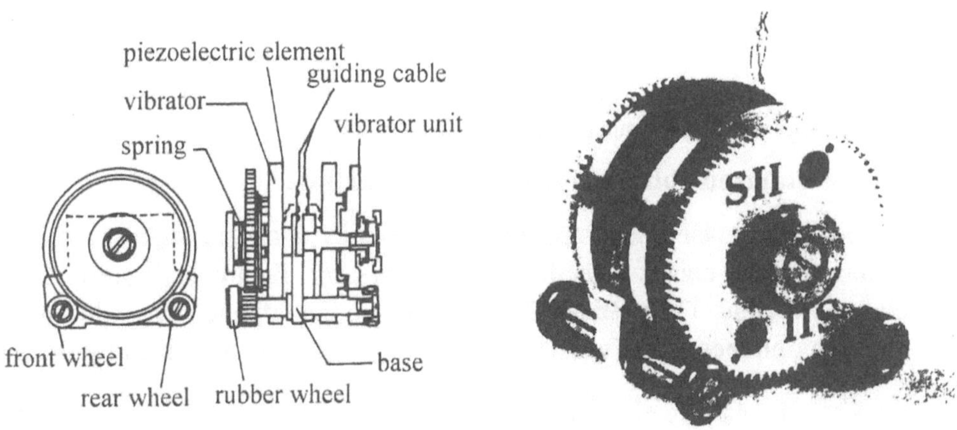

Fig.18.17. Doubled four wheel running robot "Yamakoshi Jivo" [18.6]

18.4.1.8. Robots with a piezoelectric element

Two types of mobile robots using piezoelectric vibrators with a unilateral piezoelectric element (amorphic type) and stacked type are presented here. The amorphic type solution has a U shape and consists in two piezoelectric elements (Fig.18.18a) 10*4*0,2 mm in dimension. The base has the same dimensions, but the fitness is 0,1 mm. The velvet brushes transmit the vibrations to the base in one direction only. The dimensions of the whole device are 10*10*2 mm, the mass is 0,1 g, the speed of the left and right part of the machine, as well as direction of motion, can be controlled by changes in the supply voltage and frequency of the piezoelectric element. Its maximal speed is 50 mm/s in a horizontal plane, but the machine can also move on a slope of 15 degrees.

Figure18.18b and c show different types of stacked type robots and their driven frequencies. The dimensions of the metallic frame are 36*4*0,2 mm. Piezoelectric element is 2*3*9 mm in size.

The robot is equipped with the velvet brushes. The dimensions are 10*10*10 mm and the mass, 2 g. The maximal speed experimented was 100 mm/s and the slope, 15 degrees. Further miniaturization by developing smaller actuators is planned.

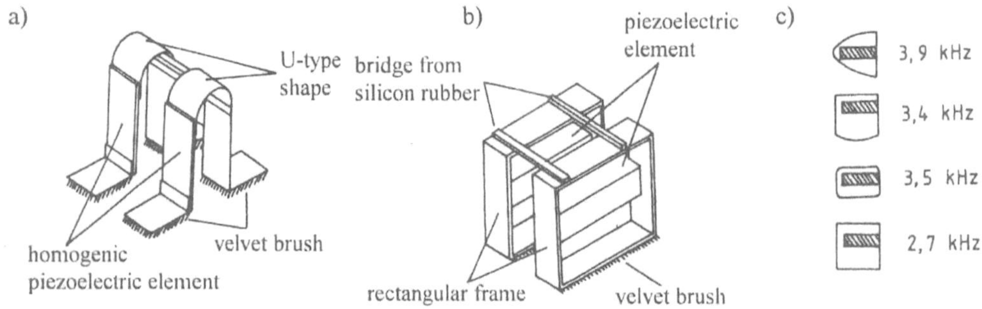

Fig.18.18. Different types of robots with a piezoelectric element [18.6]: a) amorphic type, b) stacked type, c) different types of machines studied and driven frequencies

18.4.1.9. Miniaturized robot driven by an air turbine

This robot can move on a flat terrain. It is driven by two small dentist turbines. The air comes through elastic tubes from the compressor. Figure18.19 presents the structure of this vehicle. The turbines (3) are placed on the left and right sides of the robot, and are independently controlled. The driven torque produced by the air

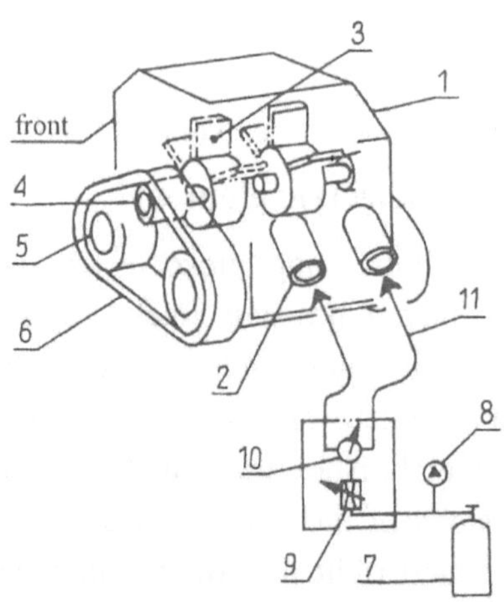

Fig.18.19. The structure of the robot driven by air turbines [18.9, 18.24]: 1 - body, 2 - air input, 3 - air turbine, 4 - driven rollers, 5 - executive roller, 6 - caterpillar, 7 - tank air, 8 - pressure sensor, 9 - velocity transducer, 10 - direction of transducer's motion, 11 - elastic tube.

turbines is transformed by two rollers with a barrel on the end (4 - active wheel) through caterpillar (6) on the front wheel (5 - passive). The turbine torque is controlled by changing the volume of the air transmitted by the 2 mm diameter elastic tubes (11) to the diameter 1mm input nozzles (2).

The speed of the turbine is controlled by the flow of the air. It allows the robot to move in all directions. The caterpillar is made of the rubber. The technical data are 9,9*4,5*6,4 mm, mass - 5 g, slope - 10°, maximal pressure 0,3 Mpa.

18.4.1.10. Micromobile robot with an oscillating electrostatic actuator

The actuator, shown in Figure18.20, consists in two constant electrodes and one moving electrode. When a constant voltage is applied to the constant electrodes, the moving electrode begins to oscillate. The frequency of oscillation is controlled by the voltage and distance between the constant electrodes. This actuator has a simple, light structure and can be used for the construction of artificial insects.

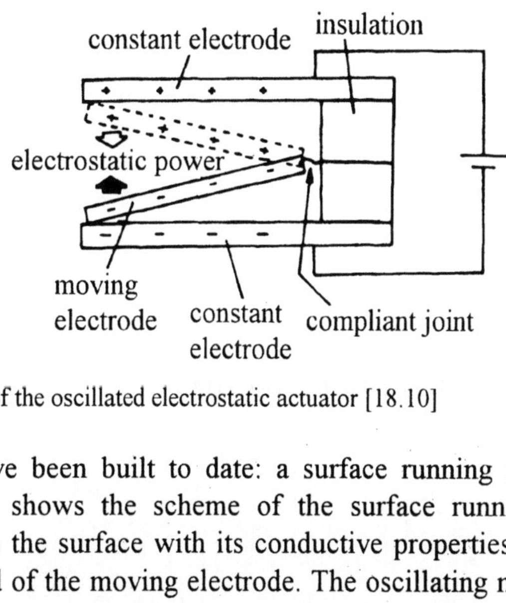

Fig.18.20. The scheme of the oscillated electrostatic actuator [18.10]

Two robots have been built to date: a surface running model, and a flying model. Figure18.21 shows the scheme of the surface running robot. Here the constant electrode is the surface with its conductive properties. Two front legs are connected to the end of the moving electrode. The oscillating motion causes the leg to jump forward. The dimensions of the robot are: 10*10*6 mm; its mass, 0,11 g; its maximal velocity, 50 mm/s on a flat surface.

Figures 18.22 and 18.23 show the diagram and overall view of the flying microrobot. The vehicle hangs on the two threads using as an input of driving strain. This vehicle was designed to move forward by oscillations of its wings. To obtain this motion, the wings were designed so that one is connected to the moving electrode, and the other to the construction, allowing the oscillation in a phase. The motion of the wings is shown in Figure18.24.

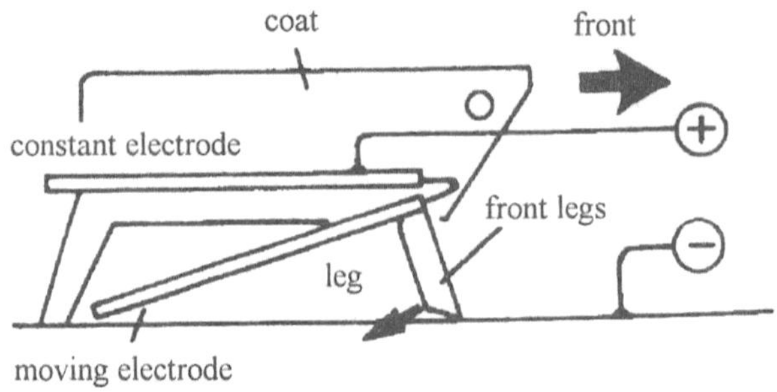

Fig. 18.21. The diagram of the running microrobot on the surface [18.10]

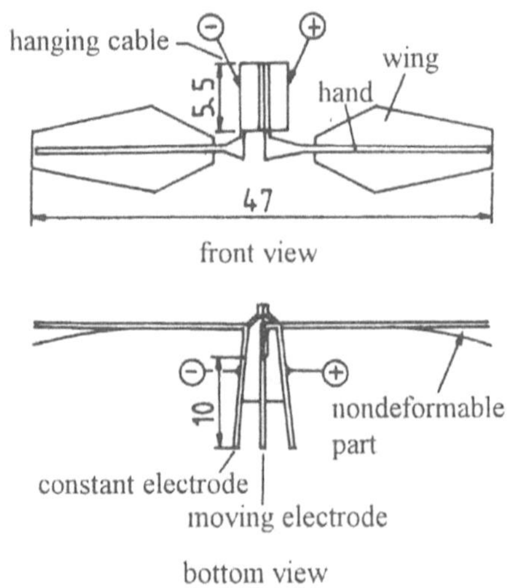

Fig. 18.22. The diagram of the flying robot (top view) [18.10]

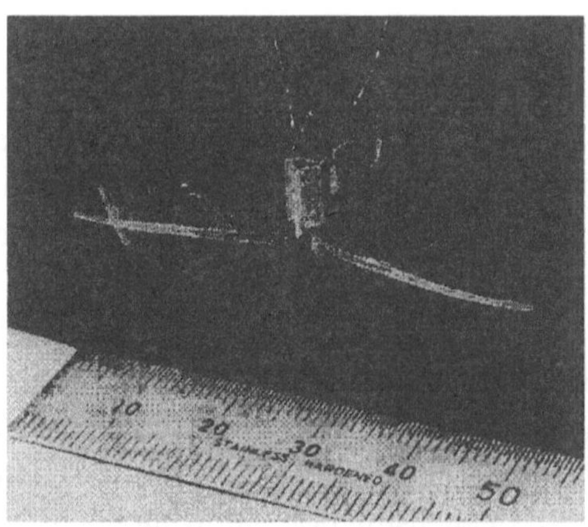

Fig. 18.23. The photograph of the flying robot [18.10]

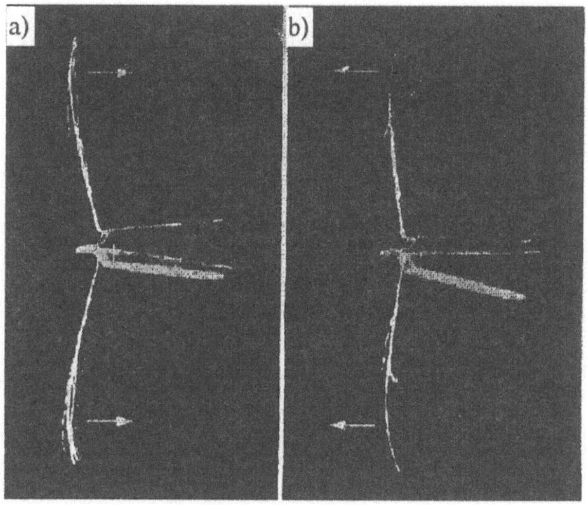

Fig.18.24. General view of the flying robot [18.10]: a - power impulse, b - backward impulse.

18.4.1.11. Swimming robot - underwater robot

The micro underwater robot shown, in different views, in Figure 18.25 is 45 mm long and has a mass of 30 g. Two DC micromotors and batteries 1,2 V, 50 mAh, Ni-Cd were used for the drive. The actuator was built from material with shape memory (SMA) and is used to lift the swim-fin. This micro vehicle is controlled by a microcomputer driven by IR - infrared and, a Pulse Code Modulation signal.

The robot can perform various tasks and investigations in places inaccessible to large machines, such as inside the human body. Figure18.26 shows the diagram of the actuator moving the swim-fin. A general view of the micro boat is given in Figure 18.27 [18.11].

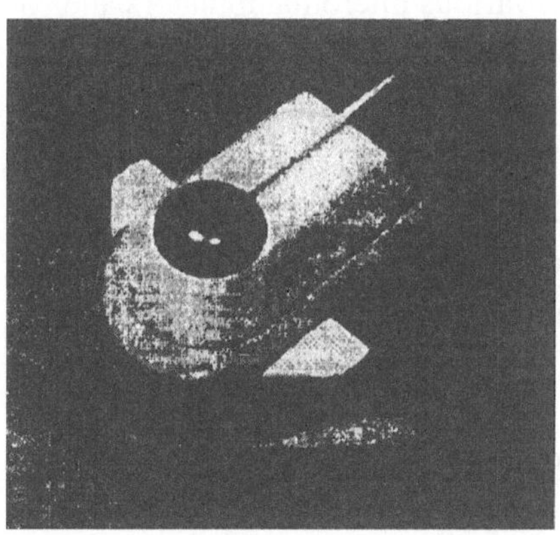

Fig.18.25. Underwater robot [18.11]

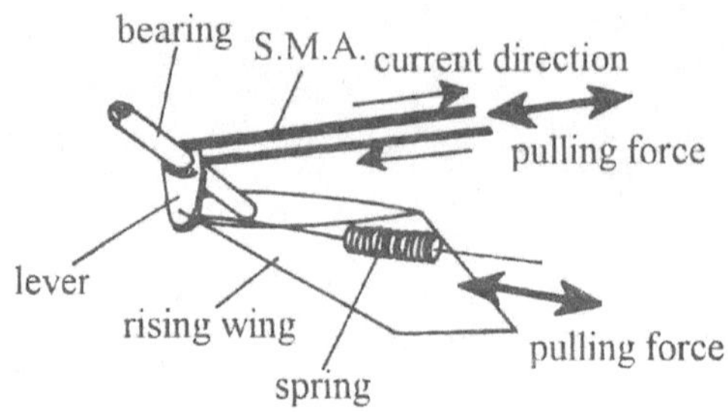

Fig.18.26. Wing actuator SMA type [18.11]

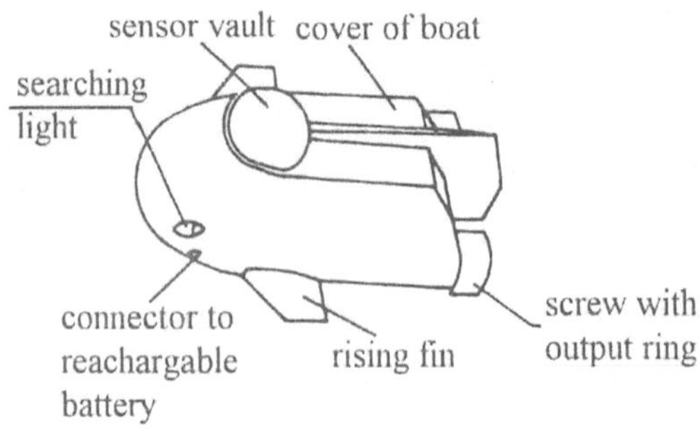

Fig.18.27. General view of the micro-boat [18.11]

18.5. Applications

An important factor, in the further evolution of MM is the presence of attractive applications. A good example of this is the micro-manipulators developed in the period from 1991 to 1993 for medical purposes (endoscope), and inspection robots [18.1]. The list of the models and devices built is long, and includes, besides actuators, motors, turbines and mobile robots, various micromanipulators and robots.

Actually various models of relatively small dimensions had already been built in Japan some centuries ago. Figure18.28a shows an illustration from "Karakuizui" (a collection of mechanical devices) published in 1796. Figure18.28b presents an example of Japanese puppet [18.15].

Let us look at a few examples of today's. applications:

- the automatic micromanipulator used for cell fusion (Fig.18.29);
- the microtelerobot for the manipulation of microscopic objects immersed in liquid (Fig.18.30);
- multi-finger gripper (Fig.18.31);
- inspection robot (Fig. 18.32);
- robot working in a group (Fig.18.33).

Fig.18.28. Mechanical dolls [18.15]: a) the illustration from "Karakuizui (collection of mechanical devices); b) example of mechanical dolls.

18.5.1. Automatic micromanipulator for cell fusion

Research is now addressing the automatization of protoplast micromanipulation under microscope, as well as microinjection. The system which provides for the displacement of the protoplast and control of its position using dielectrophoresis phenomena, is shown in Figure 18.29 [18.12].

The micromanipulator consists in a glass pipe with an external radius of 1 mm and an internal one of 0,6 mm, together with a glass electrode (needle) of 10 μm. in radius. The protoplast can be moved, held, transported and connected to another protoplast without damage and then released.

Figure18.29a outlines an application of this kind. The majority of the protoplasts concentrate close to the micromanipulator and form what is called a „pearl chain".

In Figure 18.29b the first protoplast is moved by the micromanipulator and held after applying a variable field of 300 V/cm, 0,5 Hz. The opposite electrode is made from of aluminium with a fitness of 6 mm. The next protoplast is moved and connected to the previous one. The following figures show the run of this process and fusion phenomena. Cell fusion can be performed in the condition of constant impulses with bright 50 μm [18.12].

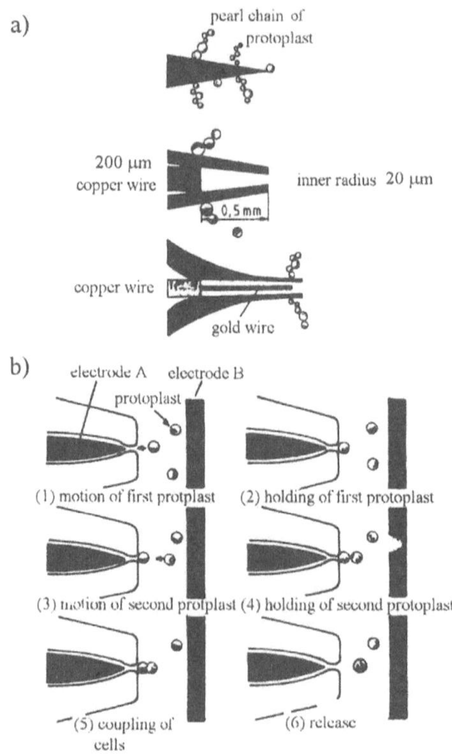

Fig.18.29. Micromanipulator used for biological reseach [18.12]: a) the experiments with the pearl chain of the protoplasts to different electrodes; b) attraction of the protoplasts.

18.5.2. Microtelerobot for manipulation of microobjects immersed in the liquid

In [18.13] Wang presents a microtelerobot for the manipulation of objects such as micromolecules on living organisms immersed in a liquid viscotic flow. The dimensions of these objects and the fitness of the liquid flow change range from a few hundred μm to a few μm. The method is based on the injection of a small number of microscopic ferromagnetic or magnetostrictive particles to control of the liquid before or after connecting the object to be manipulated. The effectors are activated and controlled by electromagnet located on outside the flow. A nonlinear mathematical model was developed to describ the dynamic properties of the microrobot.

It is important during manipulation not to break the liquid surface which is under a big surface strain because this could damage the liquid flow. Figure18.30 gives the scheme of the microtelerobot with its active end-effector for manipulation of the ferromagnetic or magnetostatic objects.

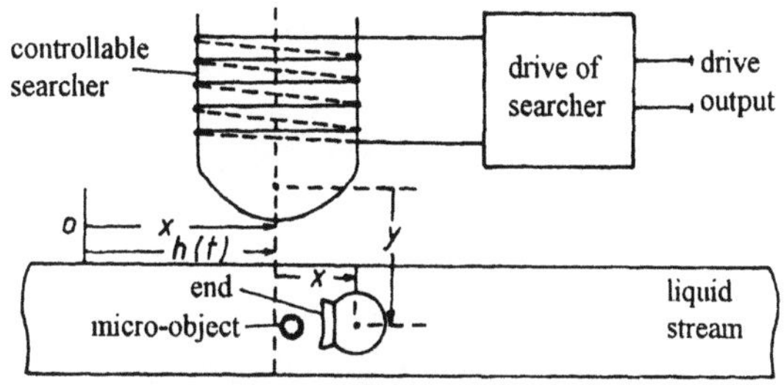

Fig.18.30. Micromanipulation in the liquid flow [18.13]

Simulation investigations of the linearized mathematical model were performed. during the study. They produced the following data: radii of gripper r_o = 0,005 cm, radii of the curvature of the searcher top r_p = 0,01 cm, distance of the searcher top from the flow surface y = 0,01 cm, revolution of the coil of the main searcher N = 1000, the current intensity in the coil 0,1 A, searcher speed V = 0,05 cm/s, liquid density q = 1 g/cm^3, effector mass 1,5708*10^{-6}g.

18.5.3. Multifinger gripper

The compliant actuator described in item 18.4.1 and situated in parallel [18.3] can be used as an multifinger gripper for manipulation of the micro objects. Figure18.31 shows it holding a contact lens with a diameter of 9 mm. The prototype of the gripper has three fingers with a diameter of 2 mm each. The gripper provides different kinds of prehension and changes in object orientation; it can also adopt to changes in object shape.

The elastic actuator can also be used in the robot for the inspection of thin pipes (Fig.18.32). This snake-like robot consists in two sets of the connected elastic microactuators and is used as a locomotion vehicle in small tubes. Figure18.32 shows the prototype of the 110 mm long robot.

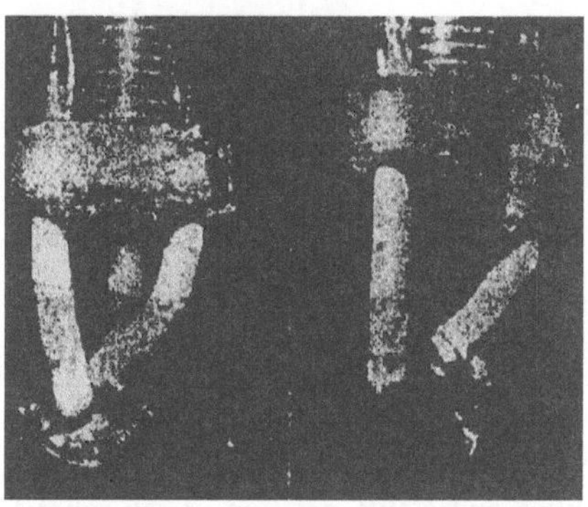

Fig.18.31. Multifinger gripper [18.3]

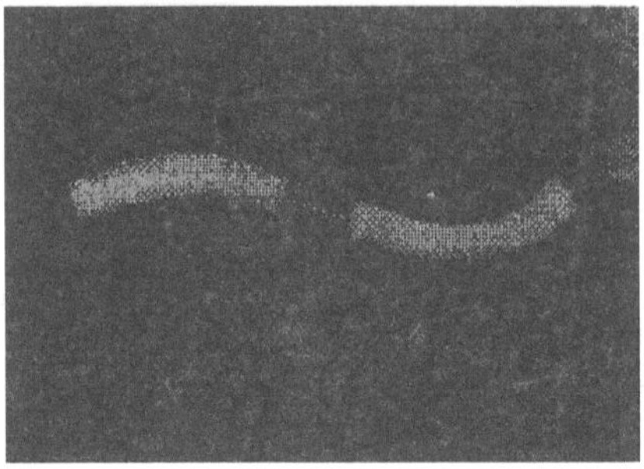

Fig.18.32. Microrobot for inspection

It has diameter of 12 mm and a mass of 15 g. Six elastic pneumatic tubes with an internal diameter of 0,5 mm and an external one of 1 mm are connected inside the robot. The pressure inside the six chambers is controlled independently. The actuators are driven periodically by analog pneumatic signals with the same frequency and phase difference 1/4. The actuators enable displacement of the robot through so called traveling waves. The robot was experimented in acrylic tube with an external diameter of 25 mm. Its maximal speed in the horizontal tube was 10 mm/s. The robot can also move in the vertical direction by loading 0,7 N [18.3].

18.5.4. Calibration of a group of micro-robots and their stable position

Figure18.33a shows the group of micro-robots, and Figure18.33b, their positioning in relation to the objects selected. The next two figures show the configurations obtained in relation to the speed of the leader [18.14].

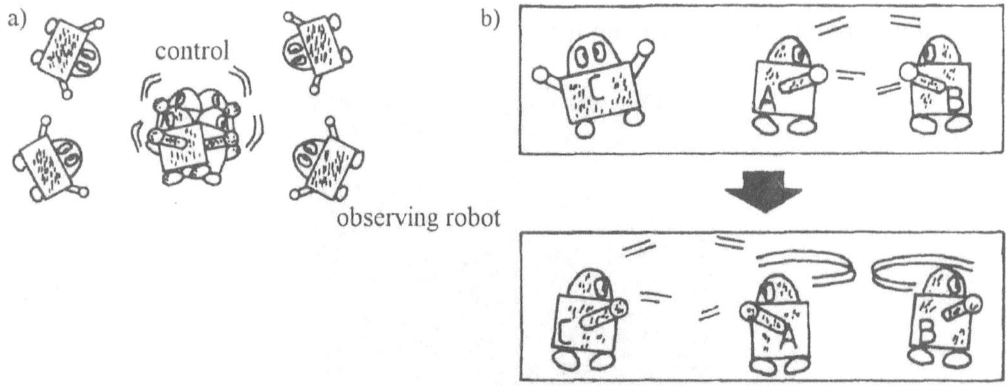

Fig.18.33: a) Potential possibilities and stable position of the robot [18.14], b) Positioning of the robot in relation to the objects selected [18.14]

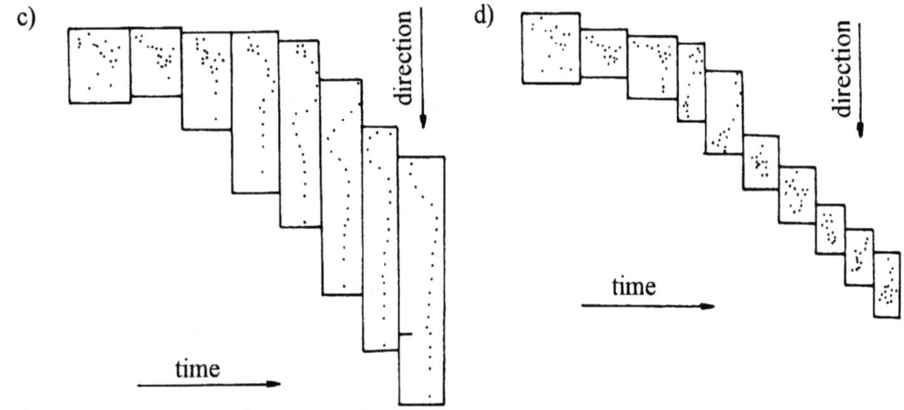

Fig.18.33 c) Configuration during team motion (the leader is moving slowly) [18.14], d) Configuration during team motion (the leader is moving quickly) [18.14]

18.6. Biomorphic micromachines (robots)

In [18.28] a new parallel type of machines, which are not robots in any traditional sense are described. These are not machine designed to perform a set of goal-oriented tasks or work, but rather to express modes of survivalist behavior The survival of a autonomous mobile machine in an unknown and possibly hostile environment. The name of this category of machines is an abbreviation of biological morphology and the authors Brosl Hausslacher and Mark T.Tilden treat them as a form of live parallels.

An ATT motor is used to build these machines, which makes it possible to design potentially useful forms of live. What distinguishes biomorphic machines from typical mobile robots is not their material base, but how they are organized. They use a dynamical, non-symbolic internal world representation and a compliant, bi-directional, interactive response, where their internal world assumes a crucial role. In this they have much in common with biological forms: in fact, these machines are designed along biological paradigms rather than on first principle notions of how such machines should be organized.

Over 70 working machines have been built. The authors consider this field an experimental science in which we both learn from the machines and are for the moment their evolutionary agents [18.28].

The following rules govern the survival of an autonomous machine:

1. the machine must protect its existence.
2. the machine must acquire more energy then it exerts.
3. the machine must exhibit (directed) motion.

These biomorphic laws differ from those propounded by Asimov.

The nervous structure of the VOG 1,2 "LOBSTER" has an almost perfectly concentric biomorphic map. The neutral layers have mapped a total of 64 possible response influences (based upon external stimuli) onto the 6 distinct patterns the nervous ring is capable of sustaining. This device is quite capable in low stimulus environments but becomes "confused" when its environment becomes complex (i.e. dynamic, rather than stable external stimulus).

The first class of thought machines, constructible with current technology, is down to dimensions of a few tens of square microns. Further shrinking of machines to the order of a square micron or slightly below will be possible in the future. A typical cell is the size of a few microns in diameter. At this scale self-assembling colonies of biomorphic machines are not only possible, but desirable. Colonies will both self-organize physically and develop collectively, or will be formed by electrical initiation. Helio batteries, ultrasonic waves, and the control loops can be used at present microelectronics achievements. Very soon it will be necessary to use submicroelectronic loops.

The authors assume that in the future we shall have machines on a nanoscale (100 nanometers and below), which will be able to enter the cell [11.28].

18.7. Selected problems

Problems now being investigated are: sensors and actuators; mechanisms of micromotion; elements and assembly; and surface technology. We shall briefly discuss some of them.

Electromagnetic actuators of the dimension of 0,5 μm shall see further development together with what we call the artificial bioactuator.

The next problem that must be addressed is that of the micromechanisms for motion transmission. Materials with a low coefficient of friction will be used. At present a finger made from Ti-Ni with a diameter 0,2 mm can open and close in 10 ms. A micromotor designed at Berkeley has a diameter of 60 μm. The first transistor (1 mm) was built in the Bell Laboratory in 1948. In recent years many components of dimension 20 nm have been built.

18.8. Possible applications and perspectives

The further development of MM depends on possible applications. Promising examples of such applications were seen in the years 1991-93. They include medical micromanipulators (endoscopes), and inspection robots. The list of possible models is rather long, and includes motors, turbines, micro-actuators, and walking machines. The development of MM in the future (mobile robots, actuators, mechanisms of micro motion, machine service, assembly etc.) will depend strongly on the research in living organisms at the mili-, micro- and nano-level. The next century may be the century of micro- and nano-engineering based on the micro structure of the organic world [18.18, 18.22].

In medicine the future may bring microsize machines to investigate and repair damage in the cardiovascular, and digestive systems. They may be used to perform urgent operations on the inner ear and nose, sectioning and suturing even individual nerve fibers. In agriculture, armies of crawling or flying insect-like micromechanisms may be used for pollination, planting seeds, targeted chemical spraying, cell implants and pest control. New elements in remote locomotion elements will be like: micro motors, micro actuators, turbines, propellers, bearings, generators, sensors, and micro optic devices. At JPL in California a new prototype accelerometer has already been built, which measures changes in acceleration smaller than 10^{-6} g. MM will become an industry, with a new philosophy of MM design, closer to microorganisms than to existing machines and mechanisms.

18.9. References

[18.1] *Hayashi T.*: On Micromechanisms and their Researches and Developments. Proc. of the Ninth CISM-IFToMM Symposium on Theory and Practice of Robots and Manipulators Ro.Man.Sy`9, Springer Verlag, London 1993. Edited by A. Morecki, G. Bianchi and K. Jaworek

[18.2] *Seireg A.*: Micromechanisms: Future Expectations and Design Methodologies. Proc. of the 1st IFToMM Int. MM Symposium, TIT, June 1-3, 1993.[18.3] *Suzumori K., Kondo F. and Tanaka H.*: Applications of the Flexible Micro-actuator to Microrobts. Proc. of the 1st IFToMM Int. MM Symposium, TIT, June 1-3, 1993.

[18.4] *Miagazawa O.*: A Self-contained Micromechanisms. Proc. of the 1st IFToMM Int. MM Symposium, TIT, June 1-3, 1993.

[18.5] *Ogasawara H. and Kojima S.*: Mobile Micromechanisms with Caterpillar-type. Proc. of the 1st IFToMM Int. MM Symposium, TIT, June 1-3, 1993.

[18.6] *Kasuga M. at all*: Development of Running Microrobot Using Ultrasonic Micromotors. Proc. of the 1st IFToMM Int. MM Symposium, TIT, June 1993.[18.7] *Fukushima H. at all*: Study for Micro Mobile Machines with Piezoelectric Element. Proc. of the 1st IFToMM Int. MM Symposium, TIT, June 1-3, 1993.

[18.8] *Mizukami M., Koyabu K. and Ohira F.*: 1 cm^3 Miniaturized Mobile Machine Driven by Electromagnetic Force. Proc. of the 1st IFToMM Int. MM Symp. TIT, June 1-3, 1993.

[18.9] *Nakada H., Sawada R. and Watabe A.*: A 10-mm Cubic Miniaturized Vehicle with Air-driven Turbines. Proc. of the 1st IFToMM Int. MM Symposium, TIT, June 1-3, 1993.

[18.10] *Inagaki H. and Saito A.*: Flapping Electrostatic Actuator for Micro-Mobile-Mechanisms. Proc. of the 1st IFToMM Int. MM Symposium, TIT, June 1993.

[18.11] *Dai Homma*: Development of Micro Submarine. Proc. of the 1st IFToMM Int. MM Symposium, TIT, June 1-3, 1993.

[18.12] *Miwa Y., Yoshimura S. and Itoi A.*: Automation of Micro-Manipulator for Cell Fusion. Jnl of Robotic Systems, April 1992.

[18.13] *Wang P.K.C.*: Micro-Telerobot for Manipulations of Microscopic Objects Immersed in Liquid Layer. Jnl of Robotic Systems 10 (3), 1993.

[18.14] *Watanabe T.*: Group Control Algorithm. Jnl of Robotics and Mechatronics, Vol. 4, No 3, 1992.

[18.15] *Suzuki K.*: Historical Observations on MM. Proc. of the 1st IFToMM Int. MM Symposium, TIT, June 1-3, 1993.

[18.16] *Dai Homma*: From the Second Mountain - Climbing, MM Contest WT. J.J.S.P. Eng., Vol.26, No 3, Sept. 1992.

[18.17] *Hayashi T. and Yoshida K.*: Traveling Micromechanisms. Proc. of the 8th World Congress on TMM, Vol.6, 1991.

[18.18] *Morecki A.*: Mili, micro and nanomechanisms, robots, manipulators and walking machines. Proc. of the 4th National Conf. on Robotics, Vol.1, Wrocław 1993.

[18.19] *Miura H.*: Robot Intelligence and Microrobot. Proc. of the 9th World Congress on the TMM, Vol. 1, Milan 1995.

[18.20] *Morecki A.*: Micromechanisms and microwalking machines. Proc. of the 14th Polish Conf. on TMM, Gdańsk/Gdynia, 1994.

[18.21] *Morecki A.*: A micromechanisms and Microwalking Machines. Proc. of the Conf.: System Modelling, Control, Vol.1, Polish Society of Medical informatics, Łódź, 1995.

[18.22] *Morecki A.*: Biological and Biomechanical Basis for Design of Microwalking Machines. Proc. of the Polish Conf. Biomechanika`95, Z.N. No 73, AWF, Cracow 1995.

[18.23] *Nieœko J.*: The construction of the walking micromechanisms family driven by electromagnetic force. Msc Diploma, (supervisor A. Morecki), Warsaw Univ. of Technology, Dec., 1994.

[18.24] *Baran K.*: Micromobile vehicle driven by the air turbines. Msc Diploma, (supervisor A. Morecki), Warsaw University of Technology, Dec., 1996.

[18.25] Proceed. of the Int. Symp. on Microsystems, Intelligent Materials and Robots, Sept. 27-29, 1995, Sendai Int. Centre, Japan.

[18.26] HERMES, the compact walking robot, 1994, IS Robotics Inc.Twin City Office Centre, Sommerville HA 02143, USA

[18.27] *Szabelak P.*: Parameters Identification of Walking Machine HERMES II, Msc Diploma, (supervisor T.Zielińska), Warsaw Univ. of Technology, June 1996.

[18.28] *Hasslacher B. and Tilden M.*: Living machines. Robotics and Autonomous Systems 448, 1995.

MIX
Papier aus verantwortungsvollen Quellen
Paper from responsible sources
FSC® C105338

If you have any concerns about our products,
you can contact us on
ProductSafety@springernature.com

In case Publisher is established outside the EU,
the EU authorized representative is:
Springer Nature Customer Service Center GmbH
Europaplatz 3, 69115 Heidelberg, Germany

Printed by Libri Plureos GmbH
in Hamburg, Germany